ThomsonNOW™
Completely integrated with *Chemistry for Today: General, Organic, and Biochemistry, Sixth Edition*

What do you need to learn now? Take charge of your learning with ThomsonNOW™

ThomsonNOW provides interactive tutorials to study and learn concepts more effectively, helping you manage and make the most of your study time. Using a diagnostic pre-test, **ThomsonNOW** gauges your unique study needs to provide you with a Personalized Study plan that optimizes your time investment by focusing your study time on the concepts you need to master.

Look for references throughout the text that lead you to **ThomsonNOW.** They direct you to the corresponding media-enhanced activities within the program. Precise page-by-page integration enables you to go beyond reading about chemistry—you'll actually experience it in action!

Easy to use

The **ThomsonNOW** system includes two powerful assessment components:

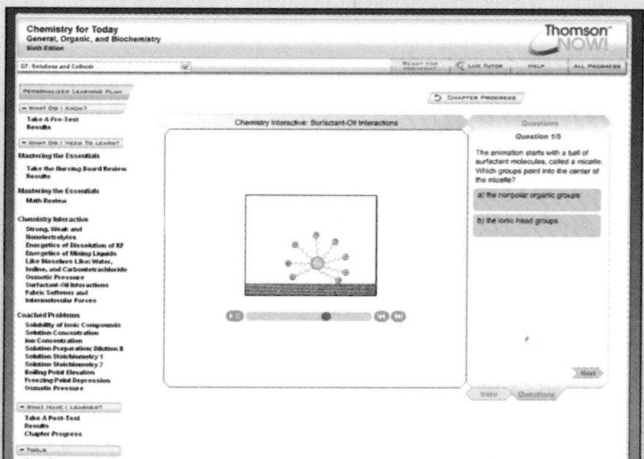

WHAT DO I KNOW?
This diagnostic Exam-Prep Quiz, based on the core concepts in each chapter, gives you an initial assessment.

WHAT DO I NEED TO LEARN?
A **Personalized Study** plan leads you to exercises, homework problems, and tutorials. With a click of the mouse, **ThomsonNOW's** unique assets allow you to:

- Create a **Personalized Study** plan or review for an exam
- Explore chemical concepts through tutorials, simulations, and animations
- View **Active Figures** and interact with text illustrations. These **Active Figures** help you master key concepts from the book. Each figure is paired with corresponding questions to help you focus on chemistry at work and ensure that you truly understand the concepts played out in the animations.

Make the most of your study time—log on to **ThomsonNOW** today!

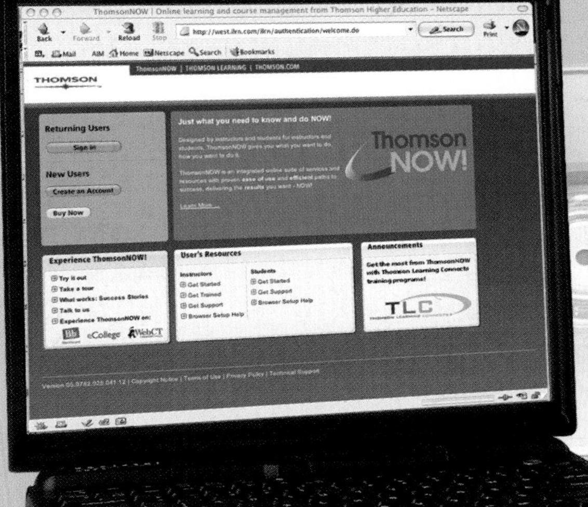

You can access **ThomsonNOW** through the URL and code on the card that accompanies your textbook, or purchase access online today at **www.thomsonedu.com/thomsonnow/buy.**

Organic and Biochemistry for Today

Sixth Edition

Spencer L. Seager
Weber State University

Michael R. Slabaugh
Weber State University

Australia • Canada • Mexico • Singapore • Spain
United Kingdom • United States

Organic and Biochemistry for Today, Sixth Edition
Spencer L. Seager, Michael R. Slabaugh

Publisher: *David Harris*
Editor: *Lisa Lockwood*
Assistant Editor: *Sylvia Krick*
Developmental Editor: *Rebecca Heider*
Art Director: *John Walker*
Creative Director: *Rob Hugel*
Technology Project Manager: *Lisa Weber*
Marketing Manager: *Amee Mosley*
Advertising Project Manager: *Bryan Vann*
Project Manager, Editorial Production: *Belinda Krohmer*
Print Buyer: *Judy Inouye*
Permissions Editor: *Roberta Broyer*
Production Service: *Lachina Publishing Services*
Text Designer: *Ellen Pettengell*
Photo Researcher: *Jane Sanders Miller*
Copy Editor: *Sara Black*
Illustrator: *Lachina Publishing Services*
OWL Producers: *Stephen Battisti, Cindy Stein, David Hart, (Center for Educational Software Development, University of Massachusetts, Amherst)*
Cover Designer: *Gia Giasulo*
Cover Image: *Getty/Stockdisc*
Cover Printer: *Courier Kendallville*
Compositor: *Lachina Publishing Services*
Printer: *Courier Kendallville*

Printed in the United States of America
1 2 3 4 5 6 7 11 10 09 08 07

For more information about our products, contact us at:
Thomson Learning Academic Resource Center
1-800-423-0563

For permission to use material from this text, contact us by:
Phone: 1-800-730-2214 **Fax:** 1-800-730-2215
Web: http://www.thomsonrights.com

Library of Congress Control Number 2006933546

Student Edition: ISBN 0-495-11280-1

Brooks/Cole—Thomson Learning
10 Davis Drive
Belmont, CA 94002
USA

Asia
Thomson Learning
5 Shenton Way #01-01
UIC Building
Singapore 068808

Australia/New Zealand
Thomson Learning
102 Dodds Street
Southbank, Victoria 3006
Australia

Canada
Nelson
1120 Birchmount Road
Toronto, Ontario M1K 5G4
Canada

Europe/Middle East/Africa
Thomson Learning
High Holborn House
50/51 Bedford Row
London WC1R 4LR
United Kingdom

To our grandchildren:
Nate and Braden Barlow, and Megan and Bradley Seager
Alexander, Elyse, Megan, and Mia Slabaugh

About the Authors

Spencer L. Seager

Spencer L. Seager is a professor of chemistry at Weber State University, where he served as chemistry department chairman from 1969 until 1993. He teaches general chemistry at the university and is also active in projects to help improve chemistry and other science education in local elementary schools. He received his B.S. degree in chemistry and Ph.D. degree in physical chemistry from the University of Utah. Other interests include making minor home repairs, reading history of science and technology, listening to classical music, and walking for exercise.

Michael R. Slabaugh

Michael R. Slabaugh is a senior fellow at Weber State University, where he teaches the year-long sequence of general, organic, and biochemistry. He received his B.S. degree in chemistry from Purdue University and his Ph.D. degree in organic chemistry from Iowa State University. His interest in plant alkaloids led to a year of postdoctoral study in biochemistry at Texas A&M University. His current professional interests are chemistry education and community involvement in science activities, particularly the State Science and Engineering Fair in Utah. He also enjoys the company of family, hiking in the mountains, and fishing the local streams.

Brief Contents

CHAPTER 1
Organic Compounds: Alkanes 1

CHAPTER 2
Unsaturated Hydrocarbons 38

CHAPTER 3
Alcohols, Phenols, and Ethers 70

CHAPTER 4
Aldehydes and Ketones 101

CHAPTER 5
Carboxylic Acids and Esters 129

CHAPTER 6
Amines and Amides 157

CHAPTER 7
Carbohydrates 185

CHAPTER 8
Lipids 218

CHAPTER 9
Proteins 245

CHAPTER 10
Enzymes 274

CHAPTER 11
Nucleic Acids and Protein Synthesis 298

CHAPTER 12
Nutrition and Energy for Life 331

CHAPTER 13
Carbohydrate Metabolism 358

CHAPTER 14
Lipid and Amino Acid Metabolism 383

CHAPTER 15
Body Fluids 409

Contents

CHAPTER 1

Organic Compounds: Alkanes 1

1.1 Carbon: The Element of Organic Compounds *2*
1.2 Organic and Inorganic Compounds Compared *3*
1.3 Bonding Characteristics and Isomerism *5*
1.4 Functional Groups: The Organization of Organic Chemistry *7*
1.5 Alkane Structures *10*
1.6 Conformations of Alkanes *13*
1.7 Alkane Nomenclature *15*
1.8 Cycloalkanes *21*
1.9 The Shape of Cycloalkanes *23*
1.10 Physical Properties of Alkanes *26*
1.11 Alkane Reactions *28*
Concept Summary *30*
Key Terms and Concepts *30*
Key Reactions *31*
Exercises *31*
Allied Health Exam Connection *36*
Chemistry for Thought *36*
STUDY SKILLS 1.1 Changing Gears for Organic Chemistry *4*
CHEMISTRY AND YOUR HEALTH 1.1 Are Organic Foods Better for You? *11*
CHEMISTRY AROUND US 1.1 Petroleum *26*
OVER THE COUNTER 1.1 Hydrating the Skin *28*

CHEMISTRY AROUND US 1.2 Carbon Monoxide: Silent and Deadly *29*

CHAPTER 2

Unsaturated Hydrocarbons 38

2.1 The Nomenclature of Alkenes *39*
2.2 The Geometry of Alkenes *42*
2.3 Properties of Alkenes *46*
2.4 Addition Polymers *51*
2.5 Alkynes *53*
2.6 Aromatic Compounds and the Benzene Structure *56*
2.7 The Nomenclature of Benzene Derivatives *58*
2.8 Properties and Uses of Aromatic Compounds *61*
Concept Summary *63*
Key Terms and Concepts *64*
Key Reactions *64*
Exercises *65*
Allied Health Exam Connection *69*
Chemistry for Thought *69*
CHEMISTRY AROUND US 2.1 Seeing the Light *41*
CHEMISTRY AROUND US 2.2 Watermelon: A Source of Lycopene *44*
STUDY SKILLS 2.1 Keeping a Reaction Card File *50*
STUDY SKILLS 2.2 A Reaction Map for Alkenes *53*
HOW REACTIONS OCCUR 2.1 The Hydration of Alkenes: An Addition Reaction *56*
CHEMISTRY AND YOUR HEALTH 2.1 Beautiful, Brown . . . and Overdone *59*
OVER THE COUNTER 2.1 Smoking: It's Quitting Time *62*

CHAPTER 3

Alcohols, Phenols, and Ethers 70

3.1 The Nomenclature of Alcohols and Phenols *72*
3.2 Classification of Alcohols *74*

3.3 Physical Properties of Alcohols *75*
3.4 Reactions of Alcohols *77*
3.5 Important Alcohols *82*
3.6 Characteristics and Uses of Phenols *85*
3.7 Ethers *88*
3.8 Properties of Ethers *89*
3.9 Thiols *90*
3.10 Polyfunctional Compounds *92*
Concept Summary *94*
Key Terms and Concepts *94*
Key Reactions *94*
Exercises *95*
Allied Health Exam Connection *100*
Chemistry for Thought *100*
HOW REACTIONS OCCUR 3.1 The Dehydration of an Alcohol *79*
STUDY SKILLS 3.1 A Reaction Map for Alcohols *82*
CHEMISTRY AROUND US 3.1 Driving on Corn Fumes *85*
OVER THE COUNTER 3.1 Outsmarting Poison Ivy *86*
CHEMISTRY AND YOUR HEALTH 3.1 Weaned from the Bottle *88*
CHEMISTRY AROUND US 3.2 General Anesthetics *91*

CHAPTER 4

Aldehydes and Ketones 101

4.1 The Nomenclature of Aldehydes and Ketones *102*
4.2 Physical Properties *105*
4.3 Chemical Properties *108*
4.4 Important Aldehydes and Ketones *119*
Concept Summary *120*
Key Terms and Concepts *121*
Key Reactions *122*
Exercises *123*
Allied Health Exam Connection *127*
Chemistry for Thought *128*
CHEMISTRY AROUND US 4.1 Faking a Tan *106*
OVER THE COUNTER 4.1 Birth Control: Progesterone Substitutes *110*
HOW REACTIONS OCCUR 4.1 Hemiacetal Formation *114*
STUDY SKILLS 4.1 A Reaction Map for Aldehydes and Ketones *115*
CHEMISTRY AROUND US 4.2 Vanilloids: Hot Relief from Pain *118*
CHEMISTRY AND YOUR HEALTH 4.1 Vitamin A and Birth Defects *119*

CHAPTER 5

Carboxylic Acids and Esters 129

5.1 The Nomenclature of Carboxylic Acids *130*
5.2 Physical Properties of Carboxylic Acids *132*
5.3 The Acidity of Carboxylic Acids *135*
5.4 Salts of Carboxylic Acids *136*
5.5 Carboxylic Esters *138*
5.6 The Nomenclature of Esters *142*
5.7 Reactions of Esters *143*
5.8 Esters of Inorganic Acids *147*
Concept Summary *149*
Key Terms and Concepts *150*
Key Reactions *150*
Exercises *151*
Allied Health Exam Connection *155*
Chemistry for Thought *155*
OVER THE COUNTER 5.1 Alpha Hydroxy Acids in Cosmetics *134*
CHEMISTRY AND YOUR HEALTH 5.1 Aspirin: Should You Take a Daily Dose? *144*
STUDY SKILLS 5.1 A Reaction Map for Carboxylic Acids *146*
HOW REACTIONS OCCUR 5.1 Ester Saponification *146*
CHEMISTRY AROUND US 5.1 Nitroglycerin in Dynamite and in Medicine *148*

CHAPTER 6

Amines and Amides 157

6.1 Classification of Amines *158*
6.2 The Nomenclature of Amines *158*
6.3 Physical Properties of Amines *160*
6.4 Chemical Properties of Amines *161*
6.5 Amines as Neurotransmitters *168*
6.6 Other Biologically Important Amines *171*
6.7 The Nomenclature of Amides *174*
6.8 Physical Properties of Amides *176*
6.9 Chemical Properties of Amides *176*
Concept Summary *179*
Key Terms and Concepts *179*
Key Reactions *179*
Exercises *180*
Allied Health Exam Connection *184*
Chemistry for Thought *184*
OVER THE COUNTER 6.1 Cough Syrup—or Are You Just Coughing Up More Money? *160*
CHEMISTRY AROUND US 6.1 Aspirin Substitutes *167*
STUDY SKILLS 6.1 A Reaction Map for Amines *170*
CHEMISTRY AND YOUR HEALTH 6.1 Chocolate: A New Health Food or Fad? *174*

CHAPTER 7

Carbohydrates 185

7.1 Classes of Carbohydrates *186*
7.2 The Stereochemistry of Carbohydrates *187*

7.3 Fischer Projections *192*
7.4 Monosaccharides *195*
7.5 Properties of Monosaccharides *196*
7.6 Important Monosaccharides *202*
7.7 Disaccharides *204*
7.8 Polysaccharides *207*
Concept Summary *212*
Key Terms and Concepts *212*
Key Reactions *213*
Exercises *213*
Allied Health Exam Connection *216*
Chemistry for Thought *217*
CHEMISTRY AROUND US 7.1 Sugar-Free Foods and Diabetes *200*
STUDY SKILLS 7.1 Biomolecules: A New Focus *205*
CHEMISTRY AND YOUR HEALTH 7.1 Sliced White Wheat Bread . . . Is It Really the Next Best Thing? *206*
OVER THE COUNTER 7.1 Dietary Fiber *210*

CHAPTER 8

Lipids 218

8.1 Classification of Lipids *219*
8.2 Fatty Acids *220*
8.3 The Structure of Fats and Oils *223*
8.4 Chemical Properties of Fats and Oils *225*
8.5 Waxes *228*
8.6 Phosphoglycerides *228*
8.7 Sphingolipids *231*
8.8 Biological Membranes *232*
8.9 Steroids *234*
8.10 Steroid Hormones *237*
8.11 Prostaglandins *239*
Concept Summary *240*
Key Terms and Concepts *241*
Key Reactions *241*
Exercises *242*
Allied Health Exam Connection *244*
Chemistry for Thought *244*
STUDY SKILLS 8.1 A Reaction Map for Triglycerides *227*
CHEMISTRY AND YOUR HEALTH 8.1 Going after Those Trans Fatty Acids *229*

CHEMISTRY AROUND US 8.1 Nuts: Good Food in Small Packages *235*

OVER THE COUNTER 8.1 Melatonin and DHEA: Hormones at Your Own Risk *236*

CHEMISTRY AROUND US 8.2 Biodiesel: A Fuel for the 21st Century? *239*

CHAPTER 9

Proteins 245

9.1 The Amino Acids *246*
9.2 Zwitterions *248*
9.3 Reactions of Amino Acids *250*
9.4 Important Peptides *253*
9.5 Characteristics of Proteins *255*
9.6 The Primary Structure of Proteins *259*
9.7 The Secondary Structure of Proteins *260*
9.8 The Tertiary Structure of Proteins *262*
9.9 The Quaternary Structure of Proteins *265*
9.10 Protein Hydrolysis and Denaturation *266*
Concept Summary *268*
Key Terms and Concepts *269*
Key Reactions *269*
Exercises *270*
Allied Health Exam Connection *272*
Chemistry for Thought *273*

OVER THE COUNTER 9.1 Medicines and Nursing Mothers *251*

CHEMISTRY AND YOUR HEALTH 9.1 C-Reactive Protein: A Message from the Heart *254*

CHEMISTRY AROUND US 9.1 Alzheimer's Disease *258*

CHEMISTRY AROUND US 9.2 Sickle-Cell Disease *263*

STUDY SKILLS 9.1 Visualizing Protein Structure *265*

CHAPTER 10

Enzymes 274

10.1 General Characteristics of Enzymes *275*
10.2 Enzyme Nomenclature and Classification *276*
10.3 Enzyme Cofactors *277*
10.4 The Mechanism of Enzyme Action *279*
10.5 Enzyme Activity *281*
10.6 Factors Affecting Enzyme Activity *282*
10.7 Enzyme Inhibition *284*
10.8 The Regulation of Enzyme Activity *289*
10.9 Medical Application of Enzymes *292*
Concept Summary *294*
Key Terms and Concepts *295*
Key Reactions *295*
Exercises *295*
Allied Health Exam Connection *297*
Chemistry for Thought *297*

CHEMISTRY AND YOUR HEALTH 10.1 Enzymes and Disease *278*

OVER THE COUNTER 10.1 Are All Vitamin Brands Created Equal? *280*

CHEMISTRY AROUND US 10.1 Enzyme Discovery Heats Up *285*

CHEMISTRY AROUND US 10.2 Mercury in Fish *286*

STUDY SKILLS 10.1 A Summary Chart of Enzyme Inhibitors *291*

CHAPTER 11

Nucleic Acids and Protein Synthesis 298

11.1 Components of Nucleic Acids *299*
11.2 The Structure of DNA *301*
11.3 DNA Replication *305*
11.4 Ribonucleic Acid (RNA) *309*
11.5 The Flow of Genetic Information *313*
11.6 Transcription: RNA Synthesis *314*
11.7 The Genetic Code *316*
11.8 Translation and Protein Synthesis *318*
11.9 Mutations *322*
11.10 Recombinant DNA *322*

Concept Summary 326
Key Terms and Concepts 327
Exercises 327
Allied Health Exam Connection 329
Chemistry for Thought 329
OVER THE COUNTER 11.1 Nucleic Acid Supplements 305
CHEMISTRY AROUND US 11.1 The Clone Wars 310
CHEMISTRY AROUND US 11.2 The Race Against Avian Flu 315
STUDY SKILLS 11.1 Remembering Key Words 317
CHEMISTRY AROUND US 11.3 Stem Cell Research 319
CHEMISTRY AROUND US 11.4 DNA and the Crime Scene 324
CHEMISTRY AND YOUR HEALTH 11.1 Genetically Modified Foods 326

CHAPTER 12

Nutrition and Energy for Life 331

12.1 Nutritional Requirements 332
12.2 The Macronutrients 333
12.3 Micronutrients I: Vitamins 337
12.4 Micronutrients II: Minerals 339
12.5 The Flow of Energy in the Biosphere 340
12.6 Metabolism and an Overview of Energy Production 343
12.7 ATP: The Primary Energy Carrier 344
12.8 Important Coenzymes in the Common Catabolic Pathway 349
Concept Summary 353
Key Terms and Concepts 354
Key Reactions 354
Exercises 355
Allied Health Exam Connection 357
Chemistry for Thought 357
CHEMISTRY AROUND US 12.1 The Ten Most Dangerous Foods to Eat While Driving 337
CHEMISTRY AND YOUR HEALTH 12.1 The Health Gauge 339
STUDY SKILLS 12.1 Bioprocesses 346
CHEMISTRY AROUND US 12.2 Eating Disorders 347
OVER THE COUNTER 12.1 Creatine Supplements: The Jury Is Still Out 351

CHAPTER 13

Carbohydrate Metabolism 358

13.1 The Digestion of Carbohydrates 359
13.2 Blood Glucose 359
13.3 Glycolysis 360
13.4 The Fates of Pyruvate 363
13.5 The Citric Acid Cycle 365
13.6 The Electron Transport Chain 368
13.7 Oxidative Phosphorylation 369
13.8 The Complete Oxidation of Glucose 370
13.9 Glycogen Metabolism 372
13.10 Gluconeogenesis 374
13.11 The Hormonal Control of Carbohydrate Metabolism 375
Concept Summary 377
Key Terms and Concepts 378
Key Reactions 378
Exercises 379
Allied Health Exam Connection 382
Chemistry for Thought 382
OVER THE COUNTER 13.1 Lactose Intolerance 360
CHEMISTRY AROUND US 13.1 Lactate Accumulation 367
STUDY SKILLS 13.1 Key Numbers for ATP Calculations 373
CHEMISTRY AROUND US 13.2 Carbohydrate Loading 376
CHEMISTRY AND YOUR HEALTH 13.1 Prediabetic . . . or Already There? 377

CHAPTER 14

Lipid and Amino Acid Metabolism 383

14.1 Blood Lipids *384*
14.2 Fat Mobilization *386*
14.3 Glycerol Metabolism *388*
14.4 The Oxidation of Fatty Acids *388*
14.5 The Energy from Fatty Acids *391*
14.6 Ketone Bodies *392*
14.7 Fatty Acid Synthesis *394*
14.8 Amino Acid Metabolism *395*
14.9 Amino Acid Catabolism: The Fate of the Nitrogen Atoms *396*
14.10 Amino Acid Catabolism: The Fate of the Carbon Skeleton *400*
14.11 Amino Acid Biosynthesis *403*
Concept Summary *404*
Key Terms and Concepts *405*
Key Reactions *405*
Exercises *406*
Allied Health Exam Connection *408*
Chemistry for Thought *408*
OVER THE COUNTER 14.1 Cholesterol-Lowering Drugs *387*
STUDY SKILLS 14.1 Key Numbers for ATP Calculations *393*
CHEMISTRY AND YOUR HEALTH 14.1 The Magic Bean *398*
CHEMISTRY AROUND US 14.1 Phenylketonuria (PKU) *401*
CHEMISTRY AROUND US 14.2 Steroids in High Schools *402*

CHAPTER 15

Body Fluids 409

15.1 A Comparison of Body Fluids *410*
15.2 Oxygen and Carbon Dioxide Transport *410*
15.3 Chemical Transport to the Cells *415*
15.4 The Constituents of Urine *416*
15.5 Fluid and Electrolyte Balance *417*
15.6 Acid–Base Balance *418*
15.7 Buffer Control of Blood pH *418*
15.8 Respiratory Control of Blood pH *420*
15.9 Urinary Control of Blood pH *420*
15.10 Acidosis and Alkalosis *421*
Concept Summary *424*
Key Terms and Concepts *425*
Key Reactions *425*
Exercises *425*
Allied Health Exam Connection *427*
Chemistry for Thought *427*
OVER THE COUNTER 15.1 Avoiding Food-Drug Interactions *413*
CHEMISTRY AND YOUR HEALTH 15.1 Exercise Beats Angioplasty *415*
CHEMISTRY AROUND US 15.1 Exercise and Altitude *419*

Appendix A The International System of Measurements A-1

Appendix B Answers to Even-Numbered End-of-Chapter Exercises B-1

Appendix C Solutions to Learning Checks C-1

Glossary G-1

Index I-1

Preface

The Image of Chemistry

We, as authors, are pleased that the acceptance of the previous five editions of this textbook by students and their teachers has made it possible to publish this sixth edition. In the earlier editions, we expressed our concern about the negative image of chemistry held by many of our students, and their genuine fear of working with chemicals in the laboratory. Unfortunately, this negative image not only persists, but seems to be intensifying. Reports in the media related to chemicals or to chemistry continue to be primarily negative, and in many cases seem to be designed to increase the fear and concern of the general public. With this edition, we continue to hope that those who use this book will gain a more positive understanding and appreciation of the important contributions that chemistry makes in their lives.

Theme and Organization

This edition continues the theme of the positive and useful contributions made by chemistry in our world. Consistent with that theme, we continue to use the chapter opening focus on health care professionals introduced in the second edition. The photos and accompanying brief descriptions of the role of chemistry in each profession continue to emphasize positive contributions of chemistry in our lives.

This text is designed to be used in either a two-semester or three-quarter course of study that provides an introduction to general chemistry, organic chemistry, and biochemistry. Most students who take such courses are majoring in nursing, other health professions, or the life sciences, and consider biochemistry to be the most relevant part of the course of study. However, an understanding of biochemistry depends upon a sound background in organic chemistry, which in turn depends upon a good foundation in general chemistry. We have attempted to present the general and organic chemistry in sufficient depth and breadth to make the biochemistry understandable.

As with previous editions, this textbook is published in a complete hardcover form and a two-volume paperback edition. One volume of the paperback edition contains all the general chemistry and the first two chapters of organic chemistry from the hardcover text. The second volume of the paperback edition contains all the organic and biochemistry of the hardcover edition. The availability of the textbook in these various forms has been a very popular feature among those who use the text because of the flexibility it affords them.

The decisions about what to include and what to omit from the text were based on our combined 65-plus years of teaching, input from numerous reviewers and adopters, and our philosophy that a textbook functions as a personal tutor to each student. In the role of a personal tutor, a text must be more than just a collection of facts, data, and exercises. It should also help students relate to the material they are studying, carefully guide them through more difficult material, provide them with interesting and relevant examples of chemistry in their lives, and become a reference and a resource that they can use in other courses or their professions.

New to This Edition

In this sixth edition of the text, we have retained features that received a positive reception from our own students, the students of other adopters, other teachers, and reviewers. The retained features are 24 *Study Skills* boxes that include 5 reaction maps; 4 *How Reactions Occur* boxes; 44 *Chemistry Around Us* boxes, including 18 new to this edition; 24 *Over the Counter* boxes with 1 new to this edition; and 22 *Chemistry and Your Health* boxes with 8 new to this edition. A new feature of this sixth edition is the *Allied Health Exam Connection* that follows the exercises of each chapter. This feature consists of examples of chemistry questions found on typical entrance examinations used to screen applicants to allied health professional programs. We have also added a section of additional exercises in each problem set, not tied to a specific chapter section, that consists of problems that often integrate several concepts. The answers to half of these questions are included in Appendix B. In addition, approximately 10% of the end-of-chapter exercises have been changed.

ALLIED HEALTH EXAM CONNECTION

Reprinted with permission from Nursing School and Allied Health Entrance Exams, COPYRIGHT 2005 Petersons.

18.64 Fats belong to the class of organic compounds represented by the general formula, RCOOR′, where R and R′ represent hydrocarbon groups. What is the name of the functional group present in fats? What functional group is common to all saponifiable lipids?

18.65 Identify each of the following characteristics as describing an unsaturated fatty acid or a saturated fatty acid:

a. Contains more hydrogen atoms

b. Is more healthy

c. More plentiful in plant sources

d. Is usually a solid at room temperature

A concept summary section is located at the end of each chapter. One or two appropriate end-of-chapter exercises are given after each summary in the section. The ability to solve these exercises will provide an approximate but quick assessment of how well the *Learning Objective* related to that concept has been understood.

Concept Summary

ThomsonNOW™ Sign in at **www.thomsonedu.com** to:

- Assess your understanding with Exercises keyed to each learning objective.
- Check your readiness for an exam by taking the **Pre-test** and exploring the modules recommended in your **Personalized Learning Plan.**

Classification of Lipids. Lipids are a family of naturally occurring compounds grouped together on the basis of their relative insolubility in water and solubility in nonpolar solvents. Lipids are energy-rich compounds that are used as waxy coatings,

Exercise 18.24. Waxes are insoluble in water and serve as protective coatings in nature.

Phosphoglycerides. Phosphoglycerides consist of glycerol esterified to two fatty acids and phosphoric acid. The phosphoric acid is further esterified to choline (in the lecithins) and to ethanolamine or serine (in the cephalins). The phosphoglycerides are particularly important in membrane formation. **OBJECTIVE 7 (Section 18.6), Exercises 18.28 and 18.30**

Sphingolipids. These complex lipids contain a backbone of sphingosine rather than glycerol and only one fatty acid component. They are abundant in brain and nerve tissue. **OBJECTIVE 8 (Section 18.7), Exercise 18.34**

Features

Each chapter has features especially designed to help students organize, study effectively, understand, and enjoy the material in the course.

Chapter Opening Photos. Each chapter opens with a photo of one of the many health care professionals that provide us with needed services. These professionals represent some of the numerous professions that require an understanding of chemistry.

CHAPTER 10

Radioactivity and Nuclear Processes

LEARNING OBJECTIVES

When you have completed your study of this chapter, you should be able to:

1. Describe and characterize the common forms of radiation emitted during radioactive decay and other nuclear processes. (Section 10.1)
2. Write balanced equations for nuclear reactions. (Section 10.2)
3. Solve problems using the half-life concept. (Section 10.3)
4. Describe the effects of radiation on health. (Section 10.4)
5. Describe and compare the units used to measure quantities of radiation. (Section 10.5)
6. Describe, with examples, medical uses of radioisotopes. (Section 10.6)
7. Describe, with examples, nonmedical uses of radioisotopes. (Section 10.7)
8. Show that you understand the concept of induced nuclear reactions. (Section 10.8)
9. Describe the differences between nuclear fission and nuclear fusion reactions. (Section 10.9)

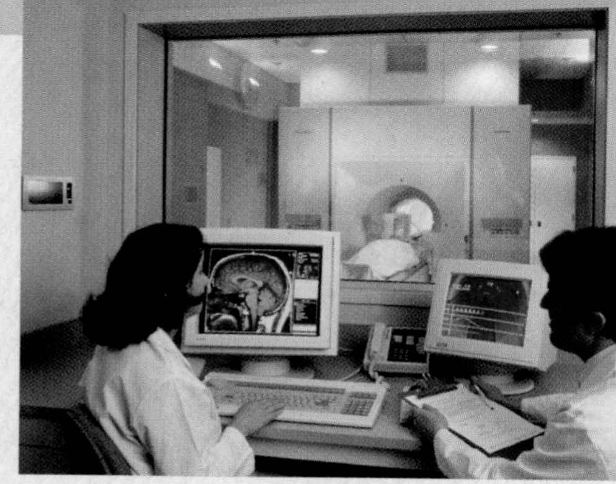

Scientists are continually developing new technology for use in the health sciences; the goal is to provide new information for improved diagnosis and better treatment of patients. Here, a medical imaging technologist uses magnetic resonance imaging (MRI) equipment to visualize soft tissues of the body. MRI is one of the special topics included in this chapter.

Chapter Outlines and Learning Objectives. At the beginning of each chapter, a list of learning objectives provides students with a convenient overview of what they should gain by studying the chapter. In order to help students navigate through each chapter and focus on key concepts, these objectives are repeated at the beginning of the section in which the applicable information is discussed. The objectives are referred to again in the concept summary of each chapter along with one or two suggested end-of-chapter exercises. By working the suggested exercises, students get a quick indication of how well they have met the stated learning objectives. Thus, students begin each chapter with a set of objectives and end with an indication of how well they satisfied the objectives.

Key Terms. Identified within the text by the use of bold type, key terms are defined in the margin near the place where they are introduced. Students reviewing a chapter can quickly identify the important concepts on each page with this marginal glossary. A full glossary of key terms and concepts appears at the end of the text.

Over the Counter. These boxed features contain useful information about health-related products that are readily available to consumers without a prescription. The information in each box provides a connection between the chemical behavior of the product and its effect on the body.

OVER THE COUNTER 2.1 Calcium Supplements

Some meanings of the word *supplement* are "to add to," "to fill up," and "to complete." When used in a nutritional context, a supplement provides an amount of a substance that is in addition to the amount obtained from the diet. An important question, then, is: Who, if anyone, should take dietary supplements? The obvious answer is that anyone who does not get enough of a particular nutrient from the diet to satisfy the needs of the body should take a supplement. How do we apply this obvious answer to the question of whether or not to take a calcium supplement?

Calcium in various forms performs numerous functions in the body. However, about 99% is used to build bones and teeth. During the body's lifetime, all bones undergo a continuous natural process of buildup and breakdown. The rate of buildup exceeds the rate of breakdown for the first 25–30 years in women and the first 30–35 years in men. After these ages, the rate of breakdown catches and exceeds the rate of buildup, resulting in a gradual decrease in bone density. As among people in these age groups are the tendency to skip meals and the substitution of soft drinks and other nondairy drinks in place of milk.

If a calcium supplement is needed, a number of factors should be considered. Vitamin D is essential for optimal calcium absorption by the body. For this reason, many calcium supplements include vitamin D in their formulation, and clearly indicate this on their labels. Calcium supplements are most efficiently absorbed when taken in individual doses of 500 mg or less. The dose per tablet or capsule is generally indicated on the label. The dosages found in the calcium supplements carried by a typical pharmacy range from a low of 333 mg to a high of 630 mg. The calcium comes in various compound forms, including calcium carbonate (often from oyster shells), calcium citrate, and calcium phosphate. Different forms can be advantageous for different body conditions. For example, a person with high gastric acid production should take calcium carbonate with food to improve absorption.

Chemistry Around Us. These boxed features present everyday applications of chemistry that emphasize in a real way the important role of chemistry in our lives. Forty percent of these are new to this edition and emphasize health-related applications of chemistry.

Chemistry and Your Health. These boxed features contain current chemistry-related health issues and questions such as safety questions surrounding genetically modified foods and the relationship between C-reactive protein and heart disease.

Examples. To reinforce students in their problem-solving skill development, carefully worked out solutions in numerous examples are included in each chapter.

Learning Checks. Short self-check exercises follow examples and discussions of key or difficult concepts. A complete set of solutions is included in Appendix C. These allow students to measure immediately their understanding and progress.

Study Skills. Most chapters contain a *Study Skills* feature in which a challenging topic, skill, or concept of the chapter is addressed. Study suggestions, analogies, and approaches are provided to help students master these ideas.

How Reactions Occur. The mechanisms of representative organic reactions are presented in four boxed inserts to help students dispel the mystery of how these reactions take place.

Concept Summary. Located at the end of each chapter, this feature provides a concise review of the concepts and includes suggested exercises to check achievement of the learning objectives related to the concepts.

STUDY SKILLS 14.1 A Reaction Map for Aldehydes and Ketones

This reaction map is designed to help you master organic reactions. Whenever you are trying to complete an organic reaction, use these two basic steps: (1) Identify the functional group that is to react, and (2) identify the reagent that is to react with the functional group. If the reacting functional group is an aldehyde or a ketone, find the reagent in the summary diagram, and use the diagram to predict the correct products.

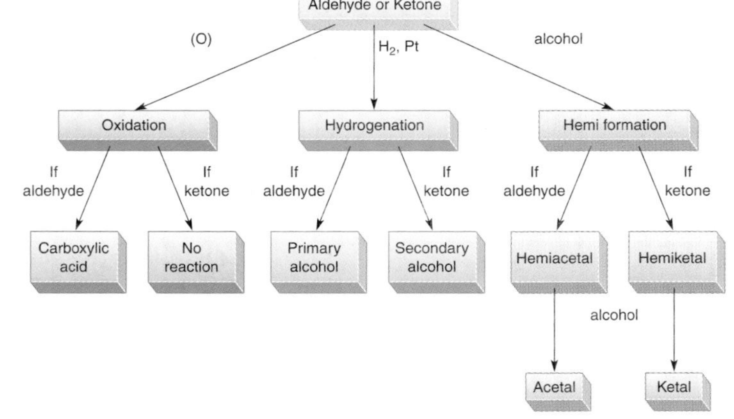

Key Terms and Concepts. These are listed at the end of the chapter for easy review, with a reference to the chapter section in which they are presented.

Key Equations. This feature provides a useful summary of general equations and reactions from the chapter. This feature is particularly helpful to students in the organic chemistry chapters.

Exercises. Nearly 1,700 end-of-chapter exercises are arranged by section. Approximately half of the exercises are answered in the back of the text. Completely worked out solutions to these answered exercises are included in the Student Study Guide. Solutions and answers to the remaining exercises are provided in the Instructor's Manual. We have included a significant number of clinical and other familiar applications of chemistry in the exercises.

Allied Health Exam Connection. These examples of chemistry questions from typical entrance exams used to screen applicants to allied health professional programs help students focus their attention on the type of chemical concepts considered important in such programs.

Chemistry for Thought. Included at the end of each chapter are special questions designed to encourage students to expand their reasoning skills. Some of these exercises are based on photographs found in the chapter, while others emphasize clinical or other useful applications of chemistry.

Possible Course Outlines

This text may be used effectively in either a two-semester or three-quarter course of study:

First semester: Chapters 1–13 (general chemistry and three chapters of organic chemistry)
Second semester: Chapters 14–25 (organic chemistry and biochemistry)

First semester: Chapters 1–10 (general chemistry)
Second semester: Chapters 11–21 (organic chemistry and some biochemistry)

First quarter: Chapters 1–10 (general chemistry)
Second quarter: Chapters 11–18 (organic chemistry)
Third quarter: Chapters 19–25 (biochemistry)

Ancillaries

The following ancillaries are available to qualified adopters. Please consult your local Brooks/Cole • Thomson sales representative for details.

Print Resources

Safety-Scale Laboratory Experiments for Chemistry for Today: General, Organic, and Biochemistry, 6th Edition. Prepared by Spencer L. Seager and Michael R. Slabaugh, this well-tested collection of experiments has been developed during more than 35 years of laboratory instruction with students at Weber State University. This manual provides a blend of training in laboratory skills and experiences illustrating concepts from the authors' textbook. The experiments are designed to use small quantities of chemicals, and emphasize safety and proper disposal of used materials. ISBN 0-495-11269-0

Instructor's Guide for Safety-Scale Laboratory Experiments. Prepared by the authors of the laboratory manual, this useful resource gives complete directions for preparing the reagents and other materials used in each experiment. It also contains useful comments concerning the experiments, answers to questions included in the experiments, and suggestions for the proper disposal of used materials. This product is available online at **www.thomsonedu.com/chemistry/seagerlab6**.

Study Guide and Solutions Manual. Prepared by Jennifer P. Harris of Portland Community College, each chapter contains a chapter outline, learning objectives, a programmed review of important topics and concepts, detailed solutions to the even-numbered exercises answered in the text, and self-test questions. ISBN 0-495-11271-2

Transparency Acetates. The publisher provides a correlation guide to 200 full-color transparencies illustrating key figures from the text for use in class. ISBN 0-495-11273-9

Media Resources

ThomsonNOW. Developed in concert with the text, ThomsonNOW is a Web-based, assessment-centered learning tool designed to assess student knowledge. In each chapter, icons with captions alert students to tutorials, animations, and coached problems, which enhance problem-solving skills and improve conceptual understanding. In ThomsonNOW, students are provided with a Personalized Learning Plan—based on a diagnostic Pre-Test—that targets their study needs. An access code is required for ThomsonNOW and may be packaged with a new copy of the text or purchased separately. Visit **www.thomsonedu.com/login** to register for ThomsonNOW.

ThomsonNOW™

OWL (Online Web-based Learning) for the GOB Course. Authored by Roberta Day, Beatrice Botch, and David Gross of the University of Massachusetts, Amherst; William Vining of the State University of New York at Oneonta; and Susan Young of Hartwick College. Class-tested by tens of thousands of students and used by more than 300 institutions, **OWL** is a reliable and customizable cross-platform online homework system and assessment tool that gives students instant analysis and feedback on homework problems such as tutors, simulations, and chemically and/or numerically parameterized short-answer questions. OWL is the only system specifically designed to support mastery learning, where students work as long as necessary to master each chemical concept and skill. A fee-based access code is required for **OWL.** OWL is only available for use within North America. Visit **http://owl.thomsonlearning.com** for a demo.

***Chemistry for Today's* e-Book in OWL** includes the complete textbook as an assignable resource that is fully linked to OWL content. This new e-Book in OWL is an exclusive option that will be available to all students if the instructor chooses it. It can be packaged with the text and/or ordered as a text replacement. Please consult your Thomson Brooks/Cole representative for pricing details.

WebCT/Blackboard. ThomsonNOW can be integrated with WebCT and Blackboard so that professors who are using one of those platforms can access all for the NOW assessments and content without an extra login. Please contact your Thomson Brooks/Cole representative for more information.

PowerLecture is a digital library and presentation tool that is available on one convenient multi-platform CD-ROM. With its easy-to-use interface, a professor can take advantage of Brooks/Cole's text-specific presentations, which consist of text art, photos, tables, and more, in a variety of e-formats that are easily exported into presentation software or used on Web-based course support materials. It allows professors to customize their own presentations by importing personal lecture slides or other material of their choosing. The result is an interactive and fluid lecture that truly engages your students. PowerLecture includes:

- **JoinIn™ on TurningPoint®.** Thomson Brooks/Cole is pleased to offer book-specific JoinIn content for Response Systems, allowing you to transform your classroom and assess your students' progress with instant in-class quizzes and polls. You can pose book-specific questions and display students' answers seamlessly within the Microsoft® PowerPoint® slides of your own lecture, and in conjunction with the "clicker" hardware of your choice. Enhance how your students interact with you, your lecture, and each other. Consult your Brooks/Cole representative for further details.
- **ExamView** allows professors to create, deliver, and customize tests and study guides (both print and online) in minutes with this easy-to-use assessment and tutorial system. ExamView offers both a Quick Test Wizard and an Online Test Wizard that guide professors step-by-step through the process of creating tests, and the unique "WYSIWYG" capability allows professors to see the test being created on the screen exactly as it will print or display online. Tests of up to 250 questions can be built using up to 12 question types. Using ExamView's complete word processing capabilities, an unlimited number of new questions can be entered or existing questions can be edited.
- **Instructor's Manual and Testbank.** Prepared by James K. Hardy of the University of Akron, each chapter contains a summary chapter outline, learning objectives, instructor resource materials, solutions to *Chemistry for Thought* questions, answers and solutions to odd-numbered exercises not answered in the text, and more than 1,300 exam questions.

Acknowledgments

We express our sincere appreciation to the following reviewers, who read and commented on the fifth edition and offered helpful advice and suggestions for improving this edition:

Jonathan T. Brockman
College of DuPage

Kathleen Brunke
Christopher Newport University

David C. Hawkinson
University of South Dakota

Margaret G. Kimble
Indiana University–Purdue University Fort Wayne

James F. Kirby
Quinnipiac University

Regan Luken
University of South Dakota

James McConaghy
Wayne College

Melvin Merken
Worcester State College

Jean Yockey
University of South Dakota

We also express appreciation to the following reviewers, who helped us revise the first five editions:

Hugh Akers
Lamar University–Beaumont

Johanne I. Artman
Del Mar College

Gabriele Backes
Portland Community College

Bruce Banks
University of North Carolina–Greensboro

David Boykin
Georgia State University

Deb Breiter
Rockford College

Lorraine C. Brewer
University of Arkansas

Martin Brock
Eastern Kentucky University

Christine Brzezowski
University of Utah

Sybil K. Burgess
University of North Carolina–Wilmington

Sharmaine S. Cady
East Stroudsburg University

Linda J. Chandler
Salt Lake Community College

Sharon Cruse
Northern Louisiana University

Thomas D. Crute
Augusta College

Jack L. Dalton
Boise State University

Lorraine Deck
University of New Mexico

Kathleen A. Donnelly
Russell Sage College

Jan Fausset
Front Range Community College

Patricia Fish
The College of St. Catherine

Harold Fisher
University of Rhode Island

John W. Francis
Columbus State Community

Wes Fritz
College of DuPage

Jean Gade
Northern Kentucky University

Galen George
Santa Rosa Junior College

Linda Thomas-Glover
Guilford Technical Community College

Jane D. Grant
Florida Community College

James K. Hardy
University of Akron

Leland Harris
University of Arizona

Robert H. Harris
University of Nebraska–Lincoln

Jack Hefley
Blinn College

Claudia Hein
Diablo Valley College

John Henderson
Jackson Community College

Mary Herrmann
University of Cincinnati

Laura Kibler-Herzog
Georgia State University

Arthur R. Hubscher
Brigham Young University–Idaho

Kenneth Hughes
University of Wisconsin–Oshkosh

Jeffrey A. Hurlbut
Metropolitan State College of Denver

Jim Johnson
Sinclair Community College

Richard. F. Jones
Sinclair Community College

Frederick Jury
Collin County Community College

Lidija Kampa
Kean College of New Jersey

James F. Kirby
Quinnipiac College

Peter J. Krieger
Palm Beach Community College

Terrie L. Lampe
De Kalb College–Central Campus

Carol Larocque
Cambrian College

Richard Lavallee
Santa Monica College

Leslie J. Lovett
Fairmont State College

Armin Mayr
El Paso Community College

Evan McHugh
Pikes Peak Community College

Trudy McKee
Thomas Jefferson University

Melvin Merken
Worcester State College

W. Robert Midden
Bowling Green State University

Pamela S. Mork
Concordia College

Phillip E. Morris, Jr.
University of Alabama–Birmingham

Robert N. Nelson
Georgia Southern University

Elva Mae Nicholson
Eastern Michigan University

H. Clyde Odom
Charleston Southern University

Howard K. Ono
California State University–Fresno

James A. Petrich
San Antonio College

Thomas G. Richmond
University of Utah

James Schreck
University of Northern Colorado

William M. Scovall
Bowling Green State University

William Scovell
Bowling Green State University

Jean M. Shankweiler
El Camino Community College

Francis X. Smith
King's College

J. Donald Smith
University of Massachusetts–Dartmouth

Malcolm P. Stevens
University of Hartford

Eric R. Taylor
University of Southwestern Louisiana

James A. Thomson
University of Waterloo

Mary Lee Trawick
Baylor University

Katherin Vafeades
University of Texas–San Antonio

Cary Willard
Grossmont College

Don Williams
Hope College

Les Wynston
California State University–Long Beach

We also give special thanks to Rebecca Heider, development editor for Brooks/Cole who guided and encouraged us in the preparation of this sixth edition. Lisa Lockwood, chemistry editor, Belinda Krohmer, production project manager, and Lisa Weber, technology project manager, were also essential to the team and contributed greatly to the success of the project. We are very grateful for the superb work of Lachina Publishing Services, especially to Sheila McGill, for outstanding work in coordinating the production. We appreciate the significant help of three associates: Mary Ann Francis, Wayne April, and Brooke Robbins who did excellent work in researching special topics, typing, working exercises, and proofreading.

Finally, we extend our love and heartfelt thanks to our families for their patience, support, encouragement, and understanding during a project that occupied much of our time and energy.

Spencer L. Seager

Michael R. Slabaugh

CHAPTER 1

Organic Compounds: Alkanes

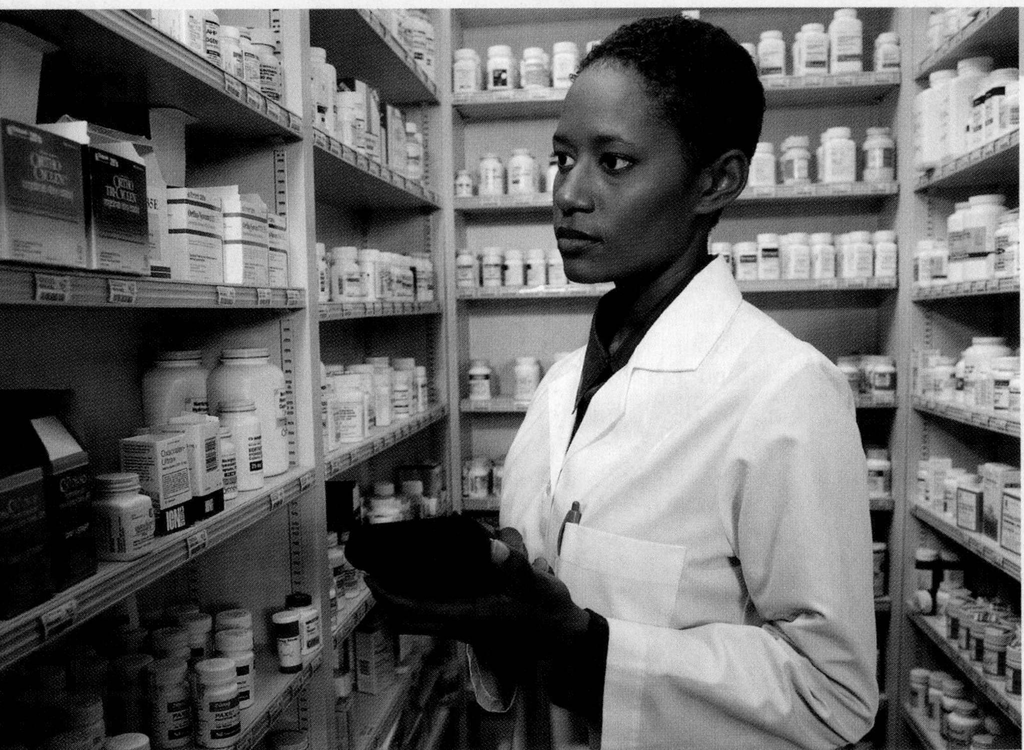

Pharmacists are responsible for the appropriate dispensing of drugs and other medications. Pharmacists must have a broad knowledge of the many drugs and medications available, as well as the effects they have on the human body when administered individually and in combinations. The great majority of drugs are organic compounds. This and the next several chapters provide the basis for understanding the structure, properties, and physiological effects of organic compounds.

LEARNING OBJECTIVES

When you have completed your study of this chapter, you should be able to:

1. Show that you understand the general importance of organic chemical compounds. (Section 1.1)
2. Be able to recognize the molecular formulas of organic and inorganic compounds. (Section 1.1)
3. Explain some general differences between inorganic and organic compounds. (Section 1.2)
4. Be able to use structural formulas to identify compounds that are isomers of each other. (Section 1.3)
5. Write condensed or expanded structural formulas for compounds. (Section 1.4)
6. Classify alkanes as normal or branched. (Section 1.5)
7. Use structural formulas to determine whether compounds are structural isomers. (Section 1.6)
8. Assign IUPAC names and draw structural formulas for alkanes. (Section 1.7)
9. Assign IUPAC names and draw structural formulas for cycloalkanes. (Section 1.8)
10. Name and draw structural formulas for geometric isomers of cycloalkanes. (Section 1.9)
11. Describe the key physical properties of alkanes. (Section 1.10)
12. Write alkane combustion reactions. (Section 1.11)

ThomsonNOW™ Throughout the chapter this icon introduces resources on the ThomsonNOW website for this text. Sign in at **www.thomsonedu.com** to:

- Evaluate your knowledge of the material
- Take an exam prep quiz
- Identify areas you need to study with a **Personalized Learning Plan**.

The word *organic* is used in several different contexts. Scientists of the 18th and 19th centuries studied compounds extracted from plants and animals and labeled them "organic" because they had been obtained from organized (living) systems. Organic fertilizer is organic in the original sense that it comes from a living organism. There is no universal definition of organic foods, but the term is generally taken to mean foods grown without the application of pesticides or synthetic fertilizers. When referring to organic chemistry, however, we mean the chemistry of carbon-containing compounds.

1.1 Carbon: The Element of Organic Compounds

LEARNING OBJECTIVES

1. Show that you understand the general importance of organic chemical compounds.
2. Be able to recognize the molecular formulas of organic and inorganic compounds.

Early chemists thought organic compounds could be produced only through the action of a "vital force," a special force active only in living organisms. This idea was central to the study of organic chemistry until 1828, because up to that time, no one had been able to synthesize an organic compound from its elements or from naturally occurring minerals. In that year, Friedrich Wöhler, a German chemist, heated an inorganic salt called ammonium cyanate and produced urea. This compound, normally found in blood and urine, was unquestionably organic, and it had come from an inorganic source. The reaction is

$$\underset{\text{ammonium cyanate}}{NH_4NCO} \xrightarrow{\text{Heat}} \underset{\text{urea}}{H-\overset{\overset{\textstyle H}{|}}{N}-\overset{\overset{\textstyle O}{\|}}{C}-\overset{\overset{\textstyle H}{|}}{N}-H} \tag{1.1}$$

Wöhler's urea synthesis discredited the "vital force" theory, and his success prompted other chemists to attempt to synthesize organic compounds. Today, organic compounds are being synthesized in thousands of laboratories, and most of the synthetics have never been isolated from natural sources.

organic compound A compound that contains the element carbon.

organic chemistry The study of carbon-containing compounds.

inorganic chemistry The study of the elements and all noncarbon compounds.

Organic compounds share one unique feature: They all contain carbon. Therefore, **organic chemistry** is defined as the study of carbon-containing compounds. There are a few exceptions to this definition; a small number of carbon compounds—such as CO, CO_2, carbonates, and cyanides—were studied before Wöhler's urea synthesis. These were classified as inorganic because they were obtained from nonliving systems, and even though they contain carbon, we still consider them to be a part of **inorganic chemistry.**

The importance of carbon compounds to life on Earth cannot be overemphasized. If all carbon compounds were removed from Earth, its surface would be somewhat like the barren surface of the moon (see ■ Figure 1.1). There would be

■ **FIGURE 1.1** Organic chemistry makes a tremendous difference between Earth and the moon.

no animals, plants, or any other form of life. If carbon-containing compounds were removed from the human body, all that would remain would be water, a very brittle skeleton, and a small residue of minerals. Many of the essential constituents of living matter—such as carbohydrates, fats, proteins, nucleic acids, enzymes, and hormones—are organic chemicals.

The essential needs of daily human life are food, fuel, shelter, and clothing. The principal components of food (with the exception of water) are organic. The fuels we use (e.g., wood, coal, and natural gas) are mixtures of organic compounds. Our homes typically involve wood construction, and our clothing, whether made of natural or synthetic fibers, is organic.

Besides the major essentials, many of the smaller everyday things often taken for granted are also derived from carbon and its compounds. Consider an ordinary pencil. The "lead" (actually graphite), the wood, the rubber eraser, and the paint on the surface are all either carbon or carbon compounds. The paper in this book, the ink on its pages, and the glue holding it all together are also made of carbon compounds.

1.2 Organic and Inorganic Compounds Compared

LEARNING OBJECTIVE

3. Explain some general differences between inorganic and organic compounds.

It is interesting that the subdivision of chemistry into its organic and inorganic branches results in one branch that deals with compounds composed mainly of one element and another branch that deals with compounds formed by the more than 100 remaining elements. However, this classification seems more reasonable when we recognize that known organic compounds are much more numerous than inorganic compounds. An estimated 250,000 inorganic compounds have been identified, but more than 6 million organic compounds are known, and thousands of new ones are synthesized or isolated each year.

One of the reasons for the large number of organic compounds is the unique ability of carbon atoms to form stable covalent bonds with other carbon atoms and with atoms of other elements. The resulting covalently bonded molecules may contain as few as one or more than a million carbon atoms.

In contrast, inorganic compounds are often characterized by the presence of ionic bonding. Covalent bonding also may be present, but it is less common. These differences generally cause organic and inorganic compounds to differ physically (see ■ Figure 1.2) and chemically, as shown in ■ Table 1.1.

TABLE 1.1 Properties of typical organic and inorganic compounds

Property	Organic compounds	Inorganic compounds
Bonding within molecules	Usually covalent	Often ionic
Forces between molecules	Generally weak	Quite strong
Normal physical state	Gases, liquids, or low-melting-point solids	Usually high-melting-point solids
Flammability	Often flammable	Usually nonflammable
Solubility in water	Often low	Often high
Conductivity of water solutions	Nonconductor	Conductor
Rate of chemical reactions	Usually slow	Usually fast

FIGURE 1.2 Many organic compounds, such as ski wax, have relatively low melting points. What does this fact reveal about the forces between organic molecules?

Learning Check 1.1 Classify each of the following compounds as organic or inorganic:

a. $NaCl$
b. CH_4
c. C_6H_6
d. $NaOH$
e. CH_3OH
f. $Mg(NO_3)_2$

Learning Check 1.2 Decide whether each of the following characteristics most likely describes an organic or inorganic compound:

a. Flammable b. Low boiling point c. Soluble in water

STUDY SKILLS 1.1 Changing Gears for Organic Chemistry

You will find that organic chemistry is very different from general or inorganic chemistry. By quickly picking up on the changes, you will help yourself prepare for quizzes and exams.

There is almost no math in these next six chapters or in the biochemistry section. Very few mathematical formulas need to be memorized. The problems you will encounter fall mainly into four categories: naming compounds and drawing structures, describing physical properties of substances, writing reactions, and identifying typical uses of compounds. This pattern holds true for all six of the organic chemistry chapters.

The naming of compounds is introduced in this chapter, and the rules developed here will serve as a starting point in the next five chapters. Therefore, it is important to master naming in this chapter. A well-developed skill in naming will help you do well on exams covering the coming chapters.

Only a few reactions are introduced in this chapter, but many more will be in future chapters. Writing organic reactions is just as important (and challenging) as naming, and Study Skills 2.1 will help you. Identifying the uses of compounds can best be handled by making a list as you read the chapter or by highlighting compounds and their uses so that they are easy to review. All four categories of problems are covered by numerous end-of-chapter exercises to give you practice.

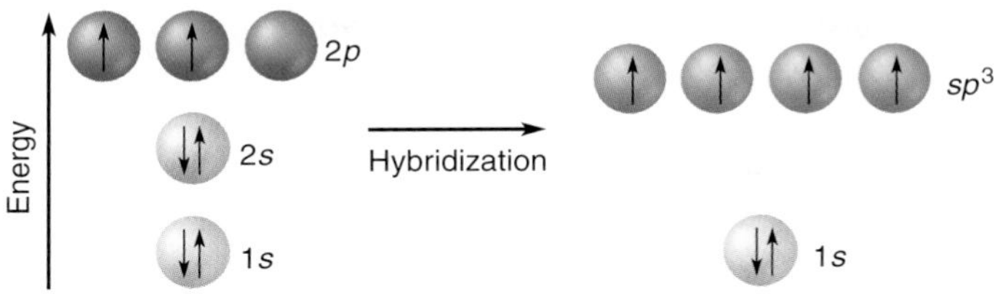

■ **FIGURE 1.3** The mixing of 2*s* and three 2*p* orbitals of a carbon atom to generate four sp^3 carbon orbitals, each with energy intermediate between 2*s* and 2*p* orbitals.

1.3 Bonding Characteristics and Isomerism

LEARNING OBJECTIVE

4. Be able to use structural formulas to identify compounds that are isomers of each other.

There are two major reasons for the astonishing number of organic compounds: the bonding characteristics of carbon atoms, and the isomerism of carbon-containing molecules. As a group IVA(14) element, a carbon atom has four valence electrons. Two of these outermost-shell electrons are in an *s* orbital, and two are in *p* orbitals (see Section 3.4):

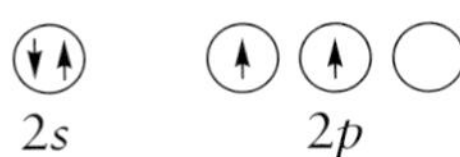

With only two unpaired electrons, we might predict that carbon would form just two covalent bonds with other atoms. Yet, we know from the formula of methane (CH_4) that carbon forms four bonds.

Linus Pauling (1901–1994), winner of the Nobel Prize in chemistry (1954) and Nobel Peace Prize (1963), developed a useful model to explain the bonding characteristics of carbon. Pauling found that a mathematical mixing of the 2*s* and three 2*p* orbitals could produce four new, equivalent orbitals (see ■ Figure 1.3). Each of these **hybrid orbitals** has the same energy and is designated sp^3. An sp^3 orbital has a two-lobed shape, similar to the shape of a *p* orbital but with different-sized lobes (see ■ Figure 1.4). Each of the four sp^3 hybrid orbitals contains a single unpaired electron available for covalent bond formation. Thus, carbon forms four bonds.

hybrid orbital An orbital produced from the combination of two or more nonequivalent orbitals of an atom.

Each carbon–hydrogen bond in methane arises from an overlap of a C (sp^3) and an H (1*s*) orbital. The sharing of two electrons in this overlap region creates a sigma (σ) bond. The four equivalent sp^3 orbitals point toward the corners of a regular tetrahedron (see ■ Figure 1.5).

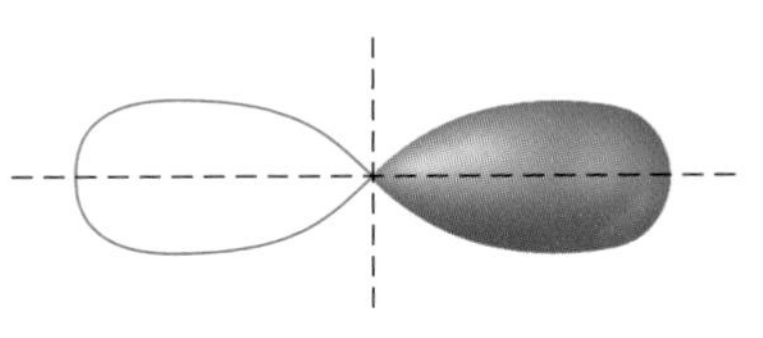

■ **FIGURE 1.4** A comparison of unhybridized *p* and sp^3 hybridized orbital shapes. The atomic nucleus is at the junction of the lobes in each case.

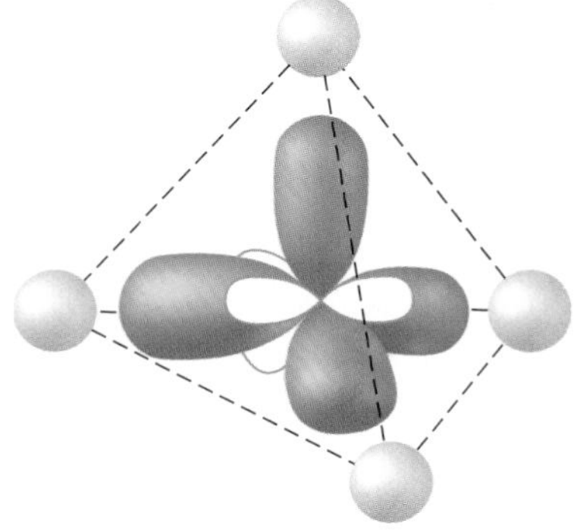

■ **FIGURE 1.5** Directional characteristics of sp^3 hybrid orbitals of carbon and the formation of C—H bonds in methane (CH_4). The hybrid orbitals point toward the corners of a regular tetrahedron. Hydrogen 1s orbitals are illustrated in position to form bonds by overlap with the major lobes of the hybrid orbitals.

Carbon atoms also have the ability to bond covalently to other carbon atoms to form chains and networks. This means that two carbon atoms can join by sharing two electrons to form a single covalent bond:

$$\cdot\dot{\underset{\cdot}{C}}\cdot + \cdot\dot{\underset{\cdot}{C}}\cdot \longrightarrow \cdot\dot{\underset{\cdot}{C}}{:}\dot{\underset{\cdot}{C}}\cdot \qquad \text{or} \qquad \cdot\dot{\underset{\cdot}{C}}-\dot{\underset{\cdot}{C}}\cdot \tag{1.2}$$

A third carbon atom can join the end of this chain:

$$\cdot\dot{\underset{\cdot}{C}}-\dot{\underset{\cdot}{C}}\cdot + \cdot\dot{\underset{\cdot}{C}}\cdot \longrightarrow \cdot\dot{\underset{\cdot}{C}}-\dot{\underset{\cdot}{C}}-\dot{\underset{\cdot}{C}}\cdot \tag{1.3}$$

This process can continue and form carbon chains of almost any length, such as

$$\cdot\dot{\underset{\cdot}{C}}-\dot{\underset{\cdot}{C}}-\dot{\underset{\cdot}{C}}-\dot{\underset{\cdot}{C}}-\dot{\underset{\cdot}{C}}-\dot{\underset{\cdot}{C}}-\dot{\underset{\cdot}{C}}-\dot{\underset{\cdot}{C}}-\dot{\underset{\cdot}{C}}\cdot$$

The electrons not involved in forming the chain can be shared with electrons of other carbon atoms (to form chain branches) or with electrons of other elements such as hydrogen, oxygen, or nitrogen. Carbon atoms may also share more than one pair of electrons to form multiple bonds:

$$\cdot\dot{\underset{\cdot}{C}}-\dot{C}=\dot{C}-\dot{\underset{\cdot}{C}}\cdot \qquad\qquad \cdot\dot{\underset{\cdot}{C}}-C\equiv C-\dot{\underset{\cdot}{C}}\cdot$$

Chain with double bond — Chain with triple bond

In principle, there is no limit to the number of carbon atoms that can bond covalently. Thus, organic molecules range from the simple molecules such as methane (CH_4) to very complicated molecules containing over a million carbon atoms.

The variety of possible carbon atom arrangements is even more important than the size range of the resulting molecules. The carbon atoms in all but the very simplest organic molecules can bond in more than one arrangement, giving rise to different compounds with different structures and properties. This property, called **isomerism,** is characterized by compounds that have identical molecular formulas but different arrangements of atoms. One type of isomerism is characterized by compounds called **structural isomers.** Other types of isomerism are covered in Chapters 2 and 7.

isomerism
A property in which two or more compounds have the same molecular formula but different arrangements of atoms.

structural isomers
Compounds that have the same molecular formula but in which the atoms bond in different patterns.

EXAMPLE 1.1

Use the usual rules for covalent bonding to show that a compound with the molecular formula C_2H_6O demonstrates the property of isomerism. Draw formulas for the isomers, showing all covalent bonds.

Solution

Carbon forms four covalent bonds by sharing its four valence-shell electrons. Similarly, oxygen should form two covalent bonds, and hydrogen a single bond. On the basis of these bonding relationships, two structural isomers are possible:

$$\begin{array}{ccccccccc} & & H & & H & & & & \\ & & | & & | & & & & \\ H & - & C & - & C & - & O & - & H \\ & & | & & | & & & & \\ & & H & & H & & & & \end{array} \qquad\qquad \begin{array}{ccccccccc} & & H & & & & H & & \\ & & | & & & & | & & \\ H & - & C & - & O & - & C & - & H \\ & & | & & & & | & & \\ & & H & & & & H & & \end{array}$$

ethyl alcohol — dimethyl ether

■ **Learning Check 1.3** Which one of the structures below represents a structural isomer of

```
     H  O  H
     |  ‖  |
  H—C—C—C—H    ?
     |     |
     H     H
```

```
     H  H  O—H                  H  H  O
     |  |  |                    |  |  ‖
a. H—C=C—C—H            c. H—C—C—C—O—H
           |                    |  |
           H                    H  H

     H  H  O
     |  |  ‖
b. H—C=C—C—H
```

The two isomers of Example 1.1 are quite different. Ethyl alcohol (grain alcohol) is a liquid at room temperature, whereas dimethyl ether is a gas. Thus, the structural differences exert a significant influence on properties. From this example, we can see that molecular formulas such as C_2H_6O provide much less information about a compound than do structural formulas. ■ Figure 1.6 shows ball-and-stick models of these two molecules.

As the number of carbon atoms in the molecular formula increases, the number of possible isomers increases dramatically. For example, 366,319 different isomers are possible for a molecular formula of $C_{20}H_{42}$. No one has prepared all these isomers or even drawn their structural formulas, but the number helps us understand why so many organic compounds have been either isolated from natural sources or synthesized.

ThomsonNOW™ Go to Chemistry Interactive to explore **conformation and isomers.**

1.4 Functional Groups: The Organization of Organic Chemistry

LEARNING OBJECTIVE

5. Write condensed or expanded structural formulas for compounds.

ThomsonNOW™ Go to Chemistry and Coached Problems to identify **functional groups.**

Because of the enormous number of possible compounds, the study of organic chemistry might appear to be hopelessly difficult. However, the arrangement of organic compounds into a relatively small number of classes can simplify the study a great deal. This organization is done on the basis of characteristic structural features called **functional groups.** For example, compounds with a carbon–carbon double bond

functional group
A unique reactive combination of atoms that differentiates molecules of organic compounds of one class from those of another.

```
 \     /
  C=C
 /     \
```

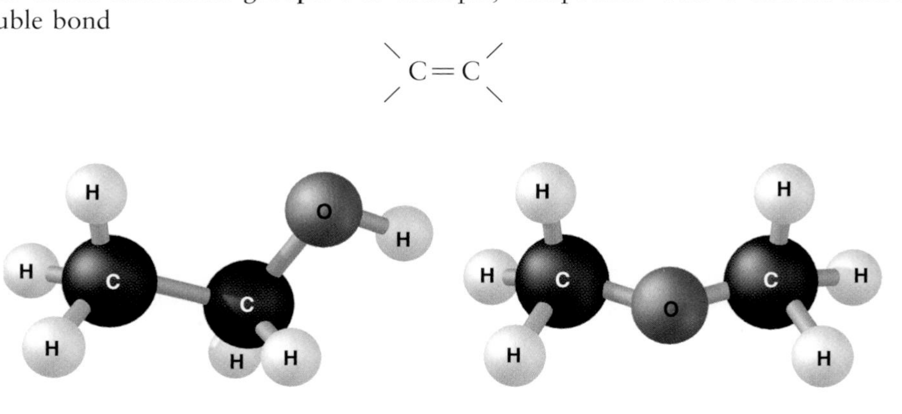

■ **FIGURE 1.6** Ball-and-stick models of the isomers of C_2H_6O. Ethyl alcohol is a liquid at room temperature and completely soluble in water, whereas dimethyl ether is a gas at room temperature and only partially soluble in water.

TABLE 1.2 Classes and functional groups of organic compounds

<table>
<tr><th>Class</th><th>Functional group</th><th>Example of expanded structural formula</th><th>Example of condensed structural formula</th><th>Common name</th></tr>
<tr><td>Alkane</td><td>None</td><td>H H
| |
H—C—C—H
| |
H H</td><td>CH_3CH_3</td><td>ethane</td></tr>
<tr><td>Alkene</td><td>\ /
C=C
/ \</td><td>H H
\ /
C=C
/ \
H H</td><td>$H_2C{=}CH_2$</td><td>ethylene</td></tr>
<tr><td>Alkyne</td><td>—C≡C—</td><td>H—C≡C—H</td><td>$HC\equiv CH$</td><td>acetylene</td></tr>
<tr><td>Aromatic</td><td>\ /
C—C
// \\
—C C—
\ /
C=C
/ \</td><td>H H
\ /
C—C
// \\
H—C C—H
\ /
C=C
/ \
H H</td><td></td><td>benzene</td></tr>
<tr><td>Alcohol</td><td>|
—C—O—H
|</td><td>H H
| |
H—C—C—O—H
| |
H H</td><td>$CH_3CH_2—OH$</td><td>ethyl alcohol</td></tr>
<tr><td>Ether</td><td>| |
—C—O—C—
| |</td><td>H H
| |
H—C—O—C—H
| |
H H</td><td>$CH_3—O—CH_3$</td><td>dimethyl ether</td></tr>
<tr><td>Amine</td><td>H
|
—N—H</td><td>H H
| |
H—C—N—H
|
H</td><td>$CH_3—NH_2$</td><td>methylamine</td></tr>
<tr><td>Aldehyde</td><td>O
‖
—C—H</td><td>H O
| ‖
H—C—C—H
|
H</td><td>O
‖
$CH_3—C—H$</td><td>acetaldehyde</td></tr>
<tr><td>Ketone</td><td>O
| ‖ |
—C—C—C—
| |</td><td>H O H
| ‖ |
H—C—C—C—H
| |
H H</td><td>O
‖
$CH_3—C—CH_3$</td><td>acetone</td></tr>
<tr><td>Carboxylic acid</td><td>O
‖
—C—O—H</td><td>H O
| ‖
H—C—C—O—H
|
H</td><td>O
‖
$CH_3—C—OH$</td><td>acetic acid</td></tr>
<tr><td>Ester</td><td>O
‖ |
—C—O—C—
|</td><td>H O H
| ‖ |
H—C—C—O—C—H
| |
H H</td><td>O
‖
$CH_3—C—O—CH_3$</td><td>methyl acetate</td></tr>
<tr><td>Amide</td><td>O H
‖ |
—C—N—H</td><td>H O H
| ‖ |
H—C—C—N—H
|
H</td><td>O
‖
$CH_3—C—NH_2$</td><td>acetamide</td></tr>
</table>

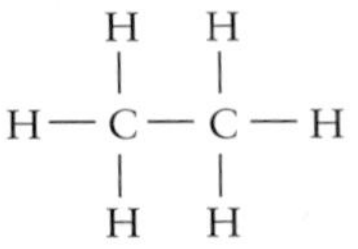

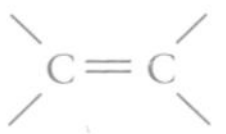

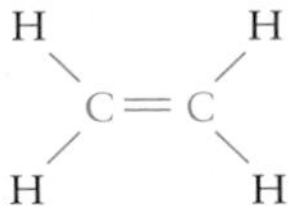

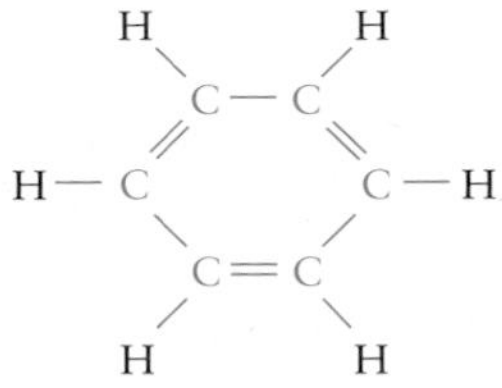

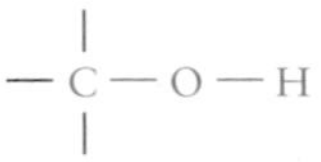

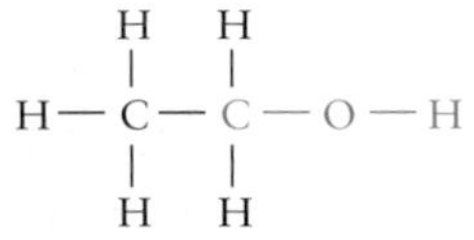

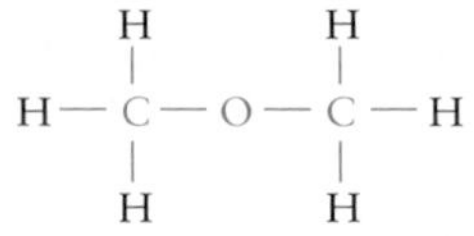

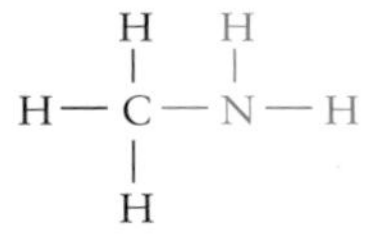

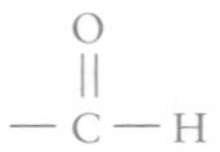

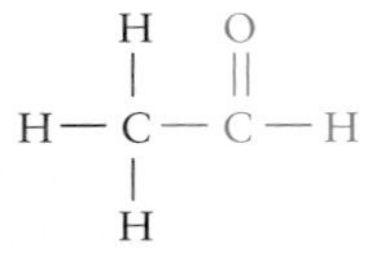

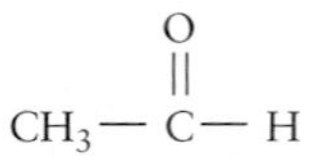

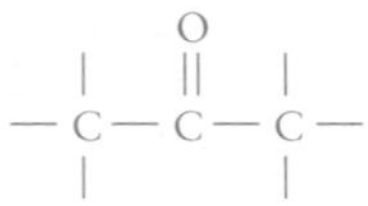

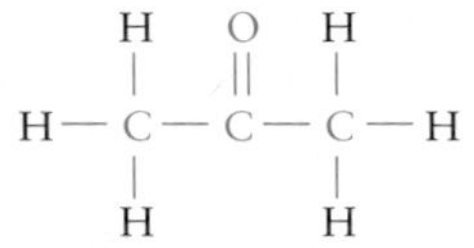

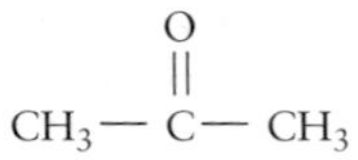

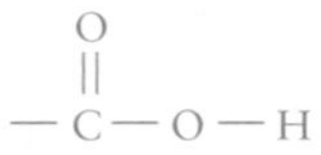

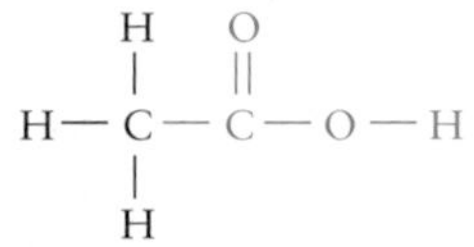

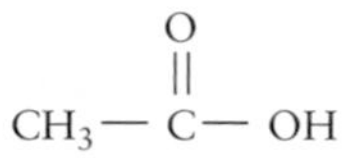

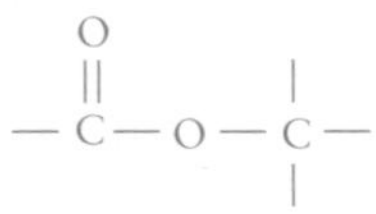

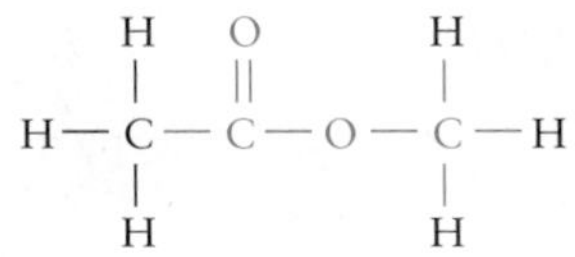

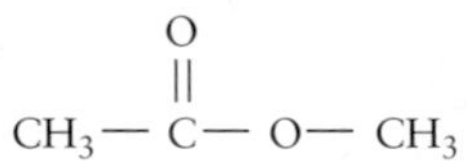

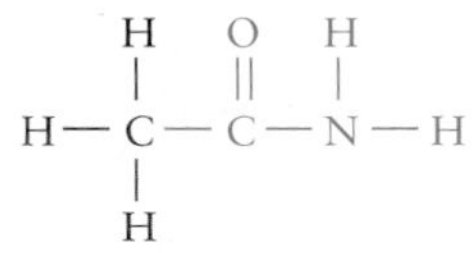

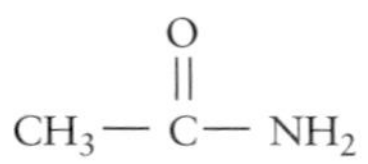

are classified as alkenes. The major classes and functional groups are given in ■ Table 1.2. Notice that each functional group in Table 1.2 (except for alkanes) contains a multiple bond or at least one oxygen or nitrogen atom.

In Table 1.2, we have used both expanded and condensed structural formulas for the compounds. **Expanded structural formulas** show all covalent bonds, whereas **condensed structural formulas** show only specific bonds. You should become familiar with both types, but especially with condensed formulas because they will be used often.

expanded structural formula A structural molecular formula showing all the covalent bonds.

condensed structural formula A structural molecular formula showing the general arrangement of atoms but without showing all the covalent bonds.

EXAMPLE 1.2

Write a condensed structural formula for each of the following compounds:

$$\text{a.}\quad \begin{array}{ccccccccc} & & H & & H & & H & & H & & \\ & & | & & | & & | & & | & & \\ H & - & C & - & C & - & C & - & C & - & H \\ & & | & & | & & | & & | & & \\ & & H & & H & & H & & H & & \end{array} \qquad \text{b.}\quad \begin{array}{ccccccccccc} & & & & H & & & & & & & & \\ & & & & | & & & & & & & & \\ & & H & & H-C-H & & H & & H & & H & & \\ & & | & & | & & | & & | & & | & & \\ H & - & C & — & C & — & C & - & C & - & C & - & H \\ & & | & & | & & | & & | & & | & & \\ & & H & & H & & H & & H & & H & & \end{array}$$

ThomsonNOW™ Go to Chemistry Interactive to explore the **structure of organic compounds.**

Solution

a. Usually the hydrogens belonging to a carbon are grouped to the right. Thus, the group

$$\begin{array}{ccc} & H & \\ & | & \\ H- & C & - \\ & | & \\ & H & \end{array}$$

condenses to CH_3—, and

$$\begin{array}{ccc} & H & \\ & | & \\ - & C & - \\ & | & \\ & H & \end{array}$$

condenses to —CH_2—. Thus, the formula condenses to

$$CH_3—CH_2—CH_2—CH_3$$

Other acceptable condensations are

$$CH_3CH_2CH_2CH_3 \qquad \text{and} \qquad CH_3(CH_2)_2CH_3$$

Parentheses are used here to denote a series of two —CH_2— groups.

b. The group

$$\begin{array}{ccc} & | & \\ - & C & - \\ & | & \\ & H & \end{array}$$

condenses to $\begin{array}{c} | \\ -CH- \end{array}$. The condensed formula is therefore

$$\begin{array}{c} CH_3 \\ | \\ CH_3-CH-CH_2-CH_2-CH_3 \end{array} \qquad \text{or} \qquad \begin{array}{c} CH_3 \\ | \\ CH_3CHCH_2CH_2CH_3 \end{array}$$

or

$$(CH_3)_2CH(CH_2)_2CH_3$$

In the last form, the first set of parentheses indicates two identical CH_3 groups attached to the same carbon atom, and the second set of parentheses denotes two CH_2 groups.

Learning Check 1.4 Write a condensed structural formula for each of the following compounds. Retain the bonds to and within the functional groups.

```
a.     H  OH    H  H  H              b.     H        H        H  O
       |  |     |  |  |                     |        |        |  ||
   H - C - C ——— C - C - C - H          H - C ——————— C ——————— C - C - OH
       |  |     |  |  |                     |        |        |
       H  H     H  H  H                     H    H - C - H    H
                                                     |
                                                     H
```

1.5 Alkane Structures

LEARNING OBJECTIVE

6. Classify alkanes as normal or branched.

hydrocarbon
An organic compound that contains only carbon and hydrogen.

saturated hydrocarbon
Another name for an alkane.

alkane
A hydrocarbon that contains only single bonds.

Hydrocarbons, the simplest of all organic compounds, contain only two elements, carbon and hydrogen. **Saturated hydrocarbons** or **alkanes** are organic compounds in which carbon is bonded to four other atoms by single bonds; there are no double or triple bonds in the molecule. Unsaturated hydrocarbons, studied later, are called alkenes, alkynes, and aromatics and contain double bonds, triple bonds, or six-carbon rings, as shown in ■ Figure 1.7.

Most life processes are based on the reactions of functional groups. Since alkanes have no functional group, they are not abundant in the human body. However, most compounds in human cells contain parts consisting solely of carbon and hydrogen that behave very much like hydrocarbons. Thus, to understand the chemical properties of the more complex biomolecules, it is useful to have some understanding of the structure, physical properties, and chemical behavior of hydrocarbons.

Another important reason for becoming familiar with the characteristics of hydrocarbons is the crucial role they play in modern industrial society. We use naturally occurring hydrocarbons as primary sources of energy and as important sources of raw materials for the manufacture of plastics, synthetic fibers, drugs, and hundreds of other compounds used daily (see ■ Figure 1.8).

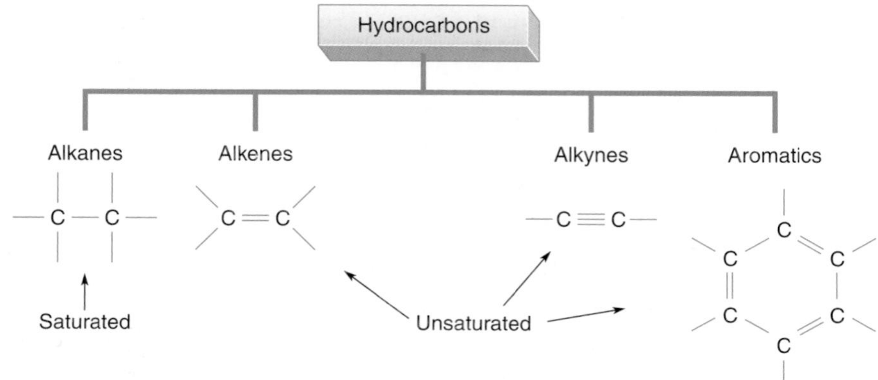

■ **FIGURE 1.7** Classification of hydrocarbons.

CHEMISTRY AND YOUR HEALTH 1.1 Are Organic Foods Better for You?

The answer to the question posed by the title is mixed. The United States Department of Agriculture (USDA) makes no claim that organically produced food is safer or more nutritious than conventionally produced food. Experts in the organic food producing industry seem to agree with the USDA. They point out that the word *organic* refers only to a method of food production. In December 2001, the USDA standardized the way the word *organic* can be used in food labeling. According to the USDA, a food product labeled *100% organic* must contain only ingredients that meet the following requirements: No genetic engineering, ionizing radiation, sewage-sludge fertilizer, or synthesized antibiotics, pesticides, hormones, or fertilizers can be used in their production. In order for a food product to be labeled *95% organic,* at least 95% of the ingredients must meet this definition, and the label *made with organic ingredients* can only be used on food products that contain a minimum of 70% organic ingredients.

Consumers of organic foods have a different answer to the question. Some say they use organic foods because of a combination of environmental and personal health concerns, while a larger number use good flavor as their primary reason. Those who are concerned about their health feel that organic products are better for them because no pesticides, growth hormones, antibiotics, or other synthesized chemicals were used in the food production and so cannot remain in the food as a residue.

While it is true that conventionally produced food may contain residues of such things as pesticides that are known to be toxic in high doses, there is no scientific evidence that they cause health problems when ingested in the quantities found on conventional food products. Some researchers feel that the concern over pesticide residues is misplaced because food-borne bacteria are a much greater health hazard than pesticide residues, and organic farming techniques that use no antibiotics are more likely to produce food carrying disease-causing organisms than are conventional techniques.

There is supporting evidence for those who say organic foods taste better. Organically grown fruits and vegetables are allowed to ripen naturally on the tree or vine, a practice generally recognized to improve flavor over produce that is picked green and ripened artificially. Also, such produce must be transported to market quickly to avoid spoiling, and so tends to be fresher when consumed. Proponents of organic food also point out that the lack of pesticide and antibiotic use in organically grown foods helps slow down the development of resistant strains of bacteria, weeds, and insects. One characteristic of organic foods on which everyone agrees is that they are generally more expensive than conventional foods.

It appears that the answer to the original question about organic foods versus conventional foods is going to continue to be based on who is answering, but it is important to note that all foods have to meet the same USDA standards of safety and quality. As a result, all consumers can be confident that they are benefitting from a safe, high-quality food supply.

Alkanes can be represented by the general formula C_nH_{2n+2}, where n is the number of carbon atoms in the molecule. The simplest alkane, methane, contains one carbon atom and therefore has the molecular formula CH_4. The carbon atom is at the center, and the four bonds of the carbon atom are directed toward the hydrogen atoms at the corners of a regular tetrahedron; each hydrogen atom is geometrically equivalent to the other three in the molecule (see ■ Figure 1.9). A tetrahedral orientation of bonds with bond angles of 109.5° is typical for carbon atoms that form four single bonds. Methane is the primary compound in natural gas. Tremendous quantities of natural gas are consumed worldwide because methane is an efficient, clean-burning fuel. It is used to heat homes, cook food, and power factories.

The next alkane is ethane, which has the molecular formula C_2H_6 and the structural formula $CH_3—CH_3$. This molecule may be thought of as a methane molecule with one hydrogen removed and a $—CH_3$ put in its place. Again, the carbon bonds have a tetrahedral geometry as shown in ■ Figure 1.10. Ethane is a minor component of natural gas.

Propane, the third alkane, has the molecular formula of C_3H_8 and the structural formula $CH_3—CH_2—CH_3$. Again, we can produce this molecule by removing a hydrogen atom from the preceding compound (ethane) and substituting a $—CH_3$ in its place (see ■ Figure 1.11). Since all six hydrogen atoms of ethane are equivalent, it makes no difference which one is replaced. Propane is used extensively as an industrial fuel, as well as for home heating and cooking (see Figure 1.11c).

■ **FIGURE 1.8** This golf club has the strength and light weight of aluminum yet is made from graphite (carbon) fibers reinforced with plastic. What other sports equipment is made from graphite fibers?

FIGURE 1.9 Structural representations of methane, CH_4.

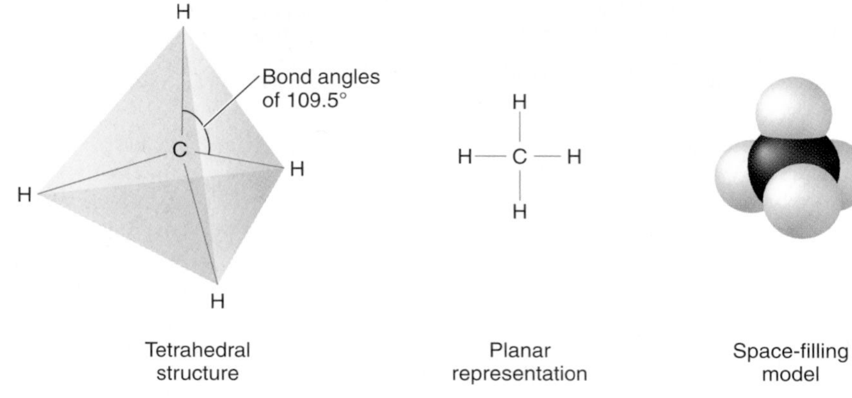

FIGURE 1.10 Perspective models of the ethane molecule CH_3CH_3.

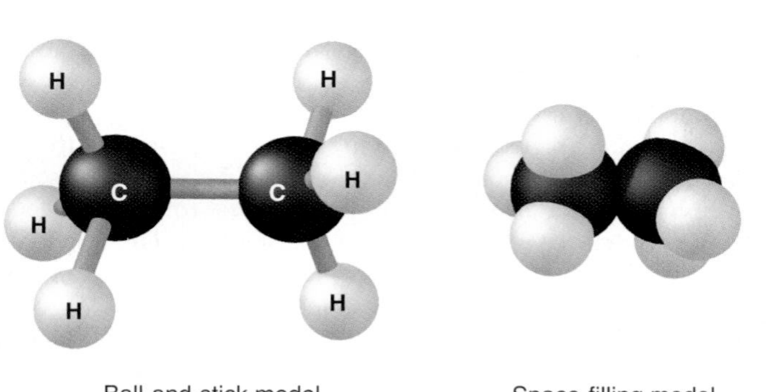

The fourth member of the series, butane, with molecular formula C_4H_{10}, can also be produced by removing a hydrogen atom (this time from propane) and adding a $—CH_3$. However, all the hydrogen atoms of propane are not geometrically equivalent, and more than one position is available for substitution. Replacing a hydrogen atom on one of the end carbons of propane with $—CH_3$ produces a butane molecule that has the structural formula $CH_3—CH_2—CH_2—CH_3$. If,

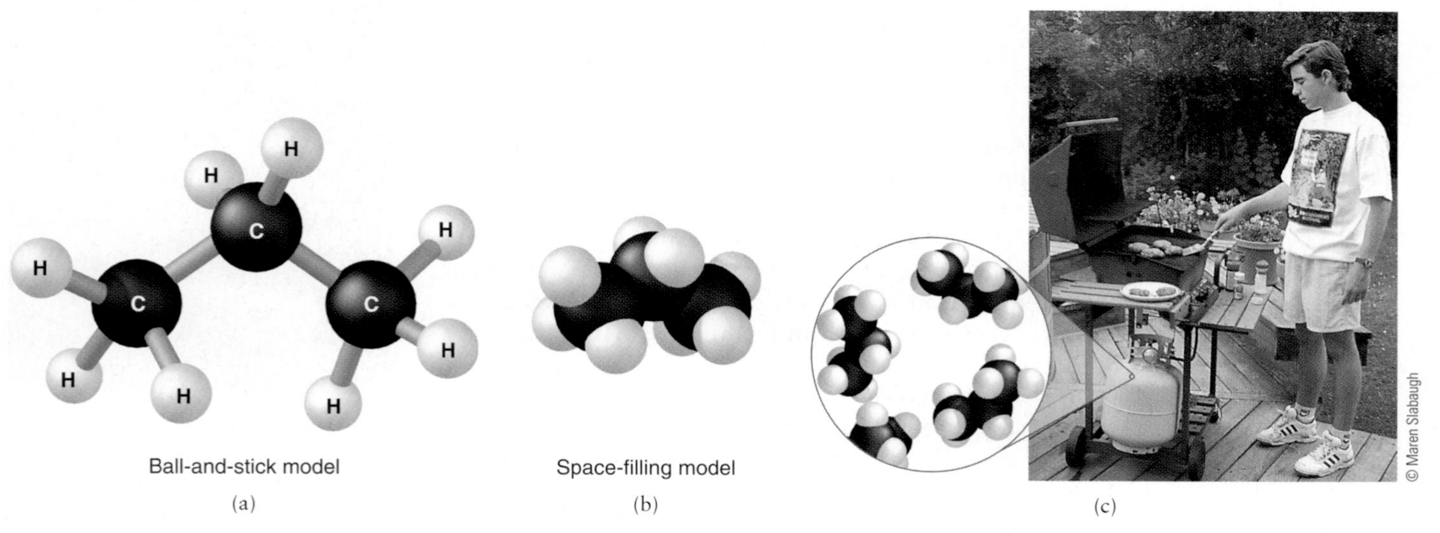

FIGURE 1.11 Perspective models (a and b) of the propane molecule $CH_3CH_2CH_3$. Propane is a common fuel for gas grills (c).

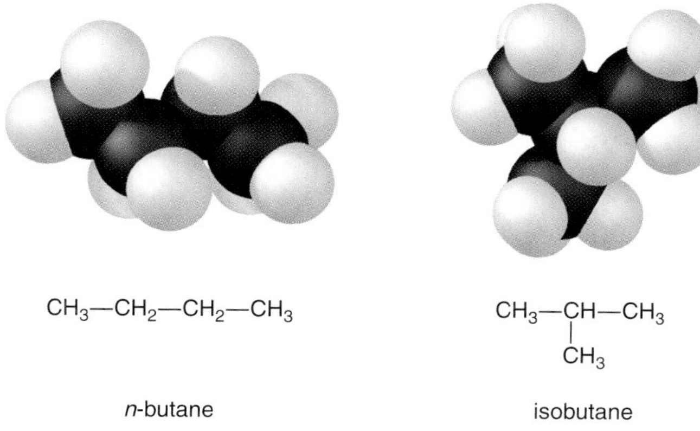

■ **FIGURE 1.12** Space-filling models of the isomeric butanes.

however, substitution is made on the central carbon atom of propane, the butane produced is

$$\begin{array}{c} CH_3-CH-CH_3 \\ | \\ CH_3 \end{array}$$

Notice that both butanes have the same molecular formula, C_4H_{10}. These two possible butanes are structural isomers because they have the same molecular formulas, but they have different atom-to-atom bonding sequences. The straight-chain isomer is called a **normal alkane** and the other is a **branched alkane** (■ Figure 1.12).

The number of possible structural isomers increases dramatically with the number of carbon atoms in an alkane, as shown in ■ Table 1.3.

TABLE 1.3 Molecular formulas and possible structural isomers of alkanes

Molecular formula	Number of possible structural isomers
C_4H_{10}	2
C_5H_{12}	3
C_6H_{14}	5
$C_{10}H_{22}$	75
$C_{20}H_{42}$	366,319
$C_{30}H_{62}$	4,111,846,763

normal alkane
Any alkane in which all the carbon atoms are aligned in a continuous chain.

branched alkane
An alkane in which at least one carbon atom is not part of a continuous chain.

■ **Learning Check 1.5**

a. Determine the molecular formula of the alkane containing eight carbon atoms.
b. Draw the condensed structural formula of the normal isomer of the compound in part a.

1.6 Conformations of Alkanes

LEARNING OBJECTIVE

7. Use structural formulas to determine whether compounds are structural isomers.

Remember that planar representations such as $CH_3—CH_2—CH_3$ or

$$\begin{array}{ccccccc} & H & & H & & H & \\ & | & & | & & | & \\ H- & C & - & C & - & C & -H \\ & | & & | & & | & \\ & H & & H & & H & \end{array}$$

are given with no attempt to accurately portray correct bond angles or molecular geometries. Structural formulas are usually written horizontally simply because it is convenient. It is also important to know that actual organic molecules are in constant motion—twisting, turning, vibrating, and bending. Groups connected by a single bond are capable of rotating about that bond much like a wheel rotates around an axle (see ■ Figure 1.13). As a result of such rotation about single

■ **FIGURE 1.13** Rotation about single bonds (n-butane molecule).

conformations
The different arrangements of atoms in space achieved by rotation about single bonds.

ThomsonNOW Go to Coached Problems to explore the **conformations of butane.**

bonds, a molecule can exist in many different orientations, called **conformations.** In a sample of butane containing billions of identical molecules, there are countless conformations present at any instant, and each conformation is rapidly changing into another. Two of the possible conformations of butane are shown in ■ Active Figure 1.14. We must be sure to recognize that these different conformations do not represent different structural isomers. In each case, the four carbon atoms are bonded in a continuous (unbranched) chain. Since the order of bonding is not changed, the conformations correspond to the same molecule. Two structures would be structural isomers only if bonds had to be broken and remade to convert one into the other.

EXAMPLE 1.3

Which of the following pairs are structural isomers, and which are simply different representations of the same molecule?

a. $CH_3-CH_2-CH_2-CH_3$ and $CH_2(CH_3)-CH_2(CH_3)$ (drawn as CH_2-CH_2 with a CH_3 below the first CH_2 and a CH_3 above the second CH_2)

b. $CH_3-CH(CH_3)-CH_2-CH_3$ and $CH_3-CH_2-CH(CH_3)-CH_3$

c. $CH_3-CH(CH_3)-CH_2-CH_2-CH_3$ and $CH_3-CH_2-CH(CH_3)-CH_2-CH_3$

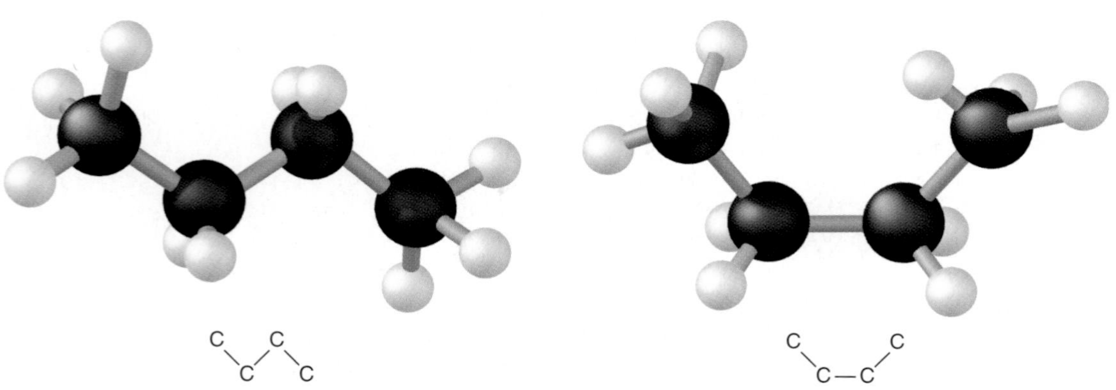

■ **ACTIVE FIGURE 1.14** Perspective models and carbon skeletons of two conformations of n-butane. **Sign in at www.thomsonedu.com to explore an interactive version of this figure.**

Solution

a. Same molecule: In both molecules, the four carbons are bonded in a continuous chain.

$$\overset{1}{CH_3}-\overset{2}{CH_2}-\overset{3}{CH_2}-\overset{4}{CH_3} \qquad \begin{array}{ccc} & & {}^{4}CH_3 \\ & & | \\ {}^{2}CH_2 & - & \underset{3}{CH_2} \\ | & & \\ {}^{1}CH_3 & & \end{array}$$

b. Same molecule: In both molecules, there is a continuous chain of four carbons with a branch at position 2. The molecule has simply been turned around.

$$\begin{array}{c} CH_3 \\ | \\ \underset{1}{CH_3}-\underset{2}{CH}-\underset{3}{CH_2}-\underset{4}{CH_3} \end{array} \qquad \begin{array}{c} CH_3 \\ | \\ \underset{4}{CH_3}-\underset{3}{CH_2}-\underset{2}{CH}-\underset{1}{CH_3} \end{array}$$

c. Structural isomers: Both molecules have a continuous chain of five carbons, but the branch is located at different positions.

$$\begin{array}{c} CH_3 \\ | \\ \underset{1}{CH_3}-\underset{2}{CH}-\underset{3}{CH_2}-\underset{4}{CH_2}-\underset{5}{CH_3} \end{array} \qquad \begin{array}{c} CH_3 \\ | \\ \underset{1}{CH_3}-\underset{2}{CH_2}-\underset{3}{CH}-\underset{4}{CH_2}-\underset{5}{CH_3} \end{array}$$

Learning Check 1.6 Which of the following pairs represent structural isomers, and which are simply the same compound?

a. $$\begin{array}{ccccc} CH_3 & - & CH_2 & - & CH_2 \\ & & & & | \\ & & & & CH_2-CH_3 \end{array} \qquad \begin{array}{ccc} CH_3-CH_2 & & \\ | & & \\ CH_2 & - & CH_2 \\ & & | \\ & & CH_3 \end{array}$$

b. $$\begin{array}{c} CH_3 \\ | \\ CH_3-CH-CH_2-CH_3 \end{array} \qquad \begin{array}{c} CH_2-CH_3 \\ | \\ CH_3-CH-CH_3 \end{array}$$

c. $$\begin{array}{c} CH_3 \\ | \\ CH_3-CH_2-CH-CH_2-CH_3 \end{array} \qquad \begin{array}{c} CH_3 \\ | \\ CH_3-CH_2-CH_2-CH-CH_3 \end{array}$$

1.7 Alkane Nomenclature

LEARNING OBJECTIVE

8. Assign IUPAC names and draw structural formulas for alkanes.

When only a relatively few organic compounds were known, chemists gave them what are today called trivial or common names, such as methane, ethane, propane, and butane. The names for the larger alkanes were derived from the Greek prefixes that indicate the number of carbon atoms in the molecule. Thus, *pent*ane contains five carbons, *hex*ane has six, *hep*tane has seven, and so forth, as shown in ■ Table 1.4.

As more compounds and isomers were discovered, however, it became increasingly difficult to devise unique names and much more difficult to commit them to memory. Obviously, a systematic method was needed. Such a method is now in use, but so are several common methods.

The names for the two isomeric butanes (*n*-butane and isobutane) illustrate the important features of the common nomenclature system used for alkanes. The

ThomsonNOW™ Go to Coached Problems to explore **naming alkanes.**

TABLE 1.4 Names of alkanes

Number of carbon atoms	Name	Molecular formula	Structure of normal isomer
1	methane	CH_4	CH_4
2	ethane	C_2H_6	CH_3CH_3
3	propane	C_3H_8	$CH_3CH_2CH_3$
4	butane	C_4H_{10}	$CH_3CH_2CH_2CH_3$
5	pentane	C_5H_{12}	$CH_3CH_2CH_2CH_2CH_3$
6	hexane	C_6H_{14}	$CH_3CH_2CH_2CH_2CH_2CH_3$
7	heptane	C_7H_{16}	$CH_3CH_2CH_2CH_2CH_2CH_2CH_3$
8	octane	C_8H_{18}	$CH_3CH_2CH_2CH_2CH_2CH_2CH_2CH_3$
9	nonane	C_9H_{20}	$CH_3CH_2CH_2CH_2CH_2CH_2CH_2CH_2CH_3$
10	decane	$C_{10}H_{22}$	$CH_3CH_2CH_2CH_2CH_2CH_2CH_2CH_2CH_2CH_3$

stem *but-* indicates that four carbons are present in the molecule. The *-ane* ending signifies the alkane family. The prefix *n-* indicates that all carbons form an unbranched chain. The prefix *iso-* refers to compounds in which all carbons except one are in a continuous chain and in which that one carbon is branched from a next-to-the-end carbon, as shown:

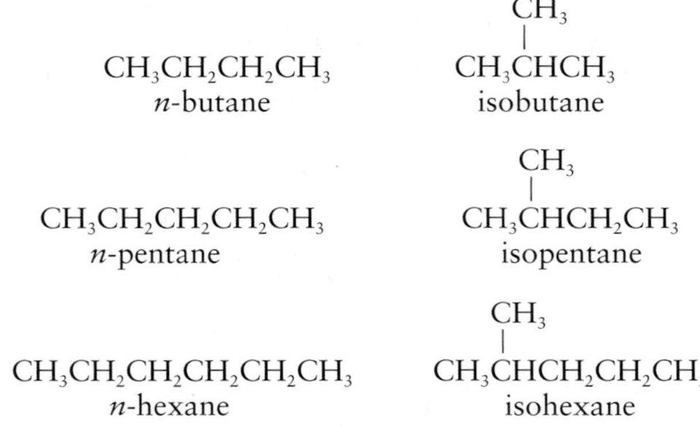

This common naming system has limitations. Pentane has three isomers, and hexane has five. The more complicated the compound, the greater the number of isomers, and the greater the number of special prefixes needed to name all the isomers. It would be extremely difficult and time-consuming to try to identify each of the 75 isomeric alkanes containing 10 carbon atoms by a unique prefix or name.

To devise a system of nomenclature that could be used for even the most complicated compounds, committees of chemists have met periodically since 1892. The system resulting from these meetings is called the IUPAC (International Union of Pure and Applied Chemistry) system. This system is much the same for all classes of organic compounds. The IUPAC name for an organic compound consists of three component parts:

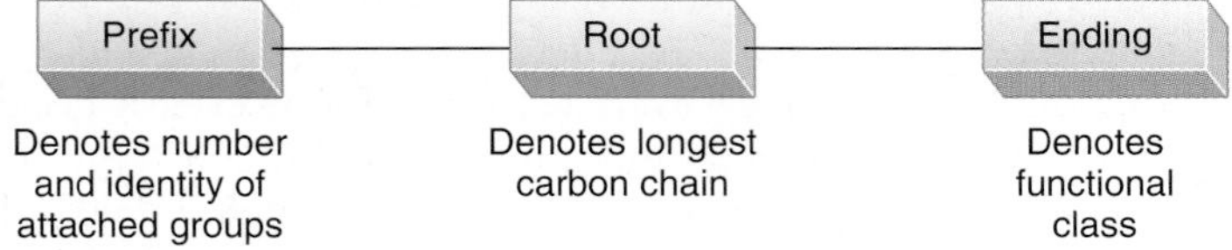

The *root* of the IUPAC name specifies the longest continuous chain of carbon atoms in the compound. The roots for the first 10 normal hydrocarbons are based on the names given in Table 1.4: C_1 *meth-*, C_2 *eth-*, C_3 *prop-*, C_4 *but-*, C_5 *pent-*, C_6 *hex-*, C_7 *hept-*, C_8 *oct-*, C_9 *non-*, C_{10} *dec-*.

The *ending* of an IUPAC name specifies the functional class or the major functional group of the compound. The ending *-ane* specifies an alkane. Each of the other functional classes has a characteristic ending; for example, *-ene* is the ending for alkenes, and the *-ol* ending designates alcohols.

Prefixes are used to specify the identity, number, and location of atoms or groups of atoms that are attached to the longest carbon chain. ■ Table 1.5 lists several common carbon-containing groups referred to as **alkyl groups.** Each alkyl group is a collection of atoms that can be thought of as an alkane minus one hydrogen atom. Alkyl groups are named simply by dropping *-ane* from the name of the corresponding alkane and replacing it with *-yl.* For example, $CH_3—$ is called a methyl group and $CH_3—CH_2—$ an ethyl group:

alkyl group
A group differing by one hydrogen from an alkane.

CH_4	$CH_3—CH_3$	$CH_3—$	$CH_3—CH_2—$
methane	ethane	methyl group	ethyl group

Two different alkyl groups can be derived from propane, depending on which hydrogen is removed. Removal of a hydrogen from an end carbon results in a propyl group:

$CH_3—CH_2—CH_3$	$CH_3—CH_2—CH_2—$
propane	propyl group

Removal of a hydrogen from the center carbon results in an isopropyl group:

$CH_3—CH_2—CH_3$	$CH_3—\overset{\vert}{C}H—CH_3$
propane	isopropyl group

TABLE 1.5 Common alkyl groups

Parent alkane	Structure of parent alkane	Structure of alkyl group	Name of alkyl group
methane	CH_4	$CH_3—$	methyl
ethane	CH_3CH_3	$CH_3CH_2—$	ethyl
propane	$CH_3CH_2CH_3$	$CH_3CH_2CH_2—$	propyl
		$CH_3\overset{\vert}{C}HCH_3$	isopropyl
n-butane	$CH_3CH_2CH_2CH_3$	$CH_3CH_2CH_2CH_2—$	butyl
		$CH_3CH_2\overset{\vert}{C}HCH_3$	*sec*-butyl (secondary-butyl)[a]
isobutane	$CH_3\overset{\overset{CH_3}{\vert}}{C}HCH_3$	$CH_3\overset{\overset{CH_3}{\vert}}{C}HCH_2—$	isobutyl
		$CH_3\underset{\vert}{\overset{\overset{CH_3}{\vert}}{C}}CH_3$	*t*-butyl (tertiary-butyl)[a]

[a]For an explanation of *secondary* and *tertiary*, see Section 3.2.

TABLE 1.6 Common nonalkyl groups

Group	Name
—F	fluoro
—Cl	chloro
—Br	bromo
—I	iodo
$—NO_2$	nitro
$—NH_2$	amino

An isopropyl group also can be represented by $(CH_3)_2CH—$. As shown in Table 1.5, four butyl groups can be derived from butane, two from the straight-chain, or normal, butane, and two from the branched-chain isobutane. A number of nonalkyl groups are also commonly used in naming organic compounds (see Table 1.6).

The following steps are useful when the IUPAC name of an alkane is written on the basis of its structural formula:

Step 1. Name the longest chain. The longest continuous carbon-atom chain is chosen as the basis for the name. The names are those given in Table 1.4.

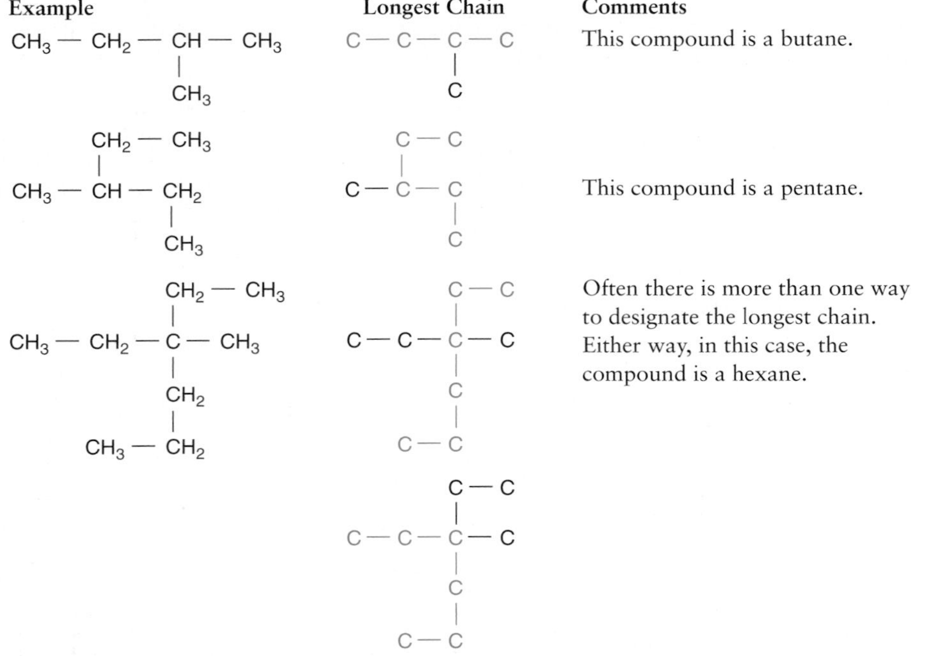

Example
Longest Chain
Comments
This compound is a butane.
This compound is a pentane.
Often there is more than one way to designate the longest chain. Either way, in this case, the compound is a hexane.

Learning Check 1.7 Identify the longest carbon chain in the following:

a. CH_3 (attached to first carbon of) $CH_2—CH_2—CH_3$

b. CH_3 on each end carbon of $CH_2—CH—CH_2$, with CH_3 on the middle CH

c. $CH_3—CH—CH_3$ bonded through CH to $CH_3—CH—CH_3$

Step 2. Number the longest chain. The carbon atoms in the longest chain are numbered consecutively from the end that will give the lowest possible number to any carbon to which a group is attached.

$\overset{4}{C}H_3—\overset{3}{C}H_2—\overset{2}{C}H(CH_3)—\overset{1}{C}H_3$ and not $\overset{1}{C}H_3—\overset{2}{C}H_2—\overset{3}{C}H(CH_3)—\overset{4}{C}H_4$

$CH_3—\overset{3}{C}H(\overset{2}{C}H_2—\overset{1}{C}H_3)—\overset{4}{C}H_2—\overset{5}{C}H_3$ the chain may also be numbered $CH_3—\overset{3}{C}H(\overset{4}{C}H_2—\overset{5}{C}H_3)—\overset{2}{C}H_2—\overset{1}{C}H_3$

If two or more alkyl groups are attached to the longest chain and more than one numbering sequence is possible, the chain is numbered to get the lowest series of numbers. An easy way to follow this rule is to number from the end of the chain nearest a branch:

```
1     2      3       4      5  6                       6     5      4       3      2       1
CH3 — CH — CH2 — CH — CH2CH3        and not        CH3 — CH — CH2 — CH — CH2 — CH3
      |              |                                   |              |
      CH3            CH2 — CH3                           CH3            CH2 — CH3
```

Groups are located at positions 2 and 4 | Groups are located at positions 3 and 5

```
                    CH3                                               CH3
5     4      3     2|   1                           1     2      3     4|    5
CH3 — CH — CH2 — C — CH3          and not           CH3 — CH — CH2 — C — CH3
      |            |                                      |            |
      CH3          CH3                                    CH3          CH3
```

Groups are located at positions 2,2,4 | Groups are located at positions 2,4,4

Here, a difference occurs with the second number in the sequence, so positions 2,2,4 is the lowest series and is used rather than positions 2,4,4.

Learning Check 1.8 Decide how to correctly number the longest chain in the following according to IUPAC rules:

a.
```
CH3 — CH — CH2 — CH2 — CH3
      |
      CH2
      |
      CH3
```

b.
```
                                 CH3
                                 |
CH3 — CH2 — CH2 — CH2 — C — CH2 — CH — CH3
                                 |          |
                                 CH3        CH3
```

Step 3. Locate and name the attached alkyl groups. Each group is located by the number of the carbon atom to which it is attached on the chain.

```
4     3      2     1
CH3 — CH2 — CH — CH3
             |
             CH3
```

The attached group is located on carbon 2 of the chain, and it is a methyl group.

```
1     2      3      4      5      6
CH3 — CH — CH2 — CH — CH2 — CH3
      |             |
      CH3           CH2 — CH3
```

The one-carbon group at position 2 is a methyl group. The two-carbon group at position 4 is an ethyl group.

Learning Check 1.9 Identify the alkyl groups attached to the dashed line, which symbolizes a long carbon chain. Refer to Table 1.5 if necessary.

```
                        CH3
                        |
                        CH — CH3
                        |
—— —┬— —— —┬— ——┴— —┬— —— —— —┬— ——
    |      |            |          |
    CH3    CH2          CH — CH3   CH2 — CH3
           |            |
           CH2          CH2
           |            |
           CH3          CH3
```

Step 4. Combine the longest chain and the branches into the name. The position and the name of the attached alkyl group are added to the name of the longest chain and written as one word:

$$\begin{array}{ccccccc} \overset{4}{CH_3} & - & \overset{3}{CH_2} & - & \overset{2}{CH} & - & \overset{1}{CH_3} \\ & & & & | & & \\ & & & & CH_3 & & \end{array}$$

2-methylbutane

Additional steps are needed when more than one alkyl group is attached to the longest chain.

Step 5. Indicate the number and position of attached alkyl groups. If two or more of the same alkyl group occur as branches, the number of them is indicated by the prefixes *di-*, *tri-*, *tetra-*, *penta-*, etc., and the location of each is again indicated by a number. These position numbers, separated by commas, are put just before the name of the group, with hyphens before and after the numbers when necessary:

$$\begin{array}{ccccccccc} & & & & & & CH_3 & & \\ & & & & & & | & & \\ \overset{1}{CH_3} & - & \overset{2}{CH} & - & \overset{3}{CH_2} & - & \overset{4}{CH} & - & \overset{5}{CH_3} \\ & & | & & & & & & \\ & & CH_3 & & & & & & \end{array}$$

2,4-dimethylpentane

$$\begin{array}{ccccccccc} & & & & CH_3 & & & & \\ & & & & | & & & & \\ \overset{1}{CH_3} & - & \overset{2}{CH_2} & - & \overset{3}{C} & - & \overset{4}{CH_2} & - & \overset{5}{CH_3} \\ & & & & | & & & & \\ & & & & CH_3 & & & & \end{array}$$

3,3-dimethylpentane

If two or more *different* alkyl groups are present, their names are alphabetized and added to the name of the basic alkane, again as one word. For purposes of alphabetizing, the prefixes *di-*, *tri-*, and so on are ignored, as are the italicized prefixes secondary *(sec)* and tertiary *(t)*. The prefix *iso-* is an exception and is used for alphabetizing:

$$\begin{array}{ccccccccccccccc} \overset{1}{CH_3} & - & \overset{2}{CH} & - & \overset{3}{CH} & - & \overset{4}{CH_2} & - & \overset{5}{CH} & - & \overset{6}{CH_2} & - & \overset{7}{CH_2} & - & \overset{8}{CH_3} \\ & & | & & | & & & & | & & & & & & \\ & & CH_3 & & CH_3 & & & & CH & - & CH_3 & & & & \\ & & & & & & & & | & & & & & & \\ & & & & & & & & CH_3 & & & & & & \end{array}$$

5-isopropyl-2,3-dimethyloctane

■ Learning Check 1.10 Give the correct IUPAC name to each of the following:

a.
$$\begin{array}{ccccccccccc} & & & & & & & & CH_3 & & \\ & & & & & & & & | & & \\ CH_3 & - & CH_2 & - & CH_2 & - & CH_2 & - & CH & - & CH_3 \end{array}$$

b.
$$\begin{array}{ccccccccccccc} & & & & & & CH_2 & - & CH_3 & & & & \\ & & & & & & | & & & & & & \\ CH_3 & - & CH_2 & - & CH_2 & - & CH & - & CH_2 & - & CH & - & CH_3 \\ & & & & & & & & & & | & & \\ & & & & & & & & & & CH_2 & - & CH_3 \end{array}$$

c.
$$\begin{array}{ccccccccccccc} & & & & & & & & CH_3 & & & & \\ & & & & & & & & | & & & & \\ CH_3 & - & CH_2 & - & CH_2 & - & CH & - & CH & - & CH & - & CH_3 \\ & & & & & & | & & & & | & & \\ & & & & & & CH & - & CH_3 & & CH_3 & & \\ & & & & & & | & & & & & & \\ & & & & & & CH_3 & & & & & & \end{array}$$

Naming compounds is a very important skill, as is the reverse process of using IUPAC nomenclature to specify a structural formula. The two processes are very similar. To obtain a formula from a name, determine the longest chain, number the chain, and add any attached groups.

EXAMPLE 1.4

Draw a condensed structural formula for 3-ethyl-2-methylhexane.

Solution

Use the last part of the name to determine the longest chain. Draw a chain of six carbons. Then, number the carbon atoms.

$$\underset{1}{C}-\underset{2}{C}-\underset{3}{C}-\underset{4}{C}-\underset{5}{C}-\underset{6}{C}$$

Attach a methyl group at position 2 and an ethyl group at position 3.

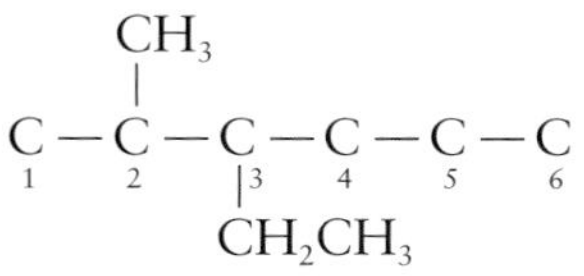

$$\underset{1}{C}-\underset{2}{\overset{\overset{CH_3}{|}}{C}}-\underset{\underset{CH_2CH_3}{|3}}{C}-\underset{4}{C}-\underset{5}{C}-\underset{6}{C}$$

Complete the structure by adding enough hydrogen atoms so that each carbon has four bonds.

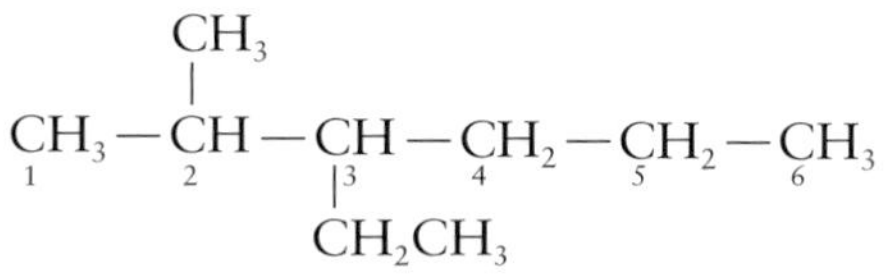

$$\underset{1}{CH_3}-\underset{2}{\overset{\overset{CH_3}{|}}{CH}}-\underset{\underset{CH_2CH_3}{|3}}{CH}-\underset{4}{CH_2}-\underset{5}{CH_2}-\underset{6}{CH_3}$$

Learning Check 1.11 Draw a condensed structural formula for each of the following compounds:

a. 2,2,4-trimethylpentane
b. 3-isopropylhexane
c. 3-ethyl-2,4-dimethylheptane

1.8 Cycloalkanes

LEARNING OBJECTIVE

9. Assign IUPAC names and draw structural formulas for cycloalkanes.

From what we have said so far, the formula C_3H_6 cannot represent an alkane. Not enough hydrogens are present to allow each carbon to form four bonds, unless there are multiple bonds. For example, the structural formula $CH_3—CH=CH_2$ fits the molecular formula but cannot represent an alkane because of the double bond. The C_3H_6 formula does become acceptable for an alkane if the carbon atoms form a ring, or cyclic, structure rather than the open-chain structure shown:

$$C-C-C \qquad \begin{array}{c} C \\ / \ \backslash \\ C-C \end{array}$$

Open chain Cyclic

The resulting saturated cyclic compound, called cyclopropane, has the structural formula

$$\begin{array}{c} CH_2 \\ / \ \backslash \\ CH_2-CH_2 \end{array}$$

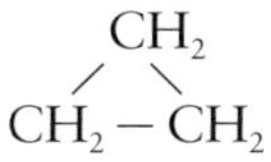

TABLE 1.7 Structural formulas and symbols for common cycloalkanes

Name	Structural formula	Condensed formula
cyclopropane	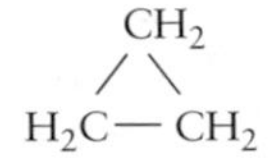 CH_2 / $H_2C—CH_2$ (three-membered ring)	 △
cyclobutane	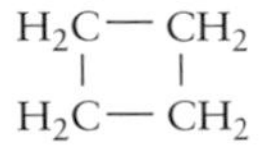 $H_2C—CH_2$ / $H_2C—CH_2$ (four-membered ring)	□
cyclopentane	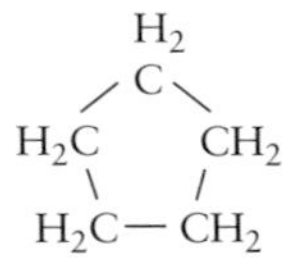 H_2C / H_2C CH_2 / $H_2C—CH_2$ (five-membered ring)	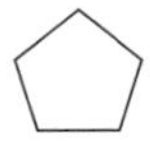 ⬠
cyclohexane	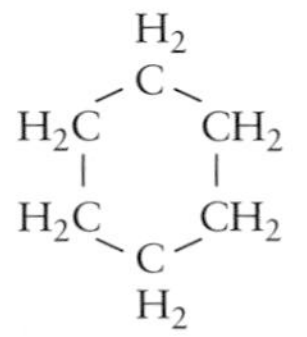 H_2C / H_2C CH_2 / H_2C CH_2 / CH_2 (six-membered ring)	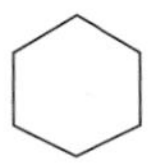 ⬡

cycloalkane
An alkane in which carbon atoms form a ring.

Alkanes containing rings of carbon atoms are called **cycloalkanes.** Like the other alkanes, cycloalkanes are not found in human cells. However, several important molecules in human cells do contain rings of five or six atoms, and the study of cycloalkanes will help you better understand the chemical behavior of these complex molecules.

According to IUPAC rules, cycloalkanes are named by placing the prefix *cyclo-* before the name of the corresponding alkane with the same number of carbon atoms. Chemists often abbreviate the structural formulas for cycloalkanes and draw them as geometric figures (triangles, squares, etc.) in which each corner represents a carbon atom. The hydrogens are omitted (see Table 1.7). It is important to remember that each carbon atom still possesses four bonds, and that hydrogen is assumed to be bonded to the carbon atoms unless something else is indicated. When substituted cycloalkanes (those with attached groups) are named, the position of a single attached group does not need to be specified in the name because all positions in the ring are equivalent. However, when two or more groups are attached, their positions of attachment are indicated by numbers, just as they were for alkanes. The ring numbering begins with the carbon attached to the first group alphabetically and proceeds around the ring in the direction that will give the lowest numbers for the locations of the other attached groups.

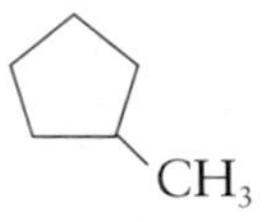

CH_3

methylcyclopentane

CH_3 (1), CH_3 (2)

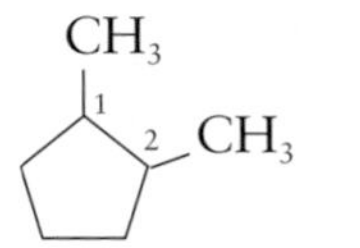

1,2-dimethylcyclopentane
not 1,5-dimethylcyclopentane

Cl (1), CH_3 (3)

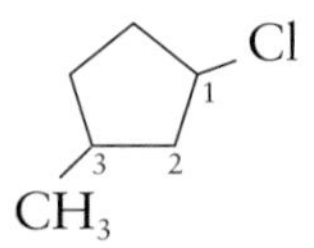

1-chloro-3-methylcyclopentane
not 1-chloro-4-methylcyclopentane
not 3-chloro-1-methylcyclopentane

EXAMPLE 1.5

Represent each of the following cycloalkanes by a geometric figure, and name each compound:

a. Condensed structural formula: five-membered ring of CH_2 groups and one CH bearing a CH_3 (ring atoms: H_2C, CH_2, H_2C, H_2C, CH; substituent CH_3)

b. Condensed structural formula: six-membered ring (H_2C, H_2C, CH_2, CH_2, CH–CH_3, CH–CH_3)

Solution

a. A pentagon represents a five-membered ring, which is called cyclopentane. This compound has a methyl group attached, so the name is methylcyclopentane. The position of a single alkyl group is not indicated by a number because the positions of all carbons in the ring are equivalent.

CH_3

b. A hexagon represents a six-carbon ring, which is called cyclohexane. Two methyl groups are attached; thus we have a dimethylcyclohexane.

CH_3

CH_3

However, the positions of the two alkyl groups must be indicated. The ring is numbered beginning with a carbon to which a methyl group is attached, counting in the direction giving the lowest numbers. The correct name is 1,3-dimethylcyclohexane. Notice that a reverse numbering beginning at the same carbon would have given 1,5-dimethylcyclohexane. The number 3 in the correct name is lower than the 5 in the incorrect name.

6 CH_3 5 1 4 2 3 CH_3

Learning Check 1.12 Give each of the following compounds the correct IUPAC name:

a. CH_3 / CH_3 (cyclohexane ring)

b. CH_2-CH_3 (cyclopropane ring)

c. CH_3 / CH_2-CH_3 (cyclopentane ring)

1.9 The Shape of Cycloalkanes

LEARNING OBJECTIVE

10. Name and draw structural formulas for geometric isomers of cycloalkanes.

Recall from Section 1.5 that a tetrahedral orientation of bonds with bond angles of 109.5° is characteristic of carbon atoms that form four single bonds. A tetrahedral

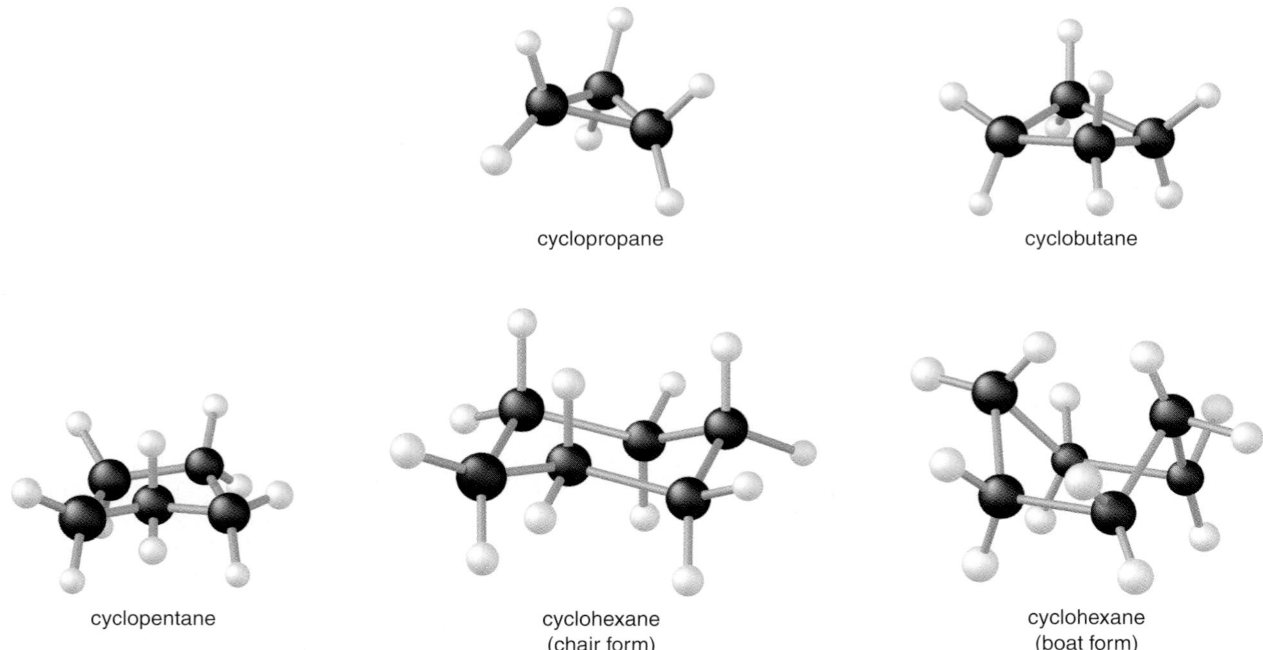

■ **FIGURE 1.15** Ball-and-stick models for common cycloalkanes.

Free rotation

Free rotation not possible

■ **FIGURE 1.16** Rotation about C—C single bonds occurs in open-chain compounds but not within rings.

arrangement is the most stable because it results in the least crowding of the atoms. In certain cycloalkanes, however, a tetrahedral arrangement for all carbon-to-carbon bonds is not possible. For example, the bond angles between adjacent carbon–carbon bonds in planar cyclopropane molecules must be 60° (see ■ Figure 1.15). In cyclobutane, they are close to 90°. As a result, cyclopropane and cyclobutane rings are much less stable than compounds with bond angles of about 109°. Both cyclobutane and cyclopentane bend slightly from a planar structure to reduce the crowding of hydrogen atoms (Figure 1.15). In larger cycloalkanes, the bonds to carbon atoms can be tetrahedrally arranged only when the carbon atoms do not lie in the same plane. For example, cyclohexane can assume several nonplanar shapes. The chair and boat forms, where all the bond angles are 109.5°, are shown in Figure 1.15.

The free rotation that can take place around C—C single bonds in alkanes (Section 1.6) is not possible for the C—C bonds of cycloalkanes. The ring structure allows bending or puckering but prevents free rotation (see ■ Figure 1.16). Any rotation of one carbon atom 180° relative to an adjacent carbon atom in a cycloalkane would require a single carbon–carbon bond to be broken somewhere in the ring. The breaking of such bonds would require a large amount of energy.

The lack of free rotation around C—C bonds in disubstituted cycloalkanes leads to an extremely important kind of isomerism called *stereoisomerism*. Two different compounds that have the same molecular formula and the same structural formula but different spatial arrangements of atoms are called **stereoisomers.** For example, consider a molecule of 1,2-dimethylcyclopentane. The cyclopentane ring is drawn in ■ Figure 1.17 as a planar pentagon with the heavy lines indicating that two of the carbons are in front as one views the structure. The groups attached to the ring project above or below the plane of the ring. Two stereoisomers are possible: Either both groups may project in the same direction from the plane, or they may project in opposite directions from the plane of the ring. Since the methyl groups cannot rotate from one side of the ring to the other, molecules of the two compounds represented in Figure 1.17 are distinct.

These two compounds have physical and chemical properties that are quite different and therefore can be separated from each other. Stereoisomers of this type, in which the spatial arrangement or geometry of their groups is maintained by rings, are called **geometric isomers** or cis-trans isomers. The prefix ***cis-*** denotes the isomer

stereoisomers
Compounds with the same structural formula but different spatial arrangements of atoms.

geometric isomers
Molecules with restricted rotation around C—C bonds that differ in the three-dimensional arrangements of their atoms in space and not in the order of linkage of atoms.

cis-
On the same side (as applied to geometric isomers).

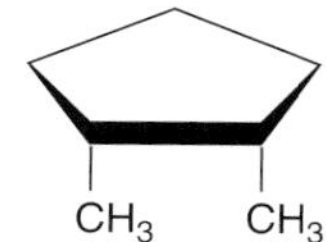

cis-1,2-dimethylcyclopentane

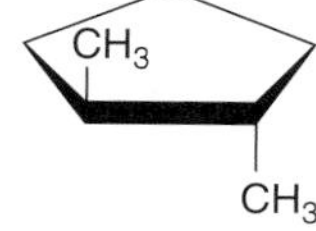

trans-1,2-dimethylcyclopentane

■ **FIGURE 1.17** Two geometric isomers of 1,2-dimethylcyclopentane.

in which both groups are on the same side of the ring, and ***trans-*** denotes the isomer in which they are on opposite sides. To exist as geometric isomers, a disubstituted cycloalkane must be bound to groups at two different carbons of the ring. For example, there are no geometric isomers of 1,1-dimethylcyclohexane:

CH_3

CH_3

trans-
On opposite sides (as applied to geometric isomers).

EXAMPLE 1.6

Name and draw structural formulas for all the isomers of dimethylcyclobutane. Indicate which ones are geometric isomers.

Solution

There are three possible locations for the two methyl groups: positions 1,1, positions 1,2, and positions 1,3.

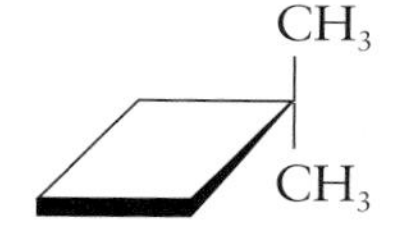

1,1-dimethylcyclobutane

Geometric isomerism is not possible in this case with the two groups bound to the same carbon of the ring

ThomsonNOW™ Go to Coached Problems to practice **identifying *cis* and *trans* isomers of cycloalkanes.**

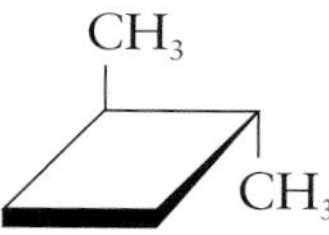

trans-1,2-dimethylcyclobutane

Two groups on opposite sides of the planar ring

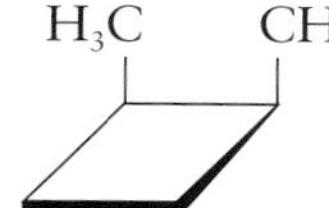

cis-1,2-dimethylcyclobutane

Two groups on the same side of the ring

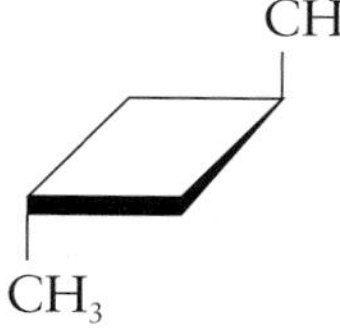

trans-1,3-dimethylcyclobutane

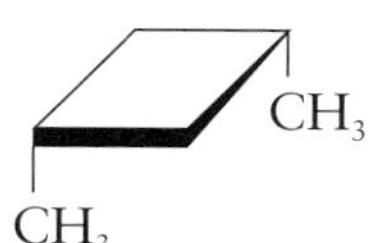

cis-1,3-dimethylcyclobutane

■ Learning Check 1.13

a. Identify each of the following cycloalkanes as a *cis-* or *trans-* compound:

(1) 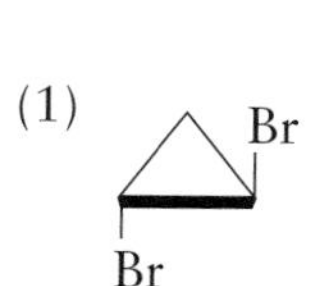

(2) 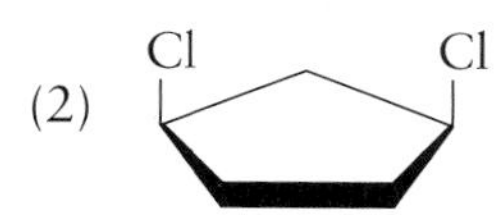

(3) 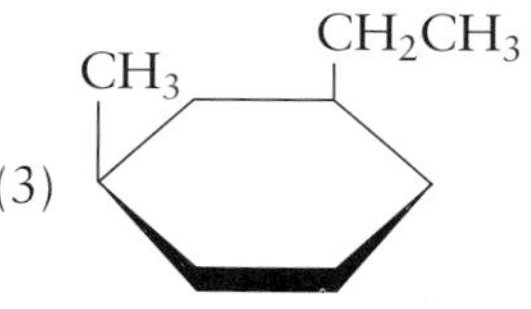

b. Draw the structural formula for *cis*-1,2-dichlorocyclobutane.

CHEMISTRY AROUND US 1.1 Petroleum

Petroleum, the most important of the fossil fuels used today, is sometimes called "black gold" in recognition of its importance in the 20th century. At times, the need for petroleum to keep society fueled has seemed second only to our need for food, shelter, and clothing.

It is generally believed that this complex mixture of hydrocarbons was formed over eons through the gradual decay of ocean-dwelling microscopic animals. The resulting crude oil, a viscous black liquid, collects in vast underground pockets in sedimentary rock. It is brought to the surface via drilling and pumping.

Useful products are obtained from crude oil by heating it to high temperatures to produce various fractions according to boiling point (see table). Most petroleum products are eventually burned as a fuel, but about 2% are used to synthesize organic compounds. This seemingly small amount is quite large in actual tonnage because of the huge volume of petroleum that is refined annually. In fact, more than half of all industrial synthetic organic compounds are made from this source. These industrial chemicals are eventually converted into dyes, drugs, plastics, artificial fibers, detergents, insecticides, and other materials deemed indispensable by many in industrialized nations.

© Michael C. Slabaugh

Asphalt for paving roads is a petroleum product.

Fraction	Boiling point range (°C)	Molecular size range	Typical uses
Gas	−164–30	C_1–C_4	Heating, cooking
Gasoline	30–200	C_5–C_{12}	Motor fuel
Kerosene	175–275	C_{12}–C_{16}	Fuel for stoves and diesel and jet engines
Heating oil	Up to 375	C_{15}–C_{18}	Furnace oil
Lubricating oils	350 and up	C_{16}–C_{20}	Lubrication, mineral oil
Greases	Semisolid	C_{18}–up	Lubrication, petroleum jelly
Paraffin (wax)	Melts at 52–57	C_{20}–up	Candles, toiletries
Pitch and tar	Residue in boiler	High	Roofing, asphalt paving

1.10 Physical Properties of Alkanes

LEARNING OBJECTIVE

11. Describe the key physical properties of alkanes.

Since alkanes are composed of nonpolar carbon–carbon and carbon–hydrogen bonds, alkanes are nonpolar molecules. Alkanes have lower melting and boiling points than other organic compounds of comparable molecular weight (see ■ Table 1.8). This is because their nonpolar molecules exert very weak attractions for each other. Alkanes are odorless compounds.

homologous series Compounds of the same functional class that differ by a $—CH_2—$ group.

ThomsonNOW Go to Coached Problems to examine the **boiling points of alkanes.**

The normal, or straight-chain, alkanes make up what is called a **homologous series.** This term describes any series of compounds in which each member differs from a previous member only by having an additional $—CH_2—$ unit. The physical and chemical properties of compounds making up a homologous series are usually closely related and vary in a systematic and predictable way. For example, the boiling points of normal alkanes increase smoothly as the length of the carbon chain increases (see ■ Figure 1.18). This pattern results from increasing dispersion forces as molecular weight increases. At ordinary temperatures and pressures, normal alkanes with 1 to 4 carbon atoms are gases, those with 5 to 20 carbon atoms are liquids, and those with more than 20 carbon atoms are waxy solids.

Because they are nonpolar, alkanes and other hydrocarbons are insoluble in water, which is a highly polar solvent. They are also less dense than water and

TABLE 1.8 Physical properties of some normal alkanes

Number of carbon atoms	IUPAC name	Condensed structural formula	Melting point (°C)	Boiling point (°C)	Density (g/mL)
1	methane	CH_4	−182.5	−164.0	0.55
2	ethane	CH_3CH_3	−183.2	−88.6	0.57
3	propane	$CH_3CH_2CH_3$	−189.7	−42.1	0.58
4	butane	$CH_3CH_2CH_2CH_3$	−133.4	−0.5	0.60
5	pentane	$CH_3CH_2CH_2CH_2CH_3$	−129.7	36.1	0.63
6	hexane	$CH_3CH_2CH_2CH_2CH_2CH_3$	−95.3	68.9	0.66
7	heptane	$CH_3CH_2CH_2CH_2CH_2CH_2CH_3$	−90.6	98.4	0.68
8	octane	$CH_3CH_2CH_2CH_2CH_2CH_2CH_2CH_3$	−56.8	125.7	0.70
9	nonane	$CH_3CH_2CH_2CH_2CH_2CH_2CH_2CH_2CH_3$	−53.5	150.8	0.72
10	decane	$CH_3CH_2CH_2CH_2CH_2CH_2CH_2CH_2CH_2CH_3$	−29.7	174.1	0.73

thus float on it. These two properties of hydrocarbons are partly responsible for the well-known serious effects of oil spills from ships (see ■ Figure 1.19).

Liquid alkanes of higher molecular weight behave as emollients (skin softeners) when applied to the skin. An alkane mixture known as mineral oil is sometimes used to replace natural skin oils washed away by frequent bathing or swimming. Petroleum jelly (Vaseline is a well-known brand name) is a semisolid mixture of alkanes that is used as both an emollient and a protective film. Water and water solutions such as urine don't dissolve or penetrate the film, and the underlying skin is protected. Many cases of diaper rash have been prevented or treated this way.

ThomsonNOW™ Go to Coached Problems to examine the **physical properties of alkanes.**

The word **hydrophobic** (literally "water fearing") is often used to refer to molecules or parts of molecules that are insoluble in water. Many biomolecules, the large organic molecules associated with living organisms, contain nonpolar (hydrophobic) parts. Thus, such molecules are not water-soluble. Palmitic acid, for example, contains a large nonpolar hydrophobic portion and is insoluble in water.

hydrophobic
Molecules or parts of molecules that repel (are insoluble in) water.

Nonpolar portion

$$\overbrace{CH_3-CH_2-CH_2-CH_2-CH_2-CH_2-CH_2-CH_2-CH_2-CH_2-CH_2-CH_2-CH_2-CH_2-CH_2}^{\text{Nonpolar portion}}-\overset{\overset{\displaystyle O}{\|}}{C}-OH$$

palmitic acid

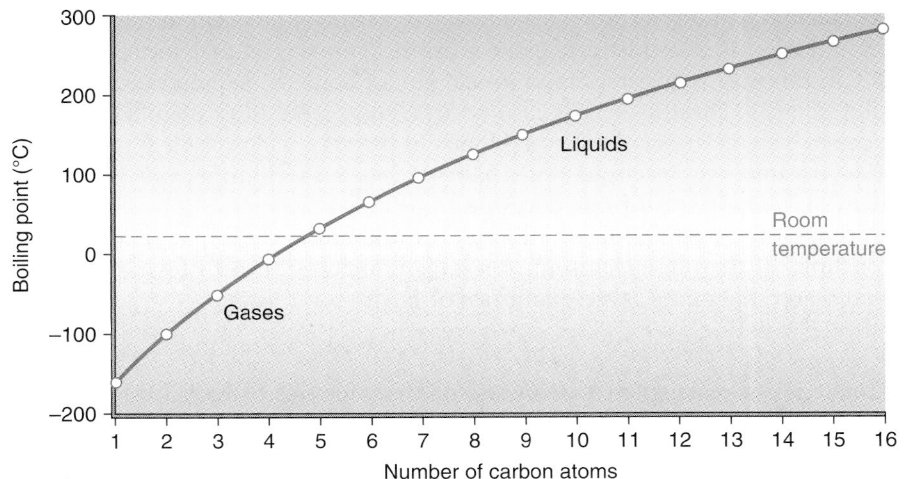

■ **FIGURE 1.18** Normal alkane boiling points depend on chain length.

■ **FIGURE 1.19** Oil spills can have serious and long-lasting effects on the environment because of the insolubility of hydrocarbons in water.

OVER THE COUNTER 1.1 Hydrating the Skin

Healthy, normal human skin is kept moist by natural body oils contained in sebum, a secretion of the skin's sebaceous glands. The oily sebum helps the epidermis, or outer layer of the skin, retain the 10–30% of water it normally contains. However, some people suffer from skin that is naturally dry, or dry because of aging or contact with materials like paint thinner that dissolve and remove the sebum. Such individuals may find some relief by using OTC products called *moisturizers.*

Two types of skin moisturizers are commonly available. One type, which behaves like natural sebum, contains nonpolar oily substances that form a barrier and prevent water from passing through and evaporating from the skin. These barrier-forming products often contain combinations of materials from various sources, including mineral oil and petroleum jelly from petroleum, vegetable oils such as apricot oil, sesame seed oil, palm kernel oil, olive oil, and safflower oil, and lanolin, an animal fat from sheep oil glands and wool. While they effectively keep water from leaving the skin, these products are somewhat messy to use and leave the skin feeling greasy.

A second, more popular type of moisturizer works by attracting water from the air and skin. These products form a water-rich layer that adheres to the skin without giving it a greasy feel. The substances that attract water are called humectants, and are compounds capable of forming hydrogen bonds with water. Some examples of humectants used in products of this type are glycerol, urea, lactic acid, and propylene glycol.

If you are shopping for a moisturizer, remember that the main characteristic you should look for in a product is the ability to form a barrier to prevent water evaporation, or the ability to act as a humectant. Some expensive products advertise that in addition to moisturizing, they also beautify the skin and even reverse aging because they contain proteins such as collagen and elastin, vitamins, hormones, or even DNA. It is unlikely that such substances can pass through the epidermis of the skin in sufficient amounts to provide the advertised benefits.

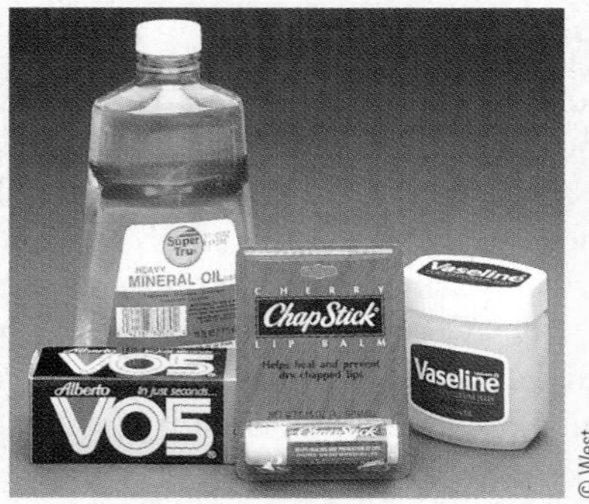

Alkane products marketed to soften and protect the skin.

1.11 Alkane Reactions

LEARNING OBJECTIVE

12. Write alkane combustion reactions.

The alkanes are the least reactive of all organic compounds. In general, they do not react with strong acids (such as sulfuric acid), strong bases (such as sodium hydroxide), most oxidizing agents (such as potassium dichromate), and most reducing agents (such as sodium metal). This unreactive nature is reflected in the name *paraffins,* sometimes used to identify alkanes (from Latin words that mean "little affinity"). Paraffin wax, sometimes used to seal jars of homemade preserves, is a mixture of solid alkanes. Paraffin wax is also used in the preparation of milk cartons and wax paper. The inertness of the compounds in the wax makes it ideal for these uses. Alkanes do undergo reactions with halogens such as chlorine and bromine, but these are not important for our purposes.

ThomsonNOW™ Go to Chemistry Interactive to examine **hydrocarbons** in more detail.

Perhaps the most significant reaction of alkanes is the rapid oxidation called combustion. In the presence of ample oxygen, alkanes burn to form carbon dioxide and water, liberating large quantities of heat:

$$CH_4 + 2O_2 \rightarrow CO_2 + 2H_2O + 212.8 \text{ kcal/mol} \quad (1.4)$$

It is this type of reaction that accounts for the wide use of hydrocarbons as fuels.

Natural gas contains methane (80–95%), some ethane, and small amounts of other hydrocarbons. Propane and butane are extracted from natural gas and sold in pressurized metal containers (bottled gas). In this form, they are used for heating and cooking in campers, trailers, boats, and rural homes.

CHEMISTRY AROUND US 1.2 Carbon Monoxide: Silent and Deadly

The incomplete combustion of alkanes and other carbon-containing fuels may produce carbon monoxide, CO. It is a common component of furnace, stove, and automobile exhaust. Normally, these fumes are emitted into the air and dispersed. However, whenever CO gas is released into closed spaces, lethal concentrations may develop. All of the following are known to have caused CO poisoning deaths: poor ventilation of a furnace or stove exhaust, running the engine of an automobile for a period of time in a closed garage, idling the engine of a snowbound car to keep the heater working, and using a barbecue grill as a source of heat in a house or camper (burning charcoal generates a large amount of carbon monoxide). Recently, the exhaust systems of snow-blocked vehicles have been identified as another cause of carbon monoxide poisoning. The problem is created when the end of a vehicle's exhaust pipe is obstructed or plugged by snow. CO-containing exhaust can leak through cracks in the exhaust system, penetrate the floorboard, and enter the passenger compartment of the vehicle. An estimated 1000 Americans die each year from unintentional CO poisoning, and as many as 10,000 require medical treatment.

Carbon monoxide is dangerous because of its ability to bind strongly with hemoglobin in red blood cells. When this occurs, the ability of the blood to transport oxygen (O_2) is reduced because the CO molecules occupy sites on hemoglobin molecules that are normally occupied by O_2 molecules.

Because it is colorless and odorless, carbon monoxide gas is a silent and stealthy potential killer. The following symptoms of CO poisoning are given in order of increasing severity and seriousness:

1. Slight headache, dizziness, drowsiness
2. Headache, throbbing temples
3. Weakness, mental confusion, nausea
4. Rapid pulse and respiration, fainting
5. Possibly fatal coma

All except the most severe cases of CO poisoning are reversible. The most important first aid treatment is to get the victim fresh air. Any person who feels ill and suspects CO poisoning might be the cause should immediately evacuate the building, get fresh air, and summon medical assistance. Victims are often treated with 100% oxygen delivered from a mask.

Consumer safety experts stress that combustion appliances should be inspected regularly for leaks or other malfunctions. Also, it is recommended that drivers should inspect exhaust pipes and clear any obstructions before starting vehicles that have been parked in snow. The Consumer Product Safety Commission recommends that CO detectors be installed near bedrooms in residences. These detectors sound an alarm before the gas reaches potentially lethal concentrations in the air.

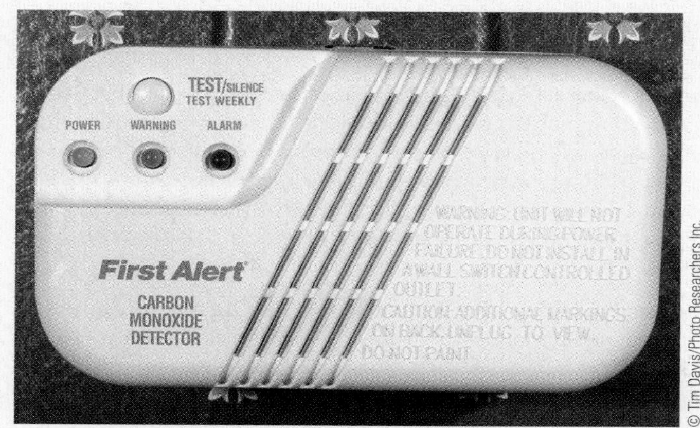

© Tim Davis/Photo Researchers Inc.

Fireplaces must be properly constructed to carry away carbon monoxide fumes.

Gasoline is a mixture of hydrocarbons (primarily alkanes) that contain 5 to 12 carbon atoms per molecule. Diesel fuel is a similar mixture, except the molecules contain 12 to 16 carbon atoms. The hot CO_2 and water vapor generated during combustion in an internal combustion engine have a much greater volume than the air and fuel mixture. It is this sudden increase in gaseous volume and pressure that pushes the pistons and delivers power to the crankshaft.

If there is not enough oxygen available, incomplete combustion of hydrocarbons occurs, and some carbon monoxide (CO) or even carbon (see ■ Figure 1.20) may be produced (Reactions 1.5 and 1.6):

$$2CH_4 + 3O_2 \rightarrow 2CO + 4H_2O \quad (1.5)$$

$$CH_4 + O_2 \rightarrow C + 2H_2O \quad (1.6)$$

These reactions are usually undesirable because CO is toxic, and carbon deposits hinder engine performance. Occasionally, however, incomplete combustion is deliberately caused; specific carbon blacks (particularly lampblack, a pigment for ink) are produced by the incomplete combustion of natural gas.

© Michael C. Slabaugh

■ **FIGURE 1.20** A luminous yellow flame from a laboratory burner produces a deposit of carbon when insufficient air (O_2) is mixed with the gaseous fuel.

Concept Summary

ThomsonNOW™ Sign in at **www.thomsonedu.com** to:

- Assess your understanding with Exercises keyed to each learning objective.
- Check your readiness for an exam by taking the **Pre-test** and exploring the modules recommended in your **Personalized Learning Plan.**

Carbon: The Element of Organic Compounds. Organic compounds contain carbon, and organic chemistry is the study of those compounds. Inorganic chemistry is the study of the elements and all noncarbon compounds. **OBJECTIVE 2, Exercise 1.4** Carbon compounds are of tremendous everyday importance to life on Earth and are the basis of all life processes. **OBJECTIVE 1, Exercise 1.2**

Organic and Inorganic Compounds Compared. The properties of organic and inorganic compounds often differ, largely as a result of bonding differences. Organic compounds contain primarily covalent bonds, whereas ionic bonding is more prevalent in inorganic compounds. **OBJECTIVE 3, Exercise 1.8**

Bonding Characteristics and Isomerism. Large numbers of organic compounds are possible because carbon atoms link to form chains and networks. An additional reason for the existence of so many organic compounds is the phenomenon of isomerism. Isomers are compounds that have the same molecular formula but different arrangements of atoms. **OBJECTIVE 4, Exercise 1.20**

Functional Groups: The Organization of Organic Chemistry. All organic compounds are grouped into classes based on characteristic features called functional groups. Compounds with their functional groups are represented by two types of structural formulas. Expanded structural formulas show all covalent bonds, whereas condensed structural formulas show no covalent bonds or only selected bonds. **OBJECTIVE 5, Exercise 1.24**

Alkane Structures. Alkanes are hydrocarbons that contain only single covalent bonds and can be represented by the formula C_nH_{2n+2}. Alkanes possess a three-dimensional geometry in which each carbon is surrounded by four bonds directed to the corners of a tetrahedron. Methane, the simplest alkane, is an important fuel (natural gas) and a chemical feedstock for the preparation of other organic compounds. The number of structural isomers possible for an alkane increases dramatically with the number of carbon atoms present in the molecule. The straight-chain isomer is called a normal alkane; others are called branched isomers. **OBJECTIVE 6, Exercise 1.28**

Conformations of Alkanes. Rotation about the single bonds between carbon atoms allows alkanes to exist in many different conformations. When an alkane is drawn using only two dimensions, the structure can be represented in a variety of ways as long as the order of bonding is not changed. **OBJECTIVE 7, Exercise 1.30**

Alkane Nomenclature. Some simple alkanes are known by common names. More complex compounds are usually named using the IUPAC system. The characteristic IUPAC ending for alkanes is *-ane*. **OBJECTIVE 8, Exercise 1.34**

Cycloalkanes. These are alkanes in which the carbon atoms form a ring. The prefix *cyclo-* is used in the names of these compounds to indicate their cyclic nature. **OBJECTIVE 9, Exercise 1.44**

The Shape of Cycloalkanes. The carbon atom rings of cycloalkanes are usually shown as planar, although only cyclopropane is planar. Because rotation about the single bonds in the ring is restricted, certain disubstituted cycloalkanes can exist as geometric (cis-trans) isomers. **OBJECTIVE 10, Exercise 1.54**

Physical Properties of Alkanes. The physical properties of alkanes are typical of all hydrocarbons: nonpolar, insoluble in water, less dense than water, and increasing melting and boiling points with increasing molecular weight. **OBJECTIVE 11, Exercise 1.56**

Alkane Reactions. Alkanes are relatively unreactive and remain unchanged by most reagents. The reaction that is most significant is combustion. **OBJECTIVE 12, Exercise 1.60**

Key Terms and Concepts

Alkane (1.5)
Alkyl group (1.7)
Branched alkane (1.5)
cis- (1.9)
Condensed structural formula (1.4)
Conformations (1.6)
Cycloalkane (1.8)
Expanded structural formula (1.4)
Functional group (1.4)
Geometric isomers (1.9)
Homologous series (1.10)
Hybrid orbital (1.3)
Hydrocarbon (1.5)
Hydrophobic (1.10)
Inorganic chemistry (1.1)
Isomerism (1.3)
Normal alkane (1.5)
Organic chemistry (1.1)
Organic compound (1.1)
Saturated hydrocarbon (1.5)
Stereoisomers (1.9)
Structural isomers (1.3)
trans- (1.9)

Key Reactions

1. Complete combustion of alkanes (Section 1.11):

$$\text{Alkane} + O_2 \rightarrow CO_2 + H_2O$$

2. Incomplete combustion of alkanes (Section 1.11):

$$\text{Alkane} + O_2 \rightarrow CO \text{ (or C)} + H_2O$$

Exercises

SYMBOL KEY

Even-numbered exercises are answered in Appendix B.

Blue-numbered exercises are more challenging.

■ denotes exercises available in ThomsonNow and assignable in OWL.

ThomsonNOW™ To assess your understanding of this chapter's topics with sample tests and other resources, sign in at **www.thomsonedu.com.**

CARBON: THE ELEMENT OF ORGANIC COMPOUNDS (SECTION 1.1)

1.1 Why were the compounds of carbon originally called organic compounds?

1.2 Name at least six items you recognize to be composed of organic compounds.

1.3 Describe what Wöhler did that made the vital force theory highly questionable.

1.4 What is the unique structural feature shared by all organic compounds?

1.5 Classify each of the following compounds as organic or inorganic:

a. KBr
b. H_2O
c. H—C≡C—H
d. LiOH
e. $CH_3—NH_2$

ORGANIC AND INORGANIC COMPOUNDS COMPARED (SECTION 1.2)

1.6 What kind of bond between atoms is most prevalent among organic compounds?

1.7 Are the majority of all compounds that are insoluble in water organic or inorganic? Why?

1.8 ■ Indicate for each of the following characteristics whether it more likely describes an inorganic or an organic compound. Give one reason for your answer.

a. This compound is a liquid that readily burns.
b. A white solid upon heating is found to melt at 735°C.
c. A liquid added to water floats on the surface and does not dissolve.
d. This compound exists as a gas at room temperature and ignites easily.
e. A solid substance melts at 65°C.

1.9 Devise a test, based on the general properties in Table 1.1, that you could use to quickly distinguish between the substances in each of the following pairs:

a. Gasoline (liquid, organic) and water (liquid, inorganic)
b. Naphthalene (solid, organic) and sodium chloride (solid, inorganic)
c. Methane (gaseous, organic) and hydrogen chloride (gaseous, inorganic)

1.10 Explain why organic compounds are nonconductors of electricity.

1.11 Explain why the rate of chemical reactions is generally slow for organic compounds and usually fast for inorganic compounds.

BONDING CHARACTERISTICS AND ISOMERISM (SECTION 1.3)

1.12 Give two reasons for the existence of the tremendous number of organic compounds.

1.13 How many of carbon's electrons are unpaired and available for bonding according to an sp^3 hybridization model?

1.14 Describe what atomic orbitals overlap to produce a carbon–hydrogen bond in CH_4.

1.15 What molecular geometry exists when a central carbon atom bonds to four other atoms?

1.16 Compare the shapes of unhybridized p and hybridized sp^3 orbitals.

1.17 Use Example 1.1 and Tables 1.2 and 1.6 to determine the number of covalent bonds formed by atoms of the following elements: carbon, hydrogen, oxygen, nitrogen, and bromine.

1.18 Complete the following structures by adding hydrogen atoms where needed:

a. C—C—C
b. C—C=C
c. C—C(=O)
d. C—C—N

1.19 Complete the following structures by adding hydrogen atoms where needed.

```
         C
         |
a.  C—C—C—C
```

b. $C-C\equiv C$

```
           O
           ||
c.    C—C—N
```

```
           O
           ||
d.   O—C—C—C
```

1.20 ■ Which of the following pairs of compounds are structural isomers?

a. $CH_3-CH=CH-CH_3$ and $CH_3-CH_2-CH_2-CH_3$

```
                                                      CH3
                                                       |
b. CH3—CH2—CH2—CH2—CH3   and   CH3—C—CH3
                                                       |
                                                      CH3
```

```
                 CH3                              O
                  |                                ||
c. CH3—CH2—CH—OH    and    CH3—C—CH2—CH3
```

```
               O                     O
               ||                     ||
d. CH3—CH2—C—H    and    CH3—C—CH3
```

e. $CH_3-CH_2-CH_2-NH_2$ and $CH_3-CH_2-NH-CH_3$

1.21 Group all the following compounds together that represent structural isomers of each other:

```
                      O
                      ||
a.  CH3—CH2—C—OH
```

```
            O
            ||
b.  CH3—C—CH2—OH
```

c. $CH_3-CH_2-CH_2-OH$

```
            O
            ||
d.  CH3—C—CH2—CH3
```

```
            OH   O
            |    ||
e.  CH3—CH—C—H
```

```
            O
            ||
f.  CH3—C—O—CH3
```

g. $HO-CH_2-CH_2-CH_2-OH$

1.22 On the basis of the number of covalent bonds possible for each atom, determine which of the following structural formulas are correct. Explain what is wrong with the incorrect structures.

```
       H        H
       |        |
a. H—C—H—C—H
       |        |
       H        H
```

```
            H    H
            |    |
b. H—N—C—C—H
       |    |    |
       H    H    H
```

```
       H          O            H
       |          ||           |
c. H—C————C————C—H
       |          |            |
       H      H—C—H        H
                  |
                  H
```

d. $CH_3-C=C-CH_3$

```
                 OH   O
                 |    ||
e. CH3—CH—CH—C—H
         |
        CH3
```

FUNCTIONAL GROUPS: THE ORGANIZATION OF ORGANIC CHEMISTRY (SECTION 1.4)

1.23 Identify each of the following as a condensed structural formula, expanded structural formula, or molecular formula:

```
       H   H   H
       |   |   |
a. H—C—C=C
       |       |
       H       H
```

b. $CH_3CH_2CH_2-OH$

c. $C_6H_{12}O_6$

```
              O
              ||
d.  H2N—C—CH2CH3
```

1.24 ■ Write a condensed structural formula for the following compounds:

```
       H   H   H           O   H   H
       |   |   |           ||  |   |
a. H—C—C—C—C=C—C—C—C—H
       |   |   |   |   |       |   |
       H   H   H   H   H       H   H
```

```
       H   H   O
       |   |   ||
b. H—C—C—C—N—H
       |   |       |
       H   H       H
```

1.25 Write a condensed structural formula for the following compounds:

```
                    H  H
                     \/
       H             N    H   H
       |             |    |   |
a. H—C———————C—C=C—H
       |             |
       H         H—C—H
                     |
                     H
```

```
       O  H        H  O
       ‖  |        |  ‖
b. H—O—C—C—O—C—C—H
          |        |
        H—C—H      H
          |
          H
```

1.26 Write an expanded structural formula for the following:

```
       O
       ‖
a. H—C—CH2—CH2—NH2
```

```
b. CH3—CH—O—CH3
          |
          CH3
```

ALKANE STRUCTURES (SECTION 1.5)

1.27 The name of the normal alkane containing 9 carbon atoms is nonane. What are the molecular and condensed structural formulas for nonane?

1.28 ■ Classify each of the following compounds as a normal alkane or a branched alkane:

```
a. CH3—CH—CH3          d. CH3—CH2—CH2
          |                         |
          CH3                CH3—CH2
```

```
                             CH3         CH3
                             |           |
b. CH3—CH2—CH2          e. CH2—CH2—CH
              |                          |
              CH2—CH3                    CH3
```

```
        CH3                          CH3
        |                            |
c. CH2—CH2             f. CH3—CH2—C—CH3
   |                                 |
   CH3                               CH3
```

CONFORMATIONS OF ALKANES (SECTION 1.6)

1.29 Why are different conformations of an alkane not considered structural isomers?

1.30 ■ Which of the following pairs represent structural isomers, and which are simply the same compound?

```
a. H3C    CH2    and   CH3CH2CH2CH3
      \  /   \
      CH2     CH3
```

```
b. CH3—CH2—CH2   and   CH3CH2CH2CH2CH2CH3
              |
   CH3—CH2—CH2
```

```
c. CH3—CH—CH3   and   CH3CH2CH2CH3
          |
          CH3
```

```
   CH3
   |
d. CH—CH2—CH3    and    CH3—CH2—CH—CH3
   |                                |
   CH3                              CH3
```

ALKANE NOMENCLATURE (SECTION 1.7)

1.31 ■ For each of the following carbon skeletons, give the number of carbon atoms in the longest continuous chain:

```
   C—C                          C—C—C
     |                          |
a. C—C—C—C             c. C—C—C—C
     |  |                       |
     C  C—C                     C—C
```

```
     C—C—C—C—C
       |
b.     C—C—C—C
             |
       C—C—C
           |
           C—C
```

1.32 For each of the following carbon skeletons, give the number of carbon atoms in the longest continuous chain:

```
       C—C
       |
a.  C—C—C—C—C
```

```
       C
       |
b.  C—C—C—C
          |
          C—C—C
```

c. 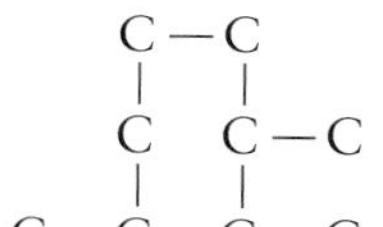

1.33 ■ Identify the following alkyl groups:

```
a. CH3—CH—
          |
          CH3
```

b. $CH_3—CH_2—CH_2—CH_2—$

```
c. CH3—CH—CH2—
          |
          CH3
```

d. $CH_3—$

1.34 ■ Give the correct IUPAC name for each of the following alkanes:

```
   CH3—CH2
        |
a. CH3—CH—CH2—CH3
```

```
         CH3
         |
b. CH3—CH
         |
         CH3
```

```
                    CH2 — CH3
                    |
c. CH3 — CH2 — CH — CH — CH2
                         |       |
                        CH3    CH — CH2 — CH2 — CH3
                                 |
                                CH2CH3
```

```
                    CH3
                    |
                    CH2
                    |
                    CH2
                    |
d. CH3 — CH2 — CH — CH2 — CH — CH3
                                 |
          CH3 — CH2 — CH2 — CH
                                 |
                                CH2 — CH3
```

```
          CH2 — CH2 — CH2 — CH3
          |
e. CH3 — CH — CH2 — CH — CH2CH3
                         |
                        CH2 — CH3
```

1.35 Give the correct IUPAC name for each of the following alkanes:

```
a. CH3 — CH2 — CH — CH3
                 |
                CH3
```

```
          CH3
          |
b. CH3 — C — CH3
          |
          CH3
```

```
          CH2 — CH3
          |
c. CH3 — CH — CH — CH3
                |
               CH3
```

```
                   CH3
                   |
  CH3              CH2
  |                |
d. CH2 — CH — C — CH2 — CH — CH2 — CH3
          |    |          |
         CH3  CH2        CH2
               |          |
              CH3        CH — CH3
                          |
                         CH3
```

```
               CH3    CH3        CH3
               |      |          |
e. CH3 — CH2 — CH — CH — CH — CH2 — CH — CH3
                    |
                   CH2 — CH2 — CH2 — CH3
```

1.36 ■ Draw a condensed structural formula for each of the following compounds:

a. 3-ethylpentane
b. 2,2-dimethylbutane
c. 4-ethyl-3,3-dimethyl-5-propyldecane
d. 5-*sec*-butyldecane

1.37 Draw a condensed structural formula for each of the following compounds:

a. 2,3-dimethylbutane
b. 3-isopropylheptane
c. 5-*t*-butyl-2,3-dimethyloctane
d. 4-ethyl-4-methyloctane

1.38 Draw the condensed structural formula for each of the three structural isomers of C_5H_{12}, and give the correct IUPAC names.

1.39 Isooctane is 2,2,4-trimethylpentane. Draw structural formulas for and name a branched heptane, hexane, pentane, and butane that are structural isomers of isooctane.

1.40 Draw structural formulas for the compounds and give correct IUPAC names for the five structural isomers of C_6H_{14}.

1.41 The following names are incorrect, according to IUPAC rules. Draw the structural formulas and tell why each name is incorrect. Write the correct name for each compound.

a. 2,2-methylbutane
b. 4-ethyl-5-methylheptane
c. 2-ethyl-1,5-dimethylhexane

1.42 The following names are incorrect, according to IUPAC rules. Draw the structural formulas and tell why each name is incorrect. Write the correct name for each compound.

a. 1,2-dimethylpropane
b. 3,4-dimethylpentane
c. 2-ethyl-4-methylpentane
d. 2-bromo-3-ethylbutane

CYCLOALKANES (SECTION 1.8)

1.43 The general formula for alkanes is C_nH_{2n+2}. Write a general formula for cycloalkanes.

1.44 ■ Write the correct IUPAC name for each of the following:

a.

c. CH3 CH3

b.

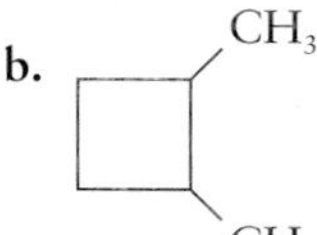

d. CH3 CH3 CH3

1.45 Write the correct IUPAC name for each of the following:

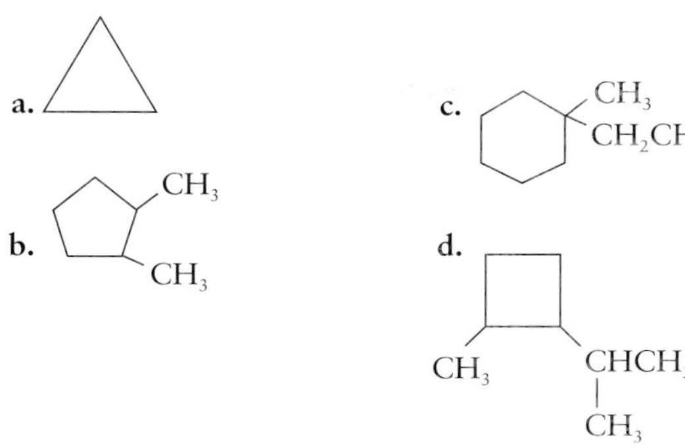

1.46 Draw the structural formulas corresponding to each of the following IUPAC names:

a. ethylcyclobutane

b. 1,1,2,5-tetramethylcyclohexane

c. 1-butyl-3-isopropylcyclopentane

1.47 Draw the structural formulas corresponding to each of the following IUPAC names:

a. 1,2-diethylcyclopentane

b. 1,2,4-trimethylcyclohexane

c. propylcyclobutane

1.48 Which of the following pairs of cycloalkanes represent structural isomers?

a. [cyclobutane with CH_3, CH_3] — [cyclobutane with CH_3, CH_3]

b. [cyclopentane with CH_2CH_3, CH_3] — [cyclopentane with CH_2CH_3, CH_3]

c. [cyclopropane with CH_2CH_3] — [cyclopropane with CH_3, CH_3]

d. [cyclohexane with CH_3, CH_3] — [cyclohexane with CH_3, CH_3]

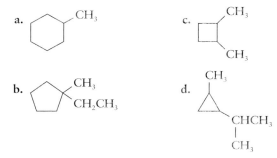

1.49 Draw structural formulas for the five structural isomers of C_5H_{10} that are cycloalkanes.

THE SHAPE OF CYCLOALKANES (SECTION 1.9)

1.50 Why does cyclohexane assume a chair form rather than a planar hexagon?

1.51 Explain the difference between geometric and structural isomers.

1.52 ■ Which of the following cycloalkanes could show geometric isomerism? For each that could, draw structural formulas, and name both the *cis*- and the *trans*- isomers.

a. [cyclohexane with CH_3]

b. [cyclopentane with CH_3, CH_2CH_3]

c. [cyclobutane with CH_3, CH_3]

d. [cyclopropane with CH_3, $CHCH_3$ / CH_3]

1.53 Draw structural formulas for *cis*- and *trans*-1,3-dibromocyclobutane.

1.54 ■ Using the prefix *cis*- or *trans*-, name each of the following:

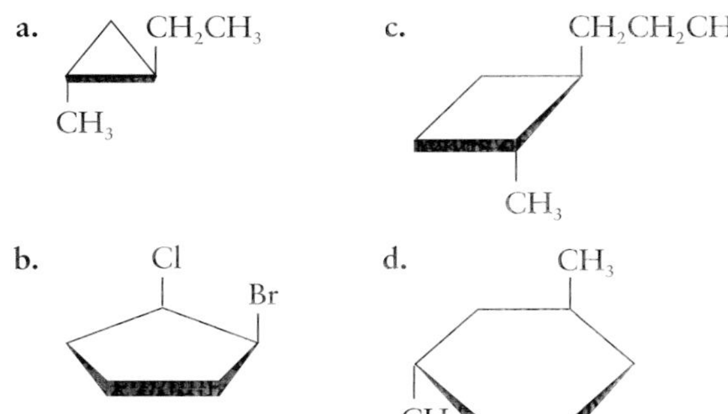

1.55 For each of the following molecular formulas, give the structural formulas requested. In most cases, there are several possible structures.

a. C_6H_{14}, a normal alkane

b. C_6H_{14}, a branched alkane

c. C_5H_{12}, a pair of conformations

d. C_5H_{12}, a pair of structural isomers

e. C_5H_{10}, a cyclic hydrocarbon

f. C_6H_{12}, two cycloalkane geometric isomers

g. C_6H_{12}, a cycloalkane that has no geometric isomers

PHYSICAL PROPERTIES OF ALKANES (SECTION 1.10)

1.56 ■ The compound decane is a straight-chain alkane. Predict the following:

a. Is decane a solid, liquid, or gas at room temperature?

b. Is it soluble in water?

c. Is it soluble in hexane?

d. Is it more or less dense than water?

1.57 Explain why alkanes of low molecular weight have lower melting and boiling points than water.

1.58 Suppose you have a sample of 2-methylhexane and a sample of 2-methylheptane. Which sample would you expect to have the higher melting point? Boiling point?

1.59 Identify (circle) the alkanelike portions of the following molecules:

a.
```
            O
            ‖
H2N — CH — C — OH
      |
      CH — CH3
      |
      CH2
      |
      CH3
```
isoleucine (an amino acid)

b.
```
         CH3—CH2—CH2—CH2—CH2—CH2—CH2—CH2
                                      |
       O                              |
       ‖                              |
Na+ -O—C—CH2—CH2—CH2—CH2—CH2—CH2—CH2
```
sodium palmitate (a soap)

c.
```
     H2                                   O
     C                                    ‖
H2C     CH — CH2 — CH2 — CH2 — CH2 — C — OH
   S—S
```
lipoic acid (a component of a coenzyme)

d. CH_3

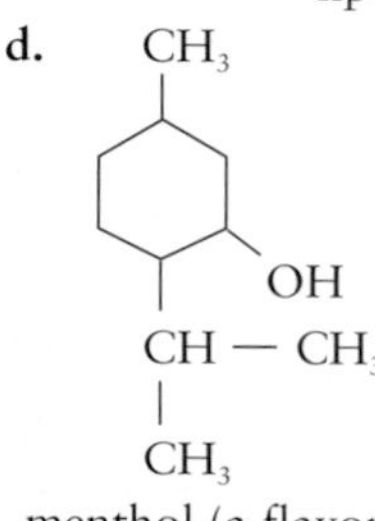

OH

CH — CH_3

|

CH_3

menthol (a flavoring)

ALKANE REACTIONS (SECTION 1.11)

1.60 ■ Write a balanced equation to represent the complete combustion of each of the following:

a. Butane

b.
```
      CH3
      |
CH3 — C — CH3
      |
      CH3
```

c. CH_3

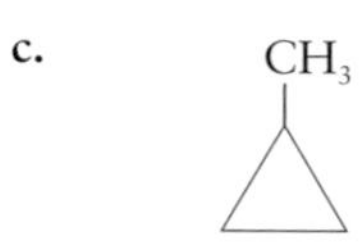

1.61 Write a balanced equation to represent the complete combustion of each of the following:

a. Ethane

b.
```
CH3 — CH — CH2 — CH3
      |
      CH3
```

c.

1.62 Write a balanced equation for the incomplete combustion of hexane, assuming the formation of carbon monoxide and water as the products.

1.63 Why is it dangerous to relight a furnace when a foul odor is present?

ADDITIONAL EXERCISES

1.64 Using the concept of dispersion forces, explain why most cycloalkanes have higher boiling points than normal alkanes with the same number of carbon atoms.

1.65 Draw a Lewis electron dot formula for ethane ($CH_3—CH_3$). Explain why ethane molecules do not hydrogen-bond.

1.66 Your sometimes inept lab technician performed experiments to determine the vapor pressure of pentane, hexane, and heptane at 20°C. He gave you back the numbers of 113.9, 37.2, and 414.5 torr without identifying the compounds. Which vapor pressure value goes with which compound?

1.67 How many grams of water will be produced by the complete combustion of 10.0 g of methane (CH_4)?

1.68 How many liters of air at STP are needed to completely combust 1.00 g of methane (CH_4)? Air is composed of about 21% v/v oxygen (O_2).

ALLIED HEALTH EXAM CONNECTION

Reprinted with permission from Nursing School and Allied Health Entrance Exams, COPYRIGHT 2005 Petersons.

1.69 ■ Which of the following are examples of organic compounds?

a. $CaCO_3$

b. NH_3

c. NaCl

d. $C_6H_{12}O_6$

1.70 Use the generic formula for alkanes (C_nH_{2n+2}) to derive molecular and condensed structural formulas for:

a. Propane, 3 carbon atoms

b. Octane, 8 carbon atoms

c. Butane, 4 carbon atoms

CHEMISTRY FOR THOUGHT

1.71 Would you expect a molecule of urea produced in the body to have any different physical or chemical properties from a molecule of urea prepared in a laboratory?

1.72 Why might the study of organic compounds be important to someone interested in the health or life sciences?

1.73 Why do very few aqueous solutions of organic compounds conduct electricity?

1.74 The ski wax being examined in Figure 1.2 has a relatively low melting point. What does that fact reveal about the forces between molecules?

1.75 Charcoal briquettes sometimes burn with incomplete combustion when the air supply is limited. Why would it be hazardous to place a charcoal grill inside a home or a camper in an attempt to keep warm?

1.76 If carbon did not form hybridized orbitals, what would you expect to be the formula of the simplest compound of carbon and hydrogen?

1.77 What types of sports equipment are made from graphite fibers besides that shown in Figure 1.8?

1.78 A semi truck loaded with cyclohexane overturns during a rainstorm, spilling its contents over the road embankment. If the rain continues, what will be the fate of the cyclohexane?

1.79 On the way home from school, you drove through a construction zone, resulting in several tar deposits on the car's fender. What substances commonly found in the kitchen might help in removing the tar deposits?

1.80 Oil spills along coastal shores can be disastrous to the environment. What physical and chemical properties of alkanes contribute to the consequences of an oil spill?

1.81 Why might some farmers hesitate to grow and sell organic produce?

CHAPTER 2

Unsaturated Hydrocarbons

LEARNING OBJECTIVES

When you have completed your study of this chapter, you should be able to:

1. Classify unsaturated hydrocarbons as alkenes, alkynes, or aromatics. (Section 2.1)
2. Write the IUPAC names of alkenes from their molecular structures. (Section 2.1)
3. Predict the existence of geometric (cis-trans) isomers from formulas of compounds. (Section 2.2)
4. Write the names and structural formulas for geometric isomers. (Section 2.2)
5. Write equations for addition reactions of alkenes, and use Markovnikov's rule to predict the major products of certain reactions. (Section 2.3)
6. Write equations for addition polymerization, and list uses for addition polymers. (Section 2.4)
7. Write the IUPAC names of alkynes from their molecular structures. (Section 2.5)
8. Classify organic compounds as aliphatic or aromatic. (Section 2.6)
9. Name and draw structural formulas for aromatic compounds. (Section 2.7)
10. Recognize uses for specific aromatic compounds. (Section 2.8)

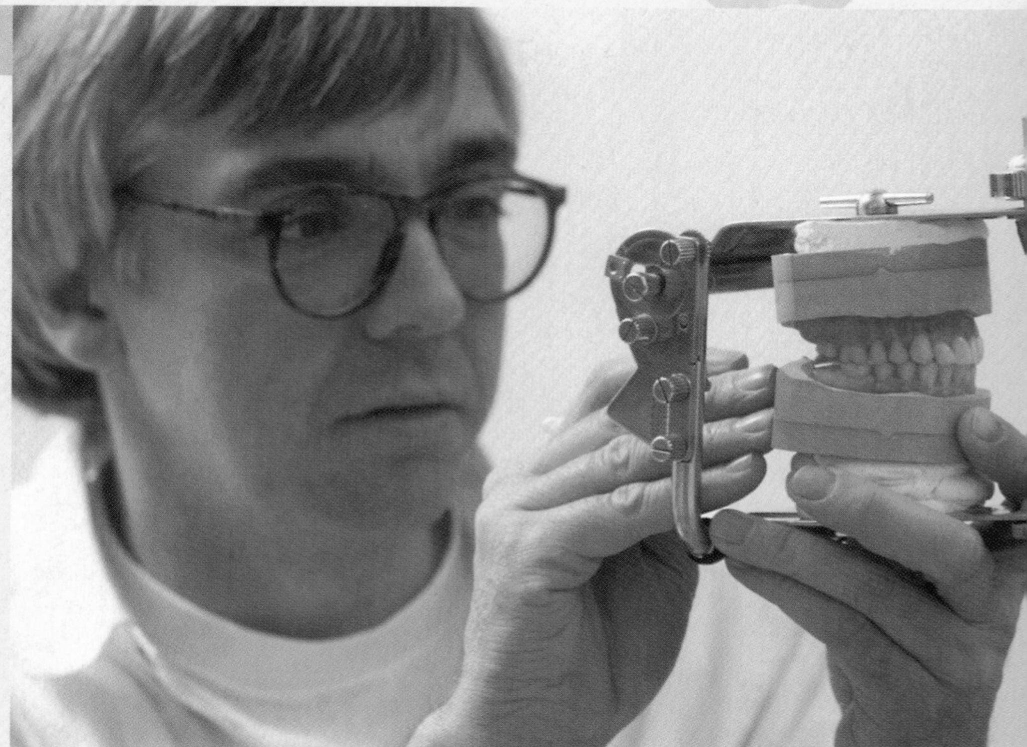

Dental technicians make dental prostheses such as false teeth, crowns, and bridges according to orders placed by dentists. In some large dental laboratories, the demand is great enough for certain types of devices that technicians become specialists in working with specific materials such as plastics, gold, or porcelain. In this chapter, you will be introduced to the molecular nature of one type of plastic and the versatility of these useful products of human ingenuity.

Unsaturated hydrocarbons contain one or more double or triple bonds between carbon atoms and belong to one of three classes: alkenes, alkynes, or aromatic hydrocarbons. **Alkenes** contain one or more double bonds, **alkynes** contain one or more triple bonds, and **aromatic hydrocarbons** contain three double bonds alternating with three single bonds in a six-carbon ring. Ethylene (the simplest alkene), acetylene (the simplest alkyne), and benzene (the simplest aromatic) are represented by the following structural formulas:

$$\underset{\text{ethylene (an alkene)}}{\mathrm{H_2C{=}CH_2}} \qquad \underset{\text{acetylene (an alkyne)}}{\mathrm{H{-}C{\equiv}C{-}H}} \qquad \underset{\text{benzene (an aromatic)}}{\mathrm{C_6H_6}}$$

Alkenes and alkynes are called *unsaturated* because more hydrogen atoms can be added in somewhat the same sense that more solute can be added to an unsaturated solution. Benzene and other aromatic hydrocarbons also react to add hydrogen atoms; in general, however, they have chemical properties very different from those of alkenes and alkynes.

ThomsonNOW Throughout the chapter this icon introduces resources on the ThomsonNOW website for this text. Sign in at **www.thomsonedu.com** to:

- Evaluate your knowledge of the material
- Take an exam prep quiz
- Identify areas you need to study with a **Personalized Learning Plan.**

unsaturated hydrocarbon
A hydrocarbon containing one or more multiple bonds.

alkene
A hydrocarbon containing one or more double bonds.

alkyne
A hydrocarbon containing one or more triple bonds.

aromatic hydrocarbons
Any organic compound that contains the characteristic benzene ring or similar feature.

2.1 The Nomenclature of Alkenes

LEARNING OBJECTIVES

1. Classify unsaturated hydrocarbons as alkenes, alkynes, or aromatics.
2. Write the IUPAC names of alkenes from their molecular structure.

The general formula for alkenes is C_nH_{2n} (the same as that for cycloalkanes). The simplest members are well known by their common names, ethylene and propylene:

$$\underset{\text{ethylene, } C_2H_4}{CH_2{=}CH_2} \qquad \underset{\text{propylene, } C_3H_6}{CH_3{-}CH{=}CH_2}$$

Three structural isomers have the formula C_4H_8:

$$CH_3{-}CH_2{-}CH{=}CH_2 \qquad CH_3{-}CH{=}CH{-}CH_3 \qquad CH_3{-}\underset{\displaystyle CH_3}{\underset{|}{C}}{=}CH_2$$

The number of structural isomers increases rapidly as the number of carbons increases because, besides variations in chain length and branching, variations occur in the position of the double bond. IUPAC nomenclature is extremely useful in differentiating among these many alkene compounds.

The IUPAC rules for naming alkenes are similar to those used for the alkanes, with a few additions to indicate the presence and location of double bonds.

ThomsonNOW Go to Coached Problems to practice **naming alkenes.**

Step 1. Name the longest chain that contains the double bond. The characteristic name ending is *-ene.*
Step 2. Number the longest chain of carbon atoms so that the carbon atoms joined by the double bond have numbers as low as possible.
Step 3. Locate the double bond by the lower-numbered carbon atom bound by the double bond.
Step 4. Locate and name attached groups.
Step 5. Combine the names for the attached groups and the longest chain into the name.

EXAMPLE 2.1

Name the following alkenes:

a. $CH_3-CH=CH-CH_3$

b. $CH_3-CH(CH_3)-CH=CH_2$

c. $CH_3-CH_2-C(=CH_2)-CH_2-CH_2-CH_3$ (drawn with CH_3-CH_2 above and $CH_3-CH_2-CH_2$ below the $C=CH_2$ carbon)

d. cyclopentene ring bearing a CH_2-CH_3 group

Solution

a. The longest chain containing a double bond has four carbon atoms. The four-carbon alkane is butane. Thus, the compound is a butene:

$$\overset{1}{C}H_3-\overset{2}{C}H=\overset{3}{C}H-\overset{4}{C}H_3$$

The chain can be numbered from either end because the double bond will be between carbons 2 and 3 either way. The position of the double bond is indicated by the lower-numbered carbon atom that is double bonded, carbon 2 in this case. The name is 2-butene.

b. To give lower numbers to the carbons bound by the double bond, the chain is numbered from the right:

$$\overset{4}{C}H_3-\overset{3}{C}H(CH_3)-\overset{2}{C}H=\overset{1}{C}H_2 \quad \text{not} \quad \overset{1}{C}H_3-\overset{2}{C}H(CH_3)-\overset{3}{C}H=\overset{4}{C}H_2$$

Thus, the compound is a 1-butene with an attached methyl group on carbon 3. Therefore, the name is 3-methyl-1-butene.

c. Care must be taken to select the longest chain containing the double bond. This compound is named as a pentene and not as a hexene because the double bond is not contained in the six-carbon chain:

$$CH_3-CH_2-\overset{2}{C}(=\overset{1}{C}H_2)-\overset{3}{C}H_2-\overset{4}{C}H_2-\overset{5}{C}H_3 \quad \text{not} \quad \overset{1}{C}H_3-\overset{2}{C}H_2-\overset{3}{C}(=CH_2)-\overset{4}{C}H_2-\overset{5}{C}H_2-\overset{6}{C}H_3$$

The compound is a 1-pentene with an ethyl group at position 2. Therefore, the name is 2-ethyl-1-pentene.

d. In cyclic alkenes, the ring is numbered so as to give the lowest possible numbers to the double-bonded carbons (they become carbons 1 and 2). The numbering direction around the ring is chosen so that attached groups are located on the lowest-numbered carbon atoms possible. Thus, the name is 3-ethylcyclopentene:

(cyclopentene ring numbered 1–5, double bond between C1 and C2, CH_2CH_3 on C3)

Notice that it is not called 3-ethyl-1-cyclopentene because the double bond is always between carbons 1 and 2, and therefore its position need not be indicated.

Learning Check 2.1 Give the IUPAC name for each of the following:

a. $\overset{\displaystyle Br}{\overset{|}{CH_2}}-CH=CH_2$

b. $CH_2=\underset{\underset{\displaystyle CH_3}{\underset{|}{\underset{\displaystyle CH_2}{|}}}}{C}-CH_2-CH_2-CH_3$

c. (cyclohexene ring with two adjacent CH_3 substituents) CH_3 CH_3

Some compounds contain more than one double bond per molecule. Molecules of this type are important components of natural and synthetic rubber and other useful materials. The nomenclature of these compounds is the same as for the alkenes with one double bond, except that the endings *-diene, -triene,* and the like are used to denote the number of double bonds. Also, the locations of all the multiple bonds must be indicated in all molecules, including those with rings:

$$\overset{1}{C}H_2=\overset{2}{C}H-\overset{3}{C}H=\overset{4}{C}H_2$$

1,3-butadiene

(cyclohexadiene ring, carbons numbered 1, 2, 3, 4)

1,3-cyclohexadiene

CHEMISTRY AROUND US 2.1 Seeing the Light

Cis-trans isomerism is important in several biological processes, one of which is vision. When light strikes the retina, a cis double bond in the compound *retinal* (structurally related to vitamin A) is converted to a trans double bond. The conversion triggers a chain of events that finally results in our being able to see.

In a series of steps, *trans*-retinal is enzymatically converted back to *cis*-retinal so that the cycle can be repeated. Bright light temporarily destroys our ability to see in dim light because large quantities of *cis*-retinal are rapidly converted to the trans isomer by the bright light. It takes time for conversion of the *trans*-retinal back to *cis*-retinal.

(Structure of *cis*-retinal, labels: CH_3, H, C, CH_3, H, cis double bond, H, C, C, C, C, C, CH_3, CH_3, H, H, C, H, CH_3, C, C=O, H)

cis-retinal

Light ↓↑ Several steps

trans-retinal

(Structure of *trans*-retinal, labels: CH_3, H, CH_3, H, CH_3, O, C, C, C, C, C—H, C, C, C, C, CH_3, CH_3, H, H, H, H)

trans double bond

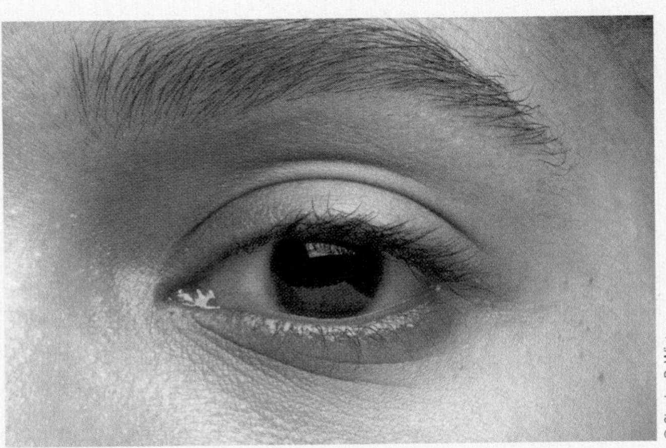

© Charles D. Winters

The vision process depends on a cis-trans reaction.

EXAMPLE 2.2

Name the following compounds:

a. 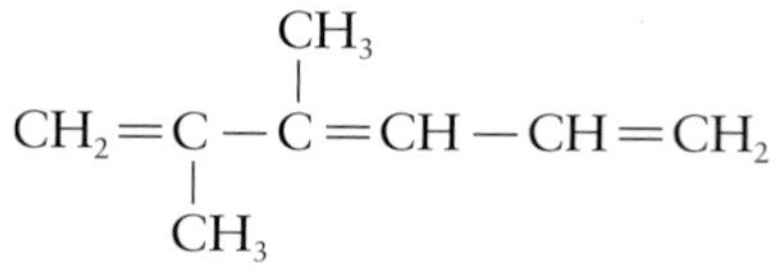 $CH_2{=}C(CH_3){-}C(CH_3){=}CH{-}CH{=}CH_2$

b. 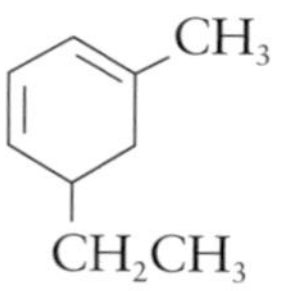 (cyclohexadiene ring bearing CH_3 and CH_2CH_3 substituents)

Solution

a. This compound is a methyl-substituted hexatriene. The chain is numbered from the end nearest the branch because the direction of numbering, again, makes no difference in locating the double bonds correctly. The name is 2,3-dimethyl-1,3,5-hexatriene:

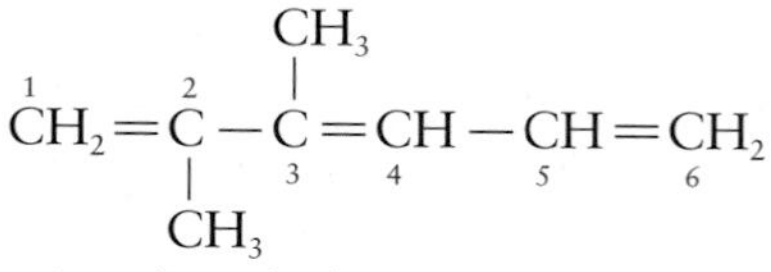

$$\overset{1}{C}H_2{=}\overset{2}{C}(CH_3){-}\overset{3}{C}(CH_3){=}\overset{4}{C}H{-}\overset{5}{C}H{=}\overset{6}{C}H_2 \quad \text{and not} \quad \overset{6}{C}H_2{=}\overset{5}{C}(CH_3){-}\overset{4}{C}(CH_3){=}\overset{3}{C}H{-}\overset{2}{C}H{=}\overset{1}{C}H_2$$

2,3-dimethyl-1,3,5-hexatriene   4,5-dimethyl-1,3,5-hexatriene

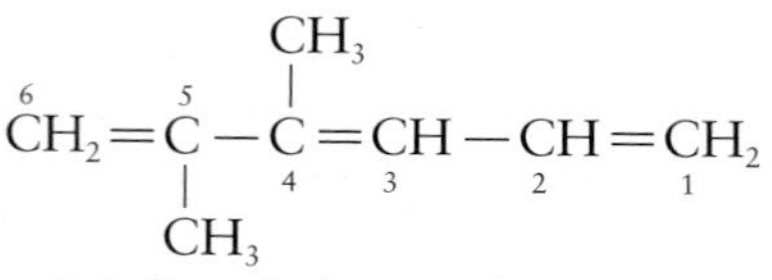

b. This compound is a substituted cyclohexadiene. The ring is numbered as shown. The name is 5-ethyl-1-methyl-1,3-cyclohexadiene:

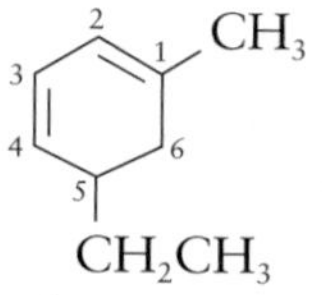

5-ethyl-1-methyl-1,3-cyclohexadiene and not 6-ethyl-4-methyl-1,3-cyclohexadiene

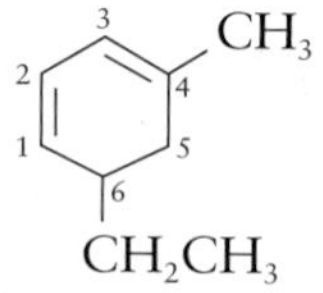

■ **Learning Check 2.2** Give the IUPAC name for each of the following:

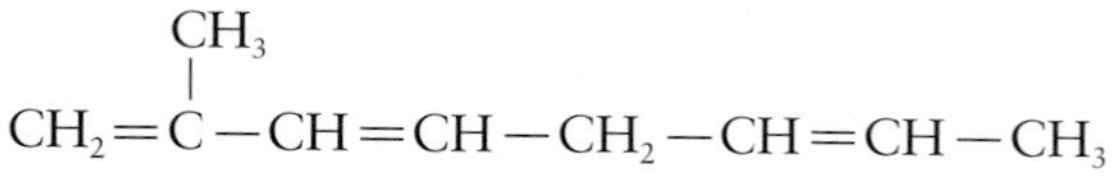

a. 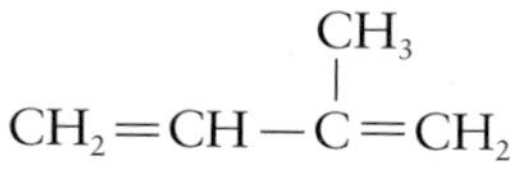 $CH_2{=}CH{-}C(CH_3){=}CH_2$

b. $CH_2{=}C(CH_3){-}CH{=}CH{-}CH_2{-}CH{=}CH{-}CH_3$

c. Br (bromo-substituted cycloheptatriene ring)

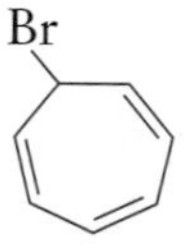

2.2 The Geometry of Alkenes

LEARNING OBJECTIVES

3. Predict the existence of geometric (cis-trans) isomers from the formulas of compounds.
4. Write the names and structural formulas for geometric isomers.

The hybridization of atomic orbitals discussed in Section 11.3 to explain the bonding characteristics of carbon atoms bonded to four other atoms can also be used to describe alkenes, compounds in which some carbon atoms are bonded to only three atoms. This hybridization involves mixing a $2s$ orbital and two $2p$ orbitals of a carbon atom to form three hybrid sp^2 orbitals (see ■ Figure 2.1).

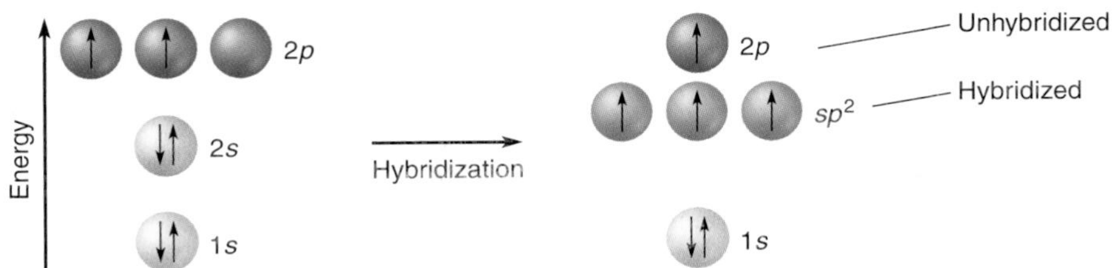

FIGURE 2.1 A representation of sp^2 hybridization of carbon. During hybridization, two of the 2*p* orbitals mix with the single 2*s* orbital to produce three sp^2 hybrid orbitals. One 2*p* orbital is not hybridized and remains unchanged.

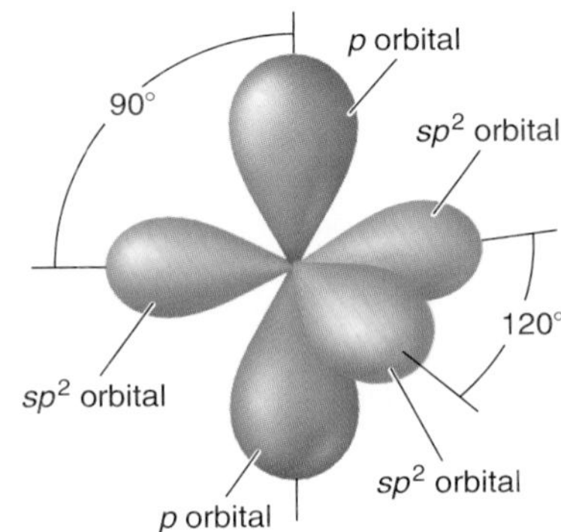

FIGURE 2.2 The unhybridized *p* orbital is perpendicular to the plane of the three sp^2 hybridized orbitals.

The three sp^2 hybrid orbitals lie in the same plane and are separated by angles of 120°. The unhybridized 2*p* orbital of the carbon atom is located perpendicular to the plane of the sp^2 hybrid orbitals (see Figure 2.2).

The bonding between the carbon atoms of ethylene results partially from the overlap of one sp^2 hybrid orbital of each carbon to form a sigma (σ) bond. The second carbon–carbon bond is formed when the unhybridized 2*p* orbitals of the carbons overlap sideways to form what is called a pi (π) bond. The remaining two sp^2 hybrid orbitals of each carbon overlap with the 1*s* orbitals of the hydrogen atoms to form sigma bonds. Thus, each ethylene molecule contains five sigma bonds (one carbon–carbon and four carbon–hydrogen) and one pi bond (carbon–carbon), as shown in Figure 2.3.

Experimental data support this hybridization model. Ethylene has been found to be a planar molecule with bond angles close to 120° between the atoms (see Figure 2.4).

ThomsonNOW Go to Chemistry Interactive to examine **bonding in alkenes and alkynes.**

In addition to geometry, alkenes also differ from open-chain alkanes in that the double bonds prevent the relatively free rotation that is characteristic of carbon atoms bonded by single bonds. As a result, alkenes can exhibit geometric isomerism, the same type of stereoisomerism seen earlier for the cycloalkanes (Section 1.9). There are two geometric isomers of 2-butene:

```
H       H          CH3      H
 \     /              \    /
  C=C                  C=C
 /     \              /    \
CH3     CH3          H      CH3
cis-2-butene        trans-2-butene
```

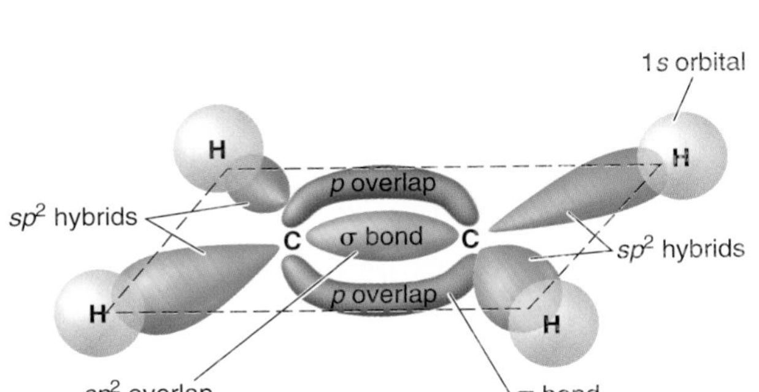

FIGURE 2.3 The bonding in ethylene is explained by combining two sp^2 hybridized carbon atoms. The C—H sigma bonds all lie in the same plane. The unhybridized p orbitals of the carbon atoms overlap to form the pi bond.

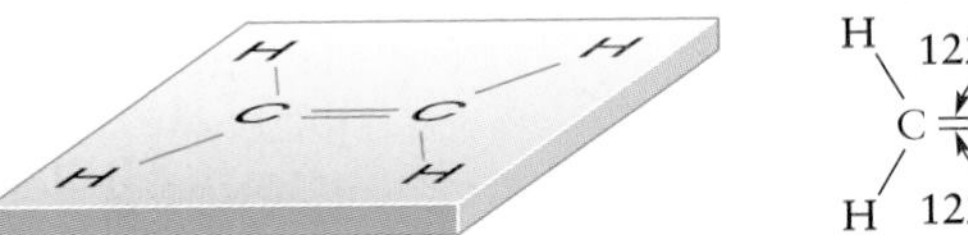

FIGURE 2.4 All six atoms of ethylene lie in the same plane.

CHEMISTRY AROUND US 2.2 Watermelon: A Source of Lycopene

Summer conjures up images of picnic tables and cool, fresh slices of watermelon. What most watermelon eaters don't realize is that in addition to enjoying a cool summer snack, they are also benefitting from a great source of lycopene, a red substance with the following highly unsaturated molecular structure:

Lycopene is known to help prevent certain types of cancer as well as heart disease, and watermelon is one of a small number of foods that contain this useful compound in large quantities. Other good food sources of lycopene are tomatoes, guava, and pink grapefruit.

The anti-cancer characteristics of lycopene are attributed to its antioxidant properties. It reacts with highly reactive oxygen-containing molecules that can oxidize cell components and cause the cells to malfunction. Research results indicate that tomatoes in the diet—especially cooked tomatoes, which contain concentrated amounts of lycopene—reduce the incidence of prostate cancer. In a study conducted at Harvard University, the incidence of prostate cancer was one-third lower in men who ate a lycopene-rich diet compared to a group who ate a low-lycopene diet.

It was long thought that heat-processed tomatoes represented the best source of lycopene in the diet. This was based on the large amounts of lycopene found in small servings of tomato juice. However, recent studies have found that red seedless watermelons contain as much lycopene as cooked tomatoes and in some cases more, depending on the variety of melon and the growing conditions.

Another factor to consider when looking for a good lycopene source is the ability of the body to digest and use the compound (bioavailability) when a lycopene-containing food is eaten. For example, the lycopene in cooked tomatoes is absorbed more readily during digestion than the lycopene in raw tomatoes. Also, it has been found that the absorption rate of lycopene goes up if the food containing it is eaten with food that contains fat. This characteristic is related to the nonpolar nature of lycopene molecules and their resulting significant solubility in fats. The level of lycopene in body fat is used as an indication of how much lycopene has been consumed in the diet. Increased lycopene levels in fat tissue have been linked to reduced risk of heart attack. The bioavailability of lycopene from raw watermelon has been found to be equal to that of lycopene obtained from cooked tomatoes. Studies are now being conducted to determine if the bioavailability of watermelon lycopene increases if the watermelon is eaten in a diet that also includes fat-containing foods.

In addition to its ability to satisfy a sweet tooth and serve as a good source of lycopene, watermelon is also an excellent source of the vitamins A, B_6, C, and thiamin. It is also fat free and low in calories, so why not include a big juicy slice of it as one of the five servings of fruits and vegetables recommended by the American Institute of Cancer Research as a way to reduce your risk of cancer? Enjoy yourself and try not to drip juice on your clothes.

Watermelon is an excellent source of lycopene.

Once again, the prefix *cis-* is used for the isomer in which the two similar or identical groups are on the same side of the double bond and *trans-* for the one in which they are on opposite sides. The two isomers *cis-* and *trans-*2-butene represent distinctly different compounds with different physical properties (see ■ Table 2.1).

Not all double-bonded compounds show cis-trans stereoisomerism. Cis-trans stereoisomerism is found only in alkenes that have two different groups attached

TABLE 2.1 Physical properties of a pair of geometric isomers

Isomer	Melting point (°C)	Boiling point (°C)	Density (g/mL)
cis-2-butene	−139.9	3.7	0.62
trans-2-butene	−105.6	0.9	0.60

to each double-bonded carbon atom. In 2-butene, the two different groups are a methyl and a hydrogen for each double-bonded carbon:

two groups are different — H, H / C=C / CH_3, CH_3 — *two groups are different*

cis-2-butene

H, CH_3 / C=C / CH_3, H — *two groups are different*

trans-2-butene

If either double-bonded carbon is attached to identical groups, no cis-trans isomers are possible. Thus, there are no geometric isomers of ethene or propene:

H, H / C=C / H, H — *two groups are the same*

ethene

CH_3, H / C=C / H, H — *two groups are the same*

propene

To see why this is so, let's try to draw geometric isomers of propene:

CH_3, H / C=C / H, H (a) is the same as H, H / C=C / CH_3, H (b)

Notice that structure (b) can be converted into (a) by just flipping it over. Thus, they are identical and not isomers.

EXAMPLE 2.3

Determine which of the following molecules can exhibit geometric isomerism, and draw structural formulas to illustrate your conclusions:

a. Cl—CH=CH—Cl
b. CH_2=CH—Cl
c. Cl—CH=CH—CH_3

Solution

a. Begin by drawing the carbon–carbon double bond with bond angles of 120° about each carbon atom:

C=C

Next, complete the structure and analyze each carbon of the double bond to see if it is attached to two different groups:

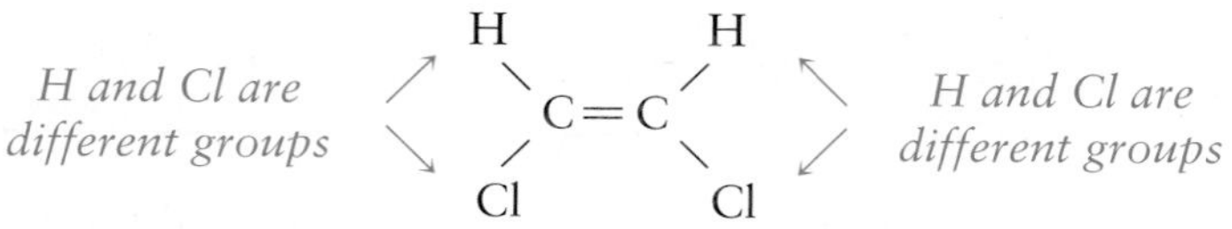

H and Cl are different groups → $\mathrm{H(Cl)C{=}C(H)Cl}$ ← *H and Cl are different groups*

In this case, each carbon is attached to two different groups and geometric isomers are possible:

$$\underset{\text{cis}}{\mathrm{H(Cl)C{=}C(H)(Cl)}} \qquad \underset{\text{trans}}{\mathrm{Cl(H)C{=}C(H)(Cl)}}$$

b. No geometric isomers are possible because one carbon contains two identical groups:

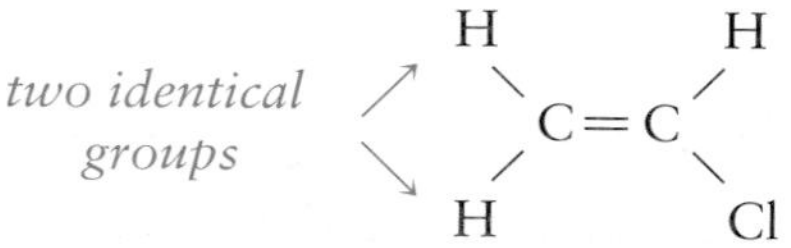

two identical groups → $\mathrm{H_2C{=}C(H)Cl}$

c. Geometric isomers are possible because there are two different groups on each carbon:

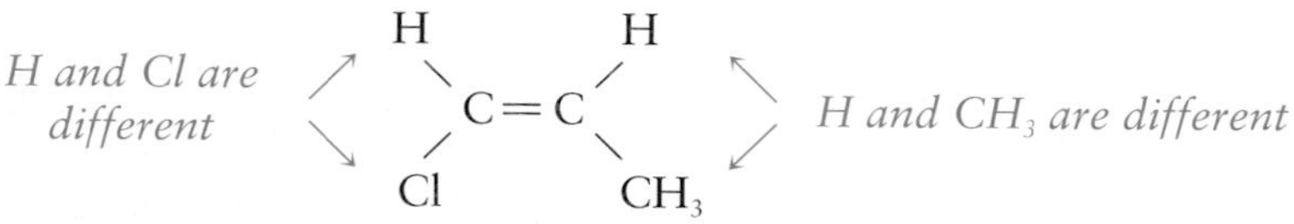

H and Cl are different → $\mathrm{H(Cl)C{=}C(H)CH_3}$ ← *H and CH_3 are different*

$$\underset{cis\text{-1-chloropropene}}{\mathrm{H(Cl)C{=}C(H)(CH_3)}} \qquad \underset{trans\text{-1-chloropropene}}{\mathrm{Cl(H)C{=}C(H)(CH_3)}}$$

Learning Check 2.3 Determine which of the following can exhibit geometric isomerism, and draw structural formulas for the cis and trans isomers of those that can:

a. $CH_2{=}C(CH_3){-}CH_2{-}CH_3$

b. $CH_3{-}CH(CH_3){-}CH{=}CH{-}CH_3$

c. $CH_3{-}C(CH_3){=}CH{-}CH_3$

2.3 Properties of Alkenes

LEARNING OBJECTIVE

5. Write equations for addition reactions of alkenes, and use Markovnikov's rule to predict the major products of certain reactions.

Physical Properties

The alkenes are similar to the alkanes in physical properties—they are nonpolar compounds that are insoluble in water, soluble in nonpolar solvents, and less

TABLE 2.2 Physical properties of some alkenes

IUPAC name	Structural formula	Boiling point (°C)	Melting point (°C)	Density (g/mL)
ethene	$CH_2{=}CH_2$	−104	−169	0.38
propene	$CH_2{=}CHCH_3$	−47	−185	0.52
1-butene	$CH_2{=}CHCH_2CH_3$	−6	−185	0.60
1-pentene	$CH_2{=}CHCH_2CH_2CH_3$	30	−138	0.64
1-hexene	$CH_2{=}CHCH_2CH_2CH_2CH_3$	63	−140	0.67
cyclohexene	(cyclohexene ring structure)	83	−104	0.81

dense than water (see ■ Table 2.2). Alkenes containing 4 carbon atoms or fewer are gases under ordinary conditions. Those containing 5 to 17 carbon atoms are liquids, and those with 18 or more carbon atoms are solids. Low molecular weight alkenes have somewhat unpleasant, gasoline-like odors.

ThomsonNOW™ Go to Chemistry Interactive and Coached Problems to examine the **reactivity of alkenes.**

Chemical Properties

In contrast to alkanes, which are inert to almost all chemical reagents (Section 1.11), alkenes are quite reactive chemically. Since the only difference between an alkane and an alkene is the double bond, it is not surprising to learn that most of the reactions of alkenes take place at the double bond. These reactions follow the pattern

$$\gt C{=}C\lt \; + \; A{-}B \longrightarrow -\overset{A}{\underset{|}{\overset{|}{C}}}-\overset{B}{\underset{|}{\overset{|}{C}}}- \qquad (2.1)$$

They are called **addition reactions** because a substance is added to the double bond. Addition reactions are characterized by two reactants that combine to form one product.

addition reaction A reaction in which a compound adds to a multiple bond.

Halogenation is one of the most common addition reactions. For example, when bromine, a halogen, is added to an alkene, the double bond reacts, and only a carbon–carbon single bond remains in the product. Without a double bond present, the product is referred to as a **haloalkane** or **alkyl halide.** Reaction 2.2 gives the general reaction, and Reaction 2.3 is a specific example, the bromination of 1-butene.

haloalkane or alkyl halide A derivative of an alkane in which one or more hydrogens are replaced by halogens.

General reaction: $$\gt C{=}C\lt \; + \; Br{-}Br \longrightarrow -\underset{Br}{\underset{|}{\overset{|}{C}}}-\underset{Br}{\underset{|}{\overset{|}{C}}}- \qquad (2.2)$$

Specific example: $$\underset{\text{1-butene}}{CH_3{-}CH_2{-}CH{=}CH_2} + Br{-}Br \longrightarrow \underset{\text{1,2-dibromobutane}}{CH_3{-}CH_2{-}\underset{Br}{\underset{|}{CH}}{-}\underset{Br}{\underset{|}{CH_2}}} \qquad (2.3)$$

The addition of Br_2 to double bonds provides a simple laboratory test for unsaturation (see ■ Active Figure 2.5). As the addition takes place, the characteristic red-brown color of the added bromine fades as it is used up, and the colorless dibromoalkane product forms.

■ **ACTIVE FIGURE 2.5** The reaction of bromine with an unsaturated hydrocarbon. **Sign in at www.thomsonedu.com to explore an interactive version of this figure.**

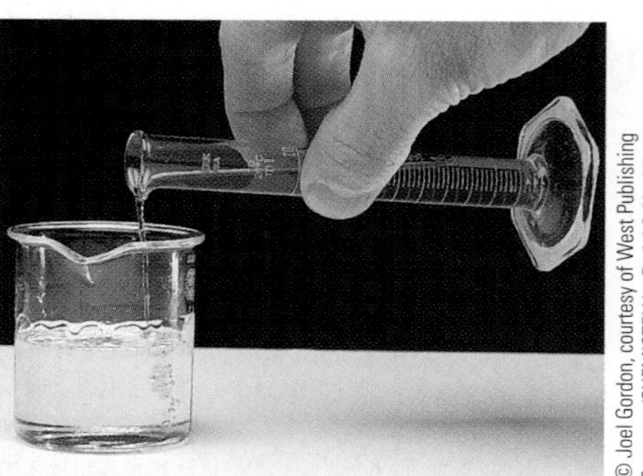

Dilute bromine solution added to 1-hexene loses its red-brown color immediately.

The remainder of the bromine solution is added. The last drops react as quickly as the first.

The addition of halogens is also used to quantitatively determine the degree of unsaturation in vegetable oils, margarines, and shortenings (Section 18.4). Chlorine reacts with alkenes to give dichloro products in an addition reaction similar to that of bromine. However, it is not used as a test for unsaturation because it is difficult to see the pale green color of chlorine in solution.

■ **Learning Check 2.4** Write the structural formula for the product of each of the following reactions:

a. $CH_3{-}\underset{\displaystyle CH_3}{\underset{|}{C}}{=}CH_2 + Br_2 \longrightarrow$

b. (cyclopentene) $+ \; Cl_2 \longrightarrow$

hydrogenation
A reaction in which the addition of hydrogen takes place.

In the presence of an appropriate catalyst (such as platinum, palladium, or nickel), hydrogen adds to alkenes and converts them into the corresponding alkanes. This reaction, which is called **hydrogenation,** is illustrated in Reactions 2.4 and 2.5.

General reaction:

$$\gt C{=}C\lt \; + \; H{-}H \xrightarrow{Pt} {-}\underset{\displaystyle H}{\underset{|}{\overset{|}{C}}}{-}\underset{\displaystyle H}{\underset{|}{\overset{|}{C}}}{-} \qquad (2.4)$$

an alkane

Specific example:

$$\underset{\text{2-butene}}{CH_3CH{=}CHCH_3} + H{-}H \xrightarrow{Pt} \underset{\text{butane}}{CH_3\underset{\displaystyle H}{\underset{|}{CH}}{-}\underset{\displaystyle H}{\underset{|}{CH}}CH_3} \qquad (2.5)$$

polyunsaturated
A term usually applied to molecules with several double bonds.

The hydrogenation of vegetable oils is a very important commercial process. Vegetable oils, such as soybean and cottonseed oil, are composed of long-chain organic molecules that contain many alkene bonds. The high degree of unsaturation characteristic of these oils gave rise to the term **polyunsaturated.** Upon hydrogenation, the melting point of the oils is raised, and the oils become low-melting-point solids. These products are used in the form of margarine and shortening.

■ **Learning Check 2.5** Write the structural formula for the product of each of the following reactions:

a. $CH_3{-}\underset{\displaystyle CH_3}{\underset{|}{C}}{=}CH{-}CH_3 + H_2 \xrightarrow{Pt}$

b. (1-methylcyclopentene, with CH_3 on the double-bonded carbon) $+ \; H_2 \xrightarrow{Pt}$

A number of acidic compounds, such as the hydrogen halides—HF, HCl, HBr, and HI—also add to alkenes to give the corresponding alkyl halide. The reaction with HCl is illustrated as follows:

General reaction: $$\rangle C{=}C\langle \; + \; H{-}Cl \longrightarrow -\overset{|}{\underset{\underset{H}{|}}{C}}-\overset{|}{\underset{\underset{Cl}{|}}{C}}- \qquad (2.6)$$

Specific example $$CH_3CH = CHCH_3 + H-Cl \longrightarrow CH_3\underset{\underset{H}{|}}{C}H\underset{\underset{Cl}{|}}{C}HCH_3 \qquad (2.7)$$

The addition reactions involving H_2, Cl_2, and Br_2 yield only one product because the same group (H and H or Br and Br) adds to each double-bonded carbon. However, with H—X, a different group adds to each carbon, and for certain alkenes, there are two possible products. For example, in the reaction of HBr with propene, two products might be expected: 1-bromopropane or 2-bromopropane,

$$CH_2{=}CH{-}CH_3 + H{-}Br \longrightarrow \underset{\text{1-bromopropane}}{\overset{\overset{Br}{|}}{C}H_2{-}\overset{\overset{H}{|}}{C}H{-}CH_3} \text{ or } \underset{\text{2-bromopropane}}{\overset{\overset{H}{|}}{C}H_2{-}\overset{\overset{Br}{|}}{C}H{-}CH_3} \qquad (2.8)$$

It turns out that only one product, 2-bromopropane, is formed in significant amounts. This fact, first reported in 1869 by Russian chemist Vladimir Markovnikov, gave rise to a rule for predicting which product will be exclusively or predominantly formed. According to **Markovnikov's rule**, when a molecule of the form H—X adds to an alkene, the hydrogen becomes attached to the carbon atom that is already bonded to more hydrogens. A phrase to help you remember this rule is "the rich get richer." Applying this rule to propene, we find

Markovnikov's rule
In the addition of H—X to an alkene, the hydrogen becomes attached to the carbon atom that is already bonded to more hydrogens.

ThomsonNOW™ Go to Coached Problems to examine **Markovnikov's rule.**

$$CH_2 = CH - CH_3$$

one hydrogen attached (CH)

two hydrogens attached (CH_2)

three hydrogens but they are not attached to the double-bonded carbons (CH_3)

Therefore, H attaches to the end carbon of the double bond and Br attaches to the second carbon, and 2-bromopropane is the major product.

EXAMPLE 2.4

Use Markovnikov's rule to predict the major product in the following reactions:

a. $$CH_3-\overset{\overset{CH_3}{|}}{C}=CH_2 + H-Cl$$

b. (1-methylcyclopentene, ring with CH_3 on the double-bonded carbon) + H — Cl ⟶

Solution

a. Analyze the C=C to see which carbon atom has more hydrogens attached:

$$CH_3 - \overset{\overset{CH_3}{|}}{C} = CH_2$$

two hydrogens attached (CH_2)

no hydrogens attached (C)

The H of H—Cl will attach to the position that has more hydrogens. Thus, 2-chloro-2-methylpropane is the major product:

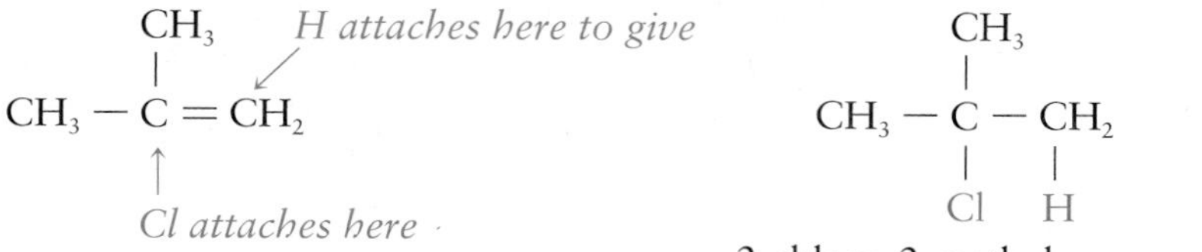

$$CH_3-\overset{\overset{\large CH_3}{|}}{C}=CH_2 \quad \longrightarrow \quad CH_3-\underset{\underset{\large Cl}{|}}{\overset{\overset{\large CH_3}{|}}{C}}-\underset{\underset{\large H}{|}}{CH_2}$$

H attaches here to give (arrow to CH_2); *Cl attaches here* (arrow to C)

2-chloro-2-methylpropane

b. The challenge with a cyclic alkene is to remember that the hydrogens are not shown. Thus,

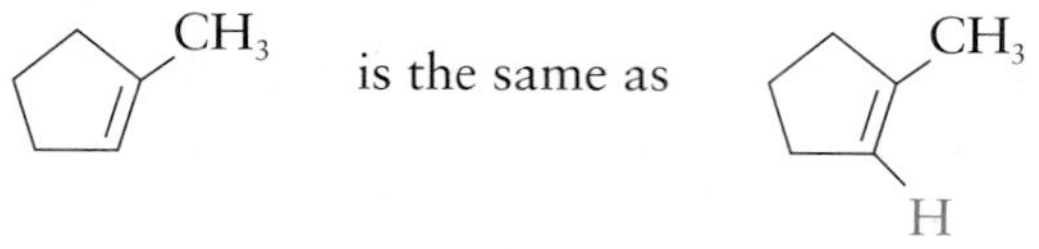

[1-methylcyclopentene, CH_3] is the same as [1-methylcyclopentene, CH_3, with H shown on the other double-bonded carbon]

As before, the H of H—Cl will attach to the double-bonded carbon that has more hydrogens:

[1-methylcyclopentene, CH_3, H] *Cl attaches here* / *H attaches here* ⟶ [cyclopentane with CH_3 and Cl on one carbon, H and H on adjacent carbon] or [cyclopentane with CH_3 and Cl on one carbon]

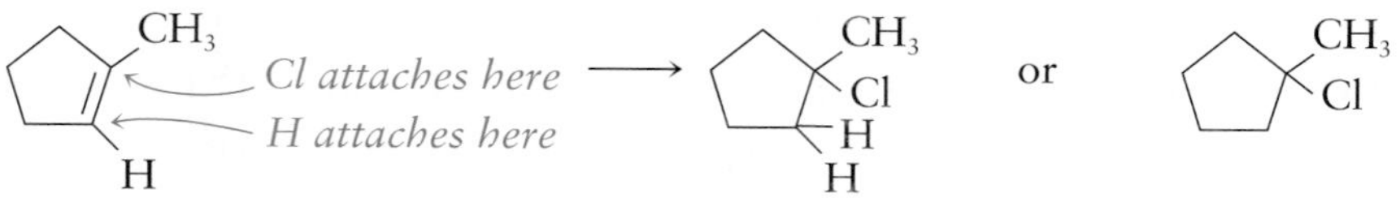

Learning Check 2.6 Use Markovnikov's rule to predict the major product in the following reactions:

a. 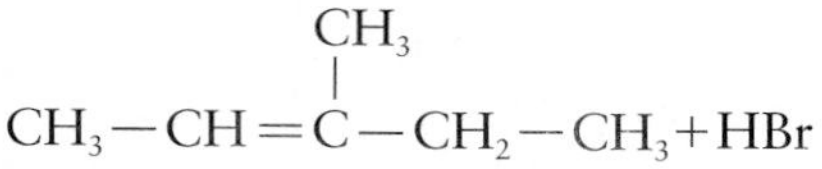 $CH_3-CH=\overset{\overset{\large CH_3}{|}}{C}-CH_2-CH_3+HBr$

b. 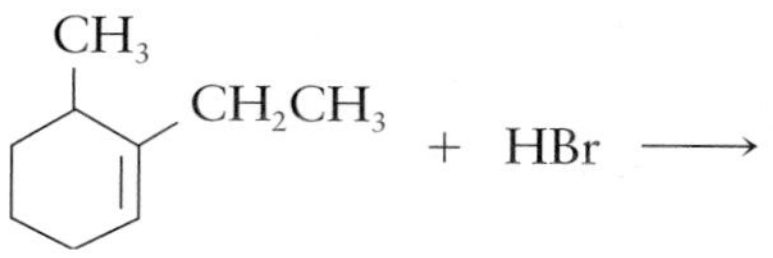 [cyclohexene ring bearing CH_3 and CH_2CH_3] $+\ HBr \longrightarrow$

STUDY SKILLS 2.1 Keeping a Reaction Card File

Remembering organic reactions for exams is challenging for most students. Because the number of reactions being studied increases rapidly, it is a good idea to develop a systematic way to organize them for easy and effective review.

One way to do this is to focus on the functional group concept. When an exam question asks you to complete a reaction by identifying the product, your first step should be to identify the functional group of the reactant. Usually, only the functional group portion of a molecule undergoes reaction; in addition, a particular functional group usually undergoes the same characteristic reactions regardless of the other features of the organic molecule to which it is bound. Thus, by remembering the behavior of a functional group under specific conditions, you can predict the reactions of many compounds, no matter how complex the structures look, as long as they contain the same functional group. For example, any structure containing a C=C will undergo reactions typical of alkenes. Other functional groups will be introduced in later chapters.

Keeping a reaction card file based on the functional group concept is a good way to organize reactions for review. Write the structures and names of the reactants on one side of an index card with an arrow showing any catalyst or special conditions. Write the product structure and name on the back of the card. We recommend that you do this for the general reaction (like those in the Key Reactions section at the end of most chapters) and for a specific example. Review your cards every day (this can even be done while waiting for a bus, etc.), and add to them as new reactions are studied. As an exam approaches, put aside the reactions you know well, and concentrate on the others in what should be a dwindling deck. This is an effective way to focus on learning what you don't know.

In the absence of a catalyst, water does not react with alkenes. But, if an acid catalyst such as sulfuric acid is added, water adds to carbon–carbon double bonds to give alcohols. In this reaction, which is called **hydration,** a water molecule is split in such a way that —H attaches to one carbon of the double bond, and —OH attaches to the other carbon. In Reactions 2.9–2.11, H_2O is written H—OH to emphasize the portions that add to the double bond. Notice that the addition follows Markovnikov's rule:

hydration
The addition of water to a multiple bond.

$$\mathrm{>C=C<} + \mathrm{H-OH} \xrightarrow{H_2SO_4} \mathrm{-\underset{\large H}{\overset{|}{\underset{|}{C}}}-\underset{\large OH}{\overset{|}{\underset{|}{C}}}-} \quad (2.9)$$

an alcohol

$$\mathrm{CH_3CH=CHCH_3} + \mathrm{H-OH} \xrightarrow{H_2SO_4} \mathrm{CH_3\underset{\large H}{\underset{|}{C}H}-\underset{\large OH}{\underset{|}{C}H}CH_3} \quad (2.10)$$

2-butene → 2-butanol

$$\mathrm{CH_3CH=CH_2} + \mathrm{H-OH} \xrightarrow{H_2SO_4} \mathrm{CH_3\underset{\large OH}{\underset{|}{C}H}-\underset{\large H}{\underset{|}{C}H_2}} \quad (2.11)$$

propene → 2-propanol

The hydration of alkenes provides a convenient method for preparing alcohols on a large scale. The reaction is also important in living organisms, but the catalyst is an enzyme rather than sulfuric acid. For example, one of the steps in the body's utilization of carbohydrates for energy involves the hydration of fumaric acid, which is catalyzed by the enzyme fumarase:

$$\mathrm{HO-\overset{\overset{\large O}{\|}}{C}-CH=CH-\overset{\overset{\large O}{\|}}{C}-OH} + \mathrm{H_2O} \xrightarrow{\text{fumarase}} \mathrm{HO-\overset{\overset{\large O}{\|}}{C}-\overset{\overset{\large H}{|}}{C}H-\overset{\overset{\large OH}{|}}{C}H-\overset{\overset{\large O}{\|}}{C}-OH} \quad (2.12)$$

fumaric acid → malic acid

Learning Check 2.7 Draw structural formulas for the major organic product of each of the following reactions:

a. $CH_3CH_2CH_2CH=CH_2 + H_2O \xrightarrow{H_2SO_4}$

b. (cyclopentene) $+ H_2O \xrightarrow{H_2SO_4}$

2.4 Addition Polymers

LEARNING OBJECTIVE

6. Write equations for addition polymerization, and list uses for addition polymers.

Certain alkenes undergo a very important reaction in the presence of specific catalysts. In this reaction, alkene molecules undergo an addition reaction with one another. The double bonds of the reacting alkenes are converted to single bonds as

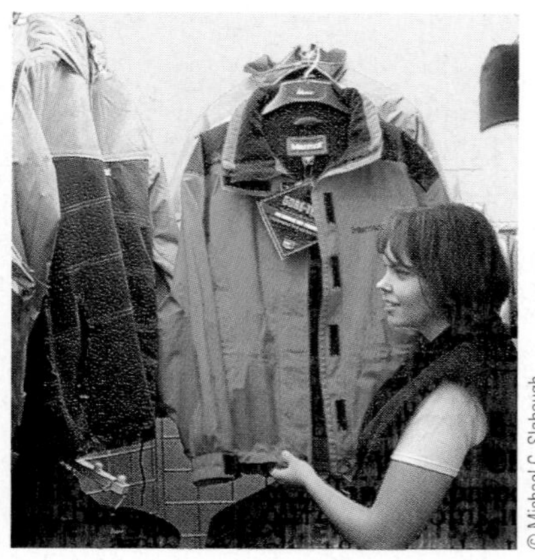
© Michael C. Slabaugh

■ **FIGURE 2.6** Gore-Tex® is a thin, membranous material made by stretching Teflon fibers. Fabrics layered with Gore-Tex repel wind and rain but allow body perspiration to escape, making it an excellent fabric for sportswear.

polymerization
A reaction that produces a polymer.

polymer
A very large molecule made up of repeating units.

addition polymer
A polymer formed by the linking together of many alkene molecules through addition reactions.

monomer
The starting material that becomes the repeating units of polymers.

hundreds or thousands of molecules bond and form long chains. For example, several ethylene molecules react as follows:

$$\underset{\text{ethylene molecules}}{CH_2{=}CH_2 + CH_2{=}CH_2 + CH_2{=}CH_2 + CH_2{=}CH_2} \xrightarrow[\text{catalysts}]{\text{Heat, pressure,}} \underset{\text{polyethylene}}{-CH_2-CH_2-CH_2-CH_2-CH_2-CH_2-CH_2-CH_2-} \quad (2.13)$$

The product is commonly called polyethylene even though there are no longer any double bonds present. The newly formed bonds in this long chain are shown in color. This type of reaction is called a **polymerization,** and the long-chain product made up of repeating units is a **polymer** (*poly* = many, *mer* = parts). The trade names of many polymers such as Orlon®, Plexiglas®, Lucite®, and Teflon® are familiar (see ■ Figure 2.6). These products are referred to as **addition polymers** because of the addition reaction between double-bonded compounds that is used to produce them. The starting materials that make up the repeating units of polymers are called **monomers** (*mono* = one, *mer* = part). Quite often, common names are used for both the polymer and the monomer.

It is not possible to give an exact formula for a polymer produced by a polymerization reaction because the individual polymer molecules vary in size. We could represent polymerization reactions as in Reaction 2.13. However, since this type of reaction is inconvenient, we adopt a commonly used approach: The polymer is represented by a simple repeating unit based on the monomer. For polyethylene, the unit is $+CH_2-CH_2+$. The large number of units making up the polymer is denoted by n, a whole number that varies from several hundred to several thousand. The polymerization reaction of ethylene is then written as

$$\underset{\text{ethylene}}{nCH_2{=}CH_2} \xrightarrow[\text{catalysts}]{\text{Heat, pressure,}} \underset{\text{polyethylene}}{+CH_2-CH_2+_n} \quad (2.14)$$

The lowercase n in Reaction 2.14 represents a large, unspecified number. From this reaction, we see that polyethylene is essentially a very long chain alkane. As a result, it has the chemical inertness of alkanes, a characteristic that makes polyethylene suitable for food storage containers, garbage bags, eating utensils, laboratory apparatus, and hospital equipment (see ■ Figure 2.7). Polymer characteristics are modified by using alkenes with different groups attached to

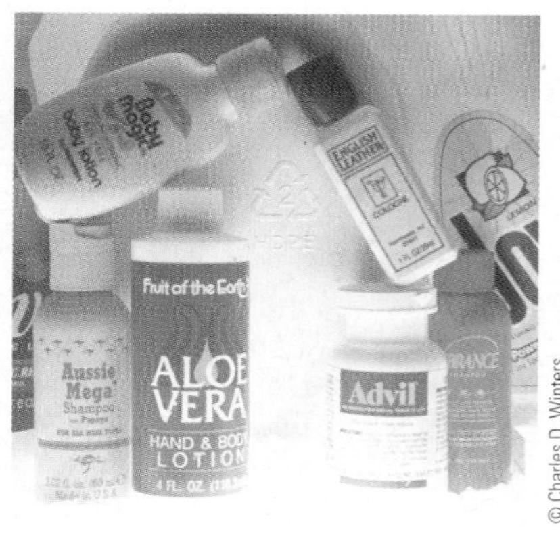

© Charles D. Winters

(a)

© Charles D. Winters

(b)

© Charles D. Winters

(c)

■ **FIGURE 2.7** Common polymer-based consumer products. (a) Packaging materials from polyethylene, (b) polystyrene, and (c) polyvinyl chloride.

STUDY SKILLS 2.2 A Reaction Map for Alkenes

A diagram may help you visualize and remember the four common addition reactions in this section. In each case, the alkene double bond reacts, and an alkane or alkane derivative is produced. The specific reagent determines the outcome of the reaction.

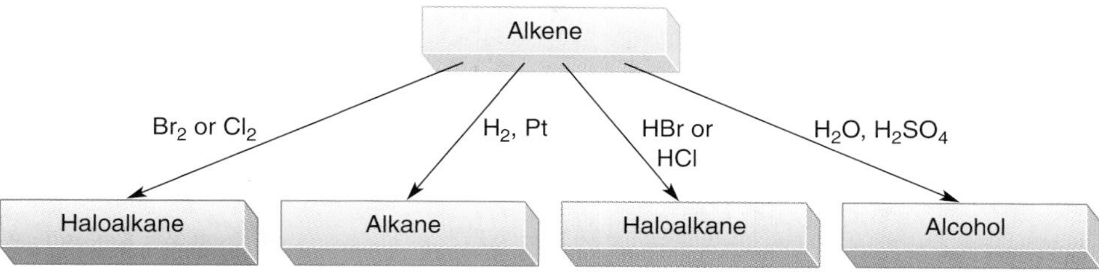

either or both of the double-bonded carbons. For example, the polymerization of vinyl chloride gives the polymer poly (vinyl chloride), PVC:

$$n\mathrm{CH_2{=}\underset{}{\overset{\overset{\displaystyle Cl}{|}}{C}}H} \xrightarrow[\text{catalysts}]{\text{Heat,}} \mathrm{+CH_2-\overset{\overset{\displaystyle Cl}{|}}{C}H+_{\mathit{n}}} \qquad (2.15)$$

vinyl chloride — poly (vinyl chloride)

The commercial product Saran Wrap is an example of a **copolymer,** which is a polymer made up of two different monomers (Reaction 2.16):

copolymer An addition polymer formed by the reaction of two different monomers.

$$n\mathrm{CH_2=\underset{\underset{\displaystyle Cl}{|}}{C}H} + n\mathrm{CH_2=\underset{\underset{\displaystyle Cl}{|}}{\overset{\overset{\displaystyle Cl}{|}}{C}}} \xrightarrow{\text{Catalyst}} \mathrm{+CH_2-\underset{\underset{\displaystyle Cl}{|}}{C}H-CH_2-\underset{\underset{\displaystyle Cl}{|}}{\overset{\overset{\displaystyle Cl}{|}}{C}}+_{\mathit{n}}} \qquad (2.16)$$

vinyl chloride — vinylidene chloride — Saran Wrap

A number of the more important addition polymers are shown in ■ Table 2.3. As you can tell by looking at some of the typical uses, addition polymers have become nearly indispensable in modern life (see ■ Figure 2.8).

■ **Learning Check 2.8** Draw the structural formula of a portion of polypropylene containing four repeating units of the monomer propylene,

$$\mathrm{CH_2{=}\overset{\overset{\displaystyle CH_3}{|}}{C}H}$$

2.5 Alkynes

LEARNING OBJECTIVE

7. Write the IUPAC names of alkynes from their molecular structures.

The characteristic feature of alkynes is the presence of a triple bond between carbon atoms. Thus, alkynes are also unsaturated hydrocarbons. Only a few compounds containing the carbon–carbon triple bond are found in nature. The simplest

TABLE 2.3 Common addition polymers

Chemical name and trade name(s)	Monomer	Polymer	Typical uses
polyethylene	$CH_2{=}CH_2$	$\left(CH_2-CH_2 \right)_n$	Bottles, plastic bags, film
polypropylene	$CH_2{=}CH$ $\vert$ CH_3	$\left(CH_2-CH \right)_n$ $\vert$ CH_3	Carpet fiber, pipes, bottles, artificial turf
poly (vinyl chloride) (PVC)	$CH_2{=}CH$ $\vert$ Cl	$\left(CH_2-CH \right)_n$ $\vert$ Cl	Synthetic leather, floor tiles, garden hoses, water pipe
polytetrafluoroethylene (Teflon®)	$CF_2{=}CF_2$	$\left(CF_2-CF_2 \right)_n$	Pan coatings, plumbers' tape, heart valves, fabrics
poly (methyl methacrylate) (Lucite®, Plexiglas®)	CH_3 $\vert$ $CH_2{=}C$ $\vert$ $C-O-CH_3$ $\Vert$ O	CH_3 $\vert$ $\left(CH_2-C \right)_n$ $\vert$ $C-O-CH_3$ $\Vert$ O	Airplane windows, paint, contact lenses, fiber optics
poly (vinyl acetate)	$CH_2{=}CH$ $\vert$ $O-C-CH_3$ $\Vert$ O	$\left(CH_2-CH \right)_n$ $\vert$ $O-C-CH_3$ $\Vert$ O	Adhesives, latex paint, chewing gum
polyacrylonitrile (Orlon®, Acrilan®)	$CH_2{=}CH$ $\vert$ CN	$\left(CH_2-CH \right)_n$ $\vert$ CN	Carpets, fabrics
polystyrene (Styrofoam®)	$CH_2{=}CH$ $\vert$ C_6H_5 (benzene ring)	$\left(CH_2-CH \right)_n$ $\vert$ C_6H_5 (benzene ring)	Food coolers, drinking cups, insulation

and most important compound of this series is ethyne, more commonly called acetylene (C_2H_2):

$$H-C{\equiv}C-H$$
acetylene

Automobile safety glass contains a sheet of poly (vinyl acetate) layered between two sheets of glass to prevent the formation of sharp fragments.

The elasticity of bubble gum comes from a copolymer of styrene and 1,3-butadiene.

■ **FIGURE 2.8** Two uses of addition polymers. What properties of addition polymers are exhibited in both of these products?

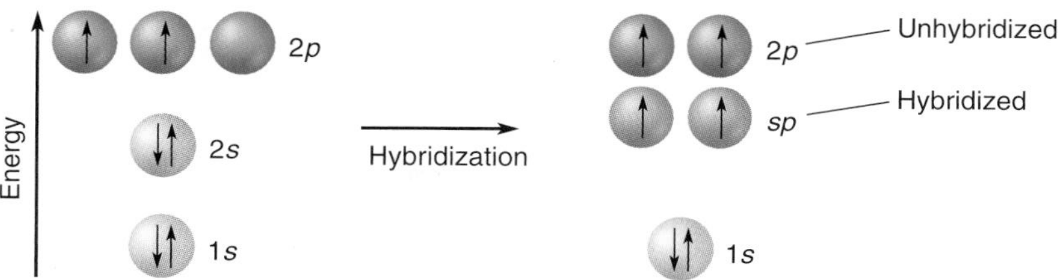

■ **FIGURE 2.9** *sp* hybridization occurs when one of the 2*p* orbitals of carbon mixes with the 2*s* orbital. Two 2*p* orbitals remain unhybridized.

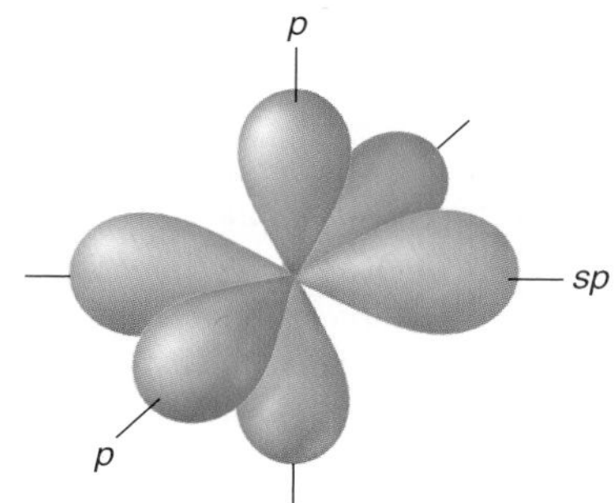

■ **FIGURE 2.10** The unhybridized *p* orbitals are perpendicular to the two *sp* hybridized orbitals.

Acetylene is used in torches for welding steel and in making plastics and synthetic fibers.

Once again, orbital hybridization provides an explanation for the bonding of the carbon atoms. Structurally, the hydrogen and carbon atoms of acetylene molecules lie in a straight line. This same linearity of the triple bond and the two atoms attached to the triple-bonded carbons is found in all alkynes. These characteristics are explained by mixing a 2*s* and a single 2*p* orbital of each carbon to form a pair of *sp* hybrid orbitals. Two of the 2*p* orbitals of each carbon are unhybridized (see ■ Figures 2.9 and 2.10).

A carbon–carbon sigma bond in acetylene forms by the overlap of one *sp* hybrid orbital of each carbon. The other *sp* hybrid orbital of each carbon overlaps with a 1*s* orbital of a hydrogen to form a carbon–hydrogen sigma bond. The remaining pair of unhybridized *p* orbitals of each carbon overlap sideways to form a pair of pi bonds between the carbon atoms. Thus, each acetylene molecule contains three sigma bonds (two carbon–hydrogen and one carbon–carbon) and two pi bonds (both are carbon–carbon). This is shown in ■ Figure 2.11.

Alkynes are named in exactly the same ways as alkenes, except the ending *-yne* is used:

$$\overset{4}{C}H_3\overset{3}{C}H_2-\overset{2}{C}\equiv\overset{1}{C}-H$$

1-butyne

$$H-\overset{1}{C}\equiv\overset{2}{C}-\overset{3}{C}H(CH_3)-\overset{4}{C}H_3$$

3-methyl-1-butyne

■ **Learning Check 2.9** Give the IUPAC name for each of the following:

a. $CH_3-CH_2-C\equiv C-CH_3$

b. $CH_3-CH(CH_3)-CH_2-C\equiv C-CH_3$

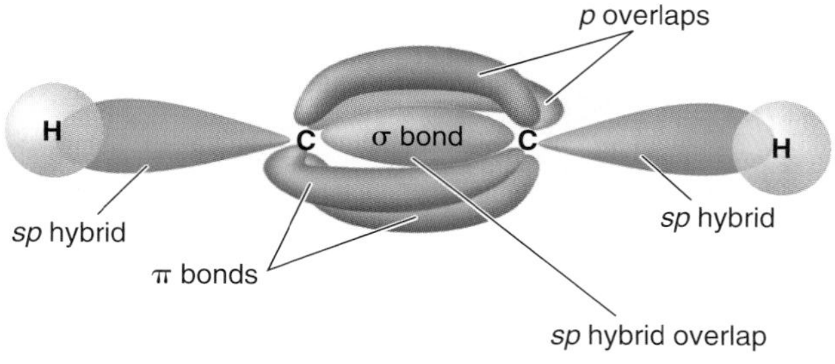

■ **FIGURE 2.11** The shape of acetylene is explained by sigma bonding between *sp* hybrid orbitals and pi bonding between unhybridized *p* orbitals.

HOW REACTIONS OCCUR 2.1 The Hydration of Alkenes: An Addition Reaction

The mechanism for the hydration of alkenes is believed to begin when H^+ from the acid catalyst is attracted to the electrons of the carbon–carbon double bond. The H^+ becomes bonded to one of the carbons by a sharing of electrons:

Step 1.

$$-\overset{|}{\underset{|}{C}}=\overset{|}{\underset{|}{C}}- + H^+ \longrightarrow -\overset{H}{\overset{|}{\underset{|}{C}}}-\overset{+}{\underset{|}{C}}-$$

a carbocation

This process leaves the second carbon with only three bonds about it and thus a positive charge. Such ions, referred to as **carbocations**, are extremely reactive.

As soon as it forms, the positive carbocation attracts any species that has readily available nonbonding electrons, whether it is an anion or a neutral molecule. In the case of water, the oxygen atom has two unshared pairs of electrons:

$$H-\ddot{\underset{\cdot\cdot}{O}}-H$$

One pair of oxygen electrons forms a covalent bond with the carbocation:

Step 2.

$$-\overset{H}{\overset{|}{\underset{|}{C}}}-\overset{+}{\underset{|}{C}}- + H-\ddot{\underset{\cdot\cdot}{O}}-H \longrightarrow -\overset{H}{\overset{|}{\underset{|}{C}}}-\overset{H_2O^+}{\overset{|}{\underset{|}{C}}}-$$

In the third step, H^+ is lost to produce the alcohol

Step 3.

$$-\overset{H}{\overset{|}{\underset{|}{C}}}-\overset{H_2O^+}{\overset{|}{\underset{|}{C}}}- \longrightarrow -\overset{H}{\overset{|}{\underset{|}{C}}}-\overset{OH}{\overset{|}{\underset{|}{C}}}- + H^+$$

an alcohol

Notice that the acid catalyst, H^+, which initiated the reaction, is recovered unchanged in the final step of the mechanism.

By applying Step 2 of the mechanism for hydration, we can understand how HCl and HBr react with alkenes:

$$-\overset{H}{\overset{|}{\underset{|}{C}}}-\overset{+}{\underset{|}{C}}- + Br^- \longrightarrow -\overset{H}{\overset{|}{\underset{|}{C}}}-\overset{Br}{\overset{|}{\underset{|}{C}}}-$$

$$-\overset{H}{\overset{|}{\underset{|}{C}}}-\overset{+}{\underset{|}{C}}- + Cl^- \longrightarrow -\overset{H}{\overset{|}{\underset{|}{C}}}-\overset{Cl}{\overset{|}{\underset{|}{C}}}-$$

carbocation
An ion of the form $-\overset{+}{\underset{|}{C}}-$.

The physical properties of the alkynes are nearly the same as those of the corresponding alkenes and alkanes: They are insoluble in water, less dense than water, and have relatively low melting and boiling points. Alkynes also resemble alkenes in their addition reactions. The same substances (Br_2, H_2, HCl, etc.) that add to double bonds also add to triple bonds. The one significant difference is that alkynes consume twice as many moles of addition reagent as alkenes in addition reactions that go on to completion.

2.6 Aromatic Compounds and the Benzene Structure

LEARNING OBJECTIVE

8. Classify organic compounds as aliphatic or aromatic.

Some of the early researchers in organic chemistry became intrigued by fragrant oils that could be extracted from certain plants. Oil of wintergreen and the flavor component of the vanilla bean are examples. The compounds responsible for the aromas had similar chemical properties. As a result, they were grouped together and called aromatic compounds.

As more and more aromatic compounds were isolated and studied, chemists gradually realized that aromatics contained at least six carbon atoms, had low hydrogen-to-carbon ratios (relative to other organic hydrocarbons), and were related to benzene (C_6H_6). For example, toluene, an aromatic compound from the bark of the South American tolu tree, has the formula C_7H_8.

Chemists also learned that the term *aromatic* was not always accurate. Many compounds that belong to the class because of chemical properties and structures are not at all fragrant. Conversely, there are many fragrant compounds that do not have aromatic compound properties or structures. Today, the old class name is used but with a different meaning. Aromatic compounds are those that contain the characteristic benzene ring or its structural relatives. Compounds that do not contain this structure (nonaromatic compounds) are referred to as **aliphatic compounds.** Alkanes, alkenes, and alkynes are, therefore, aliphatic compounds.

aliphatic compound
Any organic compound that is not aromatic.

The molecular structure of benzene presented chemists with an intriguing puzzle after the compound was discovered in 1825 by Michael Faraday. The formula C_6H_6 indicated that the molecule was highly unsaturated. However, the compound did not show the typical reactivity of unsaturated hydrocarbons. Benzene underwent relatively few reactions, and these proceeded slowly and often required heat and catalysts. This was in marked contrast to alkenes, which reacted rapidly with many reagents, in some cases almost instantaneously. This apparent discrepancy between structure and reactivity plagued chemists until 1865, when Freidrich August Kekulé von Stradonitz (see ■ Figure 2.12), a German chemist, suggested that the benzene molecule might be represented by a ring arrangement of carbon atoms with alternating single and double bonds between the carbon atoms:

or

© Stock Montage

■ **FIGURE 2.12** Friedrich August Kekulé (1829–1896).

He later suggested that the double bonds alternate in their positions between carbon atoms to give two equivalent structures:

and

Kekulé structures

A modern interpretation of the benzene structure based on hybridization enables chemists to better understand and explain the chemical properties of benzene and other aromatic compounds. Each carbon atom in a benzene ring has three sp^2 hybrid orbitals and one unhybridized p orbital.

A single sigma bond between carbons of the benzene ring is formed by the overlap of two sp^2 orbitals, one from each of the double-bonded carbons. Because each carbon forms two single bonds, two of the sp^2 hybrid orbitals of each carbon are involved. The third sp^2 hybrid orbital of each carbon forms a single sigma bond with a hydrogen by overlapping with a $1s$ orbital of hydrogen. The unhybridized p orbitals of each carbon overlap sideways above and below the plane of the carbon ring to form two delocalized pi lobes that run completely around the ring (see ■ Figure 2.13).

Go to Chemistry Interactive to explore **bonding in aromatic compounds.**

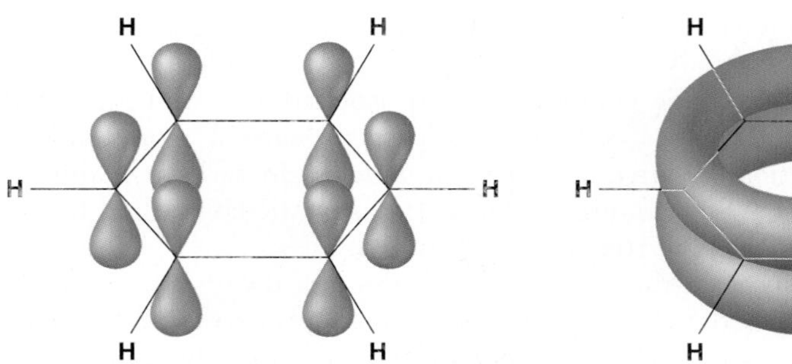

Two lobes represent the delocalized pi bonding of the six p electrons.

FIGURE 2.13 A hybrid orbital view of the benzene structure.

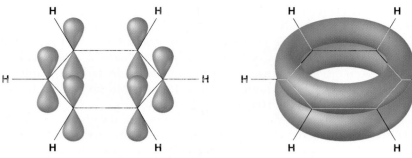

This interpretation leads to the conclusion that six bonding electrons move freely around the ring in the overlapping delocalized pi lobes. Because of this, the benzene structure is often represented by the symbol

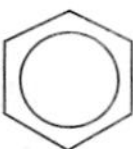

with the circle representing the evenly distributed electrons in the pi lobes. All six carbon and six hydrogen atoms in benzene molecules lie in the same plane (see Figure 2.13). Therefore, substituted aromatic compounds do not exhibit cis-trans isomerism.

As you draw the structure of aromatic compounds, remember that only one hydrogen atom or group can be attached to a particular position on the benzene ring. For example, compounds (a) and (b) below exist, but (c) does not. Examination of the Kekulé structure of compound (c) shows that the carbon has five bonds attached to it in violation of the octet rule:

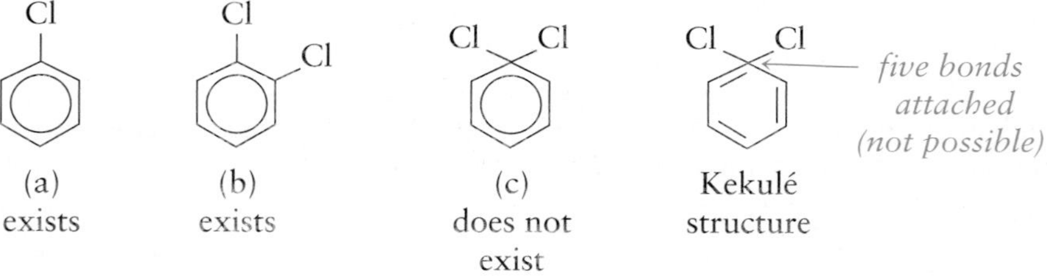

2.7 The Nomenclature of Benzene Derivatives

LEARNING OBJECTIVE

9. Name and draw structural formulas for aromatic compounds.

The chemistry of the aromatic compounds developed somewhat haphazardly for many years before systematic nomenclature schemes were developed. Some of the common names used earlier have acquired historical respectability and are used today; some have even been incorporated into the modern systematic nomenclature system.

The following guidelines are all based on the IUPAC aromatic nomenclature system. They are not complete, and you will not be able to name all aromatic compounds by using them. However, you will be able to name and recognize those used in this book.

Guideline 1. When a single hydrogen of the benzene ring is replaced, the compound can be named as a derivative of benzene:

CH_2CH_3 | NO_2 | Br | Cl

ethylbenzene nitrobenzene bromobenzene chlorobenzene

ThomsonNOW™ Go to Coached Problems to practice **naming derivatives of benzene.**

Guideline 2. A number of benzene derivatives are known by common names that are also IUPAC-accepted and are used preferentially over other possibilities. Thus, toluene is favored over methylbenzene, and aniline is used rather than aminobenzene:

CH_3 | OH | NH_2 | C(=O)—OH

toluene phenol aniline benzoic acid

Guideline 3. Compounds formed by replacing a hydrogen of benzene with a more complex hydrocarbon group can be named by designating the benzene ring as the group. When this is done, the benzene ring, C_6H_5—, or

CHEMISTRY AND YOUR HEALTH 2.1 Beautiful, Brown . . . and Overdone

For many people, a well-tanned skin is considered to be very attractive. However, the methods used to achieve this stylish appearance sometimes have some serious health drawbacks.

The body protects itself from ultraviolet (UV) radiation by producing a skin pigment called melanin. The accumulation of melanin in the skin leads to the characteristic tan color and protects the skin from burning and other damage caused by UV radiation. Melanin takes 3 to 5 days to form after initial exposure to the sun. Melanin then helps the skin to gradually tan if the exposed time is kept within reasonable bounds.

Teens are particularly susceptible to the risk of overexposure because they are still experiencing tremendous growth at the cellular level. However, people run the risk of developing skin cancer if they are exposed to the sun or to artificial sunlight in tanning salons too often or for long times.

Anyone using the facilities of a tanning salon should find out the recommended exposure times for each bed. At the beginning of developing a tan, an individual should not tan more than every other day. As the tan develops, it is wise to cut back tanning sessions to no more than twice a week. Many salon owners allow individuals to tan for greater periods of time than is safely recommended. Every tanning bed has a chart indicating what frequency and duration are approved by safety standards. If the beds in a salon do not have such charts, the salon is not following safety guidelines, and it would be a good idea to leave.

Exposure to UV rays is known to stimulate addictive behavior that results in increasing tanning frequency and duration. Although a tan is commonly thought to be beautiful, skin cancer certainly is not. Be a conscientious and perceptive consumer.

According to the Food and Drug Administration, the use of artificial tanning devices is not recommended for anyone.

phenyl group
A benzene ring with one hydrogen absent, C_6H_5—.

is called a **phenyl group:**

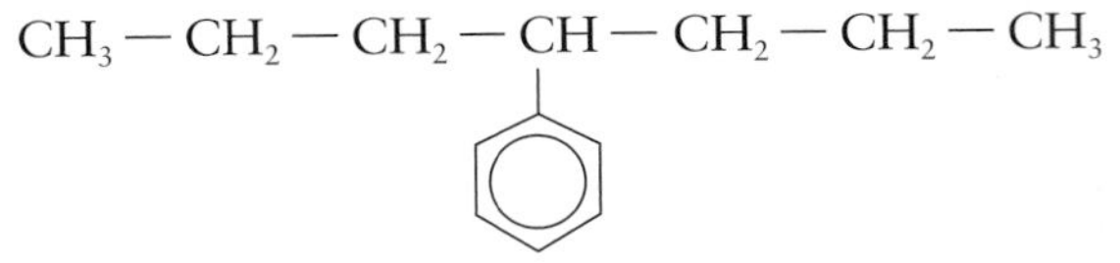

$CH_3-CH_2-CH_2-CH-CH_2-CH_2-CH_3$ (with a benzene ring on the CH)

4-phenylheptane

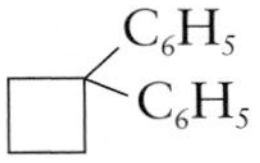

1,1-diphenylcyclobutane

It's easy to confuse the words *phenyl* and *phenol.* The key to keeping them straight is the ending: *-ol* means an alcohol (C_6H_5—OH), and *-yl* means a group (C_6H_5—).

Guideline 4. When two groups are attached to a benzene ring, three isomeric structures are possible. They can be designated by the prefixes *ortho (o), meta (m),* and *para (p):*

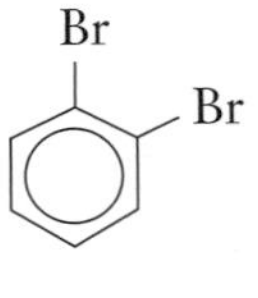

o-dibromobenzene 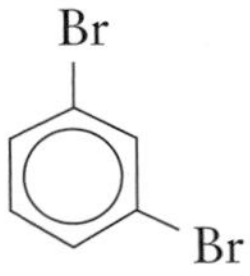 *m*-dibromobenzene  *p*-dibromobenzene

Guideline 5. When two or more groups are attached to a benzene ring, their positions can be indicated by numbering the carbon atoms of the ring so as to obtain the lowest possible numbers for the attachment positions. Groups are arranged in alphabetical order. If there is a choice of identical sets of numbers, the group that comes first in alphabetical order is given the lower number. IUPAC-acceptable common names may be used:

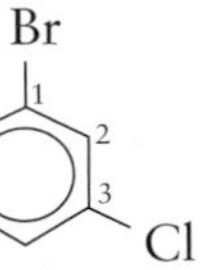

m-bromochlorobenzene or 1-bromo-3-chlorobenzene 1,2,4-trichlorobenzene 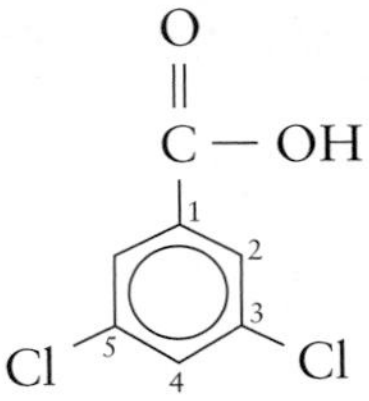 3,5-dichlorobenzoic acid

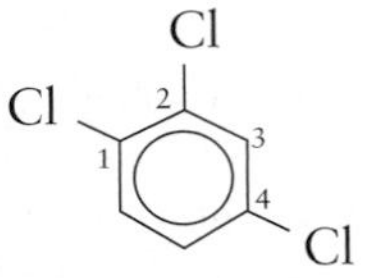

EXAMPLE 2.5

Name each of the following aromatic compounds:

a. 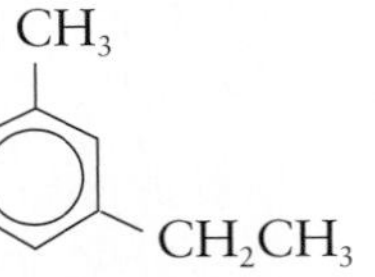 (benzene ring with CH_3 and CH_2CH_3) b. 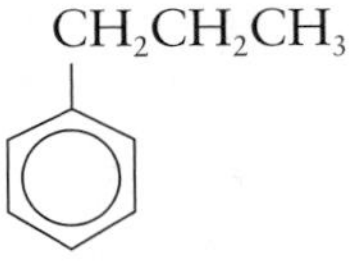 (benzene ring with $CH_2CH_2CH_3$) c. 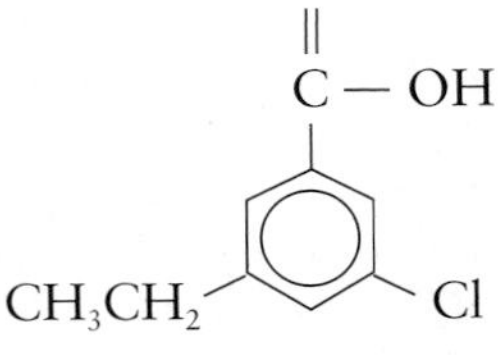 (benzene ring with C(=O)—OH, CH_3CH_2 and Cl)

Solution

a. This compound is named as a substituted toluene. Both of the following are correct: 3-ethyltoluene and *m*-ethyltoluene. Note that in 3-ethyltoluene the methyl group, which is a part of the toluene structure, must be assigned to position number one.

b. This compound may be named as a substituted benzene or a substituted propane: propylbenzene, 1-phenylpropane.

c. When three groups are involved, the ring-numbering approach must be used: 3-chloro-5-ethylbenzoic acid.

■ **Learning Check 2.10** Name the following aromatic compounds:

a. CH_2CH_3 CH_2CH_3

c. Br Br Br

b. Cl

d. CH_3 O C — OH

2.8 Properties and Uses of Aromatic Compounds

LEARNING OBJECTIVE

10. Recognize uses for specific aromatic compounds.

The physical properties of benzene and other aromatic hydrocarbons are similar to those of alkanes and alkenes. They are nonpolar and thus insoluble in water. This hydrophobic characteristic plays an important role in the chemistry of some proteins (Chapter 9).

Aromatic rings are relatively stable chemically. Because of this, benzene often reacts in such a way that the aromatic ring remains intact. Thus, benzene does not undergo the addition reactions that are characteristic of alkenes and alkynes. The predominant type of reaction of aromatic molecules is substitution, in which one of the ring hydrogens is replaced by some other group. Such aromatic reactions are of lesser importance for our purposes and are not shown here.

All aromatic compounds discussed to this point contain a single benzene ring. There are also substances called **polycyclic aromatic compounds,** which contain two or more benzene rings sharing a common side, or "fused" together. The simplest of these compounds, naphthalene, is the original active ingredient in mothballs:

polycyclic aromatic compound A derivative of benzene in which carbon atoms are shared between two or more benzene rings.

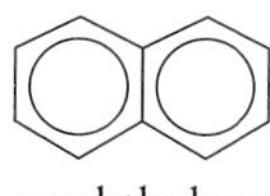

naphthalene

A number of more complex polycyclic aromatic compounds are known to be carcinogens—chemicals that cause cancer. Two of these compounds are

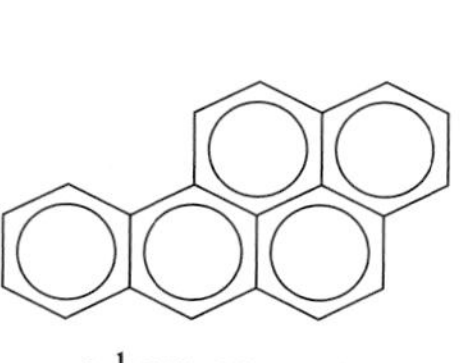

a benzopyrene

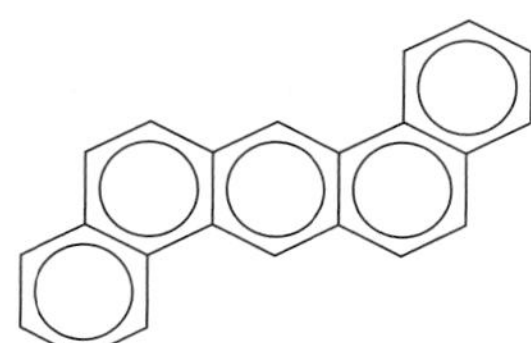

a dibenzanthracene

OVER THE COUNTER 2.1 Smoking: It's Quitting Time

Smoking is a difficult habit to break, especially if the attempt is made by stopping abruptly—going "cold turkey." Mark Twain described the more reasonable gradual approach when he said, "Habit is habit, and not to be flung out of the window by any man, but coaxed downstairs a step at a time." Smokers do not have to go "cold turkey" because there are several OTC aids available, as well as some new prescription products to help them gradually overcome the strong urge to smoke—and eventually quit.

The transdermal (absorbed through the skin) nicotine patch is available over the counter in doses of 7–22 mg. When used as directed, this method delivers a steady supply of nicotine to the bloodstream and helps minimize withdrawal symptoms. Nicotine gum helps reduce withdrawal symptoms when used correctly. The gum should be chewed briefly and then held next to the cheek, allowing the lining of the mouth to absorb the nicotine.

With a prescription, smokers can also obtain a nasal spray that provides a small dose of nicotine each time it is used. This product, called Nicotrol NS®, was approved by the FDA even though inhaling the nicotine poses a small risk that smokers will become as dependent on the mist as they are on cigarettes. The nicotine from the nasal spray gets into the bloodstream faster than nicotine from the gum or patch, providing immediate relief from cigarette craving. A squirt into each nostril gives a smoker 1 mg of nicotine, but it is not supposed to be used more than five times per hour.

The Nicotrol nicotine inhalation system has also received FDA approval. This inhaler, available only by prescription, consists of a plastic cylinder about the size of a cigarette that encloses a cartridge containing nicotine. When a smoker "puffs" on the device, nicotine vapors are absorbed through the lining of the mouth and throat. It takes about 80 puffs to deliver the amount of nicotine obtained from a single cigarette. An advantage of using the system is that a smoker still mimics the hand-to-mouth behavior of smoking, a part of the smoking habit that will be easier to break once nicotine withdrawal symptoms subside.

One of the newest prescription products designed to help break the smoking habit does not contain any nicotine. It is an antidepressant called bupropion that has been shown to be effective in the treatment of nicotine addiction. It is believed that bupropion mimics some of the action of nicotine by releasing the brain chemicals norepinephrine and dopamine, but it is not completely understood how it works. During treatment, a bupropion tablet (marketed as Zyban®) is taken once a day for 3 days and then twice daily during the week before smoking is stopped. Usually, the treatment is continued for the next 6 to 12 weeks to help curb the craving for cigarettes.

If a smoker truly wants to quit, these aids alone will not do it. A smoker must have some kind of support from a formal program, or at least informal support from family and friends. A smoker should also get rid of all tobacco products and avoid smoking triggers, such as other smokers, stress, and alcohol. Exercise can also be a distraction from smoking and can minimize the weight gain that sometimes accompanies giving up smoking.

Charles D. Winters

■ **FIGURE 2.14** Cigarette smoke contains carcinogenic polycyclic aromatic compounds.

vitamin
An organic nutrient that the body cannot produce in the small amounts needed for good health.

These cancer-producing compounds are often formed as a result of heating organic materials to high temperatures. They are present in tobacco smoke (see ■ Figure 2.14), automobile exhaust, and sometimes in burned or heavily browned food. Such compounds are believed to be at least partially responsible for the high incidence of lung and lip cancer among cigarette smokers. Those who smoke heavily face an increased risk of getting cancer. Chemists have identified more than 4000 compounds in cigarette smoke, including 43 known carcinogens. The Environmental Protection Agency (EPA) considers tobacco smoke a Class A carcinogen.

The major sources of aromatic compounds are petroleum and coal tar, a sticky, dark-colored material derived from coal. As with many classes of organic compounds, the simplest structures are the most important commercial materials.

Benzene and toluene are excellent laboratory and industrial solvents. In addition, they are the starting materials for the synthesis of hundreds of other valuable aromatic compounds that are intermediates in the manufacture of a wide variety of commercial products, including the important polymers Bakelite® and polystyrene (see ■ Table 2.4).

A number of aromatic compounds are important in another respect: They must be present in our diet for proper nutrition. Unlike plants, which have the ability to synthesize the benzene ring from simpler materials, humans must obtain any necessary aromatic rings from their diet. This helps explain why certain amino acids, the building blocks of proteins, and some **vitamins** are dietary essentials (Table 2.4).

TABLE 2.4 Some important aromatic compounds

Name	Structural formula	Use
benzene	C_6H_6 (benzene ring)	Industrial solvent and raw material
toluene	$C_6H_5-CH_3$	Industrial solvent and raw material
phenol	C_6H_5-OH	Manufacture of Bakelite® and Formica®
aniline	$C_6H_5-NH_2$	Manufacture of drugs and dyes
styrene	$C_6H_5-CH{=}CH_2$	Preparation of polystyrene products
phenylalanine	$C_6H_5-CH_2CH(NH_2)-C(=O)-OH$	An essential amino acid
riboflavin	$CH_2-CH(OH)-CH(OH)-CH(OH)-CH_2OH$ attached to N of dimethyl isoalloxazine ring (CH_3, CH_3, N, N, N, NH, O, O)	Vitamin B_2

Concept Summary

ThomsonNOW™ Sign in at **www.thomsonedu.com** to:

- Assess your understanding with Exercises keyed to each learning objective.
- Check your readiness for an exam by taking the **Pre-test** and exploring the modules recommended in your **Personalized Learning Plan.**

The Nomenclature of Alkenes. Compounds containing double or triple bonds between carbon atoms are said to be unsaturated. The alkenes contain double bonds, alkynes triple bonds, and aromatics a six-membered ring with three double bonds. **OBJECTIVE 1, Exercise 2.2** In the IUPAC nomenclature system, alkene names end in *-ene*, alkynes end in *-yne*. **OBJECTIVE 2, Exercise 2.4**

The Geometry of Alkenes. In alkenes, the double-bonded carbons and the four groups attached to these carbons lie in the same plane. Because rotation about the double bond is restricted, alkenes may exist as geometric, or cis-trans, isomers. This type of stereoisomerism is possible when each double-bonded carbon is attached to two different groups. **OBJECTIVE 3, Exercise 2.18** IUPAC names of stereoisomers contain the prefixes *cis-* or *trans-*. **OBJECTIVE 4, Assessment Exercise 2.20**

Properties of Alkenes. The physical properties of alkenes are very similar to those of the alkanes. They are nonpolar, insoluble in water, less dense than water, and soluble in nonpolar solvents. Alkenes are quite reactive, and their characteristic reaction is addition to the double bond. Three important addition reactions are bromination (an example of halogenation) to give a dibrominated alkane, hydration to produce an alcohol, and the reaction with H—X to give an alkyl halide. The addition of H_2O and H—X are governed by Markovnikov's rule. **OBJECTIVE 5, Exercise 2.26**

Addition Polymers. Addition polymers are formed from alkene monomers that undergo repeated addition reactions with each other. Many familiar and widely used materials, such as fibers and plastics, are addition polymers. **OBJECTIVE 6, Exercise 2.36**

Alkynes. The alkynes contain triple bonds and possess a linear geometry of the two carbons and the two attached groups. Alkyne names end in *-yne*. **OBJECTIVE 7, Exercise 2.44** The physical and chemical properties of alkynes are very similar to those of the alkenes.

Aromatic Compounds and the Benzene Structure. Benzene, the simplest aromatic compound, and other members of the aromatic class contain a six-membered ring with three double bonds. This aromatic ring is often drawn as a hexagon containing a circle, which represents the six electrons of the double bonds that move freely around the ring. All organic compounds that do not contain an aromatic ring are called aliphatic compounds. **OBJECTIVE 8, Exercise 2.48**

The Nomenclature of Benzene Derivatives. Several acceptable IUPAC names are possible for many benzene compounds. Some IUPAC names are based on widely used common names such as toluene and aniline. Other compounds are named as derivatives of benzene or by designating the benzene ring as a phenyl group. **OBJECTIVE 9, Exercises 2.52 and 2.54**

Properties and Uses of Aromatic Compounds. Aromatic hydrocarbons are nonpolar and have physical properties similar to those of the alkanes and alkenes. Benzene resists addition reactions typical of alkenes. Benzene and toluene are key industrial chemicals. Other important aromatics include phenol, aniline, and styrene. **OBJECTIVE 10, Exercise 2.66**

Key Terms and Concepts

Addition polymer (2.4)
Addition reaction (2.3)
Aliphatic compound (2.6)
Alkene (Introduction)
Alkyne (Introduction)
Aromatic hydrocarbon (Introduction)
Carbocation (2.5)
Copolymer (2.4)
Haloalkane or alkyl halide (2.3)
Hydration (2.3)
Hydrogenation (2.3)
Markovnikov's rule (2.3)
Monomer (2.4)
Phenyl group (2.7)
Polycyclic aromatic compound (2.8)
Polymer (2.4)
Polymerization (2.4)
Polyunsaturated (2.3)
Unsaturated hydrocarbon (Introduction)
Vitamin (2.8)

Key Reactions

1. Halogenation of an alkene (Section 2.3):

$$\mathrm{>C{=}C<} + \mathrm{Br{-}Br} \longrightarrow \mathrm{-\underset{Br}{\underset{|}{\overset{|}{C}}}-\underset{Br}{\underset{|}{\overset{|}{C}}}-}$$ Reaction 2.2

2. Hydrogenation of an alkene (Section 2.3):

$$\mathrm{>C{=}C<} + \mathrm{H{-}H} \xrightarrow{\mathrm{Pt}} \mathrm{-\underset{H}{\underset{|}{\overset{|}{C}}}-\underset{H}{\underset{|}{\overset{|}{C}}}-}$$ Reaction 2.4

3. Addition of H—X to an alkene (Section 2.3):

$$\mathrm{>C{=}C<} + \mathrm{H{-}Cl} \longrightarrow \mathrm{-\underset{H}{\underset{|}{\overset{|}{C}}}-\underset{Cl}{\underset{|}{\overset{|}{C}}}-}$$ Reaction 2.6

4. Hydration of an alkene (Section 2.3):

$$\mathrm{>C{=}C<} + \mathrm{H{-}OH} \xrightarrow{\mathrm{H_2SO_4}} \mathrm{-\underset{H}{\underset{|}{\overset{|}{C}}}-\underset{OH}{\underset{|}{\overset{|}{C}}}-}$$ Reaction 2.9

5. Addition polymerization of an alkene (Section 2.4):

$$n\mathrm{CH_2{=}CH_2} \xrightarrow[\text{catalysts}]{\text{Heat, pressure}} \mathrm{+CH_2-CH_2+}_n$$ Reaction 2.14

Exercises

SYMBOL KEY

Even-numbered exercises are answered in Appendix B.

Blue-numbered exercises are more challenging.

■ denotes exercises available in ThomsonNow and assignable in OWL.

ThomsonNOW™ To assess your understanding of this chapter's topics with sample tests and other resources, sign in at www.thomsonedu.com.

THE NOMENCLATURE OF ALKENES (SECTION 2.1) AND ALKYNES (SECTION 2.5)

2.1 What is the definition of an unsaturated hydrocarbon?

2.2 Define the terms alkene, alkyne, and aromatic hydrocarbon.

2.3 Select those compounds that can be correctly called *unsaturated* and classify each one as an *alkene* or an *alkyne:*

a. $CH_3—CH_2—CH_3$

b. $CH_3CH=CHCH_3$

c. $H—C\equiv C—CH(CH_3)—CH_3$

d. (cyclopentene)

e. (methylcyclohexane)

f. (cyclopentyl)$—CH=CH_2$

g. $CH=CH$ / $CH_2—CH_2$ (cyclobutene ring)

h. $CH_2=CHCH_2CH_3$

i. $CH_3CH(CH_3)CH_3$

2.4 ■ Give the IUPAC name for the following compounds:

a. $CH_3CH=CHCH_3$

b. $CH_3CH_2—C(CH_2CH_3)=CHCH_3$

c. $CH_3—C\equiv C—C(CH_3)_2—CH_2CH_3$

d. (cyclopentene ring with CH_3)

e. $CH_3CH(Br)CH_2—C\equiv C—CH(CH_3)—CH_3$

f. (cyclopropene ring with CH_3, CH_3, and CH_2CH_3 substituents)

g. $CH_3CH(CH_3)—CH=CHCH_2CH=CH_2$

2.5 Give the IUPAC name for the following compounds:

a. $CH_3CH(CH_3)CH=CHCH_2CH_3$

b. $CH_3CH=CHCH=CHCH(CH_3)CH_3$

c. (cyclopentene)

d. $CH_3—C\equiv C—CH_2CH_3$

e. (cyclohexene ring with CH_2CHCH_3/CH_3 and CH_3 substituents)

f. (cycloheptatriene ring with $CH_2CH_2CH_3$ substituent)

g. $CH_2=C(CH_2CH_2CH_2CH_3)—CH(CH_2CH_3)—CH=CHCH_2CH_3$

2.6 Draw structural formulas for the following compounds:

a. 3-ethyl-2-hexene

b. 3,4-dimethyl-1-pentene

c. 3-methyl-1,3-pentadiene

d. 2-isopropyl-4-methylcyclohexene

e. 1-butylcyclopropene

2.7 ■ Draw structural formulas for the following compounds:

a. 4,4,5-trimethyl-2-heptyne

b. 1,3-cyclohexadiene

c. 2-*t*-butyl-4,4-dimethyl-1-pentene

d. 4-isopropyl-3,3-dimethyl-1,5-octadiene

e. 2-methyl-1,3-cyclopentadiene

f. 3-*sec*-butyl-3-*t*-butyl-1-heptyne

2.8 A compound has the molecular formula C_5H_8. Draw a structural formula for a compound with this formula that would be classified as (a) an alkyne, (b) a diene, and (c) a cyclic alkene. Give the IUPAC name for each compound.

2.9 Draw structural formulas and give IUPAC names for the 13 alkene isomers of C_6H_{12}. Ignore geometric isomers and cyclic structures.

2.10 α-Farnesene is a constituent of the natural wax found on apples. Given that a 12-carbon chain is named as a dodecane, what is the IUPAC name of α-farnesene?

$$\begin{array}{c} \quad CH_3 \qquad\qquad\quad CH_3 \qquad\qquad\quad CH_3 \\ CH_3\overset{|}{C}{=}CHCH_2CH_2\overset{|}{C}{=}CHCH_2CH{=}\overset{|}{C}CH{=}CH_2 \end{array}$$

α-farnesene

2.11 Each of the following names is wrong. Give the structure and correct name for each compound.

a. 3-pentene

b. 3-methyl-2-butene

c. 2-ethyl-3-pentyne

2.12 Each of the following names is wrong. Give the structure and correct name for each compound.

a. 2-methyl-4-hexene

b. 3,5-heptadiene

c. 4-methylcyclobutene

THE GEOMETRY OF ALKENES (SECTION 2.2)

2.13 ■ What type of hybridized orbital is present on carbon atoms bonded by a double bond? How many of these hybrid orbitals are on each carbon atom?

2.14 What type of orbital overlaps to form a pi bond in an alkene? What symbol is used to represent a pi bond? How many electrons are in a pi bond?

2.15 Describe the geometry of the carbon–carbon double bond and the two atoms attached to each of the double-bonded carbon atoms.

2.16 Explain the difference between geometric and structural isomers of alkenes.

2.17 Draw structural formulas and give IUPAC names for all the isomeric pentenes (C_5H_{10}) that are

a. Alkenes that do not show geometric isomerism. There are four compounds.

b. Alkenes that do show geometric isomerism. There is one cis and one trans compound.

2.18 ■ Which of the following alkenes can exist as cis-trans isomers? Draw structural formulas and name the cis and trans isomers.

a. $CH_3CH_2CH_2CH_2CH{=}CH_2$

b. $CH_3CH_2CH{=}CHCH_2CH_3$

c. $CH_3\overset{\overset{CH_3}{|}}{C}{=}CHCH_2CH_3$

2.19 Which of the following alkenes can exist as cis-trans isomers? Draw structural formulas and name the cis and trans isomers.

a. $CH_3CH_2CH_2CH{=}CHCH_2CH_2CH_3$

b. $CH_3CH{=}\underset{\underset{CH_2CH_3}{|}}{C}{-}CH_2CH_3$

c. $BrCH_2CH{=}CHCH_2Br$

2.20 ■ Draw structural formulas for the following:

a. *cis*-3-hexene

b. *trans*-3-heptene

2.21 Draw structural formulas for the following:

a. *trans*-3,4-dibromo-3-heptene

b. *cis*-1,4-dichloro-2-methyl-2-butene

PROPERTIES OF ALKENES (SECTION 2.3)

2.22 In what ways are the physical properties of alkenes similar to those of alkanes?

2.23 Which of the following reactions is an addition reaction?

a. $A_2 + C_3H_6 \rightarrow C_3H_6A_2$

b. $A_2 + C_6H_6 \rightarrow C_6H_5A + HA$

c. $HA + C_4H_8 \rightarrow C_4H_9A$

d. $3O_2 + C_2H_4 \rightarrow 2CO_2 + 2H_2O$

e. $C_7H_{16} \rightarrow C_7H_8 + 4H_2$

2.24 State Markovnikov's rule, and write a reaction that illustrates its application.

2.25 Complete the following reactions. Where more than one product is possible, show only the one expected according to Markovnikov's rule.

a. $CH_2{=}CHCH_2CH_3 + H_2 \xrightarrow{Pt}$

b. $CH_2{=}CHCH_2CH_3 + Br_2 \longrightarrow$

c. 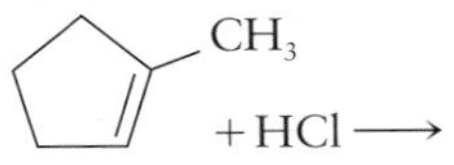 1-methylcyclopentene (CH_3 on the ring double-bond carbon) $+ HCl \longrightarrow$

d. $CH_3CH_2\overset{\overset{CH_3}{|}}{C}{=}CH_2 + H_2O \xrightarrow{H_2SO_4}$

2.26 ■ Complete the following reactions. Where more than one product is possible, show only the one expected according to Markovnikov's rule.

a. 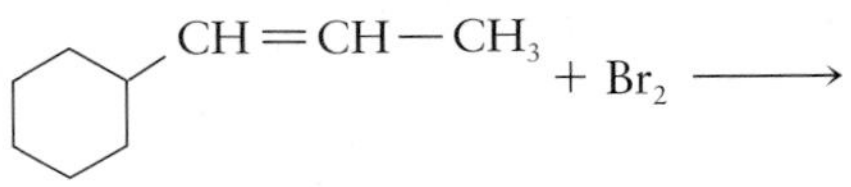 cyclohexyl–$CH{=}CH{-}CH_3$ $+ Br_2 \longrightarrow$

b. cyclohexene $+ H_2O \xrightarrow{H_2SO_4}$

c. $CH_2{=}\overset{\overset{CH_3}{|}}{C}{-}CH{=}CH_2 + 2H_2 \xrightarrow{Pt}$

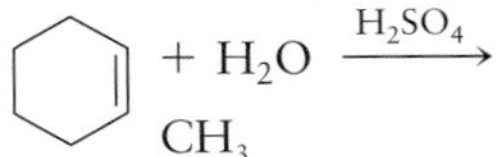

d. 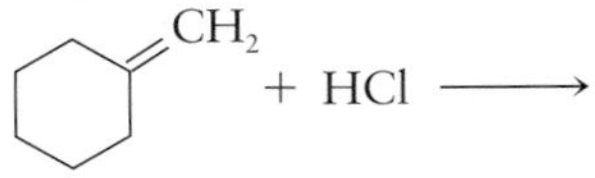 methylenecyclohexane (ring$=CH_2$) $+ HCl \longrightarrow$

2.27 Draw the structural formula for the alkenes with molecular formula C_5H_{10} that will react to give the following products. Show all correct structures if more than one starting material will react as shown.

a. $C_5H_{10} + Br_2 \longrightarrow CH_3CH_2\underset{\underset{Br}{|}}{C}H\underset{\underset{Br}{|}}{C}HCH_3$

b. $C_5H_{10} + H_2 \xrightarrow{Pt} CH_3CH_2CH_2CH_2CH_3$

c.

$$C_5H_{10} + H_2O \xrightarrow{H_2SO_4} \begin{array}{c} CH_3 \\ | \\ CH_3CCH_2CH_3 \\ | \\ OH \end{array}$$

d.

$$C_5H_{10} + HBr \longrightarrow \begin{array}{l} CH_3CHCH_2CH_2CH_3 \\ \quad\;\; | \\ \quad\;\, Br \end{array}$$

2.28 ■ What reagents would you use to prepare each of the following from 3-hexene?

a. $\begin{array}{l} \qquad\quad\;\; Br\;\; Br \\ \qquad\qquad | \quad\; | \\ CH_3CH_2CHCHCH_2CH_3 \end{array}$

b. $CH_3CH_2CH_2CH_2CH_2CH_3$

c. $\begin{array}{l} \qquad\qquad\quad\; Cl \\ \qquad\qquad\quad\;\, | \\ CH_3CH_2CH_2CHCH_2CH_3 \end{array}$

d. $\begin{array}{l} \qquad\qquad\quad\; OH \\ \qquad\qquad\quad\;\; | \\ CH_3CH_2CH_2CHCH_2CH_3 \end{array}$

2.29 What is an important commercial application of hydrogenation?

2.30 Cyclohexane and 2-hexene both have the molecular formula C_6H_{12}. Describe a simple chemical test that would distinguish one from the other.

2.31 Terpin hydrate is used medicinally as an expectorant for coughs. It is prepared by the following addition reaction. What is the structure of terpin hydrate?

CH_3

$$+ \; 2H_2O \xrightarrow{H_2SO_4} \text{terpin hydrate}$$

C

$CH_3 \quad CH_2$

ADDITION POLYMERS (SECTION 2.4)

2.32 Explain what is meant by each of the following terms: *monomer, polymer, addition polymer,* and *copolymer.*

2.33 A section of polypropylene containing three units of monomer can be shown as

$$\begin{array}{ccccccc} & CH_3 & & CH_3 & & CH_3 & \\ & | & & | & & | & \\ -CH_2- & CH & -CH_2- & CH & -CH_2- & CH & - \end{array}$$

Draw structural formulas for comparable three-unit sections of

a. Teflon

b. Orlon

c. Lucite

2.34 Identify a structural feature characteristic of all monomers listed in Table 2.3.

2.35 Rubber cement contains a polymer of 2-methylpropene (isobutylene) called polyisobutylene. Write an equation for the polymerization reaction.

2.36 ■ Much of today's plumbing in newly built homes is made from a plastic called poly (vinyl chloride), or PVC. Using Table 2.3, write a reaction for the formation of poly (vinyl chloride).

2.37 Identify a major use for each of the following addition polymers:

a. Styrofoam

b. Acrilan

c. Plexiglas

d. PVC

e. polypropylene

ALKYNES (SECTION 2.5)

2.38 ■ What type of hybridized orbital is present on carbon atoms bonded by a triple bond? How many of these hybrid orbitals are on each carbon atom?

2.39 How many sigma bonds and how many pi bonds make up the triple bond of an alkyne?

2.40 Describe the geometry in an alkyne of the carbon–carbon triple bond and the two attached atoms.

2.41 Explain why geometric isomerism is not possible in alkynes.

2.42 Give the common name and major uses of the simplest alkyne.

2.43 Describe the physical and chemical properties of alkynes.

2.44 ■ Write the structural formulas and IUPAC names for all the isomeric alkynes with the formula C_5H_8.

AROMATIC COMPOUNDS AND THE BENZENE STRUCTURE (SECTION 2.6)

2.45 What type of hybridized orbital is present on the carbon atoms of a benzene ring? How many sigma bonds are formed by each carbon atom in a benzene ring?

2.46 What type of orbital overlaps to form the pi bonding in a benzene ring?

2.47 What does the circle within the hexagon represent in the structural formula for benzene?

2.48 Define the terms *aromatic* and *aliphatic.*

2.49 Limonene, which is present in citrus peelings, has a very pleasant lemonlike fragrance. However, it is not classified as an aromatic compound. Explain.

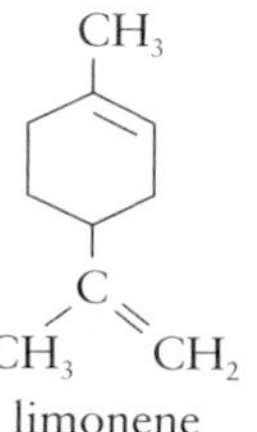

limonene

2.50 ■ A disubstituted cycloalkane such as (a) exhibits cis-trans isomerism, whereas a disubstituted benzene (b) does not. Explain.

a. 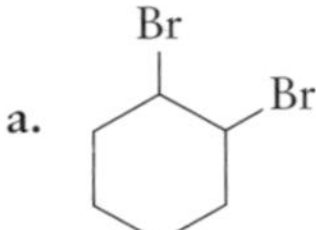

b. Br, Br

THE NOMENCLATURE OF BENZENE DERIVATIVES (SECTION 2.7)

2.51 Give an IUPAC name for each of the following hydrocarbons as a derivative of benzene:

a. 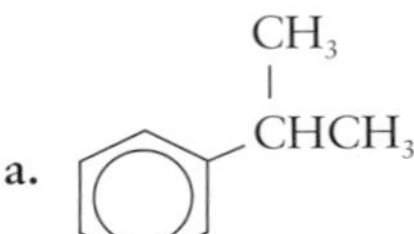

b. 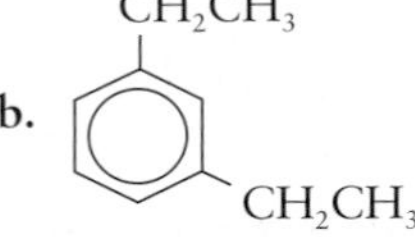

2.52 ■ Give an IUPAC name for each of the following hydrocarbons as a derivative of benzene:

a. 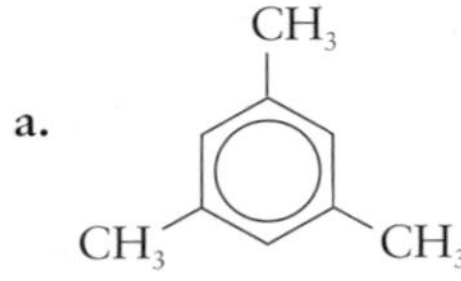

b. 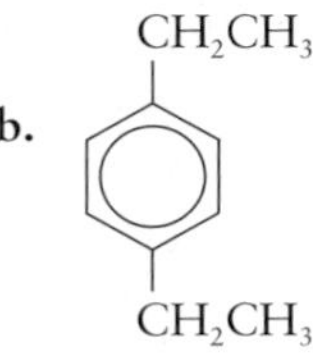

2.53 Give an IUPAC name for the following as hydrocarbons with the benzene ring as a substituent:

a. 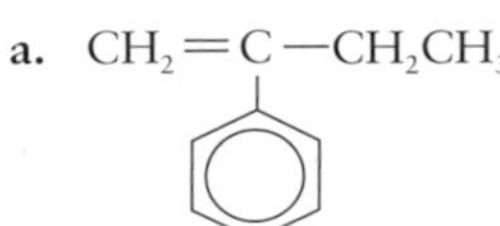

b. 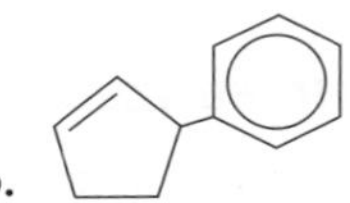

2.54 ■ Give an IUPAC name for the following as hydrocarbons with the benzene ring as a substituent:

a. $CH_3CH_2CHCH{=}CH_2$

b. 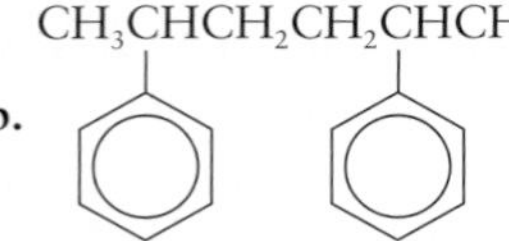

2.55 Name the following compounds, using the prefixed abbreviations for *ortho, meta,* and *para* and assigning IUPAC-acceptable common names:

a. 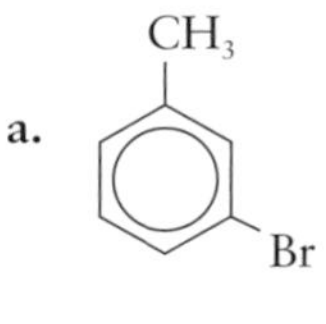

b. 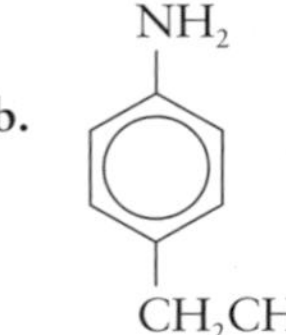

2.56 Name the following compounds, using the prefixed abbreviations for *ortho, meta,* and *para* and assigning IUPAC-acceptable common names:

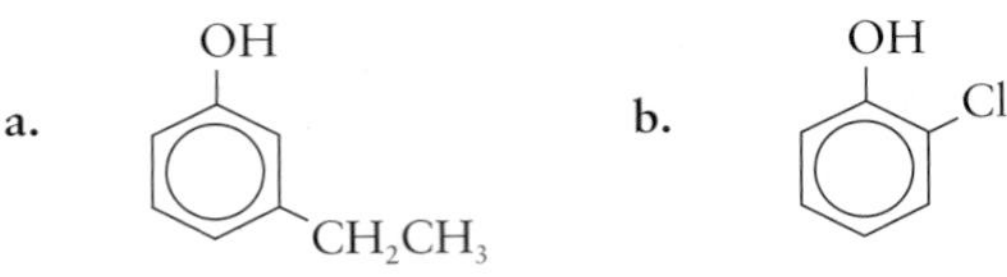

2.57 Name the following by numbering the benzene ring. IUPAC-acceptable common names may be used where appropriate:

a. Br, NH_2, Br

b. CH_2CH_3, Cl, CH_3

2.58 Name the following by numbering the benzene ring. IUPAC-acceptable common names may be used where appropriate:

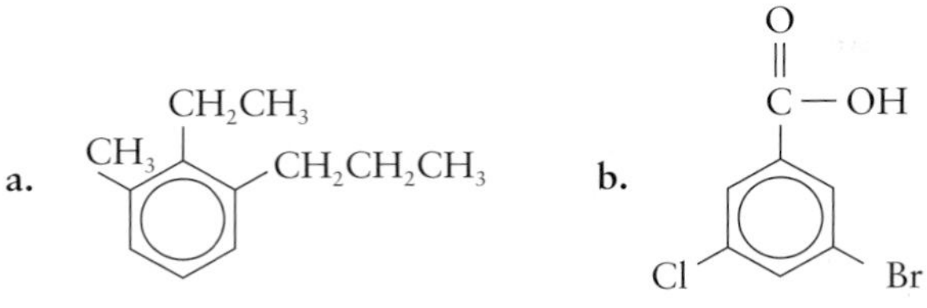

2.59 Draw structural formulas for the following:

a. 2,4-diethylaniline
b. 4-ethyltoluene
c. *p*-ethyltoluene

2.60 ■ Write structural formulas for the following:

a. *o*-ethylphenol
b. *m*-chlorobenzoic acid
c. 3-methyl-3-phenylpentane

2.61 There are three bromonitrobenzene derivatives. Draw their structures and give an IUPAC name for each one.

PROPERTIES AND USES OF AROMATIC COMPOUNDS (SECTION 2.8)

2.62 Describe the chief physical properties of aromatic hydrocarbons.

2.63 Why does benzene not readily undergo addition reactions characteristic of other unsaturated compounds?

2.64 Compare the chemical behavior of benzene and cyclohexene.

2.65 For each of the following uses, list an appropriate aromatic compound:

a. A solvent
b. A vitamin
c. An essential amino acid
d. Starting material for dyes

2.66 For each of the following uses, list an appropriate aromatic compound:

a. Used in the production of Formica
b. A starting material for polystyrene

c. Used to manufacture drugs

d. A starting material for Bakelite

ADDITIONAL EXERCISES

2.67 In general, alkynes have slightly higher boiling points and densities than structurally equivalent alkanes. What interparticle force would this be attributable to?

2.68 In Reaction 2.14, heat, pressure, and catalysts are needed to convert ethylene gas to polyethylene. Explain the effects of each of the three conditions (heat, pressure, catalysts) in terms of factors that affect reaction rates.

2.69 Propene reacts with a diatomic molecule whose atoms have the electronic configuration of $1s^2 2s^2 2p^6 3s^2 3p^5$. Draw the structure of the product formed and give its IUPAC name.

2.70 Draw a generalized energy diagram for the following reaction. Is the reaction endothermic or exothermic?

$$\text{alkene} + \text{water} \xrightarrow{H_2SO_4} \text{alcohol} + 10 \text{ kcal/mol}$$

2.71 What will be the limiting reactant when 25.0 g of 2-butene reacts with 25.0 g of iodine to produce 2,3-diiodobutane? How many moles of product could be produced?

$$CH_3CH{=}CHCH_3 + I_2 \longrightarrow CH_3\underset{}{CH(I)}CH(I)CH_3$$

ALLIED HEALTH EXAM CONNECTION

Reprinted with permission from Nursing School and Allied Health Entrance Exams, COPYRIGHT 2005 Petersons.

2.72 Which of the following are aromatic compounds?

a. Benzene

b. Ethyl alcohol

c. Methane

d. Phenol

2.73 Identify which of the following general formulas would be used to characterize (1) an alkane, (2) an alkene with one C=C bond, and (3) an alkyne with one C≡C bond.

a. C_nH_{2n}

b. C_nH_{2n+2}

c. C_nH_{2n-2}

CHEMISTRY FOR THOUGHT

2.74 Napthalene is the simplest polycyclic aromatic compound:

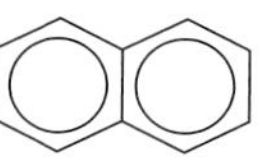

Draw a Kekulé structure for this compound like that shown for benzene in Section 2.6.

2.75 Why does propene not exhibit geometric isomerism?

2.76 Limonene is present in the rind of lemons and oranges. Based on its structure (see Exercise 2.49), would you consider it to be a solid, liquid, or gas at room temperature?

2.77 If the average molecular weight of polyethylene is 5.0×10^4 u, how many ethylene monomers ($CH_2{=}CH_2$) are contained in a molecule of the polymer?

2.78 Reactions to synthesize the benzene ring of aromatic compounds do not occur within the human body, and yet many essential body components involve the benzene structure. How does the human body get its supply of aromatic compounds?

2.79 Answer the question in the caption to Figure 2.8 pertaining to poly (vinyl acetate).

2.80 Why can't alkanes undergo addition polymerization?

2.81 Some polymers produce toxic fumes when they are burning. Which polymer in Table 2.3 produces hydrogen cyanide, HCN? Which produces hydrogen chloride, HCl?

2.82 "Super glue" is an addition polymer of the following monomer. Draw a structural formula for a three-unit section of super glue.

$$CH_2{=}\overset{\displaystyle CN}{\overset{|}{C}}{-}O{-}\overset{\displaystyle O}{\overset{\|}{C}}{-}CH_3$$

2.83 One of the fragrant components in mint plants is menthene, a compound whose IUPAC name is 1-isopropyl-4-methylcyclohexene. Draw a structural formula for menthene.

CHAPTER 3

Alcohols, Phenols, and Ethers

LEARNING OBJECTIVES

When you have completed your study of this chapter, you should be able to:

1. Name and draw structural formulas for alcohols and phenols. (Section 3.1)
2. Classify alcohols as primary, secondary, or tertiary on the basis of their structural formulas. (Section 3.2)
3. Discuss how hydrogen bonding influences the physical properties of alcohols. (Section 3.3)
4. Write equations for alcohol dehydration and oxidation reactions. (Section 3.4)
5. Recognize uses for specific alcohols. (Section 3.5)
6. Recognize uses for specific phenols. (Section 3.6)
7. Name and draw structural formulas for ethers. (Section 3.7)
8. Describe the key physical and chemical properties of ethers. (Section 3.8)
9. Write equations for a thiol reaction with heavy metal ions and production of disulfides. (Section 3.9)
10. Identify functional groups in polyfunctional compounds. (Section 3.10)

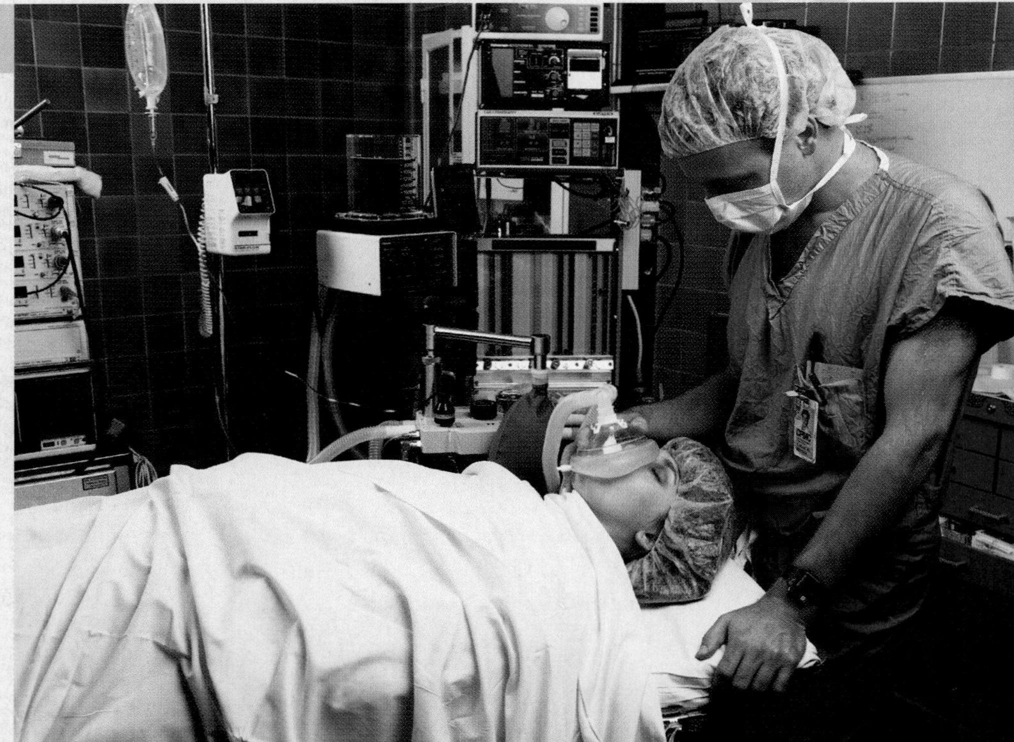

Nurse anesthetists help administer anesthesia and closely monitor patients during surgery. In large hospitals, they usually work with a physician who specializes in anesthesiology (an anesthesiologist) during the procedure. However, in small hospitals, a nurse anesthetist might have the sole responsibility for anesthesia under the direction of the surgeon who is doing the surgery. General anesthetics is one of the special topics discussed in this chapter.

Alcohols, phenols, and ethers are important organic compounds that occur naturally and are produced synthetically in significant amounts. They are used in numerous industrial and pharmaceutical applications (see ■ Figure 3.1). Ethanol, for example, is one of the simplest and best known of all organic substances; the fermentation method for producing it was known and used by ancient civilizations. Menthol, a ten-carbon alcohol obtained from peppermint oil, is widely used as a flavoring agent. Cholesterol, with its complicated multiring molecular structure, has been implicated in some forms of heart disease. The chemistry of all of these substances is similar because the alcohol functional group behaves essentially the same, regardless of the complexity of the molecule in which it is found.

■ **FIGURE 3.1** A promising anticancer drug, taxol, is found in the bark of the Pacific yew tree. Though taxol is a complex compound, what does its name suggest about its structure?

CH_3-CH_2-OH ethanol; menthol; cholesterol

Structurally, an **alcohol** is obtained from a hydrocarbon by replacing a hydrogen atom with a **hydroxy group** (—OH). A formula for an alcohol can be generalized as R—OH, where R— represents CH_3—, CH_3CH_2—, or any other singly bonded carbon–hydrogen group. Thus, R—OH can stand for CH_3—OH, CH_3CH_2—OH, and so on. If the replaced hydrogen was attached to an aromatic ring, the resulting compound is known as a **phenol:**

R—OH (an alcohol) a phenol

alcohol
A compound in which an —OH group is connected to an aliphatic carbon atom.

hydroxy group
The —OH functional group.

phenol
A compound in which an —OH group is connected to a benzene ring. The parent compound is also called phenol.

Alcohols and phenols may also be considered to be derived from water by the replacement of one of its hydrogen atoms with an alkyl group or an aromatic ring:

$$\underset{\text{water}}{H-OH} \xrightarrow{\text{Replace H with R}} \underset{\text{an alcohol}}{R-OH}$$

$$\underset{\text{water}}{H-OH} \xrightarrow{\text{Replace H with benzene ring}} \text{a phenol}$$

ThomsonNOW™ Throughout the chapter this icon introduces resources on the ThomsonNOW website for this text. Sign in at **www.thomsonedu.com** to:

- Evaluate your knowledge of the material
- Take an exam prep quiz
- Identify areas you need to study with a **Personalized Learning Plan**.

For most people, the word *ether* generates images of surgeons and operating rooms. Chemists define an **ether** as an organic compound in which two carbon atoms are bonded to an oxygen atom. This structure results when both hydrogen atoms of water are replaced by alkyl groups:

$$\underset{\text{water}}{H-O-H} \xrightarrow{\text{Replace both H's with R}} \underset{\text{an ether}}{R-O-R'}$$

R′ indicates that the two R groups can be the same or different.

ether
A compound that contains a

$$-\overset{|}{\underset{|}{C}}-O-\overset{|}{\underset{|}{C}}-$$

functional group.

3.1 The Nomenclature of Alcohols and Phenols

LEARNING OBJECTIVE

1. Name and draw structural formulas for alcohols and phenols.

The simpler alcohols are often known by common names, where the alkyl group name is followed by the word *alcohol:*

CH_3-OH	CH_3CH_2-OH	$CH_3CH_2CH_2-OH$	$CH_3CH(OH)CH_3$
methyl alcohol	ethyl alcohol	propyl alcohol	isopropyl alcohol

The IUPAC rules for naming alcohols that contain a single hydroxy group are as follows:

Step 1. Name the longest chain to which the hydroxy group is attached. The chain name is obtained by dropping the final *-e* from the name of the hydrocarbon that contains the same number of carbon atoms and adding the ending *-ol.*
Step 2. Number the longest chain to give the lower number to the carbon with the attached hydroxy group.
Step 3. Locate the position of the hydroxy group by the number of the carbon atom to which it is attached.
Step 4. Locate and name any branches attached to the chain.
Step 5. Combine the name and location for other groups, the hydroxy group location, and the longest chain into the final name.

ThomsonNOW™ Go to Coached Problems to explore the **relationship between structure and names of alcohols** as well as practice **naming alcohols.**

EXAMPLE 3.1

Name the following alcohols according to the IUPAC system:

a. CH_3-OH

b. $CH_3CH_2CH_2CHCH_2CH_3$ with CH_2-OH attached to the fourth carbon (CH)

c. Benzene ring $-CH_2CHCH_3$ with OH on the CH

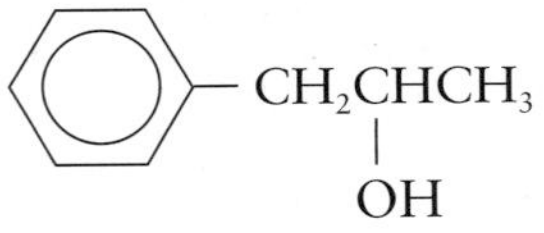

d. Cyclohexane ring bearing HO and CH_3 (1,3-positions)

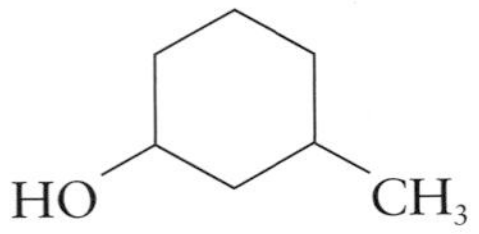

Solution

a. The longest chain has one carbon atom; thus, the alcohol will be called methanol (methane $-$ *e* + *ol*). Since the —OH group can be attached only to the single carbon atom, no location number for —OH is needed.

b. The longest chain containing the —OH group is numbered as follows:

$$\overset{5}{C}H_3\overset{4}{C}H_2\overset{3}{C}H_2\overset{2}{C}HCH_2CH_3,\quad \overset{2}{C}H-\underset{1}{C}H_2-OH$$

Note that there is a longer chain (six carbon atoms), but the —OH group is not directly attached to it. The five-carbon chain makes the alcohol a pentanol.

The —OH group is on carbon number 1; hence, the compound is a 1-pentanol. An ethyl group is attached to carbon number 2 of the chain, so the final name is 2-ethyl-1-pentanol.

c. The longest chain contains three carbon atoms, with the —OH group attached to carbon number 2. Hence, the compound is a 2-propanol. A benzene ring attached to carbon number 1 is named as a phenyl group. Therefore, the complete name is 1-phenyl-2-propanol. Note that the phenyl group is on carbon number 1 rather than number 3.

d. The —OH group is given preference, so the carbon to which it is attached is carbon number 1 of the ring. The $—CH_3$ group is located on carbon number 3 (since 3 is lower than 5, which would be obtained by counting the other way). The complete name is 3-methylcyclohexanol. Note that because the —OH group is always at the number-1 position in a ring, the position is not shown in the name.

Learning Check 3.1 Provide IUPAC names for the following alcohols:

a. $CH_3CH(CH_3)CH_2—OH$

b. $CH_3CH(CH_2CH_3)CH(OH)CH_2CH(CH_3)CH_3$

c. (cyclopentane ring bearing CH_3 and OH on adjacent carbons)

Compounds containing more than one hydroxy group are known. Alcohols containing two hydroxy groups are called *diols*. Those containing three —OH groups are called *triols*. The IUPAC nomenclature rules for these compounds are essentially the same as those for the single hydroxy alcohols, except that the ending *-diol* or *-triol* is attached to the name of the parent hydrocarbon.

EXAMPLE 3.2

Name the following alcohols according the IUPAC system:

a. $CH_2(OH)—CH_2(OH)$

b. $CH_2(OH)CH(CH_3)CH_2CH_2(OH)$

c. $CH_2(OH)—CH(OH)—CH_2(OH)$

Solution

a. The longest chain has two carbons, making this an ethanediol. The —OH groups attached to carbons 1 and 2 give the final name 1,2-ethanediol. This substance is also called ethylene glycol.

b. Because the longest chain has four carbons, the compound is a butanediol. The —OH groups are attached to carbons 1 and 4, and a methyl group is attached to carbon 2. The name is 2-methyl-1,4-butanediol.

c. The longest chain contains three carbons, so the compound is a propanetriol. The —OH groups are attached to carbons 1, 2, and 3, making the name 1,2,3-propanetriol. Common names for this substance are glycerin and glycerol.

Learning Check 3.2 Give IUPAC names to the following diols:

a. OH OH (on first and last carbons)
$CH_2CH_2CH_2CH_2$

b. OH, OH (on adjacent carbons of a cyclohexane ring)

Substituted phenols are usually named as derivatives of the parent compound phenol:

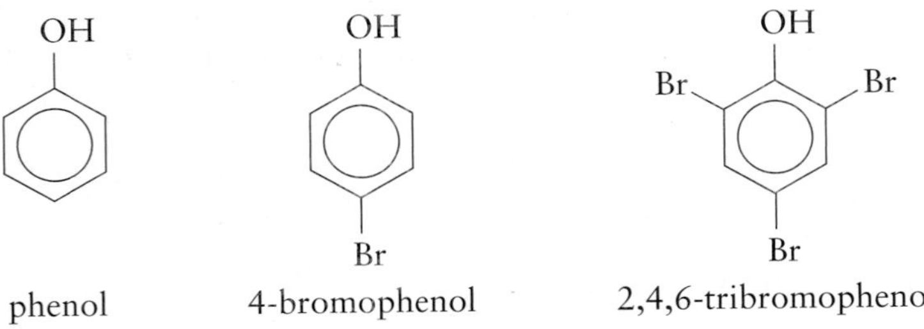

phenol 4-bromophenol 2,4,6-tribromophenol

Learning Check 3.3 Name this compound as a phenol:

CH_2CH_3
OH

3.2 Classification of Alcohols

LEARNING OBJECTIVE

2. Classify alcohols as primary, secondary, or tertiary on the basis of their structural formulas.

The characteristic chemistry of an alcohol sometimes depends on the groups bonded to the carbon atom that bears the hydroxy group. Alcohols are classified as primary, secondary, or tertiary on the basis of these attached groups (see Table 3.1). In **primary alcohols,** the hydroxy-bearing carbon atom is attached to one other carbon atom and two hydrogen atoms. The simplest alcohol, $CH_3—OH$, is also considered a primary alcohol. In **secondary alcohols,** the hydroxy-bearing carbon atom is attached to two other carbon atoms and one hydrogen atom. In **tertiary alcohols,** the hydroxy-bearing carbon atom is attached to three other carbon atoms. The use of R, R′, and R″ in Table 3.1 signifies that each of those alkyl groups may be different.

primary alcohol
An alcohol in which the —OH group is attached to CH_3— or to a carbon attached to one other carbon atom.

secondary alcohol
An alcohol in which the carbon bearing the —OH group is attached to two other carbon atoms.

tertiary alcohol
An alcohol in which the carbon bearing the —OH group is attached to three other carbon atoms.

ThomsonNOW™ Go to Coached Problems to practice **classifying alcohols and amines.**

Learning Check 3.4 Classify the following alcohols as primary, secondary, or tertiary:

a. $CH_3CH_2CH_2CH_2—OH$

b. (cyclohexane ring)—OH

c.
OH
|
CH_3CH
|
CH_2CH_3

TABLE 3.1 Primary, secondary, and tertiary alcohols

	Primary	Secondary	Tertiary
General formula	$R-CH_2-OH$	$R-CH(R')-OH$	$R'-C(R)(R'')-OH$
Specific example	$CH_3CH_2CH_2-OH$ 1-propanol (propyl alcohol)	$CH_3CH(CH_3)-OH$ 2-propanol (isopropyl alcohol)	$CH_3-C(CH_3)(CH_3)-OH$ 2-methyl-2-propanol (*t*-butyl alcohol)

3.3 Physical Properties of Alcohols

LEARNING OBJECTIVE

3. Discuss how hydrogen bonding influences the physical properties of alcohols.

The replacement of one hydrogen of water with an organic group does not cause all the waterlike properties to disappear. Thus, the lower molecular weight alcohols—methyl, ethyl, propyl, and isopropyl alcohols—are completely miscible with water. As the size of the alkyl group in an alcohol increases, the physical properties become less waterlike and more alkanelike. Long-chain alcohols are less soluble in water (see ■ Figure 3.2) and more soluble in nonpolar solvents such as benzene, carbon tetrachloride, and ether. The solubility of alcohols in water depends on the number of carbon atoms per hydroxy group in the molecule. In general, one hydroxy group can solubilize three to four carbon atoms. The high solubility of the lower molecular weight alcohols in water can be attributed to hydrogen bonding between the alcohol and water molecules (see ■ Figure 3.3).

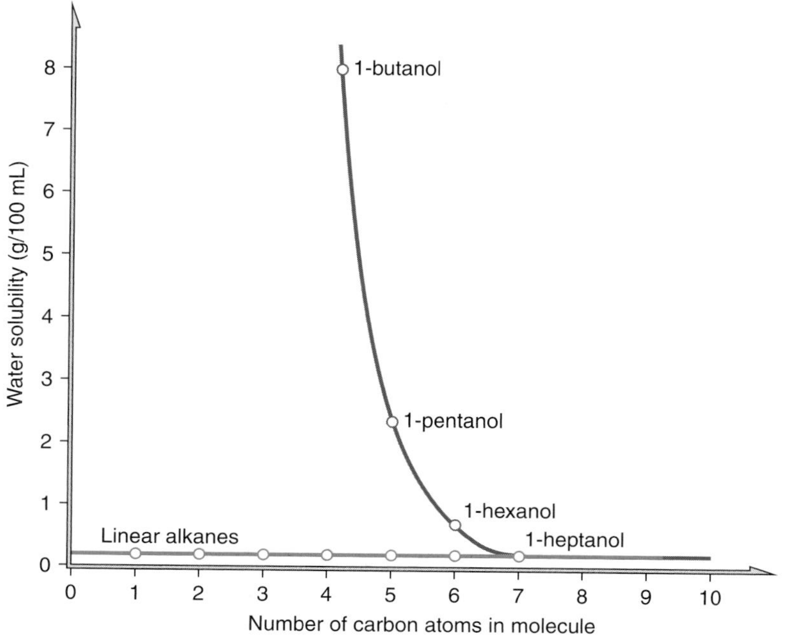

■ **FIGURE 3.2** The solubility of alcohols and linear alkanes in water.

FIGURE 3.3 Hydrogen bonding in a water–methanol solution. A three-dimensional network of molecules and hydrogen bonds is formed.

FIGURE 3.4 Water interacts only with the —OH group of 1-heptanol.

In long-chain alcohols, the hydrophobic alkyl group does not form hydrogen bonds with water. Thus, in 1-heptanol, water molecules surround and form bonds with the —OH end but do not bond to the alkyl portion (see ■ Figure 3.4). Because the alkyl portion is unsolvated, 1-heptanol is insoluble in water. Its solubility in water is comparable to that of heptane (seven carbon atoms), as shown in Figure 3.2.

FIGURE 3.5 Hydrogen bonding in pure ethanol.

Learning Check 3.5 Arrange the following compounds in order of increasing solubility in water (least soluble first, most soluble last):

a. $CH_3CH_2CH_2—OH$ b. $CH_3CH_2CH_3$ c. $CH_3(CH_2)_3CH_2—OH$

Hydrogen bonding also causes alcohols to have much higher boiling points than most other compounds of similar molecular weight. In this case, the hydrogen bonding is between alcohol molecules (see ■ Figure 3.5). Ethanol (CH_3CH_2OH) boils at 78°C, whereas methyl ether ($CH_3—O—CH_3$) (with the same molecular weight) boils at −24°C. Propane ($CH_3CH_2CH_3$) has nearly the same molecular weight but boils at −42°C (see ■ Figure 3.6).

ThomsonNOW™ Go to Coached Problems to explore the **boiling points of alcohols.**

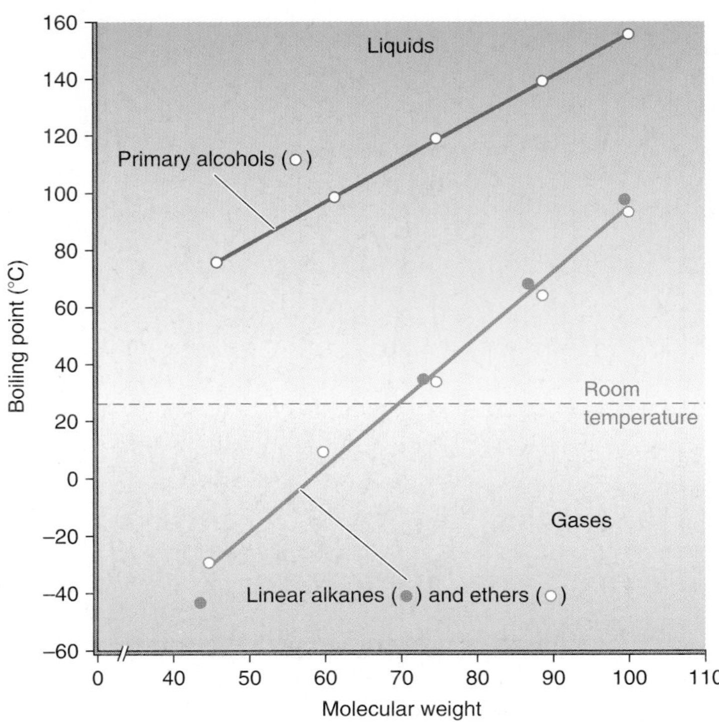

FIGURE 3.6 The boiling points of alcohols are higher than those of alkanes and ethers of similar molecular weights.

Learning Check 3.6 How can you explain the observation that methanol has a boiling point of 65°C, whereas propane, with a higher molecular weight, has a boiling point of −42°C?

3.4 Reactions of Alcohols

LEARNING OBJECTIVE

4. Write equations for alcohol dehydration and oxidation reactions.

Alcohols undergo many reactions, but we will consider only two important ones at this time—alcohol dehydration and alcohol oxidation. In alcohol **dehydration,** water is chemically removed from an alcohol. This reaction can occur in two different ways, depending on the reaction temperature. At 140°C, the main product is an ether, whereas at higher temperatures (180°C), alkenes are the predominant products.

dehydration reaction
A reaction in which water is chemically removed from a compound.

Dehydration to Produce an Alkene

When the elements of a water molecule are removed from a single alcohol molecule, an alkene is produced. A reaction of this type, in which two or more covalent bonds are broken and a new multiple bond is formed, is called an **elimination reaction.** In this case, a water molecule is eliminated from an alcohol. Reaction 3.1 gives a general reaction, and Reactions 3.2 and 3.3 are two specific examples:

elimination reaction
A reaction in which two or more covalent bonds are broken and a new multiple bond is formed.

General reaction:
$$\underset{\text{alcohol}}{-\underset{\underset{H}{|}}{\overset{|}{C}}-\underset{\underset{OH}{|}}{\overset{|}{C}}-} \xrightarrow[180°C]{H_2SO_4} \underset{\text{alkene}}{\diagdown C{=}C \diagup} + H_2O \quad (3.1)$$

Specific example:
$$\underset{\text{2-propanol}}{CH_3\underset{\underset{OH}{|}}{CH}-\underset{\underset{H}{|}}{CH_2}} \xrightarrow[180°C]{H_2SO_4} \underset{\text{propene}}{CH_3CH{=}CH_2} + H_2O \quad (3.2)$$

ThomsonNOW™ Go to Coached Problems to examine the **reactivity of alcohols.**

Specific example:
$$CH_3CH_2\underset{\underset{OH}{|}}{CH}CH_3 \xrightarrow[180°C]{H_2SO_4} CH_3CH{=}CHCH_3 + CH_3CH_2CH{=}CH_2 + H_2O \quad (3.3)$$

2-butene: Principal product (90%) (two carbon groups on double bond)

1-butene: Minor product (10%) (one carbon group on double bond)

Notice that in Reaction 3.3 two alkenes can be produced, depending on which carbon next to the hydroxy-bearing carbon loses the hydrogen atom. The principal alkene produced in such cases will be the one in which the higher number of carbon groups is bonded to the double-bonded carbon atoms.

Alcohol dehydration is an important reaction in the human body, where enzymes (rather than sulfuric acid) function as catalysts. For example, the dehydration of citrate is part of the citric acid cycle discussed in Section 13.5. Note that the dehydration of citrate, 2-propanol (Reaction 3.2), and 2-butanol (Reaction 3.3) all involve exactly the same changes in bonding. In each reaction, an H and an —OH are removed from adjacent carbons, leaving a C=C double bond. The other molecular bonds in all three of those compounds do not change:

$$\underset{\text{citrate}}{{}^{-}OOC-\underset{H}{\overset{H}{C}}-\underset{COO^-}{\overset{OH}{C}}-CH_2-COO^-} \xrightarrow{\text{Enzyme}} \underset{cis\text{-aconitate}}{(H)(^{-}OOC)C{=}C(CH_2-COO^-)(COO^-)} + H_2O \quad (3.4)$$

Learning Check 3.7 Predict the major products of the following reactions:

a. $CH_3CH_2\underset{|}{\overset{OH}{C}}HCH_2CH_3 \xrightarrow[180°C]{H_2SO_4}$

b. (cyclohexane ring bearing CH_3 and, on the adjacent carbon, OH) $\xrightarrow[180°C]{H_2SO_4}$

Dehydration to Produce an Ether

Alcohol dehydration can also occur when the elements of water are removed from two alcohol molecules. The resulting alcohol molecule fragments bond and form an ether. This ether-forming reaction is useful mainly with primary alcohols. Reaction 3.5 gives the general reaction, and the formation of diethyl ether (the anesthetic ether) is a specific example (Reaction 3.6):

General reaction:

$$\underset{\text{an alcohol}}{R-O-H} + \underset{\text{an alcohol}}{H-O-R} \xrightarrow[140°C]{H_2SO_4} \underset{\text{an ether}}{R-O-R} + H_2O \quad (3.5)$$

Specific example:

$$\underset{\substack{\text{ethanol}\\ \text{(ethyl alcohol)}}}{CH_3CH_2-OH} + \underset{\substack{\text{ethanol}\\ \text{(ethyl alcohol)}}}{HO-CH_2CH_3} \xrightarrow[140°C]{H_2SO_4} \underset{\text{diethyl ether}}{CH_3CH_2-O-CH_2CH_3} + H_2O \quad (3.6)$$

Thus, an alcohol can be dehydrated to produce either an alkene or an ether. The type of dehydration that takes place is controlled by the temperature, with sulfuric acid serving as a catalyst. Although a mixture of alkene and ether products often results, we will use these temperatures to indicate that the major dehydration product is an alkene (at 180°C) or an ether (at 140°C).

The reaction that produces ethers is an example of a dehydration synthesis in which two molecules are joined and a new functional group is generated by the removal of a water molecule. Dehydration synthesis is important in living organisms because it is part of the formation of carbohydrates, fats, proteins, and many other essential substances.

Learning Check 3.8 What catalyst and reaction temperature would you use to accomplish each of the following reactions?

a. $2CH_3CH_2CH_2-OH \rightarrow CH_3CH_2CH_2-O-CH_2CH_2CH_3 + H_2O$

b. $CH_3\underset{\underset{CH_3}{|}}{C}H-OH \rightarrow CH_3CH{=}CH_2 + H_2O$

Oxidation

An oxidation reaction occurs when a molecule gains oxygen atoms or loses hydrogen atoms. Under appropriate conditions, alcohols can be oxidized in a controlled way (not combusted) by removing two hydrogen atoms per mole-

HOW REACTIONS OCCUR 3.1 The Dehydration of an Alcohol

A strong acid such as sulfuric acid initially reacts with an alcohol in much the same way that it reacts with water in solution. Ionization of the acid produces a proton (H^+) that is attracted to an oxygen atom of the alcohol:

$$CH_3CH_2-\ddot{O}{:}-H + H_2SO_4 \rightleftarrows CH_3CH_2-\overset{+}{\ddot{O}}(H)_2 + HSO_4^-$$

protonated alcohol

The creation of a positive charge on the oxygen atom weakens the carbon–oxygen bond, which allows the following equilibrium reaction to take place:

$$CH_3CH_2-\overset{+}{O}(H)_2 \rightleftarrows CH_3\overset{+}{C}H_2 + {:}\ddot{O}(H)_2$$

a carbocation; water (one product)

The ethyl carbocation is unstable because the carbon has only three bonds (six outer electrons) around it. To achieve an octet of electrons about this carbon, the carbocation donates an H^+ to a proton acceptor, such as HSO_4^-, and forms the double bond of the alkene product:

$$H-CH_2-\overset{+}{C}H_2 \;(HSO_4^-) \longrightarrow CH_2=CH_2 + H_2SO_4$$

ethene; regenerated catalyst

cule. A number of oxidizing agents can be used, including potassium dichromate ($K_2Cr_2O_7$) and potassium permanganate ($KMnO_4$). The symbol (O) is used to represent an oxidizing agent. Reaction 3.7 gives the general reaction:

$$\underset{\text{alcohol}}{-\overset{|}{\underset{H}{C}}-O-H} + (O) \longrightarrow \underset{\text{aldehyde or ketone}}{-\overset{O}{\overset{\|}{C}}-} + H_2O \qquad (3.7)$$

The three classes of alcohols behave differently toward oxidizing agents.

Primary Alcohols. Reaction 3.8 gives the general reaction, and the formation of acetic acid is a specific example (Reaction 3.9):

General reaction:

$$\underset{\text{primary alcohol}}{R-\overset{OH}{\overset{|}{\underset{H}{C}}}-H} + (O) \longrightarrow \underset{\text{aldehyde}}{R-\overset{O}{\overset{\|}{C}}-H} + H_2O \qquad (3.8)$$

$$\text{aldehyde} \xrightarrow[(O)]{\text{further oxidation}} \underset{\text{carboxylic acid}}{R-\overset{O}{\overset{\|}{C}}-OH}$$

Specific example:

$$\underset{\text{ethanol}}{CH_3CH_2-OH} + (O) \longrightarrow \underset{\text{acetaldehyde}}{CH_3-\overset{O}{\overset{\|}{C}}-H} + H_2O \qquad (3.9)$$

$$\text{acetaldehyde} \xrightarrow[(O)]{\text{further oxidation}} \underset{\text{acetic acid}}{CH_3-\overset{O}{\overset{\|}{C}}-OH}$$

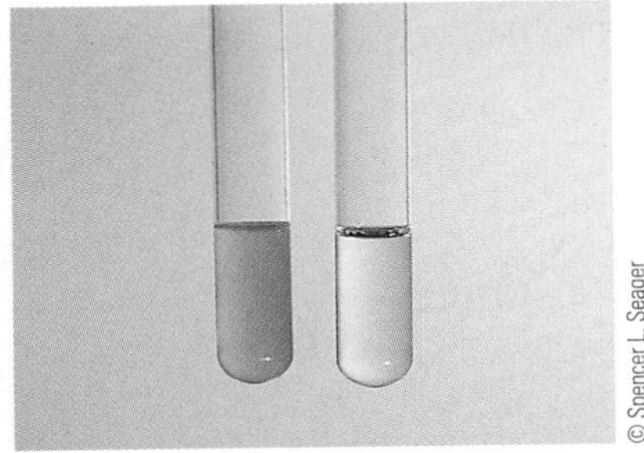

The tube on the left contains orange $K_2Cr_2O_7$ and is next to the colorless ethanol.

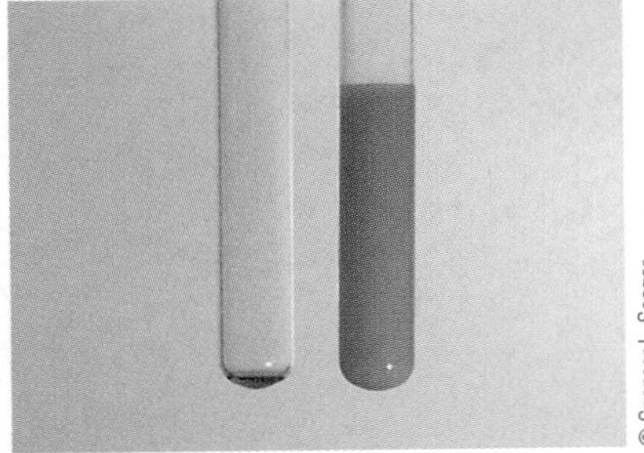

After mixing, the ethanol is oxidized, and chromium is reduced, forming a grayish-green precipitate.

FIGURE 3.7 Oxidation of ethanol by $K_2Cr_2O_7$. How might a color change like this one be used to measure alcohol concentration?

The immediate product of the oxidation of a primary alcohol is an aldehyde. However, aldehydes are readily oxidized by the same oxidizing agents (Section 14.3) to give carboxylic acids. Therefore, the oxidation of a primary alcohol normally results in the corresponding carboxylic acid as the product. The oxidation of ethanol by $K_2Cr_2O_7$ is shown in ■ Figure 3.7. Because aldehydes do not hydrogen bond (Section 14.2), they have lower boiling points than the corresponding alcohol or carboxylic acid. This makes it possible to isolate the aldehyde product before it is oxidized by maintaining the reaction temperature high enough to boil the aldehyde out of the reaction mixture before it can react.

Learning Check 3.9 Draw the structural formulas of the first and second products of the following reaction:

$$C_6H_5{-}CH_2{-}OH + (O) \longrightarrow$$

Secondary Alcohols. Reactions 3.10 and 3.11 give the general reaction and a specific example.

General reaction:
$$R{-}\underset{\displaystyle H}{\overset{\displaystyle OH}{\underset{|}{\overset{|}{C}}}}{-}R' + (O) \longrightarrow R{-}\overset{\displaystyle O}{\overset{\|}{C}}{-}R' + H_2O \qquad (3.10)$$

secondary alcohol → ketone

Specific example:
$$CH_3{-}\underset{\displaystyle H}{\overset{\displaystyle OH}{\underset{|}{\overset{|}{C}}}}{-}CH_3 + (O) \longrightarrow CH_3{-}\overset{\displaystyle O}{\overset{\|}{C}}{-}CH_3 + H_2O \qquad (3.11)$$

2-propanol → acetone

Secondary alcohols are oxidized to ketones in exactly the same way primary alcohols are oxidized to aldehydes, and by the same oxidizing agents. However, unlike aldehydes, ketones resist further oxidation, so this reaction provides an excellent way to prepare ketones.

Learning Check 3.10 Complete the following oxidation reaction:

$$C_6H_{11}{-}OH + (O) \longrightarrow$$

Tertiary Alcohols. A general reaction is given in Reaction 3.12:

General reaction:
$$R{-}\underset{\displaystyle R''}{\overset{\displaystyle OH}{\underset{|}{\overset{|}{C}}}}{-}R' + (O) \longrightarrow \text{no reaction} \qquad (3.12)$$

Tertiary alcohols do not have any hydrogen on the —OH-bearing carbon and thus do not react with oxidizing agents.

Learning Check 3.11 Which of the following alcohols will react with an oxidizing agent?

a. cyclopentyl–OH b. CH_3 OH (on the same ring carbon of cyclopentane) c. cyclopentyl–CH_2-OH

Multistep Reactions

Alcohols are important organic compounds not only for their own commercial applications but because they can be converted by chemical reactions into a wide variety of other useful products. Most alcohols do not occur naturally in commercial quantities and therefore are prepared from alkenes.

It is very common in both laboratory and industrial processes for such preparations to involve a sequence of reactions. For example, the commercial production of vinyl plastic (PVC) begins with the alkene ethene:

$$\underset{\text{ethene}}{CH_2{=}CH_2} \longrightarrow \underset{\text{vinyl chloride}}{CH_2{=}\overset{\overset{\displaystyle Cl}{|}}{C}H} \longrightarrow \underset{\text{poly (vinyl chloride) (PVC)}}{\left[CH_2-\overset{\overset{\displaystyle Cl}{|}}{C}H\right]}$$

Nature, too, employs multistep reactions to carry out processes essential to life. The glycolysis pathway used by the body to initiate the utilization of glucose for energy production involves a series of ten reactions (Section 13.3).

EXAMPLE 3.3

The following conversion requires more than one step. Show the reaction, reagents, and intermediate structures necessary to carry out the synthesis.

$$\text{cyclopentene} \longrightarrow \text{cyclopentanone}$$

Solution

The best way to initially solve a multistep synthesis is to work backward from the final product. Identify the functional group present in the product—in this case, a ketone. Think how to prepare a ketone—by oxidizing an alcohol.

$$\text{cyclopentanol (OH)} + (O) \longrightarrow \text{cyclopentanone (=O)} + H_2O$$

Now, working backward again, think how to prepare an alcohol—from an alkene through hydration.

$$\text{cyclopentene} + H_2O \xrightarrow{H_2SO_4} \text{cyclopentanol (OH)}$$

Adding these two steps together gives

$$\text{cyclopentene} + H_2O \xrightarrow{H_2SO_4} \text{cyclopentanol (OH)} \xrightarrow{(O)} \text{cyclopentanone (=O)} + H_2O$$

Learning Check 3.12 Show the reactions necessary to carry out the following conversion.

$$CH_2{=}CH_2 \longrightarrow CH_3CH_2{-}O{-}CH_2CH_3$$

3.5 Important Alcohols

LEARNING OBJECTIVE

5. Recognize uses for specific alcohols.

The simplest alcohol, methanol or methyl alcohol, is a very important industrial chemical; more than 1 billion gallons are produced and used annually. It is sometimes known as wood alcohol because the principal source for many years was the distillation of wood. Today, it is synthetically produced in large quantities by reacting hydrogen gas with carbon monoxide (Reaction 3.13):

$$CO + 2H_2 \xrightarrow[\text{heat, pressure}]{\text{Catalysts,}} CH_3{-}OH \tag{3.13}$$

The industrial importance of methanol is due to its oxidation product, formaldehyde, which is a major starting material for the production of plastics. Methanol is used as a fuel in Indy-style racing cars (see ■ Figure 3.8). It is also used as a fuel in products such as canned heat.

Methanol is highly toxic and causes permanent blindness if taken internally. Deaths and injuries have resulted as a consequence of mistakenly substituting methanol for ethyl alcohol in beverages.

STUDY SKILLS 3.1 A Reaction Map for Alcohols

To solve a test question, using a stepwise approach is often helpful. First, decide what type of question is being asked. Is it a nomenclature, typical uses, physical properties, or a reaction question? Second, if you recognize it as, say, a reaction question, look for the functional group. Third, if it's an alcohol group in the starting material, identify the reagent and use the following diagram to predict the right products.

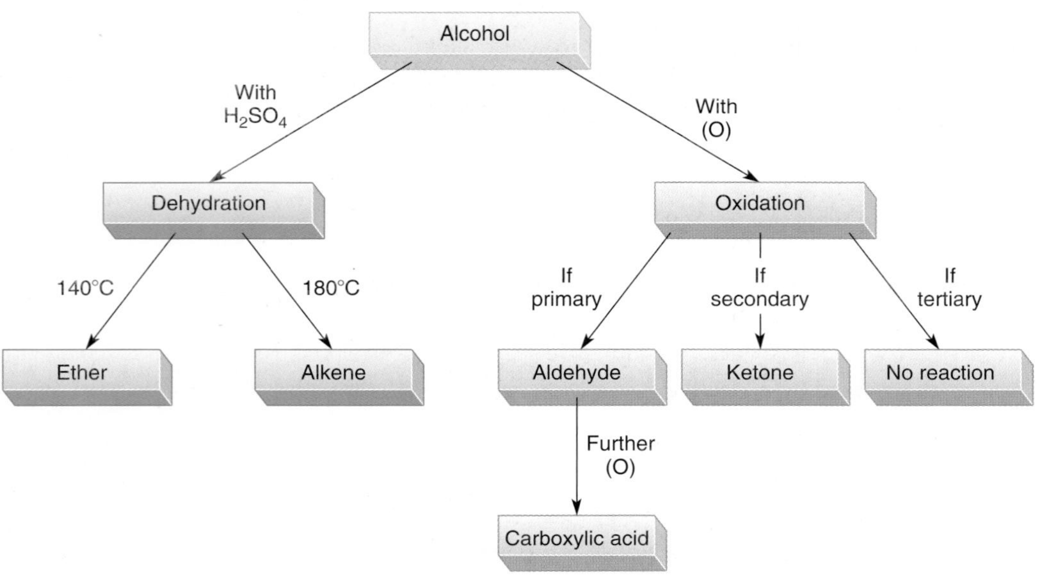

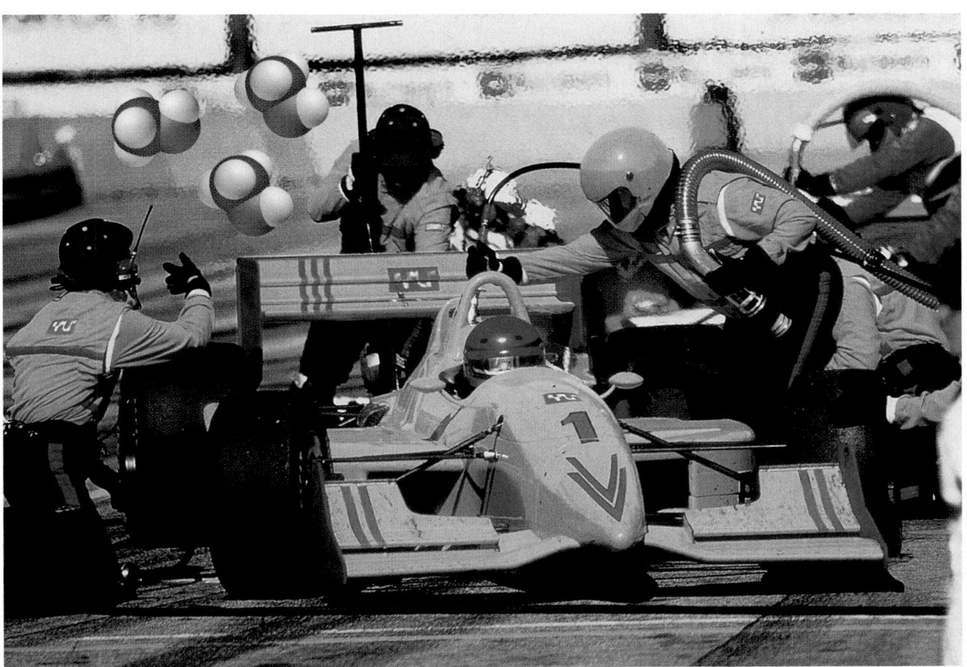
© David Madison/Tony Stone Worldwide

■ **FIGURE 3.8** Racing cars are fueled by methanol. What would be the equation for the combustion of methanol?

Ethanol, or ethyl alcohol, $CH_3CH_2—OH$, is probably the most familiar alcohol because it is the active ingredient in alcoholic beverages. It is also used in pharmaceuticals as a solvent (tinctures are ethanol solutions) and in aftershave lotions as an antiseptic and skin softener. It is an important industrial solvent as well. The gasohol now marketed at many locations around the United States is a mixture of ethanol and gasoline.

Ethanol

Most ethanol used industrially is produced on a large scale by the direct hydration of ethylene obtained from petroleum (Reaction 3.14):

$$\underset{\text{ethylene}}{H_2C{=}CH_2 + H{-}OH} \xrightarrow[300^\circ C]{70\ atm} \underset{\text{ethanol}}{\begin{matrix} H_2C - CH_2 \\ | \quad\ \ | \\ H \quad OH \end{matrix}} \tag{3.14}$$

Ethanol for beverages is produced by the yeast **fermentation** of carbohydrates such as sugars, starch, and cellulose. For example, a common method is the fermentation of glucose, grape sugar:

fermentation
A reaction of sugars, starch, or cellulose to produce ethanol and carbon dioxide.

$$\underset{\text{glucose}}{C_6H_{12}O_6} \xrightarrow{\text{Yeast}} \underset{\text{ethanol}}{2CH_3CH_2-OH} + 2CO_2$$

2-propanol (isopropyl alcohol), the main component of rubbing alcohol, acts as an astringent on the skin. It causes tissues to contract, hardens the skin, and limits secretions. Its rapid evaporation from the skin also lowers the skin temperature. It is used as an antiseptic in 70% solutions (see ■ Figure 3.9). Isopropyl alcohol is toxic and should not be taken internally.

1,2,3-propanetriol is known more commonly as glycerol or glycerin:

$$\begin{matrix} OH & & OH & & OH \\ | & & | & & | \\ CH_2 & - & CH & - & CH_2 \end{matrix}$$

1,2,3-propanetriol (glycerol)

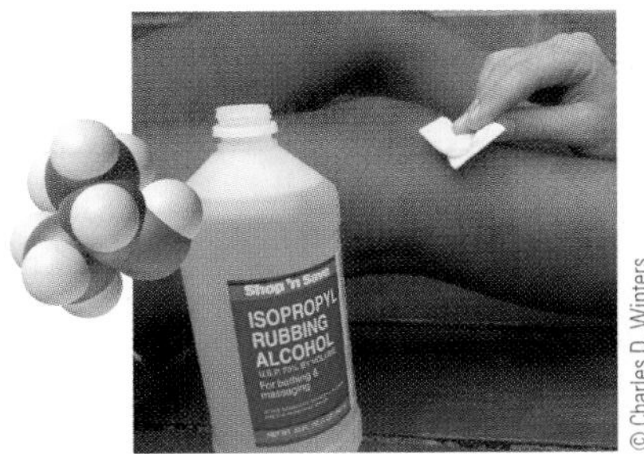

© Charles D. Winters

■ **FIGURE 3.9** Isopropyl alcohol is used as an antiseptic.

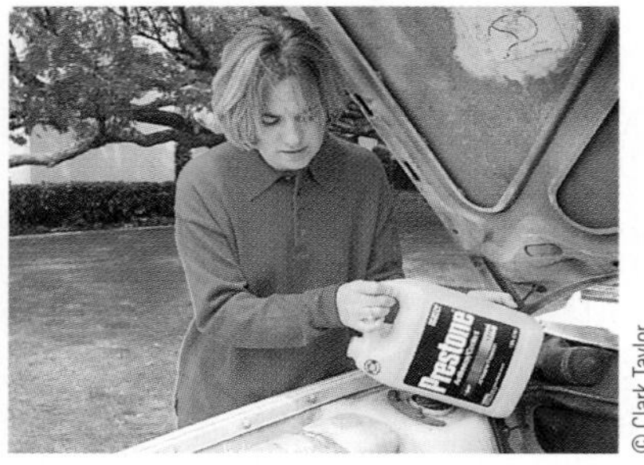

© Clark Taylor

■ **FIGURE 3.10** Ethylene glycol (HO—CH_2CH_2—OH) is used as an automotive antifreeze.

The hydrogen bonding resulting from the three hydroxy groups causes glycerol to be a syrupy, high-boiling liquid with an affinity for water. This property and its nontoxic nature make it valuable as a moistening agent in tobacco and many food products, such as candy and shredded coconut. Florists use glycerol to retain water that maintains the freshness of cut flowers. Glycerol is also used for its soothing qualities in shaving and toilet soaps and in many cosmetics. In Chapter 8 on lipids, we will see that glycerol forms part of the structure of body fats and oils.

Due to hydrogen bonding, 1,2-ethanediol (ethylene glycol) and 1,2-propanediol (propylene glycol) are both high-boiling liquids that are completely miscible with water.

$$\begin{array}{c} \text{OH} \quad \text{OH} \\ | \qquad | \\ CH_2-CH_2 \end{array} \qquad\qquad \begin{array}{l} \text{OH} \quad \text{OH} \\ | \qquad\ \ | \\ CH_2-CH-CH_3 \end{array}$$

1,2-ethanediol (ethylene glycol) — 1,2-propanediol (propylene glycol)

Their main uses are as automobile antifreeze (see ■ Figure 3.10) and as a starting material for the manufacture of polyester fibers (Reaction 5.11).

Alcohols are very important biologically because the alcohol functional group occurs in a variety of compounds associated with biological systems. For example, sugars contain several hydroxy groups, and starch and cellulose contain thousands of hydroxy groups (see Chapter 7). The structures and uses (see Figure 3.10) of some common alcohols are summarized in ■ Table 3.2.

■ **Learning Check 3.13** What is an important use for each of the following alcohols?

a. methanol
b. ethanol
c. 2-propanol
d. ethylene glycol
e. glycerol
f. menthol

TABLE 3.2 Examples of alcohols

Name	Structural formula	Typical uses
methanol (methyl alcohol)	$CH_3—OH$	Solvent, making formaldehyde
ethanol (ethyl alcohol)	$CH_3CH_2—OH$	Solvent, alcoholic beverages
2-propanol (isopropyl alcohol)	$\begin{array}{c} CH_3CHCH_3 \\ \vert \\ OH \end{array}$	Rubbing alcohol, solvent
1-butanol (butanol)	$CH_3CH_2CH_2CH_2—OH$	Solvent, hydraulic fluid
1,2-ethanediol (ethylene glycol)	$HO—CH_2CH_2—OH$	Automobile antifreeze, polyester fibers
1,2-propanediol (propylene glycol)	$\begin{array}{l} CH_3CH-CH_2 \\ \quad\ \vert \qquad \vert \\ \quad OH \quad OH \end{array}$	Moisturizer in lotions and foods, automobile antifreeze
1,2,3-propanetriol (glycerin, glycerol)	$\begin{array}{l} CH_2-CH-CH_2 \\ \vert \qquad \vert \qquad \vert \\ OH \quad OH \quad OH \end{array}$	Moisturizer in foods, tobacco, and cosmetics
menthol	cyclohexane ring bearing CH_3CHCH_3, OH, and CH_3	Cough drops, shaving lotion, mentholated tobacco

CHEMISTRY AROUND US 3.1 Driving on Corn Fumes

If you live in the Midwest, you might have heard of E85, or might even use it to fuel your car rather than gasoline. However, with most of the population of the United States living in coastal regions, this new fuel is not widely known. In his State of the Union Address in 2006, President George W. Bush cited alcohol fuel as a new resource that will help end the U.S. dependency on oil imported from other nations. If it is, you probably want to know what E85 is, and where you can find it.

E85 is a type of ethanol fuel, which is a blend of plant-derived ethanol and gasoline. In the United States, corn is usually used as the plant source of the ethanol because of its wide availability. In Brazil, sugar cane is used as the source of the ethanol used in the blend.

The two most widely used types of ethanol fuel are gasohol and E85. Gasohol is a blend containing 10% ethanol and 90% gasoline. Gasohol is popular because it is widely available and can be used by most modern cars. The E85 blend contains 85% ethanol and 15% gasoline. It is used primarily in the Midwest because most of the current E85 factories are located nearby. E85 can be used only in flexible fuel vehicles (FFVs). FFVs have been built with engines that can run on standard gasoline as well as the E85 blend. One drawback to the use of E85 is that it produces less energy per gallon than gasoline, requiring more frequent fueling stops.

So, if you're thinking about buying a car, check out what sort of fuels we might be using in a few years and determine if your dream car is compatible with them.

Ethanol-containing fuel may help the U.S. reduce oil imports.

3.6 Characteristics and Uses of Phenols

LEARNING OBJECTIVES

6. Recognize uses for specific phenols.

Phenol, C_6H_5OH, is a colorless, low-melting-point solid with a medicinal odor. The addition of a small amount of water causes the solid to liquefy at room temperature. This liquid mixture is sometimes called carbolic acid. Unlike alcohols, phenols are weak acids (Reactions 3.15 and 3.16), which can chemically burn the skin, and they must be handled with care.

$$C_6H_5OH + H_2O \longrightarrow C_6H_5O^- + H_3O^+ \qquad (3.15)$$

phenol
(a weak acid)

$$C_6H_5OH + KOH \longrightarrow C_6H_5O^-K^+ + H_2O \qquad (3.16)$$

In dilute solutions, phenol is used as an antiseptic and disinfectant. Joseph Lister, an English surgeon, introduced the use of phenol as a hospital antiseptic in the late 1800s. Before that time, antiseptics had not been used, and very few patients survived even minor surgery because of postoperative infections. Today, phenol has been largely replaced in this use by more effective derivatives that are less irritating to the skin. Two such compounds, 4-chloro-3,5-dimethylphenol and 4-hexylresorcinol, are shown here.

phenol	4-chloro-3,5-dimethylphenol	4-hexylresorcinol
OH on benzene ring	OH; CH_3, CH_3; Cl	OH; OH; $(CH_2)_5CH_3$
phenol (Lister's original disinfectant)	4-chloro-3,5-dimethylphenol (nonirritating topical antiseptic)	4-hexylresorcinol (used in mouthwashes and throat lozenges)

OVER THE COUNTER 3.1 Outsmarting Poison Ivy

Poison ivy is a woody, ropelike vine or bush recognizable by its three leaflets, which are green in the summer and red in the fall. The American Academy of Dermatology estimates that about 85% of those exposed to the plant will suffer some type of allergic reaction.

An allergic reaction usually takes the form of a rash, blisters, and itching. The culprit that causes the reaction is urushiol, an odorless oil with a phenolic component that is present in the sap of the plant. The urushiol-containing sap is released when the plant is damaged. The poison ivy plant is very fragile and can easily be damaged by passing animals, wind, and chewing insects. Urushiol that gets onto other objects can remain potent for years, depending on the environment. In a warm, moist environment, it can cause a reaction for up to a year. In a dry environment, it may remain active for decades.

The best way to avoid an adverse reaction is to avoid contact with the active oil by wearing long sleeves, long pants, gloves, boots, and a hat when moving through areas where exposure is possible. A new over-the-counter product called bentoquatam (Ivy Block®) also provides protection. This lotion is applied to the skin at least 15 minutes before any anticipated exposure to poison ivy. The lotion forms a visible claylike barrier against urushiol. However, it must be applied every 4 hours for continued protection.

If exposure to the plant occurs despite precautions, the following steps will help reduce the chances for an adverse reaction:

1. Cleanse the exposed skin with generous amounts of rubbing alcohol, and then wash with cool water. This should be done outdoors, and alcohol should not be used if a return to the site of the exposure the same day is likely. The alcohol removes the skin's protective oils along with the urushiol, and new contact will allow urushiol to penetrate the skin more rapidly.
2. Take a regular shower with soap and warm water.
3. With gloves on, wipe any clothes, tools, shoes, etc., that might have contacted the plant with alcohol and water.

If the skin has been exposed to urushiol and not cleansed quickly enough, redness and swelling will appear in 12–48 hours, often followed by itching and blisters. Some people also develop a rash within 7–10 days. The blisters, rash, and itch will normally disappear in 14–20 days without any treatment, but most sufferers want relief from the itch during that time. A number of OTC products have been approved by the FDA for the temporary relief of itching caused by poison ivy. The products are topical corticosteroids (commonly called hydrocortisones), with brand names such as Lanacort® and Cortaid®. For severe reactions, a doctor should be consulted. He or she may prescribe additional topical corticosteroid drugs or even oral corticosteroids.

Several OTC products can also be used to dry up the oozing blisters. These products include aluminum acetate, calamine lotion, zinc oxide, zinc carbonate, zinc acetate, kaolin, aluminum hydroxide gel, baking soda, and even an oatmeal bath.

Ivy Block® contains the compound bentoquatam, which protects the skin from poison ivy.

Two other derivatives of phenol, *o*-phenylphenol and 2-benzyl-4-chlorophenol, are ingredients in Lysol®, a disinfectant for walls and furniture in hospitals and homes:

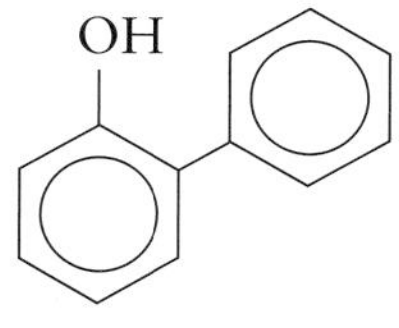

o-phenylphenol

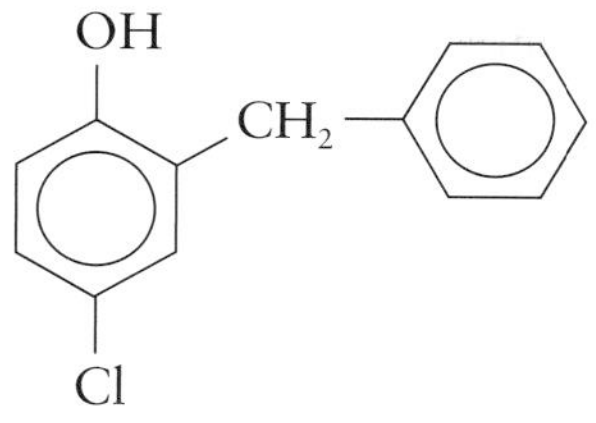

2-benzyl-4-chlorophenol

Other important phenols act as **antioxidants** by interfering with oxidizing reactions. These phenols are useful in protecting foods from spoilage and a variety of other materials from unwanted reactions. BHA and BHT are widely used as antioxidants in gasoline, lubricating oils, rubber, certain foods, and packaging materials for foods that might turn rancid (see ■ Figure 3.11). In food or containers used to package food, the amount of these antioxidants is limited to 200 ppm (parts per million) or 0.02%, based on the fat or oil content of the food. Notice that industrial nomenclature for these compounds does not follow IUPAC rules.

antioxidant
A substance that prevents another substance from being oxidized.

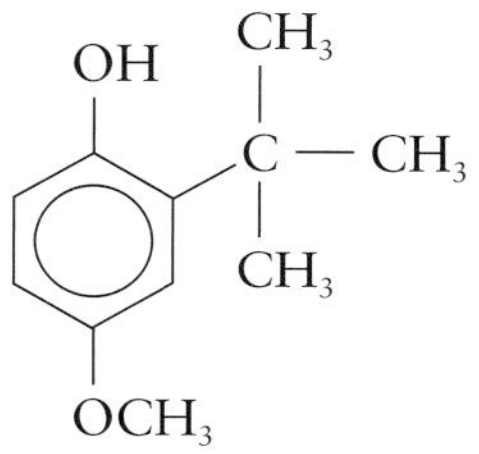

BHA (butylated hydroxy anisole)
2-*t*-butyl-4-methoxyphenol
(IUPAC name)

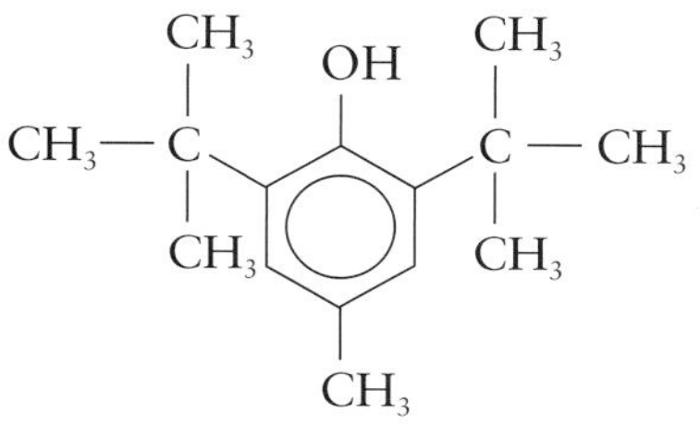

BHT (butylated hydroxy toluene)
2, 6-di-*t*-butyl-4-methylphenol
(IUPAC name)

■ **Learning Check 3.14** What is an important use for each of the following phenols?

a. BHA **b.** 4-hexylresorcinol

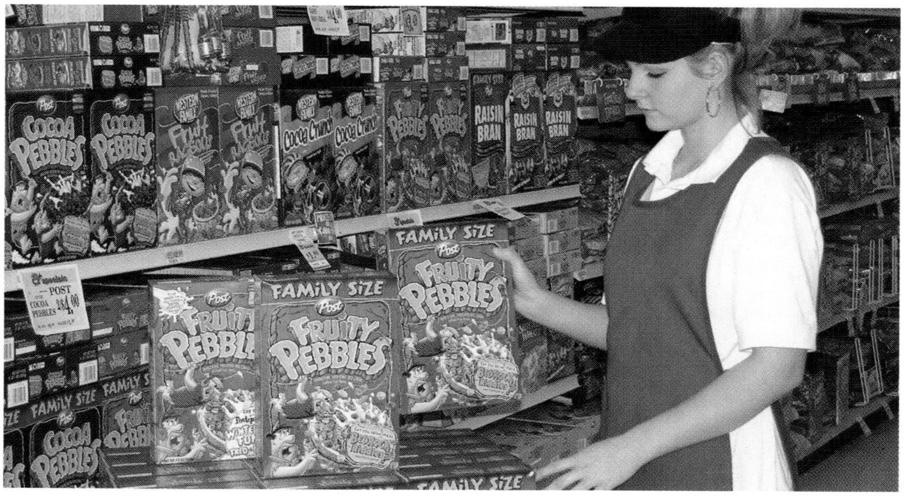

© Mark Slabaugh

■ **FIGURE 3.11** BHT is listed as a food ingredient that will "maintain freshness."

CHEMISTRY AND YOUR HEALTH 3.1 Weaned from the Bottle

Willpower sometimes isn't enough to stop drinking. As alcohol consumption continues, dependency increases, and it might be dangerous for some individuals to stop cold turkey.

Answering the following questions can help an individual answer the question: Do I have a drinking problem?

1. Have you ever felt you ought to cut down on your drinking?
2. Have people annoyed you by criticizing your drinking?
3. Have you ever felt bad or guilty about your drinking?
4. Have you ever had an "eye-opener" drink first thing in the morning to steady your nerves or help get rid of a hangover?

Answering yes to any two of these questions indicates alcoholism. Giving up alcohol is a positive life choice, but one may want to seek medical supervision before attempting to make that choice. The body of an alcoholic has become addicted to certain routines and alcoholic effects. Suddenly eliminating any alcohol intake might have adverse effects including the following withdrawal signs and symptoms that might occur within hours of stopping drinking: insomnia, vivid dreams, mild to severe anxiety, unsettled mood, agitation, irritability, tremors, appetite loss, nausea, vomiting, headache, sweating, heart palpitations, and auditory hallucinations (hearing things that aren't there). The symptoms might also include seizures and/or delirium tremens (DTs). An individual experiencing delirium tremens has an elevated blood pressure and suffers from agitation and occasionally visual hallucinations. The hallucinations are sometimes extremely vivid and can cause inappropriate and even life-threatening behavior. Delirium tremens is a dangerous condition that requires immediate medical care and hospitalization.

Giving up alcohol is an important commitment to health, and although the side effects of withdrawal seem a bit overwhelming, the long-term benefit far outweighs the immediate discomfort. It is important to remember that medical help and support are often necessary in order to successfully give up the habit.

3.7 Ethers

LEARNING OBJECTIVE

7. Name and draw structural formulas for ethers.

Dimethyl ether

Ethers were introduced at the start of this chapter as compounds whose formula is R—O—R′. Common names for ethers are obtained by first naming the two groups attached to the oxygen and then adding the word *ether.* If the groups are the same, it is appropriate to use the prefix *di-*, as in diethyl ether, but many people omit the *di-* and simply say "ethyl ether."

ThomsonNOW™ Go to Coached Problems to practice **naming ethers.**

$CH_3—O—CH_2CH_3$
ethyl methyl ether

$C_6H_5—O—CH(CH_3)—CH_3$
isopropyl phenyl ether

$CH_3—O—CH_3$
dimethyl ether
(methyl ether)

$CH_3CH_2—O—CH_2CH_3$
diethyl ether
(ethyl ether)

Learning Check 3.15 Assign a common name to the following:

a. $CH_3CH_2—O—C_6H_5$

b. $CH_3CH(CH_3)—O—CH(CH_3)CH_3$

alkoxy group
The —O—R functional group.

In the IUPAC naming system for ethers, the —O—R group is called an **alkoxy group.** The *-yl* ending of the smaller R group is replaced by *-oxy.* For example, $—O—CH_3$ is called a methoxy group, $—O—CH_2CH_3$ is called ethoxy, and so on:

$CH_3—O—CH_3$
methoxymethane

$CH_3CH_2CHCH_3$ (with $O—CH_2CH_3$ on C-2)
2-ethoxybutane

(benzene ring with $C(=O)—OH$ and para $O—CH_3$)
p-methoxybenzoic acid

■ **Learning Check 3.16** Give IUPAC names for the following:

a. (cyclopentane with two $O—CH_3$ groups on adjacent carbons)

b. $CH_3CH_2CHCH_2CH_3$ (with $O—CH—CH_3$ on C-3, the CH bearing a CH_3)

The ether linkage is also found in cyclic structures. A ring that contains elements in addition to carbon is called a **heterocyclic ring.** Two common structures with oxygen-containing heterocyclic rings are

heterocyclic ring A ring in which one or more atoms are an atom other than carbon.

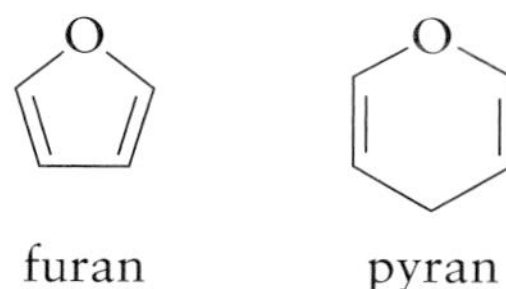

furan pyran

Many carbohydrates contain the fundamental ring systems of furan and pyran (Chapter 17).

3.8 Properties of Ethers

LEARNING OBJECTIVES

8. Describe the key physical and chemical properties of ethers.

The oxygen atom of ethers can form hydrogen bonds with water (see ■ Figure 3.12a), so ethers are slightly more soluble in water than hydrocarbons but less soluble than alcohols of comparable molecular weight. Ethers, however, cannot form hydrogen bonds with other ether molecules in the pure state (Figure 3.12b), resulting in low boiling points close to those of hydrocarbons with similar molecular weights.

The chief chemical property of ethers is that, like the alkanes, they are inert and do not react with most reagents. It is this property that makes diethyl ether

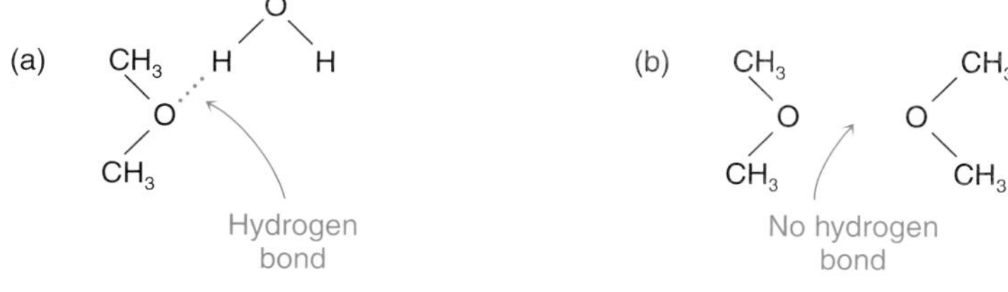

■ **FIGURE 3.12** Hydrogen bonding of dimethyl ether: (a) with water and (b) no hydrogen bonding in the pure state.

such a useful solvent. Like hydrocarbons, ethers are flammable, and diethyl ether is especially flammable because of its high volatility. Diethyl ether was the first general anesthetic used in surgery. For many years, it was the most important compound used as an anesthetic, but it has now been largely replaced by other compounds.

Learning Check 3.17

a. Explain why an ether boils at a lower temperature than an alcohol of similar molecular weight.
b. What is the most significant chemical property of ethers?

3.9 Thiols

LEARNING OBJECTIVES

9. Write equations for a thiol reaction with heavy metal ions and production of disulfides.

thiol
A compound containing an —SH group.

sulfhydryl group
The —SH functional group.

ThomsonNOW Go to Coached Problems to practice **comparing the structure and properties of thiols to the properties of alcohols.**

Sulfur and oxygen belong to the same group of the periodic table, so it is not surprising that they form similar compounds. The sulfur counterparts (analogues) of alcohols contain an —SH group rather than an —OH group and are called **thiols** (an older name is mercaptans). The —SH is known as a **sulfhydryl group.**

$$\underset{\text{an alcohol}}{R-OH} \qquad \underset{\text{a thiol}}{R-SH}$$

Perhaps the most distinguishing feature of thiols is their strong and disagreeable odors. Two different thiols and a disulfide are responsible for the odor associated with skunks (see Figure 3.13).

$$\underset{\textit{trans}\text{-2-butene-1-thiol}}{\overset{H\qquad\quad CH_3}{C=C}\ \text{with } HSCH_2 \text{ and } H} \qquad \underset{\text{3-methyl-1-butanethiol}}{HS-CH_2CH_2CH(CH_3)CH_3} \qquad \underset{\text{methyl-1-}(\textit{trans}\text{-2-butenyl})\text{disulfide}}{CH_3-S-S-CH_2CH=CHCH_3}$$

Some less-offensive odors (and flavors) are also characteristic of thiols and disulfides. Freshly chopped onions emit propanethiol, and 1-propene-3-thiol and 3,3-di-(1-propenyl)disulfide are partially responsible for the odor and flavor of garlic. It is amazing that the simple substitution of a sulfur atom for an oxygen atom can cause such a dramatic change in odor.

$$\underset{\text{propanethiol}}{CH_3CH_2CH_2-SH} \quad \underset{\text{1-propene-3-thiol}}{CH_2=CHCH_2-SH} \quad \underset{\text{3,3-di-(1-propenyl)disulfide}}{CH_2=CHCH_2-S-S-CH_2CH=CH_2}$$

© Dr. E. R. Degginger

FIGURE 3.13 Skunks use thiol chemistry as an effective defense mechanism. Thiols are more volatile than alcohols but less soluble in water. Why are these characteristics an advantage to skunks?

Gas companies make effective use of the odor of thiols. Natural gas (methane) has no odor, so the companies add a tiny amount of a thiol such as ethanethiol (CH_3CH_2—SH) to make it possible to detect leaking gas before a spark or match sets off an explosion.

Two reactions of the —SH group are important in the chemistry of proteins. When thiols are oxidized by agents such as O_2, a coupling reaction occurs (two

CHEMISTRY AROUND US 3.2 General Anesthetics

Anesthetics are compounds that induce a loss of sensation in a specific part (local anesthetic) or all (general anesthetic) of the body. A general anesthetic acts on the brain to produce unconsciousness as well as insensitivity to pain.

The word *ether* is often associated with general anesthetics because of the well-known use of ethyl ether for this purpose. Ethyl ether became generally used as an anesthetic for surgical operations in about 1850. Before that time, surgery was an agonizing procedure. The patient was strapped to a table and (if lucky) soon fainted from the pain. Divinyl ether (vinethene) is an anesthetic that acts more rapidly and is less nauseating than ethyl ether:

$$CH_2{=}CH{-}O{-}CH{=}CH_2$$
divinyl ether

Another common ether anesthetic is enflurane:

$$\begin{array}{ccccccccc} & & F & & & & F & & F & \\ & & | & & & & | & & | & \\ H & - & C & - & O & - & C & - & C & - & H \\ & & | & & & & | & & | & \\ & & H & & & & H & & Cl & \end{array}$$
enflurane

Numerous inhalation anesthetics are not ethers at all. Nitrous oxide, for example, is a simple inorganic compound with the formula N_2O. Also known as "laughing gas," it is used as a general anesthetic by some dentists because its effects wear off quickly. Halothane, currently a popular general anesthetic, is a simple halogen derivative of ethane. It is nonflammable, does not cause nausea or similar upsets, and its effects wear off quickly:

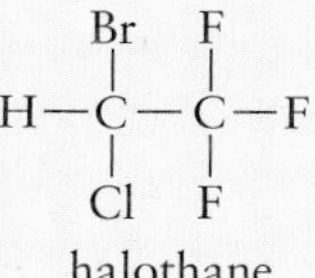
halothane

In modern surgical practice, the use of a single anesthetic has become quite rare. Usually, a patient is given an injection of a strong sedative that causes unconsciousness. A general anesthetic is then administered to provide insensitivity to pain and maintain the patient in an unconscious condition. A muscle relaxant may also be used to produce complete relaxation, while minimizing the need for, and hazards associated with, deep anesthesia.

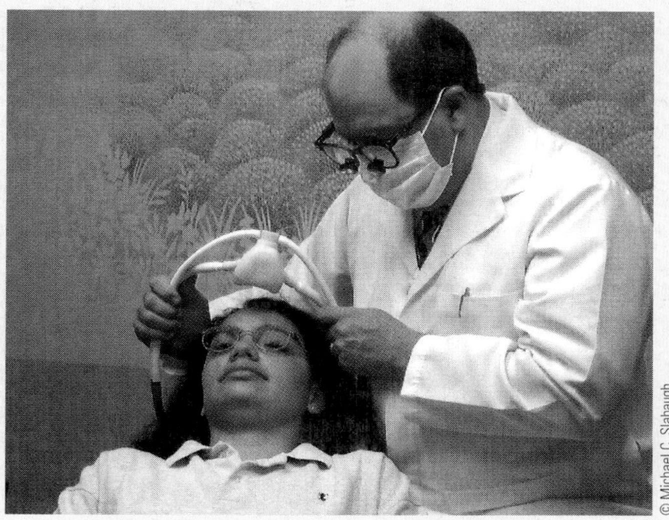

A patient receives an anesthetic prior to dental surgery.

molecules become one) to form a **disulfide,** a compound containing an —S—S— linkage:

disulfide
A compound containing an —S—S— group.

General reaction:

$$\underset{\text{a thiol}}{2R-SH}+(O)\rightarrow \underset{\text{a disulfide}}{R-S-S-R}+H_2O \qquad (3.17)$$

Specific example:

$$2CH_3-SH+(O)\rightarrow CH_3-S-S-CH_3+H_2O \qquad (3.18)$$

Learning Check 3.18 Complete the following reaction:

$$2CH_3\underset{\substack{|\\CH_3}}{CH}-SH+(O)\rightarrow$$

Disulfide linkages are important structural features of some proteins, especially those of hair (see Section 9.6). The coupling reaction can be reversed by treating disulfides with a reducing agent (H) such as H_2, which regenerates the thiol:

General reaction:

$$R{-}S{-}S{-}R + 2(H) \rightarrow 2R{-}SH \tag{3.19}$$

Specific example:

$$CH_3{-}S{-}S{-}CH_3 + 2(H) \rightarrow 2CH_3{-}SH \tag{3.20}$$

Learning Check 3.19 What products result from the reduction of the following disulfides?

a. $\underset{\displaystyle CH_3}{CH_3\underset{|}{CH}}{-}S{-}S{-}\underset{\displaystyle CH_3}{\underset{|}{CH}CH_3} + 2(H) \rightarrow$

b. (cyclopentyl)$-S-S-$(cyclopentyl) $+ 2(H) \rightarrow$

The reaction of thiols with heavy metal ions, such as those of lead and mercury, to form insoluble compounds has adverse biological results. The metal ions (M^{2+}) react with sulfhydryl groups in the following way:

General reaction:

$$2R{-}SH + M^{2+} \rightarrow R{-}S{-}M{-}S{-}R + 2H^+ \tag{3.21}$$

Specific example:

$$2CH_3{-}SH + Hg^{2+} \rightarrow CH_3{-}S{-}Hg{-}S{-}CH_3 + 2H^+ \tag{3.22}$$

Many enzymes (catalysts for biological reactions—Section 20.7) contain —SH groups. Exposure to heavy metal ions ties up these groups, through Reaction 3.21, rendering the enzyme ineffective. The loss of enzymes in this manner can lead to serious biological consequences.

Learning Check 3.20 Complete the following reaction:

$$2CH_3\underset{\displaystyle CH_3}{\underset{|}{CH}}{-}SH + Pb^{2+} \longrightarrow$$

3.10 Polyfunctional Compounds

LEARNING OBJECTIVE

10. Identify functional groups in polyfunctional compounds.

polyfunctional compound A compound with two or more functional groups.

So far our study of organic molecules has included several compounds with more than one functional group. Compounds of this type, such as dienes and diols, are said to be **polyfunctional compounds.** Substances with different functional groups are also referred to as polyfunctional. Compounds with multiple functional groups are very common in nature, and many play essential roles in life processes. For example, carbohydrate structures (Chapter 7) such as glucose typically involve a functional group at every carbon position.

$$HO{-}CH_2{-}\overset{\displaystyle OH}{\overset{|}{CH}}{-}\overset{\displaystyle OH}{\overset{|}{CH}}{-}\overset{\displaystyle OH}{\overset{|}{CH}}{-}\overset{\displaystyle OH}{\overset{|}{CH}}{-}\overset{\displaystyle O}{\overset{\|}{C}}{-}H$$

glucose

Even though some organic structures look quite complex, with multiple rings and long carbon chains, remember that it's the functional group or groups that determine the chemical properties of the compound. Thus, cholesterol, with two functional groups, should exhibit the reactions of both alkenes and alcohols.

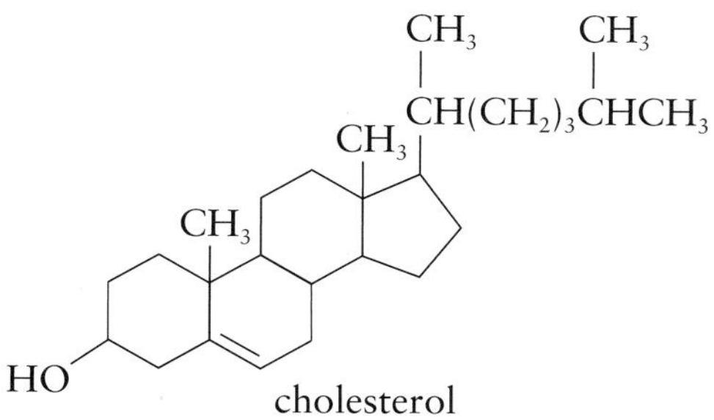

cholesterol

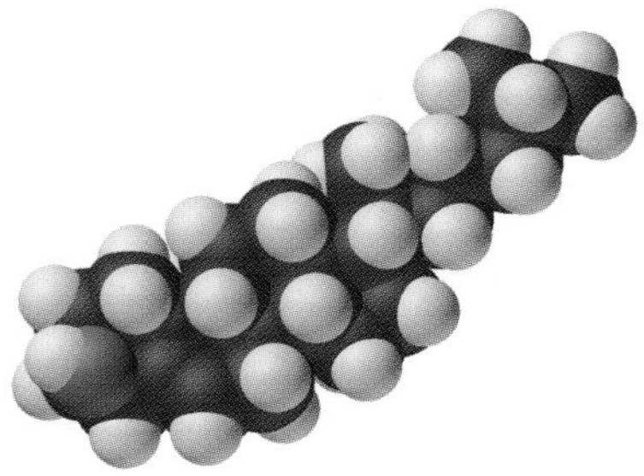

EXAMPLE 3.4

Identify the functional groups in tetrahydrocannabinol, the active ingredient in marijuana.

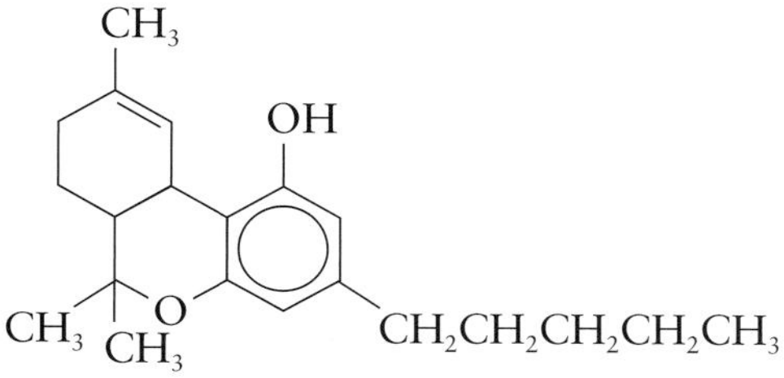

tetrahydrocannabinol

Solution

To find functional groups, look for unique atoms or collections of atoms while ignoring ordinary rings, chains, and alkyl groups. Look for multiple bonds and atoms other than carbon. The tetrahydrocannabinol molecule has four functional groups: an ether, a phenolic —OH, the benzene ring, and a carbon–carbon double bond.

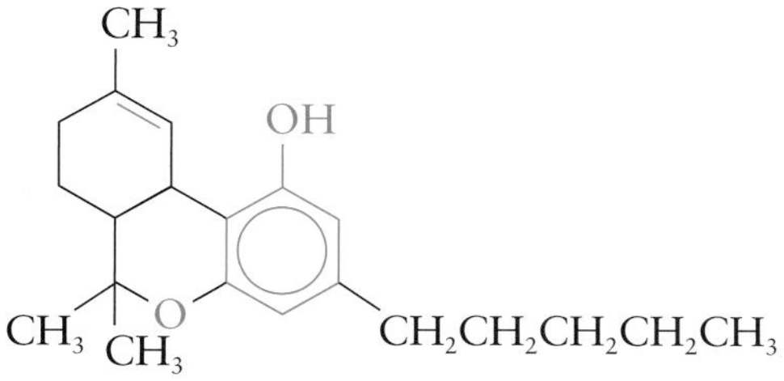

tetrahydrocannabinol

Learning Check 3.21 Identify the functional groups in vitamin E.

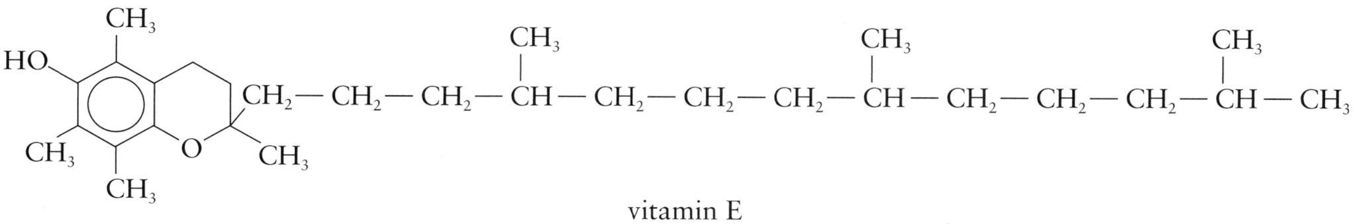

vitamin E

Concept Summary

ThomsonNOW™ Sign in at **www.thomsonedu.com** to:

- Assess your understanding with Exercises keyed to each learning objective.
- Check your readiness for an exam by taking the **Pre-test** and exploring the modules recommended in your **Personalized Learning Plan.**

The Nomenclature of Alcohols and Phenols. Alcohols are aliphatic compounds that contain the hydroxy functional group (—OH). Aromatic compounds with an —OH group attached to the ring are called phenols. Several alcohols are well known by common names. In the IUPAC system, the characteristic ending *-ol* is used to designate alcohols. Phenols are named as derivatives of the parent compound phenol. **OBJECTIVE 1, Exercises 3.4 and 3.10**

Classification of Alcohols. Alcohols are classified on the basis of the number of carbon atoms bonded to the carbon that is attached to the —OH group. In primary alcohols, the OH-bonded carbon is bonded to one other carbon. In secondary alcohols, the OH-bonded carbon is bonded to two other carbons. In tertiary alcohols, the OH-bonded carbon is bonded to three other carbons. **OBJECTIVE 2, Exercise 3.14**

Physical Properties of Alcohols. Hydrogen bonding can occur between alcohol molecules and between alcohol molecules and water. As a result, alcohols have higher boiling points than hydrocarbons of similar molecular weight, and low molecular weight alcohols are soluble in water. **OBJECTIVE 3, Exercise 3.18**

Reactions of Alcohols. Two important reactions of alcohols are dehydration and oxidation. Alcohols may be dehydrated in two different ways depending on the reaction temperature. With H_2SO_4 at 180°C, an alkene is produced, whereas at 140°C an ether is the product. The oxidation products obtained from the alcohols depend on the class of alcohol being oxidized. A primary alcohol produces an aldehyde that is further oxidized to a carboxylic acid. A secondary alcohol produces a ketone. A tertiary alcohol does not react with oxidizing agents. **OBJECTIVE 4, Exercises 3.22 and 3.26**

Important Alcohols. Some simple alcohols with commercial value are methanol (a solvent and fuel), ethanol (present in alcoholic beverages and gasohol), isopropyl alcohol (rubbing alcohol), ethylene glycol (antifreeze), and glycerol (a moistening agent). **OBJECTIVE 5, Exercise 3.36**

Characteristics and Uses of Phenols. Phenols are weak acids that are widely used for their antiseptic and disinfectant properties. Some phenols are used as antioxidants in foods and a variety of other materials. **OBJECTIVE 6, Exercise 3.38**

Ethers. Ethers contain an oxygen attached to two carbons as the characteristic functional group. In the IUPAC system, ethers are named as alkoxy derivatives of alkanes. The —O—R group is the alkoxy group that may have names such as methoxy ($—O—CH_3$) or ethoxy ($—O—CH_2CH_3$). **OBJECTIVE 7, Exercise 3.42**

Properties of Ethers. Like the alkanes, ethers are very unreactive substances. Pure ethers cannot form hydrogen bonds. As a result, their boiling points are comparable to those of hydrocarbons and lower than those of alcohols of similar molecular weight. Ethers are much less polar than alcohols but are still slightly soluble in water due to the limited formation of hydrogen bonds. **OBJECTIVE 8, Exercise 3.46**

Thiols. Thiols contain an —SH group, which often imparts a strong, disagreeable odor to the compound. Two reactions of thiols that are important in protein chemistry are their oxidation to produce a disulfide and their reaction with heavy metals such as mercury. Disulfides may be converted back to thiols by a reducing agent. **OBJECTIVE 9, Exercise 3.52**

Polyfunctional Compounds. It is common for organic compounds to contain more than one functional group. Many of the compounds that serve essential roles in life processes are polyfunctional. **OBJECTIVE 10, Exercise 3.56**

Key Terms and Concepts

Alcohol (Introduction)
Alkoxy group (3.7)
Antioxidant (3.6)
Dehydration reaction (3.4)
Disulfide (3.9)
Elimination reaction (3.4)
Ether (Introduction)
Fermentation (3.5)
Heterocyclic ring (3.7)
Hydroxy group (Introduction)
Phenol (Introduction)
Polyfunctional compound (3.10)
Primary alcohol (3.2)
Secondary alcohol (3.2)
Sulfhydryl group (3.9)
Tertiary alcohol (3.2)
Thiol (3.9)

Key Reactions

1. Dehydration of alcohols to give alkenes (Section 3.4):

$$-\underset{H}{\overset{|}{\underset{|}{C}}}-\underset{OH}{\overset{|}{\underset{|}{C}}}- \xrightarrow[180°C]{H_2SO_4} \;\rangle C=C\langle\; + H_2O \qquad \text{Reaction 3.1}$$

2. Dehydration of alcohols to give ethers (Section 3.4):

$$R-O-H + H-O-R \xrightarrow[140°C]{H_2SO_4} R-O-R + H_2O$$ Reaction 3.5

3. Oxidation of a primary alcohol to give an aldehyde and then a carboxylic acid (Section 3.4):

$$\underset{\text{primary alcohol}}{R-\overset{OH}{\underset{H}{C}}-H} + (O) \longrightarrow \underset{\text{aldehyde}}{R-\overset{O}{\overset{\|}{C}}-H} + H_2O$$ Reaction 3.8

$$\text{aldehyde} \xrightarrow[(O)]{\text{further oxidation}} \underset{\text{carboxylic acid}}{R-\overset{O}{\overset{\|}{C}}-OH}$$

4. Oxidation of a secondary alcohol to give a ketone (Section 3.4):

$$\underset{\text{secondary alcohol}}{R-\overset{OH}{\underset{H}{C}}-R'} + (O) \longrightarrow \underset{\text{ketone}}{R-\overset{O}{\overset{\|}{C}}-R'} + H_2O$$ Reaction 3.10

5. Attempted oxidation of a tertiary alcohol gives no reaction (Section 3.4):

$$R-\overset{OH}{\underset{R''}{C}}-R' + (O) \longrightarrow \text{no reaction}$$ Reaction 3.12

6. Oxidation of a thiol to give a disulfide (Section 3.9):

$$\underset{\text{a thiol}}{2R-SH} + (O) \longrightarrow \underset{\text{a disulfide}}{R-S-S-R} + H_2O$$ Reaction 3.17

7. Reduction of a disulfide to give thiols (Section 3.9):

$$R-S-S-R + 2(H) \rightarrow 2R-SH$$ Reaction 3.19

8. Reaction of thiols with heavy metals (Section 3.9):

$$2R-SH + M^{2+} \rightarrow R-S-M-S-R + 2H^+$$ Reaction 3.21

Exercises

SYMBOL KEY

Even-numbered exercises are answered in Appendix B.

Blue-numbered exercises are more challenging.

■ denotes exercises available in ThomsonNow and assignable in OWL.

ThomsonNOW To assess your understanding of this chapter's topics with sample tests and other resources, sign in at **www.thomsonedu.com.**

THE NOMENCLATURE OF ALCOHOLS AND PHENOLS (SECTION 3.1)

3.1 Draw general formulas for an alcohol and phenol, showing the functional group.

3.2 Draw a general formula for an ether, emphasizing the functional group.

3.3 ■ Assign IUPAC names to the following alcohols:

a. $CH_3CH_2CH_2-OH$

b. $C_6H_5-CH_2CH_2CH_2-OH$ (benzene ring bearing $CH_2CH_2CH_2-OH$)

c. $CH_3\underset{Cl}{C}H\overset{CH_3}{C}HCH_2-OH$

d.  $\overset{\text{OH}}{|}$ $CH_2CH_2CH_2CH_2-OH$ (OH on first carbon)

e. 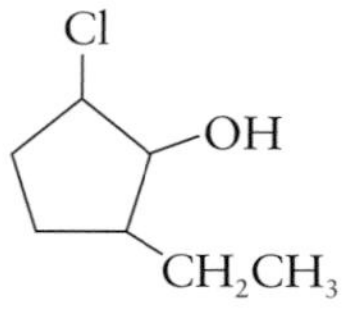 Cyclopentane ring bearing Cl, OH, and CH_2CH_3

f. 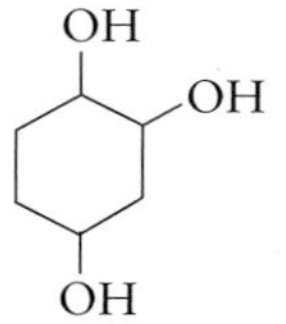 Cyclohexane ring bearing OH, OH (adjacent), and OH (para)

3.4 Assign IUPAC names to the following alcohols:

a. $CH_3CH_2CH_2CH_2-OH$

b. 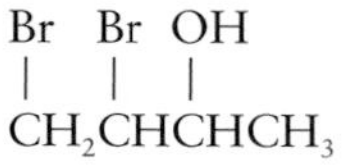 $\underset{}{CH_2}\overset{}{CH}CHCH_3$ with Br, Br, OH on the first three carbons: $CH_2(Br)CH(Br)CH(OH)CH_3$

c. 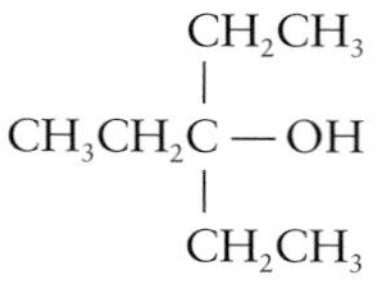 $CH_3CH_2C(CH_2CH_3)(CH_2CH_3)-OH$

d. 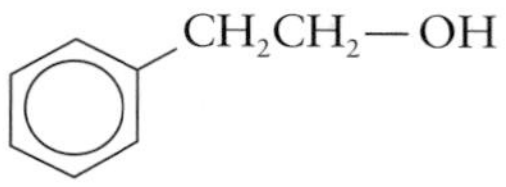 $C_6H_5-CH_2CH_2-OH$

e. 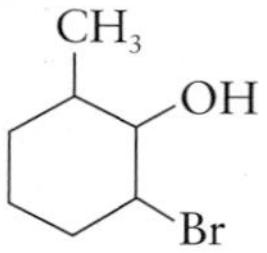 Cyclohexane ring bearing CH_3, OH, and Br on adjacent carbons

f. 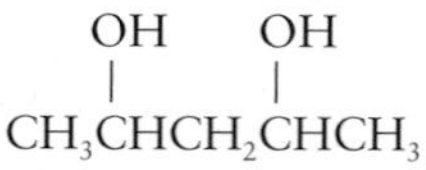 $CH_3CH(OH)CH_2CH(OH)CH_3$

3.5 ■ Several important alcohols are well known by common names. Give a common name for each of the following:

a. CH_3-OH

b. 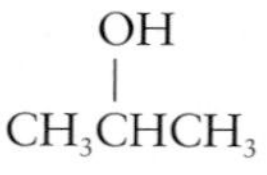 $CH_3CH(OH)CH_3$

c. CH_3CH_2-OH

d. 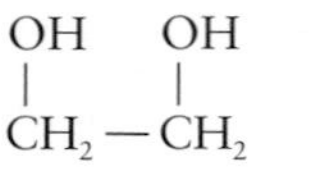 $CH_2(OH)-CH_2(OH)$

e. 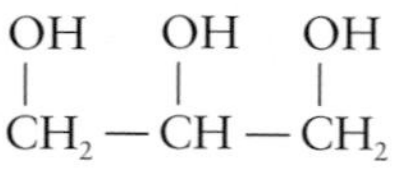 $CH_2(OH)-CH(OH)-CH_2(OH)$

3.6 Give each of the structures in Exercise 3.5 an IUPAC name.

3.7 ■ Draw structural formulas for each of the following:

a. 2-methyl-1-butanol

b. 2-bromo-3-methyl-3-pentanol

c. 1-methylcyclopentanol

3.8 Draw structural formulas for each of the following:

a. 2-methyl-2-pentanol

b. 1,3-butanediol

c. 1-ethylcyclopentanol

3.9 ■ Name each of the following as a derivative of phenol:

a. 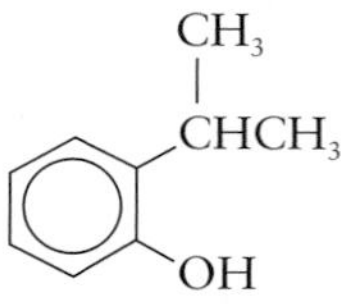 Benzene ring bearing $CHCH_3$ (with CH_3) and OH on adjacent carbons

b. 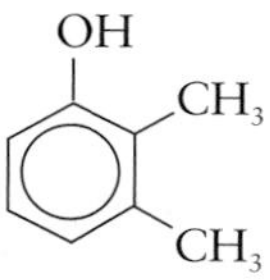 Benzene ring bearing OH, CH_3, CH_3

3.10 Name each of the following as a derivative of phenol:

a. 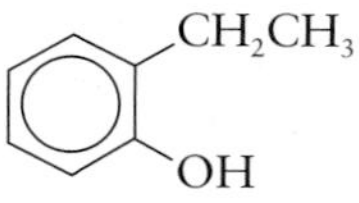 Benzene ring bearing CH_2CH_3 and OH on adjacent carbons

b. 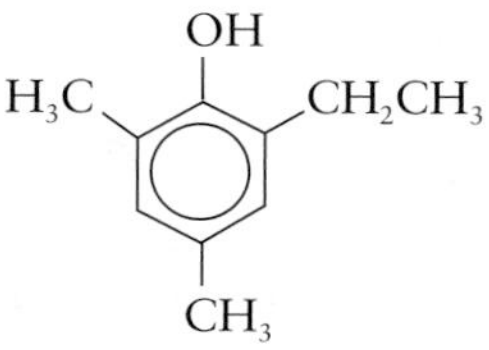 Benzene ring bearing OH, H_3C, CH_2CH_3, and CH_3

3.11 Draw structural formulas for each of the following:

a. *m*-methylphenol b. 2,3-dichlorophenol

3.12 Draw structural formulas for each of the following:

a. *o*-bromophenol b. 2,3,5-triethylphenol

CLASSIFICATION OF ALCOHOLS (SECTION 3.2)

3.13 What is the difference between a primary, secondary, and tertiary alcohol?

3.14 Classify the following alcohols as primary, secondary, or tertiary:

a. 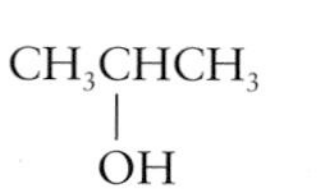 $CH_3CH(OH)CH_3$ b. CH_3CH_2-OH c. 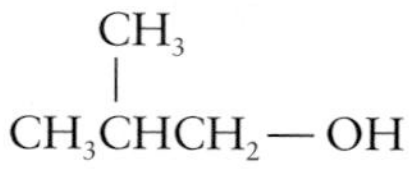 $CH_3CH(CH_3)CH_2-OH$

3.15 ■ Classify the following alcohols as primary, secondary, or tertiary:

a. 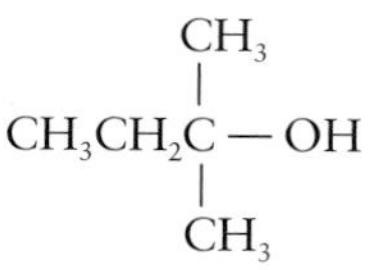 $CH_3CH_2C(CH_3)(CH_3)-OH$

b. $CH_3CH_2CH_2CH_2-OH$

c. 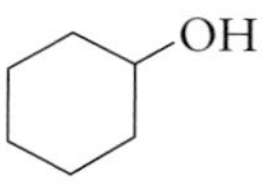 Cyclohexane ring bearing OH

3.16 Draw structural formulas for the four aliphatic alcohols with the molecular formula $C_4H_{10}O$. Name each compound using the IUPAC system and classify it as a primary, secondary, or tertiary alcohol.

PHYSICAL PROPERTIES OF ALCOHOLS (SECTION 3.3)

3.17 Why are the boiling points of alcohols much higher than the boiling points of alkanes with similar molecular weights?

3.18 ■ Arrange the compounds of each group in order of increasing boiling point.

a. ethanol, 1-propanol, methanol

b. butane, ethylene glycol, 1-propanol

3.19 ■ Which member of each of the following pairs would you expect to be more soluble in water? Briefly explain your choices.

a. butane or 2-butanol

b. 2-propanol or 2-pentanol

c. 2-butanol or 2,3-butanediol

3.20 Draw structural formulas for the following molecules and use a dotted line to show the formation of hydrogen bonds:

a. one molecule of 1-butanol and one molecule of ethanol

b. cyclohexanol and water

3.21 Explain why the use of glycerol (1,2,3-propanetriol) in lotions helps retain water and keep the skin moist.

REACTIONS OF ALCOHOLS (SECTION 3.4)

3.22 ■ Draw the structures of the chief product formed when the following alcohols are dehydrated to alkenes:

a. $CH_3CH(OH)CH_2CH_3$ b. $CH_3CH(OH)CH(CH_3)CH_3$

3.23 Draw the structures of the chief product formed when the following alcohols are dehydrated to alkenes:

a. 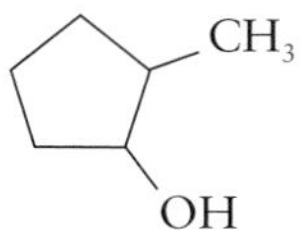 (cyclopentane ring bearing CH_3 and, on the adjacent carbon, OH)

b. $CH_3CH(OH)CH(CH_2CH_3)—CH_3$

3.24 Draw the structures of the ethers that can be produced from the following alcohols:

a. $CH_3CH_2CH_2—OH$

b. 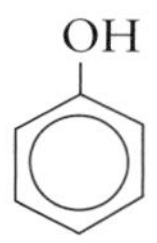 (benzene ring bearing OH)

c. 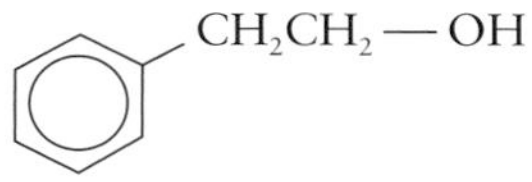 (benzene ring bearing $CH_2CH_2—OH$)

3.25 Draw the structures of the two organic compounds that can be obtained by oxidizing $CH_3CH_2CH_2—OH$.

3.26 ■ Give the structure of an alcohol that could be used to prepare each of the following compounds:

a. 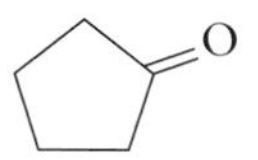 (cyclopentanone: cyclopentane ring with =O)

b. $CH_3CH_2C(=O)CH(CH_3)—CH_3$

c. $CH_3CH_2CH_2—C(=O)—OH$

3.27 Give the structure of an alcohol that could be used to prepare each of the following compounds:

a. 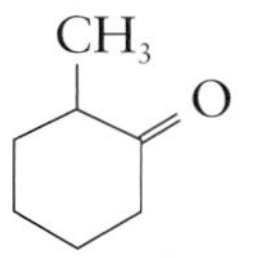 (cyclohexanone ring with =O and CH_3 on the adjacent carbon)

b. 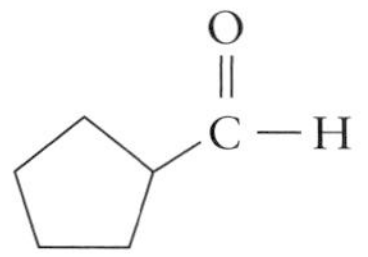 (cyclopentane ring bearing $C(=O)—H$)

c. $CH_3CH(CH_3)—C(=O)—OH$

3.28 ■ What products would result from the following processes? Write an equation for each reaction.

a. 2-Methyl-2-butanol is subjected to controlled oxidation.

b. 1-Propanol is heated to 140°C in the presence of sulfuric acid.

c. 3-Pentanol is subjected to controlled oxidation.

d. 3-Pentanol is heated to 180°C in the presence of sulfuric acid.

e. 1-Hexanol is subjected to an *excess* of oxidizing agent.

3.29 What products would result from the following processes? Write an equation for each reaction.

a. 1-Butanol is heated to 140°C in the presence of sulfuric acid.

b. 1-Butanol is subjected to an excess of oxidizing agent.

c. 2-Pentanol is subjected to controlled oxidation.

d. 2-Pentanol is heated to 180°C in the presence of sulfuric acid.

e. 2-Methyl-2-pentanol is subjected to controlled oxidation.

3.30 Each of the following conversions requires more than one step, and some reactions studied in previous chapters may be needed. Show the reagents you would use and draw structural formulas for intermediate compounds formed in each conversion.

a. $CH_3CH_2CH{=}CH_2 \rightarrow CH_3CH_2\overset{O}{\overset{\|}{C}}CH_3$

b. (2-methylcyclopentanol: ring with CH_3 and OH) $\longrightarrow$ (1-methylcyclopentanol: ring carbon bearing CH_3 and OH)

c. $CH_3CH_2CH_2\overset{OH}{\overset{|}{C}}H_2 \rightarrow CH_3CH_2\overset{O}{\overset{\|}{C}}CH_3$

3.31 Each of the following conversions requires more than one step, and some reactions studied in previous chapters may be needed. Show the reagents you would use and draw structural formulas for intermediate compounds formed in each conversion.

a. $CH_2{=}CH_2 \rightarrow CH_3CH_2{-}O{-}CH_2CH_3$

b. (cyclopentene) $\longrightarrow$ (cyclopentanone, ring with =O)

c. $CH_3CH_2\overset{OH}{\overset{|}{C}}H_2 \rightarrow CH_3\overset{OH}{\overset{|}{C}}HCH_3$

d. $CH_3CH_2CH_2\overset{OH}{\overset{|}{C}}H_2 \rightarrow CH_3CH_2\underset{Cl}{\underset{|}{C}}HCH_3$

3.32 The three-carbon diol used in antifreeze is

$$CH_3{-}\overset{OH}{\overset{|}{C}}H{-}\overset{OH}{\overset{|}{C}}H_2$$

It is nontoxic and is used as a moisturizing agent in foods. Oxidation of this substance within the liver produces pyruvic acid, which can be used by the body to supply energy. Give the structure of pyruvic acid.

IMPORTANT ALCOHOLS (SECTION 3.5)

3.33 Methanol is fairly volatile and evaporates quickly if spilled. Methanol is also absorbed quite readily through the skin. If, in the laboratory, methanol accidentally spilled on your clothing, why would it be a serious mistake to just let it evaporate?

3.34 Why is methanol such an important industrial chemical?

3.35 Suppose you are making some chocolate cordials (chocolate-coated candies with soft fruit filling). Why might you want to add a little glycerol to the filling?

3.36 Name an alcohol used in each of the following ways:

a. A moistening agent in many cosmetics

b. The solvent in solutions called tinctures

c. Automobile antifreeze

d. Rubbing alcohol

e. A flavoring in cough drops

f. Present in gasohol

CHARACTERISTICS AND USES OF PHENOLS (SECTION 3.6)

3.37 Name a phenol associated with each of the following:

a. A vitamin

b. A disinfectant cleaner

3.38 Name a phenol used in each of the following ways:

a. A disinfectant used for cleaning walls

b. An antiseptic found in some mouthwashes

c. An antioxidant used to prevent rancidity in foods

ETHERS (SECTION 3.7)

3.39 ■ Assign a common name to each of the following ethers:

a. $CH_3{-}O{-}CH_2CH_3$

b. (benzene ring)$-O{-}CH_2CH_2CH_3$

c. $CH_3CH_2\overset{CH_3}{\overset{|}{C}}H{-}O{-}\overset{CH_3}{\overset{|}{C}}HCH_2CH_3$

3.40 Assign a common name to each of the following ethers:

a. $CH_3CH_2{-}O{-}\underset{CH_3}{\underset{|}{C}}HCH_3$

b. $CH_3{-}O{-}CH_2CH_2CH_2CH_3$

c. (benzene ring)$-O-$(benzene ring)

3.41 Assign the IUPAC name to each of the following ethers. Name the smaller alkyl group as the alkoxy group.

a. $CH_3{-}O{-}CH_2CH_2CH_3$

b. $CH_3\underset{CH_2CH_3}{\underset{|}{C}}HCH_2CH_2{-}O{-}CH_2CH_3$

c. (cyclopropane ring)$-O{-}CH_2CH_2CH_3$

d. 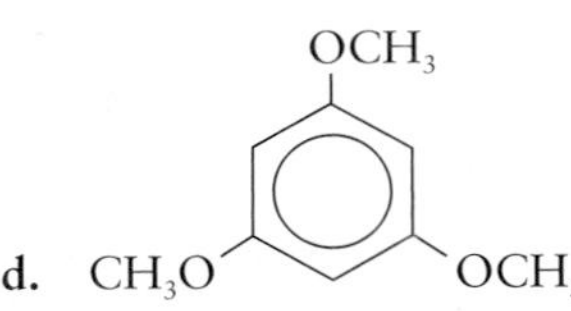

3.42 ■ Assign the IUPAC name to each of the following ethers. Name the smaller alkyl group as the alkoxy group.

a. $CH_3CH_2{-}O{-}CH_2CH_2CH_3$

b. $CH_3CH_2{-}O{-}\underset{CH_3}{\underset{|}{C}}HCH_3$

c. 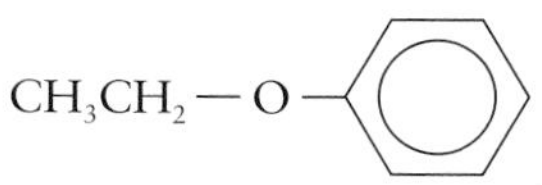 CH_3CH_2-O-(benzene ring)

d. 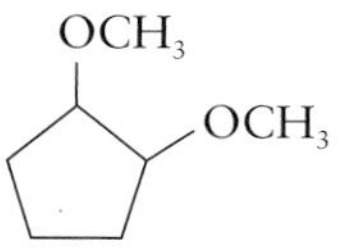 (cyclopentane with two adjacent OCH_3 groups)

3.43 Draw structural formulas for the following:
- a. butyl ethyl ether
- b. methyl phenyl ether
- c. diisopropyl ether
- d. 3-ethoxyhexane
- e. 1,2-dimethoxycyclobutane

3.44 Draw structural formulas for the following:
- a. methyl propyl ether
- b. isobutyl phenyl ether
- c. 2-methoxybutane
- d. 1,2,3-trimethoxycyclopropane
- e. 2-*t*-butoxy-2-pentene

3.45 Give the IUPAC names and draw structural formulas for the six isomeric ethers of molecular formula $C_5H_{12}O$.

PROPERTIES OF ETHERS (SECTION 3.8)

3.46 What is the chief chemical property of ethers?

3.47 Why is diethyl ether hazardous to use as an anesthetic or as a solvent in the laboratory?

3.48 Arrange the following compounds in order of decreasing solubility in water. Explain the basis for your decisions.

$CH_3CH_2CH_2-OH$ $CH_3CH_2CH_2CH_3$ $CH_3CH_2-O-CH_3$

3.49 ■ Arrange the compounds in Exercise 3.48 in order of decreasing boiling point. Explain your answer.

3.50 Draw structural formulas and use a dotted line to show hydrogen bonding between a molecule of diethyl ether and water.

THIOLS (SECTION 3.9)

3.51 Complete the following reactions:
- a. $2CH_3CH_2(SH) + Hg^{2+} \rightarrow$
- b. 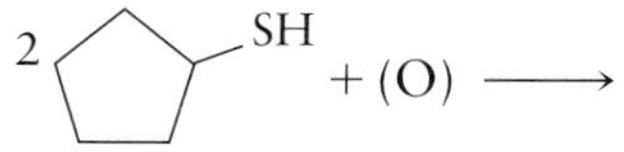 2 (cyclopentane)$-SH + (O) \longrightarrow$
- c. $CH_3-S-S-CH_2CH_2CH_3 + 2(H) \rightarrow$

3.52 ■ Complete the following reactions:
- a. $2CH_3CH_2CH_2-SH + Hg^{2+} \rightarrow$
- b. $2CH_3CH_2CH_2-SH + (O) \rightarrow$
- c. 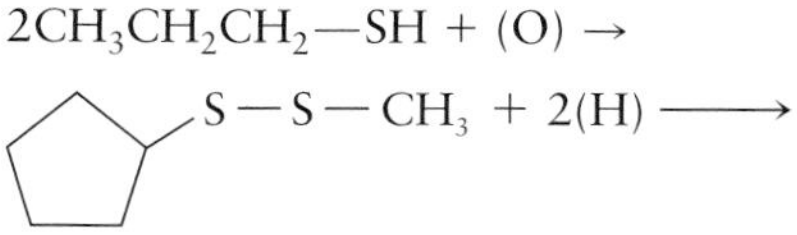 (cyclopentane)$-S-S-CH_3 + 2(H) \longrightarrow$

3.53 Lipoic acid is required by many microorganisms for proper growth. As a disulfide, it functions in the living system by catalyzing certain oxidation reactions and is reduced in the process. Write the structure of the reduction product.

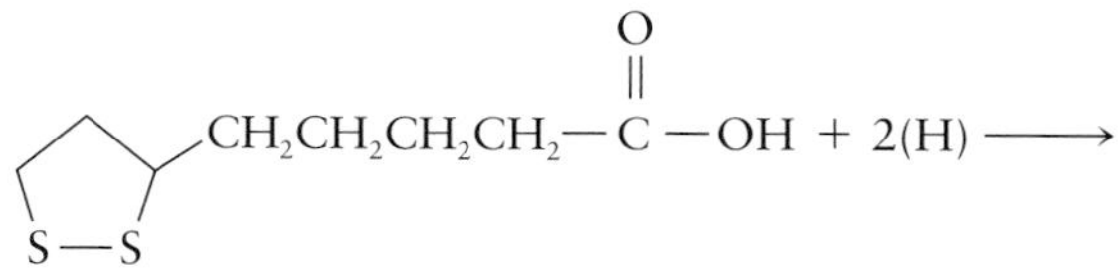

(five-membered ring containing $S-S$)$-CH_2CH_2CH_2CH_2-\overset{O}{\overset{\|}{C}}-OH + 2(H) \longrightarrow$

lipoic acid

3.54 Alcohols and thiols can both be oxidized in a controlled way. What are the differences in the products?

POLYFUNCTIONAL COMPOUNDS (SECTION 3.10)

3.55 Identify the functional groups in vanillin, the active ingredient in vanilla flavoring.

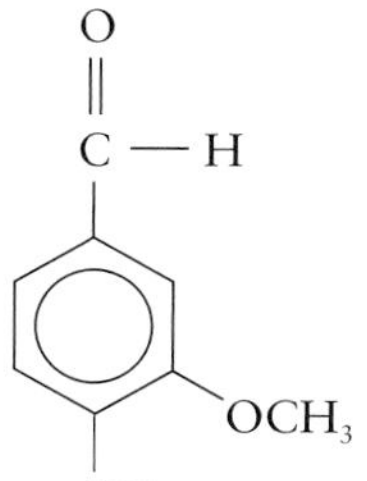

(benzene ring bearing $\overset{O}{\overset{\|}{C}}-H$, OCH_3, and OH)

3.56 Identify the functional groups in cinnamaldehyde, present in cinnamon flavoring.

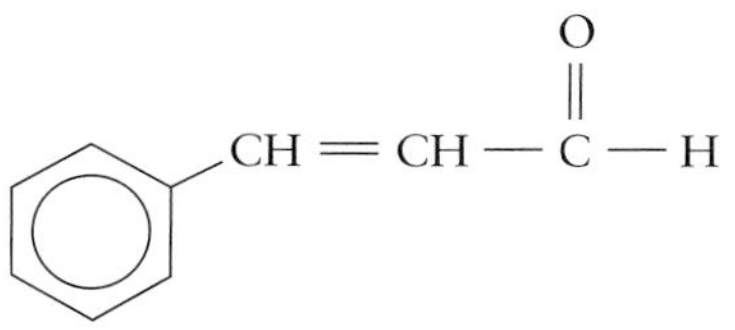

(benzene ring)$-CH=CH-\overset{O}{\overset{\|}{C}}-H$

ADDITIONAL EXERCISES

3.57
$$\text{alkene} + \text{water} \underset{}{\overset{H_2SO_4}{\rightleftharpoons}} \text{alcohol} + \text{heat}$$

In simplistic terms it can be thought that an equilibrium exists in the hydration-dehydration reactions above. Use LeChâtelier's principle to explain why the dehydration reaction is favored at 180°C.

3.58 Thiols have lower boiling points and are less water soluble than the corresponding alcohol (e.g., CH_3CH_2-OH and CH_3CH_2-SH). Explain why.

3.59 Determine the oxidation numbers on the carbon atom in the following reaction.

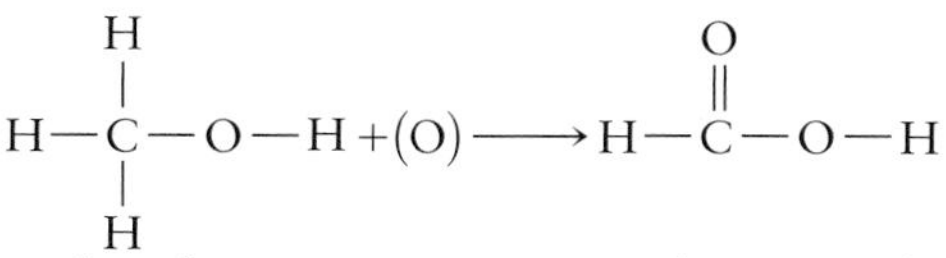

$$\underset{\text{methanol}}{H-\overset{H}{\underset{H}{C}}-O-H} + (O) \longrightarrow \underset{\text{methanoic acid}}{H-\overset{O}{\overset{\|}{C}}-O-H}$$

3.60 Using the idea of aromaticity (the electronic structure of benzene), explain why phenol is a stronger acid than cyclohexanol.

3.61 Use general reactions to show how an alcohol (R—O—H) can behave as a weak Brønsted acid when it reacts with a strong Brønsted base (X^-). Next, show how an alcohol can behave as a weak Brønsted base in the presence of a strong Brønsted acid (HA).

ALLIED HEALTH EXAM CONNECTION

Reprinted with permission from Nursing School and Allied Health Entrance Exams, COPYRIGHT 2005 Petersons.

3.62 Give IUPAC names for the following alcohols:

a. Alcohol in alcoholic beverages

b. Rubbing alcohol

c. Wood alcohol

3.63 Which of the following compounds are ethers?

a. CH_3CHO

b. $CH_3—O—CH_3$

c. $CH_3—COOH$

d. $CH_3CH_2—O—CH_3$

CHEMISTRY FOR THOUGHT

3.64 Many aftershaves are 50% ethanol. What do you think is the purpose of the ethanol?

3.65 A mixture of ethanol and 1-propanol is heated to 140°C in the presence of sulfuric acid. Careful analysis reveals that three ethers are formed. Write formulas for these three products, name each, and explain why three form, rather than a single product.

3.66 The two-carbon alcohol in some antifreeze

```
OH    OH
|     |
CH2—CH2
```

is toxic if ingested. Reactions within the liver oxidize the diol to oxalic acid, which can form insoluble salts and damage the kidneys. Give the structure of oxalic acid.

3.67 Although ethanol is only mildly toxic to adults, it poses greater risks for fetuses. Why do you think this is the case?

3.68 The medicinal compound taxol is mentioned in Figure 3.1. What does the name indicate about its possible structure?

3.69 When diethyl ether is spilled on the skin, the skin takes on a dry appearance. Explain whether this effect is more likely due to a removal of water or a removal of natural skin oils.

3.70 Figure 3.8 points out that methanol is used as a fuel in racing cars. Write a balanced equation for the combustion of methanol.

3.71 Figure 3.13 focuses on the use of thiol chemistry by skunks. If thiols are more volatile than alcohols and less soluble in water, why do these properties represent an advantage to skunks?

3.72 Why do you think a dye is added to antifreeze before it is sold in stores?

3.73 Glycerol, or glycerin, is used as a moisturizer in foods such as candy and shredded coconut. What is the molecular explanation for its utility as a moisturizer?

CHAPTER 4

Aldehydes and Ketones

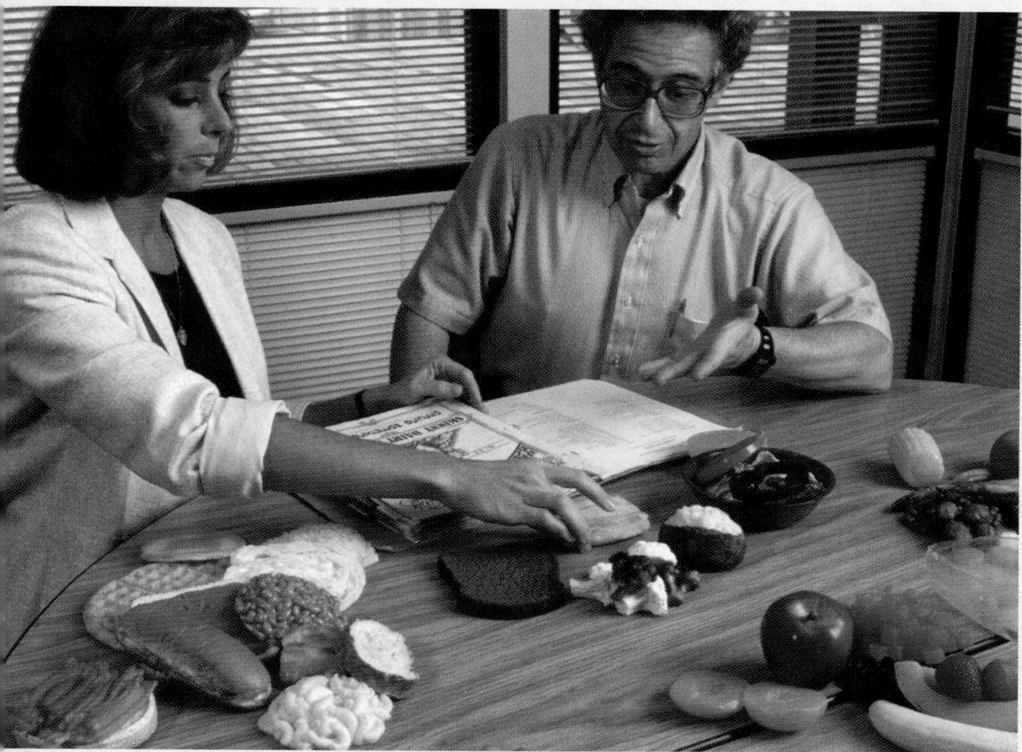

© Nathan Benn/CORBIS

LEARNING OBJECTIVES

When you have completed your study of this chapter, you should be able to:

1. Recognize the carbonyl group in compounds and classify the compounds as aldehydes or ketones. (Section 4.1)
2. Assign IUPAC names to aldehydes and ketones. (Section 4.1)
3. Compare the physical properties of aldehydes and ketones with those of compounds in other classes. (Section 4.2)
4. Write key reactions for aldehydes and ketones. (Section 4.3)
5. Give specific uses for aldehydes and ketones. (Section 4.4)

Proper nutrition and, in some cases, a carefully monitored diet play an essential role in the recovery process for hospitalized patients. Clinical dietitians also supervise the nutritional care of people in nursing homes and other settings. In this chapter, you will become familiar with aldehydes and ketones, substances whose reactions will help you understand the chemistry of carbohydrates, a nutritionally important family of compounds.

ThomsonNOW™ Throughout the chapter this icon introduces resources on the ThomsonNOW website for this text. Sign in at **www.thomsonedu.com** to:

- Evaluate your knowledge of the material
- Take an exam prep quiz
- Identify areas you need to study with a **Personalized Learning Plan**.

carbonyl group
The $-\overset{O}{\overset{\|}{C}}-$ group.

The carbonyl functional group is characteristic of both aldehydes and ketones. In a **carbonyl group,** a carbon is double bonded to an oxygen atom and single bonded to two other atoms:

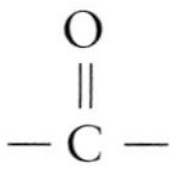

$$-\overset{O}{\overset{\|}{C}}-$$

carbonyl group

Aldehydes and ketones occur widely in nature and play important roles in living organisms. For example, the carbonyl group is found in numerous carbohydrates, including glucose and fructose. Glucose, a major source of energy in living systems, is found combined with fructose in cane sugar:

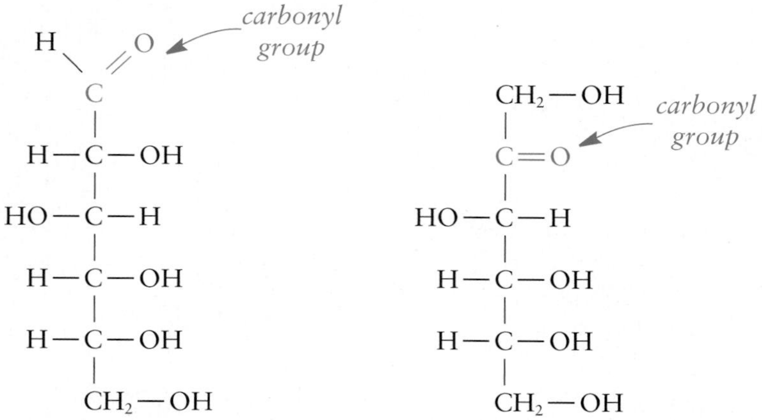

```
H     O  <-- carbonyl group        CH2—OH
  \  //                              |
   C                                 C=O  <-- carbonyl group
   |                                 |
H—C—OH                          HO—C—H
   |                                 |
HO—C—H                           H—C—OH
   |                                 |
H—C—OH                           H—C—OH
   |                                 |
H—C—OH                             CH2—OH
   |
  CH2—OH
```

glucose   fructose

In this chapter, we will study the naming of aldehydes and ketones, as well as some of their important reactions. These concepts will be very useful when we study the chemistry of carbohydrates in Chapter 7.

4.1 The Nomenclature of Aldehydes and Ketones

LEARNING OBJECTIVES

1. Recognize the carbonyl group in compounds and classify the compounds as aldehydes or ketones.
2. Assign IUPAC names to aldehydes and ketones.

aldehyde
A compound that contains the $-\overset{O}{\overset{\|}{C}}-H$ group; the general formula is $R-\overset{O}{\overset{\|}{C}}-H$.

ketone
A compound that contains the $-\overset{|}{\underset{|}{C}}-\overset{O}{\overset{\|}{C}}-\overset{|}{\underset{|}{C}}-$ group; the general formula is $R-\overset{O}{\overset{\|}{C}}-R'$.

When a carbonyl group is directly bonded to at least one hydrogen atom, the compound is an **aldehyde.** In **ketones,** two carbon atoms are directly bonded to the carbonyl group:

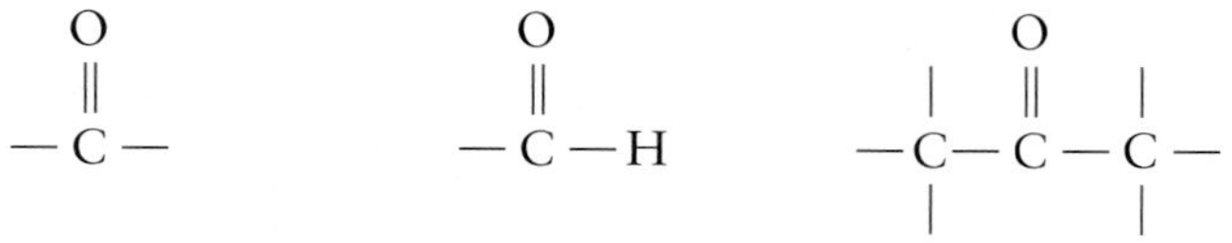

$$-\overset{O}{\overset{\|}{C}}- \qquad -\overset{O}{\overset{\|}{C}}-H \qquad -\overset{|}{\underset{|}{C}}-\overset{O}{\overset{\|}{C}}-\overset{|}{\underset{|}{C}}-$$

carbonyl group   aldehyde   ketone

Structural models of the simplest aldehyde (formaldehyde) and the simplest ketone (acetone) are shown in ■ Figure 4.1.

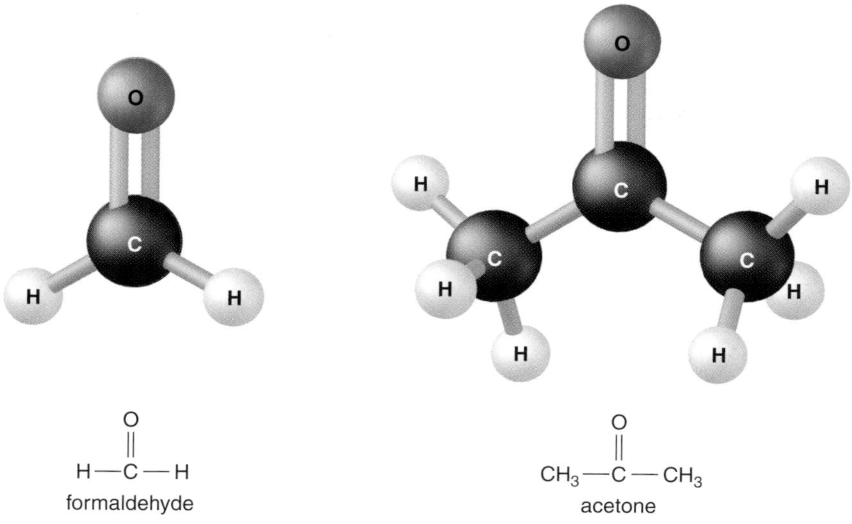

FIGURE 4.1 Ball-and-stick models of formaldehyde and acetone.

Several aldehydes and ketones, in addition to formaldehyde and acetone, are well known by common names (see Table 4.1). Notice that aldehyde and ketone common names are easily recognized by the endings *-aldehyde* and *-one*, respectively.

In the IUPAC system, aldehydes are named by selecting the longest chain that includes the carbonyl carbon. The *-e* of the alkane that has the same number of carbons is replaced by *-al*. Because the carbonyl carbon in aldehydes is always at the end of the chain, it is always designated as carbon number 1.

EXAMPLE 4.1

Name the following aldehydes according to the IUPAC system.

ThomsonNOW™ Go to Coached Problems to practice **naming aldehydes and ketones.**

a. $CH_3CH_2CH_2-\overset{\overset{\displaystyle O}{\|}}{C}-H$

b. $CH_3\overset{\overset{\displaystyle Br}{|}}{C}HCH_2-\overset{\overset{\displaystyle O}{\|}}{C}-H$

c. $CH_3CH_2\underset{}{\overset{\overset{\overset{\displaystyle O}{\|}}{C-H}}{\overset{|}{C}}}HCH_2CH_3$

Solution

a. This aldehyde contains four carbon atoms, and the name is butanal. The aldehyde group is always position number 1, so the number designating the aldehyde position is omitted from the name.

b. Carbon chain numbering begins with the carbonyl end of the molecule even though a lower number for the Br could be obtained by numbering from the left:

$$\underset{4}{CH_3}\underset{3}{\overset{\overset{\displaystyle Br}{|}}{C}H}\underset{2}{CH_2}-\underset{1}{\overset{\overset{\displaystyle O}{\|}}{C}}-H$$

The compound is 3-bromobutanal.

TABLE 4.1 Some common aldehydes and ketones

Structural formula	IUPAC name	Common name	Boiling point (°C)
Aldehydes			
$H-\overset{\overset{\displaystyle O}{\|}}{C}-H$	methanal	formaldehyde	−21
$CH_3-\overset{\overset{\displaystyle O}{\|}}{C}-H$	ethanal	acetaldehyde	21
$CH_3CH_2-\overset{\overset{\displaystyle O}{\|}}{C}-H$	propanal	propionaldehyde	49
$CH_3CH_2CH_2-\overset{\overset{\displaystyle O}{\|}}{C}-H$	butanal	butyraldehyde	76
$C_6H_5-\overset{\overset{\displaystyle O}{\|}}{C}-H$	benzaldehyde	benzaldehyde	178
Ketones			
$CH_3-\overset{\overset{\displaystyle O}{\|}}{C}-CH_3$	propanone	acetone	56
$CH_3-\overset{\overset{\displaystyle O}{\|}}{C}-CH_2CH_3$	butanone	methyl ethyl ketone (MEK)	80
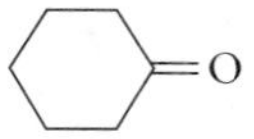 $C_6H_{10}=O$ (cyclohexane ring with =O)	cyclohexanone	cyclohexanone	156

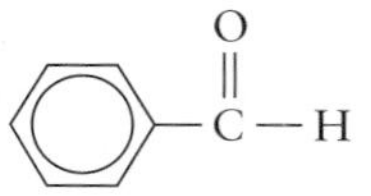

c. The carbonyl group must be position number 1 in the chain even though a longer carbon chain (pentane) exists:

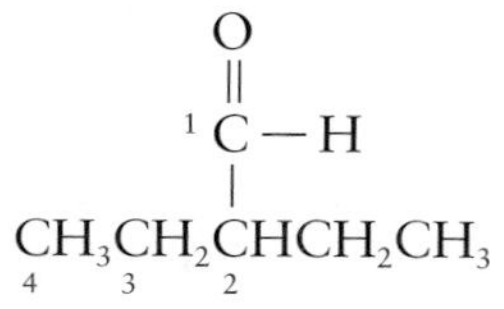

$$\underset{4}{CH_3}\underset{3}{CH_2}\underset{2}{\overset{\overset{\displaystyle {}^{1}\overset{\overset{\displaystyle O}{\|}}{C}-H}{|}}{C}H}CH_2CH_3$$

The name is 2-ethylbutanal.

Learning Check 4.1 Give IUPAC names to the following aldehydes:

a. $CH_3CH_2CH_2CH_2-\overset{\overset{\displaystyle O}{\|}}{C}-H$

b. 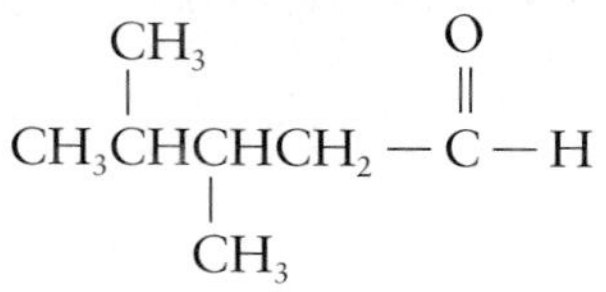 $CH_3\underset{\underset{\displaystyle CH_3}{|}}{\overset{\overset{\displaystyle CH_3}{|}}{C}}HCH_2-\overset{\overset{\displaystyle O}{\|}}{C}-H$

In naming ketones by the IUPAC system, the longest chain that includes the carbonyl group is given the name of the corresponding alkane, with the *-e* replaced by *-one*. The chain is numbered from the end closest to the carbonyl group, and the carbonyl group position is indicated by a number.

EXAMPLE 4.2

Give IUPAC names to the following ketones:

a. $CH_3-\overset{\overset{O}{\|}}{C}-CH_3$ b. $CH_3-\overset{\overset{O}{\|}}{C}-\underset{\underset{CH_3}{|}}{C}HCH_2CH_3$ c. 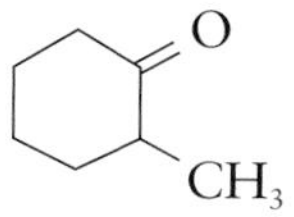 (cyclohexanone ring with =O and CH_3 on adjacent carbon)

Solution

a. There are three carbons in the chain, and the correct name is propanone. The number 2 is not used in this case because the carbonyl group in a ketone cannot be on the end of a chain. Therefore, it must be in the number 2 position.

b. The chain is numbered from the left to give a lower number for the carbonyl carbon:

$$\underset{1}{CH_3}-\underset{2}{\overset{\overset{O}{\|}}{C}}-\underset{\underset{CH_3}{|}}{\overset{3}{C}}H\overset{4}{C}H_2\overset{5}{C}H_3$$

The complete name is 3-methyl-2-pentanone.

c. In cyclic ketones, the carbonyl carbon will always be at position 1, and so the number is omitted from the name:

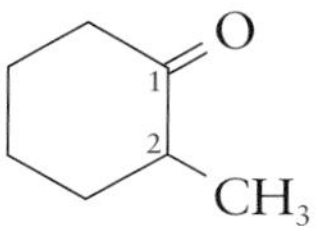

(cyclohexanone ring, carbonyl carbon numbered 1, carbon bearing CH_3 numbered 2)

The correct name is 2-methylcyclohexanone.

Learning Check 4.2 Give IUPAC names to the following ketones:

a. $CH_3\underset{\overset{|}{Br}}{C}HCH_2\underset{\overset{|}{Br}}{C}HCH_2-\overset{\overset{O}{\|}}{C}-CH_3$ b. 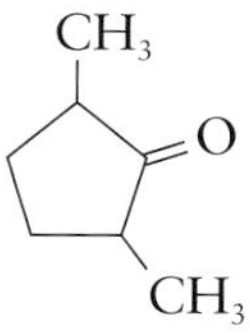 (cyclopentanone ring with =O, CH_3 on each adjacent carbon)

4.2 Physical Properties

LEARNING OBJECTIVE

3. Compare the physical properties of aldehydes and ketones with those of compounds in other classes.

The physical properties of aldehydes and ketones can be explained by an examination of their structures. First, the lack of a hydrogen on the oxygen prevents the formation of hydrogen bonds between molecules:

no H attached to the oxygen

$$R-\overset{\overset{O}{\|}}{C}-H \qquad R-\overset{\overset{O}{\|}}{C}-R'$$

CHEMISTRY AROUND US 4.1 Faking a Tan

Many people believe that a suntan makes one look healthy and attractive. Studies, however, indicate that this perception is far from the truth. According to these studies, sunbathing, especially when sunburn results, ages the skin prematurely and increases the risk of skin cancer. Cosmetic companies have developed a tanning alternative for those not willing to risk using the sun but who want to be "fashionably" tan.

Tanning lotions and creams that chemically darken the skin are now available. The active ingredient in these "bronzers" is dihydroxyacetone (DHA), a colorless compound classified by the Food and Drug Administration as a safe skin dye.

$$HO-CH_2-\overset{\overset{\displaystyle O}{\|}}{C}-CH_2-OH$$

dihydroxyacetone (DHA)

Within several hours of application, DHA produces a brown skin color by reacting with the outer layer of the skin, which consists of dead cells. Only the dead cells react with DHA, so the color gradually fades as the dead cells slough off and are replaced. This process generally leads to the fading of chemical tans within a few weeks. Another problem with chemical tans is uneven skin color. Areas of skin such as elbows and knees, which contain a thicker layer of dead cells, may absorb and react with more tanning lotion and become darker than other areas.

Perhaps the greatest problem with chemical tans is the false sense of security they might give. Some people with chemical tans think it is safe to go into the sun and get a deeper tan. This isn't true. Sunlight presents the same hazards to chemically tanned skin that it does to untanned skin.

Some DHA-containing products.

Therefore, boiling points of pure aldehydes and ketones are expected to be lower than those of alcohols with similar molecular weights. Remember, alcohols can form hydrogen bonds with one another. ■ Table 4.2 shows that the boiling points of propanal and acetone are 49°C and 56°C, respectively, whereas the alcohol of comparable molecular weight, 1-propanol, has a boiling point of 97°C.

TABLE 4.2 A comparison of physical properties

Class	Example	Formula	Molecular weight	Boiling point (°C)	Water solubility
Alkane	butane	$CH_3CH_2CH_2CH_3$	58	0	Insoluble
Aldehyde	propanal	$CH_3CH_2-\overset{\overset{\displaystyle O}{\|}}{C}-H$	58	49	Soluble
Ketone	propanone (acetone)	$CH_3-\overset{\overset{\displaystyle O}{\|}}{C}-CH_3$	58	56	Soluble
Alcohol	1-propanol	$CH_3CH_2CH_2-OH$	60	97	Soluble

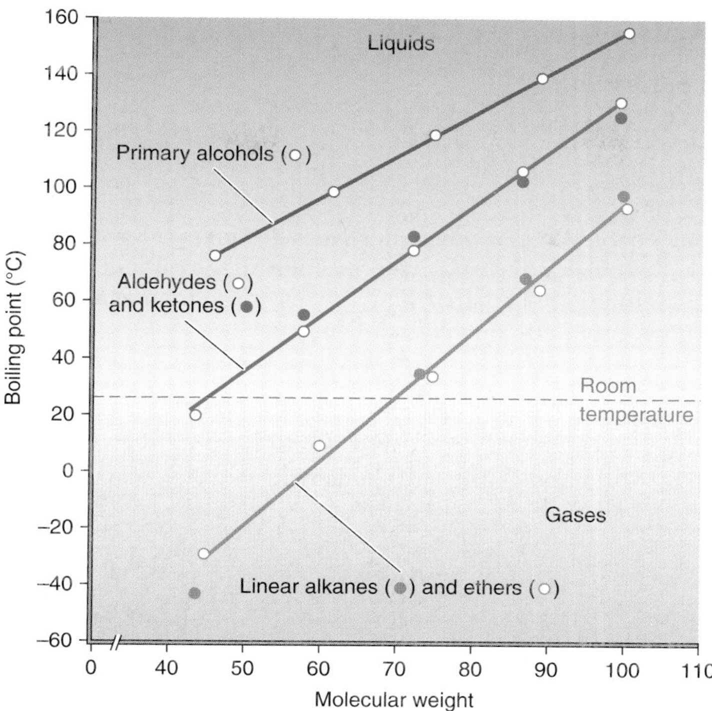

■ **FIGURE 4.2** The boiling points of aldehydes and ketones are higher than those of alkanes and ethers but lower than those of alcohols of comparable molecular weights.

Also notice in Table 4.2 that the boiling points of the aldehyde and ketone are higher than that of the comparable alkane. This can be explained by differences in polarity. Whereas alkanes are nonpolar, the carbonyl group of aldehydes and ketones is polar due to the more electronegative oxygen atom:

$$\overset{\delta-}{\mathrm{O}}\\ \| \\ -\underset{\delta+}{\mathrm{C}}-$$

The attractive forces between molecules possessing such dipoles are not as strong as hydrogen bonds, but they do cause aldehydes and ketones to boil at higher temperatures than the nonpolar alkanes. The graph in ■ Figure 4.2 summarizes the boiling point comparison for several families of compounds.

Water is a polar solvent and would be expected to be effective in dissolving compounds containing polar carbonyl groups. Besides exhibiting dipolar attractions, the carbonyl oxygen also forms hydrogen bonds with water:

ThomsonNOW Go to Coached Problems to explore the **boiling points of aldehydes and ketones.**

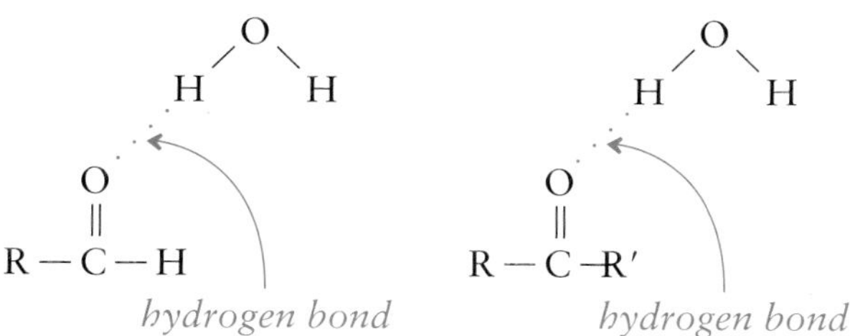

As a result, low molecular weight aldehydes and ketones are water soluble (Table 4.2).

4.3 Chemical Properties

LEARNING OBJECTIVE

4. Write key reactions for aldehydes and ketones.

Oxidation

ThomsonNOW Go to Coached Problems to explore the **reactivity of aldehydes and ketones.**

We learned in Section 3.4 that when aldehydes are prepared by oxidizing primary alcohols with $KMnO_4$ or $K_2Cr_2O_7$, the reaction may continue and produce carboxylic acids. Ketones are prepared similarly, but they are much less susceptible to further oxidation, especially by mild oxidizing agents:

$$\text{General reaction: } \underset{\substack{\text{primary}\\\text{alcohol}}}{R-CH_2-OH} \xrightarrow{(O)} \underset{\text{aldehyde}}{R-\overset{\overset{\displaystyle O}{\|}}{C}-H} \xrightarrow{(O)} \underset{\substack{\text{carboxylic}\\\text{acid}}}{R-\overset{\overset{\displaystyle O}{\|}}{C}-OH} \quad (4.1)$$

$$\text{General reaction: } \underset{\substack{\text{secondary}\\\text{alcohol}}}{R-\overset{\overset{\displaystyle OH}{|}}{CH}-R'} \xrightarrow{(O)} \underset{\text{ketone}}{R-\overset{\overset{\displaystyle O}{\|}}{C}-R'} \xrightarrow{(O)} \text{no reaction} \quad (4.2)$$

The difference in reactivity toward oxidation is the chief reason why aldehydes and ketones are classified in separate families (see ■ Figure 4.3).

Specific example:

$$\underset{\text{benzaldehyde}}{C_6H_5-\overset{\overset{\displaystyle O}{\|}}{C}-H} + (O) \longrightarrow \underset{\text{benzoic acid}}{C_6H_5-\overset{\overset{\displaystyle O}{\|}}{C}-OH} \quad (4.3)$$

Specific example:

$$\underset{\text{acetone}}{CH_3-\overset{\overset{\displaystyle O}{\|}}{C}-CH_3} + (O) \longrightarrow \text{no reaction} \quad (4.4)$$

The oxidation of aldehydes occurs so readily that even atmospheric oxygen causes the reaction. A few drops of liquid benzaldehyde (Reaction 4.3) placed on a watch glass will begin to oxidize to solid benzoic acid within an hour. In the laboratory, it is not unusual to find that bottles of aldehydes that have been opened for a length of time contain considerable amounts of carboxylic acids.

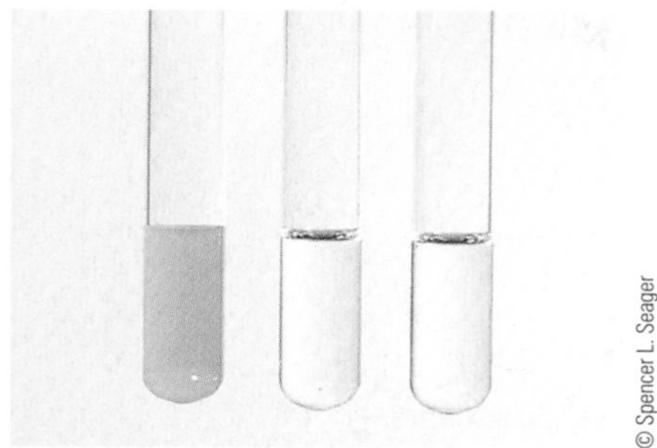

From left to right, three test tubes containing potassium dichromate ($K_2Cr_2O_7$), acetone, and benzaldehyde.

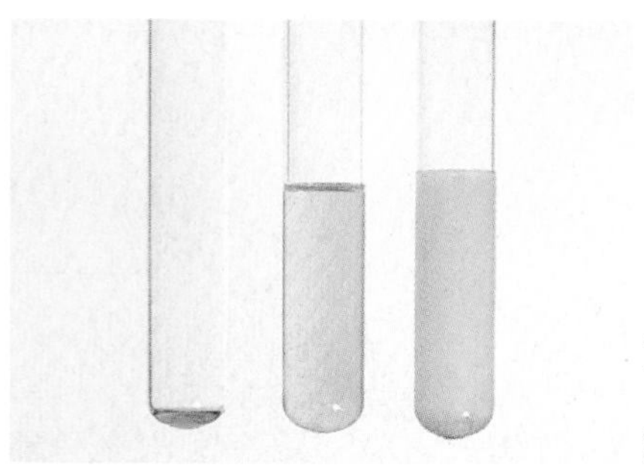

After the addition of equal amounts of $K_2Cr_2O_7$, the acetone remains unreacted, whereas the benzaldehyde is oxidized.

■ **FIGURE 4.3** Attempted oxidation of acetone and oxidation of benzaldehyde. As the reaction proceeds, chromium is reduced, forming a grayish green precipitate.

Learning Check 4.3 Draw the structural formula for each product. Write "no reaction" if none occurs.

a. $CH_3CH(CH_3)CH_2-\overset{\displaystyle O}{\overset{\|}{C}}-H + (O) \longrightarrow$

b. cyclopentanone $+ (O) \longrightarrow$

The effectiveness of two common chemical reagents used to test for the presence of aldehydes—Tollens' reagent and Benedict's reagent—depends on the ease with which aldehydes are oxidized. In general, ketones fail to react with these same reagents.

Tollens' reagent is a mild oxidizing solution containing complexed Ag^+ ions. The oxidizing agent is the silver ion, which is reduced to metallic silver in the process and precipitates out (Reaction 4.5):

Tollens' reagent
A mild oxidizing solution containing silver ions used to test for aldehydes.

$$\underset{\text{an aldehyde}}{R-\overset{\displaystyle O}{\overset{\|}{C}}-H} + \underset{\substack{\text{Tollens'}\\\text{reagent}}}{2Ag(NH_3)_2^+} + 2OH^- \longrightarrow \underset{\substack{\text{a carboxylate}\\\text{salt}}}{R-\overset{\displaystyle O}{\overset{\|}{C}}-O^-NH_4^+} + 3NH_3 + \underset{\text{silver}}{2Ag} + H_2O \qquad (4.5)$$

When the reaction is carried out in a clean, grease-free glass container, the metallic silver deposits on the glass and forms a mirror. This reaction is used commercially to silver some mirrors (see Figure 4.4).

Benedict's reagent is a mild oxidizing solution containing Cu^{2+} ions. When an aldehyde oxidation takes place in an alkaline solution containing Cu^{2+}, a red precipitate of Cu_2O is produced:

Benedict's reagent
A mild oxidizing solution containing Cu^{2+} ions used to test for the presence of aldehydes.

$$\underset{\text{certain aldehydes}}{R-\overset{\displaystyle O}{\overset{\|}{C}}-H} + \underset{\substack{\text{Benedict's}\\\text{reagent}\\\text{(blue)}}}{2Cu^{2+}} + 5OH^- \longrightarrow \underset{\text{a carboxylate ion}}{R-\overset{\displaystyle O}{\overset{\|}{C}}-O^-} + \underset{\substack{\text{red}\\\text{precipitate}}}{Cu_2O} + 3H_2O \qquad (4.6)$$

© Michael C. Slabaugh

FIGURE 4.4 This young woman can see herself because the silver coating on the back of the mirror reflects light.

OVER THE COUNTER 4.1 Birth Control: Progesterone Substitutes

The ketone *norethynodrel* is not a major industrial material, but in terms of social impact, it has had a profound influence since 1960. In that year, Enovid, with norethynodrel as the major component, became the first oral contraceptive for women approved by the FDA. Norethynodrel is effective

norethynodrel

progesterone

because it mimics the action of another ketone, *progesterone,* a hormone that causes changes in the wall of the uterus to prepare it to accept a fertilized egg and maintain the resulting pregnancy.

Norethynodrel establishes a state of false pregnancy; a woman ceases to ovulate and therefore cannot conceive. The pill (as this product is popularly called) has been used by millions of women and appears to be relatively safe even though some women experience undesirable side effects such as acne, hypertension, abnormal bleeding, or increased blood clotting. Women who suffer from these or other serious side effects should not take the pill. Women over the age of 40, particularly those who smoke or are overweight, are advised by the FDA to use contraceptive methods other than the pill.

Doctors sometimes prescribe higher doses of combined oral contraceptives for use as "morning after" pills, which are taken within 72 hours of unprotected intercourse to prevent any possibly fertilized egg from reaching the uterus. The FDA's Advisory Committee for Reproductive Health Drugs concluded that certain oral contraceptives are safe and effective for this type of use.

Another type of contraception is injectable progestin, a blend of hormones containing a compound similar to progesterone. Depo-Provera® is a product that is injected by a health care professional into the arm muscle or buttocks every 3 months. Depo-Provera prevents pregnancy in three ways. It changes the cervical mucus to help prevent sperm from reaching the egg, inhibits ovulation, and changes the uterine lining to prevent a fertilized egg from implanting in the uterus.

Long-term birth control for women became available in 1990 when the FDA approved under-the-skin implants that contain a slowly released contraceptive. In these devices, levonorgestrel (Norplant®, or the newer Norplant 2®), a synthetic compound similar to progesterone, is sealed inside small rubber rods about the size of matchsticks. The rods are implanted under the skin of the upper arm. Levonorgestrel is slowly released through the walls of the rods and prevents pregnancy for up to five years (for Norplant) or three years (for Norplant 2), with a failure rate of only 0.2%. Possible side effects from Norplant are vaginal bleeding, headaches, nausea, dizziness, and nervousness.

Stone/Getty Images

A variety of oral contraceptives contain norethynodrel.

The presence of Cu^{2+} in Benedict's reagent causes this alkaline solution to have a bright blue color. A positive Benedict's test consists of the appearance of a colored (usually red) precipitate. All aldehydes give a positive Tollens' test, but only certain aldehydes and one type of easily oxidized ketone readily give a positive Benedict's test. The necessary structural features are shown.

$$\mathrm{R-\overset{\overset{\displaystyle O}{\|}}{C}-\overset{\overset{\displaystyle O}{\|}}{C}-H}$$

(a) an aldehyde with an adjacent carbonyl group

$$\mathrm{R-\underset{\underset{\displaystyle OH}{|}}{CH}-\overset{\overset{\displaystyle O}{\|}}{C}-H}$$

(b) an aldehyde with an adjacent alcohol group

$$\mathrm{R-\underset{\underset{\displaystyle OH}{|}}{CH}-\overset{\overset{\displaystyle O}{\|}}{C}-R'}$$

(c) a ketone with an adjacent alcohol group

The structural features (b) or (c) are found in a number of sugars, including glucose. Accordingly, Benedict's reagent reacts with glucose to produce a red precipitate of Cu_2O (see ■ Figure 4.5).

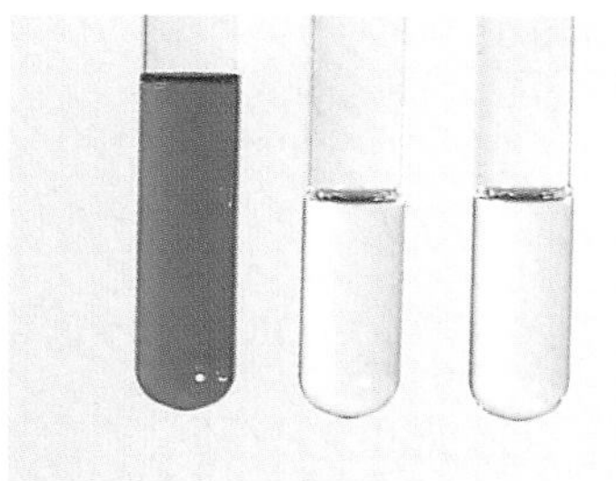

From left to right, three test tubes containing Benedict's reagent, 0.5% glucose solution, and 2.0% glucose solution.

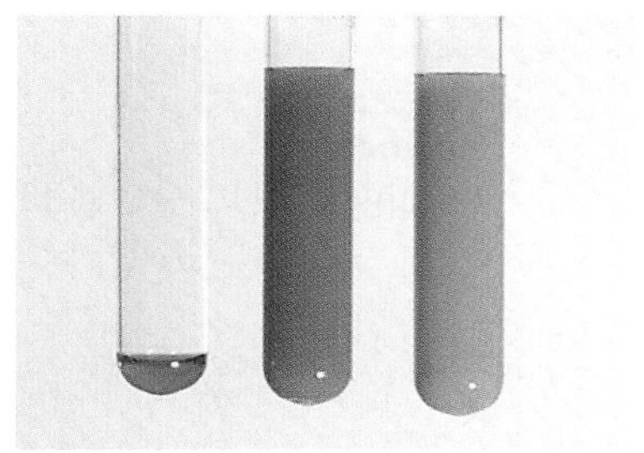

The addition of Benedict's reagent produces colors (due to the red Cu_2O) that indicate the amount of glucose present.

■ **FIGURE 4.5** The Benedict's test for glucose.

The Addition of Hydrogen

In Section 2.3, we learned that addition reactions are common for compounds containing carbon–carbon double bonds. The same is true for the carbon–oxygen double bonds of carbonyl groups. An important example is the addition of H_2 in the presence of catalysts such as nickel or platinum:

$$\text{General reaction: } \underset{\text{aldehyde}}{\mathrm{R-\overset{\overset{\displaystyle O}{\|}}{C}-H}} + \mathrm{H_2} \xrightarrow{\text{Pt}} \underset{\text{primary alcohol}}{\mathrm{R-\overset{\overset{\displaystyle OH}{|}}{\underset{\underset{\displaystyle H}{|}}{C}}-H}} \tag{4.7}$$

$$\text{General reaction: } \underset{\text{ketone}}{\mathrm{R-\overset{\overset{\displaystyle O}{\|}}{C}-R'}} + \mathrm{H_2} \xrightarrow{\text{Pt}} \underset{\text{secondary alcohol}}{\mathrm{R-\overset{\overset{\displaystyle OH}{|}}{\underset{\underset{\displaystyle H}{|}}{C}}-R'}} \tag{4.8}$$

$$\text{Specific example: } \underset{\text{propanal}}{\mathrm{CH_3CH_2-\overset{\overset{\displaystyle O}{\|}}{C}-H}} + \mathrm{H_2} \xrightarrow{\text{Pt}} \underset{\text{1-propanol}}{\mathrm{CH_3CH_2\overset{\overset{\displaystyle OH}{|}}{CH_2}}} \tag{4.9}$$

$$\text{Specific example: } \underset{\text{acetone}}{\mathrm{CH_3-\overset{\overset{\displaystyle O}{\|}}{C}-CH_3}} + \mathrm{H_2} \xrightarrow{\text{Pt}} \underset{\text{2-propanol}}{\mathrm{CH_3-\overset{\overset{\displaystyle OH}{|}}{CH}-CH_3}} \tag{4.10}$$

We see that the addition of hydrogen to an aldehyde produces a primary alcohol, whereas the addition to a ketone gives a secondary alcohol. Notice that these reactions are reductions that are essentially the reverse of the alcohol oxidations used to prepare aldehydes and ketones (Reactions 4.1 and 4.2).

■ **Learning Check 4.4** Draw the structural formula for each product of the following reactions:

a. $\mathrm{CH_3CH_2-\overset{\overset{\displaystyle O}{\|}}{C}-CH_2CH_3 + H_2 \xrightarrow{Pt}}$

b. $\mathrm{C_6H_5-\overset{\overset{\displaystyle O}{\|}}{C}-H + H_2 \xrightarrow{Pt}}$

hemiacetal
A compound that contains the $-\overset{\text{OH}}{\underset{\text{OR}}{\mathrm{C}}}-\mathrm{H}$ functional group

The Addition of Alcohols

Aldehydes react with alcohols to form **hemiacetals,** a functional group that is very important in carbohydrate chemistry (Chapter 7). Hemiacetals are usually not very stable and are difficult to isolate. With excess alcohol present and with an acid catalyst, however, a stable product called an acetal is formed:

General reaction:

$$\underset{\text{aldehyde}}{R-\overset{\overset{\displaystyle O}{\|}}{C}-H} + \underset{\text{alcohol}}{R'-OH} \rightleftharpoons \underset{\text{hemiacetal intermediate}}{R-\overset{OH}{\underset{OR'}{C}}-H} \xrightleftharpoons[]{H^+ \text{ and } R'-OH} \underset{\text{acetal}}{R-\overset{OR'}{\underset{OR'}{C}}-H} + H_2O \tag{4.11}$$

Specific example:

$$\underset{\text{acetaldehyde}}{CH_3-\overset{\overset{\displaystyle O}{\|}}{C}-H} + \underset{\text{methanol}}{CH_3-OH} \rightleftharpoons \underset{\text{hemiacetal intermediate}}{CH_3-\overset{OH}{\underset{OCH_3}{C}}-H} \xrightleftharpoons[]{H^+ \text{ and } CH_3-OH} \underset{\text{acetal}}{CH_3-\overset{OCH_3}{\underset{OCH_3}{C}}-H} + H_2O \tag{4.12}$$

acetal
A compound that contains the $-\overset{\text{OR}}{\underset{\text{OR}}{\mathrm{C}}}-\mathrm{H}$ arrangement of atoms.

Hemiacetals contain an —OH group, hydrogen, and an —OR group on the same carbon. Thus, a hemiacetal appears to have alcohol and ether functional groups. **Acetals** contain a carbon that has a hydrogen and two —OR groups attached; they look like diethers:

$$CH_3-\overset{OH}{\underset{OCH_3}{C}}-H \qquad CH_3-\overset{OCH_3}{\underset{OCH_3}{C}}-H$$

hemiacetal carbon *acetal carbon*

To better understand the changes that have occurred, let's look first at hemiacetal formation and then at acetal formation. The first molecule of alcohol adds across the double bond of the aldehyde in a manner similar to other addition reactions:

$$CH_3-\overset{\overset{\displaystyle O}{\|}}{C}-H + CH_3-O-H \rightleftharpoons \underset{\text{hemiacetal}}{CH_3-\overset{OH}{\underset{OCH_3}{C}}-H} \tag{4.13}$$

—H attaches here (to O) *—OCH_3 attaches here* (to C)

The second step, in which the hemiacetal is converted to an acetal, is like earlier substitution reactions. In this case, an $-OCH_3$ substitutes for an —OH:

$$\underset{\text{hemiacetal}}{CH_3-\overset{OH}{\underset{OCH_3}{C}}-H} + CH_3-O-H \xrightleftharpoons{H^+} \underset{\text{acetal}}{CH_3-\overset{OCH_3}{\underset{OCH_3}{C}}-H} + H-OH \tag{4.14}$$

In a very similar way, ketones can react with alcohols to form **hemiketals** and **ketals:**

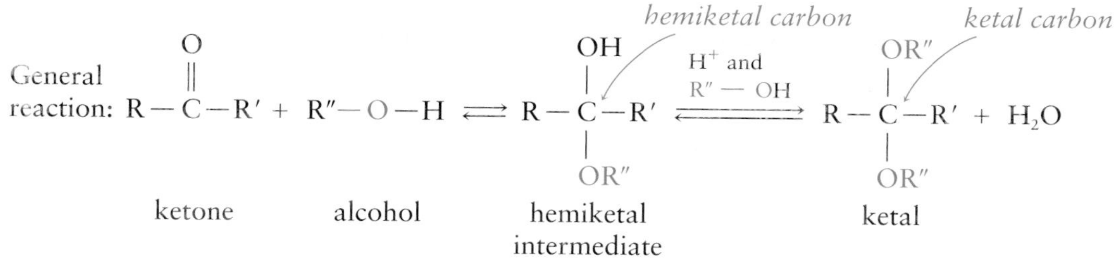

General reaction:

$$R-\overset{\overset{\displaystyle O}{\|}}{C}-R' + R''-O-H \rightleftharpoons \underset{\text{hemiketal intermediate}}{R-\overset{\overset{\displaystyle OH}{|}}{\underset{\underset{\displaystyle OR''}{|}}{C}}-R'} \xrightleftharpoons[]{H^+ \text{ and } R''-OH} \underset{\text{ketal}}{R-\overset{\overset{\displaystyle OR''}{|}}{\underset{\underset{\displaystyle OR''}{|}}{C}}-R'} + H_2O \qquad (4.15)$$

ketone, alcohol, hemiketal intermediate (*hemiketal carbon*), ketal (*ketal carbon*)

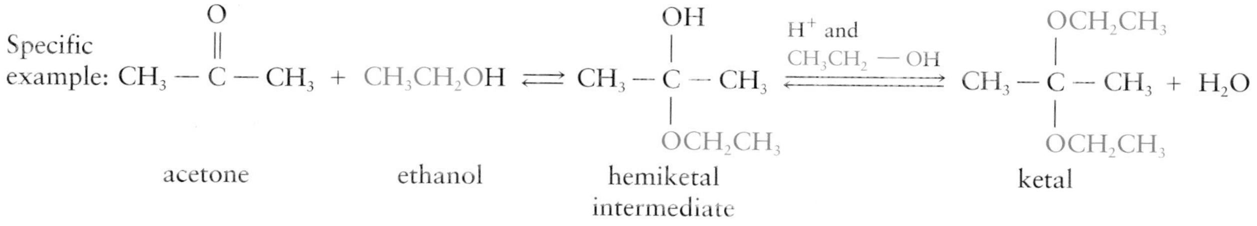

Specific example:

$$CH_3-\overset{\overset{\displaystyle O}{\|}}{C}-CH_3 + CH_3CH_2OH \rightleftharpoons CH_3-\overset{\overset{\displaystyle OH}{|}}{\underset{\underset{\displaystyle OCH_2CH_3}{|}}{C}}-CH_3 \xrightleftharpoons[]{H^+ \text{ and } CH_3CH_2-OH} CH_3-\overset{\overset{\displaystyle OCH_2CH_3}{|}}{\underset{\underset{\displaystyle OCH_2CH_3}{|}}{C}}-CH_3 + H_2O \qquad (4.16)$$

acetone, ethanol, hemiketal intermediate, ketal

hemiketal
A compound that contains the

$$-\overset{|}{\underset{|}{C}}-\overset{\overset{\displaystyle OH}{|}}{\underset{\underset{\displaystyle OR}{|}}{C}}-\overset{|}{\underset{|}{C}}-$$

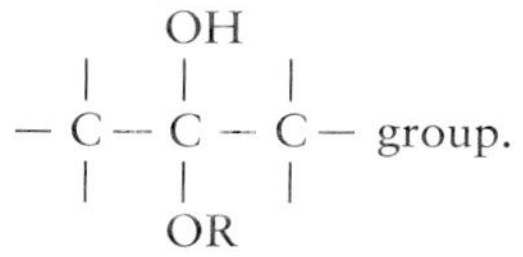

group.

ketal
A compound that contains the

$$-\overset{|}{\underset{|}{C}}-\overset{\overset{\displaystyle OR}{|}}{\underset{\underset{\displaystyle OR}{|}}{C}}-\overset{|}{\underset{|}{C}}-$$

group.

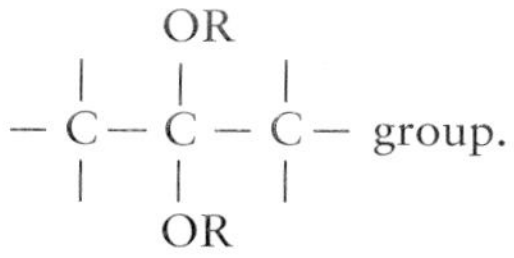

ThomsonNOW™ Go to Coached Problems to practice identifying the **structures of acetals and hemiacetals.**

EXAMPLE 4.3

Identify each of the following as a hemiacetal, hemiketal, acetal, ketal, or none of the above:

a. $CH_3-O-CH_2CH_2-OH$

b. 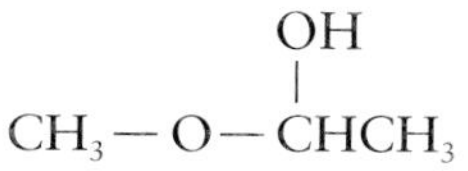 $CH_3-O-\overset{\overset{\displaystyle OH}{|}}{C}HCH_3$

c. $CH_3CH_2-O-\overset{\overset{\displaystyle OCH_3}{|}}{\underset{\underset{\displaystyle CH_3}{|}}{C}}-CH_3$

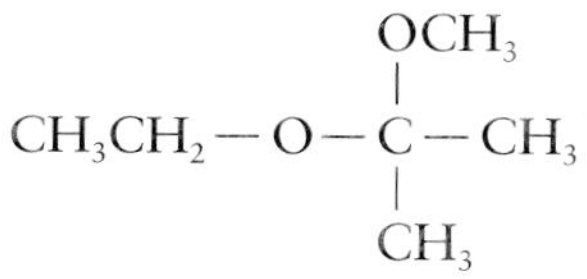

d. 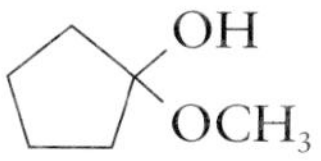 cyclopentane ring carbon bearing OH and OCH_3

Solution

a. Look first at any structure to see if there is a carbon with two oxygens attached. Hemiacetals, hemiketals, acetals, and ketals are all alike in that regard. This compound has two oxygens, but they are on different carbons; thus it is none of the possibilities.

b. After finding a carbon with two oxygens attached, next look to see if one is an —OH. If it is, that makes it a hemiacetal or hemiketal. The third step is to look at the carbon containing the two oxygens. If it is attached to an —H, the compound was originally an aldehyde and not a ketone. Thus, this compound is a hemiacetal:

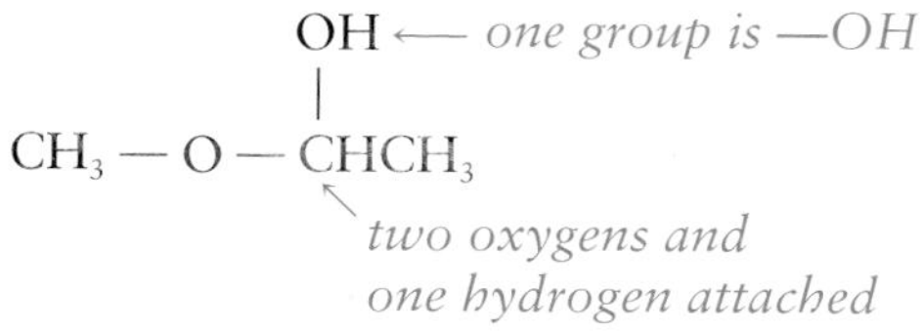

$$CH_3-O-\overset{\overset{\displaystyle OH}{|}}{C}HCH_3$$

OH ← *one group is —OH*

two oxygens and one hydrogen attached

c. There is a carbon with two oxygens attached and neither is an —OH. Thus, the compound is an acetal or ketal:

$$CH_3CH_2-O-\underset{\underset{CH_3}{|}}{\overset{\overset{OCH_3}{|}}{C}}-CH_3$$

no hydrogens attached

No hydrogens are attached to the carbon bonded to two oxygens. The compound is a ketal.

d. Two oxygens are attached to the same carbon. One group is an —OH. No hydrogens are attached to the central carbon. The compound is a hemiketal.

Learning Check 4.5 Draw the structural formula for the hemiacetal and hemiketal intermediates and for the acetal and ketal products of the following reactions. Label each structure as a hemiacetal, hemiketal, ketal, or acetal.

a. $C_6H_5-\overset{\overset{O}{||}}{C}-H + 2CH_3CH_2-OH \xrightarrow{H^+}$

b. $C_6H_5-\overset{\overset{O}{||}}{C}-CH_3 + 2CH_3-OH \xrightarrow{H^+}$

HOW REACTIONS OCCUR 4.1 Hemiacetal Formation

In the first step of the mechanism, the carbonyl oxygen is protonated:

$$R-\overset{\overset{:\overset{+}{O}-H}{||}}{C}-H \rightleftarrows R-\overset{\overset{:\ddot{O}-H}{|}}{\underset{+}{C}}-H$$

The alcohol oxygen then bonds to the carbonyl carbon:

$$R-\overset{\overset{:\overset{+}{O}-H}{||}}{C}-H + R'-\ddot{O}-H \rightleftarrows R-\overset{\overset{:\ddot{O}-H}{|}}{\underset{\underset{R'-\overset{+}{O}-H}{|}}{C}}-H$$

In the final step, a proton is given off to form the hemiacetal and regenerate the acid catalyst.

$$R-\overset{\overset{:\ddot{O}-H}{|}}{\underset{\underset{R'-\overset{+}{O}-H}{|}}{C}}-H \rightleftarrows R-\overset{\overset{:\ddot{O}H}{|}}{\underset{\underset{R'-\ddot{O}:}{|}}{C}}-H + H^+$$

hemiacetal

Some molecules contain both an —OH and a C=O group on different carbon atoms. In such cases, an intramolecular (within the molecule) reaction can occur, and a cyclic hemiacetal or hemiketal can be formed:

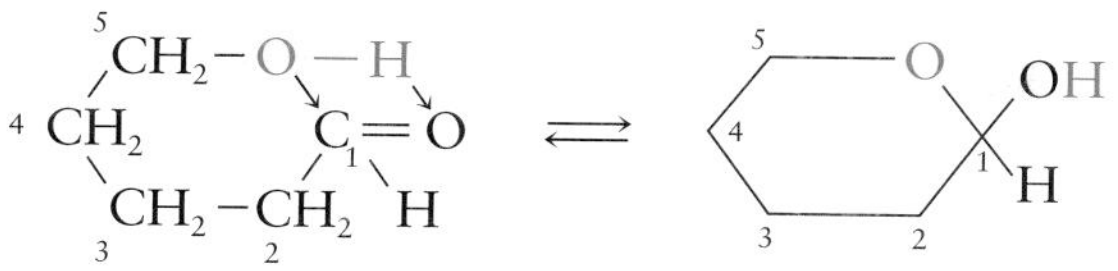

$$\overset{5}{C}H_2{-}O{-}H \quad \overset{4}{C}H_2 \quad \overset{1}{C}{=}O \quad H \quad \underset{3}{C}H_2{-}\underset{2}{C}H_2 \rightleftharpoons \text{(ring: } \overset{5}{}\text{, O, } \overset{1}{C}(OH)(H)\text{, } \overset{2}{}\text{, } \overset{3}{}\text{, } \overset{4}{}\text{)} \qquad (4.17)$$

open-chain molecule containing an aldehyde and alcohol group — cyclic structure containing a hemiacetal group

Cyclic hemiacetals and hemiketals are much more stable than the open-chain compounds discussed earlier. In Chapter 17, we'll see that sugars such as glucose and fructose exist predominantly in the form of cyclic hemiacetals and hemiketals.

In another reaction important in carbohydrate chemistry (Chapter 7), cyclic acetals and ketals may form from the hemistructures. As before, all reactions are reversible:

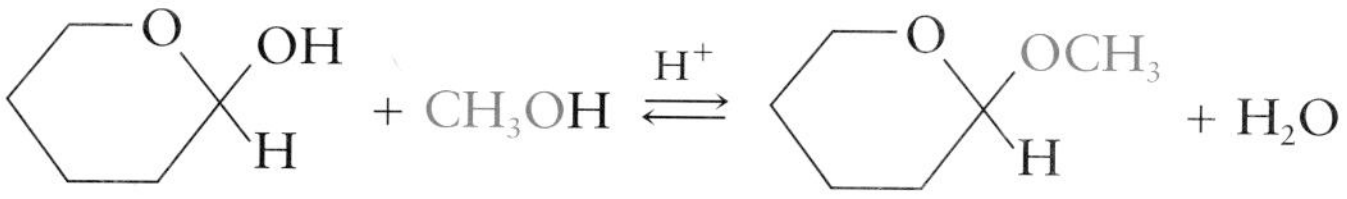

$$\text{(ring O, C(OH)(H))} + CH_3OH \overset{H^+}{\rightleftharpoons} \text{(ring O, C(OCH}_3\text{)(H))} + H_2O \qquad (4.18)$$

cyclic hemiacetal — cyclic acetal

STUDY SKILLS 4.1 A Reaction Map for Aldehydes and Ketones

This reaction map is designed to help you master organic reactions. Whenever you are trying to complete an organic reaction, use these two basic steps: (1) Identify the functional group that is to react, and (2) identify the reagent that is to react with the functional group. If the reacting functional group is an aldehyde or a ketone, find the reagent in the summary diagram, and use the diagram to predict the correct products.

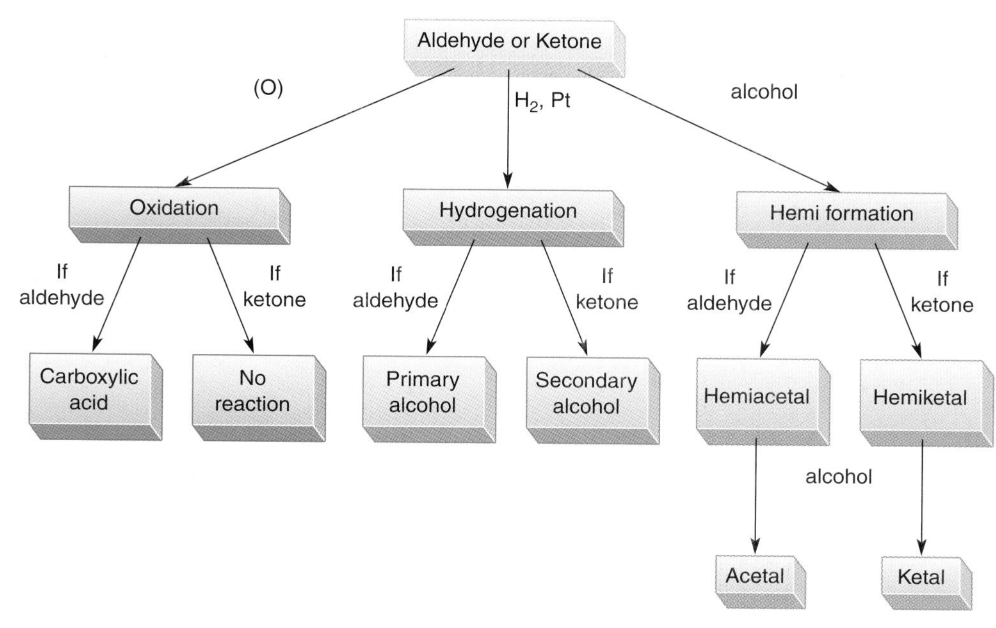

EXAMPLE 4.4

Identify each of the following as a cyclic hemiacetal, hemiketal, acetal, or ketal:

a. (five-membered ring O; ring carbon bearing OH and CH_3)

b. (five-membered ring O; ring carbon bearing OCH_3 and CH_2CH_3)

c. (six-membered ring O; ring carbon bearing OCH_3 and H)

d. (six-membered ring O; ring carbon bearing H and OH)

Solution

Use the same criteria that were applied to open-chain compounds in Example 4.3. Compound a has two oxygens attached to the same carbon. One is an —OH (hemiacetal or hemiketal). The central carbon has no hydrogens. Thus, the compound is a hemiketal. These criteria give the following results for each structure:

a. OH, CH_3 — *hemiketal carbon*

b. OCH_3, CH_2CH_3 — *ketal carbon (—OH absent, —H absent)*

c. OCH_3, H — *acetal carbon (—OH absent, —H present)*

d. H, OH — *hemiacetal carbon (—OH present, —H present)*

We have shown each of the reactions in acetal and ketal formation as being reversible. Simple hemiacetals and hemiketals are characteristically unstable and easily revert to the starting materials. Acetals and ketals are stable structures, but the reactions may be reversed by using water and an acid catalyst:

General reaction:
$$\underset{\text{acetal}}{R-\underset{OR'}{\overset{OR'}{C}}-H} + H_2O \overset{H^+}{\rightleftharpoons} \underset{\text{aldehyde}}{R-\overset{O}{\overset{\|}{C}}-H} + \underset{\text{alcohol}}{2R'-OH} \quad (4.19)$$

General reaction:
$$\underset{\text{ketal}}{R-\underset{OR''}{\overset{OR''}{C}}-R'} + H_2O \overset{H^+}{\rightleftharpoons} \underset{\text{ketone}}{R-\overset{O}{\overset{\|}{C}}-R'} + \underset{\text{alcohol}}{2R''-OH} \quad (4.20)$$

Specific example:
$$CH_3-\underset{OCH_2CH_3}{\overset{OCH_2CH_3}{C}}-CH_3 + H_2O \overset{H^+}{\rightleftharpoons} CH_3-\overset{O}{\overset{\|}{C}}-CH_3 + 2CH_3CH_2-OH \quad (4.21)$$

hydrolysis
Bond breakage by reaction with water.

This type of reaction, in which water causes a compound to split into component substances, is called **hydrolysis.** It will be an important reaction to remember as we discuss, in later chapters, the chemistry of biomolecules such as carbohydrates, lipids, and proteins.

Learning Check 4.6 Draw the structural formulas needed to complete the following hydrolysis reactions:

a. 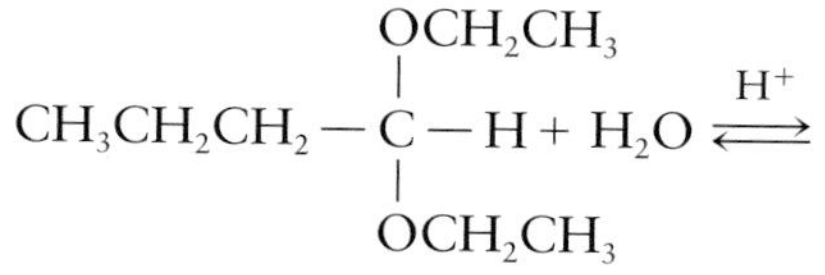 $CH_3CH_2CH_2-\underset{OCH_2CH_3}{\overset{OCH_2CH_3}{C}}-H + H_2O \overset{H^+}{\rightleftharpoons}$

b. 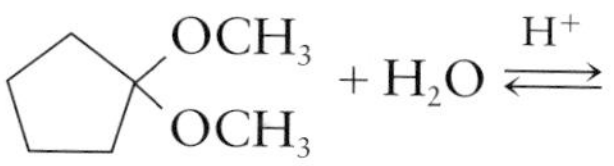 (cyclopentane ring bearing OCH_3 and OCH_3 on one carbon) $+ H_2O \overset{H^+}{\rightleftharpoons}$

EXAMPLE 4.5

Draw structural formulas for the products of the following reactions:

a. 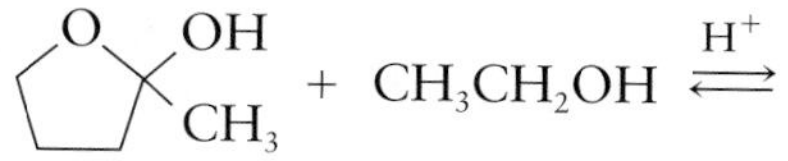 (five-membered ring with O; ring carbon bearing OH and CH_3) $+ CH_3CH_2OH \overset{H^+}{\rightleftharpoons}$

b. 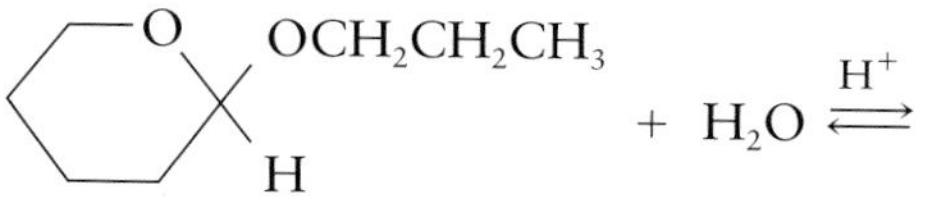 (six-membered ring with O; ring carbon bearing $OCH_2CH_2CH_3$ and H) $+ H_2O \overset{H^+}{\rightleftharpoons}$

c. 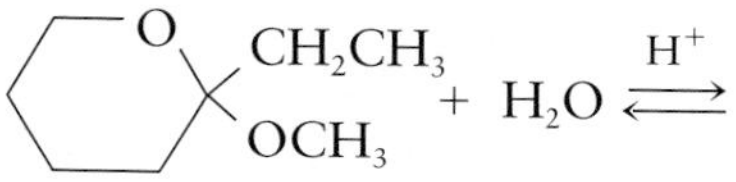 (six-membered ring with O; ring carbon bearing CH_2CH_3 and OCH_3) $+ H_2O \overset{H^+}{\rightleftharpoons}$

Solution

a. The starting material is a cyclic hemiketal. Reaction with an alcohol produces a cyclic ketal by a substitution of $—OCH_2CH_3$ for the —OH:

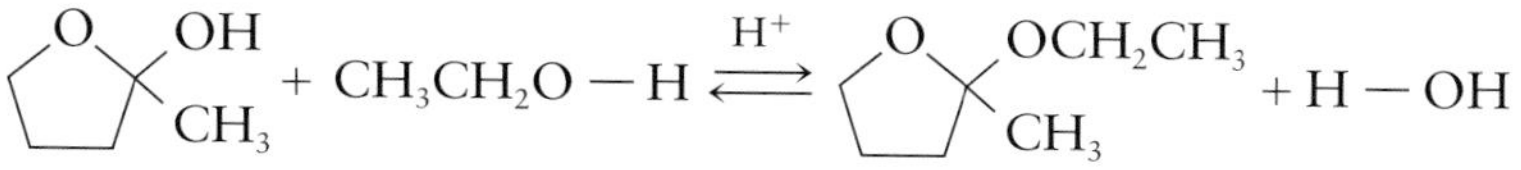

(ring O, C bearing OH and CH_3) $+ CH_3CH_2O-H \overset{H^+}{\rightleftharpoons}$ (ring O, C bearing OCH_2CH_3 and CH_3) $+ H-OH$

b. The starting compound is an acetal. Hydrolysis with water will form an aldehyde and two alcohol groups:

(ring O, C bearing $O-CH_2CH_2CH_3$ and H; ring O labelled *becomes an –OH*; C labelled *becomes* $-\overset{O}{\overset{\|}{C}}-H$) $\rightleftharpoons$ (chain with OH and $C{=}O$, H) $+ HO-CH_2CH_2CH_3$ (*alcohol groups*)

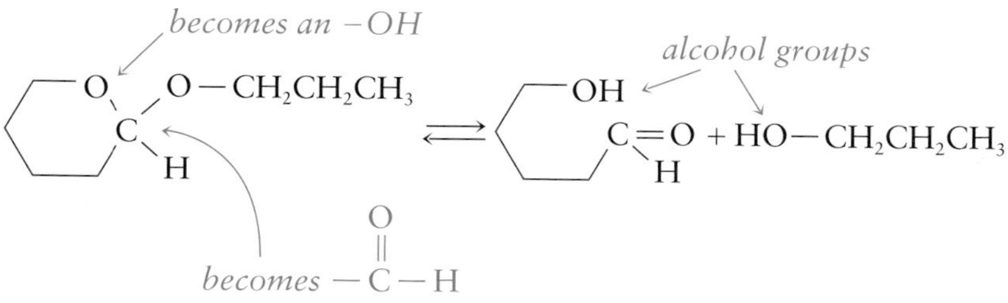

c. The starting compound is a ketal and upon hydrolysis will form a ketone and two alcohol groups:

(ring O, C bearing CH_2CH_3 and OCH_3; ring O labelled *becomes an –OH*; C labelled *becomes* $-\overset{O}{\overset{\|}{C}}-$) $\rightleftharpoons$ (chain with OH and $C{=}O$, CH_2CH_3) $+ HO-CH_3$ (*alcohol groups*)

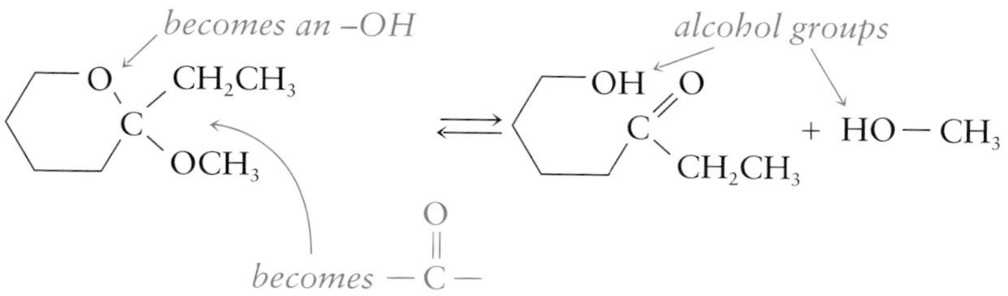

■ **Learning Check 4.7** Identify each of the following as cyclic hemiacetals, hemiketals, acetals, or ketals and show structural formulas for the hydrolysis products:

a. (six-membered ring containing O; ring carbon next to O bears OCH_3 and H)

b. (five-membered ring containing O; ring carbon next to O bears CH_3 and OCH_3)

CHEMISTRY AROUND US 4.2 Vanilloids: Hot Relief from Pain

The word *vanilloids* might bring to mind images of vanilla-flavored candies or ice cream. However, vanilloids have properties very different from those of vanilla, and their inclusion in foods would result in a hot and spicy flavor that might even be extremely irritating, rather than a smooth vanilla flavor.

The best-known vanilloids are capsaicin and resiniferatoxin (RTX). Capsaicin is the compound in various types of red peppers that gives them their characteristic hot and spicy flavor. It is used in a variety of products ranging from personal-defense pepper sprays to squirrel-proof birdseed. RTX is a powerful irritant extracted from a cactuslike plant, *Euphorbia resinifera*. The milky latex obtained from the plant is referred to in medical literature dating as far back as the first century. In these writings, it was described as a nose and skin irritant and as a treatment for chronic pain. In powdered form, RTX causes sneezing, a characteristic used by some practical jokers.

Structurally, molecules of capsaicin and RTX (not shown) are similar to those of vanillin in that they both have a homovanillyl group in their structure.

© Mark Slabaugh

Hot peppers are a source of capsaicin.

$$\text{HO}-\text{C}_6\text{H}_3(\text{OCH}_3)-\overset{\text{O}}{\overset{\|}{\text{C}}}-\text{H}$$

vanillin

$$\text{HO}-\text{C}_6\text{H}_3(\text{OCH}_3)-$$

homovanillyl group

$$\text{HO}-\text{C}_6\text{H}_3(\text{OCH}_3)-\text{CH}_2-\text{NH}-\overset{\text{O}}{\overset{\|}{\text{C}}}-(\text{CH}_2)_4-\text{CH}=\text{CH}-\overset{\text{CH}_3}{\overset{|}{\text{C}}}\text{HCH}_3$$

capsaicin

When either of the compounds is applied to the skin or mouth, a burning sensation results. This sensation is caused by the compounds bonding to receptors of small sensory neurons associated with the transmission of pain. As a result of the repeated applications of capsaicin or RTX, these neurons, or C sensory nerve fibers, become desensitized to the initial irritation or pain the compounds cause. This characteristic allows the compounds to be used topically to relieve chronic pain. Capsaicin is now used as an ingredient in a few OTC pain-relieving creams such as Capzasin-P, Heet, Menthacin, and Pain Doctor.

Both capsaicin and RTX have also been tested as treatments for a condition called *bladder hyperreflexia,* which is characterized by the urge to urinate even when there is very little urine in the bladder. Capsaicin effectively relieved the problem, but the first application caused an intense burning sensation that was intolerable and unacceptable to patients. RTX fared better in the tests. It also relieved the condition, and caused only mild discomfort and an itching sensation with the first application.

Basic research has led to some important discoveries about these compounds. The gene for the capsaicin receptor has been cloned, making available the large quantities of the receptor needed to do further research and develop new drugs. For example, it has been found that the receptor responds not only to capsaicin but also to levels of heat that can damage tissue. It is suggested that possibly the reason hot peppers feel hot in the mouth is because they stimulate the same pain receptors that heat does.

Another important development is the discovery of a way to synthesize RTX in the laboratory. The availability of synthetic RTX may reduce the cost for the material and make possible further advances in the use of RTX in medicines.

4.4 Important Aldehydes and Ketones

LEARNING OBJECTIVE

5. Give specific uses for aldehydes and ketones.

Formaldehyde: $\mathrm{H{-}\overset{\overset{\displaystyle O}{\|}}{C}{-}H}$

As seen with other functional classes, the most important compound of the class is usually the simplest member. Formaldehyde, the simplest aldehyde, is a gas at room temperature, but it is often supplied and used in the form of a 37% aqueous solution called *formalin*. This solution kills microorganisms and is effective in sterilizing surgical instruments. It is also used to embalm cadavers. Formaldehyde is a key industrial chemical in the production of plastics such as Bakelite, the first commercial polymer, and Formica.

Acetone: $\mathrm{CH_3{-}\overset{\overset{\displaystyle O}{\|}}{C}{-}CH_3}$

Judging from the quantity used, acetone is by far the most important of the ketones. Over 1 billion pounds are used annually in the United States. It is particularly useful as a solvent because it dissolves most organic compounds

vitamin international unit (IU) A measure of vitamin activity, determined by biological methods.

CHEMISTRY AND YOUR HEALTH 4.1 Vitamin A and Birth Defects

Three active forms of vitamin A are known. One is the aldehyde called retinal. In the other two forms, the aldehyde group is replaced by an alcohol group (retinol) or a carboxylic acid group (retinoic acid).

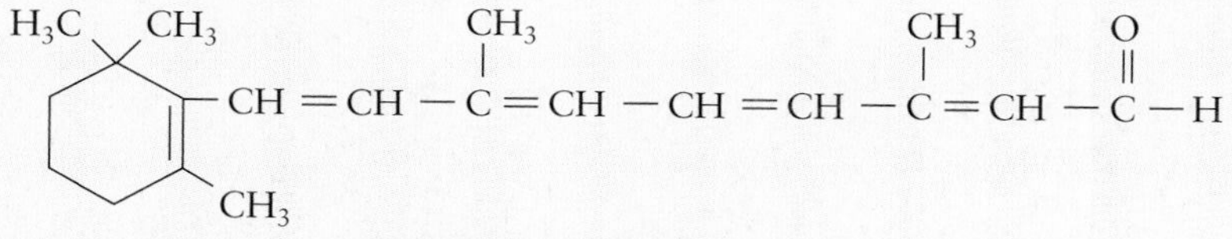

$$\text{(ring with } H_3C, CH_3, CH_3\text{)}{-}\mathrm{CH{=}CH{-}\overset{\overset{\displaystyle CH_3}{|}}{C}{=}CH{-}CH{=}CH{-}\overset{\overset{\displaystyle CH_3}{|}}{C}{=}CH{-}\overset{\overset{\displaystyle O}{\|}}{C}{-}H}$$

retinal

Vitamin A is an essential substance for healthy vision and skin, but as is sometimes true, too much of this good thing creates problems. A relationship between excessive vitamin A intake during pregnancy and birth defects has been recognized for some time. However, the minimum amount of vitamin A required to produce harmful effects is now believed to be much lower than was previously thought. According to a recent study, the consumption of vitamin A at or above 10,000 international units per day was linked to birth defects. A **vitamin international unit (IU)** is a measure of vitamin activity, determined by such biological methods as feeding a compound to vitamin-deprived animals and measuring growth. In the study, an increase in a variety of birth defects was detected in babies born to mothers who took more than 10,000 IU of vitamin A per day. The birth defects found included those of the heart, brain, limbs, kidneys, and genitals.

The active forms of vitamin A that were associated with the defects are found in fortified breakfast cereals, liver, and some vitamin supplements. Beta-carotene, a form of vitamin A that has not been linked to birth defects, is now used in many prenatal vitamin supplements. Beta-carotene is a plant-based vitamin A source that is converted into an active form of vitamin A once ingested.

A maximum dose of 5000 IU of vitamin A during pregnancy is now recommended by the American College of Obstetricians and Gynecologists. In addition, the diet should contain plenty of fruits and vegetables that are naturally rich in beta-carotene (see Chemistry Around Us 2.1).

■ **FIGURE 4.6** Acetone is an excellent solvent for many organics, including nail polish. Why does acetone-based polish remover evaporate fairly quickly when used?

■ **FIGURE 4.7** Vanilla ice cream owes its characteristic flavor to a special aldehyde.

and yet it is miscible in water. It is widely used as a solvent for coatings, which range from nail polish remover to exterior enamel paints (see ■ Figure 4.6).

A number of naturally occurring aldehydes and ketones play important roles within living systems. The carbonyl functional group is fairly common in biological compounds. For example, progesterone and testosterone, female and male sex hormones, respectively, are both ketones.

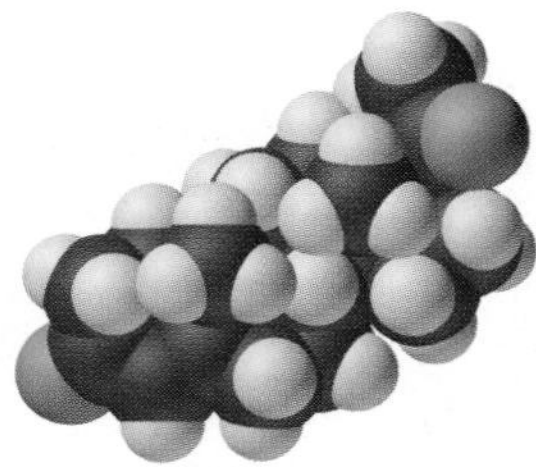

Progesterone

progesterone

testosterone

Several naturally occurring aldehydes and ketones have very fragrant odors and are used in flavorings (see ■ Figure 4.7 and ■ Table 4.3).

Concept Summary

ThomsonNOW Sign in at **www.thomsonedu.com** to:

- Assess your understanding with Exercises keyed to each learning objective.
- Check your readiness for an exam by taking the **Pre-test** and exploring the modules recommended in your **Personalized Learning Plan.**

The Nomenclature of Aldehydes and Ketones. The functional groups characteristic of aldehydes and ketones are very similar; they both contain a carbonyl group:

$$-\overset{\overset{\large O}{\|}}{C}-$$

Aldehydes have a hydrogen attached to the carbonyl carbon,

$$-\overset{\overset{\large O}{\|}}{C}-H$$

whereas ketones have two carbons attached to the carbonyl carbon:

$$-\overset{|}{\underset{|}{C}}-\overset{\overset{\large O}{\|}}{C}-\overset{|}{\underset{|}{C}}-$$

OBJECTIVE 1, Exercise 4.4

The IUPAC ending for aldehyde names is *-al,* whereas that for ketones is *-one.* **OBJECTIVE 2, Exercise 4.6**

Physical Properties. Molecules of aldehydes and ketones cannot form hydrogen bonds with each other. As a result, they have lower boiling points than alcohols of similar molecular weight. Aldehydes and ketones have higher boiling points than alkanes of similar molecular weight because of the presence of the polar carbonyl group. The polarity of the carbonyl group and the fact that aldehydes and ketones can form hydrogen bonds with water explain why the low molecular weight compounds of those kinds are water-soluble. **OBJECTIVE 3, Exercise 4.16**

TABLE 4.3 Some fragrant aldehydes and ketones

Name	Structural formula	Sources and typical uses
Vanillin	4-hydroxy-3-methoxybenzaldehyde: benzene ring bearing $-CHO$ ($C(=O)-H$), OCH_3, and OH	Vanilla bean; flavoring
Cinnamaldehyde	$C_6H_5-CH{=}CH-C(=O)-H$	Oil of cinnamon; flavoring
Citral	$CH_3C(CH_3){=}CHCH_2CH_2C(CH_3){=}CH-C(=O)-H$	Rinds of lemons, limes, and oranges; citrus flavoring
Biacetyl	$CH_3-C(=O)-C(=O)-CH_3$	Butter; flavoring for margarines
Camphor	bicyclic ketone with CH_3, CH_3, CH_3 and $=O$	Camphor tree; characteristic medicinal odor; an ingredient in some inhalants
Menthone	cyclohexanone ring with CH_3 and CH_3CHCH_3 substituents, $=O$	Mint plants; peppermint flavoring

Chemical Properties. Aldehydes and ketones are prepared by the oxidation of primary and secondary alcohols, respectively. Aldehydes can be further oxidized to carboxylic acids, but ketones resist oxidation. Thus, aldehydes are oxidized by Tollens' reagent (Ag^+) and Benedict's solution (Cu^{2+}), whereas ketones are not. A characteristic reaction of both aldehydes and ketones is the addition of hydrogen to the carbonyl double bond to form alcohols. In a reaction that is very important in sugar chemistry, an alcohol can add across the carbonyl group of an aldehyde to produce a hemiacetal. The substitution reaction of a second alcohol molecule with the hemiacetal produces an acetal. Ketones can undergo similar reactions to form hemiketals and ketals. **OBJECTIVE 4, Exercise 4.42**

Important Aldehydes and Ketones. Formaldehyde is a key industrial chemical in the production of plastics such as Bakelite and Formica. Acetone is an important solvent. Some naturally occurring aldehydes and ketones are important in living systems. Some function as sex hormones; others are used as flavorings. **OBJECTIVE 5, Exercise 4.52**

Key Terms and Concepts

Acetal (4.3)
Aldehyde (4.1)
Benedict's reagent (4.3)
Carbonyl group (Introduction)
Hemiacetal (4.3)
Hemiketal (4.3)
Hydrolysis (4.3)
Ketal (4.3)
Ketone (4.1)
Tollens' reagent (4.3)
Vitamin international unit (IU) (4.4)

Key Reactions

1. Oxidation of an aldehyde to give a carboxylic acid (Section 4.3):

$$\mathrm{R-\overset{\overset{\large O}{\|}}{C}-H + (O) \longrightarrow R-\overset{\overset{\large O}{\|}}{C}-OH}$$ Reaction 4.1

2. Attempted oxidation of a ketone (Section 4.3):

$$\mathrm{R-\overset{\overset{\large O}{\|}}{C}-R' + (O) \longrightarrow \text{no reaction}}$$ Reaction 4.2

3. Hydrogenation of an aldehyde to give a primary alcohol (Section 4.3):

$$\mathrm{R-\overset{\overset{\large O}{\|}}{C}-H + H_2 \xrightarrow{Pt} R-\underset{\underset{\large H}{|}}{\overset{\overset{\large OH}{|}}{C}}-H}$$ Reaction 4.7

4. Hydrogenation of a ketone to give a secondary alcohol (Section 4.3):

$$\mathrm{R-\overset{\overset{\large O}{\|}}{C}-R' + H_2 \xrightarrow{Pt} R-\underset{\underset{\large H}{|}}{\overset{\overset{\large OH}{|}}{C}}-R'}$$ Reaction 4.8

5. Addition of an alcohol to an aldehyde to form a hemiacetal and then an acetal (Section 4.3):

$$\mathrm{R-\overset{\overset{\large O}{\|}}{C}-H + R'-OH \rightleftarrows R-\underset{\underset{\large OR'}{|}}{\overset{\overset{\large OH}{|}}{C}}-H \xrightleftharpoons[]{H^+ \text{ and } R'-OH} R-\underset{\underset{\large OR'}{|}}{\overset{\overset{\large OR'}{|}}{C}}-H + H_2O}$$ Reaction 4.11

6. Addition of an alcohol to a ketone to form a hemiketal and a ketal (Section 4.3):

$$\mathrm{R-\overset{\overset{\large O}{\|}}{C}-R' + R''-O-H \rightleftarrows R-\underset{\underset{\large OR''}{|}}{\overset{\overset{\large OH}{|}}{C}}-R' \xrightleftharpoons[]{H^+ \text{ and } R''-OH} R-\underset{\underset{\large OR''}{|}}{\overset{\overset{\large OR''}{|}}{C}}-R' + H_2O}$$ Reaction 4.15

7. Hydrolysis of an acetal to yield an aldehyde and two moles of alcohol (Section 4.3):

$$\mathrm{R-\underset{\underset{\large OR'}{|}}{\overset{\overset{\large OR'}{|}}{C}}-H + H_2O \xrightleftharpoons{H^+} R-\overset{\overset{\large O}{\|}}{C}-H + 2R'-OH}$$ Reaction 4.19

8. Hydrolysis of a ketal to yield a ketone and two moles of alcohol (Section 4.3):

$$\mathrm{R-\underset{\underset{\large OR''}{|}}{\overset{\overset{\large OR''}{|}}{C}}-R' + H_2O \xrightleftharpoons{H^+} R-\overset{\overset{\large O}{\|}}{C}-R' + 2R''-OH}$$ Reaction 4.20

Exercises

SYMBOL KEY

Even-numbered exercises are answered in Appendix B.

Blue-numbered exercises are more challenging.

■ denotes exercises available in ThomsonNow and assignable in OWL.

ThomsonNOW To assess your understanding of this chapter's topics with sample tests and other resources, sign in at www.thomsonedu.com.

THE NOMENCLATURE OF ALDEHYDES AND KETONES (SECTION 4.1)

4.1 What is the structural difference between an aldehyde and a ketone?

4.2 Draw structural formulas for an aldehyde and a ketone that contain the fewest number of carbon atoms possible.

4.3 ■ Identify each of the following compounds as an aldehyde, a ketone, or neither:

a. $CH_3CH_2C(=O)-CH_2CH_3$

b. cyclohexane ring bearing $-C(=O)-OH$

c. six-membered ring lactone (ring O, C=O)

d. cyclobutanone

e. benzene ring $-CH_2C(=O)-H$

f. $NH_2-C(=O)-NH_2$

4.4 Identify each of the following compounds as an aldehyde, a ketone, or neither:

a. $CH_3CH_2-C(=O)-H$

b. $CH_3-C(=O)-O-CH_3$

c. cyclopentanone

d. $CH_3CH_2CH_2-C(=O)-OH$

e. $CH_3CH_2-C(=O)-CH_3$

f. $CH_3CH_2-C(=O)-NH_2$

4.5 ■ Assign IUPAC names to the following aldehydes and ketones:

a. $CH_3-C(=O)-H$

b. $CH_3CH(CH_3)CH_2CH(CH_3)-C(=O)-H$

c. $CH_3CH(CH_3)CH_2-C(=O)-CH_3$

d. $CH_3CH_2CH(C_6H_5)CH_2-C(=O)-H$

e. cyclopropanone with CH_3 on ring

4.6 Assign IUPAC names to the following aldehydes and ketones:

a. $CH_3CH_2CH_2-C(=O)-H$

b. $CH_2(Br)CH_2CH(Br)-C(=O)-H$

c. $CH_3CH(CH_3)CH_2-C(=O)-CH_3$

d. cyclopentanone with Cl on each carbon adjacent to C=O

e. benzene ring $-CH_2-C(=O)-CH(CH_3)-CH_3$

4.7 ■ Draw structural formulas for each of the following compounds:

a. methanal

b. 3-ethyl-2-pentanone

c. 3-methylcyclohexanone

d. 2,4,6-trimethylheptanal

4.8 Draw structural formulas for each of the following compounds:

a. propanal

b. 3-methyl-2-butanone

c. 2,2-dimethylcyclopentanone

d. 3-bromo-4-phenylbutanal

4.9 Draw structural formulas and give IUPAC names for all the isomeric aldehydes and ketones that have the molecular formula $C_5H_{10}O$.

4.10 Draw structural formulas and give IUPAC names for all the isomeric aldehydes and ketones that have the molecular formula C_3H_6O.

4.11 Each of the following names is wrong. Give the structure and correct name for each compound.

a. 3-ethyl-2-butanone

b. 2-ethyl-propanal

c. 4-methyl-5-propyl-3-hexanone

d. 2-methyl-2-phenylethanal

4.12 Each of the following names is wrong. Give the structure and correct name for each compound.

a. 3-ethyl-2-methylbutanal

b. 2-methyl-4-butanone

c. 4,5-dibromocyclopentanone

PHYSICAL PROPERTIES (SECTION 4.2)

4.13 Why does hydrogen bonding *not* take place between molecules of aldehydes or ketones?

4.14 Most of the remaining water in washed laboratory glassware can be removed by rinsing the glassware with acetone (propanone). Explain how this process works (acetone is much more volatile than water).

4.15 The boiling point of propanal is 49°C, whereas the boiling point of propanol is 97°C. Explain this difference in boiling points.

4.16 Explain why propane boils at −42°C, whereas ethanal, which has the same molecular weight, boils at 20°C.

4.17 Use a dotted line to show hydrogen bonding between molecules in each of the following pairs:

a. $CH_3-C(=O)-CH_3$ and $H-O-H$

b. (ring)$-C(=O)-H$ and $H-O-H$

4.18 Use a dotted line to show hydrogen bonding between molecules in each of the following pairs:

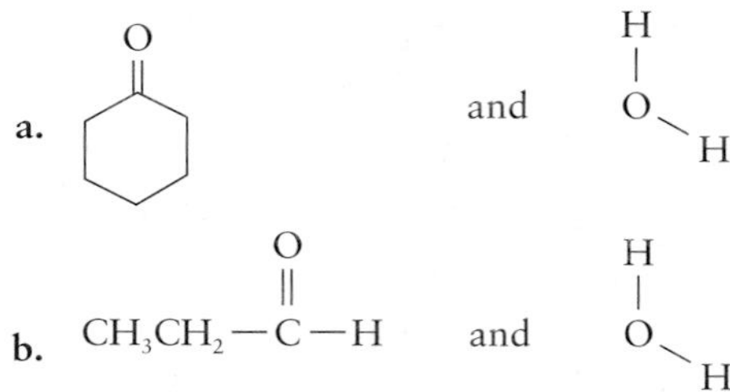

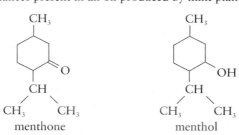

4.19 ■ Arrange the following compounds in order of increasing boiling point:

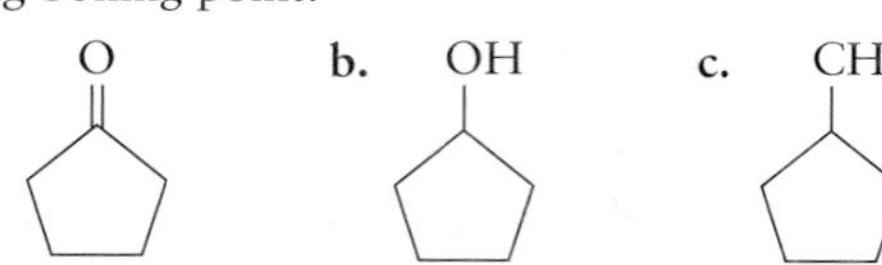

4.20 The compounds menthone and menthol are fragrant substances present in an oil produced by mint plants:

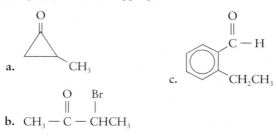

CH_3, O, CH, CH_3 CH_3 — menthone

CH_3, OH, CH, CH_3 CH_3 — menthol

When pure, one of these pleasant-smelling compounds is a liquid at room temperature; the other is a solid. Identify the solid and the liquid and explain your reasoning.

CHEMICAL PROPERTIES (SECTION 4.3)

4.21 Write an equation for the formation of the following compounds from the appropriate alcohol:

a. O, CH_3

b. $CH_3-C(=O)-CHBrCH_3$

c. $C(=O)-H$, CH_2CH_3

4.22 ■ Write an equation for the formation of the following compounds from the appropriate alcohol:

a. $CH_3-C(=O)-CH_2CH(CH_3)CH_3$

b. $C(=O)-H$

c. O, CH_2CH_3

4.23 ■ Identify the following structures as hemiacetals, hemiketals, or neither:

a. $CH_3-O-C(CH_3)_2-OH$

b. $CH_3CH_2CH(OH)-OCH_2CH_3$

c. OH, OCH_3

d. $O-CH(OH)-CH_2CH_3$

4.24 Identify the following structures as hemiacetals, hemiketals, or neither:

a. $CH_3-CH(OH)-OCH_3$

b. $CH_3CH_2C(OCH_3)(CH_3)-OH$

c. $CH_3CH(OCH_3)CH_2-OH$

d. OH, OCH_2CH_3

4.25 ■ Label each of the following as acetals, ketals, or neither:

a. OCH_3, OCH_2CH_3

b. $O-C(CH_3)_2-OCH_3$

c. $OCH_3-CH_2-OCH_2-CH_3$

4.26 Label each of the following as acetals, ketals, or neither:

a. OCH_3, OCH_2CH_3

b. $O-C(OCH_3)(CH_3)-CH_2CH_3$

c. $OCH_3-CH_2-OCH_3$

4.27 Label each of the following structures as a cyclic hemiacetal, hemiketal, acetal, ketal, or none of these:

a. 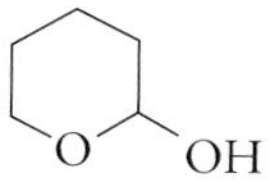 (tetrahydropyran ring with OH on carbon adjacent to O) b. 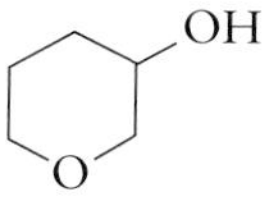 (tetrahydropyran ring with OH) c. 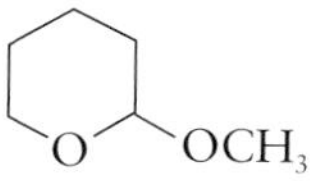 (tetrahydropyran ring with OCH_3 on carbon adjacent to O)

4.28 Label each of the following structures as a hemiacetal, hemiketal, acetal, ketal, or none of these:

a. 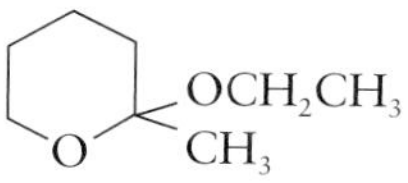 (tetrahydropyran ring with OCH_2CH_3 and CH_3 on carbon adjacent to O)

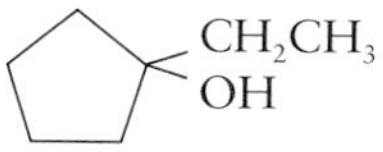

c. 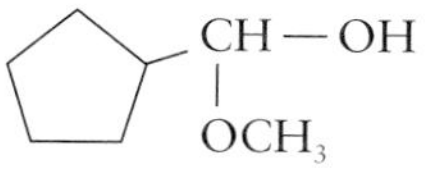 (cyclopentyl)–$\underset{\displaystyle OCH_3}{CH}$–OH

b. (cyclopentane with CH_2CH_3 and OH on the same carbon)

4.29 What two functional groups react to form the following?

a. A hemiacetal c. A ketal

b. An acetal d. A hemiketal

4.30 Hemiacetals are sometimes referred to as potential aldehydes. Explain.

4.31 ■ Complete the following statements:

a. Oxidation of a secondary alcohol produces _____.

b. Oxidation of a primary alcohol produces an aldehyde that can be further oxidized to a _____.

c. Hydrogenation of a ketone produces _____.

d. Hydrogenation of an aldehyde produces _____.

e. Hydrolysis of an acetal produces _____.

f. Hydrolysis of a ketal produces _____.

4.32 What observation characterizes a positive Tollens' test?

4.33 ■ Which of the following compounds will react with Tollens' reagent (Ag^+, NH_3, and H_2O)? For those that do react, draw a structural formula for the organic product.

a. $\overset{OH}{CH_2}-CH_2-CH_2-OCH_3$

b. $H-\overset{O}{\overset{\|}{C}}-CH_2-CH_2-CH_3$

c. 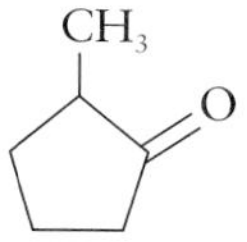 (2-methylcyclopentanone: cyclopentane ring with CH_3 and adjacent C=O)

d. 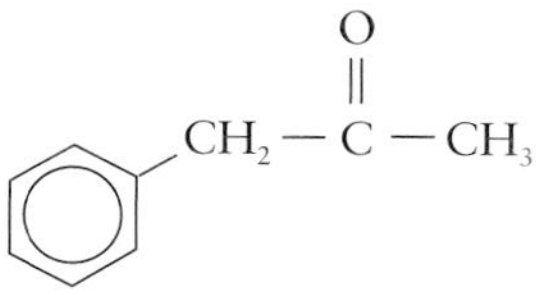 (phenyl)–$CH_2-\overset{O}{\overset{\|}{C}}-CH_3$

e. 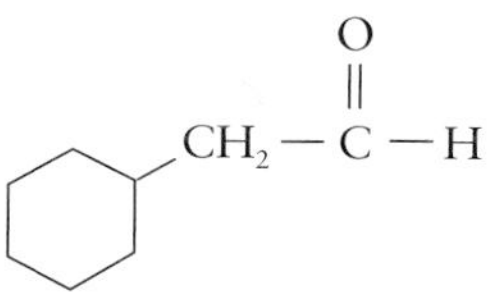 (cyclohexyl)–$CH_2-\overset{O}{\overset{\|}{C}}-H$

4.34 ■ Which of the following compounds will react with Tollens' reagent (Ag^+, NH_3, and H_2O)? For those that do react, draw a structural formula for the organic product.

a. 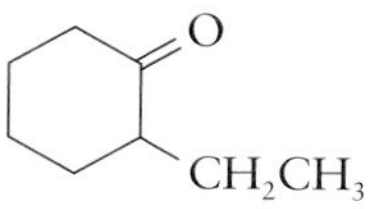 (2-ethylcyclohexanone: cyclohexane ring with =O and adjacent CH_2CH_3)

b. 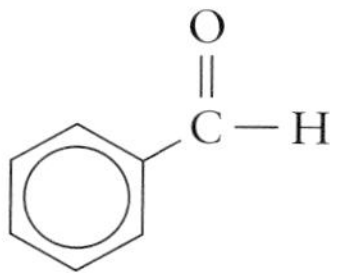 (phenyl)–$\overset{O}{\overset{\|}{C}}-H$

c. $CH_3-O-\overset{OH}{CHCH_3}$

d. $CH_3CH_2-\overset{O}{\overset{\|}{C}}-H$

e. $CH_3-\overset{O}{\overset{\|}{C}}-CH_3$

4.35 What is the formula of the precipitate that forms in a positive Benedict's test?

4.36 ■ Not all aldehydes give a positive Benedict's test. Which of the following aldehydes do?

a. $CH_3CH_2-\overset{O}{\overset{\|}{C}}-H$

b. 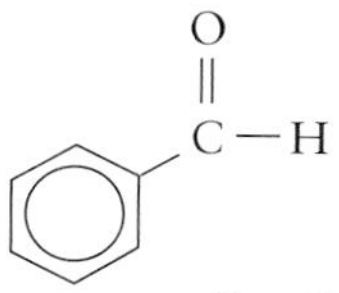 (phenyl)–$\overset{O}{\overset{\|}{C}}-H$

c. (cyclopentyl)–$\overset{O}{\overset{\|}{C}}-\overset{O}{\overset{\|}{C}}-H$

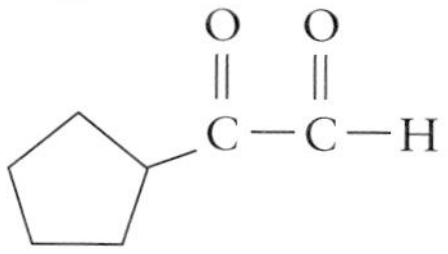

d. $CH_3\overset{OH}{CH}-\overset{O}{\overset{\|}{C}}-H$

e. $CH_3-O-CH_2-\overset{O}{\overset{\|}{C}}-H$

4.37 A stockroom assistant prepares three bottles, each containing one of the following compounds:

$CH_3CH_2-\overset{O}{\overset{\|}{C}}-CH_3$ $CH_3CH_2CH_2\overset{O}{\overset{\|}{C}}H$ $CH_3CH_2\underset{\displaystyle OH}{CH}\overset{O}{\overset{\|}{C}}H$

However, the bottles are left unlabeled. Instead of discarding the samples, another assistant runs tests on the contents, using Tollens' and Benedict's reagents. Match the compound to the bottle, based on these results.

	Tollens' reagent	Benedict's reagent
Bottle A	A dark precipitate forms	A red-orange precipitate forms
Bottle B	No reaction	No reaction
Bottle C	A dark precipitate and slight mirror form	No reaction

4.38 Glucose, the sugar present within the blood, gives a positive Benedict's test. Circle the structural features that enable glucose to react.

$\overset{OH}{CH_2}-\overset{OH}{CH}-\overset{OH}{CH}-\overset{OH}{CH}-\overset{OH}{CH}-\overset{O}{\overset{\|}{C}}-H$

4.39 Fructose, present with glucose in honey, reacts with Benedict's reagent. Circle the structural features that enable fructose to react.

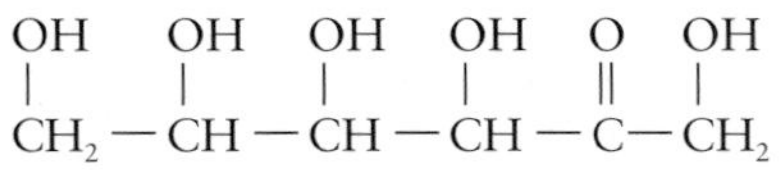

$$\underset{|}{\overset{OH}{}}CH_2-\overset{OH}{\underset{|}{C}}H-\overset{OH}{\underset{|}{C}}H-\overset{OH}{\underset{|}{C}}H-\overset{O}{\overset{\|}{C}}-\overset{OH}{\underset{|}{C}}H_2$$

4.40 Glucose can exist as a cyclic hemiacetal because one of the alcohol groups reacts with the aldehyde:

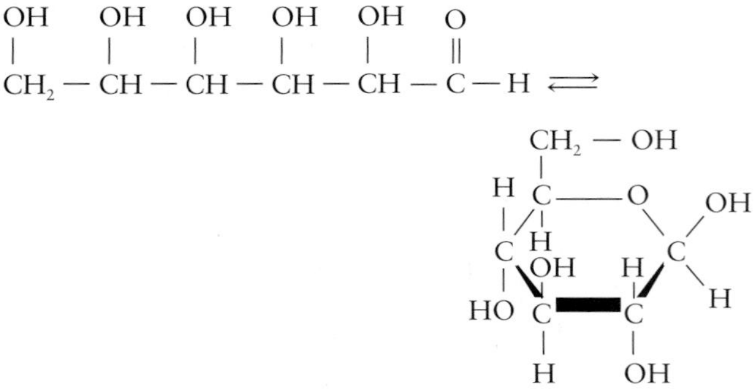

$$\overset{OH}{\overset{|}{CH_2}}-\overset{OH}{\overset{|}{CH}}-\overset{OH}{\overset{|}{CH}}-\overset{OH}{\overset{|}{CH}}-\overset{OH}{\overset{|}{CH}}-\overset{O}{\overset{\|}{C}}-H \rightleftharpoons$$

Draw an arrow to the carbon of the cyclic hemiacetal that was originally part of the aldehyde group. Circle the alcohol group in the open-chain structure that became a part of the cyclic hemiacetal.

4.41 After an intramolecular reaction fructose forms a hemiketal:

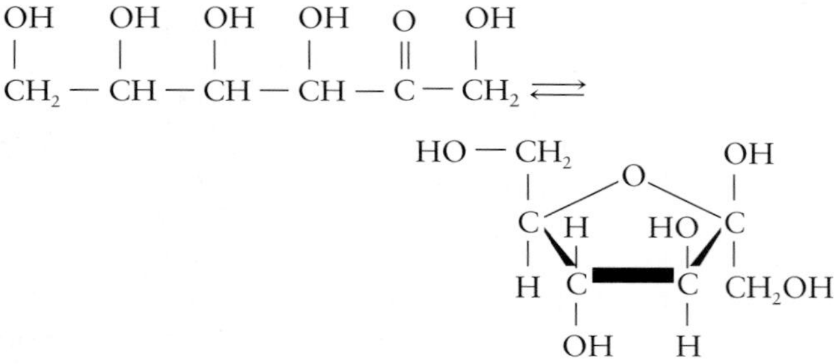

$$\overset{OH}{\overset{|}{CH_2}}-\overset{OH}{\overset{|}{CH}}-\overset{OH}{\overset{|}{CH}}-\overset{OH}{\overset{|}{CH}}-\overset{O}{\overset{\|}{C}}-\overset{OH}{\overset{|}{CH_2}} \rightleftharpoons$$

Draw an arrow to the carbon of the cyclic hemiketal that was originally part of the ketone group. Circle the alcohol group in the open-chain structure that became part of the cyclic hemiketal.

4.42 Complete the following equations. If no reaction occurs, write "no reaction."

a. (dicyclohexyl ketone) $+ H_2 \xrightarrow{Pt}$

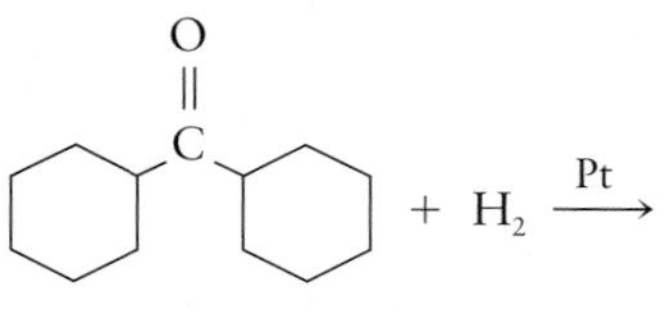

b. $CH_3CH_2-\overset{OH}{\overset{|}{\underset{CH_3}{\underset{|}{C}}}}-OCH_3 + CH_3CH_2CH_2-OH \underset{}{\overset{H^+}{\rightleftharpoons}}$

c. $CH_3CH_2-\overset{O}{\overset{\|}{C}}-H + 2CH_3\overset{OH}{\overset{|}{C}}HCH_3 \overset{H^+}{\rightleftharpoons}$

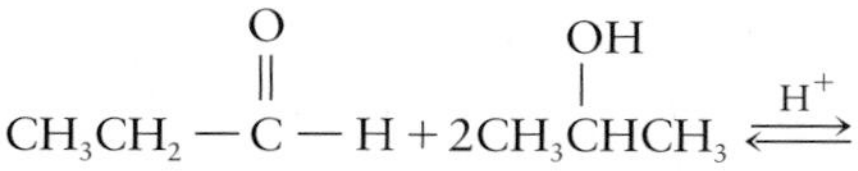

d. (cyclopentyl)$CH_2-\overset{O}{\overset{\|}{C}}-H + (O) \longrightarrow$

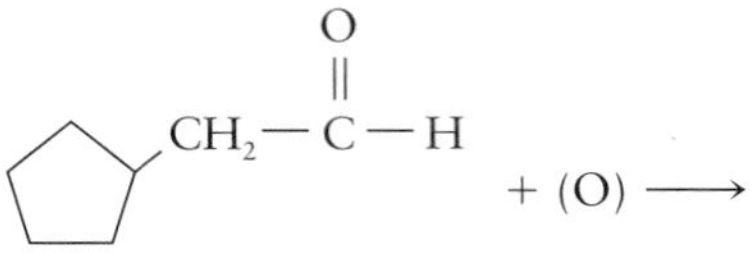

e. $CH_3CH_2\overset{CH_3}{\overset{|}{C}}H-\overset{O}{\overset{\|}{C}}-H + H_2 \xrightarrow{Pt}$

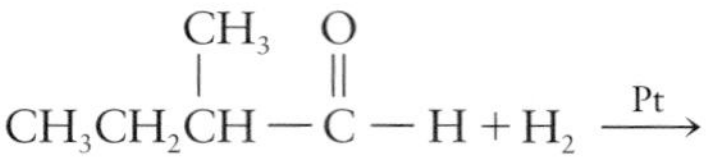

4.43 ■ Complete the following equations. If no reaction occurs, write "no reaction."

a. $CH_3CH_2-\overset{O}{\overset{\|}{C}}-CH_3 + H_2 \xrightarrow{Pt}$

b. $CH_3\underset{CH_3}{\underset{|}{C}}H\overset{Br}{\overset{|}{C}}H-\overset{O}{\overset{\|}{C}}-H + (O) \longrightarrow$

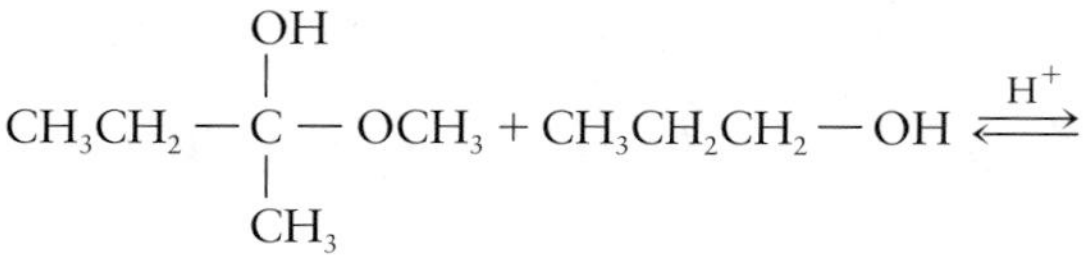

c. (cyclohexyl)$\overset{O}{\overset{\|}{C}}-H + 2CH_3-OH \overset{H^+}{\rightleftharpoons}$

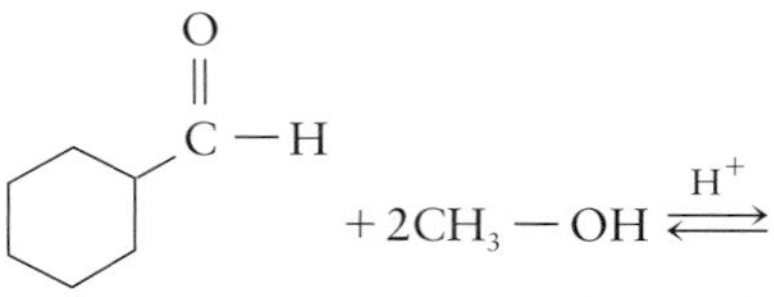

d. (cyclopentanone) $+ (O) \longrightarrow$

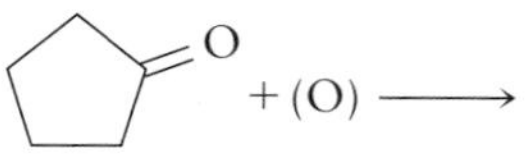

e. (2-hydroxyoxetane, OH) $+ CH_3CH_2-OH \overset{H^+}{\rightleftharpoons}$

4.44 Describe the products that result when hydrogen (H_2) is added to alkenes, alkynes, aldehydes, and ketones. You might review Chapter 12 briefly.

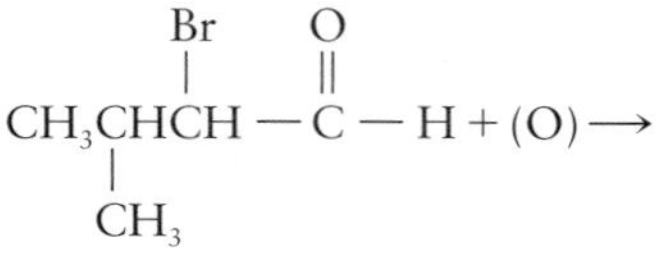

4.45 ■ Draw structural formulas for the products of the following hydrolysis reactions:

a. $CH_3\overset{OCH_2CH_3}{\overset{|}{C}}H-OCH_3 + H_2O \overset{H^+}{\rightleftharpoons}$

b. (cyclopentane with OCH_3, OCH_3) $+ H_2O \overset{H^+}{\rightleftharpoons}$

c. $CH_3-\overset{O-\text{(cyclopentyl)}}{\overset{|}{\underset{CH_3}{\underset{|}{C}}}}-O-\text{(cyclopentyl)} + H_2O \overset{H^+}{\rightleftharpoons}$

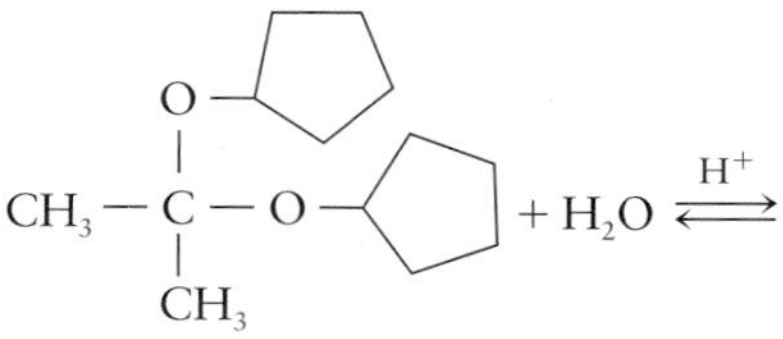

d. $\underset{OCH_2CH_3}{\underset{|}{CH_2}}-OCH_2CH_3 + H_2O \overset{H^+}{\rightleftharpoons}$

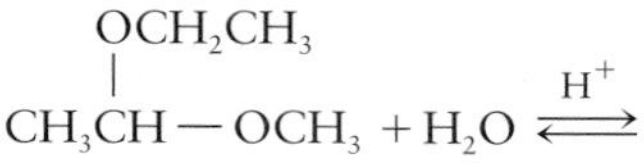

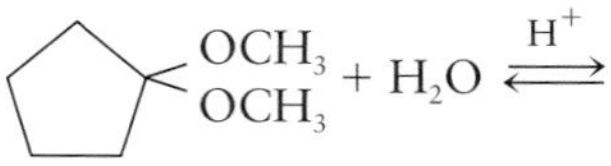

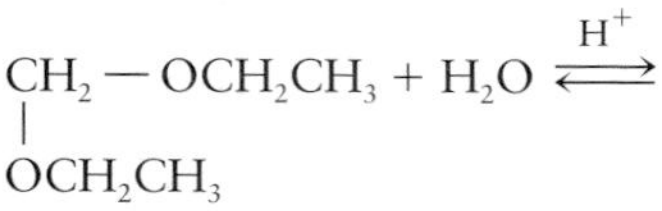

4.46 Draw structual formulas for the products of the following hydrolysis reactions:

a. (cyclohexane with OCH_2CH_3, OCH_2CH_3) $+ H_2O \overset{H^+}{\rightleftharpoons}$

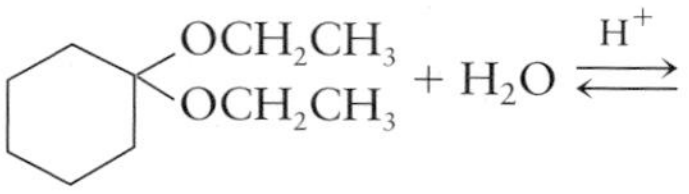

b. $$CH_3CH_2-\underset{CH_2CH_3}{\overset{OCH_3}{C}}-OCH_3 + H_2O \overset{H^+}{\rightleftharpoons}$$

c. $$CH_3CH_2CH_2O-CH_2-OCH_2CH_2CH_3 + H_2O \overset{H^+}{\rightleftharpoons}$$

d. $$CH_3O-\underset{OCH_3}{CH}-CH_2CH_3 + H_2O \overset{H^+}{\rightleftharpoons}$$

4.47 The following compounds are cyclic acetals or ketals. Write structural formulas for the hydrolysis products.

a. 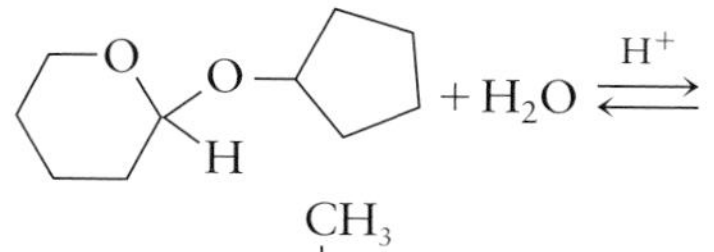 [cyclic structure: tetrahydropyran ring with O, carbon bearing H and O–cyclopentyl] $+ H_2O \overset{H^+}{\rightleftharpoons}$

b. 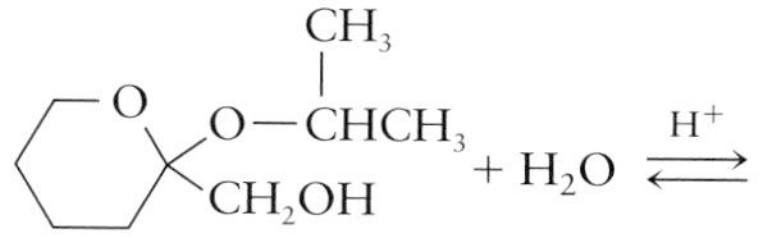 [cyclic structure: tetrahydropyran ring with O, carbon bearing $O-\overset{CH_3}{CHCH_3}$ and CH_2OH] $+ H_2O \overset{H^+}{\rightleftharpoons}$

4.48 The following compounds are cyclic acetals or ketals. Write structural formulas for the hydrolysis products.

a.  [cyclic structure: tetrahydrofuran ring with O, carbon bearing CH_3 and OCH_2CH_3] $+ H_2O \overset{H^+}{\rightleftharpoons}$

b.  [cyclic structure: tetrahydropyran ring with O, carbon bearing $OCH_2CH_2CH_3$ and H] $+ H_2O \overset{H^+}{\rightleftharpoons}$

4.49 Write equations to show how the following conversions can be achieved. More than one reaction is required, and reactions from earlier chapters may be necessary.

a. 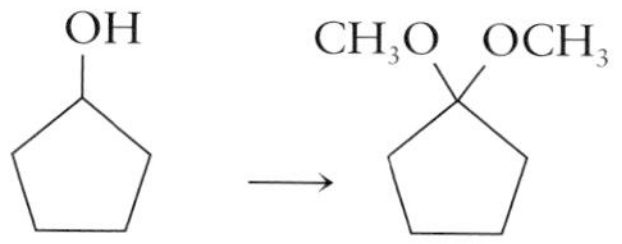 [cyclopentanol (OH)] $\longrightarrow$ [1,1-dimethoxycyclopentane (CH_3O OCH_3)]

b. 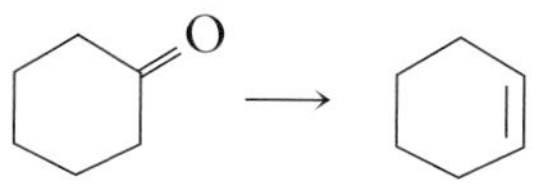 [cyclohexanone (O)] $\longrightarrow$ [cyclohexene]

4.50 Write equations to show how the following conversions can be achieved. More than one reaction is required, and reactions from earlier chapters may be necessary.

a. $$CH_3CH_2\overset{OH}{CH_2} \longrightarrow CH_3CH_2-\underset{OCH_2CH_3}{\overset{OCH_2CH_3}{CH}}$$

b. $$CH_3\overset{CH_3}{CH}-\overset{O}{\overset{\|}{C}}-CH_3 \longrightarrow CH_3\overset{CH_3}{C}=CH-CH_3$$

IMPORTANT ALDEHYDES AND KETONES (SECTION 4.4)

4.51 Identify the most important aldehyde and ketone (on the basis of amount used), and list at least one characteristic for each that contributes to its usefulness.

4.52 Using Table 4.3, name an aldehyde or ketone used in the following ways:

a. Peppermint flavoring
b. Flavoring for margarines
c. Cinnamon flavoring
d. Vanilla flavoring

ADDITIONAL EXERCISES

4.53 Arrange the following compounds starting with the carbon atom that is the most reduced and ending with the carbon that is the least reduced:

a. $R-CH_2-OH$

b. $R-\overset{O}{\overset{\|}{C}}H$

c. $R-CH_3$

4.54 $$\underset{\text{acetaldehyde}}{CH_3-\overset{O}{\overset{\|}{C}}-H} \xrightarrow{(O)} \underset{\text{acetic acid}}{CH_3-\overset{O}{\overset{\|}{C}}-OH}$$

You need to produce 500 g of acetic acid using the reaction above. The actual yield of the reaction is 79.6% of the theoretical yield. How many grams of acetaldehyde do you need to start with to produce 500 g of acetic acid?

4.55 $$CH_3-\overset{O}{\overset{\|}{C}}-CH_3 + H_2O \rightleftharpoons$$

The addition of water to aldehydes and ketones occurs rapidly, although it is not thermodynamically favored. What would be the product for the reaction above? HINT: Think of the self-ionization of water and the polarity of the carbonyl group.

4.56 Draw the product that would result when 3-pentanone reacts with an excess of 1-propanol in the presence of an acid catalyst.

4.57 Formaldehyde levels above 0.10 mg/1000 L of ambient air can cause some humans to experience difficulty breathing. How many molecules of formaldehyde would be found in 1.0 L of air at this concentration?

ALLIED HEALTH EXAM CONNECTION

Reprinted with permission from Nursing School and Allied Health Entrance Exams, COPYRIGHT 2005 Petersons.

4.58 Identify the functional group designated by each of the following:

a. RCHO
b. ROH
c. $R-\underset{\overset{\|}{O}}{C}-R'$

4.59 Which of the following would be classified as a ketone?

a. $$H-\underset{H}{\overset{H}{C}}-\underset{H}{\overset{H}{C}}-\overset{H}{C}=O$$

b.
$$\begin{array}{ccccccccc} & & \text{H} & & \text{O} & & \text{H} & & \text{H} \\ & & | & & \| & & | & & | \\ \text{H} & - & \text{C} & - & \text{C} & - & \text{C} & - & \text{C}-\text{OH} \\ & & | & & & & | & & | \\ & & \text{H} & & & & \text{H} & & \text{H} \end{array}$$

c.
$$\begin{array}{ccccccc} & & \text{H} & & \text{O} & & \text{H} \\ & & | & & \| & & | \\ \text{H} & - & \text{C} & - & \text{C} & - & \text{C}-\text{H} \\ & & | & & & & | \\ & & \text{H} & & & & \text{H} \end{array}$$

CHEMISTRY FOR THOUGHT

4.60 In the IUPAC name for the following ketone, it is not common to use a number for the position of the carbonyl group. Why not?

$$\begin{array}{c} \text{O} \\ \| \\ CH_3CH_2-\text{C}-CH_3 \end{array}$$

4.61 Why can formaldehyde (CH_2O) be prepared in the form of a 37% solution in water, whereas decanal cannot?

4.62 Other addition reactions of aldehydes occur. Water, for example, adds to the carbonyl of trichloroacetaldehyde to form chloral hydrate, a strong hypnotic and sedative known as "knock-out drops" or (when mixed with alcohol) a "Mickey Finn." Complete the reaction by drawing the structural formula of chloral hydrate:

$$\begin{array}{ccccc} & & \text{Cl} & & \text{O} \\ & & | & & \| \\ \text{Cl} & - & \text{C} & - & \text{C}-\text{H}+\text{H}-\text{OH} \longrightarrow \\ & & | & & \\ & & \text{Cl} & & \end{array}$$

trichloroacetaldehyde chloral hydrate

4.63 In the labels of some consumer products, ketone components listed with an *-one* ending can be found. However, very few aldehyde-containing (*-al* ending) products are found. How can you explain this?

4.64 In Figure 4.6, it is noted that acetone is used as a solvent in fingernail polish remover. Why do you think fingernail polish remover evaporates fairly quickly when used?

4.65 Vanilla flavoring is either extracted from a tropical orchid or synthetically produced from wood pulp byproducts. What differences would you expect in these two commercial products: vanilla extract and imitation vanilla extract?

4.66 The structure of fructose in Exercise 4.40 reveals a ketone group and several alcohol groups. If the alcohol groups are referred to as hydroxys, give an IUPAC name for fructose.

4.67 The use of acetone in laboratory experiments must be carefully monitored because of the highly flammable nature of acetone. Give an equation for the combustion of acetone.

4.68 Look ahead to Chapter 7 and locate the structure of maltose in Section 7.7. Draw the structure of maltose and circle the acetal group.

CHAPTER 5

Carboxylic Acids and Esters

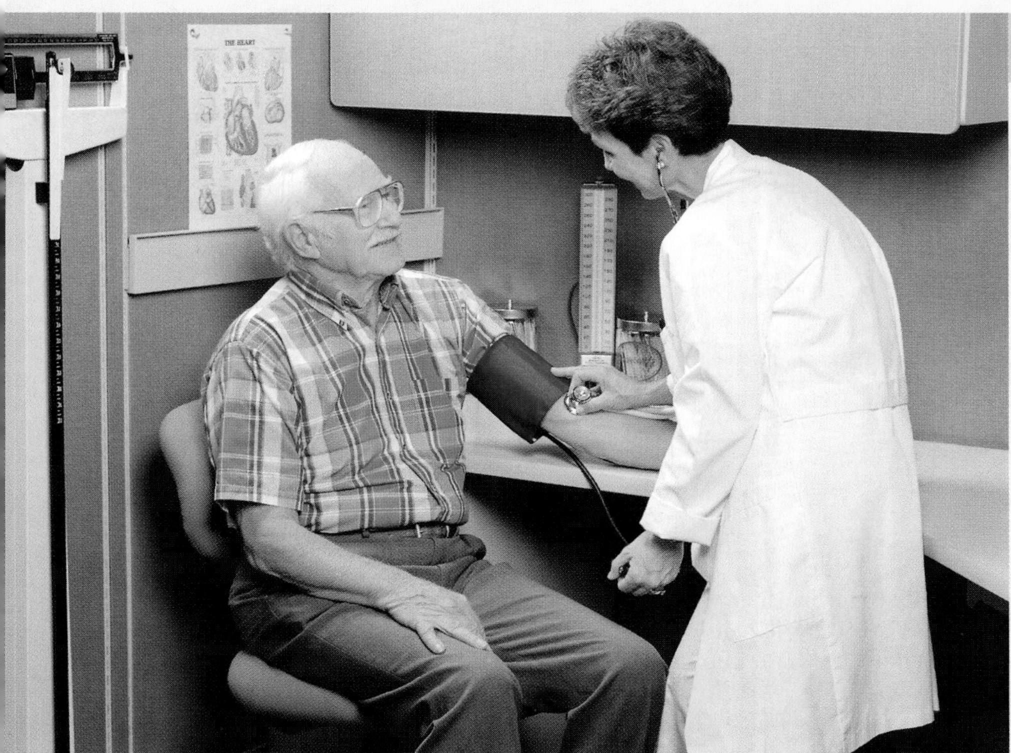

The nurse practitioner is trained to perform many tasks that traditionally were done only by doctors. These tasks include giving physical examinations, carrying out certain routine treatments of patients, suturing wounds, and prescribing medications. The Chemistry and Your Health and the Chemistry Around Us features in this chapter deal with some commonly used medications: aspirin and nitroglycerin.

LEARNING OBJECTIVES

When you have completed your study of this chapter, you should be able to:

1. Assign IUPAC names and draw structural formulas for carboxylic acids. (Section 5.1)
2. Explain how hydrogen bonding affects the physical properties of carboxylic acids. (Section 5.2)
3. Recognize and write key reactions of carboxylic acids. (Section 5.3)
4. Assign common and IUPAC names to carboxylic acid salts. (Section 5.4)
5. Describe uses for carboxylate salts. (Section 5.4)
6. Recognize and write key reactions for ester formation. (Section 5.5)
7. Assign common and IUPAC names to esters. (Section 5.6)
8. Recognize and write key reactions of esters. (Section 5.7)
9. Write reactions for the formation of phosphate esters. (Section 5.8)

ThomsonNOW™ Throughout the chapter this icon introduces resources on the ThomsonNOW website for this text. Sign in at **www.thomsonedu.com** to:

- Evaluate your knowledge of the material
- Take an exam prep quiz
- Identify areas you need to study with a **Personalized Learning Plan.**

carboxylic acid
An organic compound that contains the $-\overset{\overset{\displaystyle O}{\|}}{C}-OH$ functional group.

carboxyl group
The $-\overset{\overset{\displaystyle O}{\|}}{C}-OH$ group.

© Mark Slabaugh

FIGURE 5.1 Citrus fruits and juices are rich in citric acid. What flavor sensation is characteristic of acids?

fatty acid
A long-chain carboxylic acid found in fats.

The tart flavor of foods that taste sour is generally caused by the presence of one or more carboxylic acids. Vinegar contains acetic acid, lemons and other citrus fruits contain citric acid (see ■ Figure 5.1), and the tart taste of apples is caused by malic acid.

The characteristic functional group of **carboxylic acids** is the **carboxyl group**:

$$\underset{\text{carboxyl group}}{-\overset{\overset{\displaystyle O}{\|}}{C}-OH} \qquad \underset{\text{abbreviated forms}}{-COOH \quad \text{or} \quad -CO_2H}$$

Carboxyl-containing compounds are found abundantly in nature (see ■ Figure 5.2). Examples are lactic acid, which is formed in muscle cells during vigorous exercise, and citric acid, which is an important intermediate in cellular energy production, as well as a flavoring agent in citrus fruits (see ■ Table 5.1).

5.1 The Nomenclature of Carboxylic Acids

LEARNING OBJECTIVE

1. Assign IUPAC names and draw structural formulas for carboxylic acids.

Because of their abundance in nature, carboxylic acids were among the first organic compounds studied in detail. No systematic nomenclature system was then available, so the acids were usually named after some familiar source. Acetic acid,

$$\underset{\text{acetic acid}}{CH_3-\overset{\overset{\displaystyle O}{\|}}{C}-OH}$$

the sour constituent in vinegar (see ■ Figure 5.3), is named after *acetum*, the Latin word for vinegar. Butyric acid occurs in rancid butter (the Latin word for butter is *butyrum*). Many of these common names are still widely used today (see Table 5.1). Carboxylic acids with long hydrocarbon chains, generally 12 to 20 carbon atoms, are called **fatty acids** (Chapter 8) because they were first isolated from natural fats. These acids, such as stearic acid with 18 carbons, are known almost exclusively by their common names.

To name a carboxylic acid using the IUPAC system, the longest carbon chain including the carboxyl group is found and numbered. The numbering begins with the carboxyl carbon, so in monocarboxylic acids it is always located at the begin-

FIGURE 5.2 Rhubarb and spinach contain oxalic acid, a compound with two carboxyl groups. The nomenclature of dicarboxylic acids is not discussed here. What would you propose as the IUPAC ending for molecules with two carboxyl groups?

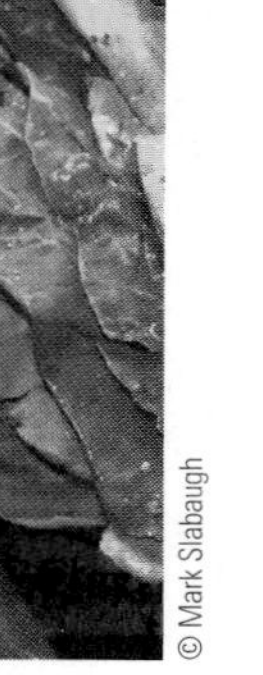

© Mark Slabaugh

© Charles D. Winters

FIGURE 5.3 Acetic acid is the active component in vinegar and a familiar laboratory weak acid.

TABLE 5.1 Examples of carboxylic acids

Common name	IUPAC name	Structural formula	Characteristics and primary uses
Formic acid	methanoic acid	$H-\overset{O}{\overset{\|\|}{C}}-OH$	Stinging agents of certain ants and nettles; used in food preservation
Acetic acid	ethanoic acid	$CH_3-\overset{O}{\overset{\|\|}{C}}-OH$	Active ingredient in vinegar; used in food preservation
Propionic acid	propanoic acid	$CH_3CH_2-\overset{O}{\overset{\|\|}{C}}-OH$	Salts used as mold inhibitors
Butyric acid	butanoic acid	$CH_3(CH_2)_2-\overset{O}{\overset{\|\|}{C}}-OH$	Odor-causing agent in rancid butter
Caproic acid	hexanoic acid	$CH_3(CH_2)_4-\overset{O}{\overset{\|\|}{C}}-OH$	Characteristic odor of Limburger cheese
Oxalic acid	ethanedioic acid	$HO-\overset{O}{\overset{\|\|}{C}}-\overset{O}{\overset{\|\|}{C}}-OH$	Present in leaves of some plants such as rhubarb and spinach; used as a cleaning agent for rust stains on fabric and porcelain
Citric acid	2-hydroxy-1,2,3-propane-tricarboxylic acid	$HO-\overset{O}{\overset{\|\|}{C}}-CH_2-\underset{\underset{\underset{O}{\|\|}}{C-OH}}{\overset{OH}{\overset{\|}{C}}}-CH_2-\overset{O}{\overset{\|\|}{C}}-OH$	Present in citrus fruits; used as a flavoring agent in foods; present in cells
Lactic acid	2-hydroxypropanoic acid	$CH_3\underset{OH}{\underset{\|}{CH}}-\overset{O}{\overset{\|\|}{C}}-OH$	Found in sour milk and sauerkraut; formed in muscles during exercise

ning of the chain, and no number is needed to locate it. The final *-e* of the parent hydrocarbon is dropped, and the ending *-oic* is added, followed by the word *acid*. Groups attached to the chain are named and located as before. Aromatic acids are given names derived from the parent compound, benzoic acid.

EXAMPLE 5.1

Use IUPAC rules to name the following carboxylic acids:

a. $CH_3CH_2-\overset{O}{\overset{\|\|}{C}}-OH$ b. $CH_3\underset{Br}{\underset{\|}{C}}HCH_2-\overset{O}{\overset{\|\|}{C}}-OH$ c. $\overset{O}{\overset{\|\|}{C}}-OH$ attached to a benzene ring bearing CH_3 (meta)

ThomsonNOW Go to Coached Problems to practice **naming carboxylic acids.**

Solution

a. The carbon chain contains three carbons. Thus, the name is based on the parent hydrocarbon, propane. The name is propanoic acid.

b. The chain contains four carbon atoms, and a bromine is attached to carbon number 3. The name is 3-bromobutanoic acid.
c. The aromatic acids are named as derivatives of benzoic acid. The carboxyl group is at position one, and the acid name is 3-methylbenzoic acid.

Learning Check 5.1 Give the IUPAC name to the following:

a.
```
      O
      ‖
      C — OH
      |
CH3CH2CHCH3
```

b.
```
        O
        ‖
        C — OH
        |
      (benzene ring)
     /      |
   Br       Br
```

5.2 Physical Properties of Carboxylic Acids

ThomsonNOW Go to Coached Problems to explore the **physical properties of carboxylic acids.**

LEARNING OBJECTIVE

2. Explain how hydrogen bonding affects the physical properties of carboxylic acids.

Carboxylic acids with low molecular weights are liquids at room temperature and have characteristically sharp or unpleasant odors (see Table 5.2). Butyric acid, for example, is a component of perspiration and is partially responsible for the odor of locker rooms and unwashed socks. As the molecular weight of carboxylic acids increases, so does the boiling point. The heavier acids (containing more than ten carbons) are waxlike solids. Stearic acid ($C_{18}H_{36}O_2$), for example, is mixed with paraffin to produce a wax used to make candles.

When boiling points of compounds with similar molecular weights are compared, the carboxylic acids have the highest boiling points of any organic compounds we have studied so far (see Table 5.3 and Figure 5.4).

TABLE 5.2 Physical properties of some carboxylic acids

Common name	Structural formula	Boiling point (°C)	Melting point (°C)	Solubility (g/100 mL H_2O)
Formic acid	H—COOH	101	8	Infinite
Acetic acid	CH_3—COOH	118	17	Infinite
Propionic acid	CH_3—CH_2—COOH	141	−21	Infinite
Butryic acid	CH_3—$(CH_2)_2$—COOH	164	−5	Infinite
Valeric acid	CH_3—$(CH_2)_3$—COOH	186	−34	5
Caproic acid	CH_3—$(CH_2)_4$—COOH	205	−3	1
Caprylic acid	CH_3—$(CH_2)_6$—COOH	239	17	Insoluble
Capric acid	CH_3—$(CH_2)_8$—COOH	270	32	Insoluble
Lauric acid	CH_3—$(CH_2)_{10}$—COOH	299	44	Insoluble
Myristic acid	CH_3—$(CH_2)_{12}$—COOH	Decomposes	58	Insoluble
Palmitic acid	CH_3—$(CH_2)_{14}$—COOH	Decomposes	63	Insoluble
Stearic acid	CH_3—$(CH_2)_{16}$—COOH	Decomposes	71	Insoluble

TABLE 5.3 Boiling points of compounds with similar molecular weight

Class	Compound	Molecular weight	Boiling point (°C)
Alkane	pentane	72	35
Ether	diethyl ether	74	35
Aldehyde	butanal	72	76
Alcohol	1-butanol	74	118
Carboxylic acid	propanoic acid	74	141

For simple aliphatic compounds, the boiling points usually increase in this order: hydrocarbons and ethers, aldehydes and ketones, alcohols, and carboxylic acids. Carboxylic acids, like alcohols, can form intermolecular hydrogen bonds. It is this bonding that causes high boiling points. Acids have higher boiling points than alcohols with similar molecular weights because hydrogen-bonded acids are held together by two hydrogen bonds rather than the one hydrogen bond that is characteristic of alcohols. A structure where two identical molecules are joined together is called a **dimer:**

dimer
Two identical molecules bonded together.

```
      O····H—O
     //        \
R—C              C—R
     \        //
      O—H····O
```

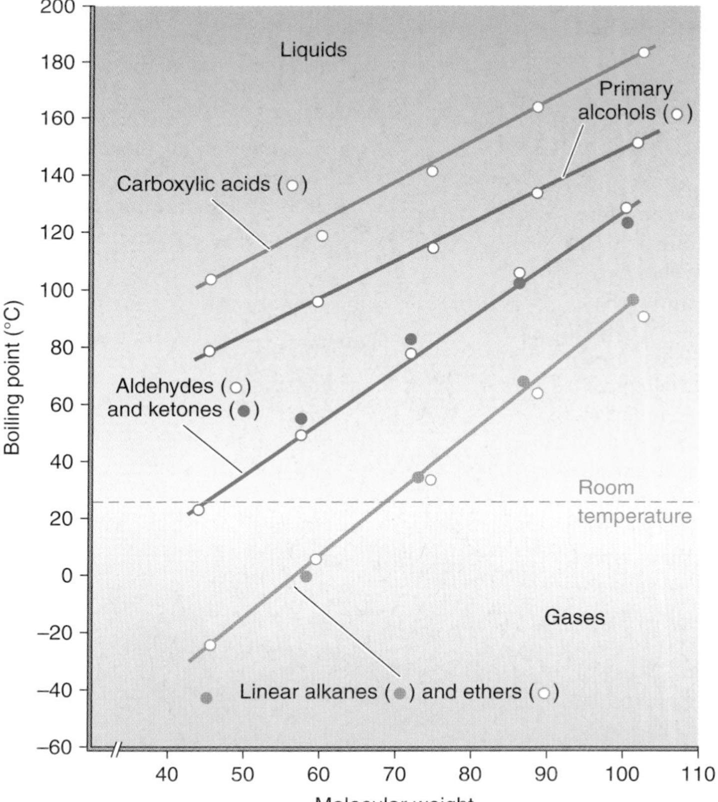

FIGURE 5.4 The boiling points of carboxylic acids are higher than those of all the other organic compounds studied thus far.

```
       O····H—O
       ‖         \H        ⁄H
R—C—O—H····O
                  \H
```

Low molecular weight acids are very soluble in water (Table 5.2) because of hydrogen bonds that form between the carboxyl group and water molecules (intermolecular hydrogen bonds). As the length of the nonpolar hydrocarbon portion (the R group) of the carboxylic acid increases, the water solubility decreases. Carboxylic acids containing eight or more carbon atoms are not appreciably soluble in water and are considered to be insoluble.

Organic compounds may be arranged according to increasing water solubility as follows: hydrocarbons, ethers, aldehydes and ketones, alcohols, and

OVER THE COUNTER 5.1 Alpha Hydroxy Acids in Cosmetics

Alpha hydroxy acids occur in soured milk as lactic acid and in many fruits. Tartaric acid occurs in grapes, malic acid is found in grapes and apples, citric acid is a common component of citrus fruits, and glycolic acid is found in sugar cane and sugar beets.

```
          O
          ‖
CH3CH — C — OH
    |
    OH
```
lactic acid

```
     O    OH    OH    O
     ‖    |     |     ‖
HO — C — CH — CH — C — OH
```
tartaric acid

```
       O
       ‖
CH2 — C — OH
 |
 OH
```
glycolic acid

```
     O    OH           O
     ‖    |            ‖
HO — C — CHCH2 — C — OH
```
malic acid

Recently, alpha hydroxy acids have been advertised as active ingredients in many cosmetics that claim to improve skin texture and tone, unblock or cleanse pores, improve oily skin or acne, and smooth fine lines and surface wrinkles. These cosmetics are most aggressively promoted for their antiwrinkle characteristics. The most commonly found alpha hydroxy acids used in cosmetics are lactic acid and glycolic acid. These weak acids are thought to work by loosening the cells of the epidermis (outer layer of skin) and enhancing cell turnover by accelerating the flaking off of dead skin (exfoliating).

However, even though exfoliation may reveal healthier looking skin, the acid-containing cosmetics that speed up exfoliation may irritate the exposed skin and cause burning and stinging. In addition, new evidence indicates that alpha hydroxy acids can increase the skin's sensitivity to sunlight and especially to the ultraviolet component of sunlight. In one industry-sponsored study, participants' skin was exposed to a 4% glycolic acid product twice daily for 12 weeks. Most of the participants developed minimal skin redness when exposed to 13% less UV radiation than normal, and three developed redness when exposed to 50% less UV light than normal. In another study, individuals who used an alpha hydroxy acid product in the presence of UV radiation experienced twice the cell damage to their skin than others who were treated similarly with a non–alpha hydroxy acid product.

Because of such test results, cosmetics companies have set guidelines suggesting that concentrations of alpha hydroxy acids in skin products should be no more than 10% for over-the-counter use and no more than 30% for professional salon use. The acid content is included on the labels of many products, and manufacturers can also provide the information.

As people age, their skin naturally becomes more wrinkled. Several factors contribute to this result of the aging process. Skin becomes drier, thinner, and less elastic with age, and people with fair skin are more predisposed to these effects than are those with darker skin. Also, smoking and long-term overexposure to sunlight dramatically accelerate the wrinkling process. Thus, one of the best ways to slow down skin wrinkling is to protect the skin from the sun by wearing long-sleeved clothing, a hat, and a sunscreen that has at least an SPF 15 rating (see Chemistry and Your Health 2.1). Because many antiwrinkling cosmetics contain alpha hydroxy acids, a sunscreen should always be worn when these products are used. Adequate protection from the sun is still the best wrinkle-prevention method available.

Cosmetics products with alpha hydroxy acids.

carboxylic acids. Because forces resulting from hydrogen bonding play important roles in determining both boiling points and water solubility, it is not surprising that the same order is generally observed for these two properties.

5.3 The Acidity of Carboxylic Acids

LEARNING OBJECTIVE

3. Recognize and write key reactions of carboxylic acids.

The most important chemical property of carboxylic acids is the acidic behavior implied by their name. The hydrogen attached to the oxygen of the carboxyl group gives carboxylic acids their acidic character. In water, this proton may leave the acid, converting it into a **carboxylate ion**. Reaction 5.1 gives the general reaction, and Reaction 5.2 gives a specific example:

carboxylate ion The $R-\overset{\overset{\displaystyle O}{\|}}{C}-O^-$ ion that results from the dissociation of a carboxylic acid.

General reaction:

$$\underset{\text{carboxylic acid}}{R-\overset{\overset{\displaystyle O}{\|}}{C}-O-H} + H_2O \rightleftharpoons \underset{\text{carboxylate ion}}{R-\overset{\overset{\displaystyle O}{\|}}{C}-O^-} + \underset{\text{hydronium ion}}{H_3O^+} \quad (5.1)$$

Generally, carboxylic acids behave as weak acids. A 1-M solution of acetic acid, for example, is only about 0.5% dissociated.

Specific example:

$$\underset{\text{acetic acid}}{CH_3-\overset{\overset{\displaystyle O}{\|}}{C}-O-H} + H_2O \rightleftharpoons \underset{\text{acetate ion}}{CH_3-\overset{\overset{\displaystyle O}{\|}}{C}-O^-} + H_3O^+ \quad (5.2)$$

It is important to realize that Reactions 5.1 and 5.2 are reversible. According to Le Châtelier's principle, the addition of H_3O^+ (low pH) should favor formation of the carboxylic acid, and removal of H_3O^+ by adding base (high pH) should favor formation of the carboxylate ion:

$$\underset{\text{carboxylic acid}}{R-\overset{\overset{\displaystyle O}{\|}}{C}-O-H} + H_2O \underset{\text{acidic conditions (low pH)}}{\overset{\text{basic conditions (high pH)}}{\rightleftharpoons}} \underset{\text{carboxylate ion}}{R-\overset{\overset{\displaystyle O}{\|}}{C}-O^-} + H_3O^+ \quad (5.3)$$

Thus, the pH of a solution determines the form in which a carboxylic acid exists in the solution. At pH 7.4, the normal pH of body fluids, the carboxylate form predominates. For example, citric acid in body fluids is often referred to as citrate. This is an important point for later chapters in biochemistry.

ThomsonNOW™ Go to Coached Problems to examine the **acidity of carboxylic acids.**

Learning Check 5.2 Pyruvic acid, an important intermediate in the energy-conversion reactions in living organisms, is usually called pyruvate because it is commonly in the carboxylate form. Draw the structure of pyruvate.

$$\underset{\text{pyruvic acid}}{CH_3-\overset{\overset{\displaystyle O}{\|}}{C}-\overset{\overset{\displaystyle O}{\|}}{C}-OH}$$

Although they are weak, carboxylic acids react readily with strong bases such as sodium hydroxide or potassium hydroxide to form salts. Reaction 5.4 gives the general reaction, and Reactions 5.5 and 5.6 are two specific examples:

ThomsonNOW™ Go to Coached Problems to practice **common reactions of carboxylic acids.**

General reaction:

$$\underset{\text{carboxylic acid}}{R-\overset{\overset{\displaystyle O}{\|}}{C}-O-H} + NaOH \longrightarrow \underset{\text{sodium carboxylate}}{R-\overset{\overset{\displaystyle O}{\|}}{C}-O^-Na^+} + H_2O \tag{5.4}$$

Specific examples:

$$\underset{\text{acetic acid}}{CH_3-\overset{\overset{\displaystyle O}{\|}}{C}-O-H} + NaOH \longrightarrow \underset{\text{sodium acetate}}{CH_3-\overset{\overset{\displaystyle O}{\|}}{C}-O^-Na^+} + H_2O \tag{5.5}$$

$$\underset{\text{benzoic acid}}{C_6H_5-\overset{\overset{\displaystyle O}{\|}}{C}-O-H} + KOH \longrightarrow \underset{\text{potassium benzoate}}{C_6H_5-\overset{\overset{\displaystyle O}{\|}}{C}-O^-K^+} + H_2O \tag{5.6}$$

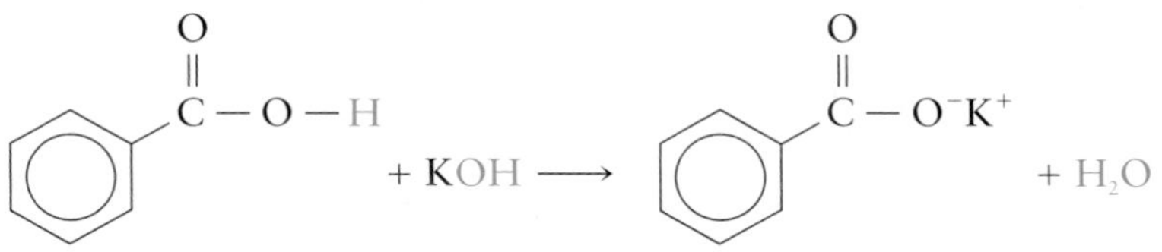

■ **Learning Check 5.3** Write the structural formulas for the products of the following reaction:

$$CH_3CH_2CH_2-\overset{\overset{\displaystyle O}{\|}}{C}-OH + NaOH \longrightarrow$$

5.4 Salts of Carboxylic Acids

LEARNING OBJECTIVES

4. Assign common and IUPAC names to carboxylic acid salts.
5. Describe uses for carboxylic salts.

Nomenclature

Both common names and IUPAC names are assigned to carboxylic acid salts by naming the metal first and changing the *-ic* ending of the acid name to *-ate*.

EXAMPLE 5.2

Give both a common name and an IUPAC name to the following salts:

a. $CH_3-\overset{\overset{\displaystyle O}{\|}}{C}-O^-Na^+$ b. 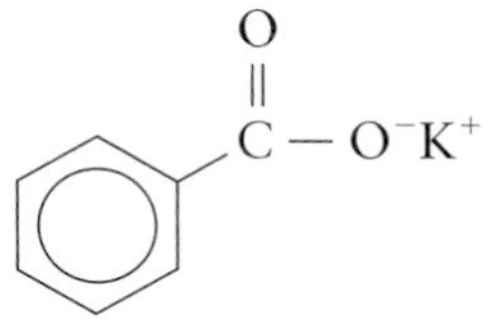 $C_6H_5-\overset{\overset{\displaystyle O}{\|}}{C}-O^-K^+$

Solution

a. This salt is formed from $CH_3-\overset{\overset{\displaystyle O}{\|}}{C}-OH$, which is called acetic acid (common name) or ethanoic acid (IUPAC name). Thus, the salt is sodium acetate (common name) or sodium ethanoate (IUPAC name).

b. This compound is the potassium salt of benzoic acid (both the common and IUPAC names). The name is potassium benzoate.

Learning Check 5.4 Give the IUPAC name for the following salts:

a. (2-methylphenyl)$-\overset{\overset{\displaystyle O}{\|}}{C}-O^{-}Li^{+}$ (ring bearing CH_3 ortho)

b. $CH_3CH_2CH_2-\overset{\overset{\displaystyle O}{\|}}{C}-O^{-}Na^{+}$

Carboxylic acid salts are solids at room temperature and are usually soluble in water because they are ionic. Even long-chain acids with an extensive nonpolar hydrocarbon portion can be solubilized by converting them into salts (Reaction 5.7):

$$\underset{\text{stearic acid (insoluble)}}{CH_3(CH_2)_{16}-\overset{\overset{\displaystyle O}{\|}}{C}-O-H} + NaOH \longrightarrow \underset{\text{sodium stearate (soluble)}}{CH_3(CH_2)_{16}-\overset{\overset{\displaystyle O}{\|}}{C}-O^{-}Na^{+}} + H_2O \qquad (5.7)$$

Useful Carboxylic Acid Salts

A number of acid salts are important around the home. Sodium stearate ($CH_3(CH_2)_{16}COO^{-}Na^{+}$) and other sodium and potassium salts of long-chain carboxylic acids are used as soaps (see Figure 5.5).

FIGURE 5.5 Soap products enable us to keep ourselves and our homes clean.

Calcium and sodium propanoate are used commercially as preservatives in bread, cakes, and cheese to prevent the growth of bacteria and molds (see Figure 5.6). The parent acid, CH_3CH_2COOH, occurs naturally in Swiss cheese. The labels for these bakery products often contain the common name propionate, rather than the IUPAC-acceptable name propanoate.

Sodium benzoate, another common food preservative, occurs naturally in many foods, especially cranberries and prunes. It is used in bakery products, ketchup, carbonated beverages, and a host of other foods:

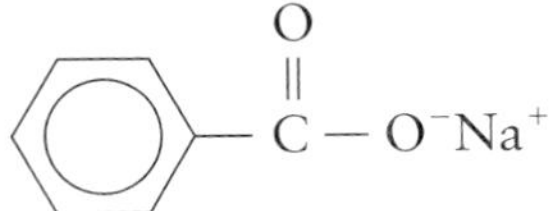

sodium benzoate

Zinc 10-undecylenate, the zinc salt of $CH_2{=}CH(CH_2)_8COOH$, is commonly used to treat athlete's foot. One commercial product that contains zinc 10-undecylenate is Desenex®.

FIGURE 5.6 Propanoates extend the shelf life of bread, preventing the formation of mold.

© Mark Slabaugh

FIGURE 5.7 Esters are partially responsible for the fragrance of oranges, pears, bananas, pineapples, and strawberries.

A mixture of sodium citrate and citric acid

$$\mathrm{HO-\overset{\overset{\displaystyle O}{\|}}{C}-CH_2-\underset{\underset{\underset{\underset{\displaystyle O}{\|}}{C-OH}}{|}}{\overset{\overset{\displaystyle OH}{|}}{C}}-CH_2-\overset{\overset{\displaystyle O}{\|}}{C}-OH}$$

citric acid

Citric acid

is widely used as a buffer to control pH. Products sold as foams or gels such as jelly, ice cream, candy, and whipped cream maintain their desirable characteristics only at certain pHs, which can be controlled by the citrate/citric acid buffer. This same buffer is used in medicines and in human blood used for transfusions. In blood, it also functions as an anticoagulant.

5.5 Carboxylic Esters

LEARNING OBJECTIVE

6. Recognize and write key reactions for ester formation.

carboxylic ester A compound with the $\mathrm{-\overset{\overset{\displaystyle O}{\|}}{C}-OR}$ functional group.

A very important reaction occurs when carboxylic acids are heated with alcohols or phenols in the presence of an acid catalyst. A molecule of water splits out and a **carboxylic ester** is formed:

General reaction:

$$\underset{\text{carboxylic acid}}{\mathrm{R-\overset{\overset{\displaystyle O}{\|}}{C}-OH}} + \underset{\text{alcohol or phenol}}{\mathrm{H-O-R'}} \xrightleftharpoons{\mathrm{H^+,\ heat}} \underset{\text{carboxylic ester}}{\mathrm{R-\overset{\overset{\displaystyle O}{\|}}{C}-O-R'}} + \mathrm{H_2O} \tag{5.8}$$

esterification The process of forming an ester.

ester linkage The carbonyl carbon–oxygen single bond of the ester group.

As before, R′ indicates that the two R groups can be the same or different. The process of ester formation is called **esterification,** and the carbonyl carbon–oxygen single bond of the ester group is called the **ester linkage:**

$$\mathrm{-\overset{\overset{\displaystyle O}{\|}}{C}-O-\overset{|}{\underset{|}{C}}-}$$

ester linkage

ester group

The ester functional group is a key structural feature in fats, oils, and other lipids (Chapter 8). Also widely found in fruits and flowers, many esters are very fragrant and represent some of nature's most pleasant odors (see ■ Figure 5.7). Because of this characteristic, esters are commonly used as flavoring agents in foods and as scents in personal products.

Reactions 5.9 and 5.10 illustrate the formation of two widely used esters:

Specific examples:

$$\underset{\text{butanoic acid}}{\mathrm{CH_3CH_2CH_2-\overset{\overset{\displaystyle O}{\|}}{C}-OH}} + \underset{\text{ethyl alcohol}}{\mathrm{HO-CH_2CH_3}} \xrightleftharpoons{\mathrm{H^+}} \underset{\text{ethyl butanoate (strawberry flavoring)}}{\mathrm{CH_3CH_2CH_2-\overset{\overset{\displaystyle O}{\|}}{C}-O-CH_2CH_3}} + \mathrm{H_2O} \tag{5.9}$$

$$\text{salicylic acid (}C_6H_4(OH)\text{—}\overset{O}{\overset{\|}{C}}\text{—OH)} + HO-CH_3 \xrightleftharpoons{H^+} \text{methyl salicylate (}C_6H_4(OH)\text{—}\overset{O}{\overset{\|}{C}}\text{—O—}CH_3\text{)} + H_2O \quad (5.10)$$

salicylic acid — methyl alcohol — methyl salicylate (oil of wintergreen)

EXAMPLE 5.3

Give the structure of the ester formed in the following reactions:

a. $CH_3-\overset{O}{\overset{\|}{C}}-OH + CH_3-OH \xrightleftharpoons{H^+,\ heat}$

b. $CH_3CH_2CH_2-\overset{O}{\overset{\|}{C}}-OH + C_6H_5-OH \xrightleftharpoons{H^+,\ heat}$

Solution

a. $CH_3-\overset{O}{\overset{\|}{C}}-OH + CH_3OH \xrightleftharpoons{H^+,\ heat} CH_3-\overset{O}{\overset{\|}{C}}-O-CH_3 + H_2O$

b. $CH_3CH_2CH_2-\overset{O}{\overset{\|}{C}}-OH + C_6H_5-OH \xrightleftharpoons{H^+,\ heat} CH_3CH_2CH_2-\overset{O}{\overset{\|}{C}}-O-C_6H_5 + H_2O$

Learning Check 5.5 Give the structure of the products formed in the following reactions:

a. $CH_3CH_2-\overset{O}{\overset{\|}{C}}-OH$ and CH_3-OH

b. $CH_3-\overset{O}{\overset{\|}{C}}-OH$ and C_6H_5-OH

c. $C_6H_5-\overset{O}{\overset{\|}{C}}-OH$ and $CH_3\underset{CH_3}{\underset{|}{CH}}-OH$

Polyesters, which are found in many consumer products, are just what the name implies—polymers made by an esterification reaction. The process used is an example of **condensation polymerization,** in which monomers combine and form a polymer; a small molecule such as water is formed as well. This is in contrast to addition polymerization, in which all atoms of the alkene monomer are incorporated into the polymer (Section 2.4). In many condensation polymerization reactions, each

condensation polymerization The process by which monomers combine together with the simultaneous elimination of a small molecule.

monomer has two functional groups so that the chain can grow from either end. A typical example is the reaction of terephthalic acid and ethylene glycol:

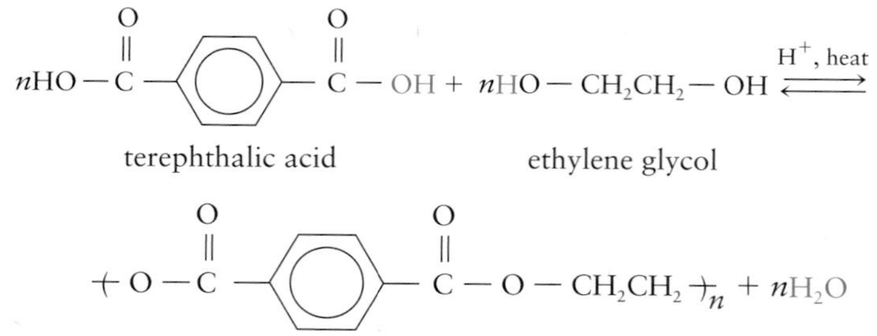

$$n\text{HO}-\overset{\overset{\displaystyle\text{O}}{\|}}{\text{C}}-\text{C}_6\text{H}_4-\overset{\overset{\displaystyle\text{O}}{\|}}{\text{C}}-\text{OH} + n\text{HO}-\text{CH}_2\text{CH}_2-\text{OH} \xrightleftharpoons{\text{H}^+,\ \text{heat}} \quad (5.11)$$

terephthalic acid ethylene glycol

$$\left(\text{O}-\overset{\overset{\displaystyle\text{O}}{\|}}{\text{C}}-\text{C}_6\text{H}_4-\overset{\overset{\displaystyle\text{O}}{\|}}{\text{C}}-\text{O}-\text{CH}_2\text{CH}_2\right)_n + n\text{H}_2\text{O}$$

polyethylene terephthalate (PET)

Molecules of product continue to react with available monomers until, finally, a polyester is formed with the general structure given above, where the n means that the unit in parentheses is repeated many times.

Over 3 billion pounds of the polyester PET are produced annually. Fibers are formed by melting the polymer and forcing the liquid through tiny holes in devices called spinnerettes. The resulting fibers are spun into thread or yarn and marketed under the trade names Dacron®, Fortrel®, and Terylene®, depending on the manufacturer. Typical uses include automobile tire cord and permanent-press clothing. In medicine, Dacron thread is used for sutures, and woven fabric is used to replace damaged or diseased sections of blood vessels and the esophagus. Dacron fibers are often blended with cotton to make a fabric that is more comfortable to wear on hot, humid days than those containing 100% polyester and that retains the latter's wrinkle resistance. Besides being forced through spinnerettes, PET melts may be forced through a narrow slit to produce thin sheets or films called Mylar® (see ■ Figure 5.8). This form of polyester is used extensively in the manufacture of magnetic tapes for tape recorders.

carboxylic acid chloride An organic compound that contains the $-\overset{\overset{\displaystyle\text{O}}{\|}}{\text{C}}-\text{Cl}$ functional group.

carboxylic acid anhydride An organic compound that contains the $-\overset{\overset{\displaystyle\text{O}}{\|}}{\text{C}}-\text{O}-\overset{\overset{\displaystyle\text{O}}{\|}}{\text{C}}-$ functional group.

Although many esters can be prepared by the method just outlined in Reaction 5.8, the reversible nature of the reaction does not always allow for good yields of product. Two compounds that are more reactive than carboxylic acids and that produce esters in good yield are **carboxylic acid chlorides** and **carboxylic acid anhydrides:**

$$\text{R}-\overset{\overset{\displaystyle\text{O}}{\|}}{\text{C}}-\text{Cl} \qquad\qquad \text{R}-\overset{\overset{\displaystyle\text{O}}{\|}}{\text{C}}-\text{O}-\overset{\overset{\displaystyle\text{O}}{\|}}{\text{C}}-\text{R}$$

carboxylic acid chloride carboxylic acid anhydride

Courtesy of DuPont deNemours Company

■ **FIGURE 5.8** Mylar, a polyester developed by E. I. DuPont deNemours Co., was used to cover the wings and pilot's compartment of the Gossamer Albatross, the first human-powered aircraft to fly across the English Channel.

These substances react vigorously with alcohols in a nonreversible manner.

General reactions:

$$\underset{\text{acid chloride}}{R-\overset{\overset{\large O}{\|}}{C}-Cl} + \underset{\text{alcohol}}{HO-R'} \longrightarrow \underset{\text{carboxylic ester}}{R-\overset{\overset{\large O}{\|}}{C}-O-R'} + HCl \qquad (5.12)$$

$$\underset{\text{carboxylic acid anhydride}}{R-\overset{\overset{\large O}{\|}}{C}-O-\overset{\overset{\large O}{\|}}{C}-R} + \underset{\text{alcohol}}{HO-R'} \longrightarrow \underset{\text{carboxylic ester}}{R-\overset{\overset{\large O}{\|}}{C}-O-R'} + \underset{\text{carboxylic acid}}{R-\overset{\overset{\large O}{\|}}{C}-OH} \qquad (5.13)$$

Reactions 5.14 and 5.15 illustrate two specific examples. In another example, the laboratory synthesis of aspirin (Reaction 5.16), using acetic anhydride, gives excellent yields of the ester.

ThomsonNOW™ Go to Coached Problems to explore the **reactivity of anhydrides.**

Specific reactions:

$$\underset{\text{acetyl chloride}}{CH_3-\overset{\overset{\large O}{\|}}{C}-Cl} + \underset{\text{ethyl alcohol}}{HO-CH_2CH_3} \longrightarrow \underset{\text{ethyl acetate}}{CH_3-\overset{\overset{\large O}{\|}}{C}-O-CH_2CH_3} + HCl \qquad (5.14)$$

$$\underset{\text{acetic anhydride}}{CH_3-\overset{\overset{\large O}{\|}}{C}-O-\overset{\overset{\large O}{\|}}{C}-CH_3} + \underset{\text{phenol}}{C_6H_5-OH} \longrightarrow \underset{\text{phenyl acetate}}{CH_3-\overset{\overset{\large O}{\|}}{C}-O-C_6H_5} + \underset{\text{acetic acid}}{CH_3-\overset{\overset{\large O}{\|}}{C}-OH} \qquad (5.15)$$

EXAMPLE 5.4

Give the structure of the ester formed in the following reactions:

a. $C_6H_5-\overset{\overset{\large O}{\|}}{C}-Cl + CH_3CH_2-OH \longrightarrow$

b. $CH_3CH_2-\overset{\overset{\large O}{\|}}{C}-O-\overset{\overset{\large O}{\|}}{C}-CH_2CH_3 + CH_3CH_2-OH \longrightarrow$

Solution

a. The products are $C_6H_5-\overset{\overset{\large O}{\|}}{C}-O-CH_2CH_3 + HCl$

b. The products are $CH_3CH_2-\overset{\overset{\large O}{\|}}{C}-O-CH_2CH_3$ $\quad CH_3CH_2-\overset{\overset{\large O}{\|}}{C}-OH$

The first product in each case is the ester.

Learning Check 5.6 Write equations to represent ester formation when the following pairs of compounds are reacted:

a. $C_6H_5-\overset{\overset{\large O}{\|}}{C}-O-\overset{\overset{\large O}{\|}}{C}-C_6H_5$ and CH_3-OH

b. $CH_3CH_2-\overset{\overset{\large O}{\|}}{C}-Cl$ and $\underset{\underset{\large CH_3}{|}}{CH_3CH}-OH$

5.6 The Nomenclature of Esters

LEARNING OBJECTIVE

7. Assign common and IUPAC names to esters.

Esters are named by a common system as well as by IUPAC rules; the common names are used most often. It is helpful with both IUPAC and common names to think of the esters as being formed from a carboxylic acid, the

$R-\overset{\overset{\large O}{\|}}{C}-$ part and an alcohol, the $-OR'$ part: $\underbrace{R-\overset{\overset{\large O}{\|}}{C}}_{\text{acid part}}-\overset{\text{alcohol part}}{OR'}$

The first word of the name of an ester is the name of the alkyl or aromatic group (R′) contributed by the alcohol. The second word is the carboxylic acid name, with the *-ic acid* ending changed to *-ate*. This is similar to the method used for naming carboxylic acid salts. Thus, an ester of acetic acid becomes an acetate, one of butyric acid becomes a butyrate, one of lactic acid becomes a lactate, and so on.

EXAMPLE 5.5

Give common and IUPAC names for the following esters:

ThomsonNOW™ Go to Coached Problems to practice **naming esters.**

a. $CH_3-\overset{\overset{\large O}{\|}}{C}-O-CH_3$

b. $CH_3CH_2CH_2-\overset{\overset{\large O}{\|}}{C}-O-C_6H_5$

c. $C_6H_5-\overset{\overset{\large O}{\|}}{C}-O-CH_2CH_3$

Solution

a. The first word is methyl, which refers to the alcohol portion:

$$\underbrace{CH_3-\overset{\overset{\displaystyle O}{\|}}{C}}_{\substack{\textit{acetate (common name)}\\ \text{ethanoate (IUPAC name)}}}-O-\overbrace{CH_3}^{\textit{methyl}}$$

The second word is derived from the name of the two-carbon carboxylic acid (acetic acid or ethanoic acid). Thus, we have methyl acetate (common name) and methyl ethanoate (IUPAC name).

b. $$\underbrace{CH_3CH_2CH_2-\overset{\overset{\displaystyle O}{\|}}{C}}_{\substack{\textit{butyrate (common name)}\\ \text{butanoate (IUPAC name)}}}-O-\overbrace{C_6H_5}^{\textit{phenyl}}$$

Thus, the two names are phenyl butyrate (common name) and phenyl butanoate (IUPAC name).

c. $$\underbrace{C_6H_5-\overset{\overset{\displaystyle O}{\|}}{C}}_{\textit{benzoate}}-O-\overbrace{CH_2CH_3}^{\textit{ethyl}}$$

Thus, we have ethyl benzoate (common name and IUPAC name).

Learning Check 5.7 Give both a common and the IUPAC names for each of the following esters:

a. $$H-\overset{\overset{\displaystyle O}{\|}}{C}-O-\underset{\underset{\displaystyle CH_3}{|}}{C}HCH_3$$

b. $$C_6H_5-\overset{\overset{\displaystyle O}{\|}}{C}-OCH_3$$

5.7 Reactions of Esters

LEARNING OBJECTIVE

8. Recognize and write key reactions of esters.

The most important reaction of carboxylic esters, both in commercial processes and in the body, involves the breaking of the ester linkage. The process is called either ester hydrolysis or saponification, depending on the reaction conditions.

CHEMISTRY AND YOUR HEALTH 5.1 Aspirin: Should You Take a Daily Dose?

Aspirin, an ester of salicylic acid first prepared in 1893, is remarkably versatile. In the body it reduces fever but does not reduce normal body temperature. It also relieves a variety of simple pains such as headache, sprain, and toothache, and is an anti-inflammatory agent often used to treat arthritis.

$$\underset{\text{salicylic acid}}{C_6H_4(OH)\overset{\overset{O}{\|}}{C}-OH} + \underset{\text{acetic anhydride}}{CH_3-\overset{\overset{O}{\|}}{C}-O-\overset{\overset{O}{\|}}{C}-CH_3} \longrightarrow \underset{\text{acetylsalicylic acid (aspirin)}}{C_6H_4(O-\underset{\underset{O}{\|}}{C}-CH_3)\overset{\overset{O}{\|}}{C}-OH} + \underset{\text{acetic acid}}{CH_3-\overset{\overset{O}{\|}}{C}-OH} \quad (5.16)$$

Americans consume an estimated 80 billion aspirin tablets each year. The *Physicians' Desk Reference* lists more than 50 OTC drugs in which aspirin is the principal ingredient. Yet, despite the fact that aspirin has been in routine use for nearly a century, scientific journals continue to publish reports about new uses for this old remedy.

In recent years, researchers have discovered one particular effect of aspirin in the body that helps explain its benefits as well as some of its side effects. Aspirin interferes with the clotting action of blood. When blood vessels are injured and bleed, cells in the blood called platelets accumulate at the injury site and clump together to form a plug that seals the opening in the blood vessel. Aspirin reduces the ability of the platelets to clump together, thereby reducing the tendency of blood to clot. In this same way, aspirin may help reduce or prevent clots from forming in arteries that are narrowed by diseases such as atherosclerosis. A heart attack results when a coronary artery that provides blood to the heart is blocked by a blood clot. One study shows that aspirin reduces the chances of having a second heart attack by more than 30%. Other studies suggest that aspirin use helps prevent a first heart attack in healthy individuals, and can also help prevent first or recurrent strokes of the type caused by blood clots that block the blood supply to the brain.

The same process that reduces blood clotting is also responsible for some of the undesirable side effects of using aspirin. It is known that aspirin use causes increased rates of gastrointestinal bleeding and hemorrhagic stroke (stroke caused by a ruptured blood vessel in the brain). Heavy aspirin use before surgery can present a danger because it increases the risk of severe bleeding after the operation.

Recent research reported in medical journals provides evidence of another important medical use of aspirin that is not related to its ability to inhibit blood clotting: Aspirin helps prevent esophageal and colorectal cancers. This indication is very significant because colorectal cancer is the second leading cancer killer after lung cancer. Researchers theorize that the aspirin exerts this effect by blocking the production of hormonelike prostaglandins, which in turn inhibits the growth of tumors. Such research may ultimately lead to the use of aspirin therapy to prevent colorectal cancer, but no official health agency has yet made such a recommendation.

Most doctors recognize that the majority of their patients with coronary artery disease benefit from daily aspirin therapy. When such therapy is recommended, a low dose of 81 mg per day (one baby aspirin) is generally used. Because the long-term, regular use of aspirin poses some risks as well as some benefits, the decision to begin such a program should be made only after consultation with a physician.

© Maren Slabaugh

A variety of aspirin products is available.

Ester Hydrolysis

Ester hydrolysis is the reaction of an ester with water to break an ester linkage and produce an alcohol and a carboxylic acid. This process is simply the reverse of the esterification reaction (Reaction 5.8). Strong acids are frequently used as catalysts to hydrolyze esters:

General reaction:

$$\underset{\text{ester}}{R-\overset{\overset{\displaystyle O}{\|}}{C}-OR'} + H-OH \underset{}{\overset{H^+}{\rightleftharpoons}} \underset{\text{carboxylic acid}}{R-\overset{\overset{\displaystyle O}{\|}}{C}-OH} + \underset{\text{alcohol or phenol}}{R'-OH} \qquad (5.17)$$

Specific example:

$$\underset{\text{ethyl acetate}}{CH_3-\overset{\overset{\displaystyle O}{\|}}{C}-O-CH_2CH_3} + H-OH \overset{H^+}{\rightleftharpoons} \underset{\text{acetic acid}}{CH_3-\overset{\overset{\displaystyle O}{\|}}{C}-OH} + \underset{\text{ethyl alcohol}}{CH_3CH_2-OH} \qquad (5.18)$$

ThomsonNOW™ Go to Coached Problems to examine the **reactivity of esters.**

Notice that when ester linkages break during hydrolysis, the elements of water (—H and —OH) are attached to the ester fragments to form the acid and alcohol (Reaction 5.17). The breaking of a bond and the attachment of the elements of water to the fragments are characteristic of all hydrolysis reactions.

The human body and other biological organisms are constantly forming and hydrolyzing a variety of esters. The catalysts used in these cellular processes are very efficient protein catalysts called enzymes (Chapter 10). Animal fats and vegetable oils are esters. Their enzyme-catalyzed hydrolysis is an important process that takes place when they are digested.

Learning Check 5.8 Give the structure of the products formed in the following hydrolysis reaction:

$$C_6H_5-\overset{\overset{\displaystyle O}{\|}}{C}-O-CH_2CH_2CH_3 + H_2O \overset{H^+}{\rightleftharpoons}$$

Saponification

Like ester hydrolysis, **saponification** results in a breaking of the ester linkage and produces an alcohol and a carboxylic acid (see Section 8.4 for more on saponification). However, unlike ester hydrolysis, saponification is done in solutions containing strong bases such as potassium or sodium hydroxide. The carboxylic acid is converted to a salt under these basic conditions.

saponification The basic cleavage of an ester linkage.

General reaction:

$$\underset{\text{ester}}{R-\overset{\overset{\displaystyle O}{\|}}{C}-O-R'} + NaOH \longrightarrow \underset{\text{carboxylic acid salt}}{R-\overset{\overset{\displaystyle O}{\|}}{C}-O^-Na^+} + \underset{\text{alcohol}}{R'-OH} \qquad (5.19)$$

Specific example:

$$\underset{\text{methyl benzoate}}{C_6H_5-\overset{\overset{\displaystyle O}{\|}}{C}-O-CH_3} + NaOH \longrightarrow \underset{\text{sodium benzoate}}{C_6H_5-\overset{\overset{\displaystyle O}{\|}}{C}-O^-Na^+} + \underset{\text{methyl alcohol}}{CH_3-OH} \qquad (5.20)$$

STUDY SKILLS 5.1 A Reaction Map for Carboxylic Acids

We hope you have become familiar with the reaction maps of Chapters 2, 3, and 4 and have found them useful in solving problems. Another reaction map is given below. It includes the key reactions of this chapter and should be useful as you answer questions and prepare for exams.

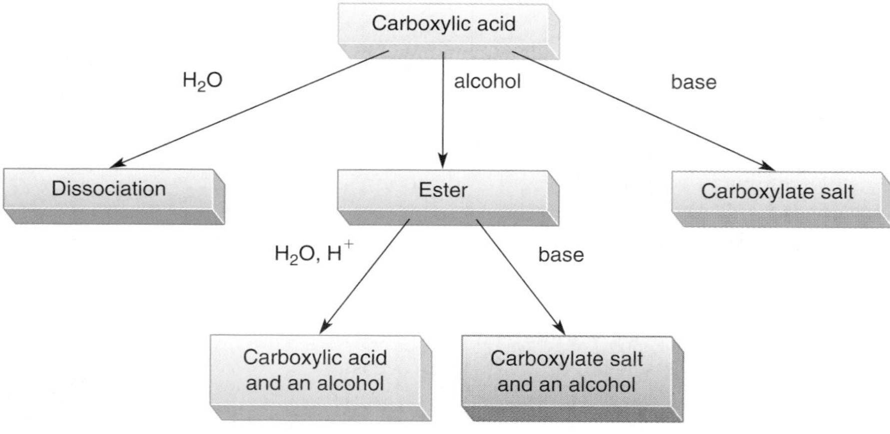

HOW REACTIONS OCCUR 5.1 Ester Saponification

The cleavage of an ester linkage under basic conditions is initiated by the attraction of the carbonyl carbon for hydroxide:

$$CH_3-\overset{\overset{\large O}{\|}}{C}-O-CH_3 \;(+\;:\ddot{O}H^-) \rightleftharpoons CH_3-\underset{\underset{\large OH}{|}}{\overset{\overset{\large :\ddot{O}:^-}{|}}{C}}-O-CH_3$$

As the —OH group bonds to the carbonyl carbon, the double bond must break (two electrons move to the carbonyl oxygen) so that carbon does not possess five bonds.

In the next step of the mechanism, the electrons reestablish the double bond. Once again, carbon will have five bonds if there is not a simultaneous bond breakage. In this case, it is the ester linkage that is cleaved:

$$CH_3-\underset{\underset{\large OH}{|}}{\overset{\overset{\large :\ddot{O}:^-}{|}}{C}}-O-CH_3 \rightleftharpoons \underset{\text{carboxylic acid}}{CH_3-\overset{\overset{\large O}{\|}}{C}-OH} + \underset{\text{strong base}}{^-:\ddot{O}-CH_3}$$

A carboxylic acid is formed momentarily but it quickly donates its proton to the strongly basic anion:

$$CH_3-\overset{\overset{\large O}{\|}}{C}-OH + {}^-:\ddot{O}-CH_3 \longrightarrow \underset{\text{carboxylate product}}{CH_3-\overset{\overset{\large O}{\|}}{C}-O^-} + \underset{\text{alcohol product}}{CH_3-OH}$$

If the source of hydroxide for the reaction was sodium hydroxide, the carboxylate product would be coupled with the positive sodium ion to form the salt.

We will see in Chapter 8 that saponification of fats and oils is important in the production of soaps.

Learning Check 5.9 Give structural formulas for the products of the following saponification reactions.

a. 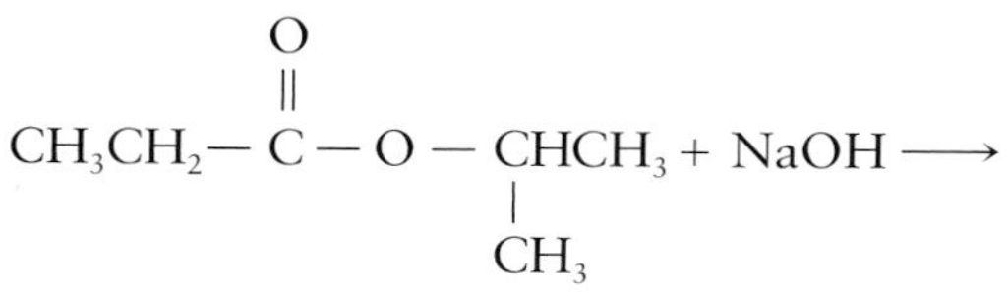 $CH_3CH_2{-}\overset{\overset{\displaystyle O}{\|}}{C}{-}O{-}\underset{\underset{\displaystyle CH_3}{|}}{C}HCH_3 + NaOH \longrightarrow$

b. 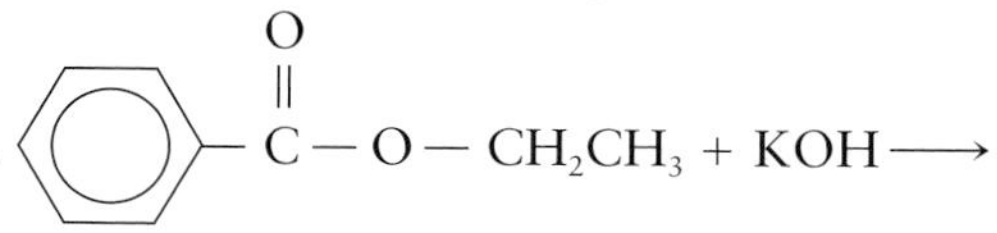 $C_6H_5{-}\overset{\overset{\displaystyle O}{\|}}{C}{-}O{-}CH_2CH_3 + KOH \longrightarrow$ (benzene ring attached to C)

5.8 Esters of Inorganic Acids

LEARNING OBJECTIVE

9. Write reactions for the formation of phosphate esters.

We have defined a carboxylic ester as the product of a reaction between a carboxylic acid and an alcohol. Alcohols can also form **esters** by reacting with inorganic acids such as sulfuric, nitric, and phosphoric acids. The most important of these in biochemistry are the esters of phosphoric acid (H_3PO_4):

ester
A compound in which the —OH of an acid is replaced by an —OR.

General reaction:

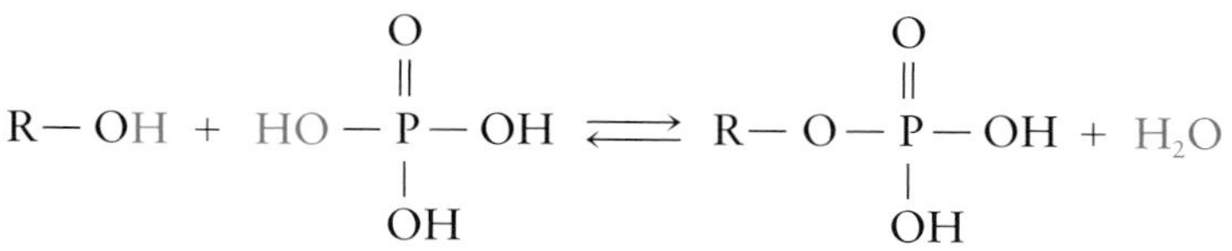

$$\underset{\text{alcohol}}{R{-}OH} + \underset{\text{phosphoric acid}}{HO{-}\overset{\overset{\displaystyle O}{\|}}{\underset{\underset{\displaystyle OH}{|}}{P}}{-}OH} \rightleftharpoons \underset{\text{phosphate ester}}{R{-}O{-}\overset{\overset{\displaystyle O}{\|}}{\underset{\underset{\displaystyle OH}{|}}{P}}{-}OH} + H_2O \qquad (5.21)$$

Specific example:

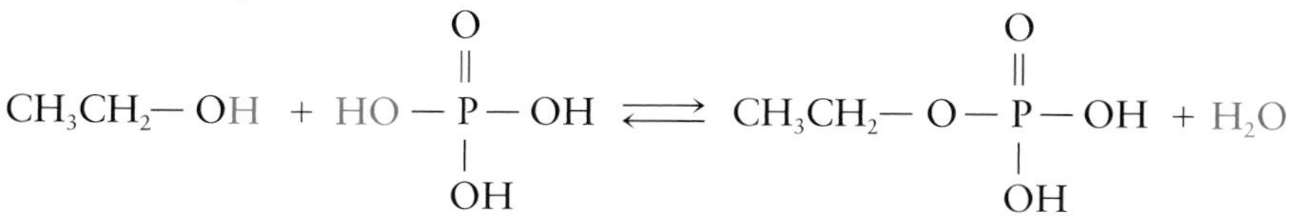

$$CH_3CH_2{-}OH + HO{-}\overset{\overset{\displaystyle O}{\|}}{\underset{\underset{\displaystyle OH}{|}}{P}}{-}OH \rightleftharpoons CH_3CH_2{-}O{-}\overset{\overset{\displaystyle O}{\|}}{\underset{\underset{\displaystyle OH}{|}}{P}}{-}OH + H_2O \qquad (5.22)$$

Because phosphoric acid has three —OH groups, it can form mono-, di-, and triesters:

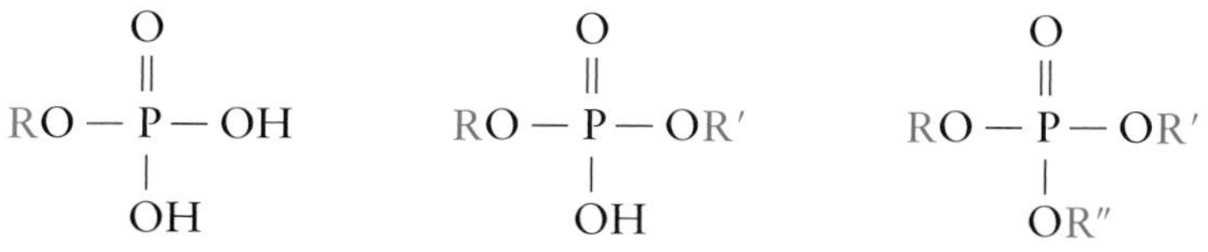

$$\underset{\text{monoester, one R group}}{RO{-}\overset{\overset{\displaystyle O}{\|}}{\underset{\underset{\displaystyle OH}{|}}{P}}{-}OH} \qquad \underset{\text{diester, two R groups}}{RO{-}\overset{\overset{\displaystyle O}{\|}}{\underset{\underset{\displaystyle OH}{|}}{P}}{-}OR'} \qquad \underset{\text{triester, three R groups}}{RO{-}\overset{\overset{\displaystyle O}{\|}}{\underset{\underset{\displaystyle OR''}{|}}{P}}{-}OR'}$$

Mono- and diesters are essential to life and represent some of the most important biological molecules. An example of an important monoester is glucose

CHEMISTRY AROUND US 5.1 Nitroglycerin in Dynamite and in Medicine

Nitroglycerin is a nitrate ester resulting from the reaction of nitric acid and glycerol:

$$\begin{array}{l} CH_2-OH \\ | \\ CH-OH \\ | \\ CH_2-OH \end{array} + 3HO-NO_2 \rightleftarrows \begin{array}{l} CH_2-O-NO_2 \\ | \\ CH-O-NO_2 \\ | \\ CH_2-O-NO_2 \end{array} + 3H_2O$$

glycerol nitric acid nitroglycerin

First made in 1846 by the Italian chemist Sobrero, nitroglycerin was discovered to be a powerful explosive with applications in both war and peace. However, as a shock-sensitive, unstable liquid, nitroglycerin proved to be extremely dangerous to manufacture and handle. The Swedish chemist Alfred Nobel later perfected its synthesis and devised a safe method for handling it. Nobel mixed nitroglycerin with a claylike absorbent material. The resulting solid, called dynamite, is an explosive that is much less sensitive to shock than liquid nitroglycerin. Dynamite is still one of the most important explosives used for mining, digging tunnels, and blasting hills for road building.

Surprising as it may seem, nitroglycerin is also an effective medicine. It is used to treat patients with angina pectoris—sharp chest pains caused by an insufficient supply of oxygen to the heart muscle. Angina pectoris is usually found in patients with coronary artery diseases, such as arteriosclerosis (hardening of the arteries). During overexertion or excitement, the partially clogged coronary arteries prevent the heart from getting an adequate supply of oxygenated blood, and pain results. Nitroglycerin relaxes cardiac muscle and causes a dilation of the arteries, thus increasing blood flow to the heart and relieving the chest pains.

Nitroglycerin (combined with other substances to render it nonexplosive) can be administered in small tablets, which are placed under the tongue during an attack of angina. The nitroglycerin is rapidly absorbed into the bloodstream and finds its way to the heart muscle within seconds. It can also be applied directly to the skin as a cream or absorbed through the skin from a transdermal patch attached to the skin. This last delivery system is a recent development that allows the nitroglycerin to be continuously absorbed for 24 hours.

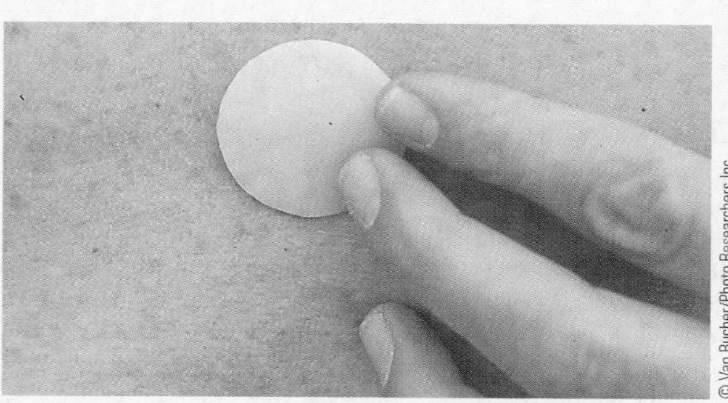

A transdermal patch allows nitroglycerin to be absorbed through the skin.

6-phosphate, which represents the first intermediate compound formed when glucose is oxidized to supply energy for the body (Chapter 13):

ester linkage

$CH_2-O-P(=O)(O^-)-O^-$ (ring with OH, HO, OH, OH)

glucose 6-phosphate

phosphoric anhydride
A compound that contains the

$$-O-\overset{\overset{O}{\|}}{\underset{\underset{O^-}{|}}{P}}-O-\overset{\overset{O}{\|}}{\underset{\underset{O^-}{|}}{P}}-O- \text{ group.}$$

At body pH, the two —OH groups of the phosphoric acid are ionized, and the phosphate group has a charge of 2^-. Examples of important diesters include the phospholipids (Chapter 8) and the nucleic acids (Chapter 11).

As we study key compounds in the storage and transfer of chemical energy in living systems (Chapter 12), we will encounter phosphate esters in which two or three phosphate groups are linked. These **phosphoric anhydrides** are referred to as diphosphates and triphosphates.

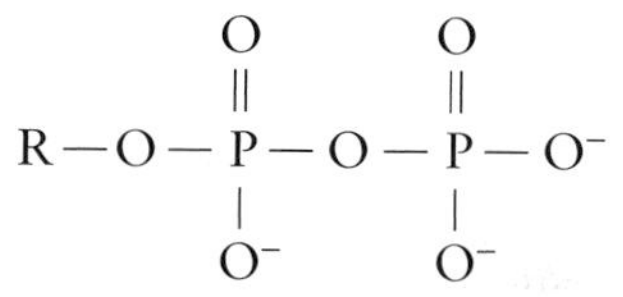

$$R-O-\overset{\overset{O}{\|}}{\underset{\underset{O^-}{|}}{P}}-O-\overset{\overset{O}{\|}}{\underset{\underset{O^-}{|}}{P}}-O^-$$

a diphosphate ester

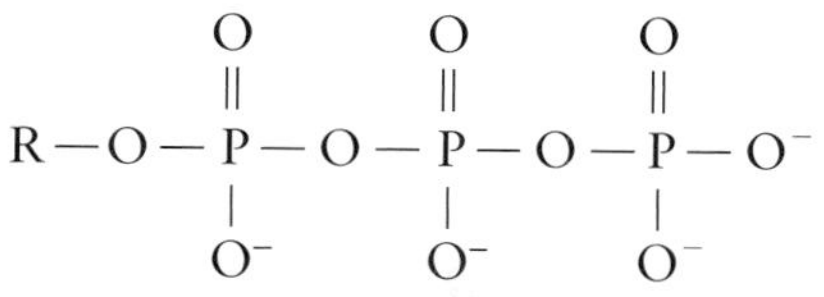

$$R-O-\overset{\overset{O}{\|}}{\underset{\underset{O^-}{|}}{P}}-O-\overset{\overset{O}{\|}}{\underset{\underset{O^-}{|}}{P}}-O-\overset{\overset{O}{\|}}{\underset{\underset{O^-}{|}}{P}}-O^-$$

a triphosphate ester

Adenosine diphosphate (ADP) and adenosine triphosphate (ATP) are the most important of these compounds. We will study their rather complex structures in more detail in Section 12.7. For now, note the ester linkages in each structure and the presence of the di- and triphosphate groups. Again, at the pH of body fluids, the —OH groups attached to the phosphorus lose H^+ and acquire negative charges.

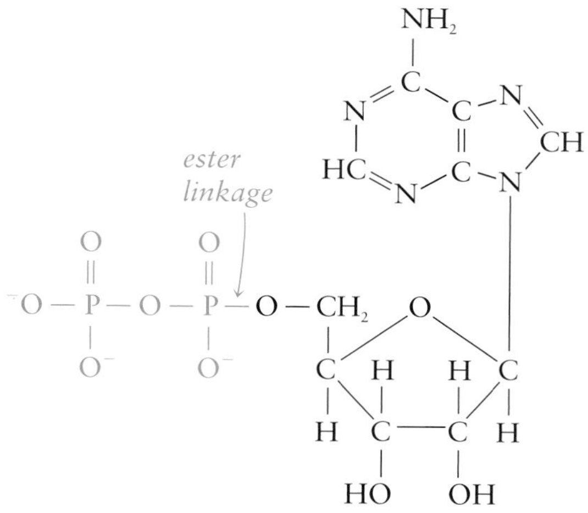

ester linkage

ADP

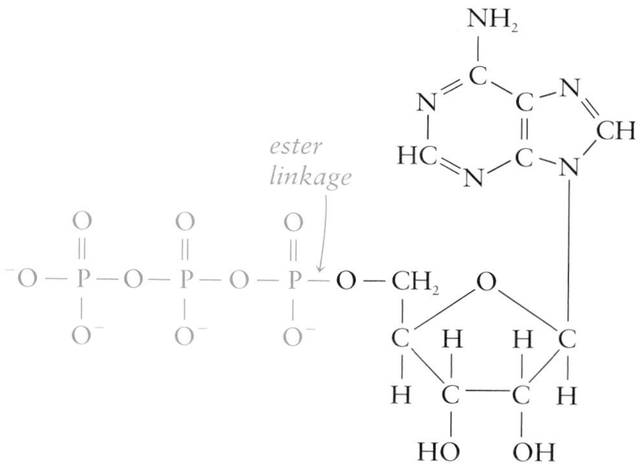

ester linkage

ATP

Concept Summary

ThomsonNOW™ Sign in at **www.thomsonedu.com** to:

- Assess your understanding with Exercises keyed to each learning objective.
- Check your readiness for an exam by taking the **Pre-test** and exploring the modules recommended in your **Personalized Learning Plan.**

The Nomenclature of Carboxylic Acids. The characteristic functional group of carboxylic acids is the carboxyl group

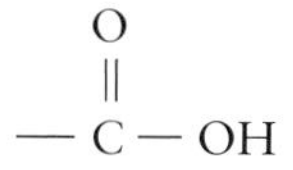

$$-\overset{\overset{O}{\|}}{C}-OH$$

Many of the simpler carboxylic acids are well known by common names. In the IUPAC system, the ending of *-oic acid* is used in the names of these compounds. Aromatic acids are named as derivatives of benzoic acid. **OBJECTIVE 1 (Section 5.1), Exercise 5.6**

Physical Properties of Carboxylic Acids. At room temperature, low molecular weight carboxylic acids are liquids with distinctively sharp or unpleasant odors. High molecular weight long-chain acids are waxlike solids. Carboxylic acids are quite effective in forming dimers, two molecules held together by hydrogen bonds. Thus, they have relatively high boiling points, and those with lower molecular weights are soluble in water. **OBJECTIVE 2 (Section 5.2), Exercise 5.10**

The Acidity of Carboxylic Acids. Soluble carboxylic acids behave as weak acids; they dissociate only slightly in water to form an equilibrium mixture with the carboxylate ion. The equilibrium concentrations of the carboxylic acid and the carboxylate ion depend on pH. At low pH, the acid form predominates, and at pH 7.4 (the pH of cellular fluids) and above, the carboxylate ion predominates. Carboxylic acids react with bases to produce carboxylate salts and water. **OBJECTIVE 3 (Section 5.3), Exercise 5.26**

Salts of Carboxylic Acids. The carboxylate salts are named by changing the *-ic* ending of the acid to *-ate.* **OBJECTIVE 4 (Section 5.4), Exercise 5.28** The ionic nature of the salts makes them water soluble. A number of carboxylate salts are useful as food preservatives, soaps, and medicines. **OBJECTIVE 5 (Section 5.4), Exercise 5.32**

Carboxylic Esters. Carboxylic acids, acid chlorides, and acid anhydrides react with alcohols to produce esters. **OBJECTIVE 6 (Section 5.5), Exercise 5.36** Polyesters result from the reaction of dicarboxylic acids and diols. Polyesters are an example of condensation polymers; these are produced when monomers react to form a polymer plus a small molecule such as water. Many esters are very fragrant and represent some of nature's most pleasant odors. Because of this characteristic, esters are widely used as flavoring agents.

The Nomenclature of Esters. Both common and IUPAC names for esters are formed by first naming the alkyl group of the alcohol portion followed by the name of the acid portion in which the *-ic acid* ending has been changed to *-ate*. **OBJECTIVE 7 (Section 5.6), Exercise 5.46**

Reactions of Esters. Esters can be converted back to carboxylic acids and alcohols under either acidic or basic conditions. Hydrolysis, the reaction with water in the presence of acid, produces the carboxylic acid and alcohol. Saponification occurs in the presence of base to produce the carboxylate salt and alcohol. **OBJECTIVE 8 (Section 5.7), Exercise 5.54**

Esters of Inorganic Acids. Alcohols can also form esters by reacting with inorganic acids such as phosphoric acid. **OBJECTIVE 9 (Section 5.8), Exercise 5.56** Phosphate esters represent some of the most important biological compounds.

Key Terms and Concepts

Carboxyl group (Introduction)
Carboxylate ion (5.3)
Carboxylic acid (Introduction)
Carboxylic acid anhydride (5.5)
Carboxylic acid chloride (5.5)
Carboxylic ester (5.5)
Condensation polymerization (5.5)
Dimer (5.2)
Ester (5.8)
Esterification (5.5)
Ester linkage (5.5)
Fatty acid (5.1)
Phosphoric anhydride (5.8)
Saponification (5.7)

Key Reactions

1. Dissociation of a carboxylic acid to give a carboxylate ion (Section 5.3):

$$R-\overset{\overset{\displaystyle O}{\|}}{C}-O-H + H_2O \rightleftharpoons R-\overset{\overset{\displaystyle O}{\|}}{C}-O^- + H_3O^+$$ Reaction 5.1

2. Reaction of a carboxylic acid with base to produce a carboxylate salt plus water (Section 5.3):

$$R-\overset{\overset{\displaystyle O}{\|}}{C}-O-H + NaOH \longrightarrow R-\overset{\overset{\displaystyle O}{\|}}{C}-O^-Na^+ + H_2O$$ Reaction 5.4

3. Reaction of a carboxylic acid with an alcohol to produce an ester plus water (Section 5.5):

$$R-\overset{\overset{\displaystyle O}{\|}}{C}-OH + H-O-R' \underset{}{\overset{H^+,\text{ heat}}{\rightleftharpoons}} R-\overset{\overset{\displaystyle O}{\|}}{C}-O-R' + H_2O$$ Reaction 5.8

4. Reaction of a carboxylic acid chloride with an alcohol to produce an ester plus hydrogen chloride (Section 5.5):

$$R-\overset{\overset{\displaystyle O}{\|}}{C}-Cl + HO-R' \longrightarrow R-\overset{\overset{\displaystyle O}{\|}}{C}-O-R' + HCl$$ Reaction 5.12

5. Reaction of a carboxylic acid anhydride with an alcohol to produce an ester plus a carboxylic acid (Section 5.5):

$$R-\overset{\overset{\displaystyle O}{\|}}{C}-O-\overset{\overset{\displaystyle O}{\|}}{C}-R + HO-R' \longrightarrow R-\overset{\overset{\displaystyle O}{\|}}{C}-O-R' + R-\overset{\overset{\displaystyle O}{\|}}{C}-OH$$ Reaction 5.13

6. Ester hydrolysis to produce a carboxylic acid and alcohol (Section 5.7):

$$R-\overset{\overset{\displaystyle O}{\|}}{C}-OR' + H-OH \overset{H^+}{\rightleftharpoons} R-\overset{\overset{\displaystyle O}{\|}}{C}-OH + R'-OH$$ Reaction 5.17

7. Ester saponification to give a carboxylate salt and alcohol (Section 5.7):

$$R-\overset{\overset{\displaystyle O}{\|}}{C}-OR' + NaOH \longrightarrow R-\overset{\overset{\displaystyle O}{\|}}{C}-O^{-}Na^{+} + R'-OH$$ Reaction 5.19

8. Phosphate ester formation (Section 5.8):

$$R-OH + HO-\underset{\underset{\displaystyle OH}{|}}{\overset{\overset{\displaystyle O}{\|}}{P}}-OH \rightleftharpoons R-O-\underset{\underset{\displaystyle OH}{|}}{\overset{\overset{\displaystyle O}{\|}}{P}}-OH + H_2O$$ Reaction 5.21

Exercises

SYMBOL KEY

Even-numbered exercises are answered in Appendix B.

Blue-numbered exercises are more challenging.

■ denotes exercises available in ThomsonNow and assignable in OWL.

ThomsonNOW™ To assess your understanding of this chapter's topics with sample tests and other resources, sign in at **www.thomsonedu.com.**

THE NOMENCLATURE OF CARBOXYLIC ACIDS (SECTION 5.1)

5.1 What is the structure of the carboxylic acid functional group? How does it differ from the structure of an alcohol and from that of an aldehyde or a ketone?

5.2 What structural features are characteristic of fatty acids? Why are fatty acids given that name?

5.3 What compound is responsible for the sour or tart taste of Italian salad dressing (vinegar and oil)?

5.4 What carboxylic acid is present in sour milk and sauerkraut?

5.5 ■ Write the correct IUPAC name for each of the following:

a. $CH_3CH_2CH_2-\overset{\overset{\displaystyle O}{\|}}{C}-OH$

b. $CH_3-O-CH_2CH_2-\overset{\overset{\displaystyle O}{\|}}{C}-OH$

c. 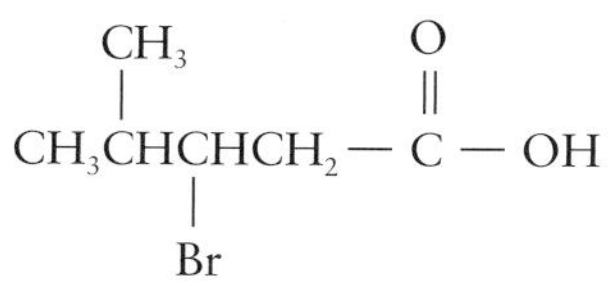 $CH_3\underset{\underset{\displaystyle Br}{|}}{C}H\overset{\overset{\displaystyle CH_3}{|}}{C}HCH_2-\overset{\overset{\displaystyle O}{\|}}{C}-OH$

d. 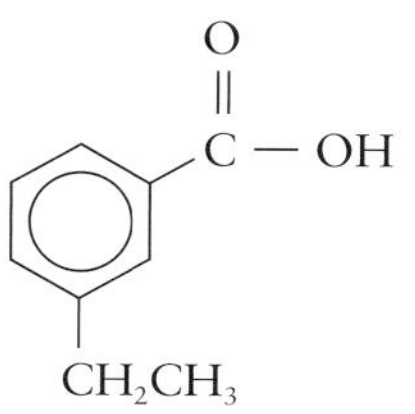 benzene ring bearing $-\overset{\overset{\displaystyle O}{\|}}{C}-OH$ and, meta to it, CH_2CH_3

e. 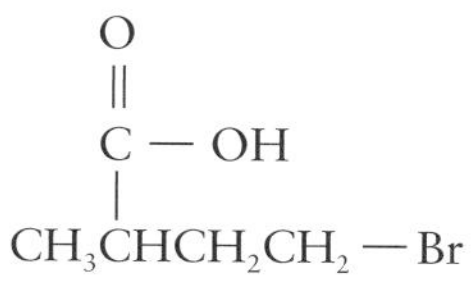 $CH_3\overset{\overset{\displaystyle \overset{\overset{\displaystyle O}{\|}}{C}-OH}{|}}{C}HCH_2CH_2-Br$

5.6 Write the correct IUPAC name for each of the following:

a. 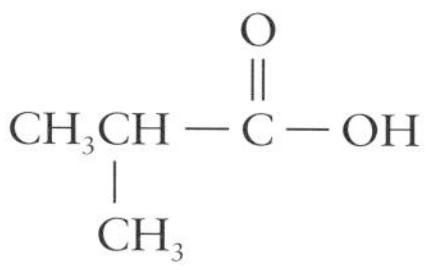 $CH_3\underset{\underset{\displaystyle CH_3}{|}}{C}H-\overset{\overset{\displaystyle O}{\|}}{C}-OH$

b. $Br-CH_2CH_2CH_2-\overset{\overset{\displaystyle O}{\|}}{C}-OH$

c. 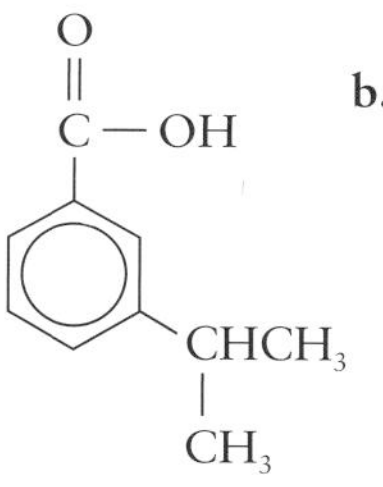 benzene ring bearing $-\overset{\overset{\displaystyle O}{\|}}{C}-OH$ and, ortho to it, $\underset{\underset{\displaystyle CH_3}{|}}{C}HCH_3$

d. 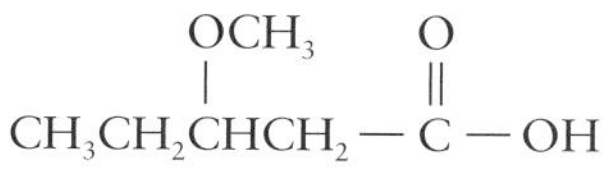 $CH_3CH_2\overset{\overset{\displaystyle OCH_3}{|}}{C}HCH_2-\overset{\overset{\displaystyle O}{\|}}{C}-OH$

e. 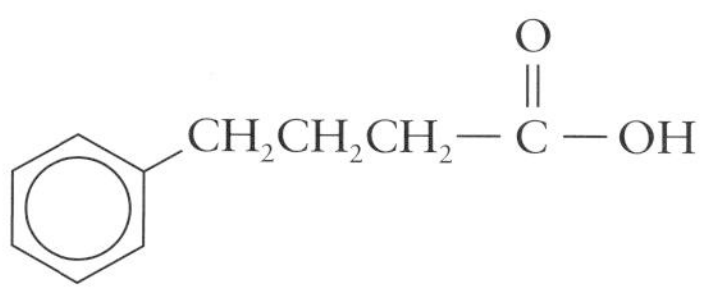 benzene ring bearing $CH_2CH_2CH_2-\overset{\overset{\displaystyle O}{\|}}{C}-OH$

5.7 ■ Write a structural formula for each of the following:

a. hexanoic acid

b. 4-bromo-3-methylpentanoic acid

c. *o*-ethylbenzoic acid

5.8 Write a structural formula for each of the following:

a. pentanoic acid

b. 2-bromo-3-methylhexanoic acid

c. 4-propylbenzoic acid

PHYSICAL PROPERTIES OF CARBOXYLIC ACIDS (SECTION 5.2)

5.9 Of the classes of organic compounds studied so far, which have particularly unpleasant odors?

5.10 ■ Which compound in each of the following pairs would you expect to have the higher boiling point? Explain your answer.

a. acetic acid or 1-propanol

b. propanoic acid or butanone

c. acetic acid or butyric acid

5.11 List the following compounds in order of increasing boiling point:

a. pentanal

b. hexane

c. 1-pentanol

d. butanoic acid

5.12 Draw the structure of the dimer formed when two molecules of propanoic acid hydrogen bond with each other.

5.13 The two isomers propanoic acid and methyl acetate are both liquids. One boils at 141°C, the other at 57°C.

$$CH_3CH_2-\overset{\overset{\displaystyle O}{\|}}{C}-OH \qquad CH_3-\overset{\overset{\displaystyle O}{\|}}{C}-OCH_3$$

propanoic acid — methyl acetate

a. Predict which compound boils at 57°C and which at 141°C. Explain your reasoning.

b. Which of these two compounds would you predict to be the more soluble in water? Explain.

5.14 Caproic acid, a six-carbon acid, has a solubility in water of 1 g/100 mL of water (Table 5.2). Which part of the structure of caproic acid is responsible for its solubility in water, and which part prevents greater solubility?

5.15 Why are acetic acid, sodium acetate, and sodium caprate all soluble in water, whereas capric acid, a 10-carbon fatty acid, is not?

5.16 ■ List the following compounds in order of increasing water solubility:

a. ethoxyethane

b. propanoic acid

c. pentane

d. 1-butanol

5.17 List the following compounds in order of increasing water solubility:

a. hexane

b. 1-pentanol

c. 2-pentanone

d. valeric acid

THE ACIDITY OF CARBOXYLIC ACIDS (SECTION 5.3)

5.18 Draw the structural formula for the carboxylate ion of propanoic acid.

5.19 What is the most important chemical property of carboxylic acids?

5.20 As we discuss the cellular importance of lactic acid in a later chapter, we will refer to this compound as lactate. Explain why.

$$CH_3-\overset{\overset{\displaystyle OH}{|}}{CH}-\overset{\overset{\displaystyle O}{\|}}{C}-OH$$

lactic acid

5.21 Write an equation to illustrate the equilibrium that is present when propanoic acid is dissolved in water. What structure predominates when OH^- is added to raise the pH to 12? What structure predominates as acid is added to lower the pH to 2?

5.22 Draw a structural formula that shows which form of butyric acid would predominate at a pH of 2.

5.23 ■ Complete each of the following reactions:

a. $CH_3(CH_2)_7-\overset{\overset{\displaystyle O}{\|}}{C}-OH + NaOH \longrightarrow$

b. (benzene ring bearing CH_3 and $-\overset{\overset{\displaystyle O}{\|}}{C}-OH$ on adjacent carbons) $+ KOH \longrightarrow$

5.24 Complete each of the following reactions:

a. $CH_3CH_2-\overset{\overset{\displaystyle O}{\|}}{C}-OH + NaOH \longrightarrow$

b. $CH_3\overset{\overset{\displaystyle OH}{|}}{CH}-\overset{\overset{\displaystyle O}{\|}}{C}-OH + KOH \longrightarrow$

5.25 Write a balanced equation for the reaction of propanoic acid with each of the following:

a. KOH

b. H_2O

c. NaOH

5.26 Write a balanced equation for the reaction of acetic acid with each of the following:

a. NaOH b. KOH c. $Ca(OH)_2$

SALTS OF CARBOXYLIC ACIDS (SECTION 5.4)

5.27 ■ Give the IUPAC name for each of the following:

a. $CH_3CH_2-\overset{\overset{\displaystyle O}{\|}}{C}-O^-Na^+$

b. $CH_3\overset{\overset{\displaystyle CH_3}{|}}{CH}-\overset{\overset{\displaystyle O}{\|}}{C}-O^-K^+$

c. 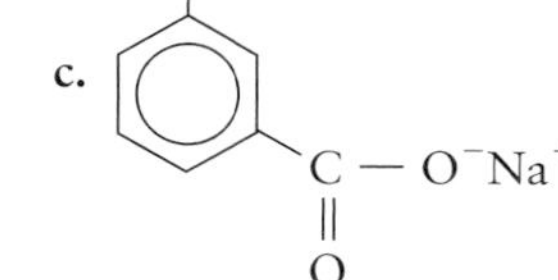

5.28 Give the IUPAC name for each of the following:

a. $CH_3\overset{\displaystyle Br}{\overset{|}{C}}HCH_2-\overset{\displaystyle O}{\overset{\|}{C}}-O^-Na^+$

b. $(H-\overset{\displaystyle O}{\overset{\|}{C}}-O^-)_2Ca^{2+}$

c. ⌬$-O-CH_2CH_2-\overset{\displaystyle O}{\overset{\|}{C}}-O^-K^+$

5.29 Draw structural formulas for the following:

a. sodium methanoate

b. calcium 3-methylbutanoate

c. potassium *p*-propylbenzoate

5.30 Draw structural formulas for the following:

a. potassium ethanoate

b. sodium *m*-methylbenzoate

c. sodium 2-methylbutanoate

5.31 Name each of the following:

a. The sodium salt of valeric acid

b. The magnesium salt of lactic acid

c. The potassium salt of citric acid

5.32 Give the name of a carboxylic acid or carboxylate salt used in each of the following ways:

a. As a soap

b. As a general food preservative used to pickle vegetables

c. As a preservative used in soft drinks

d. As a treatment for athlete's foot

e. As a mold inhibitor used in bread

f. As a food additive noted for its pH buffering ability

CARBOXYLIC ESTERS (SECTION 5.5)

5.33 ■ Which of the following compounds are esters?

a. ⌬$-CH_2-O-CH_2-OH$

b. $CH_3-\overset{\displaystyle O}{\overset{\|}{C}}-O-CH_2CH_3$

c. $CH_3CH_2-\underset{\displaystyle OCH_3}{\underset{|}{CH}}-OCH_3$

d. ⌬$-O-\overset{\displaystyle O}{\overset{\|}{C}}-CH_3$

e. $CH_3-O-CH_2-\overset{\displaystyle O}{\overset{\|}{C}}-CH_2-OH$

f. ⌬$-\overset{\displaystyle O}{\overset{\|}{C}}-O-CH_2CH_3$

5.34 For each ester in Exercise 5.33, draw a circle around the portion that came from the acid and use an arrow to point out the *ester linkage*.

5.35 ■ Complete the following reactions:

a. ⌬$-\overset{\displaystyle O}{\overset{\|}{C}}-OH + CH_3\underset{\displaystyle CH_3}{\underset{|}{C}}H-OH \xrightleftharpoons{H^+,\ heat}$

b. ⌬$-\overset{\displaystyle O}{\overset{\|}{C}}-Cl + CH_3\underset{\displaystyle CH_3}{\underset{|}{C}}H-OH \longrightarrow$

c. ⌬$-\overset{\displaystyle O}{\overset{\|}{C}}-O-\overset{\displaystyle O}{\overset{\|}{C}}-$⌬ $+ CH_3-OH \longrightarrow$

5.36 Complete the following reactions:

a. $CH_3-\overset{\displaystyle O}{\overset{\|}{C}}-O-\overset{\displaystyle O}{\overset{\|}{C}}-CH_3 + CH_3-OH \longrightarrow$

b. $CH_3\underset{\displaystyle CH_3}{\underset{|}{C}}H-\overset{\displaystyle O}{\overset{\|}{C}}-OH +$ ⌬$-CH_2-OH \xrightleftharpoons{H\ ,\ heat}$

c. $CH_3\underset{\displaystyle CH_3}{\underset{|}{C}}H-\overset{\displaystyle O}{\overset{\|}{C}}-Cl +$ ⌬$-OH \longrightarrow$

5.37 Using the alcohol $CH_3CH_2—OH$, show three different starting compounds that may be used to synthesize the ester

$$CH_3CH_2CH_2-\overset{\displaystyle O}{\overset{\|}{C}}-O-CH_2CH_3$$

5.38 ■ Give the structures of the ester that forms when propanoic acid is reacted with the following:

a. methyl alcohol

b. phenol

c. 2-methyl-1-propanol

5.39 The structures of two esters used as artificial flavorings are given below. Write the structure of the acid and the alcohol from which each ester could be synthesized.

a. Pineapple flavoring,

$$CH_3CH_2CH_2-\overset{\overset{\large O}{\|}}{C}-O-CH_2CH_2CH_2CH_3$$

b. Apple flavoring,

$$CH_3CH_2CH_2-\overset{\overset{\large O}{\|}}{C}-O-CH_3$$

5.40 Heroin is formed by reacting morphine with 2 mol of $CH_3-\overset{\overset{\large O}{\|}}{C}-Cl$ to form the diester. Show the structure of heroin.

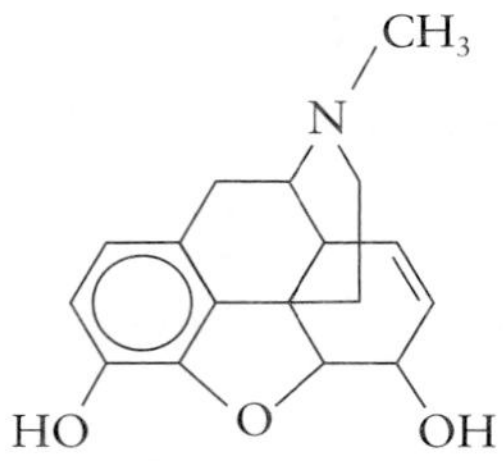

(structure of morphine: HO, O, OH, N–CH_3)

morphine

5.41 How do the acids and alcohols involved in polyester formation differ from those that commonly form simple esters?

5.42 Draw the structure of the repeating monomer unit in the polyester formed in the condensation of oxalic acid and 1,3-propanediol.

$$HO-\overset{\overset{\large O}{\|}}{C}-\overset{\overset{\large O}{\|}}{C}-OH+\overset{\overset{\large OH}{|}}{CH_2}-CH_2-\overset{\overset{\large OH}{|}}{CH_2} \rightleftharpoons$$

THE NOMENCLATURE OF ESTERS (SECTION 5.6)

5.43 ■ Assign common names to the following esters. Refer to Table 5.1 for the common names of the acids.

a. $H-\overset{\overset{\large O}{\|}}{C}-O-CH_2CH_3$

b. $CH_3(CH_2)_4-\overset{\overset{\large O}{\|}}{C}-O-\underset{\underset{\large CH_3}{|}}{CH}CH_3$

5.44 Assign common names to the following esters. Refer to Table 5.1 for the common names of the acids.

a. $CH_3\underset{\underset{\large OH}{|}}{CH}-\overset{\overset{\large O}{\|}}{C}-O-CH_2CH_3$

b. $CH_3CH_2CH_2-\overset{\overset{\large O}{\|}}{C}-O-CH_2CH_2CH_3$

5.45 ■ Give the IUPAC name for each of the following:

a. $CH_3CH_2CH_2-\overset{\overset{\large O}{\|}}{C}-O-CH_2CH_3$

b. 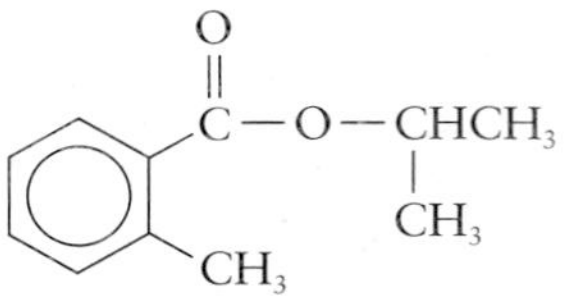 (benzene ring with $-\overset{\overset{\large O}{\|}}{C}-O-\underset{\underset{\large CH_3}{|}}{CH}CH_3$ and ortho CH_3)

5.46 Give the IUPAC name for each of the following:

a. $CH_3\underset{\underset{\large CH_3}{|}}{CH}-\overset{\overset{\large O}{\|}}{C}-OCH_3$

b. 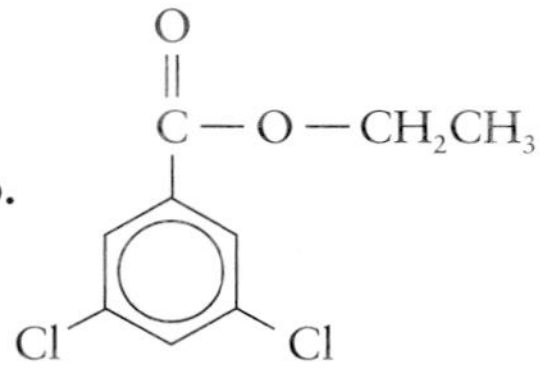 (benzene ring with $-\overset{\overset{\large O}{\|}}{C}-O-CH_2CH_3$ and two Cl at positions 3 and 5)

5.47 ■ Assign IUPAC names to the simple esters produced by a reaction between ethanoic acid and the following:

a. ethanol

b. 1-propanol

c. 1-butanol

5.48 Assign common names to the simple esters produced by a reaction between methanol and the following:

a. propionic acid

b. butyric acid

c. lactic acid

5.49 Draw structural formulas for the following:

a. methyl ethanoate

b. propyl 2-bromobenzoate

c. ethyl 3,4-dimethylpentanoate

5.50 Draw structural formulas for the following:

a. phenyl formate

b. methyl 4-nitrobenzoate

c. ethyl 2-chloropropanoate

REACTIONS OF ESTERS (SECTION 5.7)

5.51 Define and compare the terms *ester hydrolysis* and *saponification.*

5.52 ■ Write equations for the hydrolysis and saponification of ethyl acetate:

$$CH_3-\overset{\overset{\large O}{\|}}{C}-OCH_2CH_3$$

5.53 Complete the following reactions:

a. $CH_3CH_2\underset{\underset{\large CH_3}{|}}{CH}-\overset{\overset{\large O}{\|}}{C}-O-CH_2CH_3+NaOH \longrightarrow$

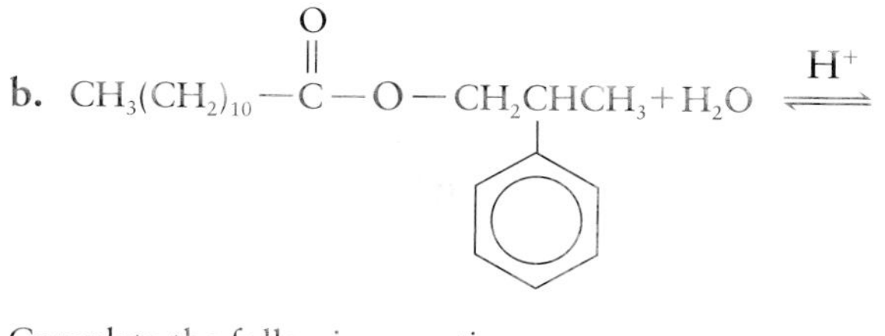

b. $CH_3(CH_2)_{10}-\overset{\overset{\displaystyle O}{\|}}{C}-O-\underset{\underset{\displaystyle C_6H_5}{|}}{CH_2CHCH_3}+H_2O \xrightleftharpoons{H^+}$

5.54 Complete the following reactions:

a. $CH_3(CH_2)_{16}-\overset{\overset{\displaystyle O}{\|}}{C}-O-CH_2CH_3 + NaOH \longrightarrow$

b. $\underset{}{\overset{\overset{\displaystyle CH_3}{|}}{CH_3CH}}-\overset{\overset{\displaystyle O}{\|}}{C}-O-C_6H_5 + H_2O \xrightleftharpoons{H^+}$

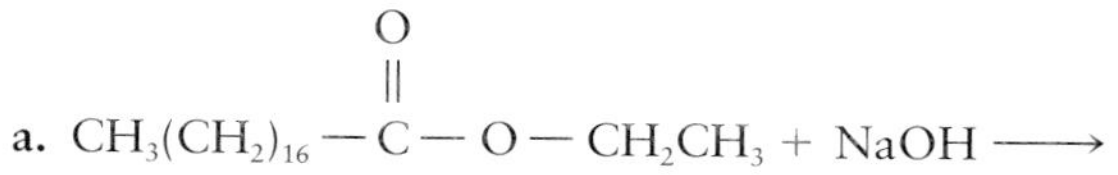

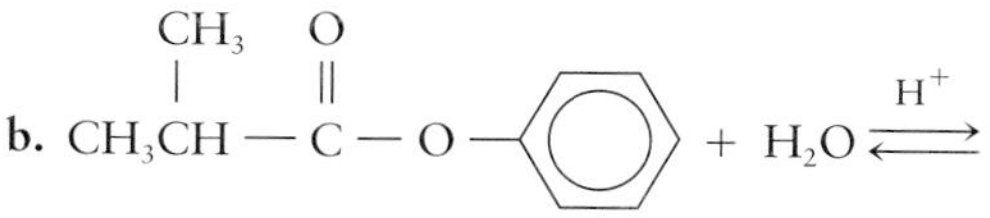

ESTERS OF INORGANIC ACIDS (SECTION 5.8)

5.55 Glyceraldehyde 3-phosphate, an important compound in the cellular oxidation of carbohydrates, is an ester of phosphoric acid and glyceraldehyde.

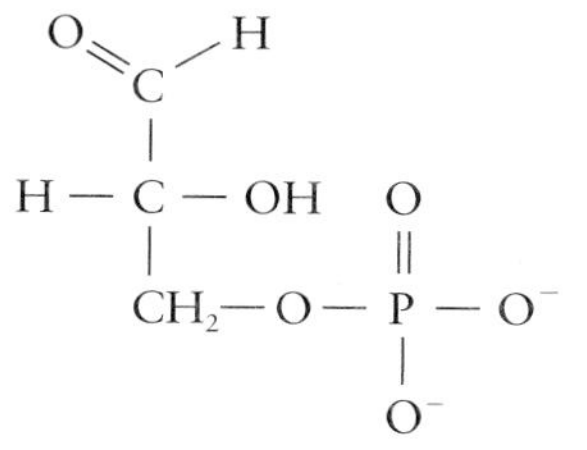

$$\begin{array}{c} O\!=\!C\!-\!H \\ | \\ H-C-OH \quad \\ | \\ CH_2-O-\overset{\overset{\displaystyle O}{\|}}{\underset{\underset{\displaystyle O^-}{|}}{P}}-O^- \end{array}$$

a. Give the structure of glyceraldehyde.

b. Why is glyceraldehyde 3-phosphate shown with a 2^- charge?

5.56 Dihydroxyacetone reacts with phosphoric acid to form the monoester called dihydroxyacetone phosphate. Complete the reaction for its formation.

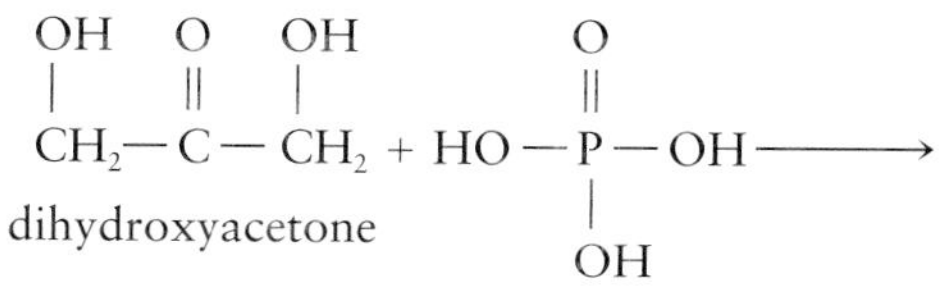

$$\underset{\text{dihydroxyacetone}}{\overset{\overset{\displaystyle OH}{|}}{CH_2}-\overset{\overset{\displaystyle O}{\|}}{C}-\overset{\overset{\displaystyle OH}{|}}{CH_2}} + HO-\overset{\overset{\displaystyle O}{\|}}{\underset{\underset{\displaystyle OH}{|}}{P}}-OH \longrightarrow$$

5.57 ■ Give the structure for the following:

a. monoethyl phosphate (with a 2^- charge)

b. monoethyl diphosphate (with a 3^- charge)

c. monoethyl triphosphate (with a 4^- charge)

ADDITIONAL EXERCISES

5.58 Explain how a sodium bicarbonate ($NaHCO_3$) solution can be used to distinguish between a carboxylic acid solution and an alcohol solution.

5.59 The following reaction requires several steps. Show the reagents you would use and draw structural formulas for intermediate compounds formed in each step.

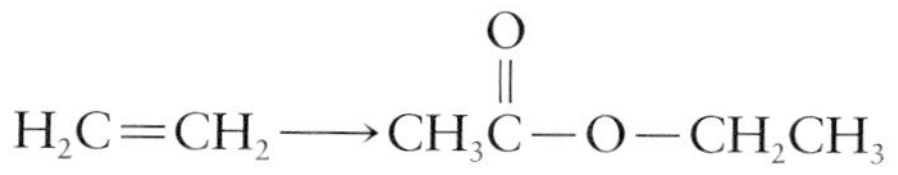

$$H_2C=CH_2 \longrightarrow CH_3\overset{\overset{\displaystyle O}{\|}}{C}-O-CH_2CH_3$$

5.60 Draw the structural formula and give the IUPAC name for the ester that results when 3-ethylhexanoic acid reacts with phenol.

5.61 How many mL of a 0.100 M NaOH solution would be needed to titrate 0.156 g of butanoic acid to a neutral pH endpoint?

5.62 Identify two things that can be done to the following equilibrium to achieve 100% conversion of butanoic acid to ethyl butanoate.

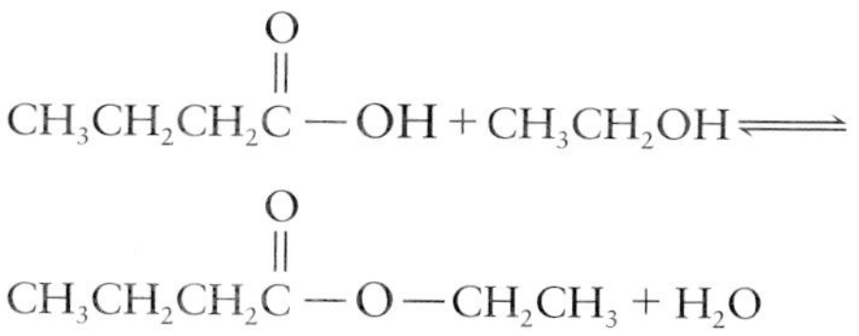

$$CH_3CH_2CH_2\overset{\overset{\displaystyle O}{\|}}{C}-OH + CH_3CH_2OH \rightleftharpoons CH_3CH_2CH_2\overset{\overset{\displaystyle O}{\|}}{C}-O-CH_2CH_3 + H_2O$$

ALLIED HEALTH EXAM CONNECTION

Reprinted with permission from Nursing School and Allied Health Entrance Exams, COPYRIGHT 2005 Petersons.

5.63 Identify the functional group designated by each of the following:

a. ROR

b. RCOOH

c. R

d. RCHO

e. 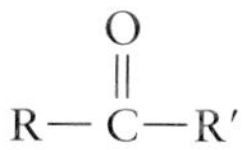 $R-\overset{\overset{\displaystyle O}{\|}}{C}-R'$

5.64 Rank the following compounds from one containing the greatest polarity to the one with the least polarity. Also, rank the compounds in order of increasing boiling points.

a. CH_3CH_2-OH

b. CH_3-O-CH_3

c. $CH_3CH_2CH_3$

d. CH_3COOH

CHEMISTRY FOR THOUGHT

5.65 Citric acid is often added to carbonated beverages as a flavoring. Why is citric acid viewed as a "safe" food additive?

5.66 Ester formation (Reaction 5.8) and ester hydrolysis (Reaction 5.17) are exactly the same reaction only written in reverse. What determines which direction the reaction proceeds and what actually forms?

5.67 Write a mechanism for the following reaction. Show the final location of the oxygen atom in the second color.

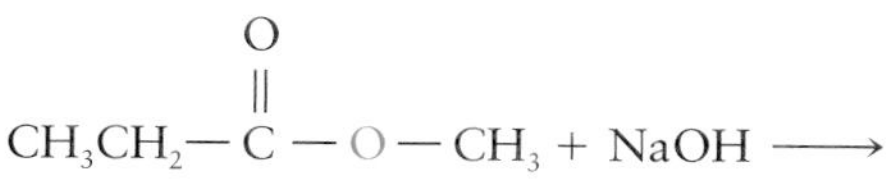

$$CH_3CH_2-\overset{\overset{\displaystyle O}{\|}}{C}-O-CH_3 + NaOH \longrightarrow$$

5.68 Answer the question in Figure 5.1. What flavor sensation is characteristic of acids?

5.69 Oxalic acid, the substance mentioned in Figure 5.2, has two carboxyl groups. If you were to give this compound an IUPAC name, what ending would be appropriate?

5.70 Why is it safe for us to consume foods like vinegar that contain acetic acids?

5.71 A Dacron patch is sometimes used in heart surgery. Why do polyesters like the Dacron patch resist hydrolysis reactions that would result in deterioration of the patch?

5.72 Citrus fruits such as grapefruit and oranges may be kept for several weeks without refrigeration. Offer some possible reasons why the spoilage rate is not higher.

5.73 In Section 8.3, locate the structure of a triglyceride. Draw its structure and circle the ester linkages.

5.74 Section 11.2 shows a tetranucleotide structure. Draw a portion of the structure and circle a portion representing a phosphate diester.

CHAPTER 6

Amines and Amides

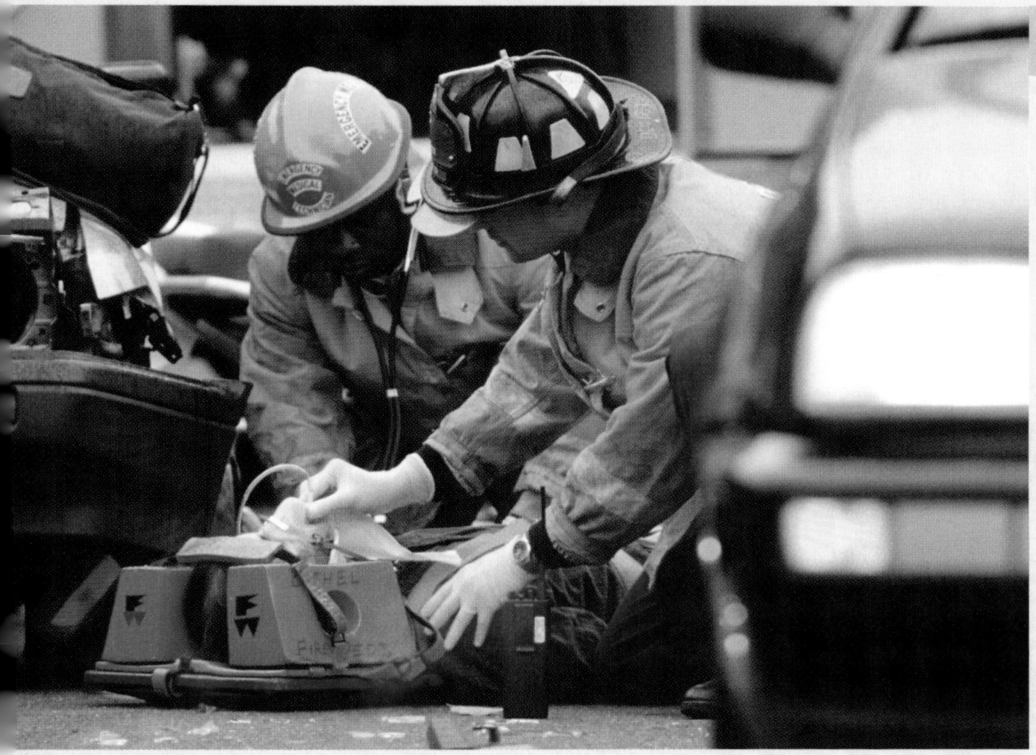

Individuals who require basic life support before arriving at a health care facility often receive it from an emergency medical technician (EMT), or paramedic. Paramedics function as extensions of an emergency room physician and must be familiar with a number of medications. More pharmaceuticals belong to the amine (or amide) family of compounds than to any other functional class.

LEARNING OBJECTIVES

When you have completed your study of this chapter, you should be able to:

1. Given structural formulas, classify amines as primary, secondary, or tertiary. (Section 6.1)
2. Assign common and IUPAC names to simple amines. (Section 6.2)
3. Discuss how hydrogen bonding influences the physical properties of amines. (Section 6.3)
4. Recognize and write key reactions for amines. (Section 6.4)
5. Name amines used as neurotransmitters. (Section 6.5)
6. Give uses for specific biological amines. (Section 6.6)
7. Assign IUPAC names for amides. (Section 6.7)
8. Show the formation of hydrogen bonds with amides. (Section 6.8)
9. Give the products of acidic and basic hydrolysis of amides. (Section 6.9)

ThomsonNOW Throughout the chapter this icon introduces resources on the ThomsonNOW website for this text. Sign in at **www.thomsonedu.com** to:

- Evaluate your knowledge of the material
- Take an exam prep quiz
- Identify areas you need to study with a **Personalized Learning Plan**.

The effectiveness of a wide variety of important medicines depends either entirely or partly on the presence of a nitrogen-containing group in their molecules. Nitrogen-containing functional groups are found in more medications than any other type of functional group. In this chapter, we will study two important classes of organic nitrogen compounds, the amines and the amides.

Both amines and amides are abundant in nature, where they play important roles in the chemistry of life. Our study of these two functional classes will help prepare us for later chapters dealing with amino acids, proteins, and nucleic acids and provide a basis for understanding the structures and chemistry of a number of medicines.

6.1 Classification of Amines

LEARNING OBJECTIVE

1. Given structural formulas, classify amines as primary, secondary, or tertiary.

amine
An organic compound derived by replacing one or more of the hydrogen atoms of ammonia with alkyl or aromatic groups, as in RNH_2, R_2NH, and R_3N.

primary amine
An amine having one alkyl or aromatic group bonded to nitrogen, as in $R{-}NH_2$.

secondary amine
An amine having two alkyl or aromatic groups bonded to nitrogen, as in R_2NH.

tertiary amine
An amine having three alkyl or aromatic groups bonded to nitrogen, as in R_3N.

Amines are organic derivatives of ammonia, NH_3, in which one or more of the hydrogens are replaced by an aromatic or alkyl group (R). Like alcohols, amines are classified as primary, secondary, or tertiary on the basis of molecular structure. However, the classification is done differently. When classifying amines, the number of groups (R) that have replaced hydrogens in the NH_3 is counted. The nitrogen atom in a **primary amine** is bonded to one R group. The nitrogen in **secondary amines** is bonded to two R groups, and that of **tertiary amines** is bonded to three R groups. These amine subclasses are summarized in ■ Table 6.1.

■ **Learning Check 6.1** Classify each of the following amines as primary, secondary, or tertiary:

a. $CH_3{-}\underset{}{\overset{H}{\overset{|}{N}}}{-}CH_2CH_3$

b. (cyclohexane ring containing N)$N{-}H$

c. $C_6H_5{-}\underset{\underset{CH_3}{|}}{N}{-}CH_3$

6.2 The Nomenclature of Amines

LEARNING OBJECTIVE

2. Assign common and IUPAC names to simple amines.

Several approaches are used in naming amines. Common names are used extensively for those with low molecular weights. These are named by alphabetically

TABLE 6.1 Subclasses of amines

Subclass	General formula	Example
Primary (1°)	$R{-}\underset{\underset{H}{\vert}}{N}{-}H$	$CH_3{-}\underset{\underset{H}{\vert}}{N}{-}H$
Secondary (2°)	$R{-}\underset{\underset{H}{\vert}}{N}{-}R'$	$CH_3{-}\underset{\underset{H}{\vert}}{N}{-}CH_3$
Tertiary (3°)	$R{-}\underset{\underset{R''}{\vert}}{N}{-}R'$	$CH_3{-}\underset{\underset{CH_3}{\vert}}{N}{-}CH_3$

linking the names of the alkyl or aromatic groups bonded to the nitrogen and attaching the suffix *-amine*. The name is written as one word, and the prefixes *di-* and *tri-* are used when identical alkyl groups are present. Some examples are

$CH_3{-}NH_2$ methylamine $CH_3{-}NH{-}CH_3$ dimethylamine $CH_3CH_2{-}NH{-}CH_3$ ethylmethylamine

ThomsonNOW Go to Coached Problems to practice **naming amines.**

Learning Check 6.2 Assign a common name to the following amines:

a. $CH_3CH_2CH_2{-}NH_2$ b. $C_6H_5{-}NH{-}CH_3$ c. $CH_3{-}N(CH_3){-}CH_3$

An IUPAC approach for naming amines is similar to that for alcohols. The root name is determined by the longest continuous chain of carbon atoms. The *-e* ending in the alkane name is changed to *–amine*. The position of the amino group on the chain is shown by a number. Numbers are also used to locate other substituents on the carbon chain. An italic capital N is used as a prefix for a substituent on nitrogen.

$CH_3CH_2CH_2CH(NH_2)CH_3$ 2-pentanamine $CH_3CH(CH_3)CH_2CH_2NH_2$ 3-methyl-1-butanamine $CH_3CH_2CH(NHCH_3)CH_3$ *N*-methyl-2-butanamine

Learning Check 6.3

Give an IUPAC name for the following amines:

a. $CH_3CH_2CH_2CH_2{-}NH_2$ b. $CH_3CH_2CH_2CH_2{-}NHCH_3$

Aromatic amines are usually given names (both common and IUPAC) based on aniline, the simplest aromatic amine. They are named as substituted anilines, and an italic capital *N* is used to indicate that an alkyl group is attached to the nitrogen and not to the ring. If two groups are attached to the nitrogen, an italic capital *N* is used for each one. Some examples are

$C_6H_5{-}NH_2$ aniline $C_6H_5{-}NH{-}CH_3$ *N*-methylaniline $C_6H_4(CH_2CH_3){-}NH{-}CH_3$ 2-ethyl-*N*-methylaniline $C_6H_5{-}N(CH_3){-}CH_3$ *N*-ethyl-*N*-methylaniline

Learning Check 6.4 Name the following amines as derivatives of aniline:

a. $C_6H_5{-}NH{-}CH_2CH_3$

b. $C_6H_5{-}N(CH_3){-}CH_2CH_3$

c. $C_6H_3(NH_2)(CH_3)(CH_2CH_3)$ (with CH_3 and CH_2CH_3 on adjacent ring carbons next to NH_2)

OVER THE COUNTER 6.1 Cough Syrup—or Are You Just Coughing Up More Money?

Coughing, the act of forcefully expelling air from the lungs, is an action with which we are all familiar. Coughs are caused by a number of conditions, with the common cold being the major culprit. In response to the widespread incidence of coughs among the population, many over-the-counter cough remedies are available, but are they effective? Most of these products contain a cough suppressant such as dextromethorphon and some contain an expectorant to thin out mucus.

A recent (2006) report from the American College of Chest Physicians casts some serious doubt on the effectiveness of OTC cough medicines. The report, based on a review of scores of studies done during the past several decades, concluded there was no conclusive scientific evidence that the ingredients in OTC cough medicines relieve coughs resulting from colds. A spokesperson for one of the OTC cough remedy manufacturers countered the report by pointing out that the Food and Drug Administration has reviewed the cough suppressant and expectorant used in the company's product and found them to be safe and effective. With an FDA stamp of approval on one side of the issue and a medical panel on the other side, it appears that the final decision will be made by the consumers who suffer from coughs.

6.3 Physical Properties of Amines

LEARNING OBJECTIVE

3. Discuss how hydrogen bonding influences the physical properties of amines.

The simple, low molecular weight amines are gases at room temperature (see ■ Table 6.2). Heavier, more complex compounds are liquids or solids.

Like alcohols, primary and secondary amines form hydrogen bonds among themselves:

$$\begin{array}{ccccc} & & & & H \\ & & & & | \\ CH_3- & N & -H\cdots & N & -CH_3 \\ & | & & | & \\ & H & & H & \end{array}$$

ThomsonNOW Go to Coached Problems to examine the **boiling points of amines.**

Because nitrogen is less electronegative than oxygen, the hydrogen bonds formed by amines are weaker than those formed by alcohols. For this reason, the boiling points of primary and secondary amines are usually somewhat lower than the boiling points of alcohols with similar molecular weights. Tertiary amines cannot form hydrogen bonds among themselves (because the nitrogen has no attached

TABLE 6.2 Properties of some amines

Name	Structure	Melting point (°C)	Boiling point (°C)
methylamine	CH_3-NH_2	−94	−6
ethylamine	$CH_3CH_2-NH_2$	−81	17
dimethylamine	$CH_3-NH-CH_3$	−93	7
diethylamine	$CH_3CH_2-NH-CH_2CH_3$	−48	56
trimethylamine	$CH_3-N(CH_3)-CH_3$	−117	3
triethylamine	$CH_3CH_2-N(CH_2CH_3)-CH_2CH_3$	−114	89

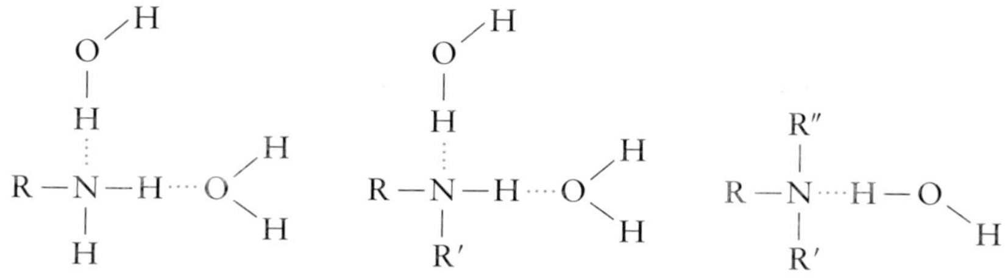

$$\begin{array}{ccc} \begin{array}{l} \quad\;\; O-H \\ \quad\;\; | \\ \quad\;\; H \\ \quad\;\; \vdots \\ R-N-H\cdots O\!\!<^{H}_{H} \\ \quad\;\; | \\ \quad\;\; H \end{array} & \begin{array}{l} \quad\;\; O-H \\ \quad\;\; | \\ \quad\;\; H \\ \quad\;\; \vdots \\ R-N-H\cdots O\!\!<^{H}_{H} \\ \quad\;\; | \\ \quad\;\; R' \end{array} & \begin{array}{l} \quad\;\; R'' \\ \quad\;\; | \\ R-N\cdots H-O-H \\ \quad\;\; | \\ \quad\;\; R' \end{array} \\ 1^\circ \text{ amine} & 2^\circ \text{ amine} & 3^\circ \text{ amine} \end{array}$$

■ **FIGURE 6.1** Hydrogen bonding between amines in water.

hydrogen atom), and their boiling points are similar to alkanes that have about the same molecular weights.

Amines with fewer than six carbon atoms are generally soluble in water as a result of hydrogen bond formation between amine functional groups and water molecules. This is true for tertiary as well as primary and secondary amines (see ■ Figure 6.1).

Low molecular weight amines have a sharp, penetrating odor similar to that of ammonia. Somewhat larger amines have an unpleasant odor reminiscent of decaying fish. In fact, some amines are partially responsible for the odor of decaying animal tissue. Two such compounds have especially descriptive common names—putrescine and cadaverine:

ThomsonNOW™ Go to Coached Problems to examine the **physical properties of amines.**

$$\underset{\substack{\text{putrescine} \\ \text{(1,4-diaminobutane)}}}{H_2N-CH_2CH_2CH_2CH_2-NH_2} \qquad \underset{\substack{\text{cadaverine} \\ \text{(1,5-diaminopentane)}}}{H_2N-CH_2CH_2CH_2CH_2CH_2-NH_2}$$

■ **Learning Check 6.5** Draw structural formulas to show how each of the following amines forms hydrogen bonds with water:

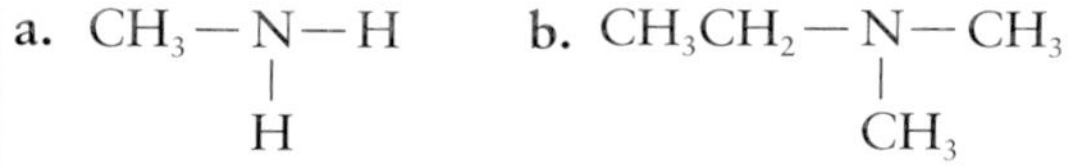

a. $\begin{array}{l} CH_3-N-H \\ \qquad\;\; | \\ \qquad\;\; H \end{array}$ b. $\begin{array}{l} CH_3CH_2-N-CH_3 \\ \qquad\qquad | \\ \qquad\qquad CH_3 \end{array}$

6.4 Chemical Properties of Amines

LEARNING OBJECTIVE

4. Recognize and write key reactions for amines.

Amines undergo many chemical reactions, but we will limit our study to the two most important properties of amines: basicity and amide formation.

Basicity and Amine Salts

The single most distinguishing feature of amines is their behavior as weak bases. They are the most common organic bases. We know from inorganic chemistry that ammonia behaves as a Brønsted base by accepting protons, and in the process becomes a conjugate acid. The reaction of NH_3 with HCl gas illustrates this point (Reaction 6.1):

ThomsonNOW™ Go to Coached Problems to explore the **relationship between structure and relative basicity for a series of amines.**

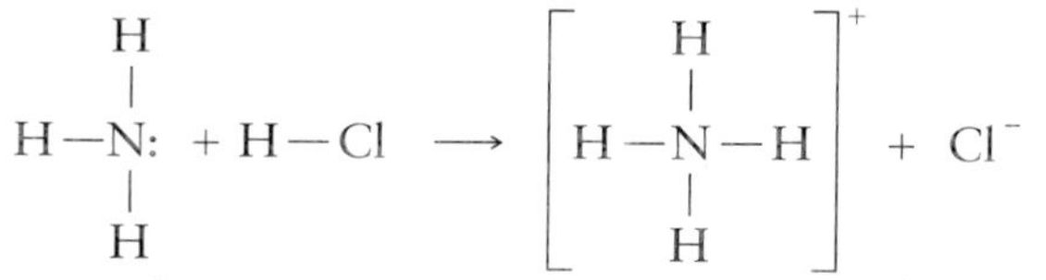

$$\underset{\text{ammonia}}{\begin{array}{c} H \\ | \\ H-\ddot{N}: \\ | \\ H \end{array}} + H-Cl \longrightarrow \underset{\text{ammonium ion}}{\left[\begin{array}{c} H \\ | \\ H-N-H \\ | \\ H \end{array}\right]^+} + \underset{\text{chloride ion}}{Cl^-} \qquad (6.1)$$

We also remember from inorganic chemistry that Brønsted bases such as ammonia react with water to liberate OH^- ions (Reaction 6.2):

$$\begin{matrix} & H & & & \\ & | & & & \\ H-&N: & + H-OH & \rightleftharpoons & \left[\begin{matrix} & H & \\ & | & \\ H-&N&-H \\ & | & \\ & H & \end{matrix}\right]^+ + OH^- \\ & | & & & \\ & H & & & \end{matrix} \quad (6.2)$$

ammonia water ammonium ion hydroxide ion

Because the amines are derivatives of ammonia, they react in similar ways. In water, they produce OH^- ions:

General reaction: $R-NH_2 + H_2O \rightleftharpoons R-NH_3^+ + OH^-$ (6.3)

Specific reaction: $CH_3-NH_2 + H_2O \rightleftharpoons CH_3-NH_3^+ + OH^-$ (6.4)
methylamine methylammonium ion

EXAMPLE 6.1

Complete the following reaction:

$$CH_3-NH-CH_3 + H_2O \rightleftharpoons$$

Solution

Primary, secondary, and tertiary amines are all basic and exhibit the same reaction in water.

$$CH_3-NH-CH_3 + H_2O \rightleftharpoons CH_3-\overset{+}{N}H_2-CH_3 + OH^-$$

■ **Learning Check 6.6** Complete the following reactions:

a. $CH_3-\underset{\underset{CH_3}{|}}{N}-CH_2CH_3 + H_2O \rightleftharpoons$

b. $C_6H_5-NH_2 + H_2O \rightleftharpoons$ (benzene ring drawn)

ThomsonNOW™ Go to Coached Problems to examine the **reactivity of amines.**

All amines behave as weak bases and form salts when they react with acids such as HCl:

General reaction: $R-NH_2 + HCl \rightarrow R-NH_3^+Cl^-$ (6.5)
amine acid amine salt

Specific example: $CH_3-NH_2 + HCl \rightarrow CH_3-NH_3^+Cl^-$ (6.6)
methylamine methylammonium chloride

Other acids, such as sulfuric, nitric, phosphoric, and carboxylic acids (Reaction 6.7), also react with amines to form salts:

$CH_3CH_2-NH_2 + CH_3COOH \rightarrow CH_3CH_2-NH_3^+CH_3COO^-$ (6.7)
ethylamine acetic acid ethylammonium acetate

■ **Learning Check 6.7** Complete the following reactions:

a. $C_6H_5-NH-CH_3 + HCl \longrightarrow$ (benzene ring drawn)

b. 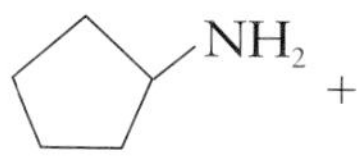 $+ \ CH_3CH_2COOH \longrightarrow$

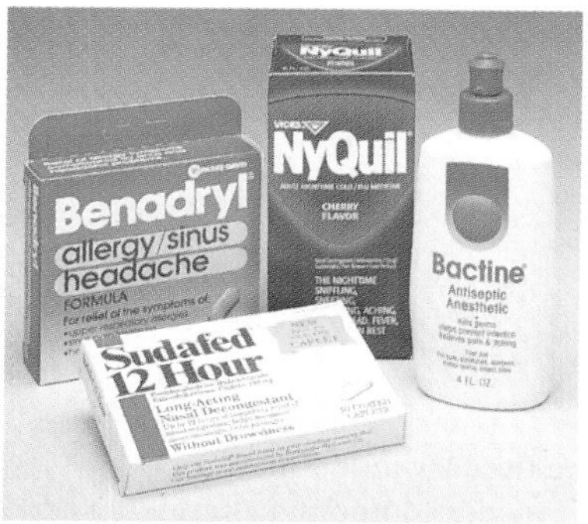

© West

■ **FIGURE 6.2** Each of these OTC medications contains an amine salt as the active ingredient.

As we have seen, amine salts may be named by changing "amine" to "ammonium" and adding the name of the negative ion derived from the acid. Some more examples are

$$\begin{array}{c} CH_3 \\ | \\ CH_3-N-H^+Br^- \\ | \\ CH_2CH_3 \end{array} \qquad\qquad (CH_3CH_2)_3NH^+CH_3COO^-$$

ethyldimethylammonium bromide — triethylammonium acetate

Amine salts have physical properties characteristic of other ionic compounds. They are white crystalline solids with high melting points. Because they are ionic, amine salts are more soluble in water than the parent amines. For this reason, amine drugs are often given in the form of salts because they will dissolve in body fluids (see ■ Figure 6.2).

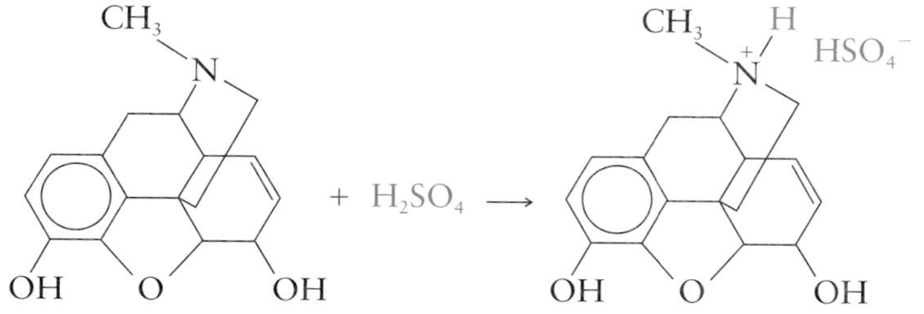
(6.8)

morphine (water insoluble) — morphine sulfate (water soluble)

Amine salts are easily converted back to amines by adding a strong base:

$$R-NH_3^+Cl^- + NaOH \rightarrow R-NH_2 + H_2O + NaCl \qquad (6.9)$$

The form in which amines occur in solutions is pH dependent, just as is the case for carboxylic acids (Section 5.3). In solutions of high H^+ (low pH), the salt form of amines predominates. In solutions of high OH^- (high pH), the amine form is favored:

$$R-NH_2 \underset{OH^-}{\overset{H^+}{\rightleftarrows}} R-NH_3^+ \qquad (6.10)$$

amine (high pH) — amine salt (low pH)

The amine cations in the salts we have discussed to this point have one, two, or three alkyl groups attached to the nitrogen:

$$R-NH_3^+ \qquad R-\overset{+}{N}H_2-R' \qquad \begin{array}{c} R-\overset{+}{N}H-R' \\ | \\ R'' \end{array}$$

primary amine cation — secondary amine cation — tertiary amine cation

It is also possible to have amine cations in which four alkyl groups or benzene rings are attached to the nitrogen. An example is

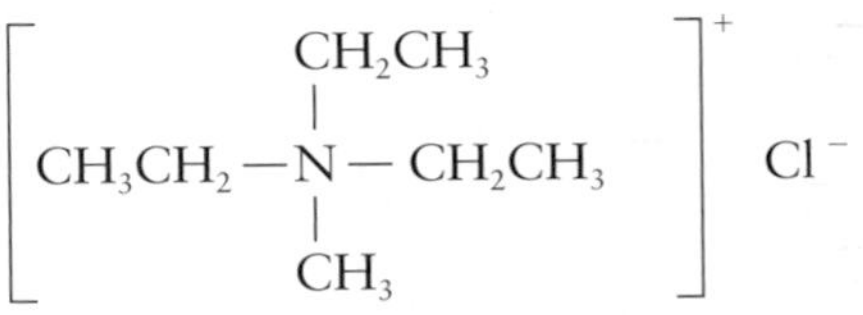

$$\left[\begin{array}{c} CH_2CH_3 \\ | \\ CH_3CH_2-N-CH_2CH_3 \\ | \\ CH_3 \end{array} \right]^+ Cl^-$$

triethylmethylammonium chloride

quaternary ammonium salt
An ionic compound containing a positively charged ion in which four alkyl or aromatic groups are bonded to nitrogen, as in R_4N^+.

These salts are called **quaternary ammonium salts** and are named the same way as other amine salts. Unlike other amine salts, quaternary salts contain no hydrogen attached to the nitrogen that can be removed by adding a base. Thus, quaternary salts are present in solution in only one form, which is independent of the pH of the solution.

Choline is an important quaternary ammonium ion that is a component of certain lipids (Section 8.6); acetylcholine is a compound involved in the transmission of nerve impulses from one cell to another (Section 6.5):

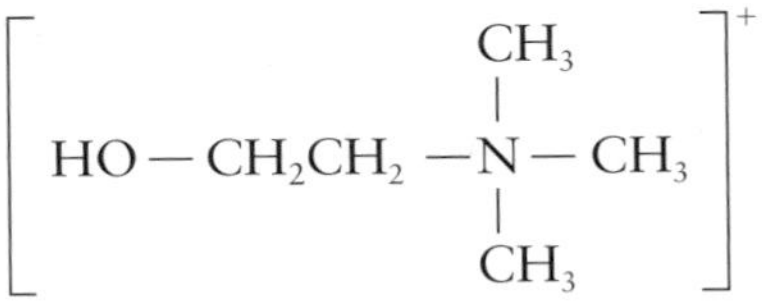

$$\left[\begin{array}{c} CH_3 \\ | \\ HO-CH_2CH_2-N-CH_3 \\ | \\ CH_3 \end{array} \right]^+ \quad \left[\begin{array}{c} O \qquad\qquad\qquad\qquad CH_3 \\ \| \qquad\qquad\qquad\qquad\quad | \\ CH_3-C-O-CH_2CH_2-N-CH_3 \\ | \\ CH_3 \end{array} \right]^+$$

choline cation 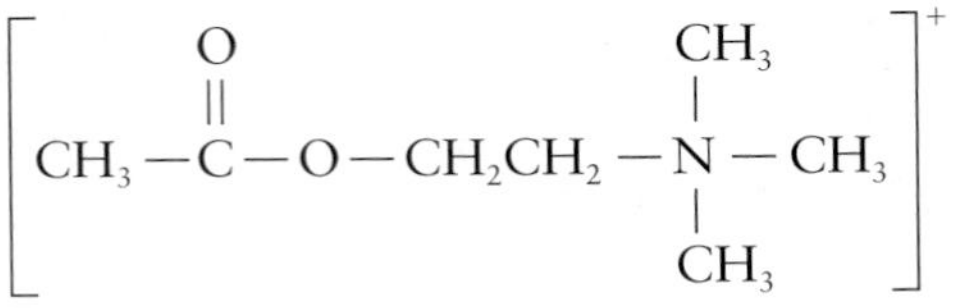 | acetylcholine cation

Some quaternary ammonium salts are important because they have disinfectant properties. For example, benzalkonium chloride (Zephiran®) is a well-known antiseptic compound that kills many pathogenic (disease-causing) bacteria and fungi on contact:

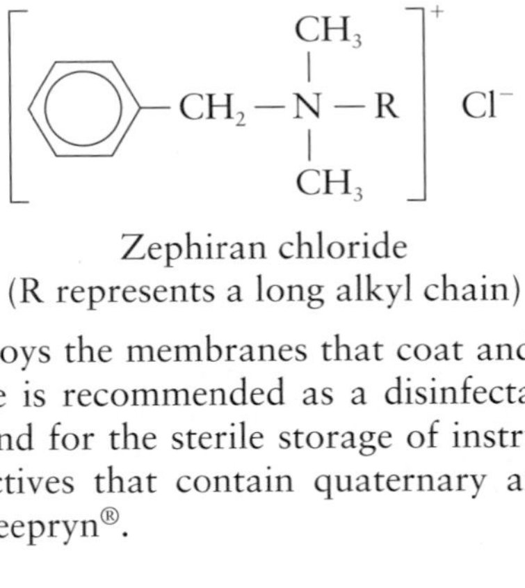

$$\left[\begin{array}{c} CH_3 \\ | \\ C_6H_5-CH_2-N-R \\ | \\ CH_3 \end{array} \right]^+ Cl^-$$

Zephiran chloride
(R represents a long alkyl chain)

Its detergent action destroys the membranes that coat and protect the microorganisms. Zephiran® chloride is recommended as a disinfectant solution for skin and hands prior to surgery and for the sterile storage of instruments. The trade names of some other anti-infectives that contain quaternary ammonium salts are Phemerol®, Bactine®, and Ceepryn®.

Amide Formation

amide
An organic compound having the functional group

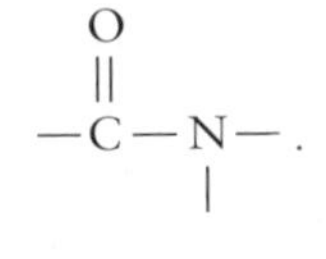

$$\begin{array}{c} O \\ \| \\ -C-N- \\ \quad | \end{array}.$$

amide linkage
The carbonyl carbon–nitrogen single bond of the amide group.

Amines react with acid chlorides or acid anhydrides to form amides. The characteristic functional group of **amides** is a carbonyl group attached to nitrogen. The single bond linking the carbonyl carbon and nitrogen atoms in the group is called an **amide linkage:**

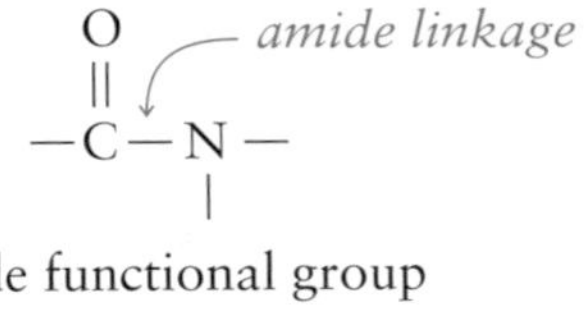

$$\begin{array}{c} O \\ \| \\ -C-N- \\ \quad | \end{array}$$

amide linkage

amide functional group

Amides may be thought of as containing an ammonia or amine portion and a portion derived from a carboxylic acid:

$$\underbrace{-\overset{\overset{\large O}{\|}}{C}}_{\text{from a carboxylic acid}}-\underbrace{\underset{|}{N}-}_{\text{from ammonia or an amine}}$$

However, we've seen that the reaction of an amine with a carboxylic acid normally produces a salt and not an amide. For this reason, the preparation of amides requires the more reactive acid chlorides or acid anhydrides.

Both primary and secondary amines can be used to prepare amides in this manner, a method similar to the formation of esters (Section 5.5). Tertiary amines, however, do not form amides because they lack a hydrogen atom on the nitrogen.

General reactions:

$$\underset{\text{acid chloride}}{R-\overset{\overset{O}{\|}}{C}-Cl} + \underset{\text{amine}}{R'-\underset{\underset{H}{|}}{N}-H} \longrightarrow \underset{\text{amide}}{R-\overset{\overset{O}{\|}}{C}-\underset{\underset{H}{|}}{N}-R'} + HCl \qquad (6.11)$$

$$\underset{\text{acid anhydride}}{R-\overset{\overset{O}{\|}}{C}-O-\overset{\overset{O}{\|}}{C}-R} + \underset{\text{amine}}{R'-\underset{\underset{H}{|}}{N}-H} \longrightarrow \underset{\text{amide}}{R-\overset{\overset{O}{\|}}{C}-\underset{\underset{H}{|}}{N}-R'} + \underset{\text{carboxylic acid}}{R-\overset{\overset{O}{\|}}{C}-OH} \qquad (6.12)$$

Specific examples:

$$\underset{\text{acetic anhydride}}{CH_3-\overset{\overset{O}{\|}}{C}-O-\overset{\overset{O}{\|}}{C}-CH_3} + \underset{\text{ammonia}}{H-\overset{\overset{H}{|}}{N}-H} \longrightarrow \underset{\text{acetamide}}{CH_3-\overset{\overset{O}{\|}}{C}-\underset{\underset{H}{|}}{N}-H} + \underset{\text{acetic acid}}{CH_3-\overset{\overset{O}{\|}}{C}-OH} \qquad (6.13)$$

$$\underset{\text{benzoyl chloride}}{C_6H_5-\overset{\overset{O}{\|}}{C}-Cl} + \underset{\text{methylamine (a 1° amine)}}{H-\overset{\overset{H}{|}}{N}-CH_3} \longrightarrow \underset{N\text{-methylbenzamide}}{C_6H_5-\overset{\overset{O}{\|}}{C}-NH-CH_3} + HCl \qquad (6.14)$$

$$\underset{\text{benzoyl chloride}}{C_6H_5-\overset{\overset{O}{\|}}{C}-Cl} + \underset{\text{dimethylamine (a 2° amine)}}{H-\overset{\overset{CH_3}{|}}{N}-CH_3} \longrightarrow \underset{N,N\text{-dimethylbenzamide}}{C_6H_5-\overset{\overset{O}{\|}}{C}-\overset{\overset{CH_3}{|}}{N}-CH_3} + HCl \qquad (6.15)$$

Learning Check 6.8 Complete the following reactions:

a. $$CH_3-\overset{\overset{O}{\|}}{C}-O-\overset{\overset{O}{\|}}{C}-CH_3 + CH_3CH_2CH_2-NH_2 \longrightarrow$$

b. $$CH_3CH_2-\overset{\overset{O}{\|}}{C}-Cl + NH_3 \longrightarrow$$

Reaction of diacid chlorides with diamines produces polyamides that, like polyesters, are condensation polymers. The repeating units in polyamides are held together by amides linkages. An example is the formation of nylon, shown in Reaction 6.16 and Figure 6.3.

$$n\text{Cl}-\overset{\overset{O}{\|}}{C}-(CH_2)_4-\overset{\overset{O}{\|}}{C}-Cl + nH_2N-(CH_2)_6-NH_2 \longrightarrow$$

adipoyl chloride hexamethylenediamine

$$---\overset{\overset{O}{\|}}{C}-(CH_2)_4-\overset{\overset{O}{\|}}{C}-NH-(CH_2)_6-NH-\overset{\overset{O}{\|}}{C}-(CH_2)_4-\overset{\overset{O}{\|}}{C}-NH-(CH_2)_6-NH--- + n\text{HCl} \quad (6.16)$$

amide linkages; amide linkage

polyamide nylon

Three billion pounds of nylon and related polyamides are produced annually. About 60% of that total goes to make the nylon fiber used in home furnishings such as carpet. The remainder is used largely as a textile fiber in clothing (shirts,

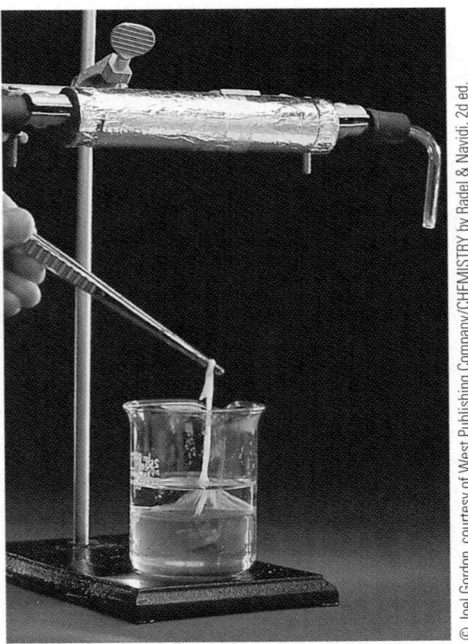

The reactants occupy separate immiscible layers. Hexamethylenediamine is dissolved in water (bottom layer), and adipoyl chloride is dissolved in hexane (top layer). The two compounds react at the interface between the two layers to form a film of nylon, which is lifted by forceps.

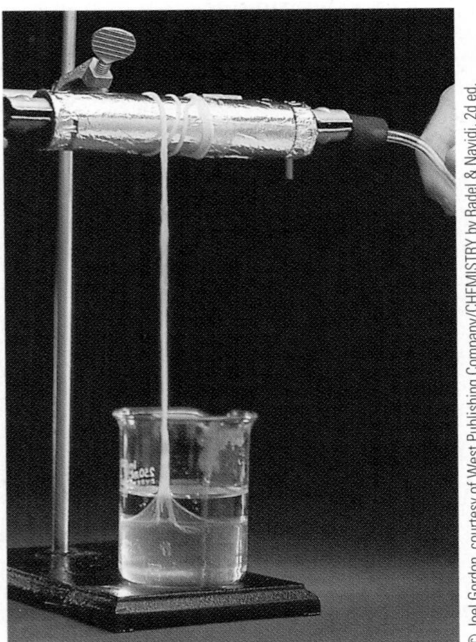

The rotating tube winds the nylon strand, which continues to form at the interface in the beaker.

FIGURE 6.3 The preparation of nylon.

CHEMISTRY AROUND US 6.1 Aspirin Substitutes

Even though aspirin is an excellent and useful drug (see Chemistry and Your Health 5.1), it occasionally produces adverse effects that cannot be tolerated by some people. These effects include gastrointestinal bleeding and such allergic reactions as skin rashes and asthmatic attacks. Children who are feverish with influenza or chickenpox should not be given aspirin because of a strong correlation with Reye's syndrome, a devastating illness characterized by liver and brain dysfunction.

The amide *acetaminophen,* marketed under trade names such as Tylenol®, Excedrin Aspirin Free®, Panadol®, and Anacin-3®, is an effective aspirin substitute available for use in such situations.

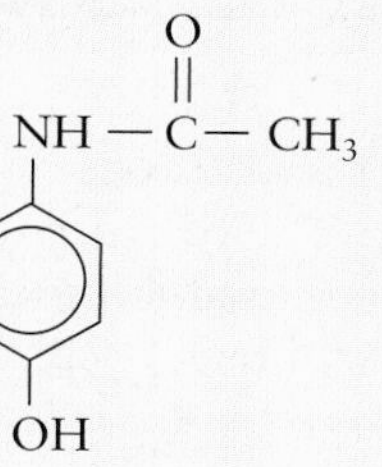

acetaminophen

Acetaminophen does not irritate the intestinal tract and yet has comparable analgesic (pain-relieving) and antipyretic (fever-reducing) effects. Acetaminophen is available in a stable liquid form suitable for administration to children and other patients who have difficulty taking tablets or capsules. Unlike aspirin, acetaminophen has virtually no anti-inflammatory effect and is therefore a poor substitute for aspirin in the treatment of inflammatory disorders such as rheumatoid arthritis. The overuse of acetaminophen has been linked to liver and kidney damage.

Ibuprofen, a carboxylic acid rather than an amide, is a drug with analgesic, antipyretic, and anti-inflammatory properties that was available for decades only as a prescription drug under the trade name Motrin®.

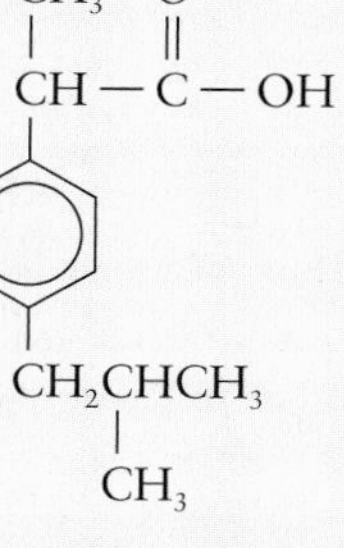

ibuprofen

In 1984, it became available in an OTC form under trade names such as Advil®, Ibuprin®, Mediprin®, and Motrin IB®. Ibuprofen is thought to be somewhat superior to aspirin as an anti-inflammatory drug, but it has about the same effectiveness as aspirin in relieving mild pain and reducing fever.

In 1994, the FDA approved another compound for sale as an over-the-counter pain reliever. *Naproxen,* marketed under the trade names Aleve® and Anaprox®, exerts its effects for a longer time in the body (8–12 hours per dose) than ibuprofen (4–6 hours per dose) or aspirin and acetaminophen (4 hours per dose). However, the chances for causing slight intestinal bleeding and stomach upset are greater with naproxen than with ibuprofen, and naproxen is not recommended for use by children under age 12.

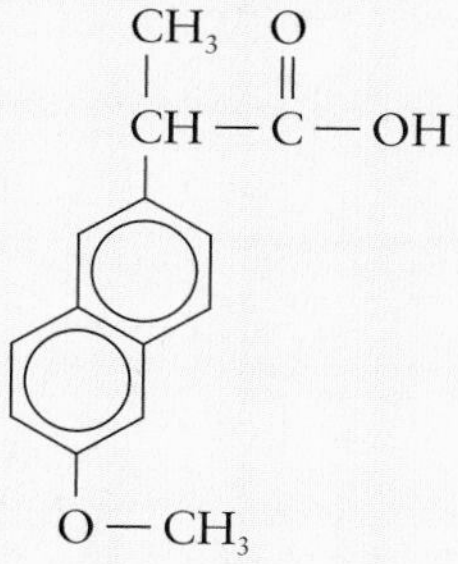

naproxen

Ketoprofen, another pain reliever, became available for OTC sales in 1996. This compound, sold under the trade names Orudis KT® and Actron®, has the advantage of requiring very small doses when compared to other pain-relieving products. A dose of ketoprofen is 12.5 mg, compared to doses of 200–500 mg for the other products. The tiny, easy-to-swallow tablet is a characteristic emphasized by the manufacturers in advertising. The side effects of ketoprofen are similar to those of aspirin and ibuprofen, so ketoprofen should not be used by people known to be allergic to either of these other two compounds.

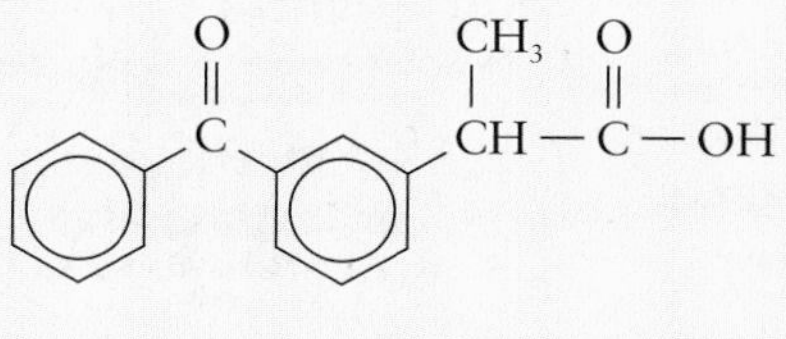

ketoprofen

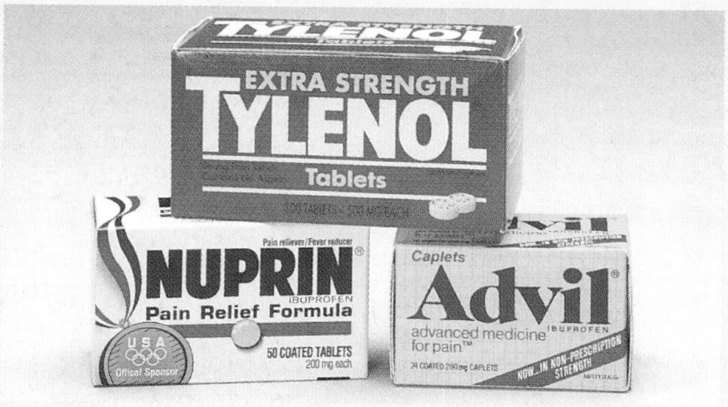

© West

Aspirin substitutes containing acetaminophen or ibuprofen.

■ **FIGURE 6.4** A parachute made of nylon fabric.

dresses, stockings, underwear, etc.) and as tire cord. Minor but important uses include fasteners, rope, parachutes, paintbrushes, and electrical parts (see ■ Figure 6.4). In medicine, nylon is used in specialized tubing. Nylon sutures were the first synthetic sutures and are still used.

Silk and wool are natural polyamides that we classify as proteins (discussed in detail in Chapter 9). For now, note the repeating units and amide linkages characteristic of proteins as shown in the following portion of a polyamide protein chain:

Go to Coached Problems to examine the **formation of polyamides.**

$$---\text{NH}-\underset{\text{R}}{\underset{|}{\text{CH}}}-\overset{\text{O}}{\overset{\|}{\text{C}}}-\text{NH}-\underset{\text{R}'}{\underset{|}{\text{CH}}}-\overset{\text{O}}{\overset{\|}{\text{C}}}-\text{NH}-\underset{\text{R}''}{\underset{|}{\text{CH}}}-\overset{\text{O}}{\overset{\|}{\text{C}}}-\text{NH}-\underset{\text{R}'''}{\underset{|}{\text{CH}}}-\overset{\text{O}}{\overset{\|}{\text{C}}}---$$

amide linkages

6.5 Amines as Neurotransmitters

LEARNING OBJECTIVE

5. Name amines used as neurotransmitters.

neurotransmitter
A substance that acts as a chemical bridge in nerve impulse transmission between nerve cells.

Neurotransmitters, the chemical messengers of the nervous system, carry nerve impulses from one nerve cell (neuron) to another. A neuron consists of a bulbous body called the soma (or cell body) attached to a long stemlike projection called an axon (see ■ Figure 6.5). Numerous short extensions called dendrites are attached to the large bulbous end of the soma, and filaments called synaptic terminals are attached to the end of the axon. At low magnification, the synaptic terminals of one neuron's axon appear to be attached to the dendrites of an adjacent neuron. However, upon further magnification, a small gap called a synapse is visible between the synaptic terminals of the axon and the dendrites of the next neuron.

Molecules of a neurotransmitter are stored in small pockets in the axon near the synapse. These pockets release neurotransmitter molecules into the synapse when an electrical current flows from the soma, along the axon to the synaptic terminals. The electrical current is generated by the exchange of positive and neg-

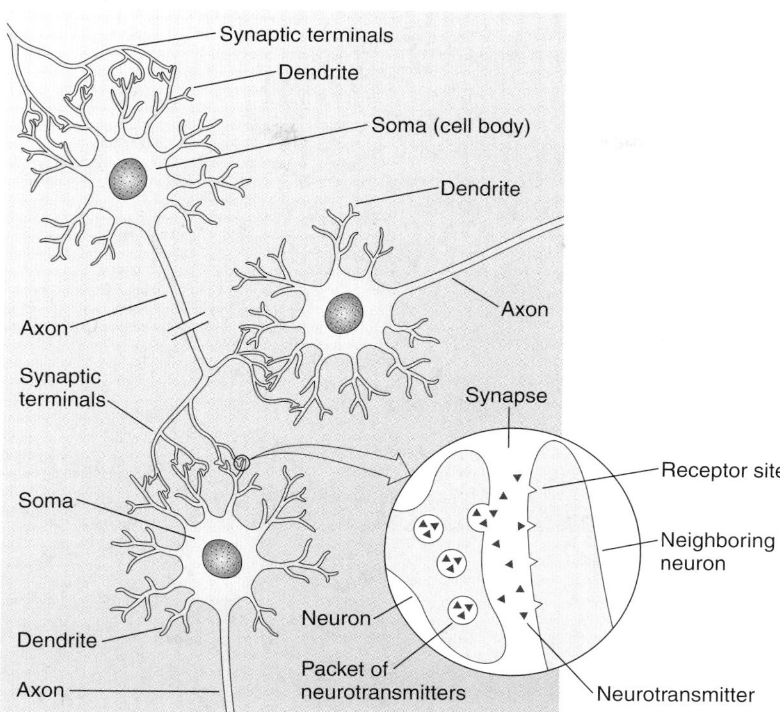

FIGURE 6.5 A nerve cell and the transmission of a nerve signal.

ative ions across membranes. The released neurotransmitter molecules diffuse across the synapse and bind to receptors on the dendrites of the next neuron. Once the neurotransmitter binds to the receptor, the message has been delivered. The receiving cell then sends an electrical current down its own axon until a synapse is reached, and a neurotransmitter delivers the message across the synapse to the next neuron. In the central nervous system—the brain and spinal cord—the most important neurotransmitters are acetylcholine and three other amines: norepinephrine, dopamine, and serotonin:

HO, HO (ring), $\overset{OH}{|}$ $CHCH_2-NH_2$

norepinephrine

HO, HO (ring), $CH_2CH_2-NH_2$

dopamine

HO (indole ring), $CH_2CH_2-NH_2$, N–H

serotonin

$$\left[CH_3-\overset{O}{\overset{\|}{C}}-O-CH_2CH_2-\underset{CH_3}{\underset{|}{\overset{CH_3}{\overset{|}{N}}}}-CH_3\right]^+$$

acetylcholine cation

Neurotransmitters are not only chemical messengers for the nervous system, they may also be partly responsible for our moods. A simplified biochemical theory of mental illness is based on two amines found in the brain. The first is norepinephrine (NE). When an excess of norepinephrine is formed in the brain, the result is a feeling of elation. Extreme excesses of NE can even induce a manic state, while low NE levels may be a cause of depression.

ThomsonNOW™ Go to Coached Problems to examine **signal transduction.**

STUDY SKILLS 6.1 A Reaction Map for Amines

We have discussed four different reactions of amines. When trying to decide what happens in a reaction, remember to focus on the functional groups. After identifying an amine as a starting material, next decide what other reagent is involved and which of the four pathways is appropriate.

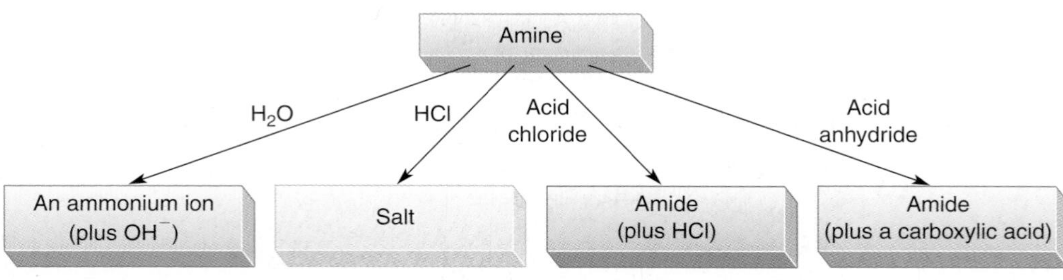

At least six different receptors in the body can be activated by norepinephrine and related compounds. A class of drugs called beta blockers reduces the stimulant action of epinephrine and norepinephrine on various kinds of cells. Some beta blockers are used to treat cardiac arrhythmias, angina, and hypertension (high blood pressure). They function by slightly decreasing the force of each heartbeat. Depression is sometimes a side effect of using such drugs.

In a complex, multistep process, the amino acid tyrosine is used in the body to synthesize norepinephrine:

$$\text{HO-C}_6\text{H}_4\text{-CH}_2\text{-CH(COOH)-NH}_2 \longrightarrow \text{(HO)}_2\text{C}_6\text{H}_3\text{-CH}_2\text{-CH(COOH)-NH}_2 \longrightarrow$$

tyrosine, dopa

$$\text{(HO)}_2\text{C}_6\text{H}_3\text{-CH}_2\text{-CH}_2\text{-NH}_2 \longrightarrow \text{(HO)}_2\text{C}_6\text{H}_3\text{-CH(OH)-CH}_2\text{-NH}_2$$

dopamine, norepinephrine

Each step in the process is catalyzed by at least one enzyme, and each of the intermediates formed in the synthesis has physiological activity. Dopa is used as a treatment for Parkinson's disease, and dopamine has been used to treat low blood pressure. Tyrosine is an essential amino acid that must be obtained from the diet, so what we eat may make a direct contribution to our mental health.

The second amine often associated with the biochemical theory of mental illness is serotonin. Like norepinephrine, it also functions as a neurotransmitter. Serotonin is produced in the body from the amino acid tryptophan:

tryptophan → serotonin

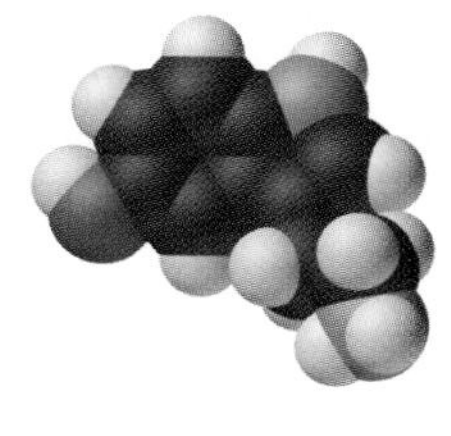

Serotonin

Serotonin has been found to influence sleeping, the regulation of body temperature, and sensory perception, but its exact role in mental illness is not yet clear. Unusually low levels of 5-hydroxyindoleacetic acid, a product of serotonin utilization, are characteristically found in the spinal fluid of victims of violent suicide. Drugs that mimic serotonin are sometimes used to treat depression, anxiety, and obsessive-compulsive disorder. Serotonin blockers are used to treat migraine headaches and relieve the nausea that accompanies cancer chemotherapy. A better understanding of the biochemistry of the brain may lead to better medications for treating various forms of mental illness.

6.6 Other Biologically Important Amines

LEARNING OBJECTIVE

6. Give uses for specific biological amines.

Epinephrine

Epinephrine, or adrenaline, has a molecular structure similar to that of norepinephrine. The difference is that epinephrine contains an *N*-methyl group:

epinephrine
(adrenaline)

Epinephrine is more important as a hormone than as a neurotransmitter. It is synthesized in the adrenal gland and acts to increase the blood level of glucose, a key source of energy for body processes. The release of epinephrine—usually in response to pain, anger, or fear—provides glucose for a sudden burst of energy. For this reason, epinephrine has been called the fight-or-flight hormone.

Epinephrine also raises blood pressure. It does this by increasing the rate and force of heart contractions and by constricting the peripheral blood vessels. Injectable local anesthetics usually contain epinephrine because it constricts blood vessels in the vicinity of the injection. This prevents the blood from rapidly "washing" the anesthetic away and prolongs the anesthetic effect in the target tissue. Epinephrine is also used to reduce hemorrhage, treat asthma attacks, and combat anaphylactic shock (collapse of the circulatory system).

Amphetamines

Amphetamine (also known as Benzedrine®) is a powerful nervous system stimulant with an amine structure similar to that of epinephrine:

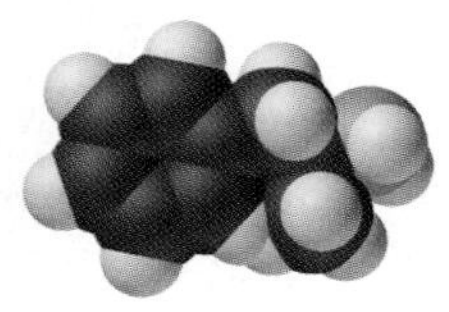

Amphetamine

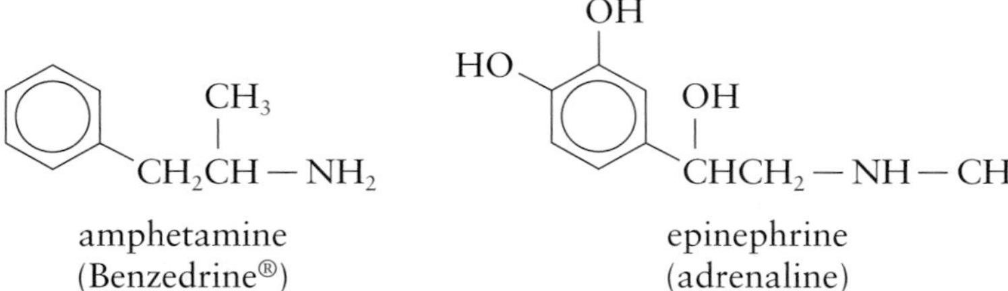

Notice that both compounds contain an aromatic ring attached to a substituted ethylamine. That structural arrangement, sometimes referred to as a phenethylamine, is apparently an essential feature of several physiologically active compounds. These structural similarities lead to similar behaviors in the body: Both compounds raise the glucose level in the blood and increase pulse rate and blood pressure. Other closely related phenethylamine compounds, such as Methedrine®, also act as powerful nervous system stimulants. Compounds of this type can be thought of as being derived from amphetamine and as a result are sometimes collectively referred to as **amphetamines:**

amphetamines
A class of drugs structurally similar to epinephrine, used to stimulate the central nervous system.

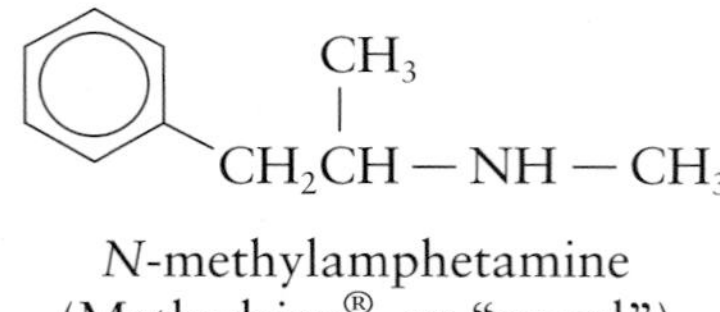

N-methylamphetamine
(Methedrine®, or "speed")

Amphetamines are used both legally (as prescribed by a physician) and illegally to elevate mood or reduce fatigue. In the drug culture, amphetamine tablets go by such names as bennies, pep pills, reds, red devils, speed, dexies, and uppers. Some amphetamines (known as STP, speed, and mescaline) cause hallucinations. The abuse of such drugs results in severe detrimental effects on both the body and the mind. They are addictive and concentrate in the brain and nervous system. Abusers often experience long periods of sleeplessness, weight loss, and paranoia. Abusers of amphetamines often compound their problems by taking other drugs in order to prevent the "crash," or deep depression, brought on by discontinuation of the drug.

Alkaloids

Plants are the sources of some of the most powerful drugs known. Numerous primitive tribes in the world possess knowledge about the physiological effects that result from eating or chewing the leaves, roots, or bark of certain plants. The effects vary from plant species to plant species. Some cure diseases; others, such as opium, are addictive drugs. Still others are deadly poisons, such as the leaves of the belladonna plant and the hemlock herb. The substances responsible for these marked physiological effects are often nitrogen-containing compounds called **alkaloids.** The name,

alkaloids
A class of nitrogen-containing organic compounds obtained from plants.

which means alkali-like, reflects the fact that these amine compounds are weakly basic.

The molecular structures of alkaloids vary from simple to complex. Two common and relatively simple alkaloids are nicotine, which is found in tobacco, and caffeine, which is present in coffee and cola drinks:

Nicotine

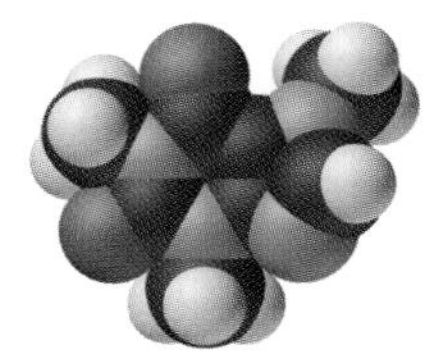

Caffeine

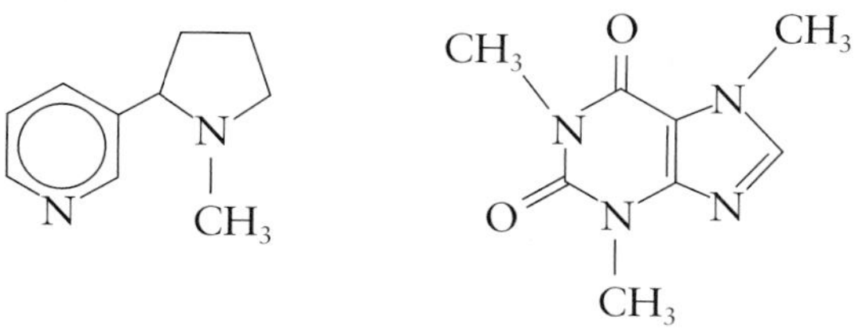

nicotine caffeine

Many people absorb nicotine into their bodies as a result of smoking or chewing tobacco. In small doses, nicotine behaves like a stimulant and is not especially harmful. However, it is habit-forming, and habitual tobacco users are exposed to other harmful substances such as tars, carbon monoxide, and polycyclic carcinogens.

Besides coffee and cola drinks, other sources of caffeine are tea, chocolate, and cocoa. Caffeine is a mild stimulant of the respiratory and central nervous systems, the reason for its well-known side effects of nervousness and insomnia. These characteristics, together with its behavior as a mild diuretic, account for the use of caffeine in a wide variety of products, including pain relievers, cold remedies, diet pills, and "stay-awake" pills (No-Doz®). Because caffeine is considered to be a drug, pregnant women should be prudent about how much caffeine they consume. Like most other drugs, caffeine enters the bloodstream, crosses the placental barrier, and reaches the fetus.

A number of alkaloids are used medicinally. Quinine, for example, is used to treat malaria. Atropine is used as a preoperative drug to relax muscles and reduce the secretion of saliva in surgical patients, and to dilate the pupil of the eye in patients undergoing eye examinations.

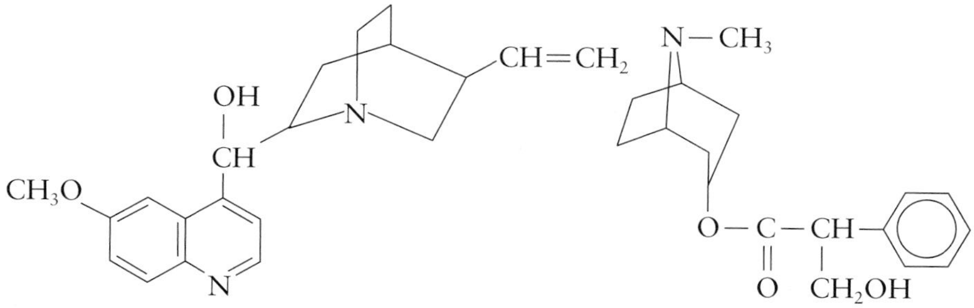

quinine atropine

Opium, the dried juice of the poppy plant, has been used for centuries as a painkilling drug. It contains numerous alkaloids, including morphine and its methyl ether, codeine. Both are central nervous system depressants, with codeine being much weaker. They exert a soothing effect on the body when administered in sufficient amounts. These properties make them useful as painkillers. Morphine is especially useful in this regard, being one of the most effective painkillers known. Codeine is used in some cough syrups to depress the action of the cough center in the brain. The major drawback to using these drugs is that they are all addictive. Heroin, a derivative of morphine, is one of the more destructive hard drugs used illegally in our society. A substantial number of addicts commit crimes to support their habits.

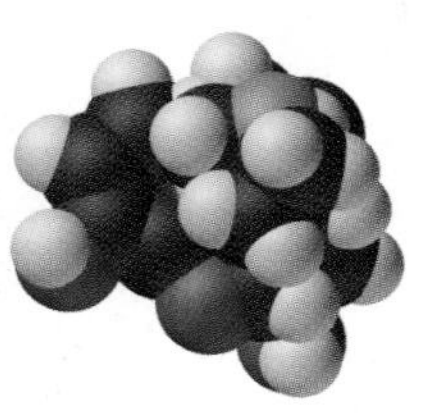

Morphine

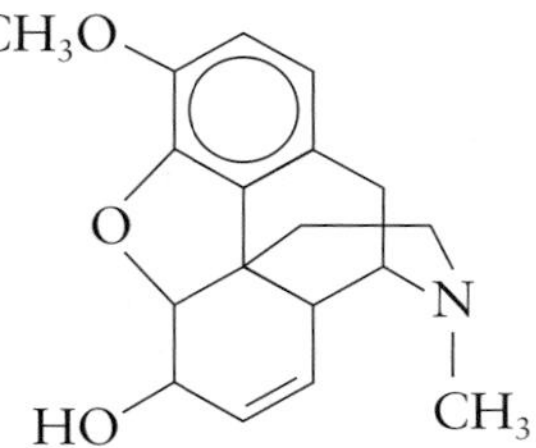

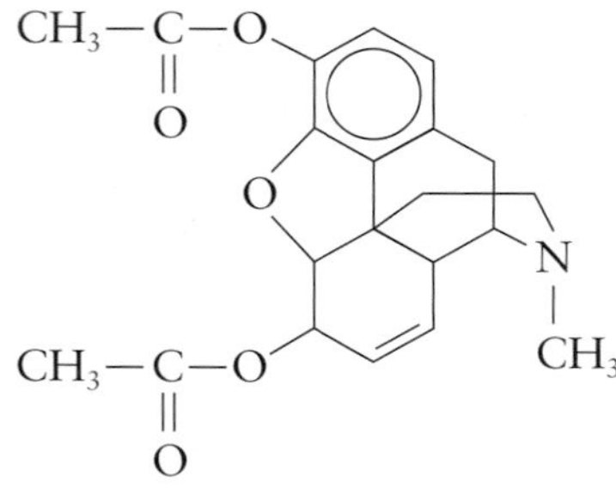

morphine codeine heroin

6.7 The Nomenclature of Amides

LEARNING OBJECTIVE

7. Assign IUPAC names for amides.

Simple amides are named after the corresponding carboxylic acids by changing the *-ic* ending (common names) or the *-oic* ending (IUPAC names) of the acid to *-amide*. Common names are used more often than IUPAC names. Some examples of the assignment of such names follow:

CHEMISTRY AND YOUR HEALTH 6.1 Chocolate: A New Health Food or Fad?

A number of candy companies are now promoting chocolate as a healthy diet supplement, a claim that needs to be carefully scrutinized. The health benefits of chocolate are determined in large part by how it was processed, as well as how much is consumed daily. To understand the chocolate of today, it is important to look at the history of this delicious little food.

Chocolate dates back to ancient America, first with the Mayans and then the Aztecs. They ground the beans of the theoboromo cacao tree into a bitter beverage that was used as a medicine. The beans were brought to Europe in the 1500s, where the raw form of cacao was used to treat anemia, fever, gout, hemorrhoids, poor digestion, depression, and heart ailments. Today's chocolate is distinctly different. It is usually a highly processed blend of chocolate liquor, cocoa butter (all fat), cocoa powder, sugar, emulsifiers, and milk. White chocolate is made from cocoa fat, sugar and flavorings, and contains no real chocolate! During processing, many of the substances thought to be responsible for the health benefits of the original cacao are lost.

The heart benefits of unprocessed cocoa are attributed in part to the presence of powerful antioxidants called flavonoids. These phytochemicals are also found in tea and red wine. Unprocessed cocoa without all the additives and fat also exerts an aspirin-like effect, which helps to prevent blood clots, a cause of heart attacks. As a bonus, chocolate also contains some plant sterols, B vitamins, magnesium, copper, and potassium, substances known to promote healthy hearts.

Unfortunately, most studies that show these benefits have been done using cocoa, or chocolate containing high levels of flavonoids, and not the chocolate candy one buys in a store. Commercial chocolates and cocoas are typically processed in ways that destroy most of their beneficial phytochemicals. The chocolate available for purchase also contains milk fat (in milk chocolate), lots of sugar, and lots of calories (about 135 to 150 calories per ounce).

Many people just read the candy company advertising or the latest headlines and then splurge on chocolate with little guilt. Once again, balance and moderation are the best policies. It's fine to buy and eat chocolate, but it shouldn't be used as a health food.

	$H-\overset{O}{\overset{\|}{C}}-OH$	$\longrightarrow$	$H-\overset{O}{\overset{\|}{C}}-NH_2$
Common name	formic acid		formamide
IUPAC name	methanoic acid		methanamide
	$CH_3-\overset{O}{\overset{\|}{C}}-OH$	$\longrightarrow$	$CH_3-\overset{O}{\overset{\|}{C}}-NH_2$
Common name	acetic acid		acetamide
IUPAC name	ethanoic acid		ethanamide
	$C_6H_5-\overset{O}{\overset{\|}{C}}-OH$	$\longrightarrow$	$C_6H_5-\overset{O}{\overset{\|}{C}}-NH_2$
Common name	benzoic acid		benzamide
IUPAC name	benzoic acid		benzamide

ThomsonNOW™ Go to Coached Problems to examine the **nomenclature of amides.**

Learning Check 6.9

a. Give the IUPAC name for the following simple amide:

$$CH_3CH_2CH_2-\overset{O}{\overset{\|}{C}}-NH_2$$

b. What is the IUPAC name for the simple amide derived from hexanoic acid?

If alkyl groups are attached to the nitrogen atom of amides (substituted amides), the name of the alkyl group precedes the name of the amide, and an italic capital *N* is used to indicate that the alkyl group is bonded to the nitrogen.

See the following examples.

	$H-\overset{O}{\overset{\|}{C}}-NH-CH_3$	$CH_3-\overset{O}{\overset{\|}{C}}-\underset{CH_3}{\underset{\|}{N}}-CH_3$	$C_6H_5-\overset{O}{\overset{\|}{C}}-\underset{CH_3}{\underset{\|}{N}}-CH_2CH_3$
Common name	*N*-methylformamide	*N*,*N*-dimethylacetamide	*N*-ethyl-*N*-methylbenzamide
IUPAC name	*N*-methylmethanamide	*N*,*N*-dimethylethanamide	*N*-ethyl-*N*-methylbenzamide

Learning Check 6.10 Give either a common name or the IUPAC name for each of the following amides:

a. $CH_3-\overset{O}{\overset{\|}{C}}-\underset{CH_2CH_3}{\underset{\|}{N}}-CH_2CH_3$

b. $CH_3CH_2CH_2-\overset{O}{\overset{\|}{C}}-NH-CH_2\overset{CH_3}{\overset{\|}{C}}HCH_3$

c. $C_6H_5-\overset{O}{\overset{\|}{C}}-\underset{CH_2CH_3}{\underset{\|}{N}}-C_6H_5$

d. $C_6H_4(CH_3)-\overset{O}{\overset{\|}{C}}-NH-CH_3$ (with CH_3 ortho on the ring)

```
         O
         ‖
H—N—C—R
   |
   H
   ⋮
   O                         H
   ‖                          |
R—C—N—H····O=C—N—H
      |                 |
      H                R
      ⋮
      O
      ‖
  R—C—N—H
          |
          H
```

■ **FIGURE 6.6** Intermolecular hydrogen bonding in an unsubstituted amide.

6.8 Physical Properties of Amides

LEARNING OBJECTIVE

8. Show the formation of hydrogen bonds with amides.

Formamide is a liquid at room temperature. It is the only unsubstituted amide that is a liquid; all others are solids.

Unsubstituted amides can form a complex network of intermolecular hydrogen bonds (see ■ Figure 6.6). It is this characteristic that causes melting points of these substances to be so high.

The substitution of alkyl or aromatic groups for hydrogen atoms on the nitrogen reduces the number of intermolecular hydrogen bonds that can form and causes the melting points to decrease. Thus, disubstituted amides often have lower melting and boiling points than monosubstituted and unsubstituted amides. We'll see in Chapter 9 that hydrogen bonding between amide groups is important in maintaining the shape of protein molecules.

Amides are rather water soluble, especially those containing fewer than six carbon atoms. This solubility results from the ability of amides to form hydrogen bonds with water. Even disubstituted amides do this because of the presence of the carbonyl oxygen (see ■ Figure 6.7).

■ **Learning Check 6.11** Show how the amide below can form the following:

a. Intermolecular hydrogen bonds with other amide molecules
b. Hydrogen bonds with water molecules

```
          O
          ‖
CH3—C—N—CH3
             |
             H
```

■ **FIGURE 6.7** Hydrogen bonding between water and (a) an unsubstituted amide and (b) a disubstituted amide.

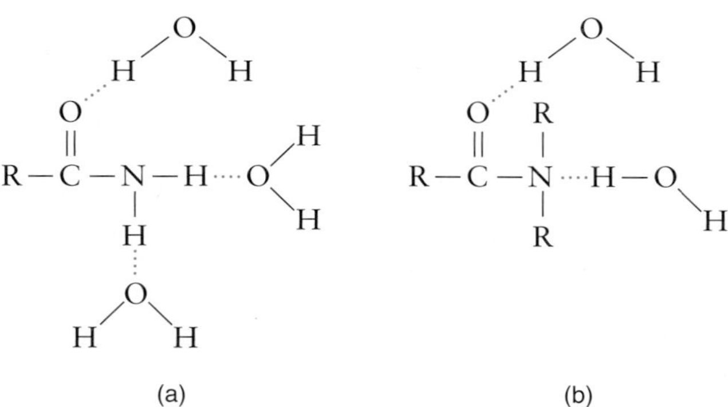

6.9 Chemical Properties of Amides

LEARNING OBJECTIVES

9. Give the products of acidic and basic hydrolysis of amides.

Neutrality

Because basicity is the most important property of amines, it is only natural to wonder if amides are also basic. The answer is no. Although they are formed from carboxylic acids and basic amines, amides are neither basic nor acidic; they are neutral. The carbonyl group bonded to the nitrogen has destroyed the basicity of the original amine, and the nitrogen of the amine has replaced the acidic —OH of the carboxylic acid.

Amide Hydrolysis

The most important reaction of amides is hydrolysis. This reaction corresponds to the reverse of amide formation; the amide is cleaved into two compounds, a carboxylic acid and an amine or ammonia:

$$\underset{\text{amide}}{R-\overset{\overset{\displaystyle O}{\|}}{C}-NH-R'} + H_2O \xrightarrow[\text{Heat}]{\text{Acid or base}} \underset{\text{carboxylic acid}}{R-\overset{\overset{\displaystyle O}{\|}}{C}-OH} + \underset{\text{amine}}{R'-NH_2} \quad (6.17)$$

(cleavage here: the C—N bond of the amide)

As with hydrolysis of carboxylic esters, amide hydrolysis requires the presence of a strong acid or base, and the nature of the final products depends on which of these catalysts is used. Notice in the following examples that under acidic conditions, the salt of the amine is produced along with the carboxylic acid, and under basic conditions, the salt of the carboxylic acid is formed along with the amine. Also, notice that heat is required for the hydrolysis of amides.

Go to Coached Problems to examine the **reactivity of amides.**

General reactions:

$$\underset{\text{amide}}{R-\overset{\overset{\displaystyle O}{\|}}{C}-NH-R'} + H_2O + HCl \xrightarrow{\text{Heat}} \underset{\text{carboxylic acid}}{R-\overset{\overset{\displaystyle O}{\|}}{C}-OH} + \underset{\text{amine salt}}{R'-NH_3^+Cl^-} \quad (6.18)$$

$$\underset{\text{amide}}{R-\overset{\overset{\displaystyle O}{\|}}{C}-NH-R'} + NaOH \xrightarrow{\text{Heat}} \underset{\text{carboxylate salt}}{R-\overset{\overset{\displaystyle O}{\|}}{C}-O^-Na^+} + \underset{\text{amine}}{R'-NH_2} \quad (6.19)$$

Specific examples:

$$\underset{\text{acetamide}}{CH_3-\overset{\overset{\displaystyle O}{\|}}{C}-NH_2} + H_2O + HCl \xrightarrow{\text{Heat}} \underset{\text{acetic acid}}{CH_3-\overset{\overset{\displaystyle O}{\|}}{C}-OH} + \underset{\text{ammonium chloride}}{NH_4^+Cl^-} \quad (6.20)$$

$$\underset{N\text{-methylacetamide}}{CH_3-\overset{\overset{\displaystyle O}{\|}}{C}-NH-CH_3} + H_2O + HCl \xrightarrow{\text{Heat}} \underset{\text{acetic acid}}{CH_3-\overset{\overset{\displaystyle O}{\|}}{C}-OH} + \underset{\text{methylammonium chloride}}{CH_3-NH_3^+Cl^-} \quad (6.21)$$

$$\underset{N\text{-methylacetamide}}{CH_3-\overset{\overset{\displaystyle O}{\|}}{C}-NH-CH_3} + NaOH \xrightarrow{\text{Heat}} \underset{\text{sodium acetate}}{CH_3-\overset{\overset{\displaystyle O}{\|}}{C}-O^-Na^+} + \underset{\text{methylamine}}{CH_3-NH_2} \quad (6.22)$$

Because the form in which carboxylic acids and amines occur in solution depends on the solution pH, one of the hydrolysis products must always be in the form of a salt.

Learning Check 6.12 Complete the following hydrolysis reactions:

a. 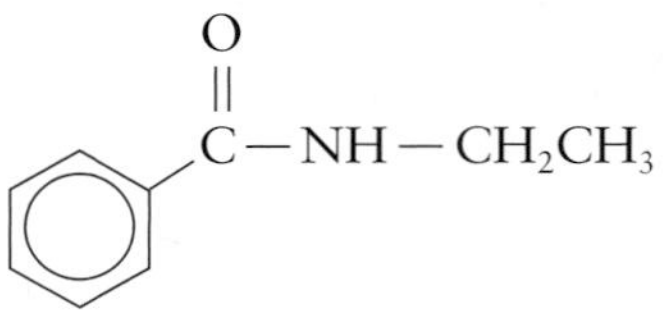 $C_6H_5-C(=O)-NH-CH_2CH_3 + H_2O + HCl \xrightarrow{\text{Heat}}$

b. 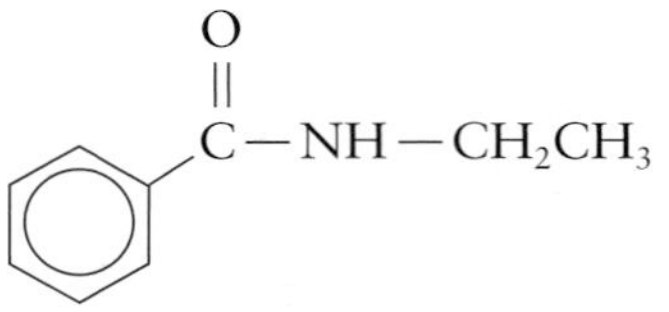 $C_6H_5-C(=O)-NH-CH_2CH_3 + NaOH \xrightarrow{\text{Heat}}$

Amide hydrolysis, a very important reaction in biochemistry, is a central reaction in the digestion of proteins and the breakdown of proteins within cells. However, most amide hydrolysis in the body is catalyzed by enzymes rather than by strong acids or bases.

Because of their physiological properties, a number of amides play valuable roles in medicine. Table 6.3 lists some of these important compounds.

TABLE 6.3 Some important amides in medicine

Structure	Generic name (trade name)	Use
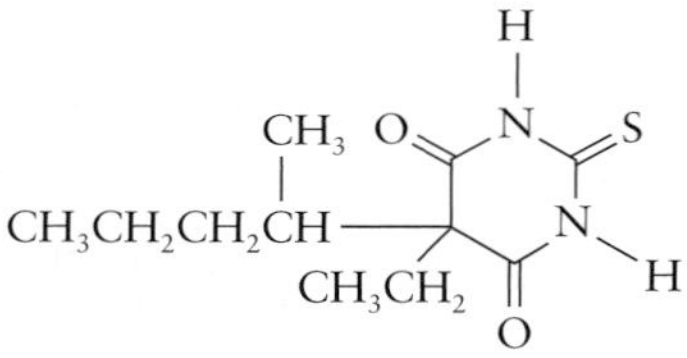 $CH_3CH_2CH_2CH(CH_3)$–, CH_3CH_2– substituted ring with N–H, C=O, C=S, N–H, C=O	thiopental (Pentothal®)	Intravenous anesthesia
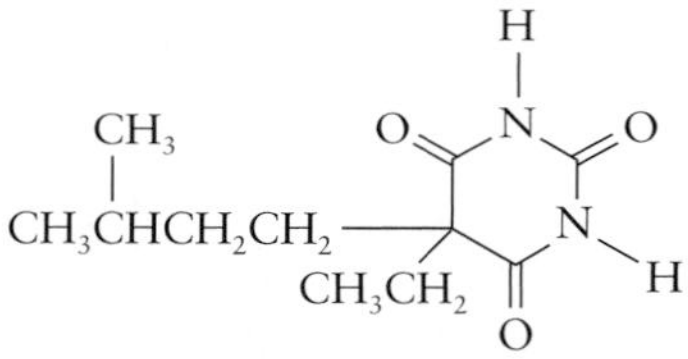 $CH_3CH(CH_3)CH_2CH_2$–, CH_3CH_2– substituted ring with N–H, C=O, C=O, N–H, C=O	amobarbital (Amytal®)	Treatment of insomnia
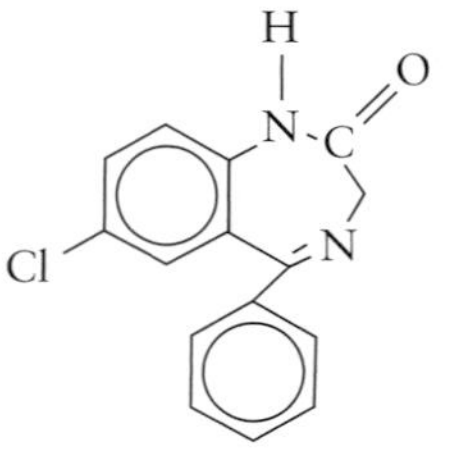 Cl-substituted benzene fused to ring with N–H, C=O, CH_2, C=N, attached phenyl	diazepam (Valium®)	Tranquilizer
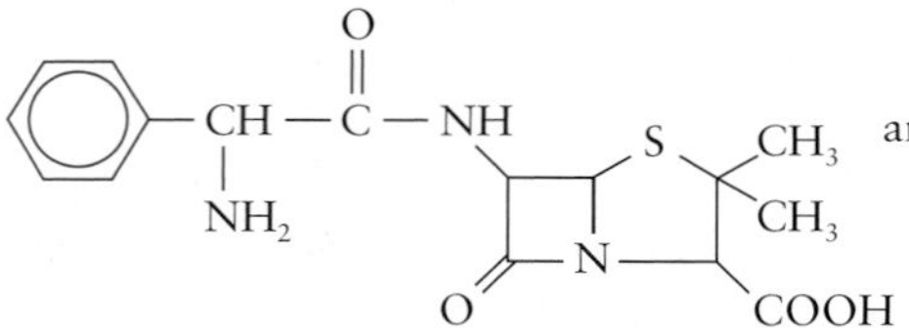 C_6H_5–CH(NH_2)–C(=O)–NH– attached to β-lactam ring (C=O, N) fused with ring containing S, C(CH_3)(CH_3), CH–COOH	ampicillin (Polycillin®)	Antibiotic

Concept Summary

ThomsonNOW™ Sign in at **www.thomsonedu.com** to:

- Assess your understanding with Exercises keyed to each learning objective.
- Check your readiness for an exam by taking the **Pre-test** and exploring the modules recommended in your **Personalized Learning Plan.**

Classification of Amines. Amines are organic derivatives of ammonia in which one or more of the ammonia hydrogens are replaced by alkyl or aromatic groups. Amines are classified as primary, secondary, or tertiary, depending on the number of groups (one, two, or three) attached to the nitrogen. **OBJECTIVE 1 (Section 6.1), Exercise 6.4**

The Nomenclature of Amines. Common names are given to simple amines by adding the ending *-amine* to the names of the alkyl groups attached to the nitrogen. In IUPAC names, the *-e* ending in the alkane name is changed to *-amine.* Aromatic amines are named as derivatives of aniline. **OBJECTIVE 2 (Section 6.2), Exercises 6.8 and 6.10**

Physical Properties of Amines. Primary and secondary amines have boiling points slightly lower than those of corresponding alcohols. Tertiary amines have boiling points similar to those of alkanes. Low molecular weight amines are water-soluble. **OBJECTIVE 3 (Section 6.3), Exercises 6.16 and 6.17**

Chemical Properties of Amines. Amines are weak bases. They react with water to liberate hydroxide ions, and they react with acids to form salts. Amines react with acid chlorides and acid anydrides to form amides. **OBJECTIVE 4 (Section 6.4), Exercise 6.25**

Amines as Neurotransmitters. Neurotransmitters are the chemical messengers of the nervous system. They carry nerve impulses from one nerve cell (neuron) to another. The most important neurotransmitters are acetylcholine and three other amines: norepinephrine, dopamine, and serotonin. **OBJECTIVE 5 (Section 6.5), Exercise 6.38**

Other Biologically Important Amines. Four neurotransmitters are acetylcholine, norepinephrine, dopamine, and serotonin. Epinephrine is also known as the fight-or-flight hormone. The amphetamines have structures similar to that of epinephrine. Alkaloids are nitrogen-containing compounds isolated from plants. They exhibit a variety of physiological effects on the body. Examples of alkaloids include nicotine, caffeine, quinine, atropine, morphine, and codeine. **OBJECTIVE 6 (Section 6.6), Exercises 6.40 and 6.44**

The Nomenclature of Amides. Amides are named by changing the *-ic acid* or *-oic acid* ending of the carboxylic acid portion of the compound to *-amide.* Groups attached to the nitrogen of the amide are denoted by an italic capital *N* that precedes the name of the attached group. **OBJECTIVE 7 (Section 6.7), Exercise 6.46**

Physical Properties of Amides. Low molecular weight amides are water-soluble due to the formation of hydrogen bonds. The melting and boiling points of unsubstituted amides are higher than those of comparable substituted amides due to intermolecular hydrogen bonding. **OBJECTIVE 8 (Section 6.8), Exercise 6.50**

Chemical Properties of Amides. Amides undergo hydrolysis in acidic conditions to yield a carboxylic acid and an amine salt. Hydrolysis under basic conditions produces a carboxylate salt and an amine. **OBJECTIVE 9 (Section 6.9), Exercise 6.52**

Key Terms and Concepts

Alkaloids (6.6)
Amide (6.4)
Amide linkage (6.4)
Amine (6.1)
Amphetamines (6.6)
Neurotransmitter (6.5)
Primary amine (6.1)
Quaternary ammonium salt (6.4)
Secondary amine (6.1)
Tertiary amine (6.1)

Key Reactions

1. Reaction of amines with water (Section 6.4):

$$R\text{—}NH_2 + H_2O \rightleftarrows R\text{—}NH_3^+ + OH^- \qquad \text{Reaction 6.3}$$

2. Reaction of amines with acids (Section 6.4):

$$R\text{—}NH_2 + HCl \rightarrow R\text{—}NH_3^+ + Cl_2 \qquad \text{Reaction 6.5}$$

3. Conversion of amine salts back to amines (Section 6.4):

$$R\text{—}NH_3^+Cl^- + NaOH \rightarrow R\text{—}NH_2 + H_2O + NaCl \qquad \text{Reaction 6.9}$$

4. Reaction of amines with acid chlorides to form amides (Section 6.4):

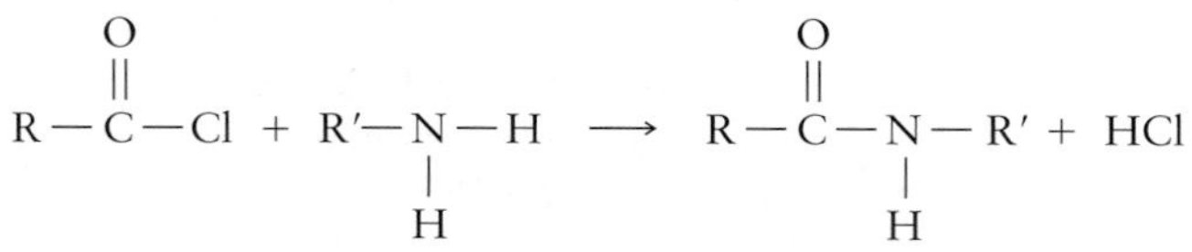

$$R-\overset{\overset{\displaystyle O}{\|}}{C}-Cl + R'-\underset{\underset{\displaystyle H}{|}}{N}-H \longrightarrow R-\overset{\overset{\displaystyle O}{\|}}{C}-\underset{\underset{\displaystyle H}{|}}{N}-R' + HCl$$ Reaction 6.11

5. Reaction of amines with acid anhydrides to form amides (Section 6.4):

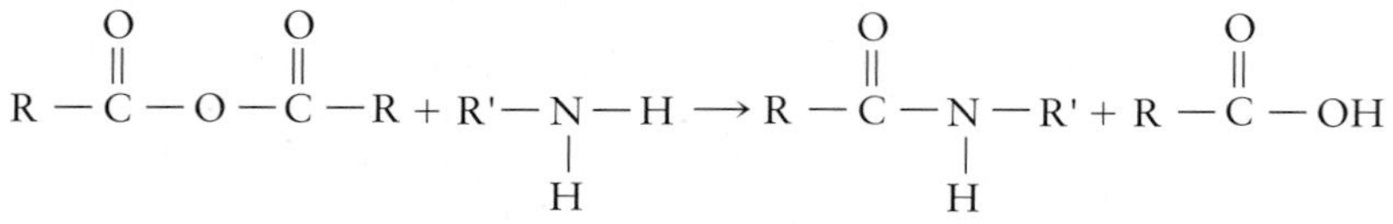

$$R-\overset{\overset{\displaystyle O}{\|}}{C}-O-\overset{\overset{\displaystyle O}{\|}}{C}-R + R'-\underset{\underset{\displaystyle H}{|}}{N}-H \longrightarrow R-\overset{\overset{\displaystyle O}{\|}}{C}-\underset{\underset{\displaystyle H}{|}}{N}-R' + R-\overset{\overset{\displaystyle O}{\|}}{C}-OH$$ Reaction 6.12

6. Acid hydrolysis of amides (Section 6.9):

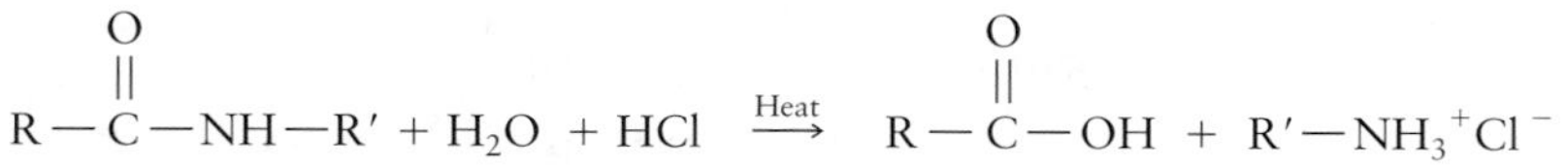

$$R-\overset{\overset{\displaystyle O}{\|}}{C}-NH-R' + H_2O + HCl \xrightarrow{\text{Heat}} R-\overset{\overset{\displaystyle O}{\|}}{C}-OH + R'-NH_3^+Cl^-$$ Reaction 6.18

7. Basic hydrolysis of amides (Section 6.9):

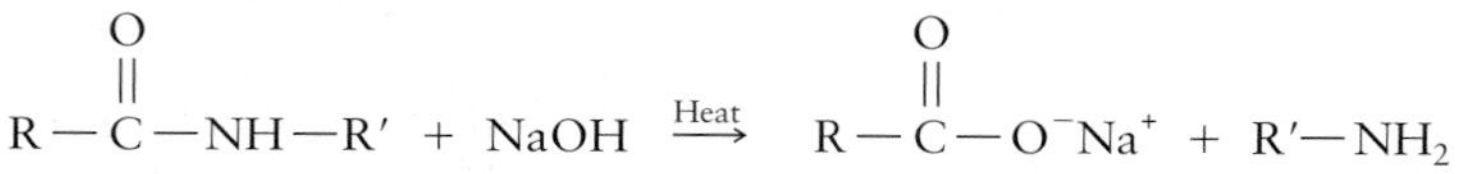

$$R-\overset{\overset{\displaystyle O}{\|}}{C}-NH-R' + NaOH \xrightarrow{\text{Heat}} R-\overset{\overset{\displaystyle O}{\|}}{C}-O^-Na^+ + R'-NH_2$$ Reaction 6.19

Exercises

SYMBOL KEY

Even-numbered exercises are answered in Appendix B.

Blue-numbered exercises are more challenging.

■ denotes exercises available in ThomsonNow and assignable in OWL.

ThomsonNOW™ To assess your understanding of this chapter's topics with sample tests and other resources, sign in at www.thomsonedu.com.

CLASSIFICATION OF AMINES (SECTION 6.1)

6.1 What is the difference among primary, secondary, and tertiary amines in terms of bonding to the nitrogen atom?

6.2 Give a general formula for a primary amine, a secondary amine, and a tertiary amine.

6.3 ■ Classify each of the following as a primary, secondary, or tertiary amine:

a. 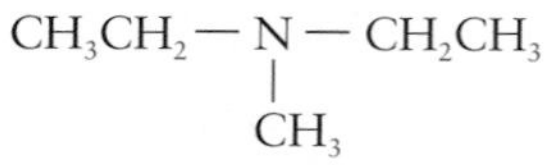 $CH_3CH_2-\underset{\underset{\displaystyle CH_3}{|}}{N}-CH_2CH_3$

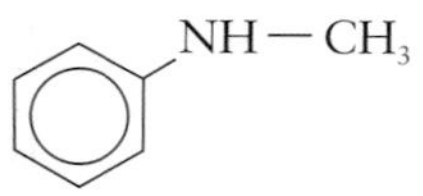

b. 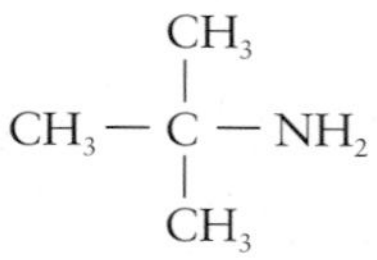 $CH_3-\underset{\underset{\displaystyle CH_3}{|}}{\overset{\overset{\displaystyle CH_3}{|}}{C}}-NH_2$

c. benzene ring $-NH-CH_3$

d. 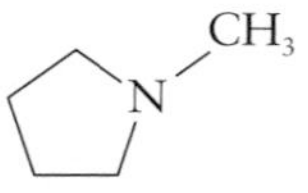 pyrrolidine ring with $N-CH_3$

6.4 Classify each of the following as a primary, secondary, or tertiary amine:

a. $CH_3CH_2-NH-CH_2CH_3$

b. 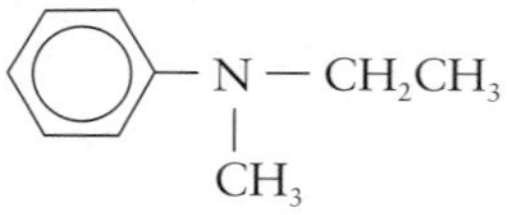 benzene ring $-\underset{\underset{\displaystyle CH_3}{|}}{N}-CH_2CH_3$

c. 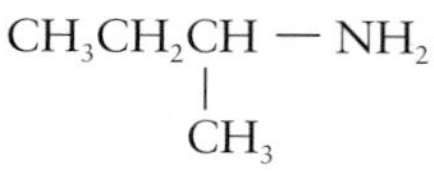 $CH_3CH_2\underset{\underset{\displaystyle CH_3}{|}}{CH}-NH_2$

d. 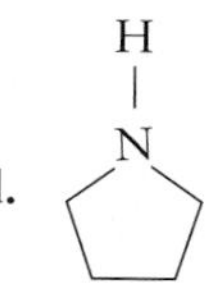 pyrrolidine ring with $N-H$

6.5 ■ Draw structural formulas for all amines that have the molecular formula $C_4H_{11}N$. Label each one as a primary, secondary, or tertiary amine.

6.6 Draw structural formulas for the four amines that have the molecular formula C_3H_9N. Label each one as primary, secondary, or tertiary.

THE NOMENCLATURE OF AMINES (SECTION 6.2)

6.7 ■ Give each of the following amines a common name by adding the ending *-amine* to alkyl group names:

a. 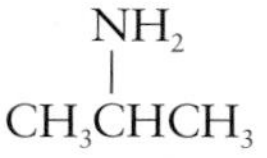 $CH_3\overset{\overset{\displaystyle NH_2}{|}}{C}HCH_3$

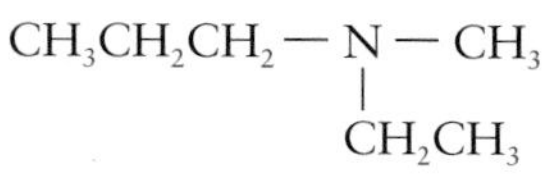

b. 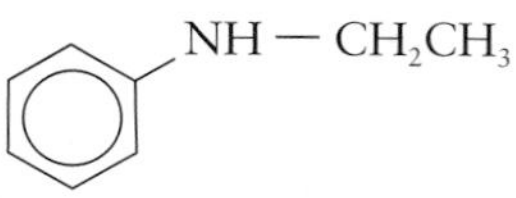 benzene ring $-NH-CH_2CH_3$

c. $CH_3CH_2CH_2-\underset{\underset{\displaystyle CH_2CH_3}{|}}{N}-CH_3$

6.8 Give each of the following amines a common name by adding the ending *-amine* to alkyl group names:

a. 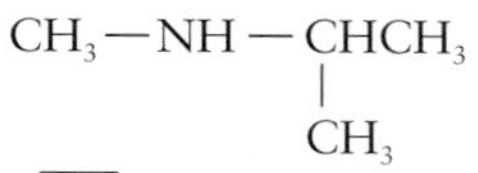 $CH_3-NH-\underset{\underset{\displaystyle CH_3}{|}}{C}HCH_3$

b. 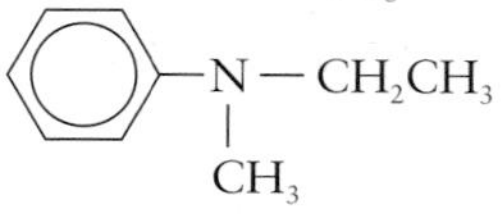 benzene ring $-\underset{\underset{\displaystyle CH_3}{|}}{N}-CH_2CH_3$

c. $CH_3CH_2CH_2CH_2-NH_2$

6.9 ■ Give each of the following amines an IUPAC name:

a. $CH_3CH(NH_2)CH_3$

b. $CH_3C(NH_2)(CH_3)CH_3$

c. $CH_3CH_2CH_2{-}NH{-}CH_2CH_3$

6.10 Give each of the following amines an IUPAC name:

a. $CH_3CH_2CH(CH_3)CH_2{-}NH_2$

b. cyclobutane ring with NH_2

c. cyclobutane ring with $NH{-}CH_3$

6.11 Name the following aromatic amines as derivatives of aniline:

a. benzene ring with $NH{-}CH(CH_3)CH_3$

b. benzene ring with $N(CH_3){-}CH_2CH_3$

6.12 Name the following aromatic amines as derivatives of aniline:

a. benzene ring with $NH{-}CH_2CH_2CH_3$

b. benzene ring with $N(CH_3){-}CH_3$

6.13 Name each of the following amines by one of the methods used in Exercises 6.7, 6.9, and 6.11:

a. $CH_3{-}C(CH_3)_2{-}NH_2$

b. $CH_3CH_2{-}N(C_6H_5){-}CH_2CH_3$

c. benzene ring with NH_2 and Cl (meta)

d. $CH_3CH(CH_3)CH_2CH(NH_2)CH_3$

6.14 ■ Draw the structural formula for each of the following amines:

a. 3-ethyl-2-pentanamine

b. *m*-ethylaniline

c. *N,N*-diphenylaniline

6.15 Draw the structural formula for each of the following amines:

a. 2,3-dimethyl-1-butanamine

b. *p*-propylaniline

c. *N,N*-dimethylaniline

PHYSICAL PROPERTIES OF AMINES (SECTION 6.3)

6.16 Explain why all classes of low molecular weight amines are water soluble.

6.17 Why are the boiling points of amines lower than those of corresponding alcohols?

6.18 ■ Why are the boiling points of tertiary amines lower than those of corresponding primary and secondary amines?

6.19 Draw diagrams similar to Figure 6.1 to illustrate hydrogen bonding between the following compounds:

a. $CH_3CH_2{-}N(H){-}CH_3$ and $CH_3CH_2{-}N(H){-}CH_3$

b. pyrrolidine (five-membered ring with N–H) and H_2O

6.20 Draw diagrams similar to Figure 6.1 to illustrate hydrogen bonding between the following compounds:

a. $CH_3CH_2{-}NH{-}CH_3$ and H_2O

b. pyrrolidine (N–H) and pyrrolidine (N–H)

6.21 Arrange the following compounds in order of increasing boiling point:

a. $CH_3CH_2CH_2{-}NH_2$

b. $CH_3{-}N(CH_3){-}CH_3$

c. $CH_3CH_2CH_2{-}OH$

6.22 Arrange the following compounds in order of increasing boiling point:

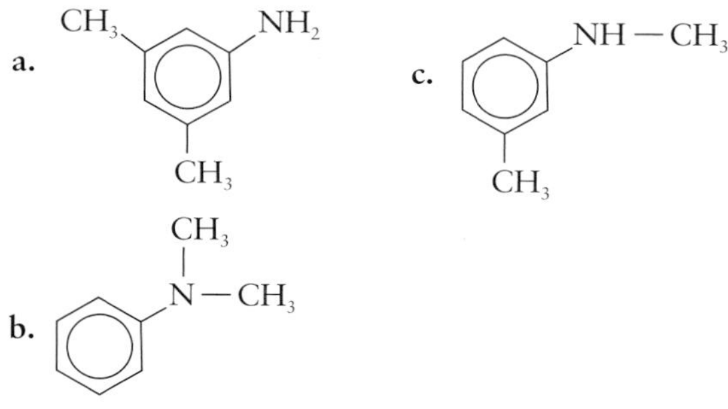

CHEMICAL PROPERTIES OF AMINES (SECTION 6.4)

6.23 Explain why $CH_3{-}NH_2$ is a Brønsted base.

6.24 ■ When diethylamine is dissolved in water, the solution becomes basic. Write an equation to account for this observation.

6.25 ■ Complete the following equations. If no reaction occurs, write "no reaction."

a. $\text{C(=O)}-\text{Cl} + NH_3 \longrightarrow$

b. $-\text{N}(CH_3)-CH_3 + CH_3\text{C(=O)}-\text{OH} \longrightarrow$

c. $CH_3CH_2-\text{N}(CH_3)-CH_3 + CH_3-\text{C(=O)}-\text{Cl} \longrightarrow$

d. $\text{N}-\text{H} + \text{HCl} \longrightarrow$

e. $\text{N}-\text{H} + H_2O \rightleftarrows$

f. $CH_3CH_2-\text{N}H_2^+(CH_3)\,Cl^- + \text{NaOH} \longrightarrow$

6.26 Complete the following equations. If no reaction occurs, write "no reaction."

a. $CH_3CH_2CH_2CH_2-NH_2 + \text{HCl} \rightarrow$

b. $NH_2 + H_2O \rightleftarrows$

c. $NH_2 + CH_3\text{C(=O)}-\text{OH} \longrightarrow$

d. $CH_3\text{CH}(NH_3^+Cl^-)CH_3 + \text{NaOH} \longrightarrow$

e. $CH_3\text{CH}(CH_3)-\text{C(=O)}-\text{Cl} + CH_3-NH_2 \longrightarrow$

f. $-\text{C(=O)}-\text{O}-\text{C(=O)}- + NH_3 \longrightarrow$

6.27 How does the structure of a quaternary ammonium salt differ from the structure of a salt of a tertiary amine?

6.28 Why are amine drugs commonly administered in the form of their salts?

6.29 Write an equation to show how the drug dextromethorphan, a cough suppressant, could be made more water soluble.

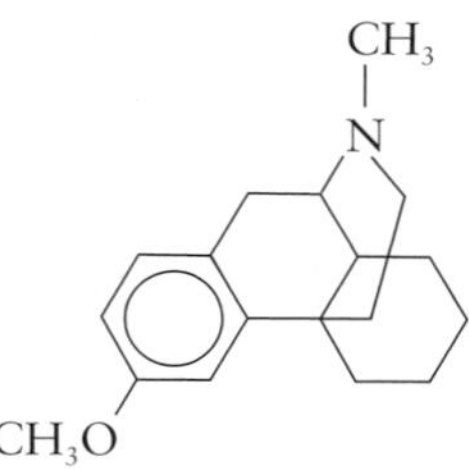

dextromethorphan

6.30 Write equations for two different methods of synthesizing the following amide:

$$CH_3-\text{C(=O)}-\text{N}(CH_3)-CH_3$$

6.31 ■ Write structures for the chemicals needed to carry out the following conversions:

a. $NH_2 \longrightarrow NH_3^+CH_3-\text{C(=O)}-\text{O}^-$

b. $CH_3-NH_2 \longrightarrow -\text{C(=O)}-\text{NH}-CH_3$

c. $CH_3\text{CH}(CH_3)-NH_2 \rightarrow CH_3\text{CH}(CH_3)-NH_3^+ + OH^-$

6.32 Write structures for the chemicals needed to carry out the following conversions:

a. $-NH_3^+Cl^- \longrightarrow -NH_2$

b. $CH_3CH_2-\text{C(=O)}-\text{Cl} \longrightarrow CH_3CH_2-\text{C(=O)}-NH_2$

c. $CH_3-\text{NH}(CH_3) \longrightarrow CH_3-NH_2^+(CH_3)\,Cl^-$

AMINES AS NEUROTRANSMITTERS (SECTION 6.5)

6.33 Describe the general structure of a neuron.

6.34 What term is used to describe the gap between one neuron and the next?

6.35 Name the two amino acids that are starting materials for the synthesis of neurotransmitters.

6.36 Name the two amines often associated with the biochemical theory of mental illness.

6.37 What role do neurotransmitters play in nerve impulse transmission?

6.38 List four neurotransmitters important in the central nervous system.

OTHER BIOLOGICALLY IMPORTANT AMINES (SECTION 6.6)

6.39 Why is epinephrine called the fight-or-flight hormone?

6.40 Describe one clinical use of epinephrine.

6.41 What are the physiological effects of amphetamines on the body?

6.42 What is the source of alkaloids?

6.43 Why are alkaloids mildly basic?

6.44 Give the name of an alkaloid for the following:

a. Found in cola drinks

b. Used to reduce saliva flow during surgery

c. Present in tobacco

d. A cough suppressant

e. Used to treat malaria

f. An effective painkiller

THE NOMENCLATURE OF AMIDES (SECTION 6.7)

6.45 ■ Assign IUPAC names to the following amides:

a. $CH_3CH_2CH(CH_2CH_3)-C(=O)-NH_2$

b. $CH_3CH_2CH_2-C(=O)-NH-C_6H_5$

c. $C_6H_5-C(=O)-NH-CH(CH_3)CH_3$

d. $CH_3CH_2-C(=O)-N(CH_3)-CH_3$

6.46 Assign IUPAC names to the following amides:

a. $CH_3CH_2CH_2-C(=O)-NH-CH_3$

b. $CH_3-C(=O)-N(CH_3)-CH_3$

c. (2-methylphenyl)$-C(=O)-NH_2$ (benzene ring with CH_3 ortho to $-C(=O)-NH_2$)

d. $CH_3CH_2CH(CH_3)-C(=O)-NH_2$

6.47 ■ Draw structural formulas for the following amides:

a. benzamide

b. *N*-methylethanamide

c. *N*-methyl-3-phenylbutanamide

6.48 Draw structural formulas for the following amides:

a. propanamide

b. *N*-ethylpentanamide

c. *N*,*N*-dimethylmethanamide

PHYSICAL PROPERTIES OF AMIDES (SECTION 6.8)

6.49 ■ Draw diagrams similar to those in Figure 6.1 to illustrate hydrogen bonding between the following molecules:

a. $CH_3-C(=O)-N(CH_3)-CH_3$ and H_2O

b. $CH_3-C(=O)-NH_2$ and $CH_3-C(=O)-NH_2$

6.50 Draw diagrams similar to those in Figure 6.1 to illustrate hydrogen bonding between the following molecules:

a. cyclopentyl$-C(=O)-NH_2$ and H_2O

b. $CH_3CH(CH_3)-C(=O)-NH_2$ and $CH_3CH(CH_3)-C(=O)-NH_2$

6.51 Explain why the boiling points of disubstituted amides are often lower than those of unsubstituted amides.

CHEMICAL PROPERTIES OF AMIDES (SECTION 6.9)

6.52 Complete the following reactions:

a. $CH_3-C(=O)-NH-CH_2CH_3 + NaOH \xrightarrow{Heat}$

b. $CH_3CH_2-C(=O)-NH_2 + H_2O + HCl \xrightarrow{Heat}$

6.53 ■ Complete the following reactions:

a. $C_6H_5-C(=O)-N(CH_3)-CH_3 + NaOH \xrightarrow{Heat}$

b. $C_6H_5-C(=O)-N(CH_3)-CH_3 + H_2O + HCl \xrightarrow{Heat}$

6.54 ■ One of the most successful mosquito repellents has the following structure and name. What are the products of the basic hydrolysis of *N,N*-diethyl-*m*-toluamide?

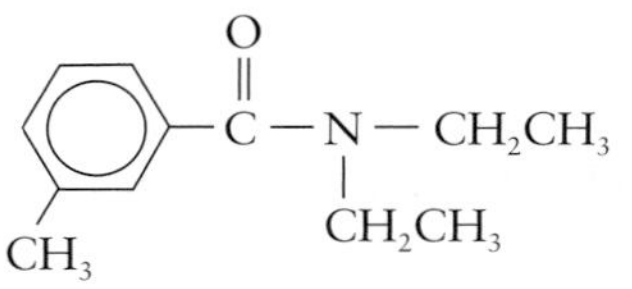

$$\text{(3-}CH_3\text{-}C_6H_4\text{)}-\overset{O}{\overset{\|}{C}}-\underset{CH_2CH_3}{\underset{|}{N}}-CH_2CH_3$$

N,N-diethyl-*m*-toluamide

6.55 What are the products of the acid hydrolysis of the local anesthetic lidocaine?

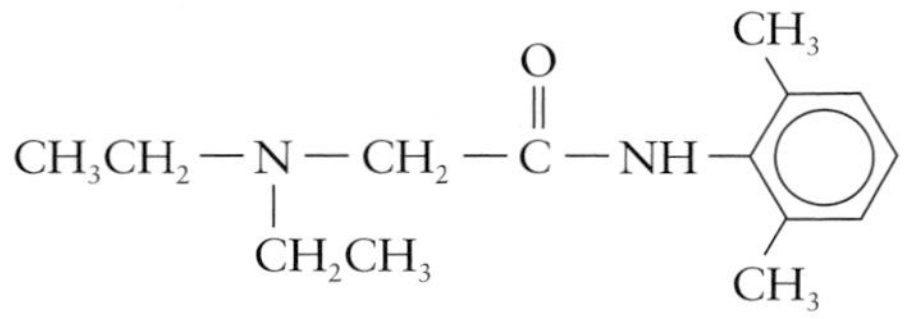

$$CH_3CH_2-\underset{CH_2CH_3}{\underset{|}{N}}-CH_2-\overset{O}{\overset{\|}{C}}-NH-\text{(2,6-}(CH_3)_2C_6H_3\text{)}$$

lidocaine (Xylocaine)

ADDITIONAL EXERCISES

6.56 ■ The following molecular formulas represent saturated amine compounds or unsaturated amine compounds that contain C=C bonds. Determine whether each compound is saturated and, if not, how many C=C bonds are present.

a. $C_4H_{12}N$

b. $C_5H_{12}N_2$

c. C_4H_7N

6.57 16.50 mL of 0.100 M HCl was needed to titrate 0.216 g of an unknown amine to a neutral pH endpoint. What is the molecular weight of the amine? Assume the compound only contains one nitrogen atom and there are no other reactive groups present.

6.58 What products would result from the hydrolysis with NaOH of nylon 66 (polymer shown here)?

$$\left[\overset{O}{\overset{\|}{C}}(CH_2)_4\overset{O}{\overset{\|}{C}}NH(CH_2)_6NH\right]_n$$

6.59 Finely ground plant material can be treated with a base to aid in the extraction of alkaloids present in the tissue. Explain why.

6.60 After a nerve impulse occurs, excess acetycholine cation molecules in the synapse are hydrolyzed in about 10^{-6} seconds by the enzyme cholinesterase. What products are formed? HINT: Identify the functional group that can undergo hydrolysis (addition of water).

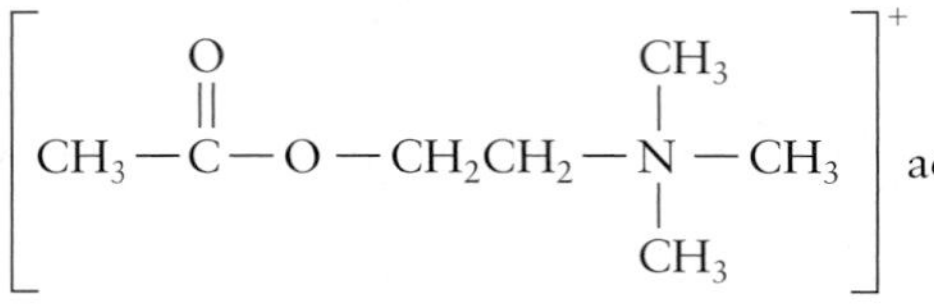

$$\left[CH_3-\overset{O}{\overset{\|}{C}}-O-CH_2CH_2-\underset{CH_3}{\underset{|}{\overset{CH_3}{\overset{|}{N}}}}-CH_3\right]^+ \quad \text{acetylcholine cation}$$

ALLIED HEALTH EXAM CONNECTION

Reprinted with permission from Nursing School and Allied Health Entrance Exams, COPYRIGHT 2005 Petersons.

6.61 Define neurotransmitter.

6.62 Give general chemical formulas for a primary, a secondary, and a tertiary amine.

CHEMISTRY FOR THOUGHT

6.63 Why might it be more practical for chemical suppliers to ship amines as amine salts?

6.64 Give structural formulas for the four-carbon diamine and the four-carbon diacid chloride that would react to form nylon.

6.65 Why might it be dangerous on a camping trip to prepare a salad using an unknown plant?

6.66 One of the active ingredients in Triaminic® cold syrup is listed as chlorpheniramine maleate. What does the word *maleate* indicate about the structure of this active ingredient?

6.67 Write reactions to show how the following conversion can be achieved. An amine reagent is required, and reactions from earlier chapters are necessary.

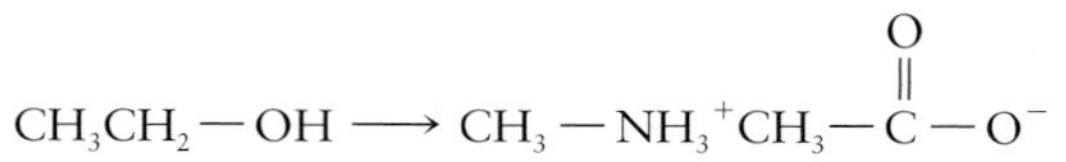

$$CH_3CH_2-OH \longrightarrow CH_3-NH_3^+CH_3-\overset{O}{\overset{\|}{C}}-O^-$$

6.68 In Figure 6.2, several products are shown that contain amine salts. Sometimes the active ingredients are given common names ending in hydrochloride. What acid do you think is used to prepare an amine hydrochloride salt?

6.69 In Section 21.1, several components of nucleic acids, DNA and RNA, are identified as "bases." What molecular feature of these substances might explain why they are referred to as bases?

6.70 The hydrolysis of amides requires rather vigorous conditions of strong acid or base and heat. What technique might be employed to achieve hydrolysis under milder conditions?

6.71 Alkaloids can be extracted from plant tissues using dilute hydrochloric acid. Why does the use of acid enhance their water solubility?

CHAPTER 7

Carbohydrates

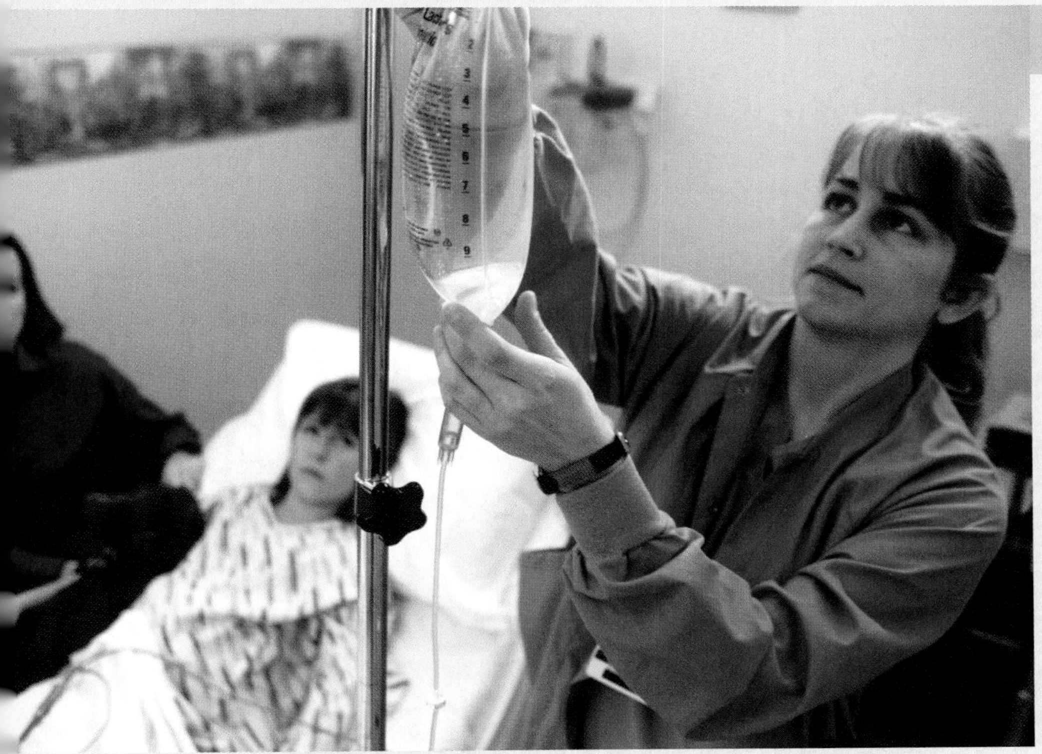

The nursing profession provides a broad range of services in a large number of different settings. This is possible because nurses specialize in one or more of a variety of patient care areas, such as emergency room, operating room, intensive care, pediatrics, and obstetrics. Here, a nurse adjusts a device that provides a patient with an intravenous infusion of a solution of glucose, a carbohydrate. In this chapter, you will learn many of the important characteristics of glucose and other carbohydrates.

LEARNING OBJECTIVES

When you have completed your study of this chapter, you should be able to:

1. Describe the four major functions of carbohydrates in living organisms. (Section 7.1)
2. Classify carbohydrates as monosaccharides, disaccharides, or polysaccharides. (Section 7.1)
3. Identify molecules possessing chiral carbon atoms. (Section 7.2)
4. Use Fischer projections to represent D and L compounds. (Section 7.3)
5. Classify monosaccharides as aldoses or ketoses, and classify them according to the number of carbon atoms they contain. (Section 7.4)
6. Write reactions for monosaccharide oxidation and glycoside formation. (Section 7.5)
7. Describe uses for important monosaccharides. (Section 7.6)
8. Draw the structures and list sources and uses for important disaccharides. (Section 7.7)
9. Write reactions for the hydrolysis of disaccharides. (Section 7.7)
10. Describe the structures and list sources and uses for important polysaccharides. (Section 7.8)

ThomsonNOW™ Throughout the chapter this icon introduces resources on the ThomsonNOW website for this text. Sign in at **www.thomsonedu.com** to:

- Evaluate your knowledge of the material
- Take an exam prep quiz
- Identify areas you need to study with a **Personalized Learning Plan**.

Carbohydrates are compounds of tremendous biological and commercial importance. Widely distributed in nature, they include such familiar substances as cellulose, table sugar, and starch (see ■ Figure 7.1). Carbohydrates have four important functions in living organisms:

- To provide energy through their oxidation.
- To supply carbon for the synthesis of cell components.
- To serve as a stored form of chemical energy.
- To form a part of the structural elements of some cells and tissues.

biomolecule
A general term referring to organic compounds essential to life.

biochemistry
A study of the compounds and processes associated with living organisms.

With carbohydrates, we begin our study of **biomolecules,** substances closely associated with living organisms. All life processes, as far as we know, involve carbohydrates and other biomolecules, including lipids, proteins, and nucleic acids. Biomolecules are organic because they are compounds of carbon, but they are also a starting point for **biochemistry,** the chemistry of living organisms. Thus, these chapters on biomolecules represent an overlap of two areas of chemistry: organic and biochemistry.

7.1 Classes of Carbohydrates

LEARNING OBJECTIVES

1. Describe the four major functions of carbohydrates in living organisms.
2. Classify carbohydrates as monosaccharides, disaccharides, or polysaccharides.

The name *carbohydrate* comes from an early observation that heating these compounds produced water and a black residue of carbon, which was incorrectly interpreted to mean that they were hydrates of carbon.

■ **FIGURE 7.1** Starches like pasta, bread, and rice are examples of carbohydrates.

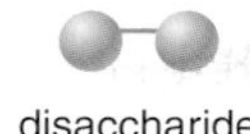

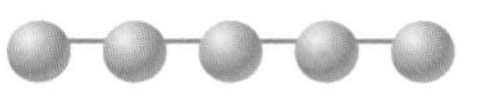

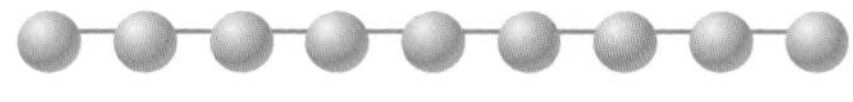

FIGURE 7.2 Carbohydrate classification.

The most striking chemical characteristic of carbohydrates is the large number of functional groups they have. For example, ribose has a functional group on every carbon atom of its five-carbon chain:

O NH S O O

saccharin

$H_3\overset{+}{N}-CH(CH_2-COO^-)-\overset{O}{\overset{\|}{C}}-NH-CH(CH_2-C_6H_5)-\overset{O}{\overset{\|}{C}}-C$

aspartame

O

ThomsonNOW™ Go to Coached Problems for practice **classifying carbohydrates.**

In addition to —OH groups, most carbohydrates have as part of their molecular structure a carbonyl group—either an aldehyde (sometimes shown as —CHO) or a ketone functional group. This is the origin of the modern definition: **Carbohydrates** are polyhydroxy aldehydes or ketones, or substances that yield such compounds on hydrolysis.

Carbohydrates can be classified according to the size of the molecules. **Monosaccharides** consist of a single polyhydroxy aldehyde or ketone unit. **Disaccharides** are carbohydrates composed of two monosaccharide units linked together chemically. Oligosaccharides (less common and of minor importance) contain from three to ten units. **Polysaccharides** consist of very long chains of linked monosaccharide units (see ■ Figure 7.2).

carbohydrate
A polyhydroxy aldehyde or ketone, or substance that yields such compounds on hydrolysis.

monosaccharide
A simple carbohydrate most commonly consisting of three to six carbon atoms.

disaccharide
A carbohydrate formed by the combination of two monosaccharides.

polysaccharide
A carbohydrate formed by the combination of many monosaccharide units.

7.2 The Stereochemistry of Carbohydrates

LEARNING OBJECTIVE

3. Identify molecules possessing chiral carbon atoms.

Glyceraldehyde, the simplest carbohydrate, exhibits a fascinating stereoisomerism shown by many organic compounds but especially characteristic of carbohydrates and many other natural substances. Glyceraldehyde can exist in two isomeric forms that are mirror images of each other (see ■ Figure 7.3). Mirror-image isomers, called **enantiomers,** have the same molecular and structural formulas but different spatial arrangements of their atoms. No amount of rotation can convert one of these structures into the other. The molecules can be compared to right and left hands, which are also mirror images of each other (see ■ Figure 7.4). Test this yourself. Put your right hand up to a mirror, and verify that its mirror image is the same as the direct image of your left hand.

enantiomers
Stereoisomers that are mirror images.

ThomsonNOW™ Go to Coached Problems for practice identifying **enantiomers.**

Compare your right hand with the right hand of a friend. Note that they can be overlapped or superimposed (see ■ Figure 7.5). Now try to rotate or twist your

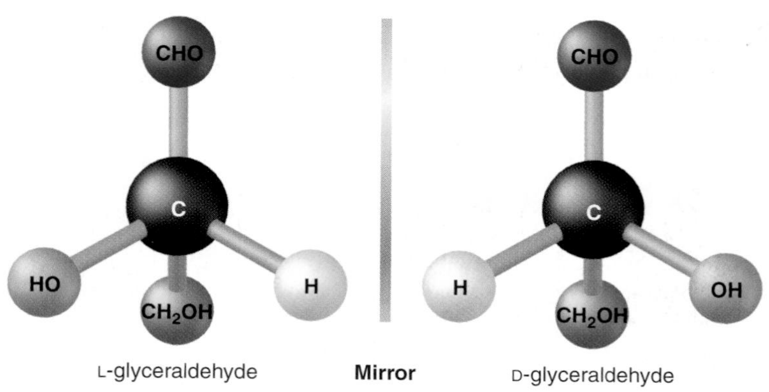

■ **FIGURE 7.3** Enantiomers of glyceraldehyde. The designations L and D are discussed in Section 7.3.

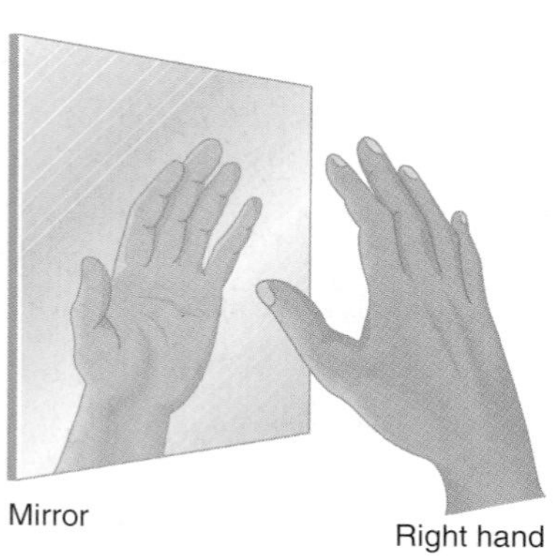

■ **FIGURE 7.4** The reflection of a right hand is a left hand.

right and left hands so that one is superimposed on the other and turned the exact same way. It cannot be done. Even if you place your hands palm-to-palm so the profiles superimpose, the nails are on the wrong sides. The same thing happens when you try to put a left-hand glove on your right hand.

chiral
A descriptive term for compounds or objects that cannot be superimposed on their mirror image.

When an object cannot be superimposed on its mirror image, the object is said to be **chiral.** Thus, a hand, a glove, and a shoe are chiral objects; they cannot be superimposed on their mirror images. However, a drinking glass, a sphere, and a cube are achiral objects because they can be superimposed on their mirror images.

Many molecules in addition to glyceraldehyde are chiral, but fortunately it is not necessary to try to superimpose them on their mirror images in order to determine whether or not they are chiral. An organic molecule that contains a carbon atom attached to four different groups will be chiral and thus will have two nonsuperimposable mirror images. A carbon atom with four different groups attached is called a **chiral carbon.**

chiral carbon
A carbon atom with four different groups attached.

The center carbon of glyceraldehyde

ThomsonNOW™ Go to Coached Problems for practice **identifying chiral centers.**

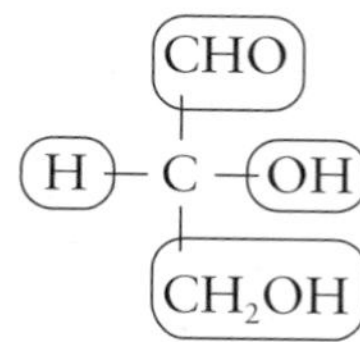

is chiral because it is attached to CHO, H, OH, and CH_2OH groups. Like your hands, the two enantiomers of glyceraldehyde are nonsuperimposable mirror images. No matter how hard we try, there is no way, short of breaking bonds, to superimpose the four groups so that they coincide.

Because the presence of a single chiral carbon atom gives rise to stereoisomerism, it is important to be able to recognize one when you see it. It is really not

■ **FIGURE 7.5** Two right hands can be superimposed; a right hand and a left hand cannot.

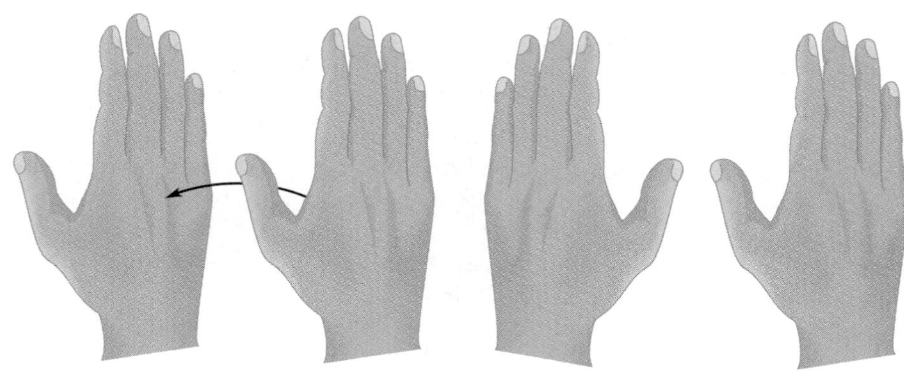

difficult. If a carbon atom is connected to four different groups, no matter how slight the differences, it is a chiral carbon. If any two groups are identical, it is not a chiral carbon.

EXAMPLE 7.1

Which of the following compounds contain a chiral carbon?

a. 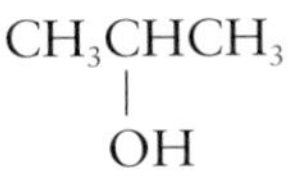 $CH_3\underset{\displaystyle OH}{\underset{|}{C}}HCH_3$

b. $CH_3\underset{\displaystyle OH}{\underset{|}{C}}HCH_2CH_3$

c. $CH_3\underset{\displaystyle O}{\underset{||}{C}}CH_2CH_3$

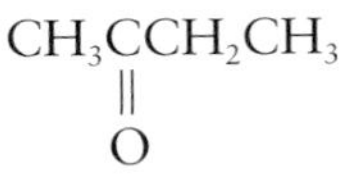

d. 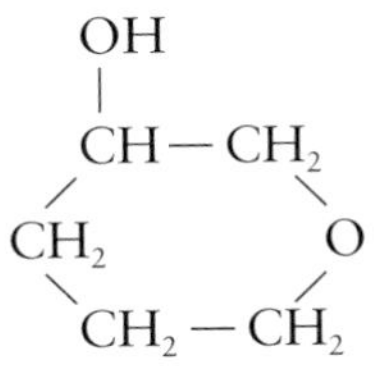 Ring: OH–CH–CH₂–O–CH₂–CH₂–CH₂– (ring closing back to CH)

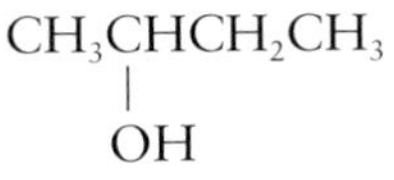

$$\begin{array}{ccc} & OH & \\ & | & \\ & CH-CH_2 & \\ / & & \backslash \\ CH_2 & & O \\ \backslash & & / \\ & CH_2-CH_2 & \end{array}$$

Solution

It may help to draw the structures in a more complete form.

a. The central carbon atom has one H, one OH, but two CH_3 groups attached. It is not chiral. Neither of the other two carbons has four different groups.

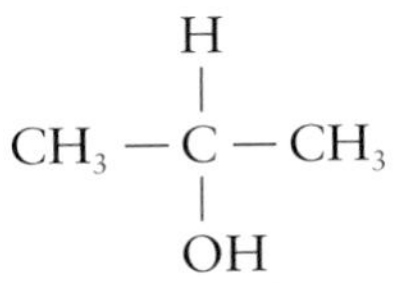

$$\begin{array}{c} H \\ | \\ CH_3-C-CH_3 \\ | \\ OH \end{array}$$

b. The central carbon atom has one H, one OH, one CH_3, and one CH_2CH_3 group attached. Both CH_3 and CH_2CH_3 are bonded to the central carbon atom through a carbon atom. However, when analyzing for chiral carbons, we look at the entire collection of atoms in a group and not just at the four atoms directly bonded. Thus, CH_3 and CH_2CH_3 groups are different (as are H and OH), so the central carbon (identified by an asterisk) is chiral. None of the other carbons are chiral.

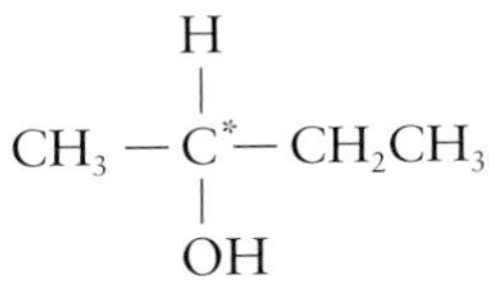

$$\begin{array}{c} H \\ | \\ CH_3-C^*-CH_2CH_3 \\ | \\ OH \end{array}$$

c. The carbonyl carbon is attached to only three groups, so it is not chiral, even though the three groups are different. None of the other carbons have four different groups attached.

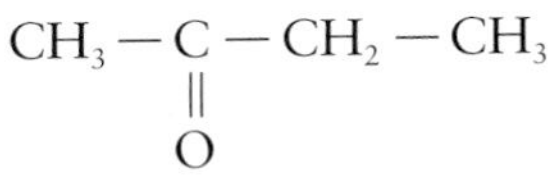

$$\begin{array}{c} CH_3-C-CH_2-CH_3 \\ || \\ O \end{array}$$

d. The rules are no different for carbons in rings. There is a chiral carbon marked with an asterisk. H and OH are two of the groups. As we proceed around the ring, —CH_2—O— (the third group) is encountered in the clockwise direction and —CH_2—CH_2— (the fourth group) in the counterclockwise direction.

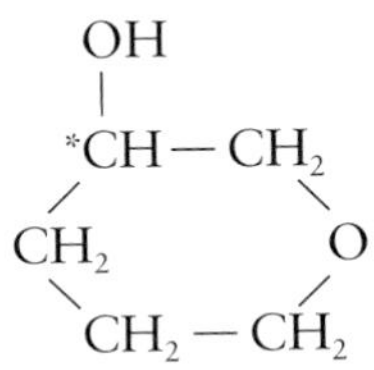

$$\begin{array}{ccc} & OH & \\ & | & \\ & {}^*CH-CH_2 & \\ / & & \backslash \\ CH_2 & & O \\ \backslash & & / \\ & CH_2-CH_2 & \end{array}$$

Learning Check 7.1 Which of the carbon atoms shown in color is chiral?

a. 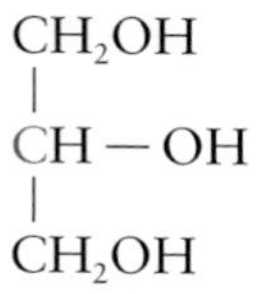
$$\begin{array}{c} CH_2OH \\ | \\ CH-OH \\ | \\ CH_2OH \end{array}$$

b.
$$\begin{array}{c} CHO \\ | \\ CH-OH \\ | \\ CH-OH \\ | \\ CH_2OH \end{array}$$

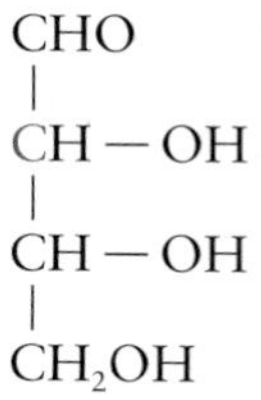

c.
$$\begin{array}{c} CH_2OH \\ | \\ C=O \\ | \\ CH-OH \\ | \\ CH_2OH \end{array}$$

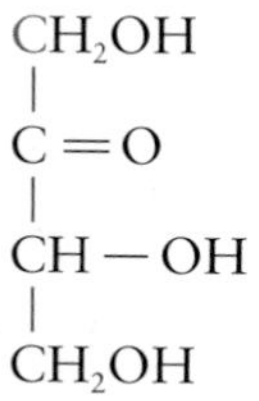

d. 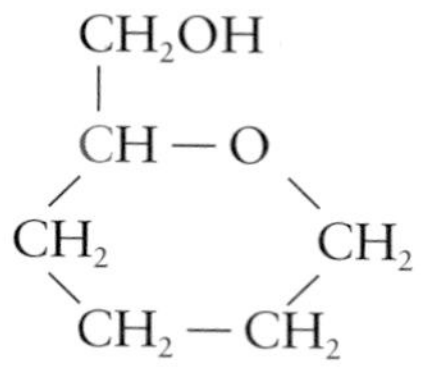
$$\begin{array}{c} CH_2OH \\ | \\ CH-O \\ / \qquad \backslash \\ CH_2 \qquad\quad CH_2 \\ \backslash \qquad / \\ CH_2-CH_2 \end{array}$$

Organic molecules (and especially carbohydrates) may contain more than one chiral carbon. Erythrose, for example, has two chiral carbon atoms:

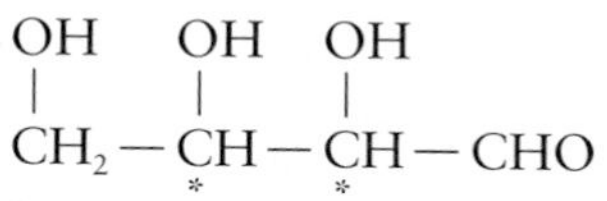

$$\begin{array}{cccc} OH & OH & OH & \\ | & | & | & \\ CH_2- & \overset{}{\underset{*}{CH}}- & \underset{*}{CH}- & CHO \end{array}$$

erythrose

EXAMPLE 7.2

Identify the chiral carbons in glucose:

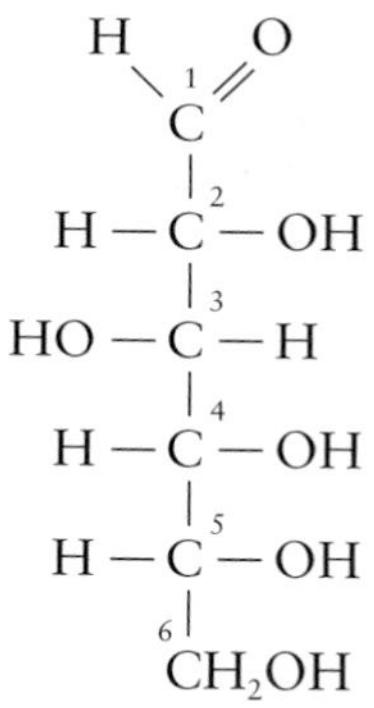

$$\begin{array}{c} H \quad O \\ \backslash \, {}^{1} \! // \\ C \\ |\,{}^{2} \\ H-C-OH \\ |\,{}^{3} \\ HO-C-H \\ |\,{}^{4} \\ H-C-OH \\ |\,{}^{5} \\ H-C-OH \\ {}^{6}| \\ CH_2OH \end{array}$$

Solution

Carbon number 1 is not chiral, because it has only three groups attached to it:

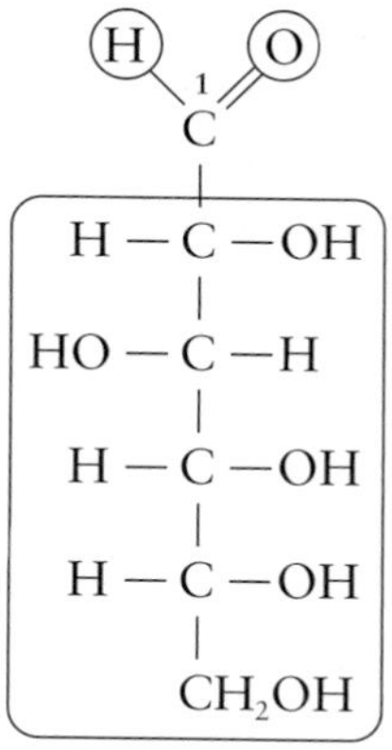

$$\begin{array}{c} (H) \quad (O) \\ \backslash \, {}^{1} \! // \\ C \\ | \\ \boxed{\begin{array}{c} H-C-OH \\ | \\ HO-C-H \\ | \\ H-C-OH \\ | \\ H-C-OH \\ | \\ CH_2OH \end{array}} \end{array}$$

Carbon atoms 2, 3, 4, and 5 are chiral because they each have four different groups attached:

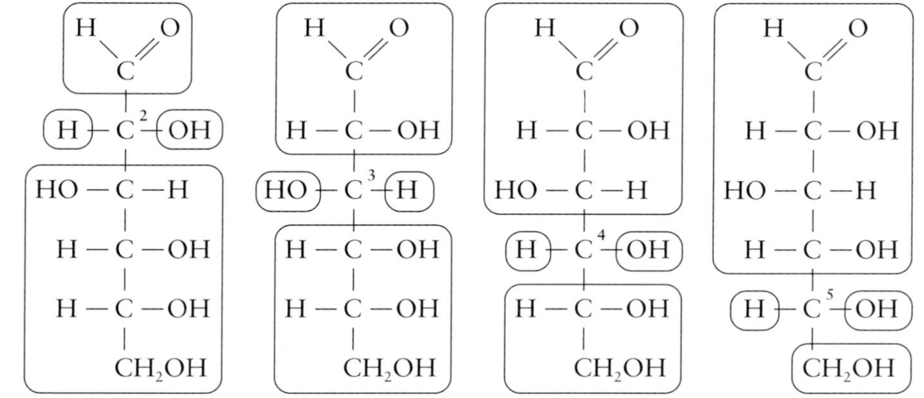

Carbon number 6 is not chiral because two of the four attached groups are the same.

■ **Learning Check 7.2** Identify the chiral carbon atoms in the following:

a.

```
      CH2OH
        |
       C=O
        |
  HO — C — H
        |
   H — C — OH
        |
      CH2OH
```

b.

```
  CHO
   |
  CH — OH
   |
  CH — OH
   |
  CH — OH
   |
  CH2OH
```

c.

```
 HOCH2   O     OH
    |  /   \   |
     H       H
  H  \_______/  H
      |     |
      OH    OH
```

When a molecule contains more than one chiral carbon, the possibility exists for two arrangements of attached groups at each chiral carbon atom. Thus, nearly all molecules having two chiral carbon atoms exist as two pairs of enantiomers, for a total of four stereoisomers. In a similar manner, when there are three chiral carbons present, there are eight stereoisomers (four pairs of enantiomers). The general formula is

$$\text{Maximum number of possible stereoisomers} = 2^n$$

where n is the number of chiral carbon atoms.

EXAMPLE 7.3

How many stereoisomers are possible for a molecule that contains four chiral carbons such as glucose?

Solution

Stereoisomers $= 2^n$

$$n = 4 \text{ chiral carbons}$$

$$2^4 = 2 \times 2 \times 2 \times 2 = 16$$

■ **Learning Check 7.3** How many stereoisomers are possible for the following?

```
 OH    OH   OH   OH
 |     |    |    |
 CH2 — CH — CH — CH — CHO
```

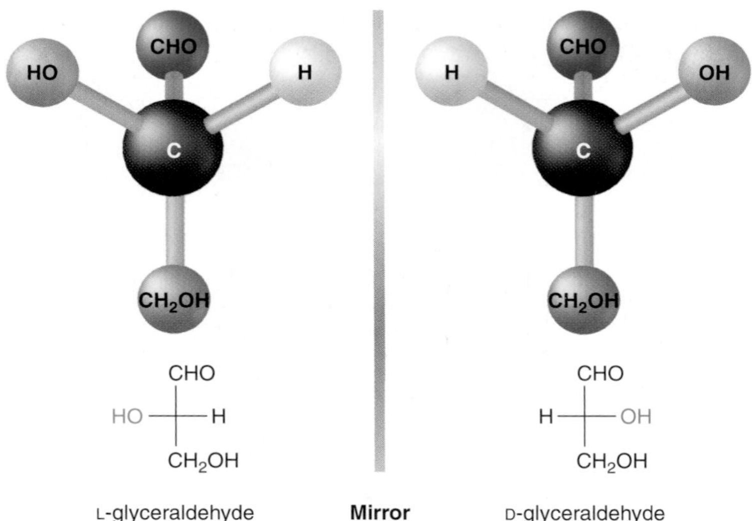

■ **FIGURE 7.6** Ball-and-stick models and Fischer projections of the two enantiomers of glyceraldehyde.

7.3 Fischer Projections

LEARNING OBJECTIVE

4. Use Fischer projections to represent D and L compounds.

It is time-consuming to draw molecules in the three-dimensional shapes shown for the two enantiomers of glyceraldehyde in Figure 7.3, but there is a way to represent these mirror images in two dimensions. Emil Fischer, a German chemist known for pioneering work in carbohydrate chemistry, introduced a method late in the 19th century. His two-dimensional structures, called **Fischer projections,** are illustrated in ■ Figure 7.6 for glyceraldehyde.

Fischer projection
A method of depicting three-dimensional shapes for chiral molecules.

In Fischer projections, the chiral carbon is represented by the intersection of two lines. If the compound is a carbohydrate, the carbonyl group is placed at or near the top. The molecule is also positioned so that the two bonds coming toward you out of the plane of the paper are drawn horizontally in the Fischer projection. The two bonds projecting away from you into the plane of the paper are drawn vertically. Thus, when you see a Fischer projection, you must realize that the molecule has a three-dimensional shape with horizontal bonds coming toward you and vertical bonds going away from you.

ThomsonNOW™ Go to Coached Problems for practice **identifying D- and L-monosaccharides.**

Fischer designated the two enantiomers of glyceraldehyde as L and D compounds. Using this system, we can represent the isomers of similar compounds by the direction of bonds about the chiral carbon. A small capital L is used to indicate that an —OH group (or another functional group) is on the left of the chiral carbon when the carbonyl is at the top. A small capital D means the —OH is on the right of the chiral carbon.

EXAMPLE 7.4

Draw Fischer projections for the D and L forms of the following:

a.
```
        OH
        |
$CH_3$ — CH — COOH
     lactic acid
```

b.
```
        NH2
        |
CH3 — CH — COOH
      alanine
```

Solution

a. The second carbon of lactic acid is chiral with four different groups attached: H, OH, CH_3, and COOH. The chiral carbon is placed at the intersection of two lines. Lactic acid is not an aldehyde, but it does contain a carbon–oxygen double bond (in the carboxyl group). This carboxyl group is placed at the top of the vertical line. The direction of the —OH groups on the chiral carbons determines the D and L notations:

L-lactic acid	Mirror	D-lactic acid
COOH HO—C—H CH_3		COOH H—C—OH CH_3

b. Similarly, the amino acid alanine has a chiral carbon and a carboxyl group. The direction of the $—NH_2$ group determines D and L notations:

L-alanine	Mirror	D-alanine
COOH H_2N—C—H CH_3		COOH H—C—NH_2 CH_3

Learning Check 7.4 Draw Fischer projections for the D and L forms of the following:

$$\begin{array}{c} \quad CH_3 \;\; NH_2 \\ CH_3CH-CH-COOH \end{array}$$

The existence of more than one chiral carbon in a carbohydrate molecule could lead to confusion of D and L designations. This is avoided by focusing on the hydroxy group attached to the chiral carbon farthest from the carbonyl group. By convention, the D family of compounds is that in which this hydroxy group projects to the right when the carbonyl is "up." In the L family of compounds, it projects to the left:

D-erythrose	L-erythrose	D-glucose	L-glucose
CHO H—C—OH H—C—OH CH_2OH	CHO HO—C—H HO—C—H CH_2OH	CHO H—C—OH HO—C—H H—C—OH H—C—OH CH_2OH	CHO HO—C—H H—C—OH HO—C—H HO—C—H CH_2OH

Notice that D and L compounds are mirror images of each other (enantiomers).

ThomsonNOW™ Go to Coached Problems for practice **relating 3-D molecular models to Fischer projections.**

■ **Learning Check 7.5** Identify each structure as D or L:

a.

$$\begin{array}{ccc} & CHO & \\ HO & - & H \\ HO & - & H \\ HO & - & H \\ & CH_2OH & \end{array}$$

b.

$$\begin{array}{ccc} & CH_2OH & \\ & C{=}O & \\ HO & - & H \\ H & - & OH \\ H & - & OH \\ & CH_2OH & \end{array}$$

c.

$$\begin{array}{ccc} & CHO & \\ H & - & OH \\ HO & - & H \\ & CH_2OH & \end{array}$$

The physical properties of the two compounds that make up a pair of D and L isomers (enantiomers) are generally the same. The only exception is the way solutions of the two compounds affect polarized light that is passed through them. Light is polarized by passing it through a special polarizing lens, such as those found in polarized sunglasses. When polarized light is passed through a solution of one enantiomer, the plane of polarization of the light is rotated to either the right or the left when viewed by looking toward the source of the light (see ■ Figure 7.7). The enantiomer that rotates it to the left is called the **levorotatory** (to the left) or (−) enantiomer. The one that rotates it to the right is the **dextrorotatory** (to the right) or (+) enantiomer.

levorotatory
Rotates plane-polarized light to the left.

dextrorotatory
Rotates plane-polarized light to the right.

The D and L designations do not represent the words *dextrorotatory* and *levorotatory* but only the spatial relationships of a Fischer projection. Thus, for example, some D compounds rotate polarized light to the right (dextrorotatory), and some rotate it to the left (levorotatory). The spatial relationships D and L and the rotation of polarized light are entirely different concepts and should not be confused with each other.

The property of rotating the plane of polarized light is called optical activity, and molecules with this property are said to be **optically active.** Measurements of optical activity are useful for differentiating between enantiomers.

optically active molecule
A molecule that rotates the plane of polarized light.

Why is it so important to be able to describe stereoisomerism and recognize it in molecules? The reason stems from the fact that living organisms, both plant and animal, consist largely of chiral substances. Most of the compounds in natural systems—including biomolecules such as carbohydrates and amino acids—are chiral. Although these molecules can in principle exist as a mixture of stereoisomers, quite often only one stereoisomer is found in nature.

In some instances, two enantiomers are found in nature, but they are not found together in the same biological system. For example, lactic acid occurs nat-

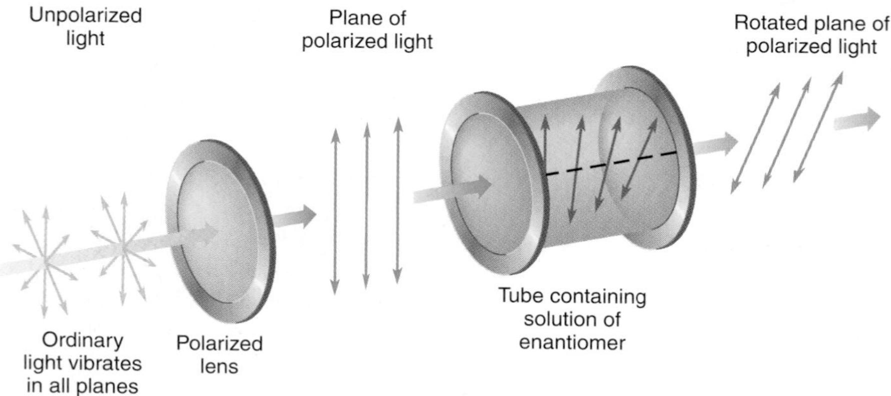

■ **FIGURE 7.7** The rotation of plane-polarized light by the solution of an enantiomer.

urally in both the forms. The L-lactic acid is found in living muscle, whereas the D-lactic acid is present in sour milk. When we realize that only one enantiomer is found in a given biological system, it is not too surprising to find that the system can usually use or assimilate only one enantiomer. Humans utilize the D-isomers of monosaccharides and are unable to metabolize the L-isomers. The D form of glucose tastes sweet, is nutritious, and is an important component of our diets. The L-isomer, on the other hand, is tasteless, and the body cannot use it. Yeast can ferment only D-glucose to produce alcohol, and most animals are able to utilize only L-amino acids in the synthesis of proteins.

TABLE 7.1 Monosaccharide classification based on the number of carbons in their chains

Number of carbon atoms	Sugar class
3	Triose
4	Tetrose
5	Pentose
6	Hexose

7.4 Monosaccharides

LEARNING OBJECTIVE

5. Classify monosaccharides as aldoses or ketoses, and classify them according to the number of carbon atoms they contain.

The simplest carbohydrates are the monosaccharides, consisting of a single polyhydroxy aldehyde or ketone unit. Monosaccharides are further classified according to the number of carbon atoms they contain (see ■ Table 7.1). Thus, simple sugars containing three, four, five, and six carbon atoms are called, respectively, trioses, tetroses, pentoses, and hexoses. The presence of an aldehyde group in a monosaccharide is indicated by the prefix *aldo-*. Similarly, a ketone group is denoted by the prefix *keto-*. Thus, glucose is an aldohexose, and ribulose is a ketopentose:

ThomsonNOW™ Go to Coached Problems for practice **identifying aldoses and ketoses.**

```
      CHO
       |
   H — C — OH
       |
  HO — C — H
       |
   H — C — OH
       |
   H — C — OH
       |
      CH2OH
glucose, an
aldohexose
```

```
      CH2OH
       |
       C = O
       |
   H — C — OH
       |
   H — C — OH
       |
      CH2OH
ribulose, a
ketopentose
```

■ **Learning Check 7.6** Classify each of the following monosaccharides by combining the aldehyde– ketone designation with terminology indicating the number of carbon atoms:

```
a.     CHO
        |
    H — C — OH
        |
   HO — C — H
        |
    H — C — OH
        |
       CH2OH
```

```
b.     CH2OH
        |
        C = O
        |
    H — C — OH
        |
       CH2OH
```

```
c.     CHO
        |
   HO — C — H
        |
    H — C — OH
        |
       CH2OH
```

Most monosaccharides are aldoses, and almost all natual monosaccharides belong to the D series. The family of D aldoses is shown in ■ Figure 7.8. D-glyceraldehyde, the smallest monosaccharide with a chiral carbon, is the standard on which the whole series is based. Notice that the bottom chiral carbon in each compound is directed to the right. The 2^n formula tells us there must be 2 trioses, 4 tetroses, 8 pentoses, and 16 hexoses. Half of those are the D compounds shown in Figure 7.8. The other half (not shown) are the enantiomers or L compounds.

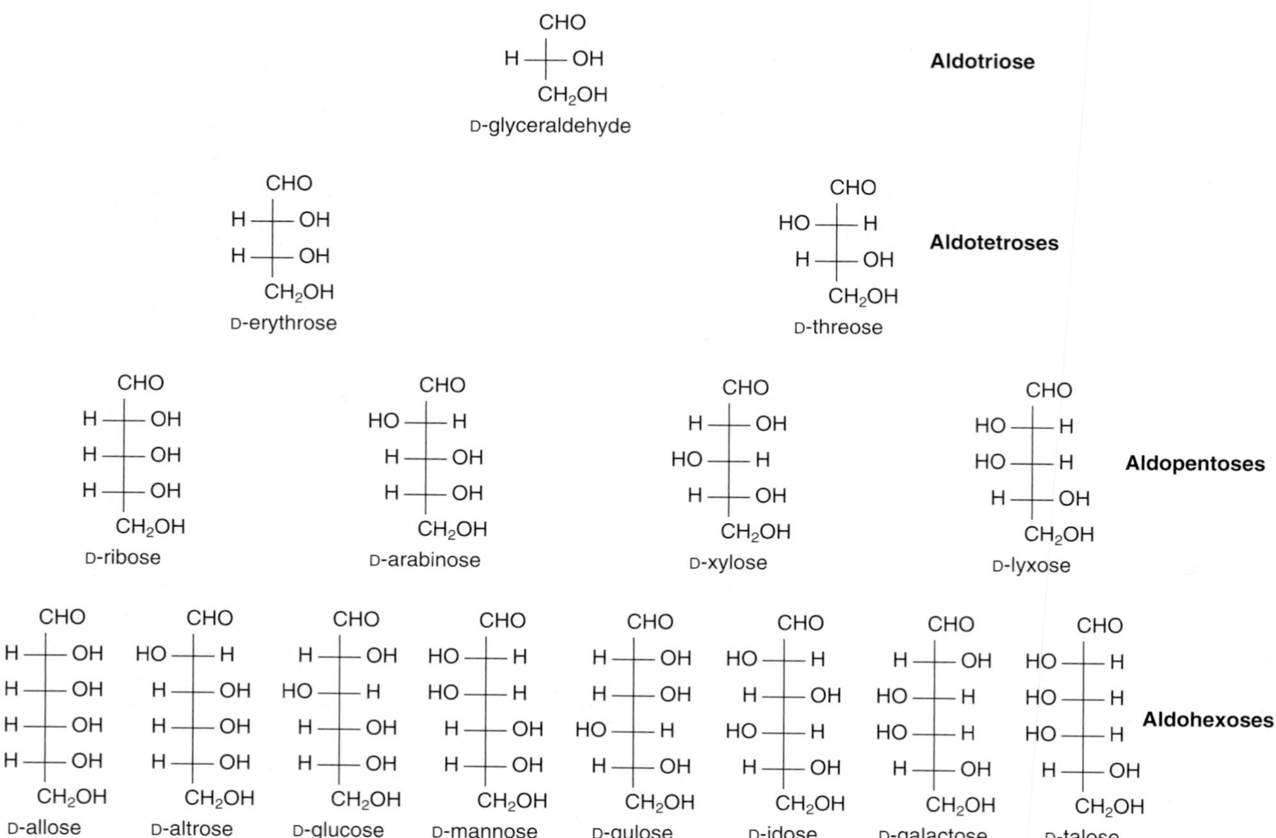

■ **FIGURE 7.8** The family of D aldoses, shown in Fischer projections.

7.5 Properties of Monosaccharides

LEARNING OBJECTIVE

6. Write reactions for monosaccharide oxidation and glycoside formation.

Physical Properties

Most monosaccharides (and disaccharides) are called sugars because they taste sweet. The degree of sweetness varies, as shown in ■ Table 7.2. Fructose, the sweetest of the common sugars, is about 73% sweeter than sucrose, ordinary table sugar.

All carbohydrates are solids at room temperature, and because of the many —OH groups present, monosaccharide carbohydrates are extremely soluble in water. The —OH groups form numerous hydrogen bonds with the surrounding water molecules. In the body, this solubility allows carbohydrates to be transported rapidly in the circulatory system.

Chemical Properties, Cyclic Forms

So far we have represented the monosaccharides as open-chain polyhydroxyaldehydes and ketones. These representations are useful in discussing the structural features and stereochemistry of monosaccharides. However, we learned in Section 4.3 that aldehydes and ketones react with alcohols to form hemiacetals and hemiketals. When these functional groups are present in the same molecule,

ThomsonNOW™ Go to Coached Problems to examine the **mutarotation of alpha and beta anomers.**

TABLE 7.2 The relative sweetness of sugars (sucrose = 1.00)

Sugar	Relative sweetness	Type
Lactose	0.16	Disaccharide
Galactose	0.22	Monosaccharide
Maltose	0.32	Disaccharide
Xylose	0.40	Monosaccharide
Glucose	0.74	Monosaccharide
Sucrose	1.00	Disaccharide
Invert sugar	1.30	Mixture of glucose and fructose
Fructose	1.73	Monosaccharide

we noted that the result of their reaction is a stable cyclic structure. You should not be surprised then to find that all monosaccharides with at least five carbon atoms exist predominantly as cyclic hemiacetals and hemiketals.

To help depict the cyclization of glucose in Reaction 7.1, the open-chain structure has been bent around to position the functional groups in closer proximity. Notice the numbering of the carbon atoms that begins at the end of the chain, giving the lowest number to the carbonyl group carbon:

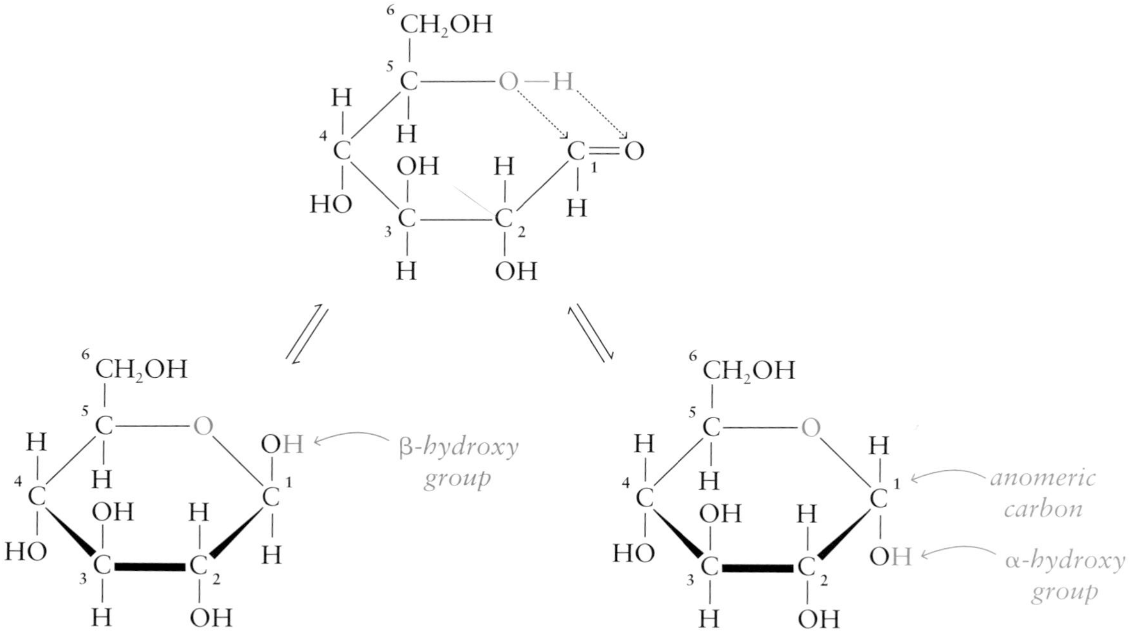

β-D-glucose α-D-glucose (7.1)

In the reaction, the alcohol group on carbon 5 adds to the aldehyde group on carbon 1. The result is a **pyranose ring,** a six-membered ring containing an oxygen atom. The attached groups have been drawn above or below the plane of the ring. This kind of drawing, called a **Haworth structure,** is used extensively in carbohydrate chemistry. Note that as the reaction occurs, a new chiral carbon is produced at position 1. Thus, two steroisomers are possible: one with the —OH group pointing down (the α form), and the other with the —OH group pointing up (the

pyranose ring
A six-membered sugar ring system containing an oxygen atom.

Haworth structure
A method of depicting three-dimensional carbohydrate structures.

anomeric carbon An acetal, ketal, hemiacetal, or hemiketal carbon atom giving rise to two stereoisomers.

anomers Stereoisomers that differ in the three-dimensional arrangement of groups at the carbon of an acetal, ketal, hemiacetal, or hemiketal group.

β form). The C-1 carbon is called an **anomeric carbon,** and the α and β forms are called **anomers.** Convenient condensed structures for the cyclic compounds omit the carbon atoms in the ring and the hydrogen atoms attached to the ring carbons, as shown in Reaction 7.2:

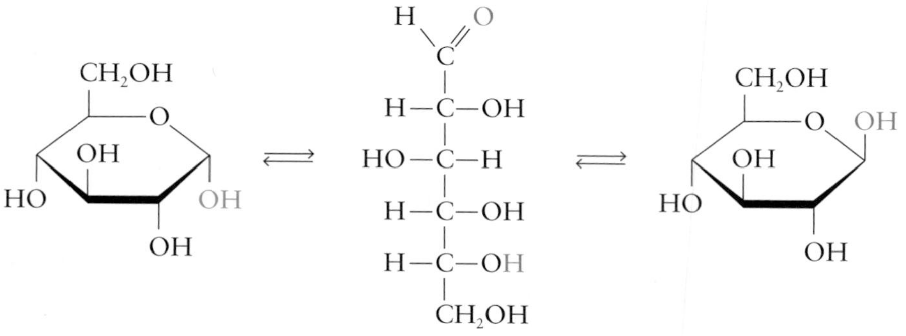

(7.2)

α-D-glucose (36%) ⇌ D-glucose (0.02%) ⇌ β-D-glucose (64%)

ThomsonNOW™ Go to Coached Problems to examine the **Haworth projections of cyclic monosaccharides.**

Thus, there are three forms of D-glucose: an open-chain structure and two ring forms. Because the reaction to form a cyclic hemiacetal is reversible, the three isomers of D-glucose are interconvertible. Studies indicate that the equilibrium distribution in water solutions is approximately 36% α-glucose, 0.02% open-chain form, and 64% β-glucose.

furanose ring A five-membered sugar ring system containing an oxygen atom.

Other monosaccharides also form cyclic structures. However, some form five-membered rings, called **furanose rings,** rather than the six-membered kind. An example is D-fructose.

The five-membered ring cyclization occurs like the ring formation in glucose. An alcohol adds across the carbonyl double bond (in this case, a ketone). The orientation of the —OH group at position 2 determines whether fructose is in the α or β form:

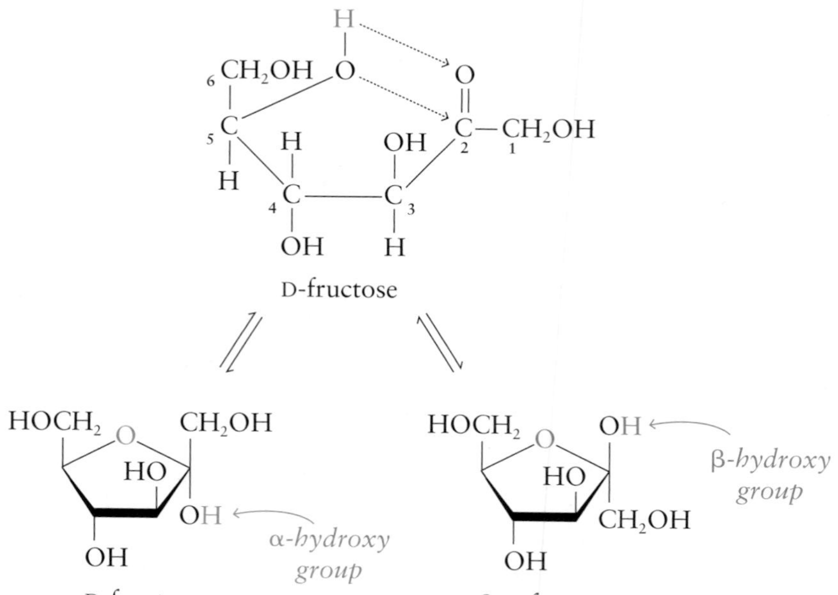

D-fructose

(7.3)

α-D-fructose (α-hydroxy group) ⇌ D-fructose ⇌ β-D-fructose (β-hydroxy group)

In drawing cyclic Haworth structures of monosaccharides, certain rules should be followed so that all our structures will be consistent. First we must draw the ring with its oxygen to the back:

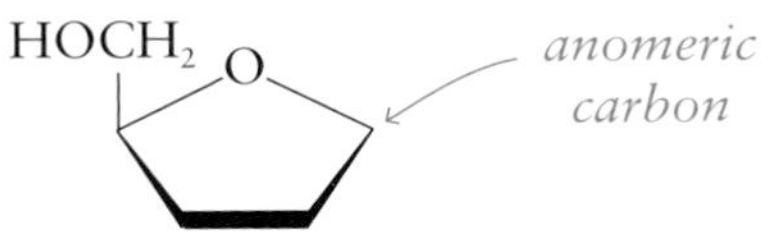

HOCH$_2$ O — anomeric carbon

furanose ring

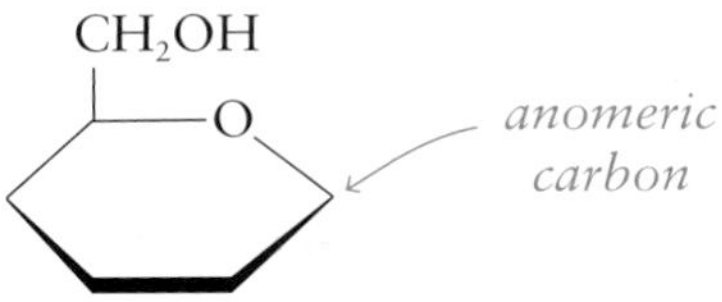

CH$_2$OH O — anomeric carbon

pyranose ring

We must also put the anomeric carbon on the right side of the ring. Then we envision the ring as planar with groups pointing up or down. The terminal —CH_2OH group (position 6) is always shown above the ring for D-monosaccharides.

EXAMPLE 7.5

The aldohexose D-galactose exists predominantly in cyclic forms. Given the structure below, draw the Haworth structure for the anomer. Label the new compound as α or β.

CH$_2$OH, HO, O, OH, OH, OH

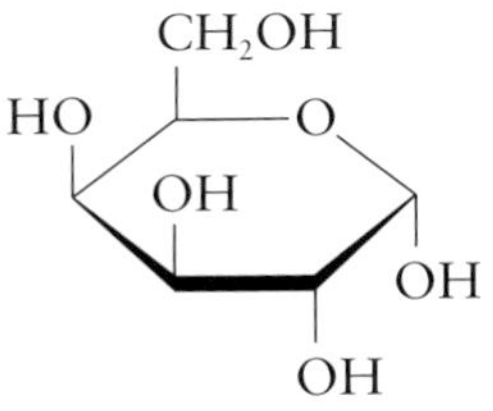

Solution

Draw the pyranose ring with the oxygen atom to the back. Number the ring starting at the right side. Position number 1 will be the anomeric carbon. Place the —OH group at position 1 in the up direction (β form) so that it is the anomer of the given compound. Place groups at the other positions exactly as they are in the given compound. Remember that anomers differ only in the position of the OH attached to the anomeric carbon.

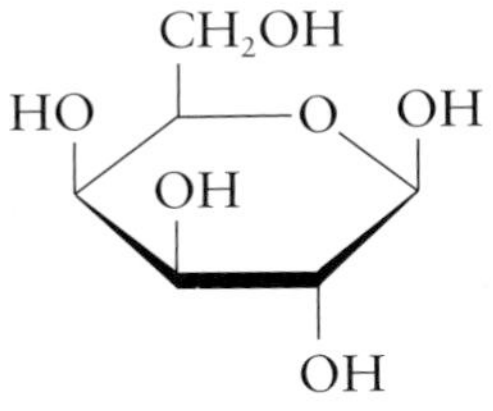

CH$_2$OH, HO, O, OH, OH, OH

β-D-galactose

Learning Check 7.7 Draw the Haworth structure for the anomer of D-ribose. Label the new compound as α or β.

HOCH$_2$, O, OH, OH, OH

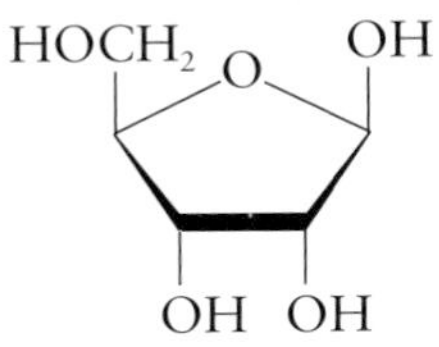

D-ribose

Oxidation

We learned in Section 4.3 that aldehydes and those ketones that contain an adjacent —OH group are readily oxidized by an alkaline solution of Cu^{2+} (Benedict's

CHEMISTRY AROUND US 7.1 Sugar-Free Foods and Diabetes

In the past, ice cream, cookies, and other sweets were considered to be off-limits to people with diabetes mellitus. However, advances in food science and a better understanding of the relationship between different foods and blood glucose levels have changed the rules, and sweet foods are now acceptable as part of the meal plan for diabetics. Diabetes is characterized by an impairment of the body's ability to utilize blood glucose.

Among the food science developments helpful to diabetics has been the discovery and commercial distribution of several synthetic noncarbohydrate sweetening agents. Three that have been approved by the American Diabetes Association are saccharin, acesulfame-K, and aspartame:

saccharin

$$\overset{+}{H_3N}-CH(CH_2COO^-)-\overset{O}{\overset{\|}{C}}-NH-CH(CH_2C_6H_5)-\overset{O}{\overset{\|}{C}}-OCH_3$$

aspartame

acesulfame-K

When compared with an equal weight of sucrose (table sugar), saccharin is about 450 times sweeter, acesulfame-K is about 200 times sweeter, and aspartame is about 180 times sweeter. Saccharin and acesulfame-K are noncaloric, whereas aspartame contributes very few calories when used in the quantities necessary to provide sweetness. Thus, these substances are useful dietary substitutes for sucrose or other calorie-laden sweeteners for people who wish to control calorie intake or for those who must avoid sugar, such as diabetics.

Until the approval of aspartame (trade name Nutrasweet®) in 1981, people with diabetes shopped for "dietetic" foods primarily in a special section of the supermarket. Aspartame vastly expanded the range of acceptable foods. Studies have verified its general safety when used in moderate amounts. The one exception is that it should not be used by individuals suffering from phenylketonuria (PKU), an inherited inability to metabolize phenylalanine properly. This condition is easily detected in newborn babies and infants, so routine screening is done shortly after birth.

In 1988, after six years of review, the FDA approved acesulfame-K under the trade name Sunette®. Pepsi-Cola® is using this artificial sweetener in a soft drink called Pepsi One®. Unlike aspartame, acesulfame-K is heat-stable and can survive the high temperatures of cooking processes.

Other sweeteners that have expanded the food options for diabetics are known as sugar alcohols. These are carbohydrate derivatives, such as sorbitol, in which the carbon–oxygen double bond of the aldehyde or ketone has been converted to an alcohol:

$$\begin{array}{c} CH_2OH \\ | \\ H-C-OH \\ | \\ HO-C-H \\ | \\ H-C-OH \\ | \\ H-C-OH \\ | \\ CH_2OH \end{array}$$

sorbitol

Sugar alcohols are incompletely absorbed during digestion and consequently contribute fewer calories than other carbohydrates. Sugar alcohols are found naturally in fruits. They are also produced commercially from carbohydrates for use in sugar-free candies, cookies, and chewing gum.

Some common artificial sweeteners.

reagent). Open-chain forms of monosaccharides exist as aldehydes or hydroxyketones and are readily oxidized. As the oxidation proceeds, the cyclic forms in equilibrium are converted to open-chain forms (Le Châtelier's principle) and also react.

Sugars that can be oxidized by weak oxidizing agents are called **reducing sugars.** Thus, all monosaccharides are reducing sugars.

reducing sugar
A sugar that can be oxidized by Cu^{2+} solutions.

General reaction:

$$\underset{\text{}}{\text{Reducing sugar}} + \underset{\substack{\text{(complex)}\\ \text{deep blue solution}}}{Cu^{2+}} \rightarrow \text{oxidized compound} + \underset{\substack{\text{red-orange}\\ \text{precipitate}}}{Cu_2O} \qquad (7.4)$$

ThomsonNOW Go to Coached Problems to examine the **reactivity of carbohydrates.**

Specific example:

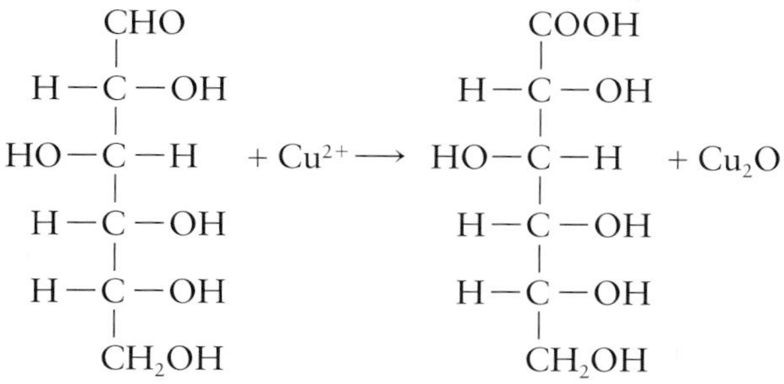

$$\underset{\text{D-glucose}}{\begin{array}{c} CHO \\ | \\ H-C-OH \\ | \\ HO-C-H \\ | \\ H-C-OH \\ | \\ H-C-OH \\ | \\ CH_2OH \end{array}} + Cu^{2+} \longrightarrow \underset{\text{D-gluconic acid}}{\begin{array}{c} COOH \\ | \\ H-C-OH \\ | \\ HO-C-H \\ | \\ H-C-OH \\ | \\ H-C-OH \\ | \\ CH_2OH \end{array}} + Cu_2O \qquad (7.5)$$

Benedict's reagent is deep blue in color. As the open-chain form of a monosaccharide is oxidized, Cu^{2+} is reduced and precipitated as Cu_2O, a red-orange solid.

When we study energy transformations in cells (Chapters 12 and 13), we will encounter many reactions that involve the stepwise oxidation of carbohydrates to CO_2 and H_2O. The controlled oxidation of carbohydrates in the body serves as an important source of heat and other forms of energy.

Phosphate Esters

The hydroxy groups of monosaccharides can behave as alcohols and react with acids to form esters. Esters formed from phosphoric acid and various monosaccharides are found in all cells, and some serve as important intermediates in carbohydrate metabolism. The structures of two representative phosphate esters are shown below:

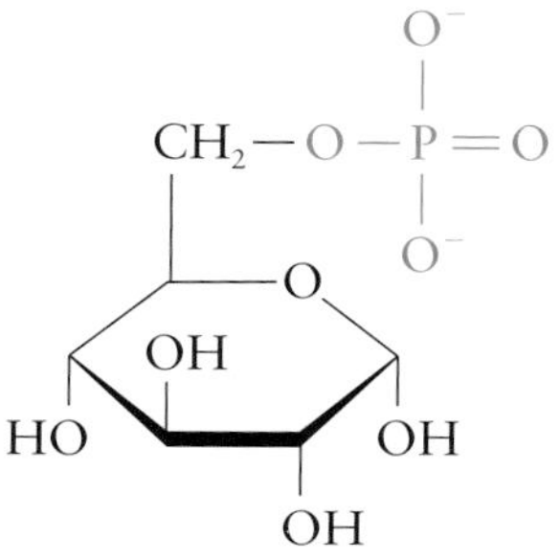

glucose 6-phosphate 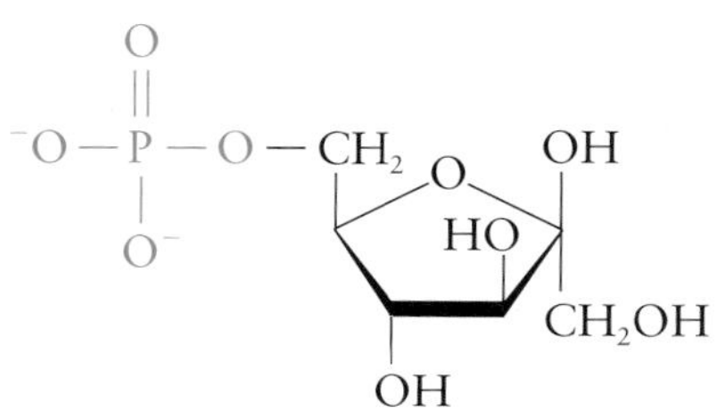 fructose 6-phosphate

Glycoside Formation

In Section 4.3, we learned that hemiacetals and hemiketals can react with alcohols in acid solutions to yield acetals and ketals, respectively. Thus, cyclic

glycoside
Another name for a carbohydrate containing an acetal or ketal group.

monosaccharides (hemiacetals and hemiketals) readily react with alcohols in the presence of acid to form acetals and ketals. The general name for these carbohydrate products is **glycosides.** The reaction of α-D-glucose with methanol is shown in Reaction 7.6.

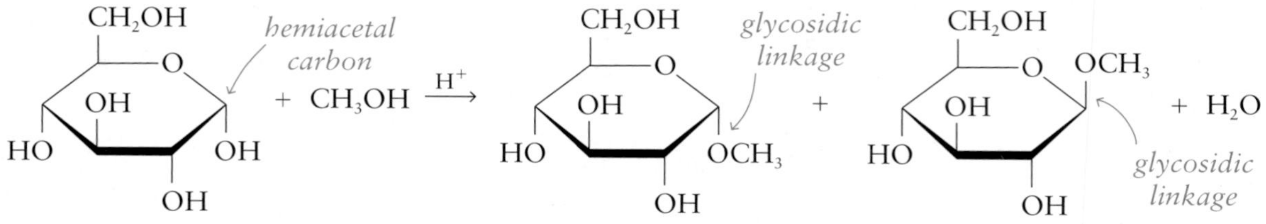

α-D-glucose (hemiacetal carbon) + CH_3OH $\xrightarrow{H^+}$ methyl α-D-glucopyranoside (glycosidic linkage) + methyl β-D-glucopyranoside (glycosidic linkage) + H_2O (7.6)

glycosides

The hemiacetal in glucose is located at position 1, and the acetal thus forms at that same position. All the other —OH groups in glucose are ordinary alcohol groups and do not react under these conditions. As the glycoside reaction takes place, a new bond is established between the pyranose ring and the $—OCH_3$ groups. The new group may point up or down from the ring. Thus, a mixture of α- and β-glycosides is formed. The new carbon–oxygen–carbon linkage that joins the components of the glycoside is called a **glycosidic linkage.**

glycosidic linkage
The carbon–oxygen–carbon linkage that joins the components of a glycoside to the ring.

Although carbohydrate cyclic hemiacetals and hemiketals are in equilibrium with open-chain forms of the monosaccharides, the glycosides (acetals and ketals) are much more stable and do not exhibit open-chain forms. Therefore, glycosides of monosaccharides are not reducing sugars.

As we'll see later in this chapter, both disaccharides and polysaccharides are examples of glycosides in which monosaccharide units are joined together by acetal (glycosidic) linkages.

Learning Check 7.8 Two glycosides are shown below. Circle any acetal or ketal groups and use an arrow to identify the glycosidic linkages.

a. [pyranose ring with CH_2OH, HO, OH, OCH_2CH_3, OH]

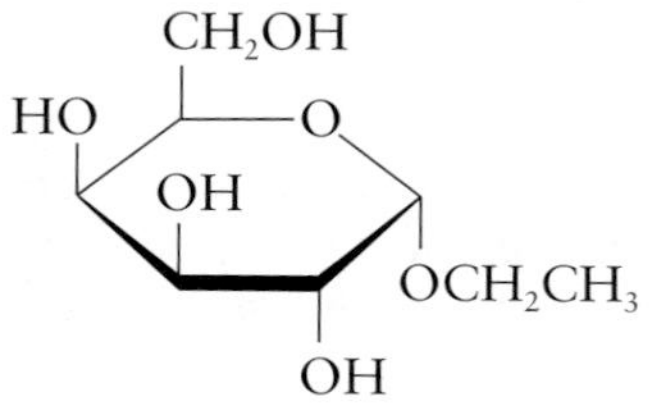

b. [furanose ring with $HOCH_2$, O, OCH_3, CH_2OH, HO, OH]

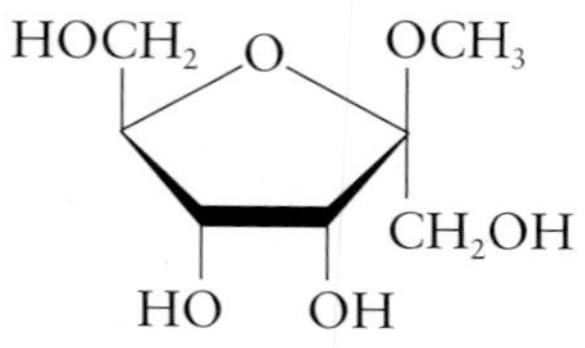

7.6 Important Monosaccharides

LEARNING OBJECTIVE

7. Describe uses for important monosaccharides.

Ribose and Deoxyribose

Two pentoses, ribose and deoxyribose, are extremely important because they are used in the synthesis of nucleic acids (DNA and RNA), substances essential in protein synthesis and the transfer of genetic material (Chapter 11). Ribose (along with phosphate groups) forms the long chains that make up ribonucleic acid (RNA). Deoxyribose (along with phosphate groups) forms the long chains of deoxyribonucleic acid (DNA). Deoxyribose differs from ribose in that the OH group on carbon 2 has been replaced by a hydrogen atom (deoxy form).

$HOCH_2$ O OH 1 2 OH OH

β-D-ribose

$HOCH_2$ O OH 1 2 OH *two hydrogens at this position*

β-D-deoxyribose

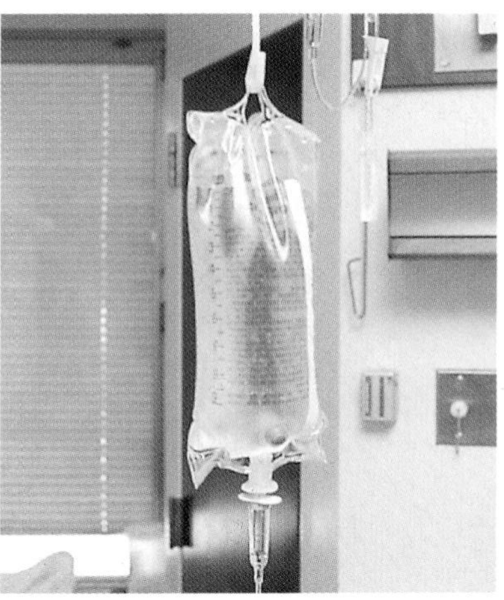

■ **FIGURE 7.9** A glucose solution prepared for intravenous use.

Glucose

Of the monosaccharides, the hexose glucose is the most important nutritionally and the most abundant in nature. Glucose is present in honey and fruits such as grapes, figs, and dates. Ripe grapes, for example, contain 20–30% glucose. Glucose is sometimes called dextrose. Glucose is also known as blood sugar because it is the sugar transported by the blood to body tissues to satisfy energy requirements (see ■ Figure 7.9). Other sugars absorbed into the body must be converted to glucose by the liver. Glucose is commonly used as a sweetener in confections and other foods, including some baby foods.

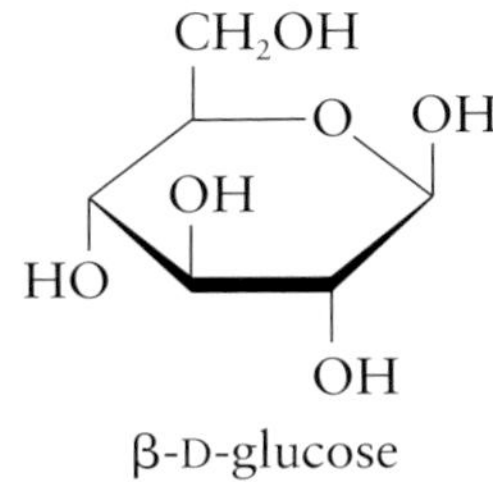

β-D-glucose

Galactose

Galactose is a hexose with a structure very similar to that of glucose. The only difference between the two is the orientation of the hydroxy group attached to carbon 4. Like glucose, galactose can exist in α, β, or open-chain forms; the β form is shown below:

ThomsonNOW™ Go to Coached Problems to examine the **common structures of monosaccharides.**

—OH group is down in glucose

$^6CH_2OH$ HO 5 O OH 1 4 OH 3 2 OH

β-D-galactose

Galactose is synthesized in the mammary glands, where it is incorporated into lactose, the sugar found in milk. It is also a component of substances present in nerve tissue.

Fructose

Fructose (Section 7.5) is the most important ketohexose. It is also known as levulose and, because of its presence in many fruits, fruit sugar. It is present in honey in a 1:1 ratio with glucose and is abundant in corn syrup. This sweetest of the common sugars is important as a food sweetener because less fructose is needed than other sugars to achieve the same degree of sweetness.

7.7 Disaccharides

LEARNING OBJECTIVES

8. Draw the structures and list sources and uses for important disaccharides.

9. Write reactions for the hydrolysis of disaccharides.

Disaccharides are sugars composed of two monosaccharide units linked together by the acetal or ketal linkages described earlier. They can be hydrolyzed to yield their monosaccharide building blocks by boiling with dilute acid or reacting them with appropriate enzymes. Nutritionally, the most important members of this group are maltose, lactose, and sucrose.

Maltose

Maltose, also called malt sugar, contains two glucose units joined by a glycosidic linkage between carbon 1 of the first glucose unit and carbon 4 of the second unit (Reaction 7.7). The configuration of the linkage on carbon 1 in the glycosidic linkage between glucose units is α, and the linkage is symbolized α(1 → 4) to indicate this fact.

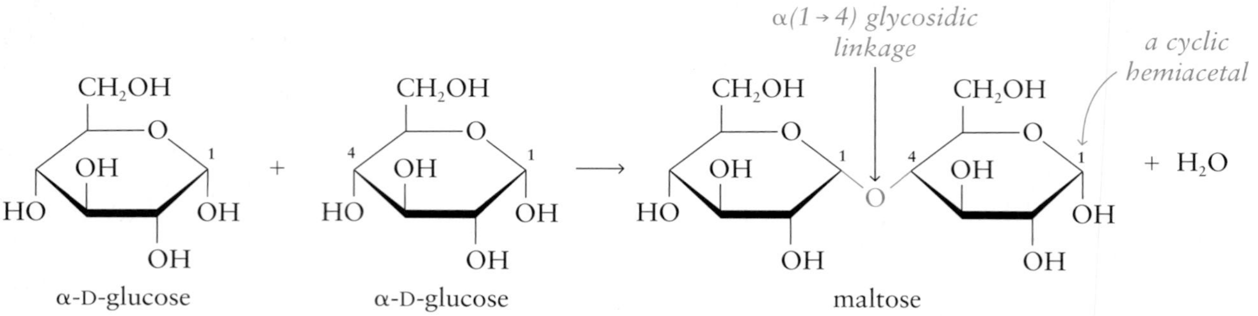

α-D-glucose + α-D-glucose ⟶ maltose + H_2O (7.7)

α(1→4) glycosidic linkage *a cyclic hemiacetal*

Notice that maltose contains both an acetal carbon (left glucose unit, position 1) and a hemiacetal carbon (right glucose unit, position 1). The —OH group at the hemiacetal position can point either up or down (the α is shown). The presence of a hemiacetal group also means that the right glucose ring can open up to expose an aldehyde group. Thus, maltose is a reducing sugar (Reaction 7.8).

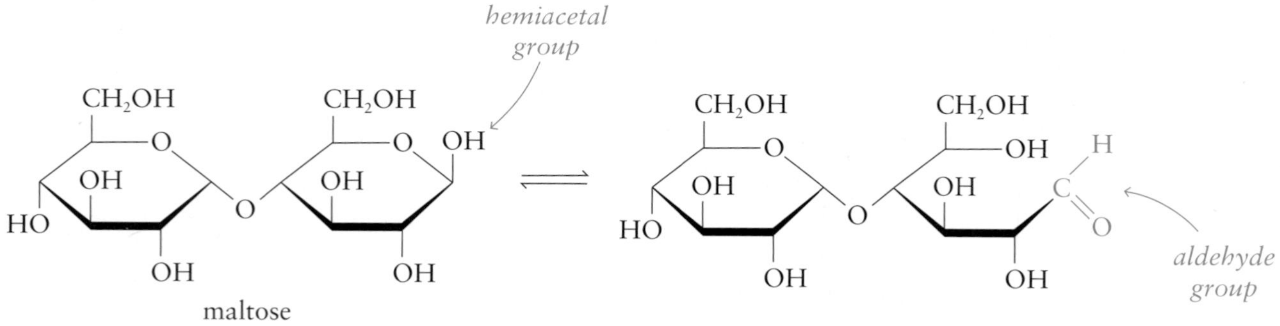

maltose ⇌ (open-chain form) (7.8)

hemiacetal group *aldehyde group*

Maltose, which is found in germinating grain, is formed during the digestion (hydrolysis) of starch to glucose. Its name is derived from the fact that during the germinating (or malting) of barley, starch is hydrolyzed, and the disaccharide is formed. On hydrolysis, maltose forms two molecules of D-glucose:

$$\text{maltose} + H_2O \xrightarrow{H^+} \text{D-glucose} + \text{D-glucose} \qquad (7.9)$$

Lactose

Lactose, or milk sugar, constitutes 5% of cow's milk and 7% of human milk by weight (see ■ Figure 7.10). Pure lactose is obtained from whey, the watery

STUDY SKILLS 7.1 Biomolecules: A New Focus

The chapters on organic chemistry (Chapters 1–6) were organized around the functional group concept. Each chapter dealt with a particular functional group or related groups. The nomenclature, properties, and uses of the compounds were discussed. The focus is different in this and the next four chapters (lipids, proteins, enzymes, and nucleic acids). Each of these chapters is devoted to a particular class of biomolecules—substances closely associated with life. Reactions and nomenclature receive much less emphasis; more time is spent describing structures and how they contribute to the processes of living organisms. Another key difference between these and earlier chapters is the reduced emphasis on mastering specific skills, such as solving numerical problems, balancing equations, and naming compounds. More attention should now be given to recognizing structures and understanding key terms and processes. It is a good time to increase your efforts to highlight important concepts in both the text and your lecture notes.

byproduct of cheese production. Lactose is composed of one molecule of D-galactose and one of D-glucose. The linkage between the two sugar units is β(1 → 4):

β-D-galactose unit *hemiacetal of α-D-glucose unit*

lactose

The presence of a hemiacetal group in the glucose unit makes lactose a reducing sugar.

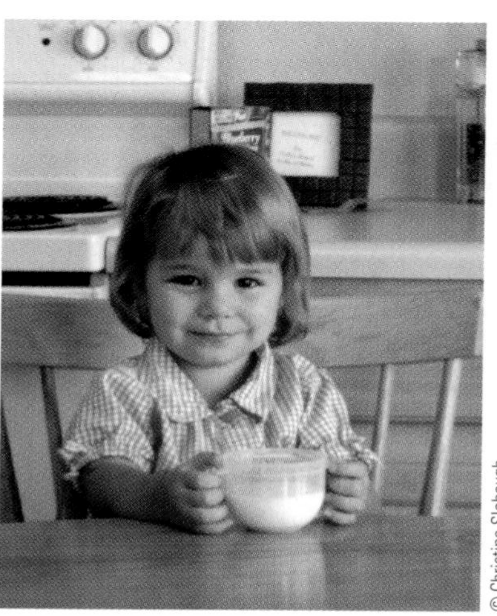

© Christina Slabaugh

FIGURE 7.10 Cow's milk is about 5% lactose. What two products form when the lactose is hydrolyzed in the child's digestive system?

Sucrose

The disaccharide sucrose, common househould sugar, is extremely abundant in the plant world. It occurs in many fruits, in the nectar of flowers (see Figure 7.11), and in the juices of many plants, especially sugar cane and sugar beets, the commercial sources. Sucrose contains two monosaccharides, glucose and fructose, joined together by a linkage that is α from carbon 1 of glucose and β from carbon 2 of fructose:

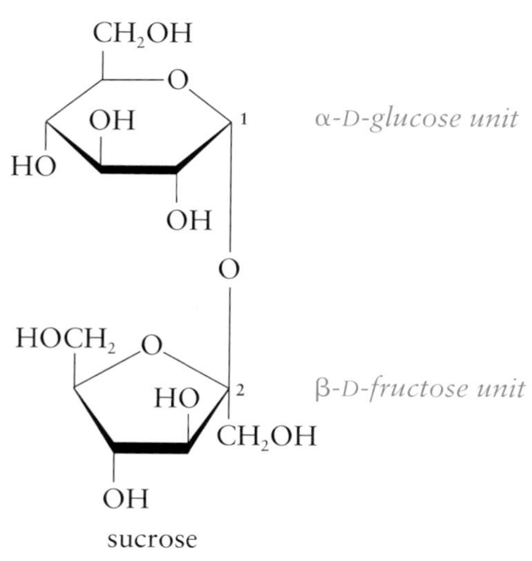

sucrose

© Neal and Mary Mishler/ Tony Stone Worldwide

FIGURE 7.11 Hummingbirds depend on the sucrose and other carbohydrates of nectar for their energy.

CHEMISTRY AND YOUR HEALTH 7.1 Sliced White Wheat Bread . . . Is It Really the Next Best Thing?

Whole grain . . . wheat . . . whole rye . . . whole wheat . . . rye . . . oats . . . white . . . wheat . . . Are you confused yet? Chances are that if you've stood in the bread aisle long enough and read the labels, you've been confused. The names and contents of the loaves sound vaguely similar, but the differences are very important and meaningful. We'll try to help you make sense of it all.

"Whole grain" means that the product is made with the entire grain kernel. A wheat-based product must say "whole wheat" or "whole grain"; otherwise, it is made from only a part of the wheat kernel. Designations of wheat, seven-grain, or multigrain all sound healthy, but such products can also contain mostly refined flour. Products bearing a label that says "whole grain" are the best options because they provide more nutrients and satisfy hunger for greater periods of time. More highly refined products use only part of the grain; the outer bran layer and inner germ of the grain are removed during processing. These removed parts contain fiber as well as a number of nutrients, which help to sustain energy and prevent the onset of hunger for greater periods of time.

To make sure you are buying products containing unrefined grains, choose foods that list one of the following whole-grain ingredients first on the label:

- Whole oats
- Whole rye
- Whole wheat
- Brown rice
- Wild rice
- Bulgur
- Graham flour
- Oatmeal
- Whole-grain corn

A new type of wheat, called "white wheat," is popular in Australia and is now being introduced in the United States. The outer bran of white wheat is lighter colored and has a milder flavor than hard red wheat, which is commonly grown and used in the United States. One disadvantage of white wheat is that a standard piece of whole grain bread contains at least 2 grams of fiber, while many of the breads made using whole grain white wheat contain only 1 to $1^1/_2$ grams of fiber.

So, the next time you are standing in the bread aisle, take time to read the labels and make thoughtful decisions.

Neither ring in the sucrose molecule contains a hemiacetal or hemiketal group necessary for ring opening. This is true because the anomeric positions in both glucose and fructose are part of the glycosidic linkage that connects the rings to each other. Thus, in contrast to maltose and lactose, both rings of sucrose are locked in the cyclic form, and sucrose is therefore not a reducing sugar (see ■ Figure 7.12).

invert sugar
A mixture of equal amounts of glucose and fructose.

The hydrolysis of sucrose (see ■ Figure 7.13) produces a 1:1 mixture of D-glucose and D-fructose called **invert sugar** (Reaction 7.10):

$$\text{sucrose} + H_2O \xrightarrow{H^+} \text{D-glucose} + \text{D-fructose} \tag{7.10}$$

This mixture is sweeter than the original sucrose (see Table 7.2).

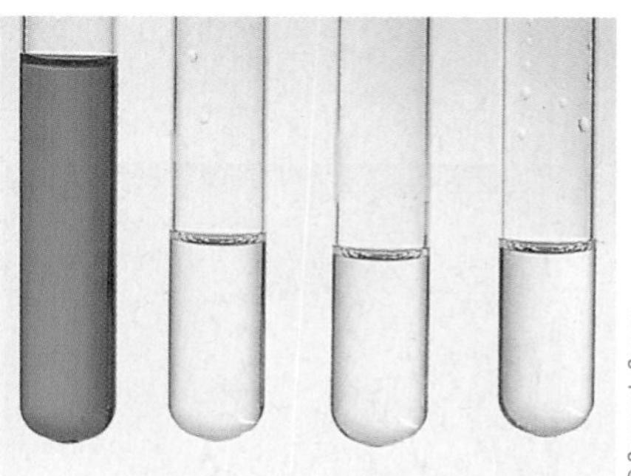
© Spencer L. Seager

From left to right, four test tubes containing Benedict's reagent, 2% maltose solution, 2% sucrose solution, and 2% lactose solution.

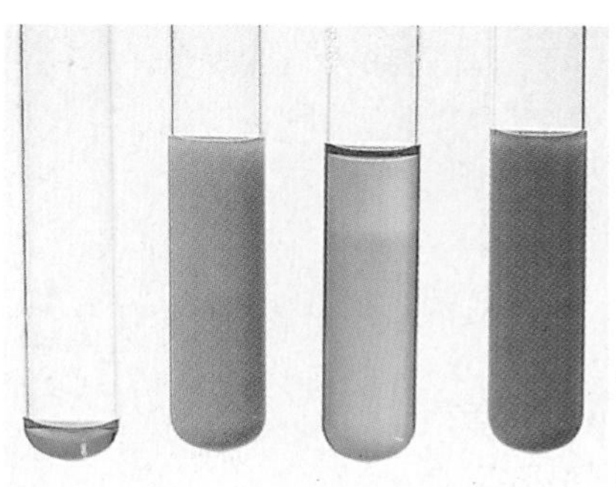
© Spencer L. Seager

Both maltose and lactose are reducing sugars. The sucrose has remained unreacted.

■ **FIGURE 7.12** Benedict's test on disaccharides.

© Christina Slabaugh

■ **FIGURE 7.13** Sucrose is hydrolyzed by bees in making honey. Why is honey sweeter than the starting sucrose?

TABLE 7.3 Some important disaccharides

Name	Monosaccharide constituents	Glycoside linkage	Source
Maltose	Two glucose units	$\alpha(1 \rightarrow 4)$	Hydrolysis of starch
Lactose	Galactose and glucose	$\beta(1 \rightarrow 4)$	Mammalian milk
Sucrose	Glucose and fructose	α-1 $\rightarrow$ β-2	Sugar cane and sugar beet juices

When sucrose is cooked with acid-containing fruits or berries, partial hydrolysis takes place, and some invert sugar is formed. The jams and jellies prepared in this manner are actually sweeter than the pure sugar originally put in them. ■ Table 7.3 summarizes some features of the three disaccharides we have discussed.

7.8 Polysaccharides

LEARNING OBJECTIVE

10. Describe the structures and list sources and uses for important polysaccharides.

Just as two sugar units are linked together to form a dissacharide, additional units can be added to form higher saccharides. Nature does just this in forming polysaccharides, which are condensation polymers containing thousands of units. Because of their size, polysaccharides are not water-soluble, but their many hydroxy groups become hydrated individually when exposed to water, and some polysaccharides form thick colloidal dispersions when heated in water. Thus, the polysaccharide known as starch can be used as a thickener in sauces, gravies, pie fillings, and other food preparations. As shown in ■ Table 7.4, the properties of polysaccharides differ markedly from those of monosaccharides and disaccharides.

ThomsonNOW™ Go to Coached Problems to examine the **structures of polysaccharides.**

Starch

Starch is a polymer consisting entirely of D-glucose units. It is the major storage form of D-glucose in plants. Two starch fractions, amylose (10–20%) and amylopectin (80–90%), can usually be isolated from plants. Amylose is made up of long unbranched chains of glucose units connected by $\alpha(1 \rightarrow 4)$ linkages (see ■ Figure 7.14). This is the same type of linkage found in maltose. The long chain,

TABLE 7.4 Properties of polysaccharides compared with those of monosaccharides and disaccharides

Property	Monosaccharides and disaccharides	Polysaccharides
Molecular weight	Low	Very high
Taste	Sweet	Tasteless
Solubility in water	Soluble	Insoluble or form colloidal dispersions
Size of particles	Pass through a membrane	Do not pass through a membrane
Test with Cu^{2+} for reducing sugars	Positive (except for sucrose)	Negative

$\alpha(1 \rightarrow 4)$ linkage

■ **FIGURE 7.14** The structure of amylose.

often containing between 1000 and 2000 glucose units, is flexible enough to allow the molecules to twist into the shape of a helix (see ■ Figure 7.15a).

An important test for the presence of starch is the reaction that occurs between iodine, I_2, and the coiled form of amylose. The product of the reaction is deep blue in color (see ■ Figure 7.16) and is thought to consist of the amylose helix filled with iodine molecules (see Figure 7.15b). This same iodine reaction is also widely used to monitor the hydrolysis of starch. The color gradually fades and finally disappears, as starch is hydrolyzed by either acid or enzymes to form dextrins (smaller polysaccharides), then maltose, and finally glucose. The disappearance of the blue iodine color is thought to be the result of the breakdown of the starch helix.

Amylopectin, the second component of starch, is not a straight-chain molecule like amylose but contains random branches. The branching point is the $\alpha(1 \rightarrow 6)$ glycosidic linkage (see ■ Figure 7.17). There are usually 24 to 30 D-glucose units, all connected by $\alpha(1 \rightarrow 4)$ linkages, between each branch point of amylopectin. These branch points give amylopectin the appearance of a bushy molecule (Figure 7.17). Amylopectin contains as many as 10^5 glucose units in one gigantic molecule.

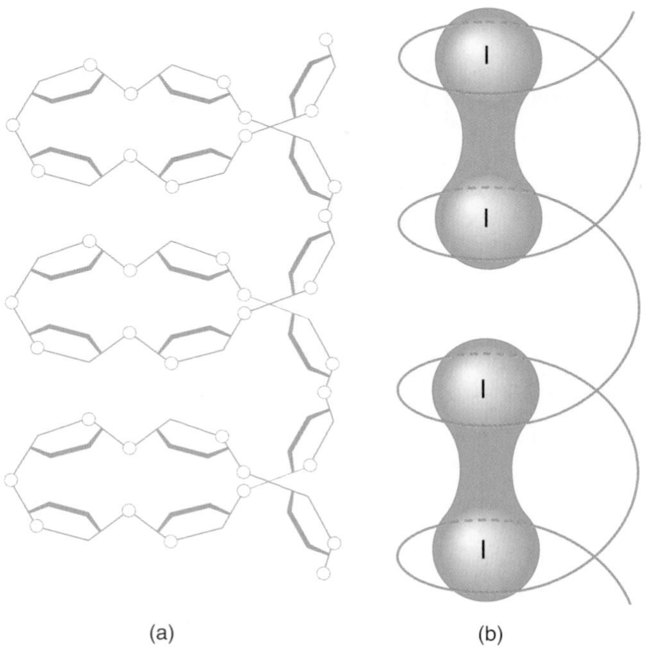

■ **FIGURE 7.15** The molecular conformation of starch and the starch–iodine complex: (a) the helical conformation of the amylose chain and (b) the starch–iodine complex.

■ **FIGURE 7.16** A dark blue color is the characteristic result when a solution of iodine encounters the starch in potatoes. What other foods would you expect to give a positive starch test?

■ **FIGURE 7.17** The partial structure of an amylopectin molecule. Glycogen has a similar structure.

Glycogen

The polysaccharide glycogen is sometimes called animal starch because it is a storage carbohydrate for animals that is analogous to the starch of plants. It is especially abundant in the liver and muscles, where excess glucose taken in by an animal is stored for future use. On hydrolysis, glycogen forms D-glucose, which helps maintain the normal blood sugar level and provide the muscles with energy. Structurally, glycogen is very similar to amylopectin, containing both $\alpha(1 \rightarrow 4)$ and $\alpha(1 \rightarrow 6)$ linkages between glucose units (see ■ Figure 7.18). The main difference between amylopectin and glycogen is that glycogen is even more highly branched; there are only 8 to 12 D-glucose units between branch points.

Cellulose

Cellulose is the most important structural polysaccharide and is the single most abundant organic compound on Earth (see ■ Figure 7.19). It is the material in plant cell walls that provides strength and rigidity. Wood is about 50% cellulose.

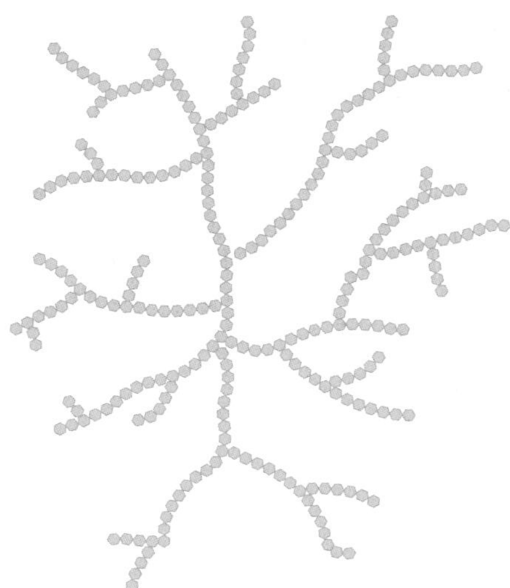

■ **FIGURE 7.18** A simplified representation of the branched polysaccharide glycogen (branches every 8–12 glucose units). Amylopectin is much less densely branched (branches every 24–30 glucose units). Each small hexagon represents a single glucose unit.

OVER THE COUNTER 7.1 Dietary Fiber

A large amount of research has been conducted on the benefits of **fiber** in the diet. Reports of the results in the mass media have often created conflicts in the minds of consumers. A few years ago, oatmeal was specifically reported to significantly reduce the chances of contracting a number of serious ailments. As a result, grocery stores had trouble keeping oatmeal on the shelves. But later reports of research findings moderated the beneficial role of oatmeal, and sales returned to normal.

For years, dietary fiber in general has been credited with helping prevent colon cancer and several other health problems. However, a prestigious new study involving more than 88,000 women has found no evidence that consuming large amounts of high-fiber food helps lower the risk of colon cancer. No research has yet disproved the results of previous studies indicating that dietary fiber may help prevent digestive disorders, some types of diabetes, heart disease, high blood pressure, and obesity.

Dietary fiber comes from plants; it consists of complex carbohydrates, such as cellulose, and other substances that make up the cell walls and structural parts of plants. Some examples of good sources of dietary fiber are cereal grains (including oatmeal), fresh fruits and vegetables, and grain products such as breads.

Dietary fiber is classified as soluble or insoluble on the basis of its solubility in water. Soluble fiber, such as pectin, has a lower molecular weight and structural features that make it more water soluble than insoluble fiber, such as cellulose. Perhaps the most important function of insoluble fiber in the diet is providing bulk to the stool. This helps the body eliminate solid wastes and in particular helps reduce the risk of *diverticulosis,* a condition in which small pouches form in the colon wall. Increased fiber consumption also helps alleviate the symptoms of diarrhea, abdominal pain, and flatulence. Also, because insoluble fiber is indigestible and passes through the body without providing many calories, it causes people to feel full and therefore eat less, thereby helping them avoid weight gain. Insoluble fiber may also contribute to weight control by inhibiting the absorption of some calorie-dense dietary fats from the digestive system.

Soluble fiber traps carbohydrates and thereby slows their digestion and absorption. This has the effect of leveling out blood sugar levels during the day. Studies indicate that a diet high in sugar and low in fiber more than doubles the chances for a woman to develop Type II diabetes. Clinical studies have also shown that a diet low in saturated fat and cholesterol, and high in fruits, vegetables, and grain products that contain soluble fiber, can lower blood cholesterol levels. Recently, the FDA approved specific labels for breakfast cereal and other foods that contain soluble fiber from psyllium seed husk. The labels state that the food (especially oats) may reduce the risk of coronary heart disease if the food is included as part of a diet that is also low in saturated fat and cholesterol. Maybe the sales of oatmeal will again increase. The soluble fiber lowers blood cholesterol levels by binding to dietary cholesterol and thus helping the body eliminate it.

If dietary fiber looks good to you and you want to include more in your daily diet, start by reading food labels and choosing those that contain more fiber. However, if you follow the current dietary recommendations and each day consume 2–4 servings of fruit, 3–5 servings of vegetables, and 6–11 servings of cereal and grain foods (including oatmeal), you will get enough fiber. Here are some suggestions to help you increase your fiber intake:

- Start each day with a whole-grain cereal.
- Try eating vegetables raw, instead of cooked.
- Don't peel fruits and vegetables before eating them.
- Add beans to soups, stews, and salads as a meat replacement.

As you make the transition to a high-fiber diet, there are some precautions to consider. First, fiber supplements have not been shown to be safe or effective for weight loss. A low-fat diet, combined with an increase in exercise, is still the most effective method. Second, don't increase the fiber content of your diet too rapidly. A gradual increase of a few grams each day will allow your digestive system to adjust. The unpleasant, common results of a sudden dietary fiber increase include abdominal cramps, gas, bloating, and diarrhea or constipation. Third, drink plenty of liquid (preferably water) daily. Finally, the use of an enzyme product, such as Beano®, will help minimize unpleasant symptoms that may accompany your change in diet.

A variety of high-fiber foods.

fiber
Indigestible plant material composed primarily of cellulose.

■ **FIGURE 7.19** Cellulose and the plant kingdom are responsible for much of the world's natural beauty.

Like amylose, cellulose is a linear polymer consisting of D-glucose units joined by the $1 \rightarrow 4$ linkage. It may contain from 300 to 3000 glucose units in one molecule. The main structural difference between amylose and cellulose is that all the $1 \rightarrow 4$ glycosidic linkages in cellulose are β instead of α (see ■ Figure 7.20). This seemingly small difference causes tremendous differences between amylose and cellulose. The shapes of the molecules are quite different. The $\alpha(1 \rightarrow 4)$-linked amylose tends to form loose spiral structures (Figure 7.15), whereas $\beta(1 \rightarrow 4)$-linked cellulose tends to form extended straight chains. These chains become aligned side by side to form well-organized, water-insoluble fibers in which the hydroxy groups form numerous hydrogen bonds with the neighboring chains. These parallel chains of cellulose confer rigidity and strength to the molecules.

Second, although starch with its α linkages is readily digestible, humans and other animals lack the necessary enzymes to hydrolyze the β linkages of cellulose. Thus, cellulose passes unchanged through the digestive tract and does not contribute to the caloric value of food. However, it still serves a useful purpose in digestion. Cellulose, a common constituent of dietary fiber, is the roughage that provides bulk, stimulates contraction of the intestines, and aids the passage of food through the digestive system.

Animals that use cellulose as a food do so only with the help of bacteria that possess the necessary enzymes. Herbivores such as cows, sheep, and horses are animals in this category. Each has a colony of such bacteria somewhere in the

■ **FIGURE 7.20** The structure of cellulose.

digestive system and uses the simple carbohydrates resulting from the bacterial hydrolysis of cellulose. Fortunately, the soil also contains organisms with appropriate enzymes. Otherwise, debris from dead plants would accumulate rather than disappear through biodegradation.

Concept Summary

Classes of Carbohydrates. Carbohydrates are polyhydroxy aldehydes or ketones, or substances that yield such compounds on hydrolysis. Carbohydrates are used as energy sources, biosynthetic intermediates, energy storage, and structural elements in organisms. **OBJECTIVE 1 (Section 7.1), Exercise 7.2.** Carbohydrates can exist either as single units (monosaccharides) or joined together in molecules ranging from two units (disaccharides) to hundreds of units (polysaccharides). **OBJECTIVE 2 (Section 7.1), Exercise 7.4**

The Stereochemistry of Carbohydrates. Carbohydrates, along with many other natural substances, exhibit a type of isomerism in which two isomers are mirror images of each other. The two isomers are called enantiomers and contain a chiral carbon atom. **OBJECTIVE 3 (Section 7.2), Exercise 7.8.** When a molecule has more than one chiral carbon, the maximum number of stereoisomers possible is 2^n, where n is the number of chiral carbons.

Fischer Projections. A useful way of depicting the structure of chiral molecules employs crossed lines (Fischer projections) to represent chiral carbon atoms. The prefixes D- and L- are used to distinguish between enantiomers. **OBJECTIVE 4 (Section 7.3), Exercise 7.12.** Signs indicating the rotation of plane-polarized light to the right (+) or to the left (−) may also be used to designate enantiomers.

Monosaccharides. Monosaccharides that contain an aldehyde group are called aldoses, whereas those containing a ketone group are ketoses. Monosaccharides are also classified by the number of carbon atoms as trioses, tetroses, etc. **OBJECTIVE 5 (Section 7.4), Exercise 7.22.** Most natural monosaccharides belong to the D family.

Properties of Monosaccharides. Monosaccharides are sweet-tasting solids that are very soluble in water. Noncarbohydrate low-calorie sweeteners such as aspartame have been developed as sugar substitutes. Pentoses and hexoses form cyclic hemiacetals or hemiketals whose structures can be represented by Haworth structures. Two isomers referred to as anomers (the α and β forms) are produced in the cyclization reaction. All monosaccharides are oxidized by Benedict's reagent and are called reducing sugars. Monosaccharides can react with alcohols to produce acetals or ketals that are called glycosides. **OBJECTIVE 6 (Section 7.5), Exercise 7.34**

Important Monosaccharides. Ribose and deoxyribose are important as components of nucleic acids. The hexoses glucose, galactose, and fructose are the most important nutritionally and the most abundant in nature. Glucose, also known as blood sugar, is transported within the bloodstream to body tissues, where it supplies energy. **OBJECTIVE 7 (Section 7.6), 7.37**

Disaccharides. Glycosidic linkages join monosaccharide units together to form disaccharides. Three important disaccharides are maltose (two glucose units $\alpha(1 \rightarrow 4)$-linked), lactose (a galactose linked to glucose by a $\beta(1 \rightarrow 4)$ glycosidic linkage), and sucrose (α-glucose joined to β-fructose). **OBJECTIVE 8, (Section 7.7), Exercise 7.44.** Hydrolysis of disaccharides produces monosaccharides. **OBJECTIVE 9 (Section 7.7), Exercise 7.52**

Polysaccharides. Cellulose, starch, and glycogen are three important polysaccharides. Starch is the major storage form of glucose in plants, whereas glycogen is the storage form of glucose in animals. Cellulose is the structural material of plants. **OBJECTIVE 10 (Section 7.8), Exercise 7.54**

Key Terms and Concepts

Anomeric carbon (7.5)
Anomers (7.5)
Biochemistry (Introduction)
Biomolecule (Introduction)
Carbohydrate (7.1)
Chiral (7.2)
Chiral carbon (7.2)
Dextrorotatory (7.3)
Disaccharide (7.1)
Enantiomers (7.2)
Fiber (7.8)
Fischer projection (7.3)
Furanose ring (7.5)
Glycoside (7.5)
Glycosidic linkage (7.5)
Haworth structure (7.5)
Invert sugar (7.7)
Levorotatory (7.3)
Monosaccharide (7.1)
Optically active molecule (7.3)
Polysaccharide (7.1)
Pyranose ring (7.5)
Reducing sugar (7.5)

Key Reactions

1. Oxidation of a sugar (Section 7.5):

$$\text{reducing sugar} + Cu^{2+} \rightarrow \text{oxidized compound} + Cu_2O$$ Reaction 7.4

2. Glycoside formation (Section 7.5):

$$\text{monosaccharide} + \text{alcohol} \xrightarrow{H^+} \text{acetals or ketals}$$

3. Hydrolysis of disaccharides (Section 7.7):

$$\text{disaccharide} + H_2O \xrightarrow[\text{enzymes}]{H^+ \text{ or}} \text{two monosaccharides}$$

Exercises

SYMBOL KEY

Even-numbered exercises are answered in Appendix B.

Blue-numbered exercises are more challenging.

■ denotes exercises available in ThomsonNow and assignable in OWL.

ThomsonNOW™ To assess your understanding of this chapter's topics with sample tests and other resources, sign in at www.thomsonedu.com.

CLASSES OF CARBOHYDRATES (SECTION 7.1)

7.1 What are the four important roles of carbohydrates in living organisms?

7.2 Describe whether each of the following substances serves primarily as an energy source, a form of stored energy, or a structural material (some serve as more than one):

a. cellulose
b. sucrose, table sugar
c. glycogen
d. starch

7.3 What are the structural differences among monosaccharides, disaccharides, and polysaccharides?

7.4 ■ Match the terms *carbohydrate, monosaccharide, disaccharide,* and *polysaccharide* to each of the following (more than one term may fit):

a. table sugar

b. [ring structure: CH_2OH, O, OH, OH, HO, OH]

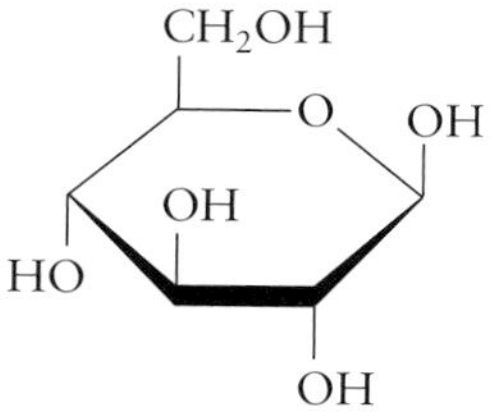

c. starch

d. fructose

e. cellulose

f. [two linked ring structures: CH_2OH, O, OH, OH, HO, OH; O; CH_2OH, O, OH, OH, OH]

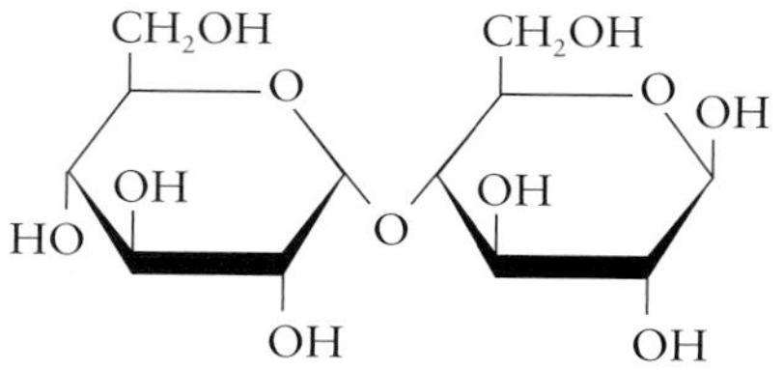

g. glycogen

h. amylose

7.5 Define *carbohydrate* in terms of the functional groups present.

THE STEREOCHEMISTRY OF CARBOHYDRATES (SECTION 7.2)

7.6 Why are carbon atoms 1 and 3 of glyceraldehyde not considered chiral?

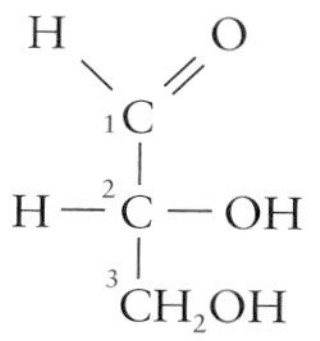

[structure: H–$_1$C=O; H–$_2$C–OH; $_3CH_2OH$]

7.7 Locate the chiral carbon in amphetamine and identify the four different groups attached to it.

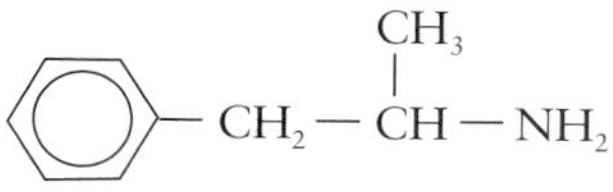

[structure: $C_6H_5-CH_2-CH(CH_3)-NH_2$]

7.8 ■ Which of the following molecules can have enantiomers? Identify any chiral carbon atoms.

a. 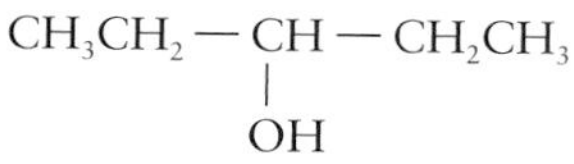 $CH_3CH_2-CH(OH)-CH_2CH_3$

b. 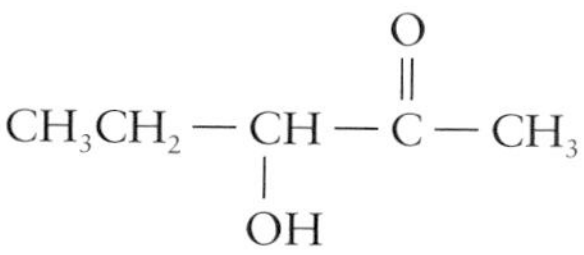 $CH_3CH_2-CH(OH)-C(=O)-CH_3$

c. cyclopentyl$-C(OH)(CH_3)-CH_2CH_3$

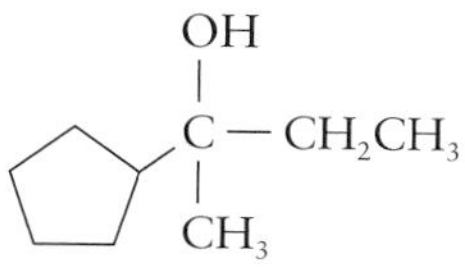

7.9 Which of the following molecules can have enantiomers? Identify any chiral carbon atoms.

a. 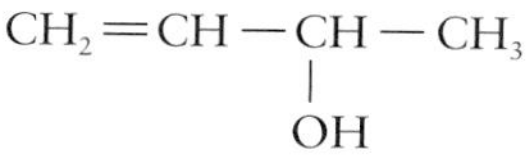 $CH_2=CH-CH(OH)-CH_3$

b. $CH_3-\underset{\overset{|}{OH}}{CH}-CH_2-\overset{\overset{O}{\|}}{C}-H$ (OH on the CH; =O on the C)

c. cyclopentane ring bearing Br

FISCHER PROJECTIONS (SECTION 7.3)

7.10 Explain what the following Fischer projection denotes about the three-dimensional structure of the compound:

$$\begin{array}{c} CHO \\ H - \!\!+\!\! - OH \\ CH_2CH_3 \end{array}$$

7.11 ■ Identify each of the following as a D or an L form and draw the structural formula of the enantiomer:

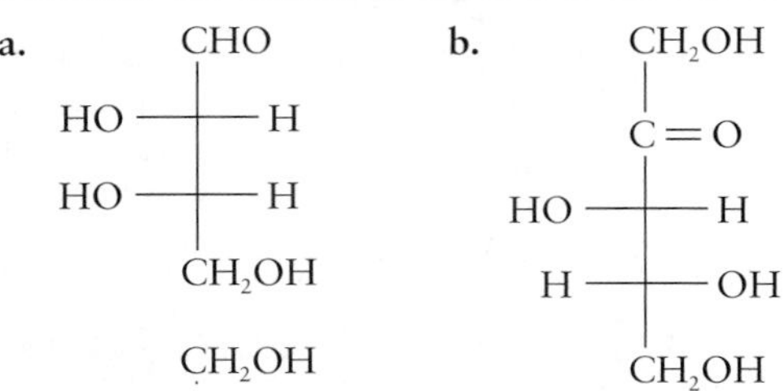

7.12 Identify each of the following as a D or an L form and draw the structural formula of the enantiomer:

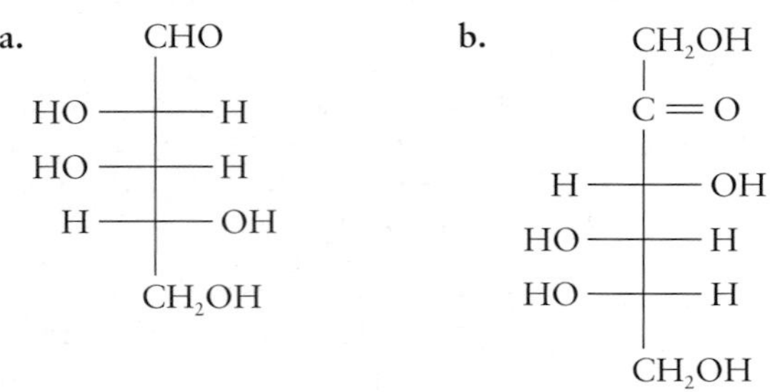

7.13 ■ Draw Fischer projections for both the D and L isomers of the following:

a. 2,3-dihydroxypropanoic acid

b. 2-chloro-3-hydroxypropanoic acid

7.14 Draw Fischer projections for both the D and L isomers of the following amino acids:

a. Alanine, $H_2N-CH(CH_3)-COOH$

b. Leucine, $H_2N-CH(CH_2-CH(CH_3)_2)-COOH$

7.15 How many chiral carbon atoms are there in each of the following sugars? How many stereoisomers exist for each compound?

a. $CH_2(OH)-CH(OH)-CH(OH)-C(=O)-H$

b. $CH_2(OH)-C(=O)-CH(OH)-CH(OH)-CH(OH)-CH_3$

7.16 ■ How many chiral carbon atoms are there in each of the following sugars? How many stereoisomers exist for each compound?

a. $CH_2(OH)-CH(OH)-CH(OH)-C(=O)-CH_2-OH$

b. $CH_2(OH)-CH(OH)-CH(OH)-C(OH)-CHO$

7.17 How many aldoheptoses are possible? How many are the D form and how many the L form?

7.18 How many aldopentoses are possible? How many are the D form and how many the L form?

7.19 Why is the study of chiral molecules important in biochemistry?

7.20 What physical property is characteristic of optically active molecules?

MONOSACCHARIDES (SECTION 7.4)

7.21 ■ Classify each of the following monosaccharides as an aldo- or keto- triose, tetrose, pentose, or hexose:

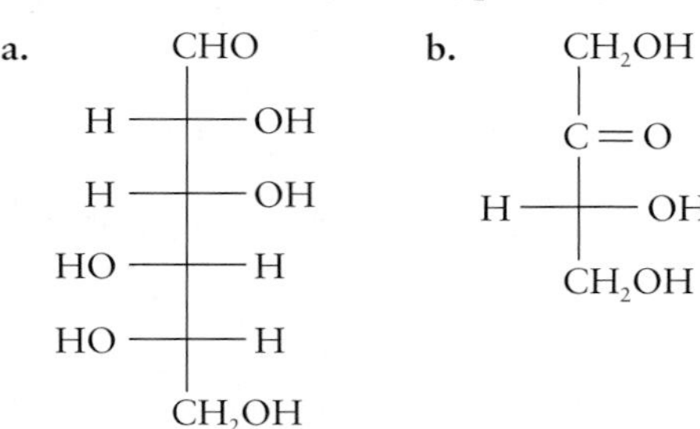

7.22 Classify each of the following monosaccharides as an aldo- or keto- triose, tetrose, pentose, or hexose:

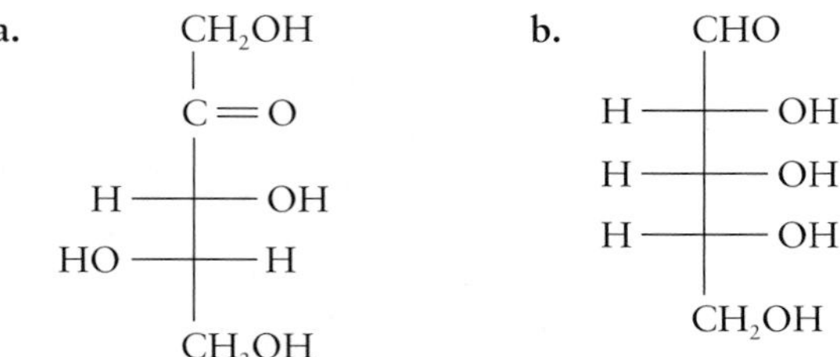

7.23 Draw Fischer projections of any aldotetrose and any ketopentose.

PROPERTIES OF MONOSACCHARIDES (SECTION 7.5)

7.24 Explain why certain carbohydrates are called sugars.

7.25 Explain why monosaccharides are soluble in water.

7.26 ■ Using a Fischer projection of a glucose molecule, identify the site(s) on the molecule where water can hydrogen-bond.

7.27 Identify each of the following as an α or a β form and draw the structural formula of the other anomer:

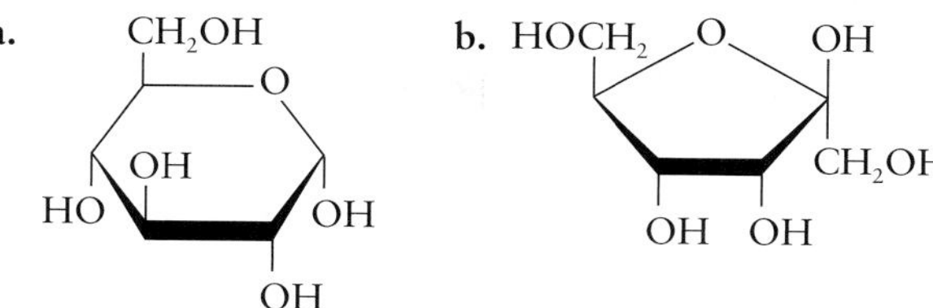

7.28 ■ Identify each of the following as an α or a β form and draw the structural formula of the other anomer:

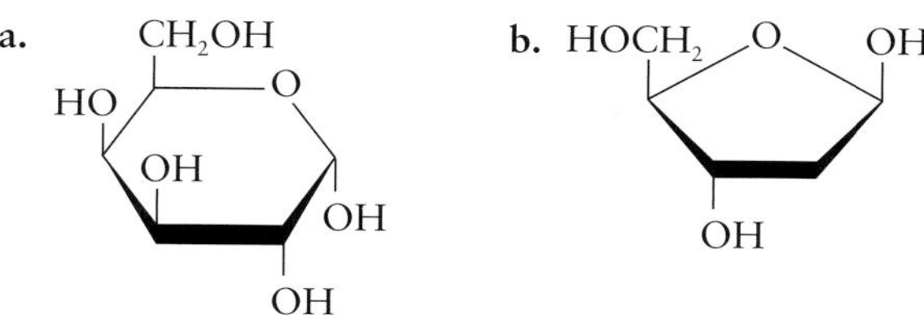

7.29 The structure of talose differs from galactose only in the direction of the —OH group at position 2. Draw and label Haworth structures for α- and β-talose.

7.30 The structure of mannose differs from glucose only in the direction of the —OH group at position 2. Draw and label Haworth structures for α- and β-mannose.

7.31 What is the difference between pyranose and furanose rings?

7.32 An unknown sugar failed to react with a solution of Cu^{2+}. Classify the compound as a reducing or nonreducing sugar.

7.33 Explain how the cyclic compound β-D-galactose can react with Cu^{2+} and be classified as a reducing sugar.

7.34 ■ Complete the following reactions:

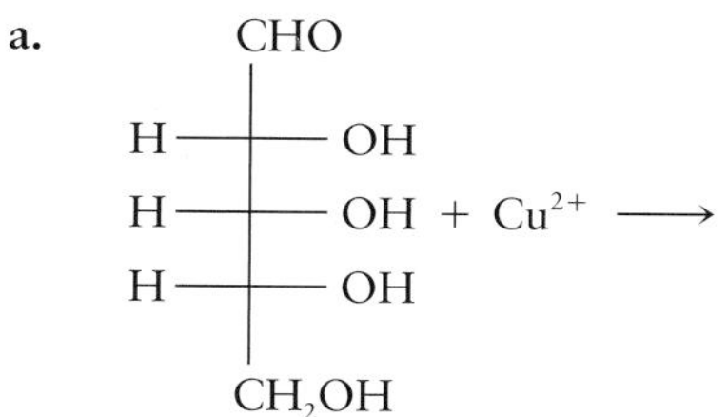

b. CH_2OH, O, OH, HO, OH, OH $+ CH_3CH_2—OH \xrightarrow{H^+}$

c. $HOCH_2$, O, OH, HO, CH_2OH, OH $+ CH_3—OH \xrightarrow{H^+}$

7.35 ■ Use an arrow to identify the glycosidic linkage in each of the following:

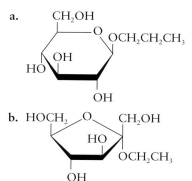

a. CH_2OH, O, $OCH_2CH_2CH_3$, OH, HO, OH

b. $HOCH_2$, O, CH_2OH, HO, OCH_2CH_3, OH

IMPORTANT MONOSACCHARIDES (SECTION 7.6)

7.36 What monosaccharides are used in the synthesis of nucleic acids?

7.37 Explain why D-glucose can be injected directly into the bloodstream to serve as an energy source.

7.38 Give two other names for D-glucose.

7.39 ■ Which of the following are ketohexoses?

a. glucose b. fructose c. galactose

7.40 How do the hexoses glucose and galactose differ structurally?

7.41 Give the natural sources for each of the following:

a. glucose b. fructose c. galactose

7.42 Explain why fructose can be used as a low-calorie sweetener.

DISACCHARIDES (SECTION 7.7)

7.43 What one monosaccharide is a component of all three of the disaccharides sucrose, maltose, and lactose?

7.44 ■ Identify a disaccharide that fits each of the following:

a. The most common household sugar
b. Formed during the digestion of starch
c. An ingredient of human milk
d. Found in germinating grain
e. Hydrolyzes when cooked with acidic foods to give invert sugar
f. Found in high concentrations in sugar cane

7.45 What type of linkage is broken when disaccharides are hydrolyzed to monosaccharides?

7.46 Explain the process of how the hemiacetal group of a lactose molecule is able to react with Benedict's reagent.

7.47 Why is invert sugar sweeter than sucrose?

7.48 Sucrose and honey are commonly used sweeteners. Suppose you had a sweet-tasting water solution that contained either honey or sucrose. How would you chemically determine which sweetener was present?

7.49 Draw Haworth projection formulas of the disaccharides maltose and sucrose. Label the hemiacetal, hemiketal, acetal, or ketal carbons.

7.50 Using structural characteristics, show why

a. lactose is a reducing sugar.
b. sucrose is not a reducing sugar.

7.51 ■ What type of linkage, for example, $\alpha(1 \rightarrow 4)$, is present in the following?

a. lactose b. maltose c. sucrose

7.52 ■ The disaccharide melibiose is present in some plant juices.

a. What two monosaccharides are formed on hydrolysis of melibiose?

b. Is melibiose a reducing sugar? Explain.

c. Describe the glycosidic linkage between the two monosaccharide units.

POLYSACCHARIDES (SECTION 7.8)

7.53 Match the following structural characteristics to the polysaccharides amylose, amylopectin, glycogen, and cellulose (a characteristic may fit more than one):

a. Contains both $\alpha(1 \rightarrow 4)$ and $\alpha(1 \rightarrow 6)$ glycosidic linkages

b. Is composed of glucose monosaccharide units

c. Contains acetal linkages between monosaccharide units

d. Is composed of highly branched molecular chains

e. Is composed of unbranched molecular chains

f. Contains only $\alpha(1 \rightarrow 4)$ glycosidic linkages

g. Contains only $\beta(1 \rightarrow 4)$ glycosidic linkages

7.54 ■ Name a polysaccharide that fits each of the following:

a. The unbranched polysaccharide in starch

b. A polysaccharide widely used as a textile fiber

c. The most abundant polysaccharide in starch

d. The primary constituent of paper

e. A storage form of carbohydrates in animals

7.55 Polysaccharides are abundant in celery, and yet celery is a good snack for people on a diet. Explain why.

ADDITIONAL EXERCISES

7.56 There are two solutions:

(1) a 10% (w/v) water–glucose solution

(2) a 10% (w/v) water–maltose solution

Can you differentiate between the two solutions based on boiling points? Explain your answer.

7.57 Why does one mole of D-glucose only react with one mole of methanol (instead of two) to form an acetal?

7.58 If 100 g of sucrose is completely hydrolyzed to invert sugar, how many grams of glucose will be present?

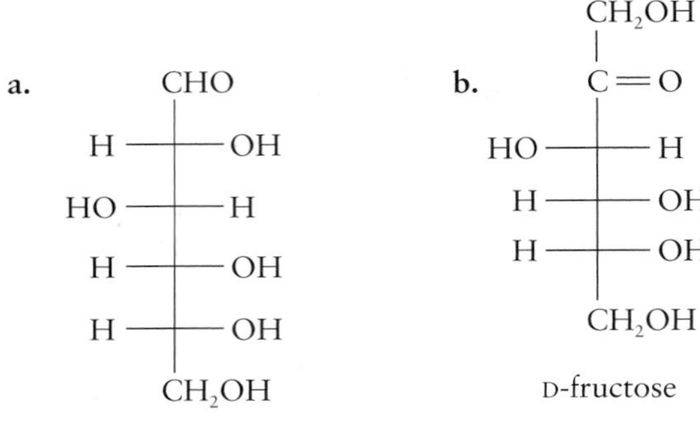

7.59 Hexanal and 1-hexanol are liquids at room temperature, but glucose (a hexose) is a solid. Give two reasons that may explain this phenomenon.

7.60 Look at Figure 7.15 "Osmosis through carrot membranes." Suppose maltose was used instead of molasses in the demonstration and the solution level rose 5 cm above the carrot in the tube. Next, maltase, an enzyme that hydrolyzes the glycosidic linkage, was added to the solution. What would happen to the solution level in the tube? Explain your reasoning.

ALLIED HEALTH EXAM CONNECTION

Reprinted with permission from Nursing School and Allied Health Entrance Exams, COPYRIGHT 2005 Petersons.

7.61 The molecular formula for glucose is $C_6H_{12}O_6$, and the molecular formula for fructose is $C_6H_{12}O_6$. Which of the following describe glucose and fructose?

a. stereoisomers

b. isomers

c. hexoses

d. anomers

7.62 Which of the following are composed of long chains of glucose molecules?

a. cellulose

b. starch

c. cholesterol

d. glycogen

7.63 Identify each of the following as a monosaccharide or a disaccharide. If it is a disaccharide, identify the individual monosaccharide components.

a. dextrose

b. fructose

c. sucrose

d. maltose

e. lactose

f. ribose

7.64 What is the general type of reaction used to decompose sucrose to glucose + fructose?

7.65 Glucose is a reducing sugar, which if boiling in Benedict's reagent, produces an orange to brick-red color. What chemical species is being reduced during the reaction?

CHEMISTRY FOR THOUGHT

7.66 Suppose human intestinal bacteria were genetically altered so they could hydrolyze the β linkages of cellulose. Would the results be beneficial? Explain.

7.67 A sample of starch is found to have a molecular weight of 2.80×10^5 u. How many glucose units are present in a molecule of the starch?

7.68 The open-chain form of glucose constitutes only a small fraction of any glucose sample. Yet, when Cu^{2+} is used to oxidize the open-chain form, nearly all the glucose in a sample reacts. Use Le Châtelier's principle to explain this observation.

7.69 Aspartame (Nutrasweet) contains calories and yet is used in diet drinks. Explain how a drink can contain aspartame as a sweetener and yet be low in calories.

7.70 From Figure 7.10, what two products form when the lactose in cow's milk is hydrolyzed in a child's digestive system?

7.71 Answer the question raised in Figure 7.16. What foods in addition to potatoes would you expect to give a positive starch test?

7.72 Amylopectin is a component of starch, yet it does not give a positive iodine test (turn bluish-black) for the presence of starch when it is tested in its pure form. Why?

7.73 Amylose is a straight-chain glucose polymer similar to cellulose. What would happen to the longevity of paper manufactured with amylose instead of cellulose?

7.74 Raffinose, a trisaccharide found in some plants, contains three monosaccharide components: galactose, glucose, and fructose. Disregarding the types of linkages, how many possible structures are there for raffinose?

CHAPTER 8

Lipids

LEARNING OBJECTIVES

When you have completed your study of this chapter, you should be able to:

1. Classify lipids as saponifiable or nonsaponifiable and list five major functions of lipids. (Section 8.1)
2. Describe four general characteristics of fatty acids. (Section 8.2)
3. Draw structural formulas of triglycerides given the formulas of the component parts. (Section 8.3)
4. Describe the structural similarities and differences of fats and oils. (Section 8.3)
5. Write key reactions for fats and oils. (Section 8.4)
6. Compare the structures of fats and waxes. (Section 8.5)
7. Draw structural formulas and describe uses for phosphoglycerides. (Section 8.6)
8. Draw structural formulas and describe uses for sphingolipids. (Section 8.7)
9. Describe the major features of cell membrane structure. (Section 8.8)
10. Identify the structural characteristic typical of steroids and list important groups of steroids in the body. (Section 8.9)
11. Name the major categories of steroid hormones. (Section 8.10)
12. Describe the biological importance and therapeutic uses of the prostaglandins. (Section 8.11)

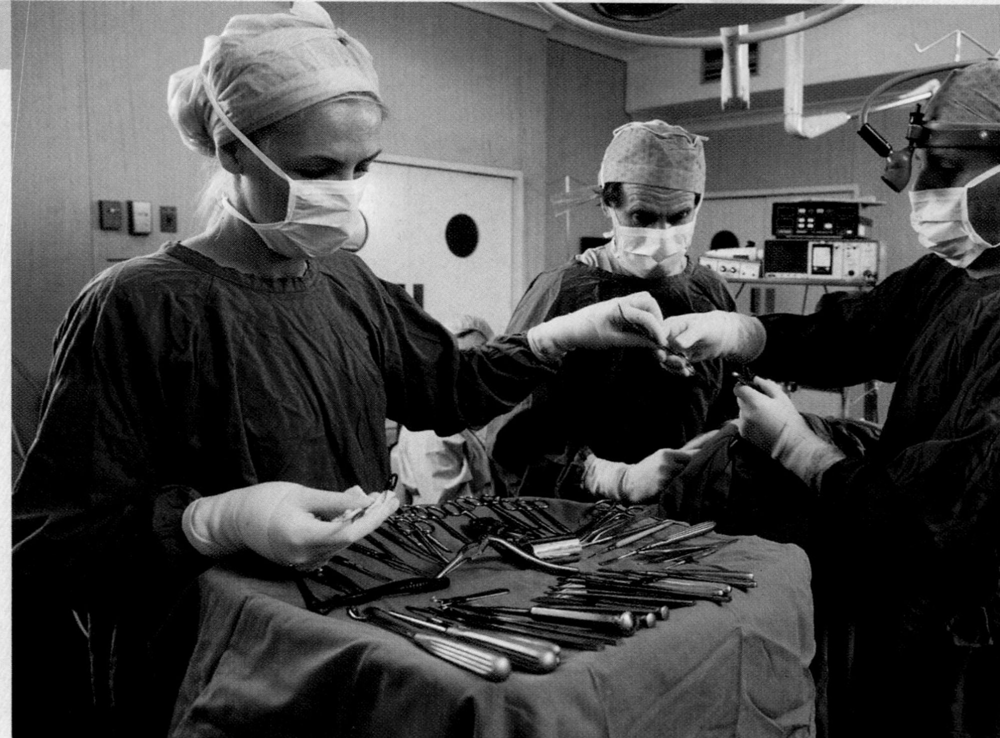

A certified surgical technologist (CST), a member of the operating room surgical team, is responsible for the preparation of supplies and equipment used in surgery. In addition, a CST usually scrubs the patients in preparation for surgery and may also assist the surgeon in other ways. Thus, a CST must be familiar with basic surgical procedures and cleansing techniques. One of the topics in this chapter on lipids describes the structural characteristics of soaps.

The group of compounds called lipids is made up of substances with widely different compositions and structures. Unlike the carbohydrates that are defined in terms of structure, lipids are defined in terms of a physical property—solubility. Lipids are biological molecules that are insoluble in water but soluble in nonpolar solvents. Lipids are the waxy, greasy, or oily compounds found in plants and animals (see ■ Figure 8.1 and ■ Figure 8.2). Lipids repel water, a useful characteristic of protective wax coatings found on some plants. Fats and oils are energy-rich and have relatively low densities. These properties account for their use as storage forms of energy in plants and animals. Still other lipids are used as structural components, especially in the formation of cellular membranes.

ThomsonNOW Throughout the chapter this icon introduces resources on the ThomsonNOW website for this text. Sign in at **www.thomsonedu.com** to:

- Evaluate your knowledge of the material
- Take an exam prep quiz
- Identify areas you need to study with a **Personalized Learning Plan**.

8.1 Classification of Lipids

LEARNING OBJECTIVE

1. Classify lipids as saponifiable or nonsaponifiable and list the major functions of lipids.

Lipids are grouped into two main classes: saponifiable lipids and nonsaponifiable lipids. Saponification refers to the process in which esters are hydrolyzed under basic conditions (Section 5.7). Triglycerides, waxes, phospholipids, and sphingolipids are esters and thus all belong to the first class. Nonsaponifiable lipids are not esters and cannot be hydrolyzed. Steroids and prostaglandins belong to this class. ■ Figure 8.3 summarizes the classification of different types of lipids. Notice from the figure that saponifiable lipids are further classified into categories of simple and complex, based on the number of components in their structure. **Simple lipids** contain just two types of components (fatty acids and an alcohol), whereas **complex lipids** contain more than two (fatty acids, an alcohol, plus other components).

lipid
A biological compound that is soluble only in nonpolar solvents.

simple lipid
An ester-containing lipid with just two types of components: an alcohol and one or more fatty acids.

complex lipid
An ester-containing lipid with more than two types of components: an alcohol, fatty acids—plus others.

■ **FIGURE 8.1** Waxes form a protective coating on these leaves.

■ **FIGURE 8.2** Thick layers of fat help insulate penguins against low temperatures.

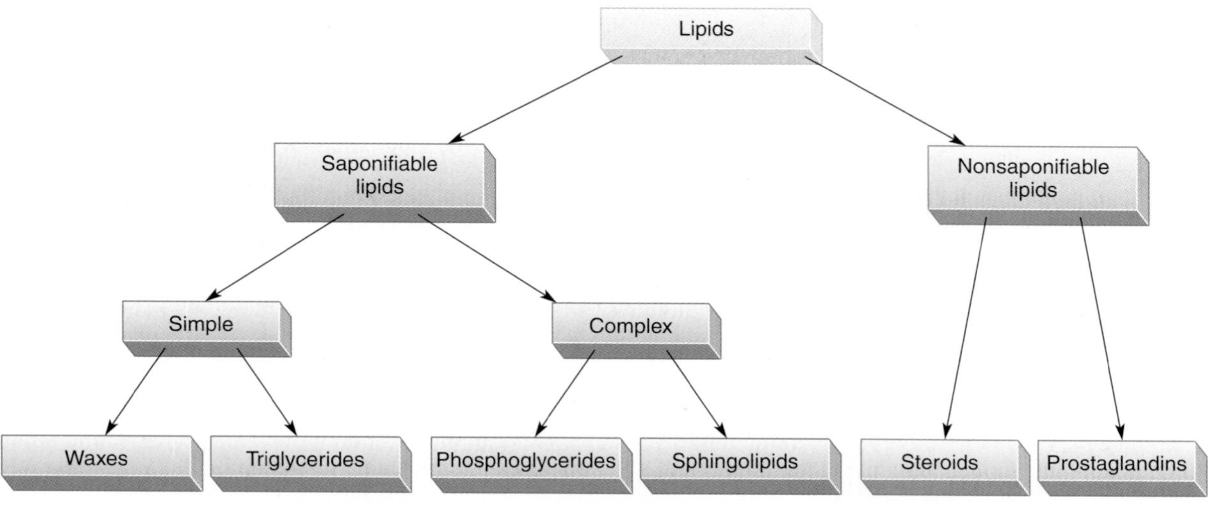

■ **FIGURE 8.3** The major types of lipids.

8.2 Fatty Acids

LEARNING OBJECTIVE

2. Describe four general characteristics of fatty acids.

Prior to discussing the chemical structures and properties of the saponifiable lipids, it is useful to first describe the chemistry of fatty acids, the fundamental building blocks of many lipids. Fatty acids (defined in Section 5.1) are long-chain carboxylic acids, as shown in ■ Figure 8.4. It is the long nonpolar tails of fatty acids that are responsible for most of the fatty or oily characteristics of fats. The carboxyl group, or polar head of fatty acids, is very hydrophilic under conditions of physiological pH, where it exists as the carboxylate anion $—COO^-$.

In aqueous solution, the ions of fatty acids associate with one another in a rather unique way. The ions form spherical clusters, called **micelles,** in which the nonpolar chains extend toward the interior of the structure away from water, and the polar carboxylate groups face outward in contact with the water. Some micelles are large on a molecular scale and contain hundreds or even thousands of fatty acid molecules. The nonpolar chains in a micelle are held together by weak dispersion forces. The structure of a micelle in a radial cross section is shown in ■ Figure 8.5. Micelle formation and structure are important in a number of biological functions, such as the transport of insoluble lipids in the blood (Section 14.1).

micelle
A spherical cluster of molecules in which the polar portions of the molecules are on the surface and the nonpolar portions are located in the interior.

ThomsonNOW™ Go to Coached Problems to investigate **micelles.**

$$CH_3—CH_2—CH_2—CH_2—CH_2—CH_2—CH_2—CH_2—CH_2—CH_2—CH_2—\overset{O}{\overset{\|}{C}}—OH$$

Nonpolar, hydrophobic tail (water insoluble)

Polar, hydrophilic head (water soluble)

(a)

Nonpolar tail COOH Polar head

(b)

■ **FIGURE 8.4** The molecular structure of fatty acids: (a) lauric acid and (b) a simplified diagram of a fatty acid with a nonpolar tail and a polar head.

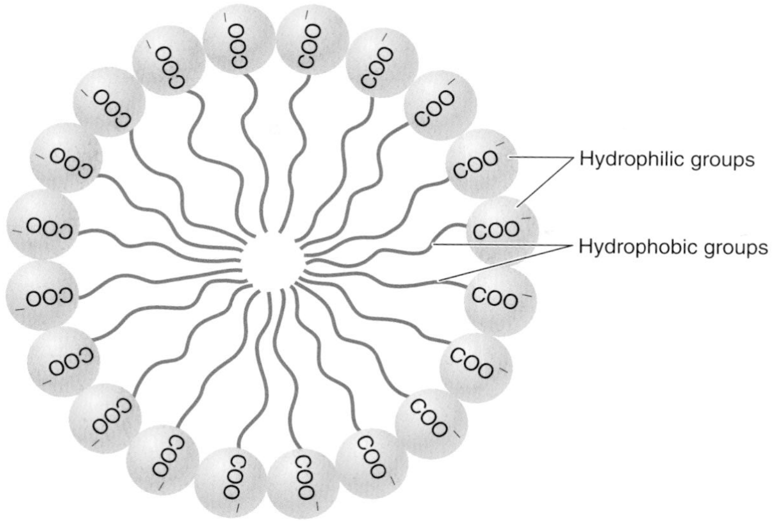

■ **FIGURE 8.5** A cross section of a fatty acid micelle.

The fatty acids found in natural lipids have several characteristics in common:

1. They are usually straight-chain carboxylic acids (no branching).
2. The sizes of most common fatty acids range from 10 to 20 carbons.
3. Fatty acids usually have an even number of carbon atoms (including the carboxyl group carbon).
4. Fatty acids can be saturated (containing no double bonds between carbons) or unsaturated (containing one or more double bonds between carbons). Apart from the carboxyl group and double bonds, there are usually no other functional groups present.

■ Table 8.1 lists some important fatty acids, along with their formulas and melting points.

Unsaturated fatty acids usually contain double bonds in the cis configuration.

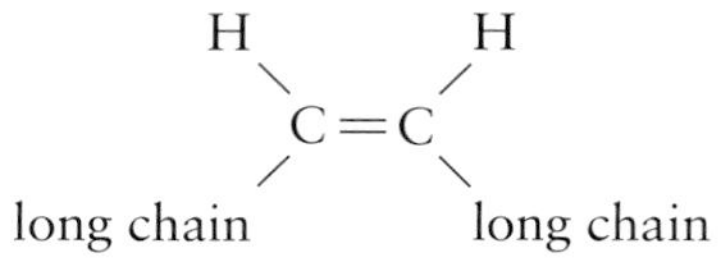

```
          H           H
           \         /
             C  =  C
           /         \
 long chain           long chain
```

This configuration creates a long characteristic bend, or kink, in the fatty acid chain that is not found in saturated fatty acids.

These kinks prevent unsaturated fatty acid chains from packing together as closely as do the chains of saturated acids (see ■ Figure 8.6). As a result, the intermolecular forces are weaker, and unsaturated fatty acids have lower melting points and are usually liquids at room temperature. For example, the melting point of stearic acid (18 carbons) is 71°C, whereas that of oleic acid (18 carbons with one cis double bond) is 13°C. The melting point of linoleic acid (18 carbons and two double bonds) is even lower (−5°C). Chain length also affects the melting point, as illustrated by the fact that palmitic acid (16 carbons) melts at 7°C lower than stearic acid (18 carbons). The presence of double bonds and the length of fatty acid chains in membrane lipids partly explain the fluidity of biological membranes, an important feature discussed in Section 8.8.

The human body can synthesize all except two of the fatty acids it needs. These two, linoleic acid and linolenic acid, are polyunsaturated fatty acids that contain 18 carbon atoms (Table 8.1). Because they are not synthesized within the

TABLE 8.1 Some important fatty acids

Compound type and number of carbons	Name	Formula	Melting point (°C)	Common sources
Saturated				
14	Myristic acid	$CH_3(CH_2)_{12}—COOH$	54	Butterfat, coconut oil, nutmeg oil
16	Palmitic acid	$CH_3(CH_2)_{14}—COOH$	63	Lard, beef fat, butterfat, cottonseed oil
18	Stearic acid	$CH_3(CH_2)_{16}—COOH$	70	Lard, beef fat, butterfat, cottonseed oil
20	Arachidic acid	$CH_3(CH_2)_{18}—COOH$	76	Peanut oil
Monounsaturated				
16	Palmitoleic acid	$CH_3(CH_2)_5CH{=}CH(CH_2)_7—COOH$	−1	Cod liver oil, butterfat
18	Oleic acid	$CH_3(CH_2)_7CH{=}CH(CH_2)_7—COOH$	13	Lard, beef fat, olive oil, peanut oil
Polyunsaturated				
18	Linoleic acid[a]	$CH_3(CH_2)_4(CH{=}CHCH_2)_2(CH_2)_6—COOH$	−5	Cottonseed oil, soybean oil, corn oil, linseed oil
18	Linolenic acid[b]	$CH_3CH_2(CH{=}CHCH_2)_3(CH_2)_6—COOH$	−11	Linseed oil, corn oil
20	Arachidonic acid[a]	$CH_3(CH_2)_4(CH{=}CHCH_2)_4(CH_2)_2—COOH$	−50	Corn oil, linseed oil, animal tissues
20	Eicosapentaenoic acid[b]	$CH_3CH_2(CH{=}CHCH_2)_5(CH_2)_2—COOH$		Fish oil, seafoods
22	Docosahexaenoic acid[b]	$CH_3CH_2(CH{=}CHCH_2)_6CH_2—COOH$		Fish oil, seafoods

[a]Omega-6 fatty acid.

[b]Omega-3 fatty acid.

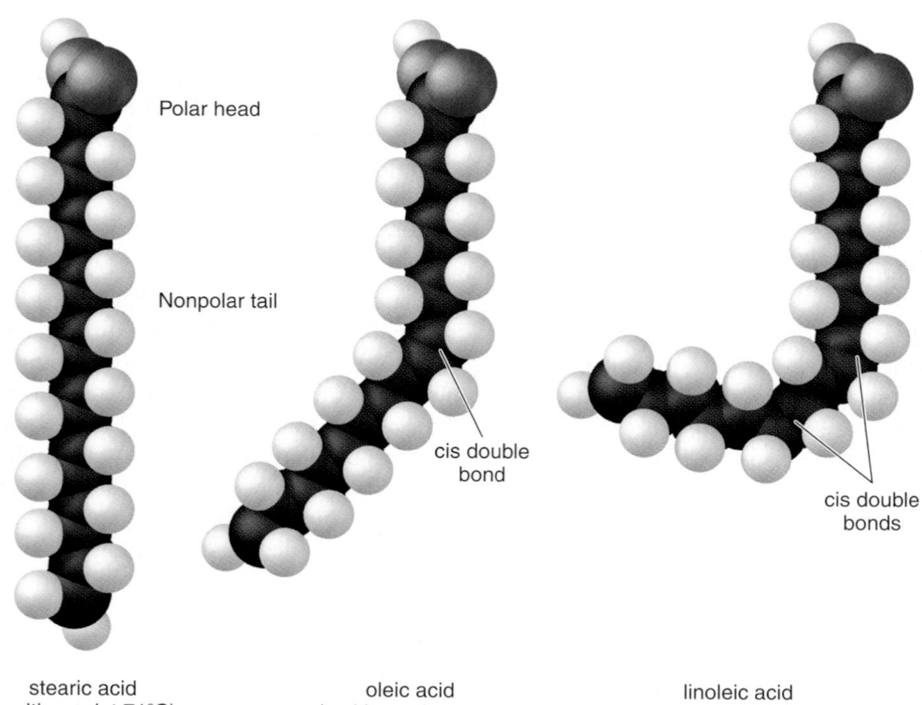

FIGURE 8.6 Space-filling models of fatty acids. Unsaturated fatty acids do not pack as tightly as saturated acids, and their melting points are lower than those of saturated acids.

body and must be obtained from the diet, they are called **essential fatty acids.** Both are widely distributed in plant and fish oils. In the body, both acids are used to produce hormonelike substances that regulate a wide range of functions and characteristics, including blood pressure, blood clotting, blood lipid levels, the immune response, and the inflammation response to injury and infection.

essential fatty acid A fatty acid needed by the body but not synthesized within the body.

Linolenic acid is called an omega-3 fatty acid, which means that the endmost double bond is three carbons from the methyl end of the chain:

$$\underset{\text{linolenic acid}}{CH_3—CH_2—(CH{=}CHCH_2)_3—(CH_2)_6—COOH}$$

In linoleic acid, an omega-6 fatty acid, the endmost double bond is located six carbons from the methyl end of the chain. Both linoleic and linolenic acids can be converted to other omega-3 and omega-6 fatty acids.

Omega-3 fatty acids have been a topic of interest in medicine since 1985. In that year, researchers reported the results of a study involving natives of Greenland. These individuals have a very low death rate from heart disease, even though their diet is very high in fat. Studies led researchers to conclude that the abundance of fish in the diet of the Greenland natives was involved. Continuing studies led to a possible involvement of the omega-3 fatty acids in the oil of the fish.

A recent Harvard University study indicated that male adults who ate just a couple of servings of fish per month were 25% less likely to die of heart disease than those who ate it less than once per month or not at all.

8.3 The Structure of Fats and Oils

LEARNING OBJECTIVES

3. Draw structural formulas of triglycerides given the formulas of the component parts.
4. Describe the structural similarities and differences of fats and oils.

Animal fats and vegetable oils are the most widely occurring lipids. Chemically, fats and oils are both esters, and as we learned in Chapter 15, esters consist of an alcohol portion and an acid portion. In fats and oils, the alcohol portion is always derived from glycerol, and the acid portion is furnished by fatty acids. Because glycerol has three —OH groups, a single molecule of glycerol can be attached to three different acid molecules. An example is the esterification reaction of stearic acid and glycerol:

$$\begin{array}{llll} CH_2—OH & HO—\overset{O}{\overset{\|}{C}}—C_{17}H_{35} & & CH_2—O—\overset{O}{\overset{\|}{C}}—C_{17}H_{35} \\ | & & & | \\ CH—OH & +\ HO—\overset{O}{\overset{\|}{C}}—C_{17}H_{35} & \longrightarrow & CH—O—\overset{O}{\overset{\|}{C}}—C_{17}H_{35} + 3H_2O \\ | & & & | \\ CH_2—OH & HO—\overset{O}{\overset{\|}{C}}—C_{17}H_{35} & & CH_2—O—\overset{O}{\overset{\|}{C}}—C_{17}H_{35} \\ \text{glycerol} & \text{stearic acid} & & \text{glyceryl tristearate} \\ \text{(1 molecule)} & \text{(3 molecules)} & & \text{(a triglyceride — 1 molecule)} \end{array} \quad (8.1)$$

The following diagram may help you remember the components of a triglyceride.

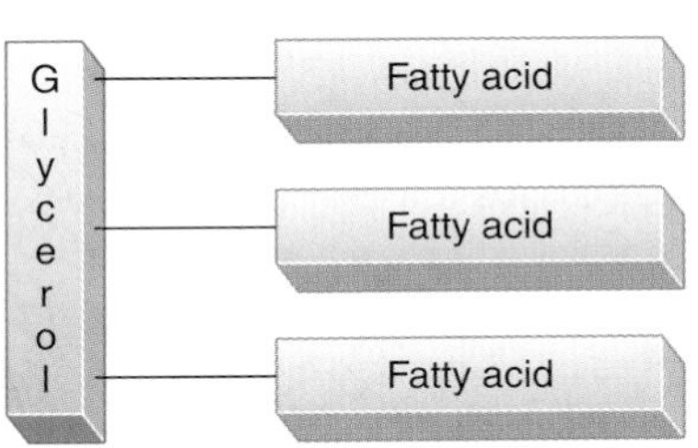

The resulting esters are called **triglycerides** or **triacylglycerols.**

triglyceride or triacylglycerol A triester of glycerol in which all three alcohol groups are esterified.

The fatty acid components in naturally occurring triglyceride molecules are rarely identical (as in the case of glyceryl tristearate). In addition, natural triglycerides (fats and oils) are usually mixtures of different triglyceride molecules. Butterfat, for example, contains at least 14 different fatty acid components.

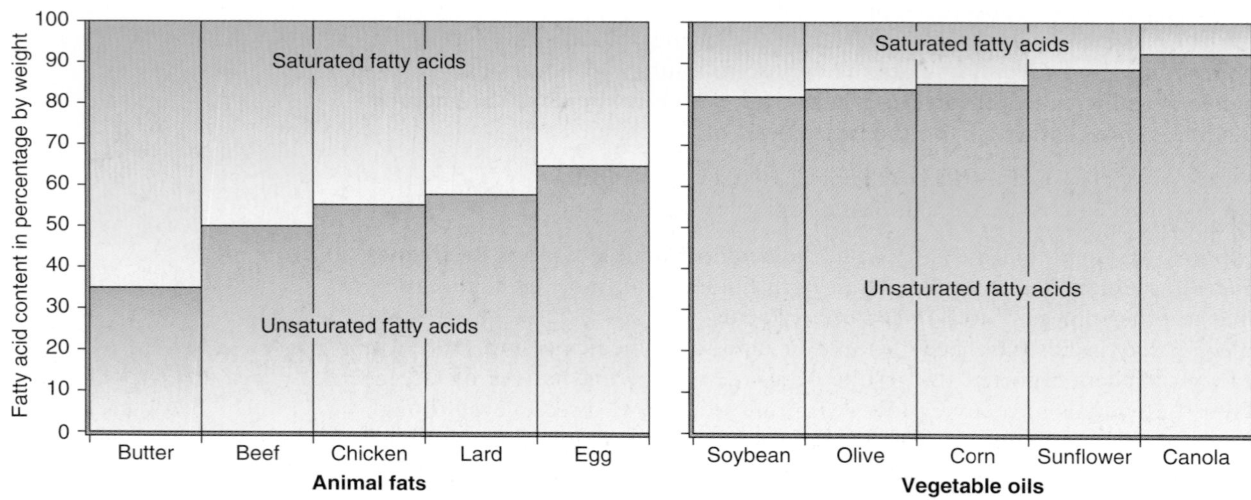

FIGURE 8.7 A comparison of saturated and unsaturated fatty acids in some foods.

Learning Check 8.1 Write one structure (several possibilities exist) for a triglyceride derived from stearic acid, oleic acid, and palmitic acid.

fat
A triglyceride that is a solid at room temperature.

oil
A triglyceride that is a liquid at room temperature.

ThomsonNOW™ Go to Coached Problems to explore **triglycerides.**

With some exceptions, triglycerides that come from animals other than fish are solids at room temperature and are called **fats.** Triglycerides from plants or fish are liquids at room temperature and are usually referred to as **oils.** Although both fats and oils are triglycerides, they differ structurally in one important respect—the degree of unsaturation. As shown in Figure 8.7, animal fats contain primarily triglycerides of long-chain saturated fatty acids (higher melting points). In contrast, vegetable oils, such as corn oil and sunflower oil, consist of triglycerides containing unsaturated fatty acids (lower melting points) (see Figure 8.8). Thus, we see that the structural and physical properties of fatty acids determine the properties of the triglycerides derived from them.

Excessive fat in the diet is recognized as one risk factor influencing the development of chronic disease. The main concern about excessive dietary fat, especially saturated fat, centers on its role in raising blood cholesterol levels (Section 8.9). High blood cholesterol is a recognized risk factor in the development of coronary heart disease, the leading cause of death in Americans every year.

FIGURE 8.8 Sunflowers are an important crop in some regions of the country because the seeds are a good source of unsaturated oils. Why do you think seeds are rich in oils?

8.4 Chemical Properties of Fats and Oils

LEARNING OBJECTIVE

5. Write key reactions for fats and oils.

The chemical properties of triglycerides are typical of esters and alkenes because these are the two functional groups present.

Hydrolysis

Hydrolysis is one of the most important reactions of fats and oils, just as it was for carbohydrates. The treatment of fats or oils with water and an acid catalyst causes them to hydrolyze to form glycerol and fatty acids. This reaction is simply the reverse of ester formation:

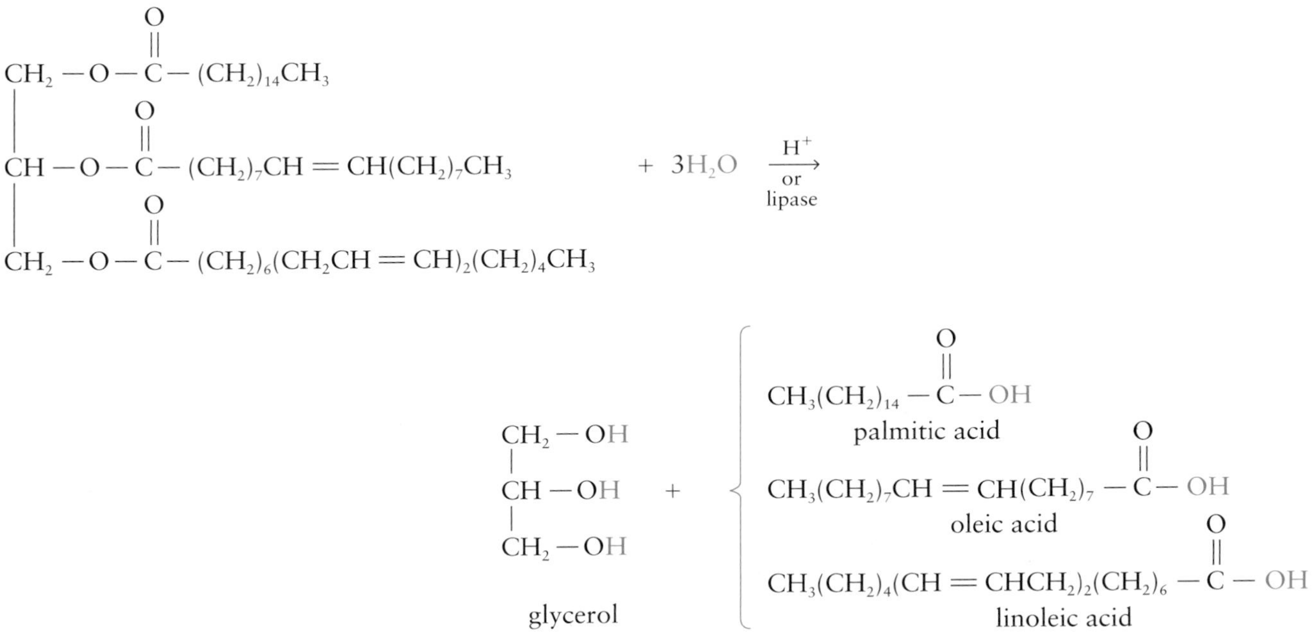

$$\begin{array}{l} CH_2-O-\overset{\overset{\large O}{\|}}{C}-(CH_2)_{14}CH_3 \\ | \\ CH-O-\overset{\overset{\large O}{\|}}{C}-(CH_2)_7CH=CH(CH_2)_7CH_3 \\ | \\ CH_2-O-\overset{\overset{\large O}{\|}}{C}-(CH_2)_6(CH_2CH=CH)_2(CH_2)_4CH_3 \end{array} + 3H_2O \xrightarrow[\text{or lipase}]{H^+} \tag{8.2}$$

$$\begin{array}{c} CH_2-OH \\ | \\ CH-OH \\ | \\ CH_2-OH \\ \text{glycerol} \end{array} + \left\{ \begin{array}{l} CH_3(CH_2)_{14}-\overset{\overset{\large O}{\|}}{C}-OH \quad \text{palmitic acid} \\ CH_3(CH_2)_7CH=CH(CH_2)_7-\overset{\overset{\large O}{\|}}{C}-OH \quad \text{oleic acid} \\ CH_3(CH_2)_4(CH=CHCH_2)_2(CH_2)_6-\overset{\overset{\large O}{\|}}{C}-OH \quad \text{linoleic acid} \end{array} \right.$$

Enzymes (lipases) of the digestive system also catalyze the hydrolysis process. This reaction represents the only important change that takes place in fats and oils during digestion. The breakdown of cellular fat deposits to supply energy also begins with the lipase-catalyzed hydrolysis reaction.

Learning Check 8.2 Write the structures of the hydrolysis products of the following reaction:

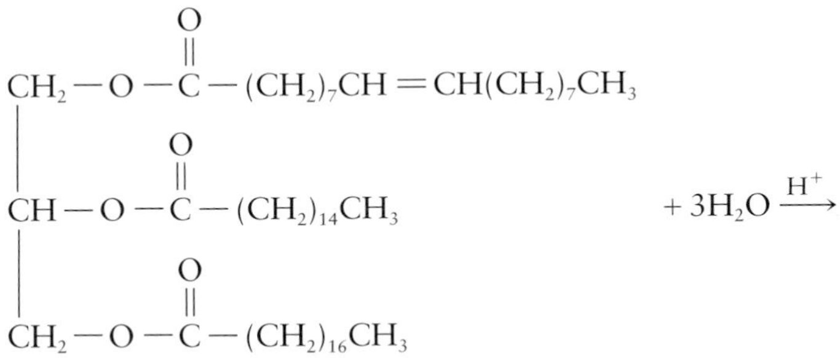

$$\begin{array}{l} CH_2-O-\overset{\overset{\large O}{\|}}{C}-(CH_2)_7CH=CH(CH_2)_7CH_3 \\ | \\ CH-O-\overset{\overset{\large O}{\|}}{C}-(CH_2)_{14}CH_3 \\ | \\ CH_2-O-\overset{\overset{\large O}{\|}}{C}-(CH_2)_{16}CH_3 \end{array} + 3H_2O \xrightarrow{H^+}$$

Saponification

soap
A salt of a fatty acid often used as a cleaning agent.

When triglycerides are reacted with a strong base, the process of saponification (soap making) occurs. In this commercially important reaction, the products are glycerol and the salts of fatty acids, which are also called **soaps** (Reaction 8.3).

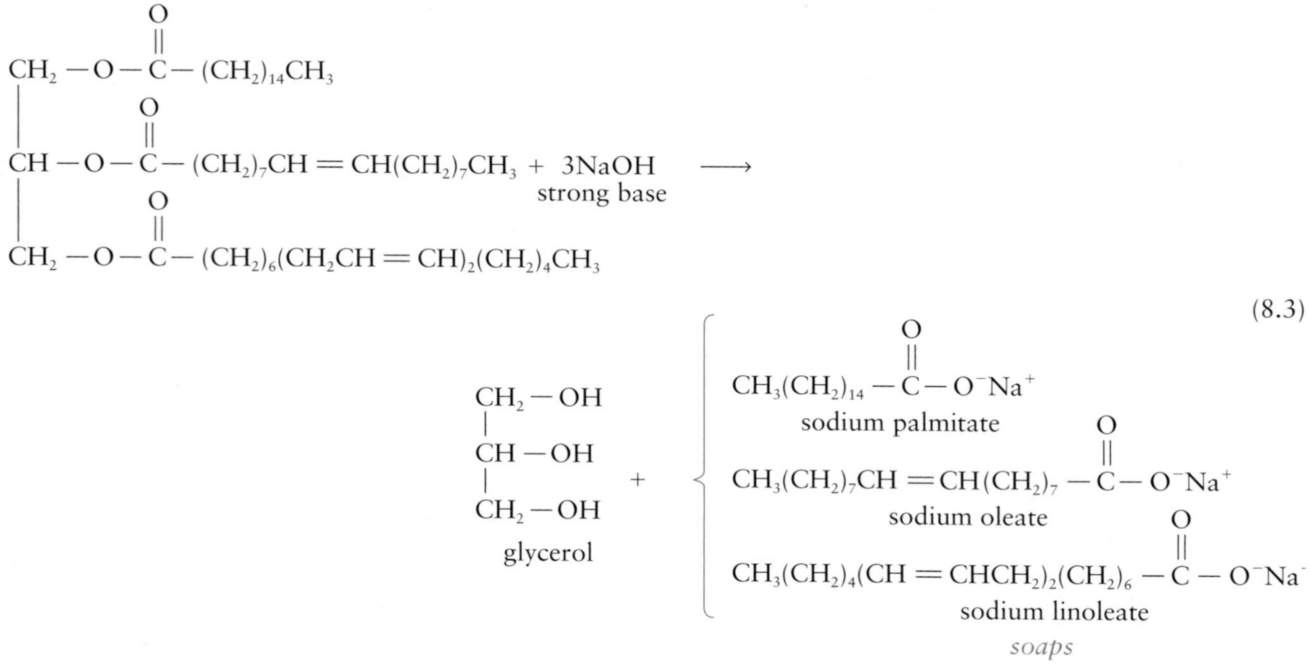

$$\begin{array}{l} CH_2-O-\overset{\overset{\Large O}{\|}}{C}-(CH_2)_{14}CH_3 \\ | \\ CH-O-\overset{\overset{\Large O}{\|}}{C}-(CH_2)_7CH=CH(CH_2)_7CH_3 + \underset{\text{strong base}}{3NaOH} \longrightarrow \\ | \\ CH_2-O-\overset{\overset{\Large O}{\|}}{C}-(CH_2)_6(CH_2CH=CH)_2(CH_2)_4CH_3 \end{array}$$

$$\underset{\text{glycerol}}{\begin{array}{l} CH_2-OH \\ | \\ CH-OH \\ | \\ CH_2-OH \end{array}} + \underset{\textit{soaps}}{\left\{ \begin{array}{l} \underset{\text{sodium palmitate}}{CH_3(CH_2)_{14}-\overset{\overset{\Large O}{\|}}{C}-O^-Na^+} \\ \underset{\text{sodium oleate}}{CH_3(CH_2)_7CH=CH(CH_2)_7-\overset{\overset{\Large O}{\|}}{C}-O^-Na^+} \\ \underset{\text{sodium linoleate}}{CH_3(CH_2)_4(CH=CHCH_2)_2(CH_2)_6-\overset{\overset{\Large O}{\|}}{C}-O^-Na^-} \end{array} \right.} \tag{8.3}$$

The salts obtained from saponification depend on the base used. Sodium salts, known as hard salts, are found in most cake soap used in the home. Potassium salts, soft soaps, are used in some shaving creams and liquid soap preparations.

In traditional soap making, animal fat is the source of triglycerides, and lye (crude NaOH) or an aqueous extract of wood ashes is the source of the base. The importance of soap making was amply proved after the soap maker's art was lost with the fall of the Roman Empire. The soapless centuries between A.D. 500 and 1500 were notorious for the devastating plagues that nearly depopulated an unsanitary western Europe.

■ **Learning Check 8.3** Write structures for the products formed when the following triglyceride is saponified using NaOH:

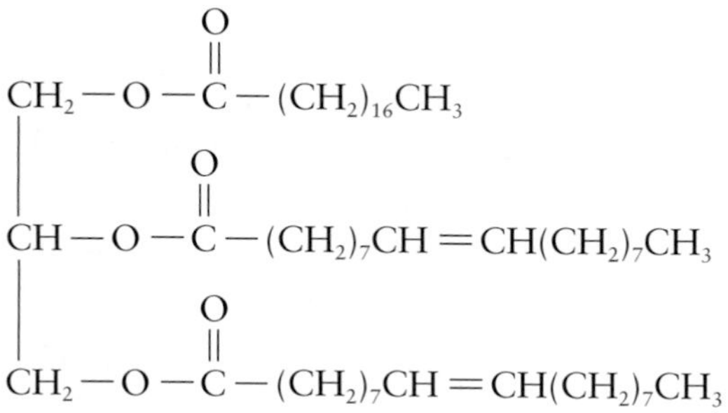

$$\begin{array}{l} CH_2-O-\overset{\overset{\Large O}{\|}}{C}-(CH_2)_{16}CH_3 \\ | \\ CH-O-\overset{\overset{\Large O}{\|}}{C}-(CH_2)_7CH=CH(CH_2)_7CH_3 \\ | \\ CH_2-O-\overset{\overset{\Large O}{\|}}{C}-(CH_2)_7CH=CH(CH_2)_7CH_3 \end{array}$$

Hydrogenation

In Chapter 2, we learned that double bonds can be reduced to single bonds by treatment with hydrogen (H_2) in the presence of a catalyst. An important commercial reaction of fats and oils uses this same hydrogenation process, in which some of the fatty acid double bonds are converted to single bonds. The result is a decrease in the degree of unsaturation and a corresponding increase in the melting point of the fat or oil (Figure 8.6). The peanut oil in many popular brands of peanut butter has been partially hydrogenated to convert the oil to a semisolid that does not separate. Hydrogenation is most often used in the production of semisolid cooking shortenings (such as Crisco®, Spry®, and Fluffo®) or margarines from liquid vegetable oils (see ■ Figure 8.9).

Partially hydrogenated vegetable oils were developed in part to help displace highly saturated animal fats used in frying, baking, and spreads. It is important not to complete the reaction and totally saturate all the double bonds. If this is done, the product is hard and waxy—not the smooth, creamy product desired by consumers. Equation 8.4 gives an example in which some of the double bonds react, showing one of the possible products.

■ **FIGURE 8.9** Hydrogenation of vegetable oils is used to prepare both shortenings and margarines. How do oils used in automobiles differ chemically from these vegetable oils?

$$
\begin{array}{l}
CH_2-O-\overset{O}{\overset{\|}{C}}-(CH_2)_{14}CH_3 \\
| \\
CH-O-\overset{O}{\overset{\|}{C}}-(CH_2)_7CH=CH(CH_2)_7CH_3 \\
| \\
CH_2-O-\overset{O}{\overset{\|}{C}}-(CH_2)_6(CH_2CH=CH)_2(CH_2)_4CH_3
\end{array}
+ 2H_2 \xrightarrow{Ni}
\begin{array}{l}
CH_2-O-\overset{O}{\overset{\|}{C}}-(CH_2)_{14}CH_3 \\
| \\
CH-O-\overset{O}{\overset{\|}{C}}-(CH_2)_7CH=CH(CH_2)_7CH_3 \\
| \\
CH_2-O-\overset{O}{\overset{\|}{C}}-(CH_2)_6(CH_2CH_2CH_2)_2(CH_2)_4CH_3
\end{array}
\quad (8.4)
$$

■ **Learning Check 8.4** If an oil with the following structure is completely hydrogenated, what is the structure of the product?

$$
\begin{array}{l}
CH_2-O-\overset{O}{\overset{\|}{C}}-(CH_2)_7CH=CH(CH_2)_7CH_3 \\
| \\
CH-O-\overset{O}{\overset{\|}{C}}-(CH_2)_7CH=CHCH_2CH=CHCH_2CH=CHCH_2CH_3 \\
| \\
CH_2-O-\overset{O}{\overset{\|}{C}}-(CH_2)_7CH=CHCH_2CH=CH(CH_2)_4CH_3
\end{array}
$$

STUDY SKILLS 8.1 A Reaction Map for Triglycerides

We've discussed three different reactions of triglycerides. Remember, when you're trying to decide what happens in a reaction, focus attention on the functional groups. After identifying a triglyceride as a starting material, next decide what other reagent is involved and which of the three pathways is appropriate.

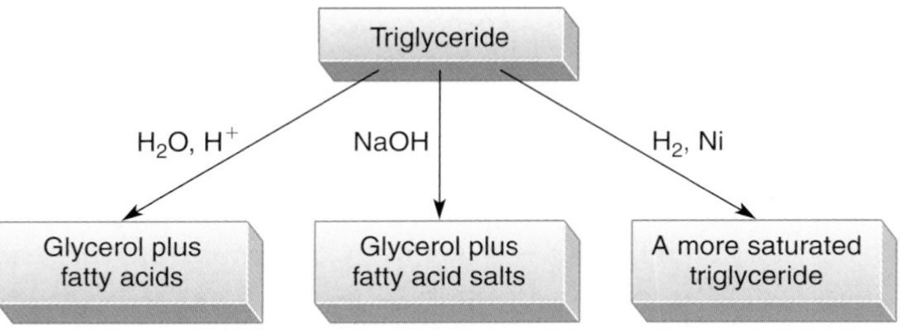

During hydrogenation, some fatty acid molecules in the common cis configuration (Section 8.2) are isomerized into trans fatty acids. The main sources of trans fatty acids in the diet are stick margarine, shortening, and high-fat baked goods. Recent clinical studies have shown that blood cholesterol levels may be raised by the consumption of trans fatty acids (see Chemistry and Your Health 8.1). Current dietary advice is to reduce the consumption of saturated fatty acids and products that contain significant amounts of trans fatty acids.

8.5 Waxes

LEARNING OBJECTIVE

6. Compare the structures of fats and waxes.

wax
An ester of a long-chain fatty acid and a long-chain alcohol.

Waxes represent a second group of simple lipids that, like fats and oils, are esters of fatty acids. However, the alcohol portion of waxes is derived from long-chain alcohols (12–32 carbons) rather than glycerol. Beeswax, for example, contains a wax with the following structure:

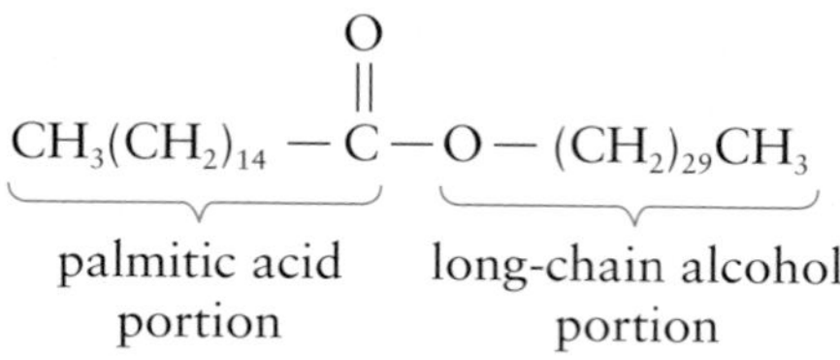

A wax.

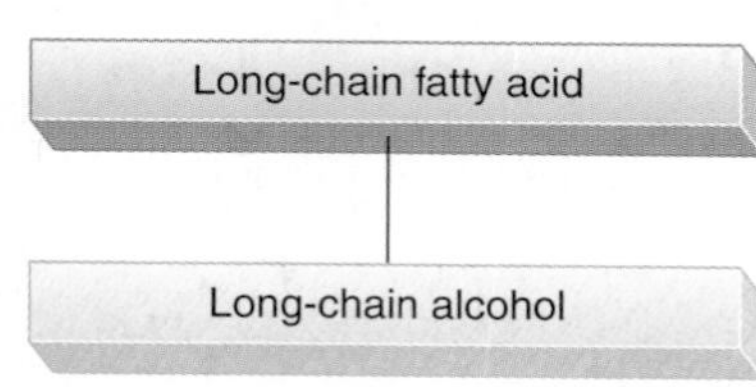

Waxes are water-insoluble and not as easily hydrolyzed as the fats and oils; consequently, they often occur in nature as protective coatings on feathers, fur, skin, leaves, and fruits. Sebum, a secretion of the sebaceous glands of the skin, contains many different waxes. It keeps the skin soft and prevents dehydration. Waxes are used commercially to make cosmetics, candles, ointments, and protective polishes.

8.6 Phosphoglycerides

LEARNING OBJECTIVES

7. Draw structural formulas and describe uses for phosphoglycerides.

phosphoglyceride
A complex lipid containing glycerol, fatty acids, phosphoric acid, and an aminoalcohol component.

phospholipid
A phosphorus-containing lipid.

Phosphoglycerides are complex lipids that serve as the major components of cell membranes. Phosphoglycerides and related compounds are also referred to as **phospholipids.** In these compounds, one of the —OH groups of glycerol is joined by an ester linkage to phosphoric acid, which in turn is linked to another alcohol (usually an aminoalcohol). The other two —OH groups of glycerol are linked to fatty acids, resulting in the following general structure:

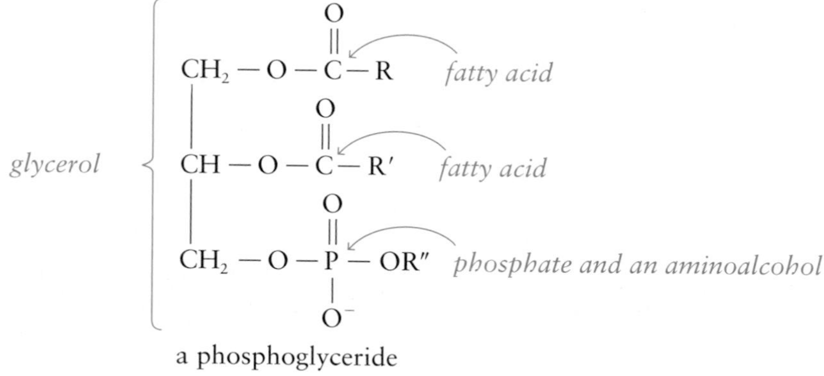

a phosphoglyceride

A phosphoglyceride.

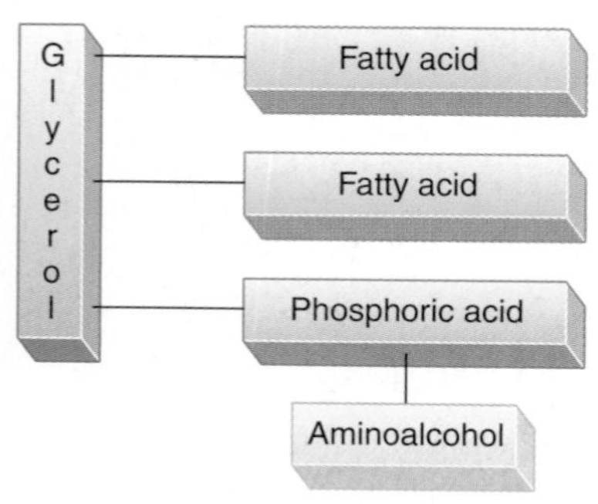

CHEMISTRY AND YOUR HEALTH 8.1 Going after Those Trans Fatty Acids

Nutritional labels are supposed to be helpful, but sometimes the average consumer is only concerned with price and quantity. A recent labeling change mandated by the FDA on January 1, 2006, requires all food products to list materials called trans-fats on the nutritional label. Many people don't distinguish between the types of fat they are consuming and consider all fats to be "bad." However, some fats promote health, while others should be largely avoided. In order to make good decisions about dietary fats, it is necessary to understand some terminology. Unsaturated fats such as those found in olives, avacados, nuts, and salmon are generally considered to promote good health. These fats can be eaten in moderate amounts with healthy benefits and have even been shown to reduce the risk of coronary heart disease. Unsaturated fats are liquids at room temperature. Saturated fats are solids at room temperature and are the type found mostly in animal products such as butter and lard. These should be eaten only in small quantities as they have been shown to have some negative health effects such as increasing the level of LDL (bad) blood lipids.

The melting point of liquid, unsaturated fats can be increased by reacting the fats with hydrogen is a process called hydrogenation. The resulting partially hydrogenated fats are solids or semisolids at room temperature. Margarines are produced this way. Fats produced by hydrogenation are still partially unsaturated, have long shelf lives, and have long useful lives when used to deep fry foods. Because they also give foods desirable taste and textures, they are used extensively in fast foods and snack foods.

On the negative side, however, the partial hydrogenation of unsaturated fats generates materials called trans-fats. These are partially unsaturated fats that are isomers of naturally occurring unsaturated fats and have become a serious enough health concern to cause the FDA to invoke the labeling requirement mentioned earlier. As far as dietary fats are concerned, it is prudent to read labels and try to make healthy choices.

The most abundant phosphoglycerides have one of the alcohols choline, ethanolamine, or serine attached to the phosphate group. These aminoalcohols are shown below in charged forms:

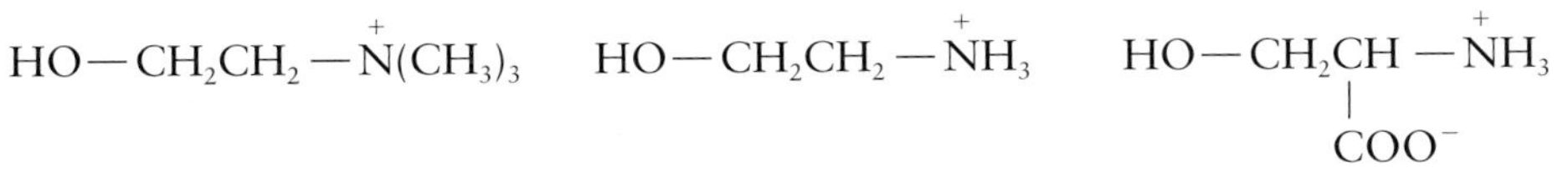

$$\mathrm{HO{-}CH_2CH_2{-}\overset{+}{N}(CH_3)_3} \qquad \mathrm{HO{-}CH_2CH_2{-}\overset{+}{N}H_3} \qquad \mathrm{HO{-}CH_2\underset{\displaystyle COO^-}{\underset{|}{C}H}{-}\overset{+}{N}H_3}$$

choline (a quaternary ammonium cation) | ethanolamine (cation form) | serine (two ionic groups present)

A typical phosphoglyceride is phosphatidylcholine, which is commonly called **lecithin:**

lecithin
A phosphoglyceride containing choline.

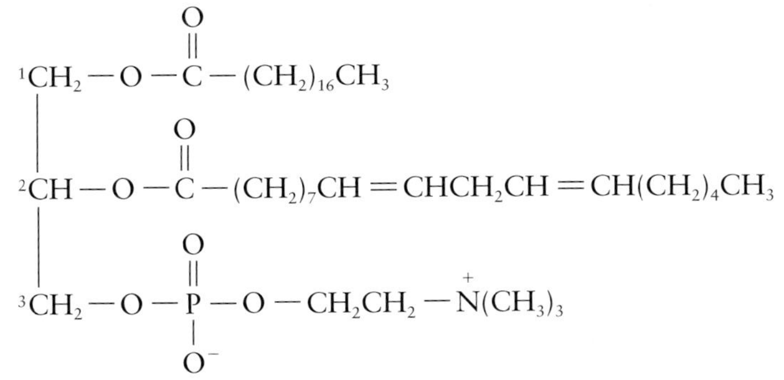

$$\begin{array}{l}
{}^{1}\mathrm{CH_2{-}O{-}\overset{\displaystyle O}{\overset{\|}{C}}{-}(CH_2)_{16}CH_3} \\
\quad | \\
{}^{2}\mathrm{CH{-}O{-}\overset{\displaystyle O}{\overset{\|}{C}}{-}(CH_2)_7CH{=}CHCH_2CH{=}CH(CH_2)_4CH_3} \\
\quad | \\
{}^{3}\mathrm{CH_2{-}O{-}\underset{\displaystyle O^-}{\underset{|}{\overset{\displaystyle O}{\overset{\|}{P}}}}{-}O{-}CH_2CH_2{-}\overset{+}{N}(CH_3)_3}
\end{array}$$

phosphatidylcholine

All phosphoglycerides that contain choline are classified as lecithins. Because different fatty acids may be bonded at positions 1 and 2 of the glycerol, a number of different lecithins are possible. Note that the lecithin shown contains a negatively charged phosphate group and a positively charged quaternary nitrogen.

■ **FIGURE 8.10** Some common foods containing lecithins as emulsifying agents.

These charges make that end of the molecule strongly hydrophilic, whereas the rest of the molecule is hydrophobic. This structure of lecithins enables them to function as very important structural components of most cell membranes (Section 8.8). It also allows lecithins to function as emulsifying and micelle-forming agents. Such micelles play an important role in the transport of lipids in the blood (Chapter 14). Lecithin (phosphatidylcholine) is commercially extracted from soybeans for use as an emulsifying agent in food. It gives a smooth texture to such products as margarine and chocolate candies (see ■ Figure 8.10).

cephalin A phosphoglyceride containing ethanolamine or serine.

Phosphoglycerides in which the alcohol is ethanolamine or serine, rather than choline, are called **cephalins.** They are found in most cell membranes and are particularly abundant in brain tissue. Cephalins are also found in blood platelets, where they play an important role in the blood-clotting process. A typical cephalin is represented as

$$\begin{array}{l} CH_2-O-\overset{\overset{\large O}{\|}}{C}-(CH_2)_{14}CH_3 \\ | \\ CH-O-\overset{\overset{\large O}{\|}}{C}-(CH_2)_7CH=CH(CH_2)_7CH_3 \\ | \\ CH_2-O-\underset{\underset{\large O^-}{|}}{\overset{\overset{\large O}{\|}}{P}}-O-CH_2\underset{\underset{\large COO^-}{|}}{CH}-\overset{+}{N}H_3 \end{array}$$

Note the presence of the negatively charged phosphate and carboxyl groups and positively charged nitrogen. This characteristic allows cephalins, like the lecithins, to function as emulsifiers.

■ **Learning Check 8.5** Draw a typical structure for a cephalin containing the cation of ethanolamine:

$$HO-CH_2CH_2-\overset{+}{N}H_3$$

8.7 Sphingolipids

LEARNING OBJECTIVE

8. Draw structural formulas and describe uses for sphingolipids.

Sphingolipids, a second type of complex lipid found in cell membranes, do not contain glycerol. Instead, they contain sphingosine, a long-chain unsaturated aminoalcohol. This substance and a typical sphingolipid, called a sphingomyelin, are represented below:

sphingolipid
A complex lipid containing the aminoalcohol sphingosine.

$$\begin{array}{l} CH_3(CH_2)_{12}CH{=}CH{-}CH{-}OH \\ \qquad\qquad\qquad\qquad\quad | \\ \qquad\qquad\qquad\qquad\quad CH{-}NH_2 \\ \qquad\qquad\qquad\qquad\quad | \\ \qquad\qquad\qquad\qquad\quad CH_2OH \end{array}$$

sphingosine

$$\begin{array}{l} CH_3(CH_2)_{12}CH{=}CH{-}CH{-}OH \\ \qquad\qquad\qquad\qquad\quad | \\ \qquad\qquad\qquad\qquad\quad CH{-}NH{-}\overbrace{\overset{\overset{\displaystyle O}{\|}}{C}{-}(CH_2)_7CH{=}CH(CH_2)_7CH_3}^{\textit{fatty acid}} \\ \qquad\qquad\qquad\qquad\quad | \\ \qquad\qquad\qquad\qquad\quad CH_2{-}\underbrace{O{-}\overset{\overset{\displaystyle O}{\|}}{\underset{\underset{\displaystyle O^-}{|}}{P}}}_{\textit{phosphate}}{-}O{-}\underbrace{CH_2CH_2\overset{+}{N}(CH_3)_3}_{\textit{choline}} \end{array}$$

sphingomyelin

A sphingomyelin.

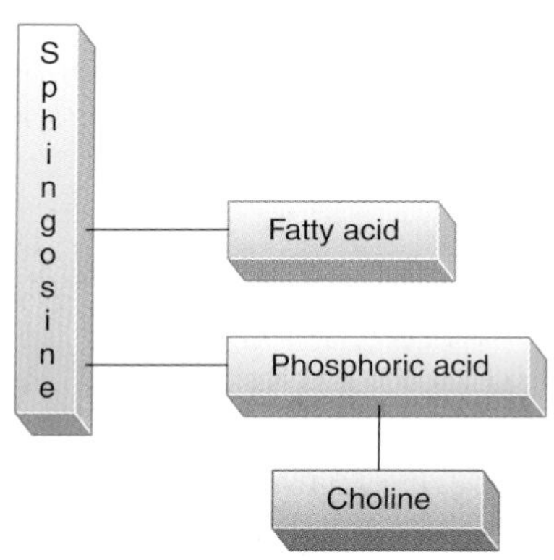

In sphingomyelin, choline is attached to sphingosine through a phosphate group. A single fatty acid is also attached to the sphingosine, but an amide linkage is involved instead of an ester linkage. A number of sphingomyelins are known; they differ only in the fatty acid component. Large amounts of sphingomyelins are found in brain and nerve tissue and in the protective myelin sheath that surrounds nerves.

Glycolipids are another type of sphingolipid. Unlike sphingomyelins, however, these complex lipids contain carbohydrates (usually monosaccharides such as glucose or galactose). Glycolipids are often called cerebrosides because of their abundance in brain tissue. The structure of a typical cerebroside is

glycolipid
A complex lipid containing a sphingosine, a fatty acid, and a carbohydrate.

$$\begin{array}{l} CH_3(CH_2)_{12}CH{=}CH{-}CH{-}OH \\ \qquad\qquad\qquad\qquad\quad | \\ \qquad\qquad\qquad\qquad\quad CH{-}NH{-}\overbrace{\overset{\overset{\displaystyle O}{\|}}{C}{-}(CH_2)_7CH{=}CH(CH_2)_7CH_3}^{\textit{fatty acid}} \\ \qquad\qquad\qquad\qquad\quad | \\ \qquad\qquad\qquad\qquad\quad CH_2 \\ \qquad\qquad\qquad\qquad\quad | \\ \qquad\qquad\qquad\qquad\quad O{-}(\text{D-galactose ring: } CH_2OH,\ HO,\ OH,\ OH) \end{array}$$

D-galactose

a cerebroside

A glycolipid.

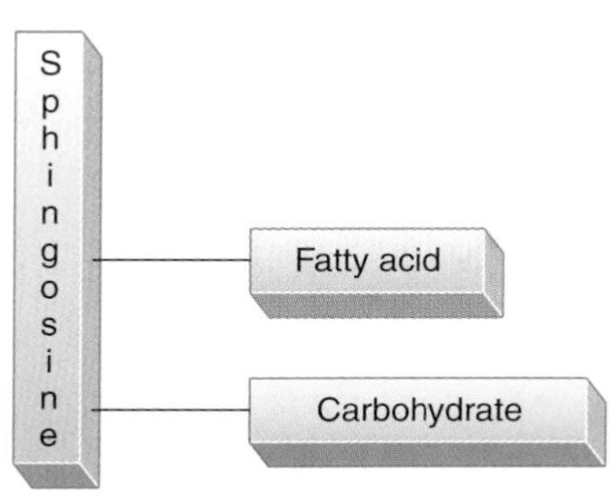

Notice that, unlike the sphingomyelins, cerebrosides do not contain phosphate.

TABLE 8.2 Diseases originating from abnormal metabolism and accumulation of glycolipids and sphingomyelins

Disease	Organs(s) affected	Type of lipid accumulated
Tay-Sachs	Brain	Glycolipid (ganglioside G_{M2})
Gaucher's	Spleen, liver	Cerebrosides containing glucose
Niemann-Pick	Several, particularly liver and spleen	Sphingomyelins

Several human diseases are known to result from an abnormal accumulation of sphingomyelins and glycolipids in the body (see ■ Table 8.2). Research has shown that each of these diseases is the result of an inherited absence of an enzyme needed to break down these complex lipids.

8.8 Biological Membranes

LEARNING OBJECTIVE

9. Describe the major features of cell membrane structure.

Cell Structure

prokaryotic cell
A simple unicellular organism that contains no nucleus and no membrane-enclosed organelles.

eukaryotic cell
A cell containing membrane-enclosed organelles, particularly a nucleus.

organelle
A specialized structure within a cell that performs a specific function.

Two cell types are found in living organisms: prokaryotic and eukaryotic. **Prokaryotic cells,** comprising bacteria and cyanobacteria (blue-green algae), are smaller and less complex than eukaryotic cells. The more complex **eukaryotic cells** make up the tissues of all other organisms, including humans. Both types of cells are essentially tiny membrane-enclosed units of fluid containing the chemicals involved in life processes. In addition, eukaryotic cells contain small bodies called **organelles,** which are are suspended in the cellular fluid, or cytoplasm. Organelles are the sites of many specialized functions in eukaryotes. The most prominent organelles and their functions are listed in ■ Table 8.3. Membranes perform two vital functions in living organisms: The external cell membrane acts as a selective barrier between the living cell and its environment, and internal membranes surround some organelles, creating cellular compartments that have separate organization and functions.

Membrane Structure

ThomsonNOW™ Go to Coached Problems to explore **cell membrane structure.**

Most cell membranes contain about 60% lipid and 40% protein. Phosphoglycerides (such as lecithin and cephalin), sphingomyelin, and cholesterol are the pre-

TABLE 8.3 The functions of some cellular organelles

Organelle	Function
Endoplasmic reticulum	Synthesis of proteins, lipids, and other substances
Lysosome	Digestion of substances taken into cells
Microfilaments and microtubules	Cellular movements
Mitochondrion	Cellular respiration and energy production
Nucleus	Contains hereditary material (DNA), which directs protein synthesis
Plastids	Contains plant pigments such as chlorophyll (photosynthesis)
Ribosome	Protein synthesis

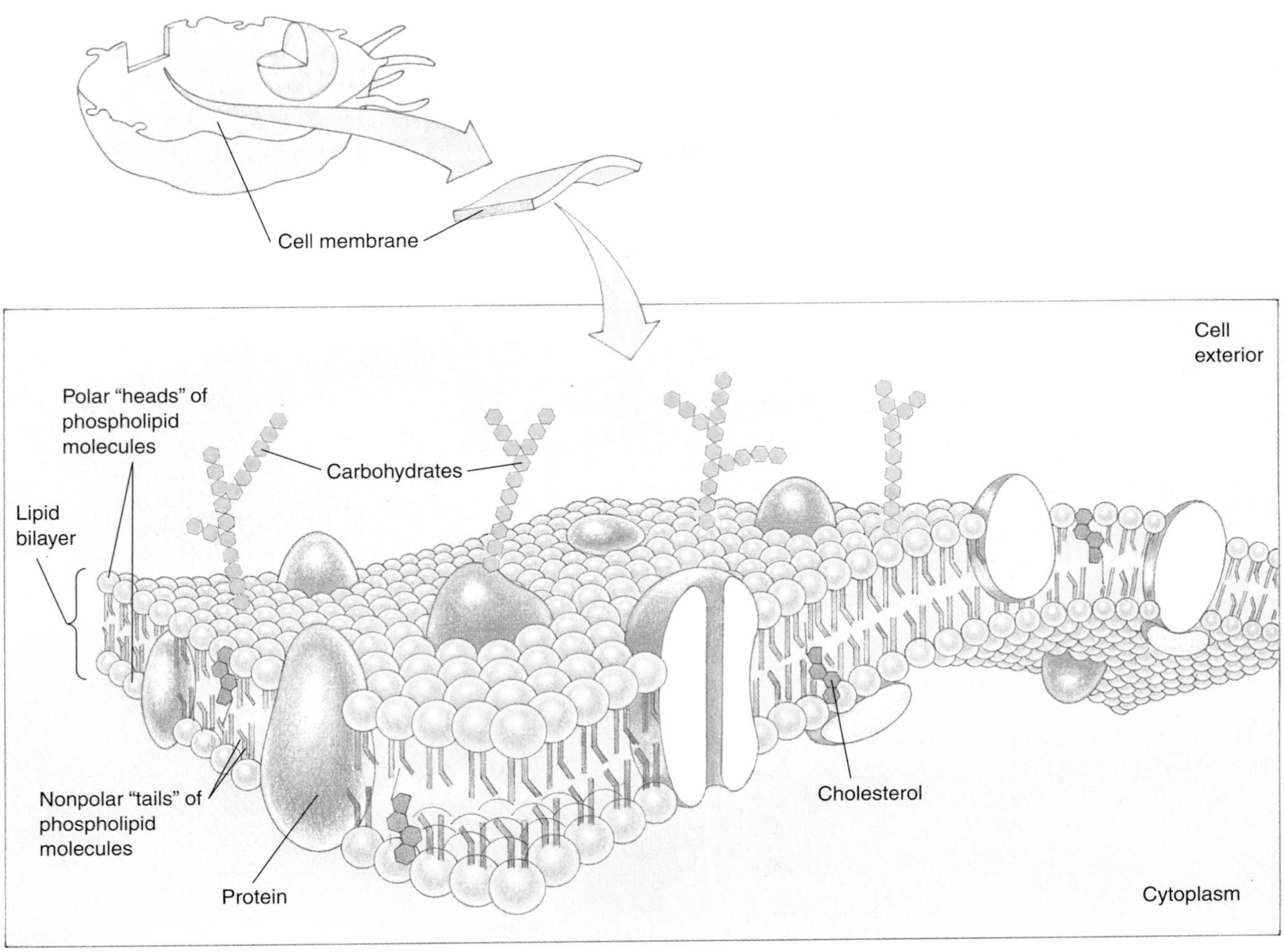

■ **FIGURE 8.11** The fluid-mosaic model of membrane structure. Phosphoglycerides are the chief lipid component. They are arranged in a bilayer. Proteins float like icebergs in a sea of lipid.

dominant types of lipids found in most membranes. Precisely how the lipids and proteins are organized to form membranes has been the subject of a great deal of research. A widely accepted model called the **fluid-mosaic model** is diagrammed in ■ Figure 8.11. The lipids are organized in a **bilayer** in which the hydrophobic chains extend toward the inside of the bilayer, and the hydrophilic groups (the phosphate groups and other polar groups) are oriented toward the outside, where they come in contact with water (see ■ Figure 8.12). Like the micelle structure discussed in Section 8.2, a lipid bilayer is a very stable arrangement where "like" associates with "like." The hydrophobic tails of the lipids are protected from water, and the hydrophilic heads are in a position to interact with water. When a

fluid-mosaic model
A model of membrane structure in which proteins are embedded in a flexible lipid bilayer.

lipid bilayer
A structure found in membranes, consisting of two sheets of lipid molecules arranged so that the hydrophobic portions are facing each other.

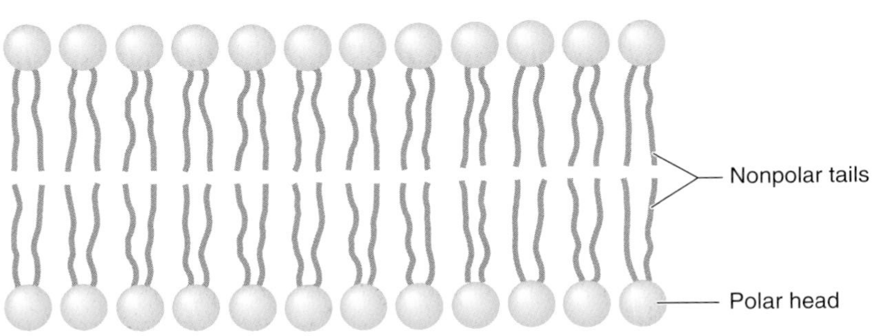

■ **FIGURE 8.12** The lipid bilayer model. Circles represent the polar portion of phosphoglycerides. The hydrocarbon "tails" are within the interior of the bilayer.

ThomsonNOW™ Go to Coached Problems to explore the **fluid mosaic model.**

lipid bilayer is broken and the interior hydrocarbon tails are exposed to water, the resulting repulsion causes the bilayer to re-form, and the break seals spontaneously.

Membrane lipids usually contain unsaturated fatty acid chains that fit into bilayers more loosely than do saturated fatty acids (Section 8.2). This increases the flexibility or fluidity of the membrane. Some of the proteins in the membrane float in the lipid bilayer like icebergs in the sea, whereas others extend completely through the bilayer. The lipid molecules are free to move laterally within the bilayer like dancers on a crowded dance floor—hence the term "fluid mosaic."

ThomsonNOW™ Go to Coached Problems to explore **lipid membrane structure.**

8.9 Steroids

LEARNING OBJECTIVE

10. Identify the structural characteristic typical of steroids and list important groups of steroids in the body.

Steroids exhibit the distinguishing feature of other lipids: They are soluble in nonpolar solvents. Structurally, however, they are completely different from the lipids already discussed. **Steroids** have as their basic structure a set of three six-membered rings and a single five-membered ring fused together:

steroid
A compound containing four rings fused in a particular pattern.

steroid ring system

Cholesterol

The most abundant steroid in the human body, and the most important, is cholesterol:

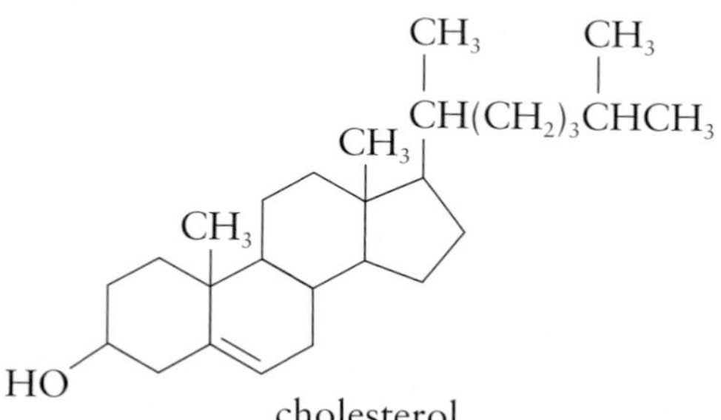

cholesterol

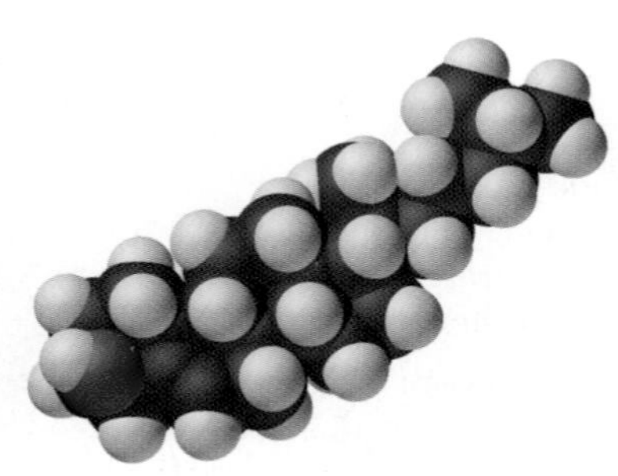
Cholesterol

Cholesterol is an essential component of cell membranes and is a precursor for other important steroids, including bile salts, male and female sex hormones, vitamin D, and the adrenocorticoid hormones (Section 8.10). Cholesterol for use in these important functions is synthesized in the liver. Additional amounts of cholesterol are present in the foods we eat.

On the other side of the coin, cholesterol has received much attention because of a correlation between its levels in the blood and the disease known as atherosclerosis, or hardening of the arteries (Section 14.1). Although our knowledge of the role played by cholesterol in atherosclerosis is incomplete, it is now considered advisable to reduce the amount of cholesterol in the foods we eat. In addition, reducing the amount of saturated fatty acids in the diet appears to lower cholesterol production by the body.

CHEMISTRY AROUND US 8.1 Nuts: Good Food in Small Packages

Many people avoid eating nuts because of a recent diet craze that branded them as containing "bad fats." However, this label, like many from fad diets, is not entirely true. The truth is, nuts contain a large amount of nutrients in relatively small packages, and should be added to the standard diet.

According to some dieticians, nuts represent a good nutritional investment. You get protein, minerals, and healthy fat and fiber all in a single food item. As a bonus, nuts are fun to eat and give you the feeling of cheating on your diet even though you are not.

However, like many good things, they are best when used in moderation. In addition, nuts are often combined with high fat and caloric ingredients in foods such as cookies, pies, and ice cream sundaes. For a healthier approach, nuts can be included in healthier foods such as oatmeal, yogurt, salads, or stir fry.

Nuts are a good source of protein, minerals, and fiber.

Bile Salts

Bile is a yellowish brown or green liver secretion that is stored and concentrated in the gallbladder. The entry of partially digested fatty food into the small intestine causes the gallbladder to contract and empty bile into the intestine. Bile does not catalyze any digestive reactions, but it is important in lipid digestion. Fats are more difficult to hydrolyze than starch and proteins because they are not soluble in water. In the watery medium of the digestive system, fats tend to form large globules with a limited surface area exposed to attack by the digestive juices. The chief constituents of bile are the bile salts and two waste components: cholesterol and bile pigments. One of the principal bile salts is sodium glycocholate:

Structure of sodium glycocholate: steroid ring system bearing HO, OH, OH and three CH_3 groups, with side chain $\text{CH(CH}_3\text{)CH}_2\text{CH}_2\text{—C(=O)—NHCH}_2\text{COO}^-\text{Na}^+$

sodium glycocholate

The bile salts act much like soaps; they emulsify the lipids and break the large globules into many smaller droplets. After such action, the total amount of lipid is still the same, but a much larger surface area is available for the hydrolysis reactions. Some researchers feel that bile salts also remove fatty coatings from particles of other types of food and thus aid in their digestion.

A second function of bile salts is to emulsify cholesterol found in the bile. Excess body cholesterol is concentrated in bile and is passed into the small intestine for excretion. If cholesterol levels become too high in the bile, or if the concentration of bile salts is too low, the cholesterol precipitates and forms gallstones (see ■ Figure 8.13). The most common gallstones contain about 80% cholesterol and are colored by entrapped bile pigments.

■ **FIGURE 8.13** Three gallstones (green) in a gallbladder (red).

OVER THE COUNTER 8.1 Melatonin and DHEA: Hormones at Your Own Risk

The search for the fountain of youth continues. Melatonin and dehydroepiandrosterone (DHEA) have both been given the image of wonder drugs with the ability to reduce the chances of contracting several diseases or conditions, including cancer, high blood pressure, Alzheimer's disease, AIDS, and coronary heart disease. In addition, these drugs are popularly believed to improve sleep, increase sexual vitality, and increase longevity. Most of these claims are based on animal studies involving mice, or on short-term human clinical studies. No results of long-term studies about the effects of these compounds are yet available. Research is progressing slowly because both compounds are natural substances that are classified as food supplements and are readily available over the counter. Since the effects or uses for natural substances cannot be patented, there is no economic incentive for expensive research.

The basis for many of the claims made for the compounds relies on the fact that as humans age, smaller amounts of both of them are found in the body. Therefore, some conclude that increasing the levels of melatonin and DHEA in older adults will restore many characteristics of their youthful bodies.

Melatonin is a hormone produced by the pea-sized pineal gland located in the center of the brain. The amount of melatonin released by the pineal gland depends on the time of day. During daylight hours, the rate of release is lower than during the night. Because of this characteristic, it is thought that melatonin levels may help the body determine the time of day. Melatonin influences biological rhythms, such as tiredness in the evening, hormone production, and internal clocks. These natural rhythms have been delayed by melatonin injections administered in the early morning or at noon. Injections given in the late afternoon or early evening have been found to speed up the onset of the rhythms. This effect accounts for the use of melatonin in some sleep-aid products, but it must be taken at the appropriate time of day in order to be effective.

CH_3O — $CH_2CH_2—NH—C(=O)—CH_3$ — N—H

melatonin

Melatonin is generally considered safe when taken in low doses for short periods of time. The FDA has not received many complaints about its effects, yet most of the previously mentioned claims have yet to be verified. Therefore, it would not be wise to take melatonin in large doses, or for long periods of time, until more is learned about its effects.

DHEA is not itself a hormone, but a precursor to both testosterone, the male sex hormone, and estrogen, the female sex hormone. The results of research on DHEA have thus far been contradictory. Some short-term studies have shown that DHEA has a tendency to increase resistance to infection, prevent or inhibit some types of cancerous growths, reverse the disease known as lupus, and provide benefits in the treatment of age-related diabetes. However, other studies have indicated that DHEA increases the risk of endometrial cancer in women, and may also increase the incidence of prostate cancer in men.

As yet, not enough long-term studies have been done to completely evaluate the safety and usefulness of DHEA. In fact, the FDA banned the use of DHEA in 1985 because of its side effects. However, the compound was put back on the market as a result of the 1994 Dietary Supplements Health and Education Act. Because of the contradictory nature of the available research results, and the lack of results from long-term studies, people who take either melatonin or DHEA should be aware that they are doing it at their own risk.

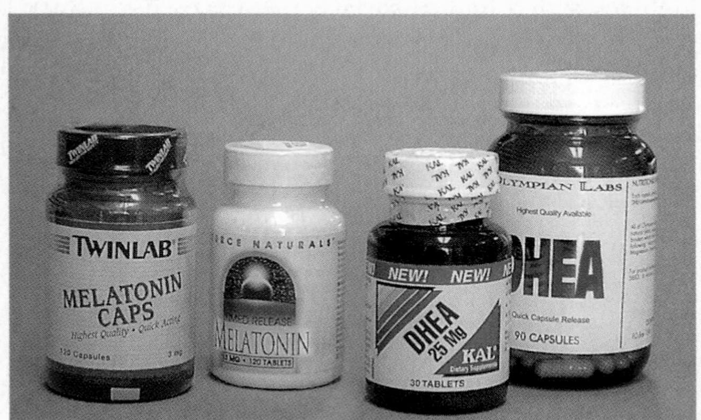

© Maren Slabaugh

A number of companies market melatonin and DHEA.

The passage of a gallstone from the gallbladder down the common bile duct to the intestine causes excruciating pain. Sometimes one or more stones become lodged in the duct and prevent bile from passing into the duodenum, and fats can no longer be digested normally. A person afflicted with this condition feels great pain and becomes quite nauseated and ill. Bile pigments are absorbed into the blood, the skin takes on a yellow coloration, and the stool becomes gray-colored because of lack of excreted bile pigments. If the condition becomes serious, both the gallbladder and the stones can be surgically removed.

8.10 Steroid Hormones

LEARNING OBJECTIVE

11. Name the major categories of steroid hormones.

A number of steroids in the body serve important roles as **hormones.** The two major categories of steroid hormones are the adrenocorticoid hormones and the sex hormones.

hormone
A chemical messenger secreted by specific glands and carried by the blood to a target tissue, where it triggers a particular response.

Adrenocorticoid Hormones

Adrenal glands are small mounds of tissue located at the top of each kidney. An outer layer of the gland, the adrenal cortex, produces a number of potent steroid hormones, the adrenocorticoids. They are classified into two groups according to function: Mineralocorticoids regulate the concentration of ions (mainly Na^+) in body fluids, and glucocorticoids enhance carbohydrate metabolism.

Cortisol, the major glucocorticoid, functions to increase the glucose and glycogen concentrations in the body. This is accomplished by the conversion of lactate and amino acids from body proteins into glucose and glycogen. The reactions take place in the liver under the influence of cortisol.

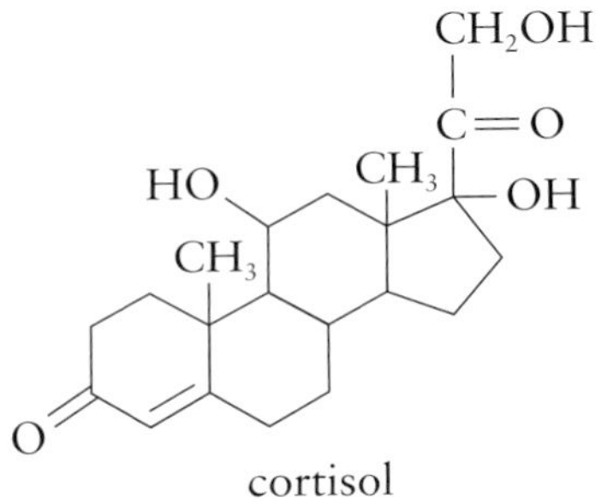

CH_2OH | C=O; HO; CH_3; OH; CH_3; O

cortisol

Cortisol and its ketone derivative, cortisone, exert powerful anti-inflammatory effects in the body. These, or similar synthetic derivatives such as prednisolone, are used to treat inflammatory diseases such as rheumatoid arthritis and bronchial asthma.

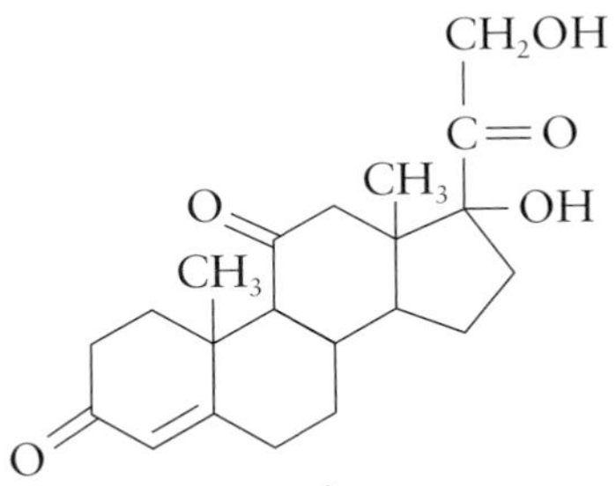

CH_2OH | C=O; O; CH_3; OH; CH_3; O

cortisone

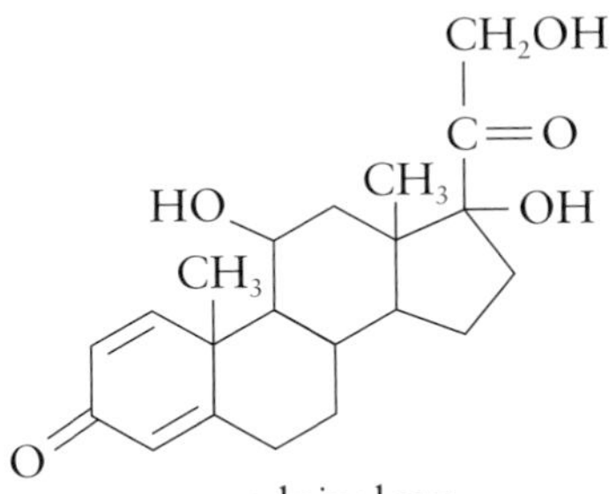

CH_2OH | C=O; HO; CH_3; OH; CH_3; O

prednisolone

By far the most important mineralocorticoid is aldosterone, which influences the absorption of Na^+ and Cl^- in kidney tubules. An increase in the level of the hormone results in a corresponding increase in absorption of Na^+ and Cl^-.

Because the concentration of Na^+ in tissues influences the retention of water, aldosterone is involved in water balance in the body.

aldosterone

Sex Hormones

The testes and ovaries produce steroids that function as sex hormones. Secondary sex characteristics that appear at puberty, including a deep voice, beard, and increased muscular development in males and a higher voice, increased breast size, and lack of facial hair in females, develop under the influence of these sex hormones.

The testes in males perform two functions: One is to produce sperm; the other is to produce male sex hormones (androgens), the most important of which is testosterone. Testosterone promotes the normal growth of the male genital organs and aids in the development of secondary sex characteristics.

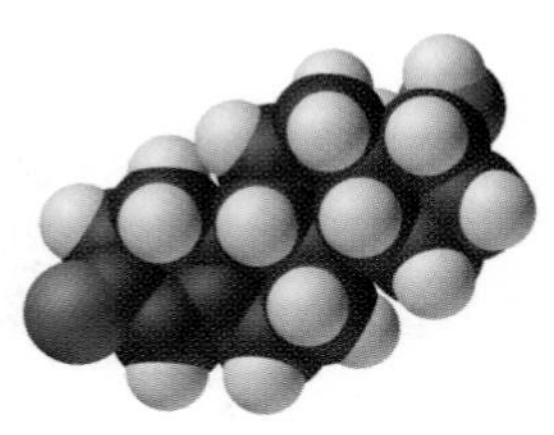

Testosterone

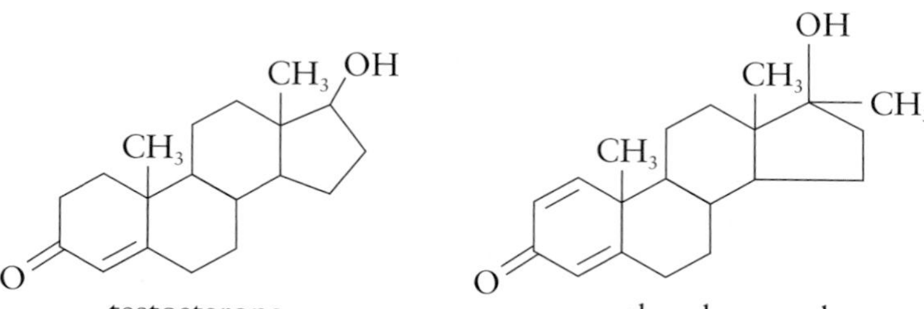

testosterone

methandrostenolone

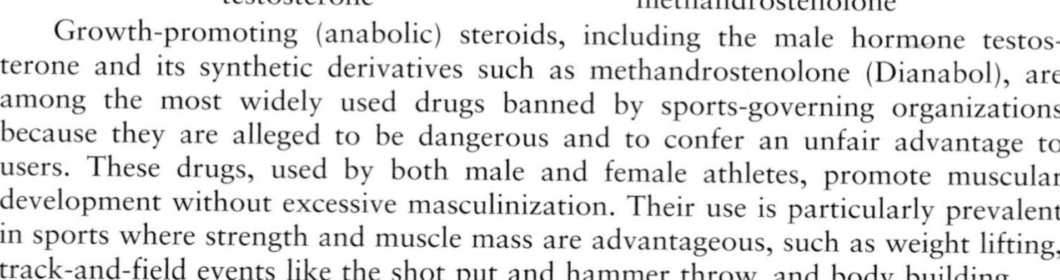

Growth-promoting (anabolic) steroids, including the male hormone testosterone and its synthetic derivatives such as methandrostenolone (Dianabol), are among the most widely used drugs banned by sports-governing organizations because they are alleged to be dangerous and to confer an unfair advantage to users. These drugs, used by both male and female athletes, promote muscular development without excessive masculinization. Their use is particularly prevalent in sports where strength and muscle mass are advantageous, such as weight lifting, track-and-field events like the shot put and hammer throw, and body building.

The side effects of anabolic steroids range from acne to deadly liver tumors. Effects on the male reproductive system include testicular atrophy, a decrease in sperm count, and, occasionally, temporary infertility.

The primary female sex hormones are the estrogens estradiol and estrone, and progesterone. Estrogens and progesterone are important in the reproductive process. Estrogens are involved in egg (ovum) development in the ovaries, and progesterone causes changes in the wall of the uterus to prepare it to accept a fertilized egg and maintain the resulting pregnancy.

estradiol

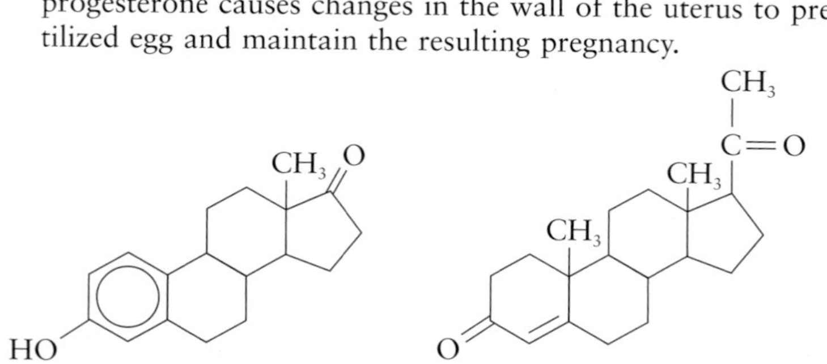

estrone

progesterone

CHEMISTRY AROUND US 8.2 Biodiesel: A Fuel for the 21st Century?

Crude oil prices seem to just keep rising with occasional plateaus. The increase in crude oil prices results in price increases for gasoline and diesel fuel, two of the main products derived from crude oil. Their increases caused many individuals and companies to begin considering alternative fuels, including biodiesel. Biodiesel is produced from plant oils such as soybean oil, and even from recycled oils used to deep fry foods.

There are currently some obstacles that prevent biodiesel from becoming a general fuel used by many consumers. Most importantly, relatively few automobiles in the United States have diesel fuel engines. A second factor is the higher price of biodiesel compared to diesel fuel from crude oil. However, it is likely that diminishing supplies of crude oil in the future will increase the prices of both gasoline and diesel fuel, which will make the biodiesel alternative much more attractive. The fact that a regular diesel engine can run on biodiesel without any mechanical modification to the engine adds to the incentives for the motoring and United States public to switch to diesel-powered vehicles.

The biodiesel industry has been developing in Europe for decades, but it only began to gain momentum in the United States in the 1990s. Most current uses of U.S. biodiesel are large vehicles fleets run by cities, schools, or businesses whose vehicles were formerly fueled with crude-oil-derived diesel.

Perhaps one of the most important factors that will encourage the future use of biodiesel in the United States is that the raw materials can be grown entirely in the United States, thus decreasing the dependence on importing crude oil from unpredictable foreign sources.

8.11 Prostaglandins

LEARNING OBJECTIVE

12. Describe the biological importance and therapeutic uses of the prostaglandins.

This group of compounds was given its name because the first prostaglandins were identified from the secretions of the male prostate gland. Recent research has identified as many as 20 prostaglandins in a variety of tissues within both males and females.

Prostaglandins are cyclic compounds synthesized in the body from the 20-carbon unsaturated fatty acid arachidonic acid. Prostaglandins are designated by codes that refer to the ring substituents and the number of side-chain double bonds. For example, prostaglandin E_2 (PGE_2) has a carbonyl group on the ring and a side chain containing two double bonds:

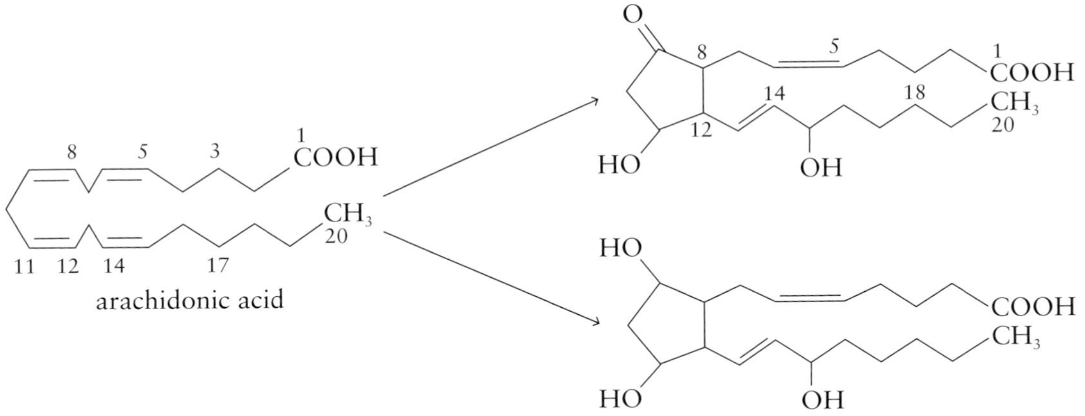

prostaglandin
A substance derived from unsaturated fatty acids, with hormonelike effects on a number of body tissues.

Prostaglandins are similar to hormones in the sense that they are intimately involved in a host of body processes. Clinically, it has been found that prostaglandins are involved in almost every phase of the reproductive process; they can act to regulate menstruation, prevent conception, and induce uterine contractions during childbirth. Certain prostaglandins stimulate blood clotting. They also lead to inflammation and fever. It has been found that aspirin inhibits prostaglandin production. This explains in part why aspirin is such a powerful drug in the treatment of inflammatory diseases such as arthritis.

From a medical viewpoint, prostaglandins have enormous therapeutic potential. For example, PGE_2 and PGF_2 induce labor and are used for therapeutic abortion in early pregnancy. PGE_2 in aerosol form is used to treat asthma; it opens up the bronchial tubes by relaxing the surrounding muscles. Other prostaglandins inhibit gastric secretions and are used in treating peptic ulcers. Many researchers believe that when prostaglandins are fully understood, they will be found useful for treating a much wider variety of ailments.

Concept Summary

Classification of Lipids. Lipids are a family of naturally occurring compounds grouped together on the basis of their relative insolubility in water and solubility in nonpolar solvents. Lipids are energy-rich compounds that are used as waxy coatings, energy storage compounds, and as structural components by organisms. Lipids are classified as saponifiable (ester-containing) or nonsaponifiable. Saponifiable lipids are further classified as simple or complex, depending on the number of structural components. **OBJECTIVE 1 (Section 8.1), Exercises 8.2 and 8.4**

Fatty Acids. A fatty acid consists of a long nonpolar chain of carbon atoms, with a polar carboxylic acid group at one end. Most natural fatty acids contain an even number of carbon atoms. They may be saturated, unsaturated, or polyunsaturated (containing two or more double bonds). **OBJECTIVE 2 (Section 8.2), Exercise 8.6**

The Structure of Fats and Oils. Triglycerides or triacylglycerols in the form of fats and oils are the most abundant lipids. Fats and oils are simple lipids that are esters of glycerol and fatty acids. **OBJECTIVE 3 (Section 8.3), Exercise 8.14.** The difference between fats and oils is the melting point, which is essentially a function of the fatty acids in the compound. **OBJECTIVE 4 (Section 8.3), Exercise 8.12**

Chemical Properties of Fats and Oils. Fats and oils can be hydrolyzed in the presence of acid to produce glycerol and fatty acids. When the hydrolysis reaction is carried out in the presence of a strong base, salts of the fatty acids (soaps) are produced. During hydrogenation, some multiple bonds of unsaturated fatty acids contained in fats or oils are reacted with hydrogen and converted to single bonds. **OBJECTIVE 5 (Section 8.4), Exercise 8.18**

Waxes. Waxes are simple lipids composed of a fatty acid esterified with a long-chain alcohol. **OBJECTIVE 6 (Section 8.5), Exercise 8.24.** Waxes are insoluble in water and serve as protective coatings in nature.

Phosphoglycerides. Phosphoglycerides consist of glycerol esterified to two fatty acids and phosphoric acid. The phosphoric acid is further esterified to choline (in the lecithins) and to ethanolamine or serine (in the cephalins). The phosphoglycerides are particularly important in membrane formation. **OBJECTIVE 7 (Section 8.6), Exercises 8.28 and 8.30**

Sphingolipids. These complex lipids contain a backbone of sphingosine rather than glycerol and only one fatty acid component. They are abundant in brain and nerve tissue. **OBJECTIVE 8 (Section 8.7), Exercise 8.34**

Biological Membranes. Membranes surround tissue cells as a selective barrier, and they encase the organelles found in eukaryotic cells. Membranes contain both proteins and lipids. According to the fluid-mosaic model, the lipids are arranged in a bilayer fashion with the hydrophobic portions on the inside of the bilayer. Proteins float in the bilayer. **OBJECTIVE 9 (Section 8.8), Exercise 8.42**

Steroids. Steroids are compounds that have four rings fused together in a specific way. The most abundant steroid in humans is cholesterol, which serves as a starting material for other important steroids such as bile salts, adrenocorticoid hormones, and sex hormones. **OBJECTIVE 10 (Section 8.9), Exercises 8.44 and 8.46**

Steroid Hormones. Hormones are chemical messengers, synthesized by specific glands, that affect various target tissues in the body. The adrenal cortex produces a number of steroid hormones that regulate carbohydrate utilization (the glucocorticoids) and electrolyte balance (the mineralocorticoids). The testes and ovaries produce steroid hormones that determine secondary sex characteristics and regulate the reproductive cycle in females. **OBJECTIVE 11 (Section 8.10), Exercise 8.50**

Prostaglandins. These compounds are synthesized from the 20-carbon fatty acid arachiodonic acid. They exert many hormone-like effects on the body and are used therapeutically to induce labor, treat asthma, and control gastric secretions. **OBJECTIVE 12 (Section 8.11), Exercise 8.58**

Key Terms and Concepts

Cephalin (8.6)
Complex lipid (8.1)
Essential fatty acid (8.2)
Eukaryotic cell (8.8)
Fat (8.3)
Fluid-mosaic model (8.8)
Glycolipid (8.7)
Hormone (8.10)
Lecithin (8.6)
Lipid (8.1)
Lipid bilayer (8.8)
Micelle (8.2)
Oil (8.3)
Organelle (8.8)
Phosphoglyceride (8.6)
Phospholipid (8.6)
Prokaryotic cell (8.8)
Prostaglandin (8.11)
Simple lipid (8.1)
Soap (8.4)
Sphingolipid (8.7)
Steroid (8.9)
Triglyceride or triacylglycerol (8.3)
Wax (8.5)

Key Reactions

1. Hydrolysis of a triglyceride to glycerol and fatty acids—general reaction (Section 8.4):

$$
\begin{array}{l}
CH_2-O-\overset{\displaystyle O}{\overset{\|}{C}}-R \\
| \\
CH-O-\overset{\displaystyle O}{\overset{\|}{C}}-R \\
| \\
CH_2-O-\overset{\displaystyle O}{\overset{\|}{C}}-R
\end{array}
+ 3H_2O \xrightarrow[\text{or lipase}]{H^+}
\begin{array}{l}
CH_2-OH \\
| \\
CH-OH \\
| \\
CH_2-OH
\end{array}
+ 3R-\overset{\displaystyle O}{\overset{\|}{C}}-OH
$$

2. Saponification of a triglyceride to glycerol and fatty acid salts—general reaction (Section 8.4):

$$
\begin{array}{l}
CH_2-O-\overset{\displaystyle O}{\overset{\|}{C}}-R \\
| \\
CH-O-\overset{\displaystyle O}{\overset{\|}{C}}-R \\
| \\
CH_2-O-\overset{\displaystyle O}{\overset{\|}{C}}-R
\end{array}
+ 3NaOH \longrightarrow
\begin{array}{l}
CH_2-OH \\
| \\
CH-OH \\
| \\
CH_2-OH
\end{array}
+ 3R-\overset{\displaystyle O}{\overset{\|}{C}}-O^-Na^+
$$

3. Hydrogenation of a triglyceride—general reaction (Section 8.4):

$$
\begin{array}{l}
CH_2-O-\overset{\displaystyle O}{\overset{\|}{C}}-R \\
| \\
CH-O-\overset{\displaystyle O}{\overset{\|}{C}}-(CH_2)_7CH=CH(CH_2)_7CH_3 \\
| \\
CH_2-O-\overset{\displaystyle O}{\overset{\|}{C}}-R
\end{array}
+ H_2 \xrightarrow{Ni}
\begin{array}{l}
CH_2-O-\overset{\displaystyle O}{\overset{\|}{C}}-R \\
| \\
CH-O-\overset{\displaystyle O}{\overset{\|}{C}}-(CH_2)_{16}CH_3 \\
| \\
CH_2-O-\overset{\displaystyle O}{\overset{\|}{C}}-R
\end{array}
$$

Exercises

SYMBOL KEY

Even-numbered exercises are answered in Appendix B.

Blue-numbered exercises are more challenging.

■ denotes exercises available in ThomsonNow and assignable in OWL.

ThomsonNOW™ To assess your understanding of this chapter's topics with sample tests and other resources, sign in at www.thomsonedu.com.

INTRODUCTION AND CLASSIFICATION OF LIPIDS (SECTION 8.1)

8.1 What is the basis for deciding if a substance is a lipid?

8.2 List two major functions of lipids in the human body.

8.3 What functional group is common to all saponifiable lipids?

8.4 ■ Classify the following as saponifiable or nonsaponifiable lipids:

a. A steroid
b. A wax
c. A triglyceride
d. A phosphoglyceride
e. A glycolipid
f. A prostaglandin

FATTY ACIDS (SECTION 8.2)

8.5 ■ Draw the structure of a typical saturated fatty acid. Label the polar and nonpolar portions of the molecule. Which portion is hydrophobic? Hydrophilic?

8.6 Describe four structural characteristics exhibited by most fatty acids.

8.7 Describe the structure of a micelle formed by the association of fatty acid molecules in water. What forces hold the micelle together?

8.8 ■ Name two essential fatty acids, and explain why they are called essential.

8.9 ■ Indicate whether each of the following fatty acids is saturated or unsaturated. Which of them are solids and which are liquids at room temperature?

a. $CH_3(CH_2)_{14}COOH$
b. $CH_3(CH_2)_4CH{=}CHCH_2CH{=}CH(CH_2)_7COOH$
c. $CH_3(C_{14}H_{24})COOH$
d. $CH_3(C_{10}H_{20})COOH$

8.10 ■ Explain why the melting points of unsaturated fatty acids are lower than those of saturated fatty acids.

8.11 What structural feature of a fatty acid is responsible for the name omega-3 fatty acid?

THE STRUCTURE OF FATS AND OILS (SECTION 8.3)

8.12 How are fats and oils structurally similar? How are they different?

8.13 From Figure 8.7, arrange the following substances in order of increasing percentage of unsaturated fatty acids: chicken fat, beef fat, corn oil, butter, and sunflower oil.

8.14 ■ Draw the structure of a triglyceride that contains one myristic acid, one palmitoleic acid, and one linoleic acid. Identify the ester bonds.

8.15 From what general source do triglycerides tend to have more saturated fatty acids? More unsaturated fatty acids?

8.16 ■ The percentage of fatty acid composition of two triglycerides is reported below. Predict which triglyceride has the lower melting point.

	Palmitic acid	Stearic acid	Oleic acid	Linoleic acid
Triglyceride A	20.4	28.8	38.6	10.2
Triglyceride B	9.6	7.2	27.5	55.7

8.17 Why is the amount of saturated fat in the diet a health concern?

CHEMICAL PROPERTIES OF FATS AND OILS (SECTION 8.4)

8.18 What process is used to prepare a number of useful products such as margarines and cooking shortenings from vegetable oils?

8.19 Write equations for the following reactions using the oil shown below:

$$\begin{array}{l} CH_2-O-\overset{O}{\overset{\|}{C}}-(CH_2)_7CH=CH(CH_2)_7CH_3 \\ | \\ CH-O-\overset{O}{\overset{\|}{C}}-(CH_2)_7CH=CH(CH_2)_7CH_3 \\ | \\ CH_2-O-\overset{O}{\overset{\|}{C}}-(CH_2)_{16}CH_3 \end{array}$$

a. Lipase-catalyzed hydrolysis
b. Saponification
c. Hydrogenation

8.20 Why is the hydrogenation of vegetable oils of great commercial importance?

8.21 Palmolive soap is mostly sodium palmitate. Write the structure of this compound.

8.22 Write reactions to show how each of the following products might be prepared from a typical triglyceride present in the source given. Use Table 8.1 as an aid:

a. Glycerol from beef fat
b. Stearic acid from beef fat
c. A margarine from corn oil
d. Soaps from lard

WAXES (SECTION 8.5)

8.23 Draw the structure of a wax formed from oleic acid and cetyl alcohol ($CH_3(CH_2)_{14}CH_2-OH$).

8.24 Like fats, waxes are esters of long-chain fatty acids. What structural difference exists between them that warrants placing waxes in a separate category?

8.25 ■ Draw the structure of a wax formed from stearic acid and cetyl alcohol.

8.26 What role do waxes play in nature?

PHOSPHOGLYCERIDES (SECTION 8.6)

8.27 How do phosphoglycerides differ structurally from triglycerides?

8.28 Draw the general block diagram structure of a phosphoglyceride.

8.29 Draw the structure of a phosphoglyceride-containing ethanolamine:

$$HO-CH_2CH_2-\overset{+}{N}H_3$$

Is it a lecithin or a cephalin?

8.30 Describe two biological roles served by the lecithins.

8.31 Draw the structure of a lecithin. What structural features make lecithin an important commercial emulsifying agent in certain food products?

8.32 ■ What is the structural difference between a lecithin and a cephalin?

8.33 Where are cephalins found in the human body?

SPHINGOLIPIDS (SECTION 8.7)

8.34 Draw the general block diagram structure of a sphingolipid.

8.35 ■ List two structural differences between sphingolipids and phosphoglycerides.

8.36 List three diseases caused by abnormal metabolism and accumulation of sphingolipids.

8.37 Describe the structural similarities and differences between the sphingomyelins and the glycolipids.

8.38 Give another name for glycolipids. In what tissues are they found?

BIOLOGICAL MEMBRANES (SECTION 8.8)

8.39 Where would you find membranes in a prokaryotic cell? Where would you find membranes in a eukaryotic cell? What functions do the membranes serve?

8.40 ■ What three classes of lipids are found in membranes?

8.41 How does the polarity of the phosphoglycerides contribute to their function of forming cell membranes?

8.42 Describe the major features of the fluid-mosaic model of cell membrane structure.

STEROIDS (SECTION 8.9)

8.43 Why is it suggested that some people restrict cholesterol intake in their diet?

8.44 Draw the characteristic chemical structure that applies to all steroid molecules.

8.45 Explain how bile salts aid in the digestion of lipids.

8.46 ■ List three important groups of compounds the body synthesizes from cholesterol. Give the location in bodily tissues where cholesterol is an essential component.

8.47 What symptoms might indicate the presence of gallstones in the gallbladder?

8.48 What is the major component in gallstones?

STEROID HORMONES (SECTION 8.10)

8.49 What is a hormone? What are the two major categories of steroid hormones?

8.50 Name the two groups of adrenocorticoid hormones, give a specific example of each group, and explain the function of those compounds in the body.

8.51 ■ How are testosterone and progesterone structurally similar? How are they different?

8.52 ■ Name the primary male sex hormone and the three principal female sex hormones.

8.53 Why do athletes use anabolic steroids? What side effects are associated with their use?

8.54 What role do the estrogens and progesterone serve in preparation for pregnancy?

PROSTAGLANDINS (SECTION 8.11)

8.55 How are prostaglandins differentiated from each other?

8.56 ■ What compound serves as a starting material for prostaglandin synthesis?

8.57 What body processes appear to be regulated in part by prostaglandins?

8.58 Name three therapeutic uses of prostaglandins.

ADDITIONAL EXERCISES

8.59 Unsaturated fatty acids are susceptible to oxidation, which causes rancidity in food products. Suggest a way to decrease the amount of oxidative rancidity occurring in a food product.

8.60 A red-brown solution of bromine (Br_2) was added to a lipid, and the characteristic red-brown bromine color disappeared. What can be deduced about the lipid structure from this information?

8.61 Which of the following waxes would have the lowest melting point? Explain your answer.

a. $CH_3(CH_2)_{14}-\overset{\overset{O}{\|}}{C}-O-(CH_2)_{29}CH_3$

b. $CH_3(CH_2)_7CH=CH(CH_2)_5-\overset{\overset{O}{\|}}{C}-O-(CH_2)_{29}CH_3$

8.62 Suggest a reason why complex lipids, such as phosphoglycerides and sphingomyelin, are more predominant in cell membranes than simple lipids.

8.63 Many esters are pleasantly fragrant. Examples include esters that are partially responsible for the smell of oranges and strawberries. What is the main difference between a fragrant ester and a wax? Explain why some esters are fragrant and waxes are not especially aromatic.

ALLIED HEALTH EXAM CONNECTION

Reprinted with permission from Nursing School and Allied Health Entrance Exams, COPYRIGHT 2005 Petersons.

8.64 Fats belong to the class of organic compounds represented by the general formula, RCOOR′, where R and R′ represent hydrocarbon groups. What is the name of the functional group present in fats? What functional group is common to all saponifiable lipids?

8.65 Identify each of the following characteristics as describing an unsaturated fatty acid or a saturated fatty acid:

a. Contains more hydrogen atoms

b. Is more healthy

c. More plentiful in plant sources

d. Is usually a solid at room temperature

8.66 Identify which sex hormones (testosterone, estrogens, or progesterone) are produced in

a. The ovaries

b. The testes

8.67 Substance X passes through a cell membrane easily. Substance X would best be described as

a. Hydrophilic or hydrophobic

b. Polar or nonpolar

c. A lipid or a protein

8.68 In diseases of the gallbladder, which of the following nutrients is limited in its digestibility? Explain your choice(s).

a. Starches

b. Proteins

c. Fats

CHEMISTRY FOR THOUGHT

8.69 Gasoline is soluble in nonpolar solvents. Why is gasoline not classified as a lipid?

8.70 The structure of cellular membranes is such that ruptures are closed naturally. Describe the molecular forces that cause the closing to occur.

8.71 In Figure 8.7, the five vegetable oils listed are derived from seeds. Why do you think seeds are rich in oils?

8.72 Answer the question asked in Figure 8.9. How do oils used in automobiles differ chemically from vegetable oils?

8.73 Why doesn't honey dissolve beeswax?

8.74 In Figure 8.5 (where the nonpolar tails are together), why is the structure more stable than a structure in which the polar heads are together?

8.75 Why are there so many structurally different kinds of lipids?

8.76 In what ways are the structural features of lecithin similar to those of a soap?

8.77 When a doughnut is placed on a napkin, the napkin will often absorb a liquid and appear "moist." What is most likely the nature of the liquid?

CHAPTER 9

Proteins

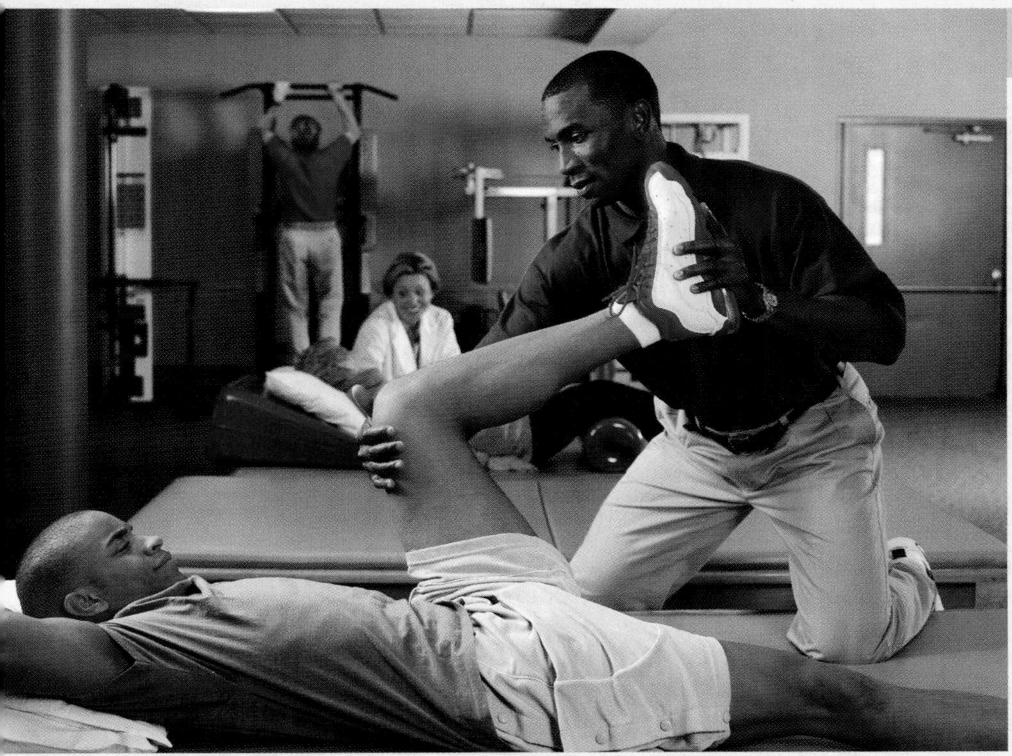

A physical therapist assesses the limitations, and potential for improvement, of individuals with physical disabilities, then selects or develops rehabilitation methods. Physical therapists use such methods as body massage, hydrotherapy, heat, light, and ultrasound to help people with sensory or motor deficiencies return to normal. Physical therapists need a basic knowledge of the protein structure of muscles and connective tissue to help them understand human physiology. Protein structure is one of the topics of this chapter.

LEARNING OBJECTIVES

When you have completed your study of this chapter, you should be able to:

1. Identify the characteristic parts of alpha-amino acids. (Section 9.1)
2. Draw structural formulas to illustrate the various ionic forms assumed by amino acids. (Section 9.2)
3. Write reactions to represent the formation of peptides and proteins. (Section 9.3)
4. Describe uses for important peptides. (Section 9.4)
5. Describe proteins in terms of the following characteristics: size, function, classification as fibrous or globular, and classification as simple or conjugated. (Section 9.5)
6. Explain what is meant by the primary structure of proteins. (Section 9.6)
7. Describe the role of hydrogen bonding in the secondary structure of proteins. (Section 9.7)
8. Describe the role of side-chain interactions in the tertiary structure of proteins. (Section 9.8)
9. Explain what is meant by the quaternary structure of proteins. (Section 9.9)
10. Describe the conditions that can cause proteins to hydrolyze or become denatured. (Section 9.10)

ThomsonNOW™ Throughout the chapter this icon introduces resources on the ThomsonNOW website for this text. Sign in at **www.thomsonedu.com** to:

- Evaluate your knowledge of the material
- Take an exam prep quiz
- Identify areas you need to study with a **Personalized Learning Plan**.

The name *protein,* coined more than 100 years ago, is derived from the Greek *proteios,* which means "of first importance," and is appropriate for these most important of all biological compounds. Proteins are indispensable components of all living things, where they play crucial roles in all biological processes. In this chapter, we begin with a discussion of the structures and chemistry of the common amino acids, then progress to show how combinations of amino acids form large molecules called peptides and larger molecules called proteins.

9.1 The Amino Acids

LEARNING OBJECTIVE

1. Identify the characteristic parts of alpha-amino acids.

Structure

alpha-amino acid
An organic compound containing both an amino group and a carboxylate group, with the amino group attached to the carbon next to the carboxylate group.

All proteins, regardless of their origin or biological function, are structurally similar. They are polymers consisting of chains of amino acids chemically bound to each other. As the name implies, an amino acid is an organic compound that contains both an amino group and a carboxylate group in ionic form (Section 9.2). Hundreds of different amino acids, both synthetic and naturally occurring, are known, but only 20 are commonly found in natural proteins. The amino acids in proteins are called **alpha (α)-amino acids** because the amino group is attached to the α carbon (the carbon next to the carboxylate group), as shown in ■ Figure 9.1. Each amino acid has a characteristic side chain, or R group, that imparts chemical individuality to the molecule. These R side chains contain different structural features, such as aromatic rings, —OH groups, $—NH_3^+$ groups, and $—COO^-$ groups. This variety in side chains causes differences in the properties of the individual amino acids and the proteins containing different combinations of them.

The polarity of the R groups is an important characteristic and is the basis for classifying amino acids into the four groups shown in ■ Table 9.1: neutral and nonpolar side chains, neutral but polar side chains, basic side chains, and acidic side chains. Note that acidic side chains contain a carboxylate group and that basic side chains contain an additional amino group. The three-letter and one-letter abbreviations that are used to represent amino acids are given after the names:

The structure of proline differs slightly from the general formula because the amino group and R group are part of a ring.

ThomsonNOW™ Go to Chemistry Interactive to explore the **amino acid structure.**

$$H_3\overset{+}{N}-CH(CH_2SH)-C(=O)-O^-$$

cysteine (Cys) C

$$H_3\overset{+}{N}-CH(CH_2CH_2C(=O)NH_2)-C(=O)-O^-$$

glutamine (Gln) Q

$$H_3\overset{+}{N}-CH(CH_2C(=O)NH_2)-C(=O)-O^-$$

asparagine (Asn) N

$$H_3\overset{+}{N}-CH(R)-C(=O)-O^-$$

amino group; α carbon; carboxylate group; side chain (different for each amino acid)

■ **FIGURE 9.1** The general structure of α-amino acids shown in ionic form.

TABLE 9.1 The common amino acids of proteins. Below each amino acid are its name, its three-letter abbreviation, and its one-letter abbreviation

Neutral, nonpolar side chains

$H_3\overset{+}{N}-CH-COO^-$ | H	$H_3\overset{+}{N}-CH-COO^-$ | CH_3	$H_3\overset{+}{N}-CH-COO^-$ | CH / \\ CH_3 CH_3	$H_3\overset{+}{N}-CH-COO^-$ | CH_2 | CH / \\ CH_3 CH_3
Glycine (Gly) G	Alanine (Ala) A	Valine (Val) V	Leucine (Leu) L
$H_3\overset{+}{N}-CH-COO^-$ | $CH-CH_3$ | CH_2 | CH_3	$H_3\overset{+}{N}-CH-COO^-$ | CH_2 | (phenyl ring)	$H_2\overset{+}{N}-CH-COO^-$ | | CH_2 CH_2 \\ / CH_2	$H_3\overset{+}{N}-CH-COO^-$ | CH_2 | CH_2 | $S-CH_3$
Isoleucine (Ile) I	Phenylalanine (Phe) F	Proline (Pro) P	Methionine (Met) M

Neutral, polar side chains

$H_3\overset{+}{N}-CH-COO^-$ | CH_2 | OH	$H_3\overset{+}{N}-CH-COO^-$ | $CH-CH_3$ | OH	$H_3\overset{+}{N}-CH-COO^-$ | CH_2 | (benzene ring) | OH	$H_3\overset{+}{N}-CH-COO^-$ | CH_2 | (indole ring) N | H
Serine (Ser) S	Threonine (Thr) T	Tyrosine (Tyr) Y	Tryptophan (Trp) W
$H_3\overset{+}{N}-CH-COO^-$ | CH_2 | SH	$H_3\overset{+}{N}-CH-COO^-$ | CH_2 | $C=O$ | NH_2	$H_3\overset{+}{N}-CH-COO^-$ | CH_2 | CH_2 | $C=O$ | NH_2	
Cysteine (Cys) C	Asparagine (Asn) N	Glutamine (Gln) Q	

Basic, polar side chains			Acidic, polar side chains	
$H_3\overset{+}{N}-CH-COO^-$ | CH_2 | (imidazole ring) $H\overset{+}{N}$ NH	$H_3\overset{+}{N}-CH-COO^-$ | CH_2 | CH_2 | CH_2 | CH_2 | NH_3^+	$H_3\overset{+}{N}-CH-COO^-$ | CH_2 | CH_2 | CH_2 | NH | $C=NH_2^+$ | NH_2	$H_3\overset{+}{N}-CH-COO^-$ | CH_2 | COO^-	$H_3\overset{+}{N}-CH-COO^-$ | CH_2 | CH_2 | COO^-
Histidine (His) H	Lysine (Lys) K	Arginine (Arg) R	Aspartate (Asp) D	Glutamate (Glu) E

Stereochemistry

A glance at the general formula for the 20 amino acids

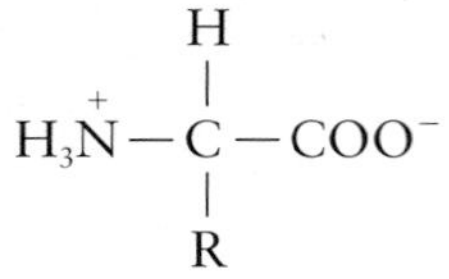

$$H_3\overset{+}{N}-\underset{R}{\overset{H}{\underset{|}{\overset{|}{C}}}}-COO^-$$

ThomsonNOW™ Go to Coached Problems to practice **naming amino acids.**

shows that with the exception of glycine, in which the R group is —H, there are four different groups attached to the α carbon. Thus, 19 of the 20 amino acids contain a chiral α carbon atom and can exist as two enantiomers; these are the L and D forms:

$\overset{+}{N}H_3$ on the left → $H_3N^+-\underset{R}{\overset{COO^-}{\underset{|}{\overset{|}{C}}}}-H$ (an L-amino acid)

$H-\underset{R}{\overset{COO^-}{\underset{|}{\overset{|}{C}}}}-\overset{+}{N}H_3$ ← *$\overset{+}{N}H_3$ on the right* (a D-amino acid)

As we saw when we studied carbohydrates, one of the two possible forms usually predominates in natural systems. With few exceptions, the amino acids found in living systems exist in the L form.

9.2 Zwitterions

LEARNING OBJECTIVE

2. Draw structural formulas to illustrate the various ionic forms assumed by amino acids.

ThomsonNOW™ Go to Coached Problems to examine **zwitterions.**

zwitterion
A dipolar ion that carries both a positive and a negative charge as a result of an internal acid–base reaction in an amino acid molecule.

The structural formulas used to this point have shown amino acids in an ionized form. The fact that amino acids are white crystalline solids with relatively high melting points and high water solubilities suggests that they exist in an ionic form. Studies of amino acids confirm that this is true both in the solid state and in solution. The presence of a carboxyl group and a basic amino group in the same molecule makes possible the transfer of a hydrogen ion in a kind of internal acid–base reaction (Reaction 9.1). The product of this reaction is a dipolar ion called a **zwitterion.**

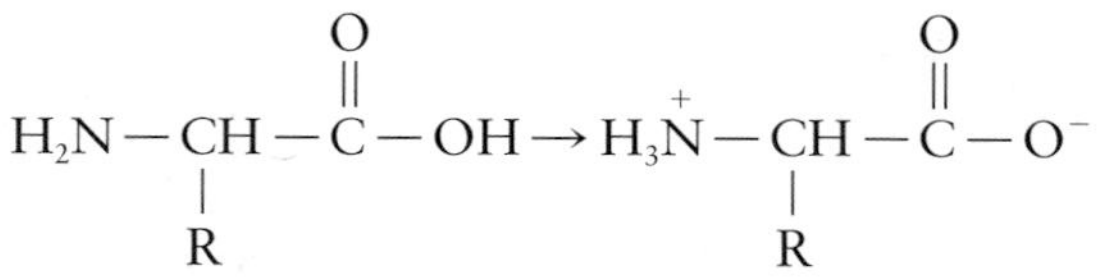

$$\underset{\substack{\text{nonionized form}\\\text{(does not exist)}}}{H_2N-\underset{R}{\underset{|}{CH}}-\overset{O}{\overset{\|}{C}}-OH} \rightarrow \underset{\substack{\text{zwitterion}\\\text{(present in solids}\\\text{and solutions)}}}{H_3\overset{+}{N}-\underset{R}{\underset{|}{CH}}-\overset{O}{\overset{\|}{C}}-O^-} \quad (9.1)$$

The structure of an amino acid in solution varies with the pH of the solution. For example, if the pH of a solution is lowered by adding a source of H_3O^+, such as hydrochloric acid, the carboxylate group (—COO^-) of the zwitterion can pick up a proton to form —COOH (Reaction 9.2):

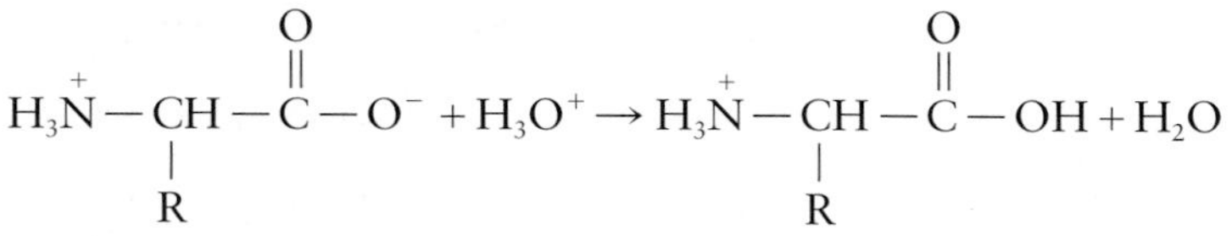

$$\underset{\substack{\text{zwitterion}\\\text{(0 net charge)}}}{H_3\overset{+}{N}-\underset{R}{\underset{|}{CH}}-\overset{O}{\overset{\|}{C}}-O^-} + H_3O^+ \rightarrow \underset{\text{(positive net charge)}}{H_3\overset{+}{N}-\underset{R}{\underset{|}{CH}}-\overset{O}{\overset{\|}{C}}-OH} + H_2O \quad (9.2)$$

The zwitterion form has a net charge of 0, but the form in acid solution has a net positive charge.

When the pH of the solution is increased by adding OH^-, the $—NH_3^+$ of the zwitterion can lose a proton, and the zwitterion is converted into a negatively charged form (Reaction 9.3):

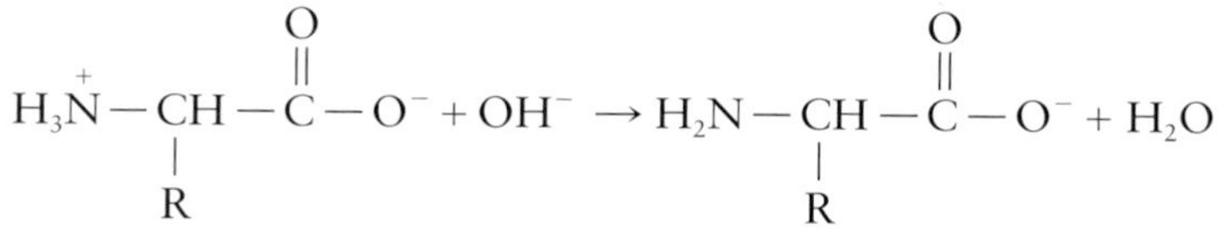

$$\underset{\substack{\text{zwitterion}\\\text{(0 net charge)}}}{H_3\overset{+}{N}-\underset{\substack{|\\R}}{CH}-\overset{\substack{O\\\|}}{C}-O^-} + OH^- \rightarrow \underset{\text{(negative net charge)}}{H_2N-\underset{\substack{|\\R}}{CH}-\overset{\substack{O\\\|}}{C}-O^-} + H_2O \qquad (9.3)$$

Changes in pH also affect acidic and basic side chains and those proteins in which they occur. Thus, all amino acids exist in solution in ionic forms, but the actual form (and charge) is determined by the solution pH.

Because amino acids assume a positively charged form in acidic solutions and a negatively charged form in basic solutions, it seems reasonable to assume that at some pH between the acidic and basic extremes the amino acid will have no net charge (it will be a zwitterion). This assumption is correct, and the pH at which the zwitterion forms is called the **isoelectric point.** Each amino acid has a unique and characteristic isoelectric point; those with neutral R groups are all near a pH of 6 (that of glycine, e.g., is at pH 6.06), those with basic R groups have higher values (that of lysine is at pH 9.47), and the two acidic amino acids have lower values (that of aspartate is at pH 2.98). Proteins, which are composed of amino acids, also have characteristic isoelectric points.

isoelectric point
The characteristic solution pH at which an amino acid has a net charge of 0.

The ability of amino acids and some proteins to react with both H_3O^+ and OH^-, as shown by Reactions 9.2 and 9.3, allows solutions of amino acids or proteins to behave as buffers. The buffering action of blood proteins is one of their most important functions.

EXAMPLE 9.1

Draw the structure of the amino acid leucine

$$H_3\overset{+}{N}-\underset{\substack{|\\CH_2\\|\\CH\\/\ \backslash\\CH_3\quad CH_3}}{CH}-\overset{\substack{O\\\|}}{C}-O^-$$

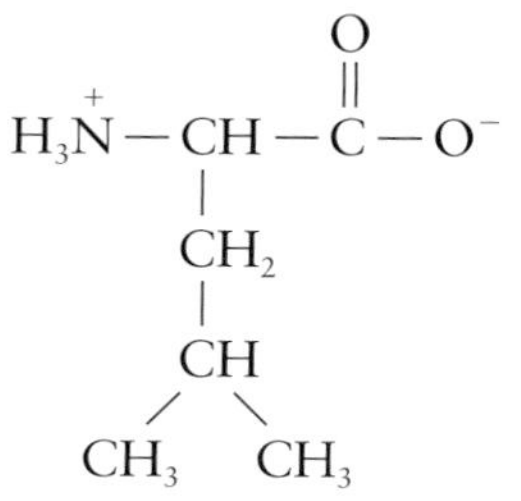

a. in acidic solution at a pH below the isoelectric point;
b. in basic solution at a pH above the isoelectric point.

ThomsonNOW Go to Coached Problems to explore the **acid–base forms of amino acids.**

Solution

a. In acidic solution, the carboxylate group of the zwitterion picks up a proton.

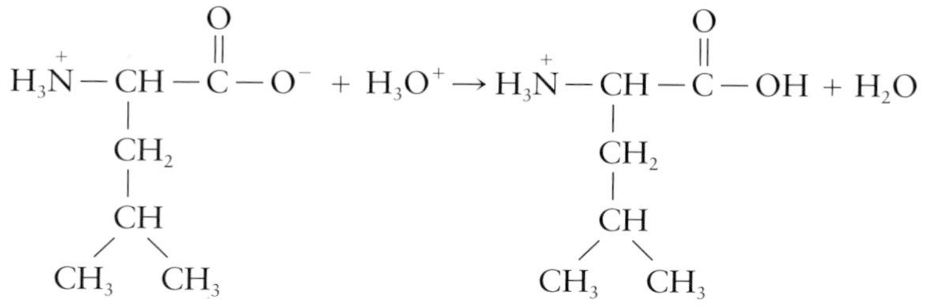

$$H_3\overset{+}{N}-\underset{\substack{|\\CH_2\\|\\CH\\/\ \backslash\\CH_3\quad CH_3}}{CH}-\overset{\substack{O\\\|}}{C}-O^- + H_3O^+ \rightarrow H_3\overset{+}{N}-\underset{\substack{|\\CH_2\\|\\CH\\/\ \backslash\\CH_3\quad CH_3}}{CH}-\overset{\substack{O\\\|}}{C}-OH + H_2O$$

b. In basic solution, the $-H_3\overset{+}{N}$ group of the zwitterion loses a proton.

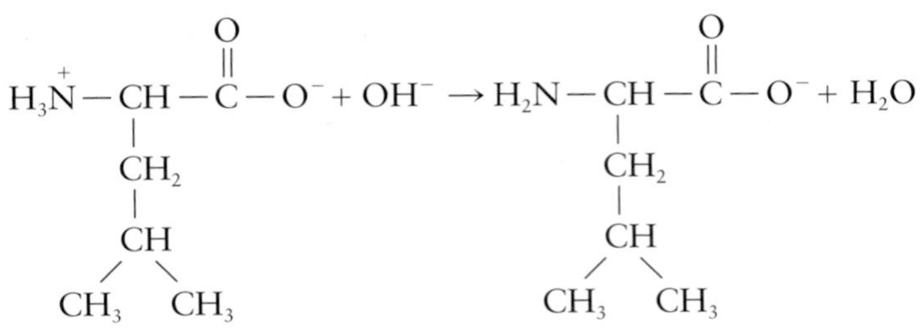

$$H_3\overset{+}{N}-\underset{\underset{\underset{CH_3\ \ CH_3}{CH}}{|}}{\underset{|}{\underset{CH_2}{CH}}}-\overset{\overset{O}{\|}}{C}-O^- + OH^- \rightarrow H_2N-\underset{\underset{\underset{CH_3\ \ CH_3}{CH}}{|}}{\underset{|}{\underset{CH_2}{CH}}}-\overset{\overset{O}{\|}}{C}-O^- + H_2O$$

Learning Check 9.1 Draw the structure of the amino acid serine

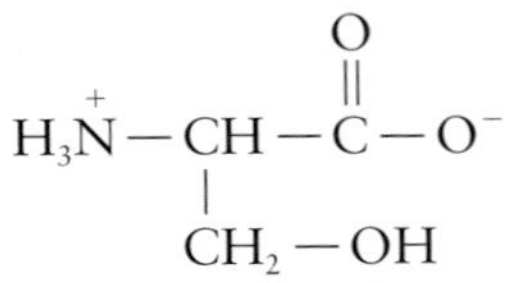

$$H_3\overset{+}{N}-\underset{\underset{CH_2-OH}{|}}{CH}-\overset{\overset{O}{\|}}{C}-O^-$$

a. in acidic solution at a pH below the isoelectric point;
b. in basic solution at a pH above the isoelectric point.

9.3 Reactions of Amino Acids

LEARNING OBJECTIVE

3. Write reactions to represent the formation of peptides and proteins.

Amino acids can undergo reactions characteristic of any functional group in the molecule. This includes reactions of the carboxylate group, the amino group, and any other functional groups attached to the R side chain. However, two reactions are of special interest because of their influence in determining protein structure.

Oxidation of Cysteine

Cysteine, the only sulfhydryl (—SH)-containing amino acid of the 20 commonly found in proteins, has a chemical property not shared by the other 19. We learned in Section 3.9 that a sulfhydryl group can easily be oxidized to form a disulfide bond (—S—S—). Thus, two cysteine molecules react readily to form a disulfide compound called cystine. The disulfide, in turn, is easily converted to —SH groups by the action of reducing agents:

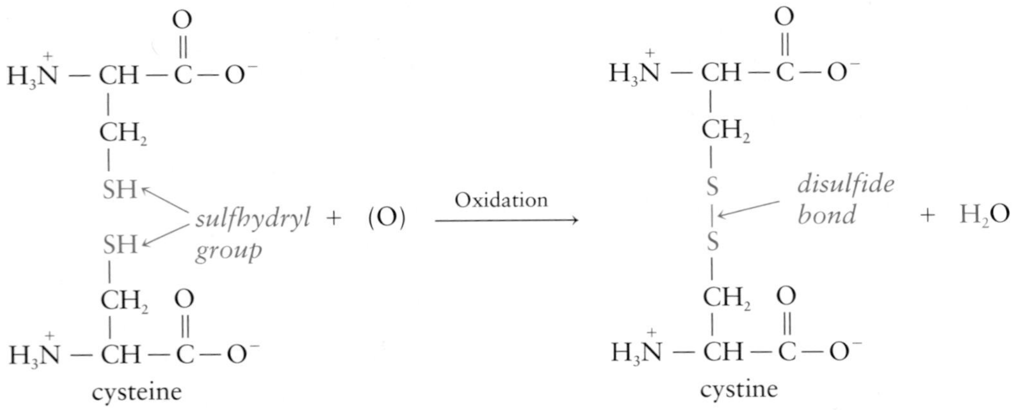

$$\begin{matrix} H_3\overset{+}{N}-CH-\overset{\overset{O}{\|}}{C}-O^- \\ | \\ CH_2 \\ | \\ SH \\ SH \\ | \\ CH_2 \\ | \\ H_3\overset{+}{N}-CH-\overset{\overset{O}{\|}}{C}-O^- \\ \text{cysteine} \end{matrix} \text{ (sulfhydryl group)} + (O) \xrightarrow{\text{Oxidation}} \begin{matrix} H_3\overset{+}{N}-CH-\overset{\overset{O}{\|}}{C}-O^- \\ | \\ CH_2 \\ | \\ S \\ | \\ S \\ | \\ CH_2 \\ | \\ H_3\overset{+}{N}-CH-\overset{\overset{O}{\|}}{C}-O^- \\ \text{cystine} \end{matrix} \text{ (disulfide bond)} + H_2O \quad (9.4)$$

As will be seen, this reaction is important in establishing the structure of some proteins.

OVER THE COUNTER 9.1 Medicines and Nursing Mothers

A concern for all mothers who are breastfeeding a baby should be the effect on the baby of substances, including medicines, that are used by the mother. The effects on breastfed infants of drugs taken by the mother have been determined for only a limited number of medications. However, because very few problems have been reported, most OTC and prescription drugs that are taken by nursing mothers, as needed and according to directions, are considered to be safe for their infants. For example, a nursing mother who takes daily medication for epilepsy, high blood pressure, or diabetes does not put her infant at risk. However, it is a good precaution to check with a pediatrician to determine whether a specific medication might cause problems. In order to minimize a baby's exposure to any medication taken by the mother, the drug should be taken just after the baby finishes nursing, or just before the baby goes to sleep for an extended time.

The following medications are generally considered to present minimal problems for nursing infants when taken by mothers:

- Acetaminophen
- Antibiotics
- Antiepileptics (use Primidone with caution)
- Antihistamines
- Antihypertensives
- Aspirin (use with caution)
- Codeine
- Decongestants
- Ibuprofen
- Insulin
- Quinine
- Thyroid medications

The drugs in the following list have been identified as hazardous to a nursing child, and should never be taken by a nursing mother:

- Bromocriptine (Parlodel): This is a drug for Parkinson's disease; it decreases a woman's milk supply.
- Cancer chemotherapy drugs: These drugs kill body cells in the mother and thus could harm the baby as well.
- Ergotamine (for migraine headaches): This drug causes vomiting, diarrhea, and convulsions in infants.
- Lithium (for manic depression): This substance is secreted in the mother's milk.
- Methotrexate (for arthritis): This drug can depress the baby's immune system.
- Tobacco smoke: Nicotine can cause vomiting, diarrhea, and restlessness in a baby, as well as decreased milk production in the mother. There is also an increased risk of sudden infant death syndrome (SIDS) and an increased frequency of respiratory and ear infections in nursing infants of smoking mothers.
- Illegal drugs: Some drugs, such as cocaine and PCP, intoxicate a baby when used by the mother. Others, such as amphetamines, heroin, and marijuana, cause a variety of symptoms in the baby, including irritability, tremors, vomiting, and poor sleeping patterns. Worst of all, the infants become addicted to the drugs.

Peptide Formation

In Section 6.4 we noted that amides could be thought of as being derived from a carboxylic acid and an amine (Reaction 9.5), although in actual practice the acid chlorides or anhydrides are generally used.

ThomsonNOW™ Go to Coached Problems to explore the **peptide bond formation.**

$$\underset{\text{a carboxylic acid}}{R-\overset{\overset{\displaystyle O}{\|}}{C}-OH} + \underset{\text{an amine}}{R'-NH_2} \xrightarrow{\text{Heat}} \underset{\text{an amide}}{R-\overset{\overset{\displaystyle O}{\|}}{C}-NH-R'} + H_2O \quad \textit{(amide linkage: C–NH)} \tag{9.5}$$

In the same hypothetical way, it is possible to envision the carboxylate group of one amino acid reacting with the amino group of a second amino acid. For example, glycine could combine with alanine:

$$\underset{\text{glycine}}{H_3\overset{+}{N}-CH_2-\overset{\overset{\displaystyle O}{\|}}{C}-O^-} + \underset{\text{alanine}}{H_3\overset{+}{N}-\underset{\underset{\displaystyle CH_3}{|}}{CH}-\overset{\overset{\displaystyle O}{\|}}{C}-O^-} \longrightarrow \underset{\text{glycylalanine (a dipeptide)}}{H_3\overset{+}{N}-CH_2-\overset{\overset{\displaystyle O}{\|}}{C}-NH-\underset{\underset{\displaystyle CH_3}{|}}{CH}-\overset{\overset{\displaystyle O}{\|}}{C}-O^-} + H_2O \quad \textit{(peptide linkage: C–NH)} \tag{9.6}$$

dipeptide
A compound formed when two amino acids are bonded by an amide linkage.

Compounds of this type made up of two amino acids are called **dipeptides,** and the amide linkage that holds them together is called a **peptide linkage** or **peptide bond.**

It is important to realize that a different dipeptide could also form by linking glycine and alanine the other way (alanine on the left in this case):

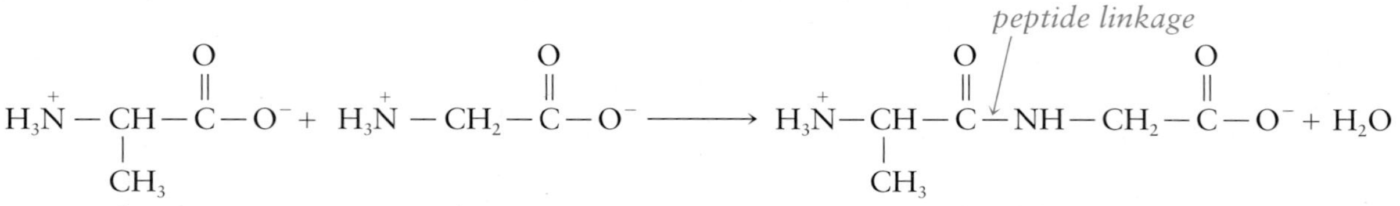

$$\underset{\text{alanine}}{H_3\overset{+}{N}-\underset{\underset{CH_3}{|}}{CH}-\overset{\overset{O}{\|}}{C}-O^-} + \underset{\text{glycine}}{H_3\overset{+}{N}-CH_2-\overset{\overset{O}{\|}}{C}-O^-} \longrightarrow \underset{\text{alanylglycine}}{H_3\overset{+}{N}-\underset{\underset{CH_3}{|}}{CH}-\overset{\overset{O}{\|}}{C}\underset{\text{peptide linkage}}{-}NH-CH_2-\overset{\overset{O}{\|}}{C}-O^-} + H_2O \qquad (9.7)$$

peptide linkage or peptide bond
The amide linkage between amino acids that results when the amino group of one acid reacts with the carboxylate group of another.

Glycylalanine and alanylglycine are structural isomers of each other, and each has a unique set of properties. The reactions to form these and other dipeptides are more complex than Reactions 9.6 and 9.7 imply. In nature, these processes are carried out according to genetic information using a mechanism we shall examine in Section 11.8.

The presence of amino and carboxylate groups on the ends of dipeptides allows for the attachment of a third amino acid to form a tripeptide. For example, valine could be attached to alanylglycine:

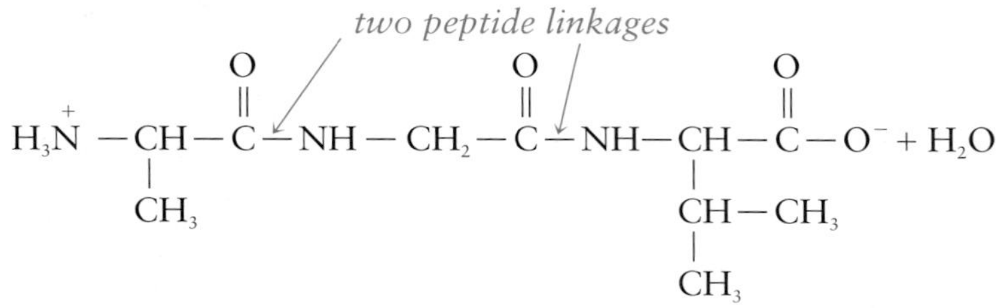

$$H_3\overset{+}{N}-\underset{\underset{CH_3}{|}}{CH}-\overset{\overset{O}{\|}}{C}-NH-CH_2-\overset{\overset{O}{\|}}{C}-NH-\underset{\underset{\underset{CH_3}{|}}{CH-CH_3}}{\underset{|}{CH}}-\overset{\overset{O}{\|}}{C}-O^- + H_2O \qquad (9.8)$$

two peptide linkages

alanylglycylvaline, a tripeptide

peptide
An amino acid polymer of short chain length.

polypeptide
An amino acid polymer of intermediate chain length containing up to 50 amino acid residues.

protein
An amino acid polymer made up of more than 50 amino acids.

amino acid residue
An amino acid that is a part of a peptide, polypeptide, or protein chain.

N-terminal residue
An amino acid on the end of a chain that has an unreacted or free amino group.

C-terminal residue
An amino acid on the end of a chain that has an unreacted or free carboxylate group.

More amino acids can react in the same way to form a tetrapeptide, a pentapeptide, and so on, until a chain of hundreds or even thousands of amino acids has formed. The compounds with shortest chains are often simply called **peptides,** those with longer chains are called **polypeptides,** and those with still longer chains are called **proteins.** Chemists differ about where to draw the lines in the use of these names, but generally polypeptide chains with more than 50 amino acids are called proteins. However, the terms *protein* and *polypeptide* are often used interchangeably.

Amino acids that have been incorporated into chains are called **amino acid residues.** The amino acid residue with an unreacted or free amino group at the end of the chain is designated the **N-terminal residue,** and, according to convention, it is written on the left end of the peptide chain. Similarly, the residue with a free carboxylate group at the other end of the chain is the **C-terminal residue;** it is written on the right end of the chain. Thus, alanine is the N-terminal residue and valine is the C-terminal residue in the tripeptide alanylglycylvaline:

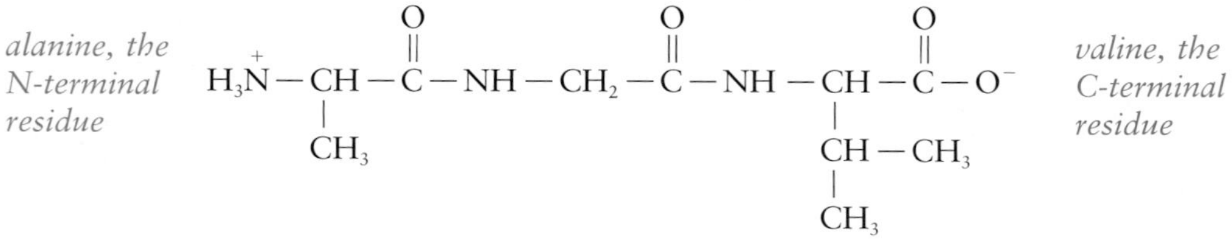

alanine, the N-terminal residue

$$H_3\overset{+}{N}-\underset{\underset{CH_3}{|}}{CH}-\overset{\overset{O}{\|}}{C}-NH-CH_2-\overset{\overset{O}{\|}}{C}-NH-\underset{\underset{\underset{CH_3}{|}}{CH-CH_3}}{\underset{|}{CH}}-\overset{\overset{O}{\|}}{C}-O^-$$

valine, the C-terminal residue

Peptides are named by starting at the N-terminal end of the chain and listing the amino acid residues in order from left to right. Structural formulas and even full names for large peptides become very unwieldy and time-consuming to write. This is simplified by representing peptide and protein structures in terms of three-letter abbreviations for the amino acid residues with dashes to show peptide linkages. Thus, the structure of glycylalanine is represented by Gly-Ala. The isomeric dipeptide, alanylglycine, is Ala-Gly. In this case, alanine is the N-terminal residue, and glycine is the C-terminal residue.

The tripeptide alanylglycylvaline can be represented by Ala-Gly-Val. In addition to this sequence, five other arrangements of these three components are possible, and each one represents an isomeric tripeptide with different properties. These sequences are Ala-Val-Gly, Val-Ala-Gly, Val-Gly-Ala, Gly-Val-Ala, and Gly-Ala-Val.

Insulin, with 51 amino acids, has 1.55×10^{66} sequences possible! From these possibilities, the body reliably produces only one, which illustrates the remarkable precision of the life process. From the simplest bacterium to the human brain cell, only the amino acid sequences needed to form essential cell peptides are produced.

Learning Check 9.2 Use Table 9.1 to draw the full structure of the following tetrapeptide. Label the N-terminal and C-terminal residues.

Phe-Cys-Ser-Ile

9.4 Important Peptides

LEARNING OBJECTIVE

4. Describe uses for important peptides.

More than 200 peptides have been isolated and identified as essential to the proper functioning of the human body. The identity and sequence of amino acid residues in these peptides is extremely important to proper physiological function. Two well-known examples are vasopressin and oxytocin, which are hormones released by the pituitary gland. Each hormone is a nonapeptide (nine amino acid residues) with six of the residues held in the form of a loop by a disulfide bond (see Figure 9.2). The disulfide bond in these hormones is formed by the oxidation of cysteine residues in the first and sixth positions (counting from the N-terminal end) of these peptide chains. In peptides and proteins, disulfide bonds such as these that draw a single peptide chain into a loop or hold two peptide chains together are called **disulfide bridges.**

disulfide bridge
A bond produced by the oxidation of —SH groups on two cysteine residues. The bond loops or holds two peptide chains together.

Figure 9.2 shows that the structures of vasopressin and oxytocin differ only in the third and eighth amino acid residues of their peptide chains. The result of

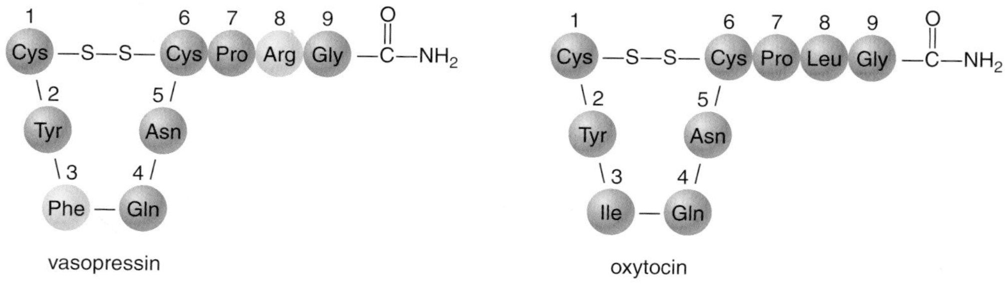

FIGURE 9.2 The structures of vasopressin and oxytocin. Differences are in amino acid residues 3 and 8. The normal carboxylate groups of the C-terminal residues (Gly) have been replaced by amide groups.

CHEMISTRY AND YOUR HEALTH 9.1 C-Reactive Protein: A Message from the Heart

The letters LDL and HDL followed by the word cholesterol are familiar to most health-conscious individuals. The letters stand for low-density lipoprotein and high-density lipoprotein, respectively, and it is also common knowledge that high LDL levels in the blood are dangerous and indicate possible heart disease. However, one of every four heart attack or stroke victims has a low LDL level, low blood pressure, does not smoke, is not diabetic, and has no other indication of cardiovascular disease—at least until recently. A protein discovered in the 1930s is gaining new prominence as a test for heart disease. The protein, called *C-Reactive Protein* (CRP) is produced in the liver and passed into the bloodstream in response to any type of inflammation in the body. Many researchers now believe that inflammation is involved in cardiovascular disease.

An inexpensive test that measures the amount of CRP in the blood can detect inflammation when no other symptoms of cardiovascular disease exist. It is estimated that the routine use of this test as a part of physical examinations could help prevent up to 25% of all heart attacks. It should be stressed that the test is only an indicator of cardiovascular disease and not a cure. Individuals who desire a healthy heart still need to eat wisely, avoid tobacco and other known harmful products, and exercise regularly. Regular exercise has been shown to reduce the level of CRP in the blood, which indicates a reduction in inflammation. It has also been shown that aspirin and similar drugs reduce inflammation, which provides an explanation for the accepted practice of taking a low dose of aspirin on a regular basis to help prevent heart attacks.

National guidelines are being developed to standardize CRP testing techniques and interpretation of the results. A paper published in the *New England Journal of Medicine* specifies three levels of risk that will probably be used as guidelines. Levels of CRP below one milligram per liter of blood are considered to be low. Levels between 1 and 3 mg/L are considered moderate, and levels above 3 mg/L are considered high.

Some factors suggest that CRP levels should not be used as the sole indicator of cardiovascular disease. For example, women tend to have higher CRP levels than men, but men have a higher incidence of cardiovascular disease. Also, women on hormone therapy seem to have abnormally high levels of CRP even in the absence of inflammation. Some researchers even suggest that CRP is just an indicator for some other factor, such as microbes that might be the real culprits, citing the atherosclerosis caused by syphilis as an example. However, it is beginning to appear that for both men and women, lowering CRP levels through exercise and diet might be as important to good heart health as lowering the level of LDL cholesterol. In any case, the low cost of CRP testing coupled with the fact that many insurance plans will pay for it makes such testing a good idea.

these variations is a significant difference in their biological functions. Vasopressin is known as antidiuretic hormone (ADH) because of its action in reducing the volume of urine formed, thus conserving the body's water. It also raises blood pressure. Oxytocin causes the smooth muscles of the uterus to contract and is often administered to induce labor. Oxytocin also acts on the smooth muscles of lactating mammary glands to stimulate milk ejection.

Another important peptide hormone is adrenocorticotropic hormone (ACTH). It is synthesized by the pituitary gland and consists of a single polypeptide chain of 39 amino acid residues with no disulfide bridges (see ■ Figure 9.3). The major function of ACTH is to regulate the production of steroid hormones in the cortex of the adrenal gland. ■ Table 9.2 lists these and some other important peptide or protein hormones.

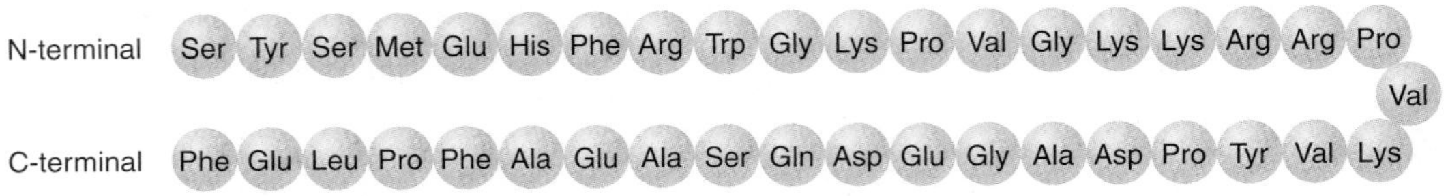

N-terminal Ser Tyr Ser Met Glu His Phe Arg Trp Gly Lys Pro Val Gly Lys Lys Arg Arg Pro Val Lys Val Tyr Pro Asp Ala Gly Glu Asp Gln Ser Ala Glu Ala Phe Pro Leu Glu Phe C-terminal

■ **FIGURE 9.3** The amino acid sequence of ACTH.

TABLE 9.2 Examples of peptide and protein hormones

Name	Origin	Action
Adrenocorticotropic hormone (ACTH)	Pituitary	Stimulates production of adrenal hormones
Angiotensin II	Blood plasma	Causes blood vessels to constrict
Follicle-stimulating hormone (FSH)	Pituitary	Stimulates sperm production and follicle maturation
Gastrin	Stomach	Stimulates stomach to secrete acid
Glucagon	Pancreas	Stimulates glycogen metabolism in liver
Human growth hormone (HGH)	Pituitary	General effects; bone growth
Insulin	Pancreas	Controls metabolism of carbohydrates
Oxytocin	Pituitary	Stimulates contraction of uterus and other smooth muscles
Prolactin	Pituitary	Stimulates lactation
Somatostatin	Hypothalamus	Inhibits production of HGH
Vasopressin	Pituitary	Decreases volume of urine excreted

9.5 Characteristics of Proteins

LEARNING OBJECTIVE

5. Describe proteins in terms of the following characteristics: size, function, classification as fibrous or globular, and classification as simple or conjugated.

Size

Proteins are extremely large natural polymers of amino acids with molecular weights that vary from about 6000 to several million u. The immense size of proteins (in the molecular sense) can be appreciated by comparing glucose with hemoglobin, a relatively small protein. Glucose has a molecular weight of 180 u, whereas the molecular weight of hemoglobin is 65,000 u (see ■ Table 9.3). The molecular formula of glucose is $C_6H_{12}O_6$ and that of hemoglobin is $C_{2952}H_{4664}O_{832}N_{812}S_8Fe_4$.

Protein molecules are too large to pass through cell membranes and are contained inside the normal cells where they were formed. However, if cells become damaged by disease or trauma, the protein contents can leak out. Thus, persistent

TABLE 9.3 Molecular weights of some common proteins

Protein	Molecular weight (u)	Number of amino acid residues
Insulin	6,000	51
Cytochrome c	16,000	104
Growth hormone	49,000	191
Hemoglobin	65,000	574
Hexokinase	96,000	730
Gamma globulin	176,000	1320
Myosin	800,000	6100

excessive amounts of proteins in the urine are indicative of damaged kidney cells. A routine urinalysis usually includes a test for protein. Similarly, a heart attack can be confirmed by the presence in the blood of certain proteins (enzymes) that are normally confined to cells in heart tissue (Section 10.9).

Acid–Base Properties

The 20 different R groups of amino acids are important factors in determining the physical and chemical properties of proteins that contain the amino acids. Acid–base behavior is one of the most important of these properties. Proteins, like amino acids, take the form of zwitterions. The R groups of glutamate and aspartate contain $—COO^-$ groups and provide H_3O^+ ions, whereas the R groups of arginine and histidine are basic (see Table 9.1). Thus, proteins, like amino acids, have characteristic isoelectric points and can behave as buffers in solutions.

The tendency for large molecules such as proteins to remain in solution or form a stable colloidal dispersion depends to a large extent on the repulsive forces acting between molecules with like charges on their surfaces. When protein molecules are at a pH where they take forms that have a net positive or negative charge, the presence of these like charges causes the molecules to repel one another and remain dispersed. These repulsive forces are smallest at the isoelectric point, when the net molecular charges are essentially zero. Thus, protein molecules tend to clump together and precipitate from solutions in which the pH is equal to or close to the isoelectric point.

Protein Function

A large part of the importance of proteins results from the crucial roles they serve in all biological processes. The various functions of proteins are discussed below and summarized in ■ Table 9.4.

1. **Catalytic function:** Nearly all the reactions that take place in living organisms are catalyzed by proteins functioning as enzymes. Without these catalytic proteins, biological reactions would take place too slowly to support life. Enzyme-catalyzed reactions range from relatively simple processes to the very complex, such as the duplication of hereditary material in cell nuclei. Enzymes are discussed in more detail in the next chapter.
2. **Structural function:** The main structural material for plants is cellulose. In the animal kingdom, structural materials other than the inorganic components of the skeleton are composed of protein. Collagen, a fiberlike protein, is responsible for the mechanical strength of skin and bone. Keratin, the chief constituent of hair, skin, and fingernails, is another example.
3. **Storage function:** Some proteins provide a way to store small molecules or ions. Ovalbumin, for example, is a stored form of amino acids that is used by embryos developing in bird eggs. Casein, a milk protein, and gliadin, in wheat seeds, are also stored forms of protein intended to nourish animals and plants, respectively. Ferritin, a liver protein, attaches to iron ions and forms a storage complex in humans and other animals.
4. **Protective function: Antibodies** are tremendously important proteins that protect the body from disease. These highly specific proteins combine with and help destroy viruses, bacteria, and other foreign substances that get into the blood or tissue of the body. Blood clotting, another protective process, is carried out by the proteins thrombin and fibrinogen. Without this process, even small wounds would result in life-threatening bleeding.

antibody
A substance that helps protect the body from invasion by viruses, bacteria, and other foreign substances.

5. **Regulatory function:** Numerous body processes are regulated by hormones, many of which are proteins. Examples are growth hormone, which regulates the growth rate of young animals, and thyrotropin, which stimulates the activity of the thyroid gland.

TABLE 9.4 Biological functions of proteins

Function	Examples	Occurrence or role
Catalysis	lactate dehydrogenase	Oxidizes lactic acid
	cytochrome c	Transfers electrons
	DNA polymerase	Replicates and repairs DNA
Structure	viral-coat proteins	Sheath around nucleic acid of viruses
	glycoproteins	Cell coats and walls
	α-keratin	Skin, hair, feathers, nails, and hooves
	β-keratin	Silk of cocoons and spiderwebs
	collagen	Fibrous connective tissue
	elastin	Elastic connective tissue
Storage	ovalbumin	Egg-white protein
	casein	A milk protein
	ferritin	Stores iron in the spleen
	gliadin	Stores amino acids in wheat
	zein	Stores amino acids in corn
Protection	antibodies	Form complexes with foreign proteins
	fibrinogen	Involved in blood clotting
	thrombin	Involved in blood clotting
Regulation	insulin	Regulates glucose metabolism
	growth hormone	Stimulates growth of bone
Nerve impulse transmission	rhodopsin	Involved in vision
	acetylcholine receptor protein	Impulse transmission in nerve cells
Movement	myosin	Thick filaments in muscle fiber
	actin	Thin filaments in muscle fiber
	dynein	Movement of cilia and flagella
Transport	hemoglobin	Transports O_2 in blood
	myoglobin	Transports O_2 in muscle cells
	serum albumin	Transports fatty acids in blood
	transferrin	Transports iron in blood
	ceruloplasmin	Transports copper in blood

6. **Nerve impulse transmission:** Some proteins behave as receptors of small molecules that pass between gaps (synapses) separating nerve cells. In this way, they transmit nerve impulses from one nerve to another. Rhodopsin, a protein found in the rod cells of the retina of the eye, functions this way in the vision process.
7. **Movement function:** Every time we climb stairs, push a button, or blink an eye, we use muscles that have proteins as their major components. The proteins actin and myosin are particularly important in processes involving movement. They are long-filament proteins that slide along each other during muscle contraction.
8. **Transport function:** Numerous small molecules and ions are transported effectively through the body only after binding to proteins. For example, fatty acids are carried between fat (adipose) tissue and other tissues or organs by serum albumin, a blood protein. Hemoglobin, a well-known example, carries oxygen from the lungs to other body tissues, and transferrin is a carrier of iron in blood plasma.

Proteins perform other functions, but these are among the most important. It is easy to see that properly functioning cells of living organisms must contain

CHEMISTRY AROUND US 9.1 Alzheimer's Disease

Alzheimer's disease is a progressive, degenerative disease that attacks the brain, causing impaired thinking, behavior, and memory. As the disease progresses, nerve cells in the brain degenerate, and the size of the brain actually decreases. About 4 million Americans suffer from this malady; it afflicts 1 in 10 people over age 65, and nearly half of all people over age 85. The disease, first identified by Alois Alzheimer in 1906, is best described by the words of one of its victims: "I have lost myself."

Some warning signs of Alzheimer's disease include difficulty in performing familiar tasks, language problems, disorientation related to time or place, poor or decreased quality of judgment, memory loss affecting job skills, problems with abstract thinking, misplacing things or putting them in inappropriate places, passivity and loss of initiative, and changes in mood, personality, or behavior. If several of these symptoms are present in an older person, a medical evaluation for Alzheimer's disease is in order.

The causes of the disease are under debate. Some researchers feel it is caused by the formation in the brain of fibrils (strands) and plaque deposits consisting largely of amyloid beta-protein. Others think the formation of tangles in a microtubule-associated protein called Tau is the culprit. Still others think both of these explanations are correct, and that the formation of abnormal amyloid beta strands causes the abnormal killer forms of the Tau protein to form. This position is supported by a recent study of the brains of aging rhesus monkeys. This study showed that amyloid beta-protein caused the same damage to the monkey brains that was found in the brains of Alzheimer's victims. In addition, the amyloid beta-protein caused changes in the Tau protein of the monkey brains that made it capable of forming tangles.

Currently, two drugs have been approved for use in treating Alzheimer's disease, Tacrine and Aricept, which treat some of the associated memory loss. Both drugs function by inhibiting the reuptake of acetylcholine in neural synapses. Unfortunately, they neither cure the disease nor slow its progress.

Current research is focused on preventing or reversing the process of changing normal amyloid protein into the deformed amyloid protein found in the brains of Alzheimer's patients. The α-helices of normal proteins are changed to β-pleated sheets in the abnormal proteins characteristic of the disease. In a recent promising study, an experimental peptide containing five amino acids was shown to mimic a region of amyloid beta-protein that regulates folding. However, the experimental peptide is so constructed that it prevents other amino acids of amyloid beta-protein from folding into the β-pleated sheet conformation necessary to form amyloid beta-protein fibrils. In cell culture studies, the experimental peptide inhibited the formation of amyloid fibrils, broke down existing fibrils, and prevented the death of neurons. This type of promising discovery may eventually lead to a drug that can stop the progress—or even reverse the effects—of Alzheimer's disease.

many proteins. It has been estimated that a typical human cell contains 9000 different proteins, and that a human body contains about 100,000 different proteins.

Classification

fibrous protein
A protein made up of long rod-shaped or stringlike molecules that intertwine to form fibers.

globular protein
A spherical protein that usually forms stable suspensions in water or dissolves in water.

simple protein
A protein made up entirely of amino acid residues.

conjugated protein
A protein made up of amino acid residues and other organic or inorganic components.

prosthetic group
The non–amino acid parts of conjugated proteins.

Proteins can be classified by the functions just discussed. They can also be classified into two major types, fibrous and globular, on the basis of their structural shape. **Fibrous proteins** are made up of long rod-shaped or stringlike molecules that can intertwine with one another and form strong fibers. They are water-insoluble and are usually found as major components of connective tissue, elastic tissue, hair, and skin. Examples are collagen, elastin, and keratin (see ■ Figure 9.4).

Globular proteins are more spherical and either dissolve in water or form stable suspensions in water. They are not found in structural tissue but are transport proteins, or proteins that may be moved easily through the body by the circulatory system. Examples of these are hemoglobin and transferrin.

Another classification scheme for proteins is based on composition; proteins are either simple or conjugated. **Simple proteins** are those that contain only amino acid residues. **Conjugated proteins** contain amino acid residues and other organic or inorganic components. The non–amino acid parts of conjugated proteins are called **prosthetic groups.** The type of prosthetic group present is often used to further classify conjugated proteins. For example, proteins containing lipids, carbohydrates, or metal ions are called lipoproteins, glycoproteins, and metalloproteins, respectively. Thus, the iron-containing proteins hemoglobin and myoglobin are metalloproteins. ■ Table 9.5 lists the common classes of conjugated proteins.

TABLE 9.5 Conjugated proteins

Class	Prosthetic groups	Examples
Nucleoproteins	nucleic acids (DNA, RNA)	Viruses
Lipoproteins	lipids	Fibrin in blood, serum lipoproteins
Glycoproteins	carbohydrates	Gamma globulin in blood, mucin in saliva
Phosphoproteins	phosphate groups	Casein in milk
Hemoproteins	heme	Hemoglobin, myoglobin, cytochromes
Metalloproteins	iron	Ferritin, hemoglobin
	zinc	Alcohol dehydrogenase

© Christina Slabaugh

■ **FIGURE 9.4** Animal hair and spiderwebs are constructed of the fibrous proteins α-keratin and β-keratin, respectively. What properties of fibrous proteins make hair, fur, and spiderwebs useful?

ThomsonNOW™ Go to Coached Problems to explore **protein structure.**

Protein Structure

As you might expect, the structures of extremely large molecules, such as proteins, are much more complex than those of simple organic compounds. Many protein molecules consist of a chain of amino acids twisted and folded into a complex three-dimensional structure. This structural complexity imparts unique features to proteins that allow them to function in the diverse ways required by biological systems. To better understand protein function, we must look at four levels of organization in their structure. These levels are referred to as primary, secondary, tertiary, and quaternary.

9.6 The Primary Structure of Proteins

LEARNING OBJECTIVE

6. Explain what is meant by the primary structure of proteins.

Every protein, be it hemoglobin, keratin, or insulin, has a similar backbone of carbon and nitrogen atoms held together by peptide bonds:

$$-NH-\underset{\displaystyle R}{\underset{|}{CH}}-\overset{\displaystyle O}{\overset{\|}{C}}-NH-\underset{\displaystyle R'}{\underset{|}{CH}}-\overset{\displaystyle O}{\overset{\|}{C}}-NH-\underset{\displaystyle R''}{\underset{|}{CH}}-\overset{\displaystyle O}{\overset{\|}{C}}-NH-\underset{\displaystyle R'''}{\underset{|}{CH}}-\overset{\displaystyle O}{\overset{\|}{C}}-NH-\underset{\displaystyle R''''}{\underset{|}{CH}}-\overset{\displaystyle O}{\overset{\|}{C}}-$$

protein backbone (in color)

The difference in proteins is the length of the backbone and the sequence of the side chains (R groups) that are attached to the backbone. The order in which amino acid residues are linked together in a protein is called the **primary structure.** Each different protein in a biological organism has a unique sequence of amino acid residues. It is this sequence that causes a protein chain to fold and curl into the distinctive shape that, in turn, enables the protein to function properly. As we will see in Section 9.10, a protein molecule that loses its characteristic three-dimensional shape cannot function.

primary protein structure The linear sequence of amino acid residues in a protein chain.

Biochemists have devised techniques for finding the order in which the residues are linked together in protein chains. A few of the proteins whose primary structures are known are listed in ■ Table 9.6.

■ Figure 9.5 shows the primary structure of human insulin. Notice that the individual molecules of this hormone consist of two chains having a total of 51

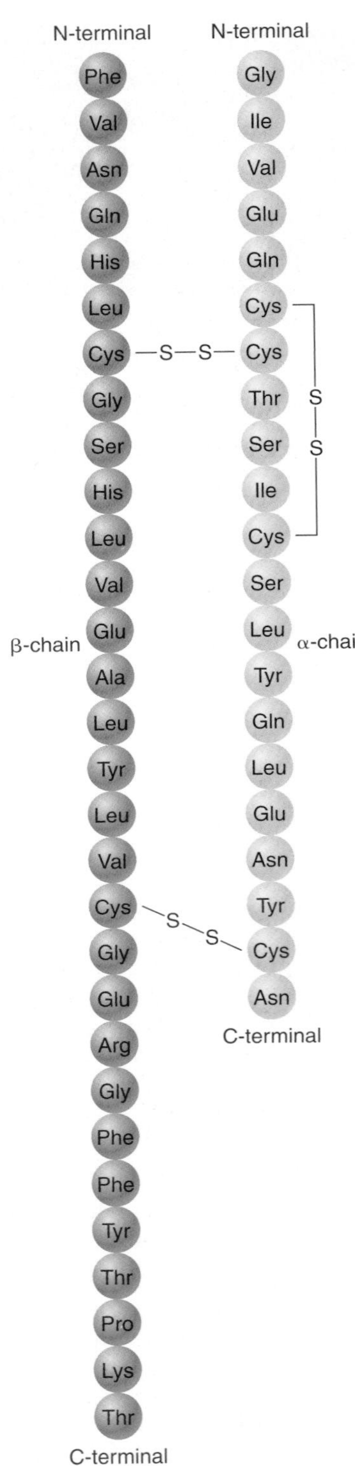

■ **FIGURE 9.5** The amino acid sequence (primary structure) of human insulin.

TABLE 9.6 Some proteins whose sequence of amino acids is known

Protein	Number of amino acid residues
Insulin	51
Hemoglobin	
α-chain	141
β-chain	146
Myoglobin	153
Human growth hormone	191
Trypsinogen	229
Carboxypeptidase A	307
Gamma globulin	1320

amino acid residues. In the molecule, two disulfide bridges hold the chains together, and there is one disulfide linkage within a chain. Insulin plays an essential role in regulating the use of glucose by cells. Inadequate production of insulin leads to diabetes mellitus, and people with severe diabetes must take insulin shots.

As we saw in the case of vasopressin and oxytocin (Section 9.4), small changes in amino acid sequence can cause profound differences in the functioning of proteins. For example, a minor change in sequence in the blood protein hemoglobin causes the fatal sickle-cell disease.

9.7 The Secondary Structure of Proteins

LEARNING OBJECTIVE

7. Describe the role of hydrogen bonding in the secondary structure of proteins.

If the only structural characteristics of proteins were their amino acid sequences (primary structures), all protein molecules would consist of long chains arranged in random fashion. However, protein chains fold and become aligned in such a way that certain orderly patterns result. These orderly patterns, referred to as **secondary structures,** result from hydrogen bonding and include the α-helix (alpha-helix) and the β-pleated (beta-pleated) sheet.

The α-Helix

In 1951, American chemists Linus Pauling and Robert Corey suggested that proteins could exist in the shape of an **α-helix,** a form in which a single protein chain twists so that it resembles a coiled helical spring (see ■ Figure 9.6). The chain is held in the helical shape by numerous intramolecular hydrogen bonds between carbonyl oxygens and amide hydrogens in adjacent turns of the helical backbone:

$$\begin{array}{c} -\underset{|}{C}{=}O \cdots H{-}\underset{|}{N}- \end{array}$$

The carbonyl group of each amino acid residue is hydrogen bonded to the amide hydrogen of the amino acid four residues away in the chain; thus, all amide groups in the helix are hydrogen bonded (Figure 9.6). The protein backbone forms the coil, and the side chains (R groups) extend outward from the coil.

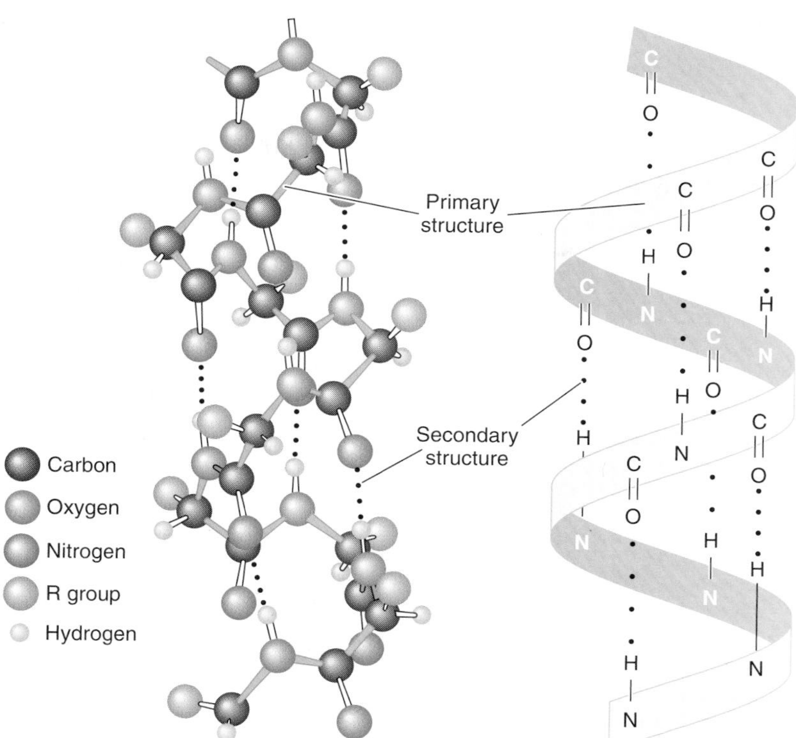

■ **FIGURE 9.6** Two representations of the α-helix, showing hydrogen bonds between amide groups.

Six years after Pauling and Corey proposed the α-helix, its presence in proteins was detected. Since that time it has been found that the amount of α-helical content is quite variable in proteins. In some it is the major structural component, whereas in others a random coil predominates and there is little or no α-helix coiling.

In the proteins α-keratin (found in hair), myosin (found in muscle), epidermin (found in skin), and fibrin (found in blood clots), two or more helices interact (supracoiling) to form a cable (see ■ Figure 9.7). These cables make up bundles of fibers that lend strength to the tissue in which they are found.

secondary protein structure
The arrangement of protein chains into patterns as a result of hydrogen bonds between amide groups of amino acid residues in the chain. The common secondary structures are the α-helix and the β-pleated sheet.

α-helix
The helical structure in proteins that is maintained by hydrogen bonds.

β-pleated sheet
A secondary protein structure in which protein chains are aligned side by side in a sheetlike array held together by hydrogen bonds.

The β-Pleated Sheet

A less common type of secondary structure is found in some proteins. This type, called a **β-pleated sheet,** results when several protein chains lie side by side and are held in position by hydrogen bonds between the amide carbonyl oxygens of one chain and the amide hydrogens of an adjacent chain (see ■ Figure 9.8).

The β-pleated sheet is found extensively only in the protein of silk. However, a number of proteins in which a single polypeptide chain forms sections of a β-pleated sheet structure by folding back on itself are known. ■ Figure 9.9 depicts a protein chain exhibiting the β-pleated sheet, the α-helix, and random structural conformations. Note the position of the hydrogen bonds in the two secondary structures.

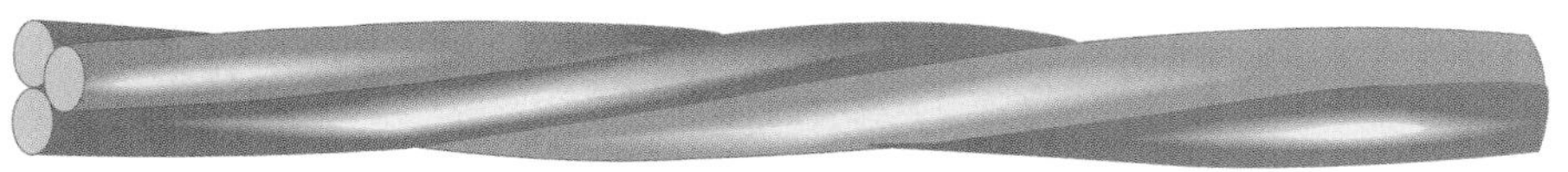

■ **FIGURE 9.7** The supracoiling of three α-helices in the keratins of hair and wool.

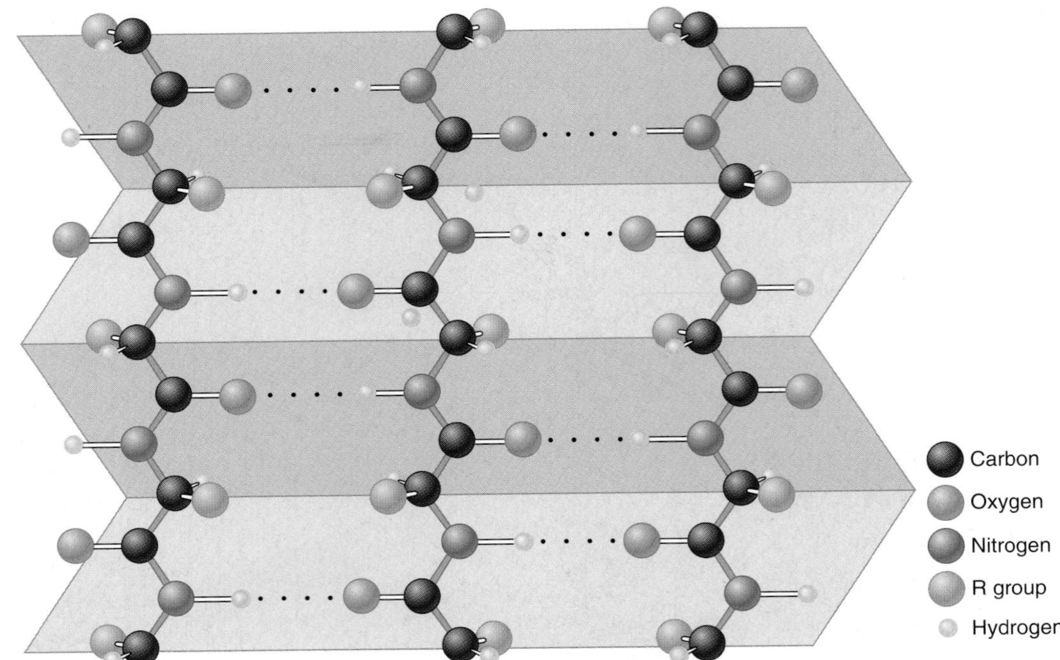

■ **FIGURE 9.8** The β-pleated sheet. The four dots show hydrogen bonds between adjacent protein chains.

9.8 The Tertiary Structure of Proteins

LEARNING OBJECTIVE

8. Describe the role of side-chain interactions in the tertiary structure of proteins.

tertiary protein structure A specific three-dimensional shape of a protein resulting from interactions between R groups of the amino acid residues in the protein.

Tertiary structure, the next higher level of complexity in protein structure, refers to the bending and folding of the protein into a specific three-dimensional shape. This bending and folding may seem rather disorganized, but nevertheless it results in a favored arrangement for a given protein.

The tertiary structure results from interactions between the R side chains of the amino acid residues. These R-group interactions are of four types:

1. **Disulfide bridges:** As in the structure of insulin (Figure 9.5), a disulfide linkage can form between two cysteine residues that are close to each other in the

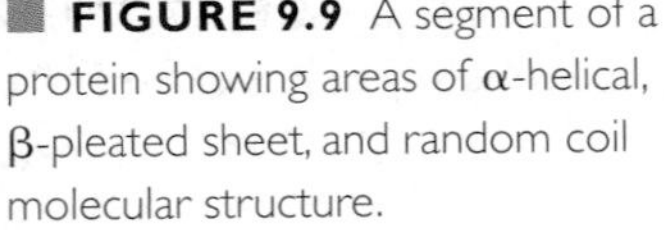

■ **FIGURE 9.9** A segment of a protein showing areas of α-helical, β-pleated sheet, and random coil molecular structure.

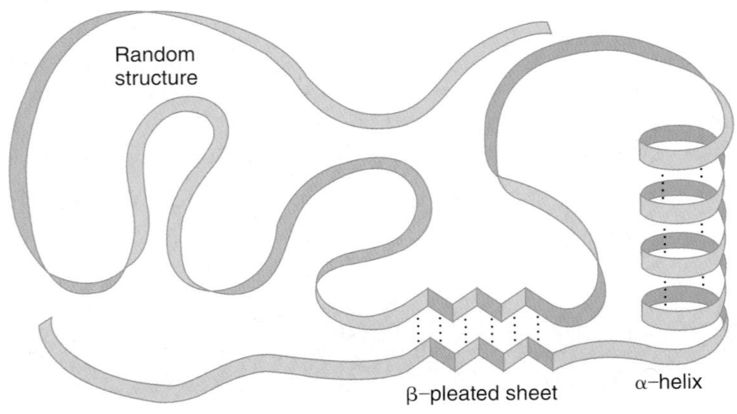

CHEMISTRY AROUND US 9.2 Sickle-Cell Disease

Hemoglobin is composed of four polypeptide chains: two α-chains, each containing 141 amino acid residues, and two β-chains, each containing 146 amino acid residues (Figure 9.12). Hemoglobin is located in red blood cells and performs the life-sustaining function of transporting oxygen from the lungs to the cells and carbon dioxide from the cells to the lungs. In some individuals, the hemoglobin molecules have a slightly different sequence of amino acid residues. It is the so-called abnormal human hemoglobins that have attracted particular attention because of the diseases associated with them. One of the best-known abnormal hemoglobins (designated HbS) differs from the normal type (HbA) by only one amino acid residue in each of the two β-chains. In HbS, the glutamate in the sixth position of normal HbA is replaced by a valine residue:

	4	5	6	7	8	9
Normal HbA	—Thr	—Pro	—Glu	—Glu	—Lys	—Ala—
Sickle-cell HbS	—Thr	—Pro	—Val	—Glu	—Lys	—Ala—

This difference affects only two positions in a molecule containing 574 amino acid residues, and yet it is enough to result in a very serious disease, sickle-cell disease.

When combined with oxygen, the red blood cells of people with sickle-cell disease have the flat, disklike shape of normal red blood cells. However, after they give up their oxygen, red blood cells containing HbS hemoglobin become distorted into a characteristic sickle shape. These sickled cells tend to clump together and wedge in capillaries, particularly in the spleen, and cause excruciating pain. Cells blocking capillaries are rapidly destroyed, and the loss of red blood cells causes anemia. Children with sickle-cell disease have an 80% lower chance of surviving to adulthood than unafflicted children.

Sickle-cell anemia is a genetic disease that can be identified by screening a blood sample for abnormal hemoglobin. A person with an HbS gene from one parent and a normal HbA gene from another parent (a heterozygote) is said to have sickle-cell trait. About 40% of the hemoglobin in these individuals is HbS, but there are generally no ill effects. A person with HbS genes from both parents (a homozygote) is said to have sickle-cell disease, and all of his or her hemoglobin is HbS. Laboratory screening tests to separate HbA and HbS proteins can easily determine whether prospective parents carry the sickle-cell trait.

There is an intriguing relationship between the sickle-cell trait and malaria resistance. In some parts of the world, up to 20% of the population has the sickle-cell trait. The trait produces an increased resistance to one type of malaria because the malarial parasite cannot feed on the hemoglobin in sickled red blood cells. Individuals with sickle-cell disease die young, while those without the sickle-cell trait also have a high probability of dying prematurely, but from malaria. Occupying the middle ground are those with the sickle-cell trait who do not suffer much from the effects of sickle-cell disease and also avoid the ravages of malaria.

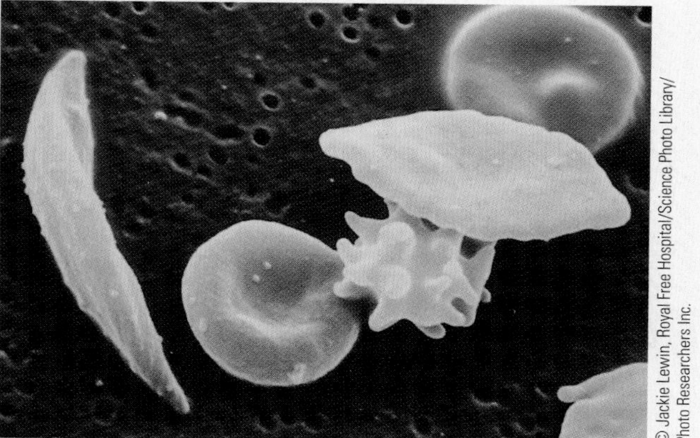

Scanning electron micrograph of normal red blood cells (round shape) in the midst of sickled cells (crescent shape).

same chain or between cysteine residues in different chains. The existence and location of these disulfide bridges is a part of the tertiary structure because these interactions hold the protein chain in a loop or some other three-dimensional shape.

2. **Salt bridges:** These interactions are a result of ionic bonds that form between the ionized side chain of an acidic amino acid ($—COO^-$) and the side chain of a basic amino acid ($—NH_3^+$).
3. **Hydrogen bonds:** Hydrogen bonds can form between a variety of side chains, especially those that possess the following functional groups:

$$—OH \qquad —NH_2 \qquad —\overset{\displaystyle O}{\overset{\|}{C}}—NH_2$$

ThomsonNOW™ Go to Coached Problems to examine the **control of protein structure.**

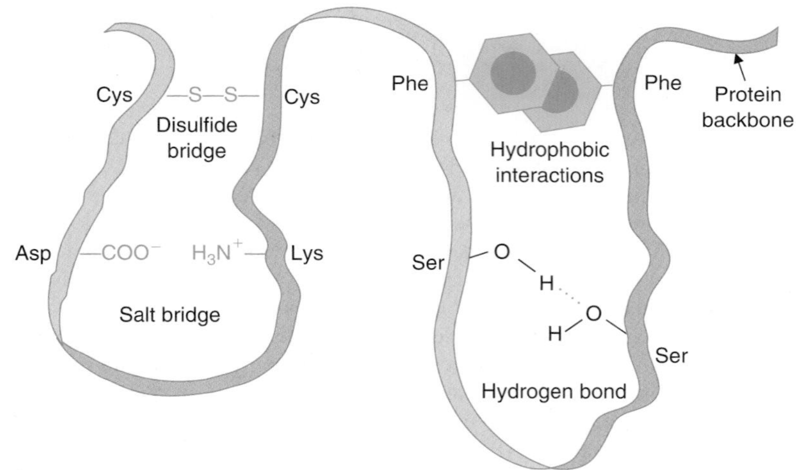

FIGURE 9.10 R-group interactions leading to tertiary protein structure.

One possible interaction, between the —OH groups of two serine residues, is shown in ■ Figure 9.10. We saw in Section 9.7 that hydrogen bonding also determines the secondary structure or proteins. The distinction is that the tertiary hydrogen bonding occurs between R groups, whereas in secondary structures, the hydrogen bonding is between backbone —C=O and —N groups.

4. **Hydrophobic interactions:** These result when nonpolar groups are either attracted to one another or forced together by their mutual repulsion of aqueous solvent. Interactions of this type are common between R groups such as the nonpolar phenyl rings of phenylalanine residues (Figure 9.10). This is illustrated by the compact structure of globular proteins in aqueous solutions (see ■ Figure 9.11). Much like in the fatty acid micelles in Section 8.2, the globular shape results because polar groups are pointed outward toward the aqueous solvent, and nonpolar groups are pointed inward, away from the water molecules. This type of interaction is weaker than the other three, but it usually acts over large surface areas so that the net effect is an interaction strong enough to stabilize the tertiary structure.

The interactions leading to tertiary protein structures are summarized in Figure 9.10.

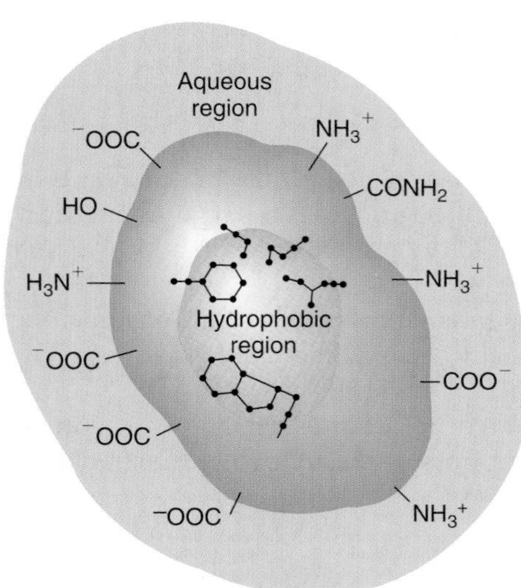

FIGURE 9.11 A globular protein with a hydrophobic region on the inside and polar groups on the outside extending into the aqueous surroundings.

Learning Check 9.3 What kind of R-group interaction might be expected if the following side chains were in close proximity?

a. $-\underset{|}{\overset{CH_3}{}}CHCH_2CH_3$ and $-\underset{|}{\overset{CH_3}{}}CHCH_2CH_3$

b. $-\overset{O}{\overset{\|}{C}}-NH_2$ and $-C_6H_4-OH$

c. $-CH_2-\overset{O}{\overset{\|}{C}}-O^-$ and $-CH_2CH_2CH_2CH_2-NH_3^+$

9.9 The Quaternary Structure of Proteins

LEARNING OBJECTIVE

9. Explain what is meant by the quaternary structure of proteins.

ThomsonNOW Go to Coached Problems to explore the **structural parts of a protein.**

Many functional proteins contain two or more polypeptide chains held together by forces such as ionic attractions, disulfide bridges, hydrogen bonds, and hydrophobic forces. Each of the polypeptide chains, called **subunits,** has its own primary, secondary, and tertiary structure. The arrangement of subunits to form the larger protein is called the **quaternary structure** of the protein.

subunit
A polypeptide chain having primary, secondary, and tertiary structural features that is a part of a larger protein.

quaternary protein structure
The arrangement of subunits that form a larger protein.

Hemoglobin is a well-known example of a protein exhibiting quaternary structure. Hemoglobin is made of four chains (subunits): two identical chains (called alpha) containing 141 amino acid residues each and two other identical chains (beta) containing 146 residues each. Each of these four hemoglobin subunits contains a heme group (a planar ring structure centered around an iron

STUDY SKILLS 9.1 Visualizing Protein Structure

A visual aid to help you "see" protein structure and remember the difference between primary, secondary, and tertiary structure is the cord of a telephone receiver. The chainlike primary structure is represented by the long straight cord. The coiling of the cord into a helical arrangement represents the secondary structure, and the tangled arrangement the cord adopts after the receiver has been used several times is the tertiary structure.

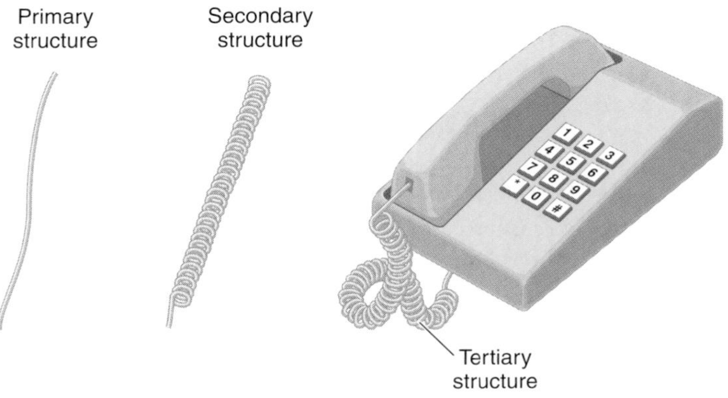

Three levels of structure of a telephone cord.

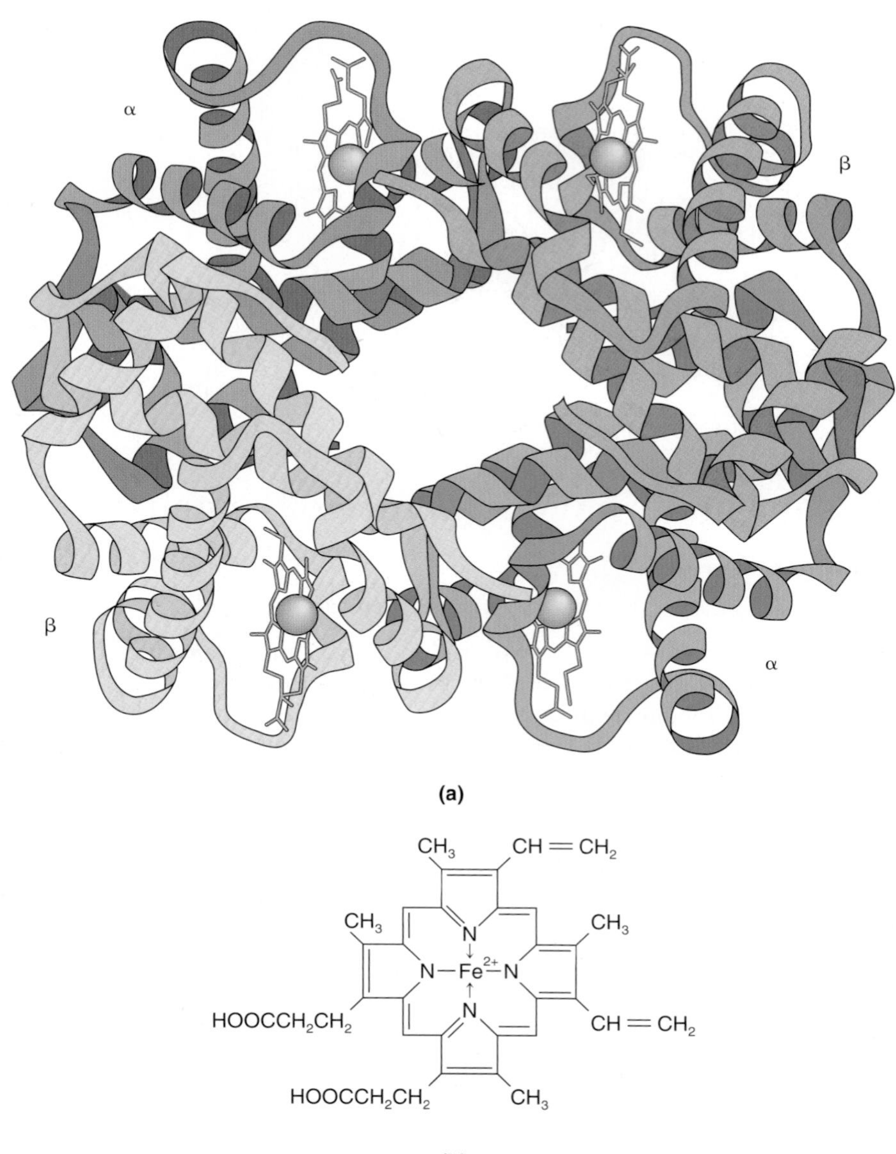

■ **FIGURE 9.12** The structure of hemoglobin. (a) Hemoglobin exhibits quaternary structure with two α-chains and two β-chains. The purple disks are heme groups. (b) A heme group.

atom, shown in ■ Figure 9.12b) located in crevices near the exterior of the molecule. The hemoglobin molecule is nearly spherical with the four subunits held together rather tightly by hydrophobic forces. A schematic drawing of the quaternary structure of hemoglobin is shown in Figure 9.12a.

9.10 Protein Hydrolysis and Denaturation

LEARNING OBJECTIVE

10. Describe the conditions that can cause proteins to hydrolyze or become denatured.

Hydrolysis

We learned in Section 6.8 that amides can be hydrolyzed by aqueous acid or base. The amide linkages of peptides and proteins also show this characteristic. As a result, the heating of a peptide or protein in the presence of acid or base causes it to break into smaller peptides, or even amino acids, depending on the hydrolysis time, temperature, and pH.

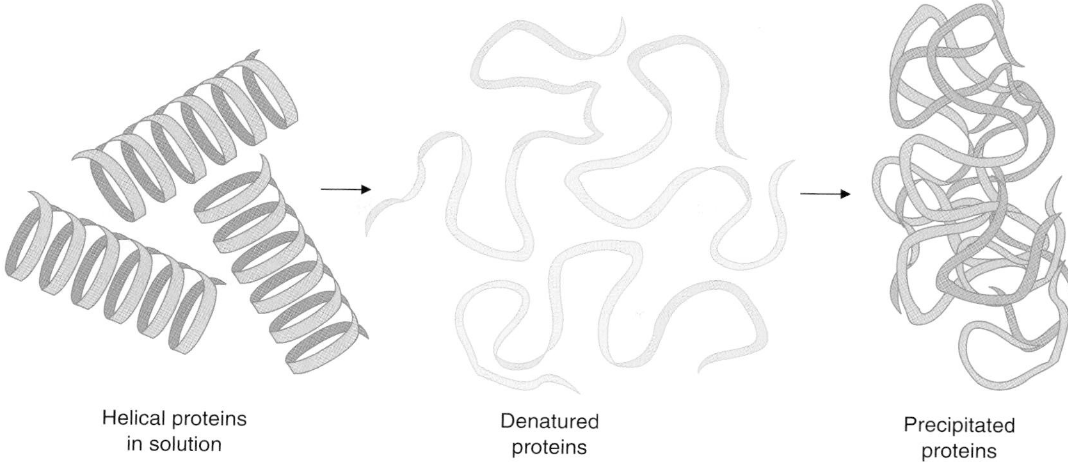

■ **FIGURE 9.13** Protein denaturation and coagulation of the proteins.

$$\text{protein} + H_2O \xrightarrow[OH^-]{H^+ \text{ or}} \text{smaller peptides} \xrightarrow[OH^-]{H^+ \text{ or}} \text{amino acids} \quad (9.9)$$

The process of protein digestion (Section 12.6) involves hydrolysis reactions that are catalyzed by enzymes in the digestive tract. In Chapter 14, we'll also see that the hydrolysis of cellular proteins to amino acids is an ongoing process as the body resynthesizes needed molecules and tissues.

Denaturation

Proteins are maintained in their natural three-dimensional conformation, or **native state,** by stable secondary and tertiary structures and through the aggregation of subunits in quaternary structures. If proteins are exposed to physical or chemical conditions, such as extreme temperatures or pH values, that disrupt these stabilizing structures, the folded native structure breaks down and the protein takes on a random, disorganized conformation (see ■ Figure 9.13). This process, called **denaturation,** causes the protein to become inactive, and it may precipitate because hydrophobic residues are likely to be exposed to the aqueous surroundings. The change of egg white from a clear, jellylike material to a white solid when heated is an example of this process (see ■ Figure 9.14). Denaturation also explains why most proteins are biologically active only over a narrow temperature range, typically 0°C–40°C. Temperatures higher than those normal for a living organism can cause intramolecular vibrations to become so intense that hydrogen bonds or other stabilizing forces are disrupted and denaturation results.

native state
The natural three-dimensional conformation of a functional protein.

denaturation
The process by which a protein loses its characteristic native structure and function.

The correlation between denaturation and loss of biological activity is clearly illustrated by the common methods used to kill microorganisms and deactivate their toxins. Most bacterial toxins, which are proteins, lose their ability to cause disease when they are heated. For example, the toxic protein excreted by *Clostridium botulinum* is responsible for botulism, an often fatal form of food poisoning. This dangerous substance loses its toxic properties after being heated at 100°C for a few minutes. Heating also deactivates the toxins of the bacteria that cause diphtheria and tetanus. Heat denaturation is used to prepare vaccines against these particular diseases. The deactivated (denatured) toxin can no longer cause the symptoms of the disease to appear, but it can stimulate the body to produce substances that induce immunity to the functional toxins. It is well known that high temperatures have been used for many years in hospitals and laboratories to sterilize

■ **FIGURE 9.14** Denaturing the protein in eggs.

TABLE 9.7 Substances and conditions that denature proteins

Substance or condition	Effect on proteins
Heat and ultraviolet light	Disrupt hydrogen bonds and ionic attractions by making molecules vibrate too violently; produce coagulation, as in cooking an egg
Organic solvents (ethanol and others miscible with water)	Disrupt hydrogen bonds in proteins and probably form new ones with the proteins
Strong acids or bases	Disrupt hydrogen bonds and ionic attractions; prolonged exposure results in hydrolysis of protein
Detergents	Disrupt hydrogen bonds, hydrophobic interactions, and ionic attractions
Heavy-metal ions (Hg^{2+}, Ag^{+}, Pb^{2+})	Form bonds to thiol groups and precipitate proteins as insoluble heavy-metal salts

surgical instruments and glassware that must be germ-free. The high temperatures denature and deactivate proteins needed within the germ cells for survival.

Similar changes in protein conformation resulting in denaturation can be brought about by heavy-metal ions such as Hg^{2+}, Ag^{+}, and Pb^{2+} that interact with —SH groups and carboxylate ($—COO^{-}$) groups. Organic materials containing mercury (mercurochrome and merthiolate) were once commonly used as antiseptics in the home to treat minor skin abrasions because they denature proteins in bacteria. Mercury, silver, and lead salts are also poisonous to humans for the same reason: They deactivate critical proteins we need to live. Large doses of raw egg white and milk are often given to victims of heavy-metal poisoning. The proteins in the egg and milk bind to the metal ions and form a precipitate. The metal-containing precipitate is removed from the stomach by pumping or is ejected by induced vomiting. If the precipitated protein is not removed, it will be digested, and the toxic metal ions will be released again.

Some of the various substances and conditions that will denature proteins are summarized in ■ Table 9.7.

Concept Summary

ThomsonNOW™ Sign in at www.thomsonedu.com to:

- Assess your understanding with Exercises keyed to each learning objective.
- Check your readiness for an exam by taking the **Pre-test** and exploring the modules recommended in your **Personalized Learning Plan.**

The Amino Acids. All proteins are polymers of amino acids, which are bifunctional organic compounds that contain both an amino group and a carboxylate group. Differences in the R groups of amino acids cause differences in the properties of amino acids and proteins. **OBJECTIVE 1 (Section 9.1), Exercise 9.2**

Zwitterions. The presence of both amino groups and carboxyl groups in amino acids makes it possible for amino acids to exist in several ionic forms, including the form of a zwitterion. **OBJECTIVE 2 (Section 9.2), Exercise 9.12** The zwitterion is a dipolar form in which the net charge on the ion is zero.

Reactions of Amino Acids. Amino acids can undergo reactions characteristic of any functional group in the molecule. Two important reactions are the reaction of two cysteine molecules to form a disulfide, and the reaction of amino groups and carboxylate groups of different molecules to form peptide (amide) linkages. **OBJECTIVE 3 (Section 9.3), Exercise 9.16**

Important Peptides. More than 200 peptides have been shown to be essential to the proper functioning of the human body. Hormones are among the peptides for which functions have been identified. **OBJECTIVE 4 (Section 9.4), Exercise 9.22**

Characteristics of Proteins. Proteins are large polymers of amino acids. Acidic and basic properties of proteins are determined by the acidic or basic character of the R groups of the amino acids constituting the protein. Proteins perform numerous important functions in the body. Proteins are classified structurally as fibrous or globular. They are classified on the basis of composition as simple or conjugated. **OBJECTIVE 5 (Section 9.5), Exercises 9.30 and 9.32**

The Primary Structure of Proteins. The primary structure of a protein is the sequence of amino acids in the polymeric chain. This gives all proteins an identical backbone of carbon and nitrogen atoms held together by peptide linkages. The difference in proteins is the sequence of R groups attached to the backbone. **OBJECTIVE 6 (Section 9.6), Exercise 9.34**

The Secondary Structure of Proteins. Protein chains are held in characteristic shapes called secondary structures by hydrogen bonds. Two specific structures that have been identified are the α-helix and the β-pleated sheet. **OBJECTIVE 7 (Section 9.7), Exercise 9.38**

The Tertiary Structure of Proteins. A third level of complexity in protein structure results from interactions between the R groups of protein chains. These interactions include disulfide bridges, salt bridges, hydrogen bonds, and hydrophobic attractions. **OBJECTIVE 8 (Section 9.8), Exercise 9.42**

The Quaternary Structure of Proteins. Some functional proteins consist of two or more polypeptide chains held together by forces such as ionic attractions, disulfide bridges, hydrogen bonds, and hydrophobic forces. The arrangement of these polypeptides to form the functional protein is called the quaternary structure of the protein. **OBJECTIVE 9 (Section 9.9), Exercise 9.46**

Protein Hydrolysis and Denaturation. The peptide (amide) linkages of peptides and proteins can be hydrolyzed under appropriate conditions. This destroys the primary structure and produces smaller peptides or amino acids. The characteristic secondary, tertiary, and quaternary structures of proteins can also be disrupted by certain physical or chemical conditions such as extreme temperatures or pH values. The disruption of these structures is called denaturation and causes the protein to become nonfunctional and, in some cases, to precipitate. **OBJECTIVE 10 (Section 9.10), Exercise 9.50**

Key Terms and Concepts

Alpha-amino acid (9.1)
α-helix (9.7)
Amino acid residue (9.3)
Antibody (9.5)
β-pleated sheet (9.7)
Conjugated protein (9.5)
C-terminal residue (9.3)
Denaturation (9.10)
Dipeptide (9.3)
Disulfide bridge (9.4)
Fibrous protein (9.5)
Globular protein (9.5)
Isoelectric point (9.2)
Native state (9.10)
N-terminal residue (9.3)
Peptide (9.3)
Peptide linkage (peptide bond) (9.3)
Polypeptide (9.3)
Primary protein structure (9.6)
Prosthetic group (9.5)
Protein (9.3)
Quaternary protein structure (9.9)
Secondary protein structure (9.7)
Simple protein (9.5)
Subunit (9.9)
Tertiary protein structure (9.8)
Zwitterion (9.2)

Key Reactions

1. Formation of a zwitterion (Section 9.2):

$$H_2N-\underset{\displaystyle R}{\underset{|}{CH}}-\overset{\displaystyle O}{\overset{||}{C}}-OH \rightarrow H_3\overset{+}{N}-\underset{\displaystyle R}{\underset{|}{CH}}-\overset{\displaystyle O}{\overset{||}{C}}-O^- \qquad \text{Reaction 9.1}$$

2. Conversion of a zwitterion to a cation in an acidic solution (Section 9.2):

$$H_3\overset{+}{N}-\underset{\displaystyle R}{\underset{|}{CH}}-\overset{\displaystyle O}{\overset{||}{C}}-O^- + H_3O^+ \rightarrow H_3\overset{+}{N}-\underset{\displaystyle R}{\underset{|}{CH}}-\overset{\displaystyle O}{\overset{||}{C}}-OH + H_2O \qquad \text{Reaction 9.2}$$

3. Conversion of a zwitterion to an anion in a basic solution (Section 9.2):

$$H_3\overset{+}{N}-\underset{\displaystyle R}{\underset{|}{CH}}-\overset{\displaystyle O}{\overset{||}{C}}-O^- + OH^- \rightarrow H_2N-\underset{\displaystyle R}{\underset{|}{CH}}-\overset{\displaystyle O}{\overset{||}{C}}-O^- + H_2O \qquad \text{Reaction 9.3}$$

4. Oxidation of cysteine to cystine (Section 9.3):

$$\begin{array}{c} \overset{+}{H_3N}-CH-\overset{O}{\overset{\|}{C}}-O^- \\ | \\ CH_2 \\ | \\ SH \\ SH \\ | \\ CH_2 \\ | \\ \overset{+}{H_3N}-CH-\overset{O}{\overset{\|}{C}}-O^- \end{array} + (O) \xrightarrow{\text{Oxidation}} \begin{array}{c} \overset{+}{H_3N}-CH-\overset{O}{\overset{\|}{C}}-O^- \\ | \\ CH_2 \\ | \\ S \\ | \\ S \\ | \\ CH_2 \\ | \\ \overset{+}{H_3N}-CH-\overset{O}{\overset{\|}{C}}-O^- \end{array} + H_2O \qquad \text{Reaction 9.4}$$

5. Formation of a peptide linkage—general reaction (Section 9.3):

$$\overset{+}{H_3N}-\underset{R}{\underset{|}{CH}}-\overset{O}{\overset{\|}{C}}-O^- + \overset{+}{H_3N}-\underset{R'}{\underset{|}{CH}}-\overset{O}{\overset{\|}{C}}-O^- \rightarrow \overset{+}{H_3N}-\underset{R}{\underset{|}{CH}}-\overset{O}{\overset{\|}{C}}-NH-\underset{R'}{\underset{|}{CH}}-\overset{O}{\overset{\|}{C}}-O^- + H_2O \qquad \text{Reaction 9.6}$$

6. Hydrolysis of proteins in acid or base (Section 9.10):

$$\text{protein} + H_2O \xrightarrow[OH^-]{H^+ \text{ or}} \text{smaller peptides} \xrightarrow[OH^-]{H^+ \text{ or}} \text{amino acids} \qquad \text{Reaction 9.9}$$

Exercises

SYMBOL KEY

Even-numbered exercises are answered in Appendix B.

Blue-numbered exercises are more challenging.

■ denotes exercises available in ThomsonNow and assignable in OWL.

ThomsonNOW To assess your understanding of this chapter's topics with sample tests and other resources, sign in at www.thomsonedu.com.

THE AMINO ACIDS (SECTION 9.1)

9.1 Draw the structure of hexanoic acid. Label the alpha carbon.

9.2 What functional groups are found in all amino acids?

9.3 ■ Identify the R group of the side chain in the following amino acids that results in the side-chain classification indicated in parentheses (see Table 9.1):

a. tyrosine (neutral, polar)
b. glutamate (acidic, polar)
c. methionine (neutral, nonpolar)
d. histidine (basic, polar)
e. cysteine (neutral, polar)
f. valine (neutral, nonpolar)

9.4 ■ Draw structural formulas for the following amino acids, identify the chiral carbon atom in each one, and circle the four different groups attached to the chiral carbon.

a. threonine
b. aspartate
c. serine
d. phenylalanine

9.5 Draw structural formulas for the following amino acids, identify the chiral carbon atom in each one, and circle the four different groups attached to the chiral carbon.

a. alanine
b. tryptophan
c. histidine
d. glutamate

9.6 Isoleucine contains two chiral carbon atoms. Draw the structural formula for isoleucine twice. In the first, identify one of the chiral carbons and circle the four groups attached to it. In the second, identify the other chiral carbon and circle the four groups attached to it.

9.7 ■ Draw Fischer projections representing the D and L forms of the following:

a. aspartate
b. phenylalanine

9.8 Draw Fischer projections representing the D and L forms of the following:

a. cysteine
b. glutamate

ZWITTERIONS (SECTION 9.2)

9.9 What is meant by the term *zwitterion?*

9.10 What characteristics indicate that amino acids exist as zwitterions?

9.11 What is meant by the term *isoelectric point?*

9.12 ■ Write structural formulas to show the form the following amino acids would have in a solution with a pH higher than the amino acid isoelectric point:

a. cysteine

b. alanine

9.13 Write structural formulas to show the form the following amino acids would have in a solution with a pH lower than the amino acid isoelectric point:

a. phenylalanine b. leucine

9.14 Write ionic equations to show how serine acts as a buffer against the following added ions:

a. OH^- b. H_3O^+

REACTIONS OF AMINO ACIDS (SECTION 9.3)

9.15 Write two reactions to represent the formation of the two dipeptides that form when valine and serine react.

9.16 Write a complete structural formula and an abbreviated formula for the tripeptide formed from aspartate, cysteine, and valine in which the C-terminal residue is cysteine and the N-terminal residue is valine.

9.17 Write a complete structural formula and an abbreviated formula for the tripeptide formed from tyrosine, serine, and phenylalanine in which the C-terminal residue is serine and the N-terminal residue is phenylalanine.

9.18 Write abbreviated formulas for the six isomeric tripeptides of alanine, phenylalanine, and arginine.

9.19 ■ Write abbreviated formulas for the six isomeric tripeptides of asparagine, glutamate, and proline.

9.20 How many tripeptide isomers that contain one residue each of valine, phenylalanine, and lysine are possible?

9.21 ■ How many tripeptide isomers that contain one glycine residue and two alanine residues are possible?

IMPORTANT PEPTIDES (SECTION 9.4)

9.22 What special role does the amino acid cysteine have in the peptides vasopressin and oxytocin?

9.23 Summarize the physiological effects of oxytocin and vasopressin.

CHARACTERISTICS OF PROTEINS (SECTION 9.5)

9.24 Explain why the presence of certain proteins in body fluids such as urine or blood can indicate that cellular damage has occurred in the body.

9.25 Write equations to show how a protein containing both lysine and aspartate could act as a buffer against the following added ions:

a. OH^- b. H_3O^+

9.26 Explain why a protein is least soluble in an aqueous medium that has a pH equal to the isoelectric point of the protein.

9.27 Wool is a protein used as a textile fiber. Classify the protein of wool into a category of Table 9.4.

9.28 List the eight principal functions of proteins.

9.29 ■ Classify each of the following proteins into one of the eight functional categories of proteins:

a. insulin

b. collagen

c. hemoglobin

d. cytochrome c

e. α-keratin

f. serum albumin

9.30 For each of the following two proteins listed in Table 9.4, predict whether it is more likely to be a globular or a fibrous protein. Explain your reasoning.

a. collagen

b. lactate dehydrogenase

9.31 For each of the following two proteins listed in Table 9.4, predict whether it is more likely to be a globular or a fibrous protein. Explain your reasoning.

a. elastin

b. ceruloplasmin

9.32 Differentiate between simple and conjugated proteins.

9.33 Give one example of a conjugated protein that contains the following prosthetic group:

a. iron

b. lipid

c. phosphate group

d. carbohydrate

e. heme

THE PRIMARY STRUCTURE OF PROTEINS (SECTION 9.6)

9.34 Describe what is meant by the term *primary structure of proteins.*

9.35 ■ What type of bonding is present to account for the primary structure of proteins?

9.36 Write the structure for a protein backbone. Make the backbone long enough to attach four R groups symbolizing amino acid side chains.

THE SECONDARY STRUCTURE OF PROTEINS (SECTION 9.7)

9.37 ■ Describe the differences between alpha and beta secondary structures of proteins.

9.38 What type of bonding between amino acid residues is most important in holding a protein or polypeptide in a specific secondary configuration?

9.39 Describe what is meant by "supracoiling" in proteins.

THE TERTIARY STRUCTURE OF PROTEINS (SECTION 9.8)

9.40 How do hydrogen bonds involved in tertiary protein structures differ from those involved in secondary structures?

9.41 Which amino acids have side-chain groups that can form salt bridges?

9.42 ■ Refer to Table 9.1 and list the type of side-chain interaction expected between the side chains of the following pairs of amino acid residues.

a. tyrosine and glutamine

b. aspartate and lysine

c. leucine and isoleucine

d. phenylalanine and valine

9.43 ■ Which amino acids have side-chain groups that can participate in hydrogen bonding?

9.44 ■ A globular protein in aqueous surroundings contains the following amino acid residues: phenylalanine, methionine, glutamate, lysine, and alanine. Which amino acid side chains would be directed toward the inside of the protein and which would be directed toward the aqueous surroundings?

THE QUATERNARY STRUCTURE OF PROTEINS (SECTION 9.9)

9.45 Will all proteins have a quaternary structure? Why or why not?

9.46 What types of forces give rise to quaternary structure?

9.47 Describe the quaternary protein structure of hemoglobin.

9.48 What is meant by the term *subunit?*

PROTEIN HYDROLYSIS AND DENATURATION (SECTION 9.10)

9.49 ■ Is a protein undergoing hydrolysis or denaturation if its peptide linkages are being broken?

9.50 Suppose a sample of a protein is completely hydrolyzed and another sample of the same protein is denatured. Compare the final products of each process.

9.51 As fish is cooked, the tissue changes from a soft consistency to a firm one. A similar change takes place when fish is pickled by soaking it in vinegar (acetic acid). In fact, some people prepare fish for eating by pickling instead of cooking. Propose an explanation of why the two processes give somewhat similar results.

9.52 In what way is the protein in a raw egg the same as that in a cooked egg?

9.53 In addition to emulsifying greasy materials, what other useful function might be served by a detergent used to wash dishes?

9.54 Once cooked, egg whites remain in a solid form. However, egg whites that are beaten to form meringue will partially change back to a jellylike form if allowed to stand for a while. Explain these behaviors using the concept of reversible protein denaturation.

9.55 Lead is a toxic material. What form of the lead is actually the harmful substance?

9.56 Explain how egg white can serve as an emergency antidote for heavy-metal poisoning.

ADDITIONAL EXERCISES

9.57 A protein has a molecular weight of about 12,000 u. This protein contains 6.4% (w/w) sulfur in the form of cystine (the cysteine disulfide). How many cystine units are present in the protein?

9.58 ■ Which amino acid could be referred to as 2-amino-3-methylbutanoic acid?

9.59 A Tyndall effect is observed as light is scattered as it passes through a small polypeptide–water mixture with a pH of 6.7. What can be deduced about the polypeptide's isoelectric point from this information?

9.60 ■ What is the conjugate acid for the following zwitterion? What is the conjugate base for it?

$$H_3\overset{+}{N}-\underset{\substack{|\\ R}}{\overset{\substack{H\\ |}}{C}}-COO^-$$

9.61 The K_a values recorded for alanine are 5.0×10^{-3} and 2.0×10^{-10}. Why are two values recorded? What functional groups do the values correspond to? How many K_a values would be recorded for glutamate?

ALLIED HEALTH EXAM CONNECTION

Reprinted with permission from Nursing School and Allied Health Entrance Exams, COPYRIGHT 2005 Petersons.

9.62 _______ stimulates the smooth muscle of the uterine wall during the labor and delivery process. After delivery, it promotes the ejection of milk.

9.63 What functional groups are found in all amino acids? How many differenct amino acids are found in naturally occurring proteins?

9.64 Which of the following are true concerning the chemical bond that forms between the carboxyl (RCOOH) group of one amino acid and the amino ($RC—NH_2$) group of another?

a. The bond is called a peptide bond.

b. It is formed by inserting a water molecule between them.

c. It is formed by a dehydration reaction.

d. A polypeptide has more of these bonds than a protein.

9.65 Rank the following components of hemoglobin in decreasing amounts that they are found in a hemoglobin molecule:

a. Iron atoms

b. Amino acids

c. Heme groups

9.66 Describe the quaternary protein structure of hemoglobin.

9.67 Complete degradation of a protein into individual amino acids involves (choose all that are correct):

a. Removal of a water molecule from between two amino acids

b. Addition of a water molecule between two amino acids

c. A hydrolysis reaction

d. The breaking of peptide linkage

CHEMISTRY FOR THOUGHT

9.68 Some researchers feel that the "high" experienced by runners may be due to the brain's synthesis of peptides. Explain why this theory might be a reasonable explanation.

9.69 Suggest a reason why individuals with the sickle-cell trait might sometimes encounter breathing problems on high-altitude flights.

9.70 In Section 9.2, you learned that amino acids like alanine are crystalline solids with high melting points. However, the ethyl ester of alanine, which is a free NH_2 group, melts at only 87°C. Explain why the ethyl ester melts 200 degrees below the melting point of alanine.

9.71 Why do health care workers wipe a patient's skin with solutions of rubbing alcohol before giving injections?

9.72 Would you expect plasma proteins to be fibrous or globular? Explain.

9.73 Why must the protein drug insulin be given by injection rather than taken by mouth?

9.74 In Figure 9.4, it is noted that animal hair is made of fibrous proteins. What properties of fibrous proteins make hair, fur, and spiderwebs useful?

9.75 Why is it necessary that protein molecules be enormous?

9.76 Name two elements that are found in proteins but are not present in fats, oils, or carbohydrates.

CHAPTER 10

Enzymes

LEARNING OBJECTIVES

When you have completed your study of this chapter, you should be able to:

1. Describe the general characteristics of enzymes and explain why enzymes are vital in body chemistry. (Section 10.1)
2. Determine the function and/or substrate of an enzyme on the basis of its name. (Section 10.2)
3. Identify the general function of cofactors. (Section 10.3)
4. Use the lock-and-key theory to explain specificity in enzyme action. (Section 10.4)
5. List two ways of describing enzyme activity. (Section 10.5)
6. Identify the factors that affect enzyme activity. (Section 10.6)
7. Compare the mechanisms of competitive and noncompetitive enzyme inhibition. (Section 10.7)
8. Describe the three methods of cellular control over enzyme activity. (Section 10.8)
9. Discuss the importance of measuring enzyme levels in the diagnosis of disease. (Section 10.9)

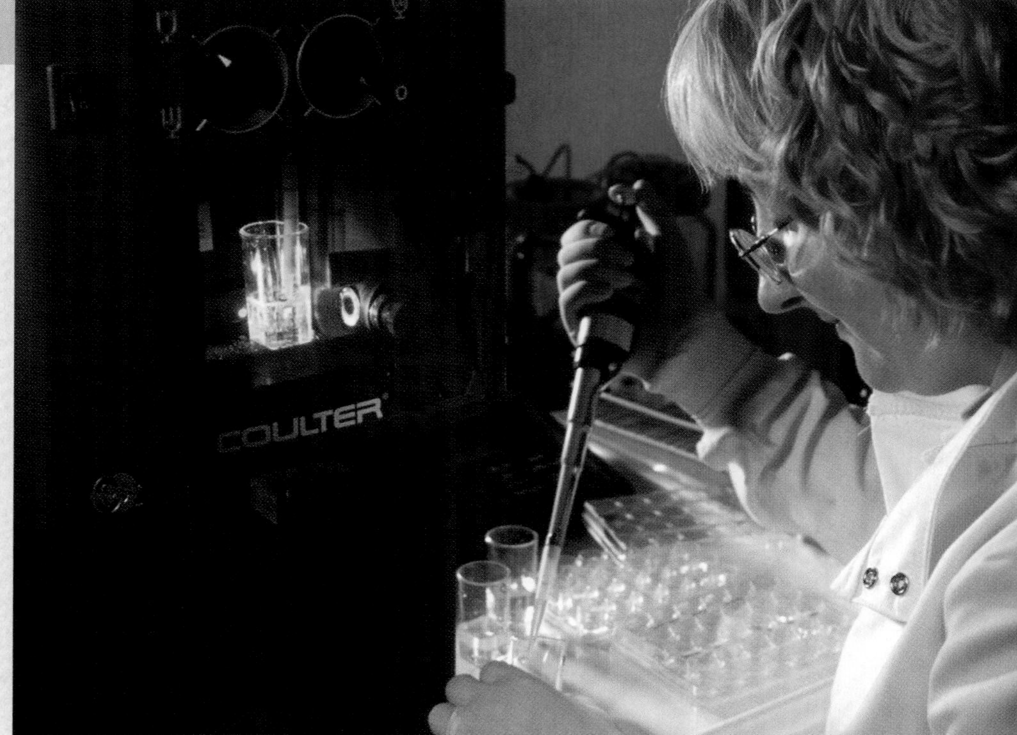

Clinical laboratory scientists often use sophisticated automated equipment to perform tests that can detect the failure or abnormal behavior of most organs of the body. The tests can identify such problems as diabetes, heart attacks, and liver and kidney damage. Many of the tests involve enzymes, the topic of this chapter.

The catalytic behavior of proteins acting as enzymes is one of the most important functions performed by cellular proteins. Without catalysts, most cellular reactions would take place much too slowly to support life (see ■ Figure 10.1). With the exception of a few recently discovered RNA molecules (discussed in Chapter 11) that catalyze their own reactions, all enzymes are globular proteins. Enzymes are the most efficient of all known catalysts; some can increase reaction rates by 10^{20} times that of uncatalyzed reactions.

ThomsonNOW™ Throughout the chapter this icon introduces resources on the ThomsonNOW website for this text. Sign in at **www.thomsonedu.com** to:

- Evaluate your knowledge of the material
- Take an exam prep quiz
- Identify areas you need to study with a **Personalized Learning Plan**.

10.1 General Characteristics of Enzymes

LEARNING OBJECTIVE

1. Describe the general characteristics of enzymes and explain why enzymes are vital in body chemistry.

Enzymes are well suited to their essential roles in living organisms in three major ways: They have enormous catalytic power, they are highly specific in the reactions they catalyze, and their activity as catalysts can be regulated.

enzyme
A biomolecule that catalyzes chemical reactions.

Catalytic Efficiency

As we know, catalysts increase the rate of chemical reactions but are not used up in the process. Although a catalyst actually participates in a chemical reaction, it is not permanently modified and may be used again and again. Enzymes are true catalysts that speed chemical reactions by lowering activation energies (see ■ Figure 10.2) and allowing reactions to achieve equilibrium more rapidly.

Many of the important enzyme-catalyzed reactions are similar to the reactions we studied in the chapters on organic chemistry: ester hydrolysis, alcohol oxidation, and so on. However, laboratory conditions cannot match what happens when these reactions are carried out in the body; enzymes cause such reactions to proceed under mild pH and temperature conditions. In addition, enzyme catalysis within the body can accomplish in seconds reactions that ordinarily take weeks or even months under laboratory conditions.

The influence of enzymes on the rates of reactions essential to life is truly amazing. A good example is the need to move carbon dioxide, a waste product of cellular respiration, out of the body. Before this can be accomplished, the carbon dioxide must be combined with water to form carbonic acid:

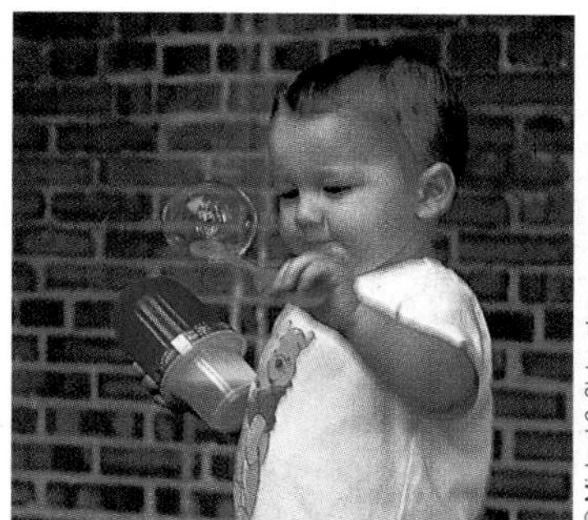

■ **FIGURE 10.1** One miracle of life is that hundreds of chemical reactions take place simultaneously in each living cell; the activity of enzymes in the cells makes it all possible.

$$CO_2 + H_2O \xrightarrow{\text{carbonic anhydrase}} H_2CO_3 \tag{10.1}$$

In the absence of the appropriate enzyme, carbonic anhydrase, the formation of carbonic acid takes place much too slowly to support the required exchange of carbon dioxide between the blood and the lungs. But in the presence of carbonic anhydrase, this vital reaction proceeds rapidly. Each molecule of the enzyme can catalyze the formation of carbonic acid at the rate of 36 million molecules per minute.

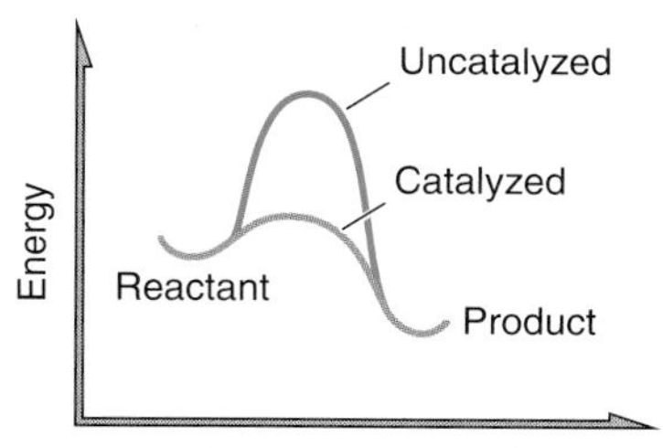

■ **FIGURE 10.2** An energy diagram for a reaction when uncatalyzed and catalyzed.

Specificity

Enzyme specificity is a second characteristic that is important in life processes. Unlike other catalysts, enzymes are often quite specific in the type of reaction they catalyze and even the particular substance that will be involved in the reaction. For example, strong acids catalyze the hydrolysis of any amide, the dehydration of any

FIGURE 10.3 The hydrolysis of urea catalyzed by urease.

The enzyme urease is added to a urea solution containing the indicator phenolphthalein.

As urea is hydrolyzed, forming ammonia, the indicator turns pink. How else might the presence of ammonia be detected?

alcohol, and a variety of other processes. However, the enzyme urease catalyzes only the hydrolysis of a single amide, urea (Reaction 10.2 and Figure 10.3):

$$H_2N-\overset{\overset{\displaystyle O}{\|}}{C}-NH_2 + H_2O \xrightarrow{\text{urease}} CO_2 + 2NH_3 \qquad (10.2)$$

absolute specificity
The characteristic of an enzyme that it acts on one and only one substance.

relative specificity
The characteristic of an enzyme that it acts on several structurally related substances.

stereochemical specificity
The characteristic of an enzyme that it is able to distinguish between stereoisomers.

This is an example of **absolute specificity,** the catalyzing of the reaction of one and only one substance. Other enzymes display **relative specificity** by catalyzing the reaction of structurally related substances. For example, the lipases hydrolyze lipids, proteases split up proteins, and phosphatases hydrolyze phosphate esters. **Stereochemical specificity** extends even to enantiomers: D-amino acid oxidase will not catalyze the reactions of L-amino acids.

Regulation

A third significant property of enzymes is that their catalytic behavior can be regulated. Even though each living cell contains thousands of different molecules that could react with each other in an almost unlimited number of ways, only a relatively small number of these possible reactions take place because of the enzymes present. The cell controls the rates of these reactions and the amount of any given product formed by regulating the action of the enzymes.

10.2 Enzyme Nomenclature and Classification

LEARNING OBJECTIVE

2. Determine the function and/or substrate of an enzyme on the basis of its name.

Some of the earliest discovered enzymes were given names ending with *-in* to indicate their protein composition. For example, three of the digestive enzymes that catalyze protein hydrolysis are named pepsin, trypsin, and chymotrypsin.

substrate
The substance that undergoes a chemical change catalyzed by an enzyme.

The large number of enzymes now known has made it desirable to adopt a systematic nomenclature system known as the Enzyme Commission (EC) system. Enzymes are grouped into six major classes on the basis of the reaction catalyzed (see Table 10.1). In the EC system, each enzyme has an unambiguous (and often long) systematic name that specifies the **substrate** (substance acted on), the func-

TABLE 10.1 The EC classification of enzymes

Group name	Type of reaction catalyzed
Oxidoreductases	Oxidation–reduction reactions
Transferases	Transfer of functional groups
Hydrolases	Hydrolysis reactions
Lyases	Addition to double bonds or the reverse of that reaction
Isomerases	Isomerization reactions
Ligases	Formation of bonds with ATP cleavage[a]

[a]ATP is discussed in Chapter 22.

tional group acted on, and the type of reaction catalyzed. All EC names end in *-ase*. The hydrolysis of urea provides a typical example:

$$H_2N-\overset{\overset{\displaystyle O}{\|}}{C}-NH_2 + H_2O \xrightarrow{\text{enzyme}} CO_2 + 2NH_3 \qquad (10.3)$$

IEC name: urea amidohydrolase
Substrate: urea
Functional group: amide
Type of reaction: hydrolysis

Enzymes are also assigned common names, which are usually shorter and more convenient. Common names are derived by adding *-ase* to the name of the substrate or to a combination of substrate name and type of reaction. For example, urea amidohydrolase is assigned the common name urease:

Substrate: urea
Common name: urea + ase = urease

The enzyme name *alcohol dehydrogenase* is an example of a common name derived from both the name of the substrate and the type of reaction:

Substrate: alcohol (ethyl alcohol)
Reaction type: dehydrogenation (removal of hydrogen)
Common name: alcohol dehydrogenation + ase = alcohol dehydrogenase

Throughout this text, we will refer to most enzymes by their accepted common names.

Learning Check 10.1 Predict the substrates for the following enzymes:

a. maltase
b. peptidase
c. glucose 6-phosphate isomerase

10.3 Enzyme Cofactors

LEARNING OBJECTIVE

3. Identify the general function of cofactors.

In Section 9.5, we learned that proteins may be simple (containing only amino acid residues) or conjugated with a prosthetic group present. Many enzymes are simple proteins, whereas many others function only in the presence of specific nonprotein molecules or metal ions. If these nonprotein components are

CHEMISTRY AND YOUR HEALTH 10.1 Enzymes and Disease

A number of hereditary diseases appear to result from the absence of enzymes or from the presence of altered enzymes. These diseases are often referred to as **inborn errors of metabolism** (the term *metabolism* refers to all reactions occurring within a living organism).

An example of an inborn error of lipid metabolism is Gaucher's disease, characterized by symptoms that include an enlarged spleen and an enlarged liver. A large amount of glucocerebroside is found in fat deposits in these organs. Normally, these organs contain β-glucosidase, an enzyme that catalyzes the degradation of glucocerebrosides (Section 18.7), so that only small amounts of these lipids are normally present. However, the enzyme activity is impaired in Gaucher's disease, and the glucocerebrosides accumulate. Most people with the adult form of the disease do not require treatment; however, the infantile form produces mental retardation and often death.

The following table lists some of the diseases caused by enzyme deficiencies.

Disease	Deficient enzyme
Albinism	tyrosinase
Galactosemia	galactose 1-phosphate uridyltransferase
Gaucher's disease	β-glucosidase
Homocystinuria	cystathionine synthetase
Maple syrup urine disease	amino acid decarboxylase
Methemoglobinemia	methemoglobin reductase
Niemann-Pick disease	sphingomyelinase
Phenylketonuria (PKU)	phenylalanine hydroxylase
Tay-Sachs disease	hexosaminidase A

inborn error of metabolism
A disease in which a genetic change causes a deficiency of a particular protein, often an enzyme.

cofactor
A nonprotein molecule or ion required by an enzyme for catalytic activity.

coenzyme
An organic molecule required by an enzyme for catalytic activity.

apoenzyme
A catalytically inactive protein formed by removal of the cofactor from an active enzyme.

tightly bound to and form an integral part of the enzyme structure, they are true prosthetic groups. Often, however, a nonprotein component is only weakly bound to the enzyme and is easily separated from the protein structure. This type of nonprotein component is referred to as a **cofactor.** When the cofactor is an organic substance, it is called a **coenzyme.** The cofactor may also be an inorganic ion (usually a metal ion). The protein portion of enzymes requiring a cofactor is called the **apoenzyme.** Thus, the combination of an apoenzyme and a cofactor produces an active enzyme:

$$\text{apoenzyme} + \text{cofactor (coenzyme or inorganic ion)} \rightarrow \text{active enzyme} \qquad (10.4)$$

Typical inorganic ions are metal ions such as Mg^{2+}, Zn^{2+}, and Fe^{2+}. For example, the enzyme carbonic anhydrase functions only when Zn^{2+} is present, and rennin needs Ca^{2+} in order to curdle milk. Numerous other metal ions are essential for proper enzyme function in humans; hence, they are required for good health.

Like metal ions, the small organic molecules that act as coenzymes bind reversibly to an enzyme and are essential for its activity. An interesting feature of coenzymes is that many of them are formed in the body from vitamins (see ■ Table 10.2), which explains why it is necessary to have certain vitamins in the diet for good health. For example, the coenzyme nicotinamide adenine dinucleotide (NAD^+), which is a necessary part of some enzyme-catalyzed oxidation–reduction reactions, is formed from the vitamin precursor nicotinamide. Reaction 10.5 shows the participation of NAD^+ in the oxidation of lactate by the enzyme lactate dehydrogenase (LDH). Like other cofactors, NAD^+ is written separately from the enzyme so that the change in its structure may be shown and to emphasize that the enzyme and cofactor are easily separated.

$$\underset{\text{lactate}}{CH_3-\overset{\overset{\displaystyle OH}{|}}{CH}-COO^-} + NAD^+ \xrightleftharpoons[]{\text{lactate dehydrogenase}} \underset{\text{pyruvate}}{CH_3-\overset{\overset{\displaystyle O}{\|}}{C}-COO^-} + NADH + H^+ \qquad (10.5)$$

We can see from the reaction that the coenzyme NAD^+ is essential because it is the actual oxidizing agent. NAD^+ is typical of coenzymes that often aid in the

TABLE 10.2 Vitamins and their coenzyme forms

Vitamin	Coenzyme form	Function
biotin	biocytin	Carboxyl group removal or transfer
folacin	tetrahydrofolic acid	One-carbon group transfer
lipoic acid	lipoamide	Acyl group transfer
niacin	nicotinamide adenine dinucleotide (NAD^+)	Hydrogen transfer
	nicotinamide adenine dinucleotide phosphate ($NADP^+$)	Hydrogen transfer
pantothenic acid	coenzyme A (CoA)	Acyl group carrier
pyridoxal, pyridoxamine, pyridoxine (B_6 group)	pyridoxal phosphate	Amino group transfer
riboflavin	flavin mononucleotide (FMN) flavin adenine dinucleotide (FAD)	Hydrogen transfer Hydrogen transfer
thiamin (B_1)	thiamin pyrophosphate (TPP)	Aldehyde group transfer
vitamin B_{12}	coenzyme B_{12}	Shift of hydrogen atoms between adjacent carbon atoms; methyl group transfer

transfer of chemical groups from one compound to another. In Reaction 10.5, NAD^+ accepts hydrogen from lactate and will transfer it to other compounds in subsequent reactions. In Chapter 12, we will discuss in greater detail several coenzymes that act as shuttle systems in the exchange of chemical substances among various biochemical pathways.

10.4 The Mechanism of Enzyme Action

LEARNING OBJECTIVE

4. Use the lock-and-key theory to explain specificity in enzyme action.

Although enzymes differ widely in structure and specificity, a general theory has been proposed to account for their catalytic behavior. Because enzyme molecules are large compared with the molecules whose reactions they catalyze (substrate molecules), it has been proposed that substrate and enzyme molecules come in contact and interact over only a small region of the enzyme surface. This region of interaction is called the **active site.**

active site
The location on an enzyme where a substrate is bound and catalysis occurs.

The binding of a substrate molecule to the active site of an enzyme may occur through hydrophobic attraction, hydrogen bonding, and/or ionic bonding. The complex formed when a substrate and an enzyme bond is called the enzyme–substrate (ES) complex. Once this complex is formed, the conversion of substrate (S) to product (P) may take place:

General reaction:

$$\underset{\text{enzyme}}{E} + \underset{\text{substrate}}{S} \rightleftarrows \underset{\text{enzyme–substrate complex}}{ES} \longrightarrow \underset{\text{enzyme}}{E} + \underset{\text{product}}{P} \qquad (10.6)$$

ThomsonNOW™ Go to Coached Problems to explore the **induced fit model.**

Specific example:

$$\underset{\text{enzyme}}{\text{Sucrase}} + \underset{\text{substrate}}{\text{sucrose}} \rightleftarrows \left\{\begin{matrix}\text{sucrase–}\\ \text{sucrose}\\ \text{complex}\end{matrix}\right\} \longrightarrow \underset{\text{enzyme}}{\text{sucrase}} + \underbrace{\text{glucose} + \text{fructose}}_{\text{products}} \quad (10.7)$$

The chemical transformation of the substrate occurs at the active site, usually aided by enzyme functional groups that participate directly in the making and breaking of chemical bonds. After chemical conversion has occurred, the product is released from the active site, and the enzyme is free for another round of catalysis.

lock-and-key theory A theory of enzyme specificity proposing that a substrate has a shape fitting that of the enzyme's active site, as a key fits a lock.

An extension of the theory of ES complex formation is used to explain the high specificity of enzyme activity. According to the **lock-and-key theory,** enzyme surfaces will accommodate only those substrates having specific shapes and sizes. Thus, only specific substrates "fit" a given enzyme and can form complexes with it (see ■ Figure 10.4a).

induced-fit theory A theory of enzyme action proposing that the conformation of an enzyme changes to accommodate an incoming substrate.

A limitation of the lock-and-key theory is the implication that enzyme conformations are fixed or rigid. However, research results suggest that the active sites of some enzymes are not rigid. This is taken into account in a modification of the lock-and-key theory known as the **induced-fit theory,** which proposes that enzymes have somewhat flexible conformations that may adapt to incoming substrates. The active site has a shape that becomes complementary to that of the substrate only after the substrate is bound (see Figure 10.4b).

OVER THE COUNTER 10.1 Are All Vitamin Brands Created Equal?

Most people realize that a flashy label on a product doesn't verify quality, but how many pay attention to the words and symbols on the label? Some do, but chances are most people might be mining information in the form of a symbol called a seal of approval. Seals of approval do not come from the FDA, but from independent companies that test products for quality. This testing is especially important for vitamins and herbals that were classified by the FDA as foods or supplements before 1994. The Dietary Supplement Health and Education Act states that a company with a product containing a new dietary ingredient not present as a food or supplement before 1994 must provide the FDA with evidence that the product is safe. A loophole in this law is that no evidence of product safety is required for products available in some form before 1994.

To complicate matters, many consumers are not really sure of what they are searching for in a "good" vitamin or supplement product. Significant help is available in the form of seals of approval from several independent organizations that require and/or perform testing to verify product quality and safety. Four organizations that provide this service for foods and supplements are the National Sanitation Foundation (NSF) International, the United States Pharmacopeia (USP), the ConsumerLab.com (CL), and Good Housekeeping (GH).

The NSF, founded in 1944, requires manufacturers to request and pay for product testing. NSF purchases the test product from a retail outlet and performs the tests in NSF laboratories. The tests are concerned with quality and quantity of all listed ingredients, the presence of any unlisted ingredients, and the consistency and accuracy between different batches. The product must pass twice yearly to receive the seal of approval.

The oldest association is USP, which was established in 1920 but did not take part in quality assurance testing until 2001. Similar to NSF, the USP tests for good manufacturing practices, purity, and quantity standards. It also assures that a supplement will dissolve after ingestion. The USP accepts requests for free product testing and uses outside laboratories to complete the process. The test must be passed early to keep the USP Verified Seal.

The third seal, the CL was established in 1999 and varies from both the NSF and the USP. The organization tests quantity, purity, and availability by randomly selecting products. If a product passes, it may print the CL seal of approval on the label. If a product fails, the manufacturer can choose to use CL consulting services to improve its product. CL requires fees for any testing and a fee to put the seal of approval on a product label.

GH does not perform any testing; instead, it awards its seal to products that pay for advertising in the *Good Housekeeping* magazine. GH requires the manufacturer to submit product-testing evidence that has been published in a peer-reviewed journal.

So the next time you pick up some vitamins, pay attention to the seal of approval rather than the fancy colors and catchy design of the label.

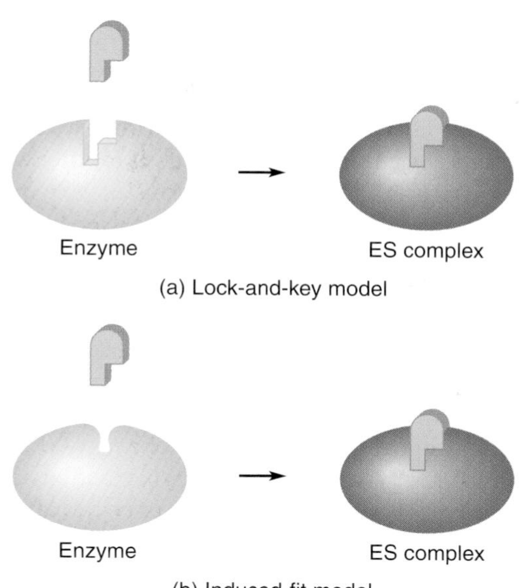

FIGURE 10.4 Models representing enzyme action: (a) In the lock-and-key model, the rigid enzyme and substrate have matching shapes. (b) In the induced-fit model, the flexible enzyme changes shape to match the substrate.

10.5 Enzyme Activity

LEARNING OBJECTIVE

5. List two ways of describing enzyme activity.

Enzyme activity refers in general to the catalytic ability of an enzyme to increase the rate of a reaction. The amazing rate (36 million molecules per minute) at which one molecule of carbonic anhydrase converts carbon dioxide to carbonic acid was mentioned earlier. This rate, called the **turnover number,** is one of the highest known for enzyme systems. More common turnover numbers for enzymes are closer to 10^3/min, or 1000 reactions per minute. Nevertheless, even such low numbers dramatize the speed with which a small number of enzyme molecules can transform a large number of substrate molecules. ■ Table 10.3 gives the turnover numbers of several enzymes.

enzyme activity The rate at which an enzyme catalyzes a reaction.

turnover number The number of molecules of substrate acted on by one molecule of enzyme per minute.

Experiments that measure enzyme activity are called *enzyme assays*. Assays for blood enzymes are routinely performed in clinical laboratories. Such assays are often done by determining how fast the characteristic color of a product forms or

TABLE 10.3 Examples of enzyme turnover numbers

Enzyme	Turnover number (per minute)	Reaction catalyzed
carbonic anhydrase	36,000,000	$CO_2 + H_2O \rightleftharpoons H_2CO_3$
catalase	5,600,000	$2H_2O_2 \rightleftharpoons 2H_2O + O_2$
cholinesterase	1,500,000	Hydrolysis of acetylcholine
chymotrypsin	6,000	Hydrolysis of specific peptide bonds
DNA polymerase I	900	Addition of nucleotide monomers to DNA chains
lactate dehydrogenase	60,000	pyruvate + NADH + $H^+ \rightleftharpoons$ lactate + NAD^+
penicillinase	120,000	Hydrolysis of penicillin

the color of a substrate decreases. Reactions in which protons (H^+) are produced or used up can be followed by measuring how fast the pH of the reacting mixture changes with time.

Because some clinical assays are done many times a day, the procedures for running the assays have been automated and computerized. The determined enzyme activity levels are usually reported in terms of **enzyme international units (IU)**, which define enzyme activity as the amount of enzyme that will convert a specified amount of substrate to a product within a certain time.

enzyme international unit (IU) A quantity of enzyme that catalyzes the conversion of 1 μmol of substrate per minute under specified conditions.

One standard international unit (1 IU) is the quantity of enzyme that catalyzes the conversion of 1 micromole (μmol) of substrate per minute under specified reaction conditions. Thus, unlike the turnover number, which is a constant characteristic for one molecule of a particular enzyme, international units measure how much enzyme is present. For example, an enzyme preparation with an activity corresponding to 40 IU contains a concentration of enzyme 40 times greater than the standard. This is a useful way to measure enzyme activity because the level of enzyme activity compared with normal activity is significant in the diagnosis of many diseases (Section 10.9).

■ **Learning Check 10.2** Differentiate between the terms *turnover number* and *enzyme international unit.*

10.6 Factors Affecting Enzyme Activity

LEARNING OBJECTIVE

6. Identify the factors that affect enzyme activity.

ThomsonNOW Go to Coached Problems to explore the **rate effects for enzymatic reactions.**

Several factors affect the rate of enzyme-catalyzed reactions. The most important factors are enzyme concentration, substrate concentration, temperature, and pH. In this section, we'll look at each of these factors in some detail. In Section 10.7, we'll consider another very important factor, the presence of enzyme inhibitors.

Enzyme Concentration

In an enzyme-catalyzed reaction, the concentration of enzyme is normally very low compared with the concentration of substrate. When the enzyme concentration is increased, the concentration of ES also increases in compliance with reaction rate theory:

$$E + S \rightleftarrows ES$$

increased [E] gives more [ES]

Thus, the availability of more enzyme molecules to catalyze a reaction leads to the formation of more ES and a higher reaction rate. The effect of enzyme concentration on the rate of a reaction is shown in ■ Figure 10.5. As the figure indicates, the rate of a reaction is directly proportional to the concentration of the enzyme—that is, if the enzyme concentration is doubled, the rate of conversion of substrate to product is also doubled.

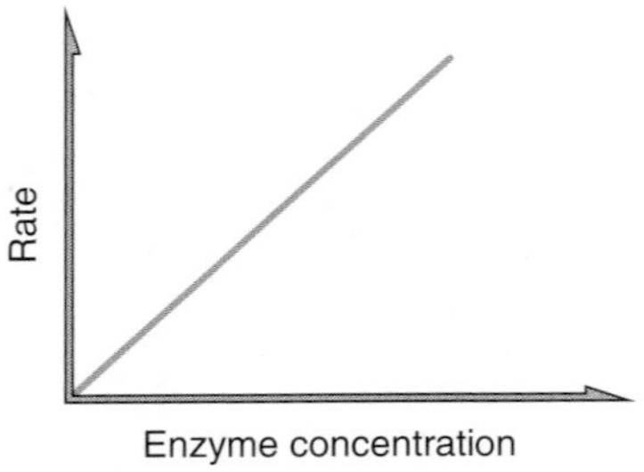

Rate

Enzyme concentration

■ **FIGURE 10.5** The dependence of reaction rate (initial velocity) on enzyme concentration.

Substrate Concentration

As shown in ■ Figure 10.6, the concentration of substrate significantly influences the rate of an enzyme-catalyzed reaction. Initially, the rate is responsive to increases in substrate concentration. However, at a certain concentration, the rate levels out and remains constant. This maximum rate (symbolized by V_{max}) occurs because the enzyme is saturated with substrate and cannot work any faster under the conditions imposed.

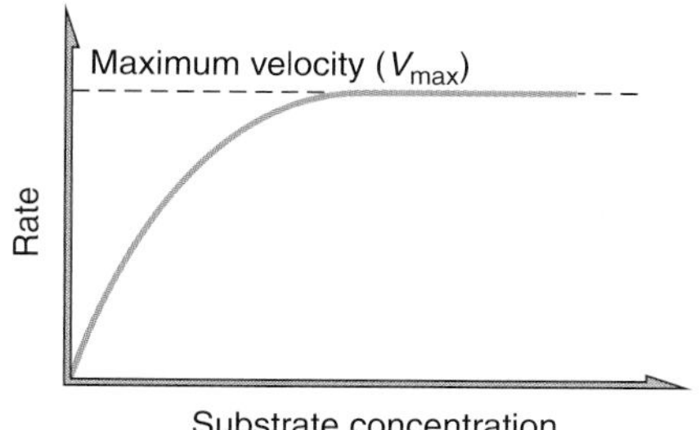

■ **FIGURE 10.6** The dependence of reaction rate (initial velocity) on substrate concentration.

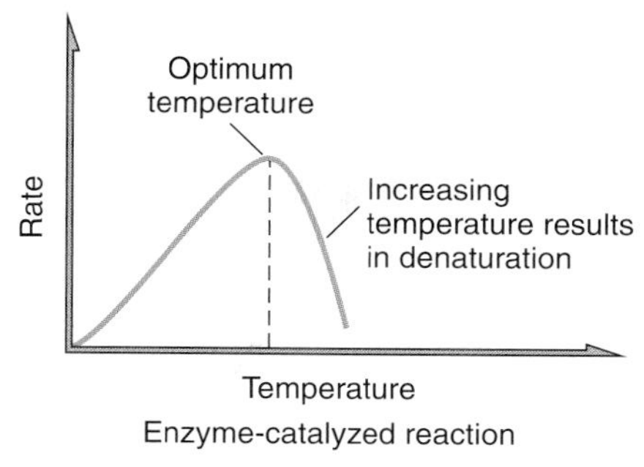

■ **FIGURE 10.7** The effect of temperature on enzyme-catalyzed reaction rates.

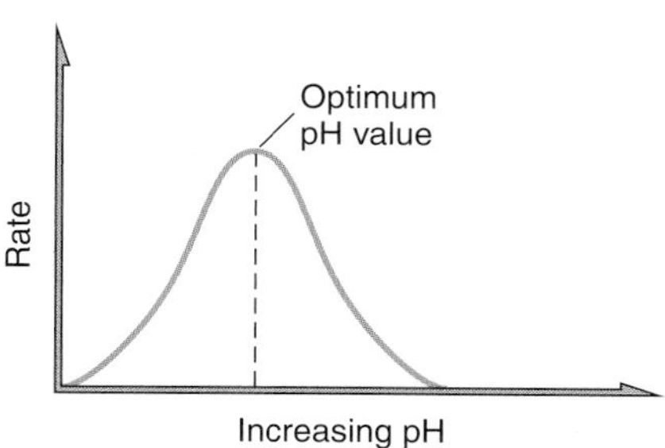

■ **FIGURE 10.8** A typical plot of the effect of pH on reaction rate.

Temperature

Enzyme-catalyzed reactions, like all chemical reactions, have rates that increase with temperature (see ■ Figure 10.7). However, because enzymes are proteins, there is a temperature limit beyond which the enzyme becomes vulnerable to denaturation. Thus, every enzyme-catalyzed reaction has an **optimum temperature,** usually in the range 25°C–40°C. Above or below that value, the reaction rate will be lower. This effect is illustrated in Figure 10.7.

optimum temperature
The temperature at which enzyme activity is highest.

The Effect of pH

The graph of enzyme activity as a function of pH is somewhat similar to the behavior as a function of temperature (see ■ Figure 10.8). Notice in Figure 10.8 that an enzyme is most effective in a narrow pH range and is less active at pH values lower or higher than this optimum. This variation in enzyme activity with changing pH may be due to the influence of pH on acidic and basic side chains within the active site. In addition, most enzymes are denatured by pH extremes. The resistance of pickled foods (see ■ Figure 10.9) to spoilage is due to the enzymes of microorganisms being less active under acidic conditions.

Many enzymes have an **optimum pH** near 7, the pH of most biological fluids. However, the optimum pH of a few is considerably higher or lower than 7. For example, pepsin, a digestive enzyme of the stomach, shows maximum activity at a pH of about 1.5, the pH of gastric fluids. ■ Table 10.4 lists optimum pH values for several enzymes.

optimum pH
The pH at which enzyme activity is highest.

TABLE 10.4 Examples of optimum pH for enzyme activity

Enzyme	Source	Optimum pH
pepsin	Gastric mucosa	1.5
β-glucosidase	Almond	4.5
sucrase	Intestine	6.2
urease	Soybean	6.8
catalase	Liver	7.3
succinate dehydrogenase	Beef heart	7.6
arginase	Beef liver	9.0
alkaline phosphatase	Bone	9.5

■ **FIGURE 10.9** Pickles are prepared under acidic conditions. Why do they resist spoilage?

Learning Check 10.3 Indicate how each of the following affects the rate of an enzyme-catalyzed reaction:

a. Increase in enzyme concentration
b. Increase in substrate concentration
c. Increase in temperature
d. Increase in pH

10.7 Enzyme Inhibition

LEARNING OBJECTIVE

7. Compare the mechanisms of competitive and noncompetitive enzyme inhibition.

enzyme inhibitor
A substance that decreases the activity of an enzyme.

An **enzyme inhibitor** is any substance that can decrease the rate of an enzyme-catalyzed reaction. An understanding of enzyme inhibition is important for several reasons. First, the characteristic function of many poisons and some medicines is to inhibit one or more enzymes and to decrease the rates of the reactions they catalyze. Second, some substances normally found in cells inhibit specific enzyme-catalyzed reactions and thereby provide a means for the internal regulation of cellular metabolism.

Enzyme inhibitors are classified into two categories, reversible and irreversible, on the basis of how they behave at the molecular level.

Irreversible Inhibition

An irreversible inhibitor forms a covalent bond with a specific functional group of the enzyme and as a result renders the enzyme inactive. A number of very deadly poisons act as irreversible inhibitors.

The cyanide ion (CN^-) is an example of an irreversible enzyme inhibitor. It is extremely toxic and acts very rapidly. The cyanide ion interferes with the operation of an iron-containing enzyme called cytochrome oxidase. The ability of cells to use oxygen depends on the action of cytochrome oxidase. When the cyanide ion reacts with the iron of this enzyme, it forms a very stable complex (Reaction 10.8), and the enzyme can no longer function properly. As a result, cell respiration stops, causing death in a matter of minutes.

$$\underset{\text{cytochrome oxidase}}{\text{Cyt—Fe}^{3+}} + \underset{\text{cyanide ion}}{\text{CN}^-} \rightarrow \underset{\text{stable complex}}{\text{Cyt—Fe—CN}^{2+}} \qquad (10.8)$$

Any antidote for cyanide poisoning must be administered quickly. One antidote is sodium thiosulfate (the "hypo" used in developing photographic film). This substance converts the cyanide ion to a thiocyanate ion that does not bind to the iron of cytochrome oxidase.

$$\underset{\text{cyanide}}{\text{CN}^-} + \underset{\text{thiosulfate}}{\text{S}_2\text{O}_3{}^{2-}} \rightarrow \underset{\text{thiocyanate}}{\text{SCN}^-} + \underset{\text{sulfite}}{\text{SO}_3{}^{2-}} \qquad (10.9)$$

The toxicity of heavy metals such as mercury and lead is due to their ability to render the protein part of enzymes ineffective. These metals act by combining with the —SH groups found on many enzymes (Reaction 10.10). In addition, as we learned in Chapter 9, metals can cause nonspecific protein denaturation. Both mercury and lead poisoning can cause permanent neurological damage.

$$\underset{\text{active enzyme}}{\text{Enz}(\text{—SH})_2} + \text{Hg}^{2+} \longrightarrow \underset{\text{inactive enzyme}}{\text{Enz}(\text{—S—})_2\text{Hg}} + 2\text{H}^+ \qquad (10.10)$$

CHEMISTRY AROUND US 10.1 Enzyme Discovery Heats Up

Enzymes provide the same advantages to a commercial process as they do to a living organism. They are highly specific and extremely fast-acting catalysts. The use of enzymes in commercial industrial processes is developing rapidly and has a potentially bright future because of the existence of enzymes in microorganisms that thrive in extreme environments. Such enzymes have been called **extremozymes** because of the extreme conditions under which they function. These conditions include temperatures greater than 100°C in hot springs and deep-sea thermal vents, or below 0°C in Antarctic waters. Some enzymes function under immense pressure on the ocean floor, whereas others are unaffected by extremes of pH and salt concentrations in solution.

One goal of current research is to obtain useful enzymes from extremophiles, the microbes that function in these extreme environments. The process used by one company to obtain commercially useful enzymes consists of the following steps:

1. Biomass samples are obtained from the extreme environment.
2. DNA is extracted from the biomass and purified.
3. Fragments of the purified DNA are cloned to produce large libraries of genes.
4. The cloned genes are screened to find any that are coded to produce enzymes.
5. The DNA sequence is determined for any genes that are coded to produce enzymes.
6. The DNA sequence for enzyme-coded genes is cloned to produce large amounts of the gene, which in turn is inserted into organisms and used to produce large amounts of the enzyme.
7. The enzyme is optimized by random DNA mutations.

The last step is particularly interesting. An organism's DNA is randomly mutated over several generations. Offspring from each generation are selected that produce a specific enzyme with activity greater than or different from that of the enzyme from the previous generation. This technique, known as *mutagenesis,* is used to alter enzymes so they will perform new tasks or do old ones better.

Because extremozymes function under conditions that are so far removed from normal, research into their behavior is intense. One of the research goals is to develop ways to use enzyme catalysis in industrial processes that require extreme temperatures or pH. One example is the current use of heat-stable DNA polymerase in the DNA replication technique called polymerase chain reaction (Section 11.3). Another research goal is to develop enzyme-catalyzed industrial processes that don't generate large amounts of undesirable waste material. For example, an esterase that catalyzes the hydrolysis of the antibiotic intermediate *p*-nitrobenzyl ester was developed. That hydrolysis reaction was usually done using catalytic zinc and organic solvents in a process that generated a lot of industrial waste. Some other applications being considered are the use of hydrolases at high temperature in the food-processing industry, and the use of heat-stable enzymes in the production of corn syrup.

Extremozymes enable microorganisms to survive in the harsh environment of deep-sea thermal vents.

extremozyme
A nickname for certain enzymes isolated from microorganisms that thrive in extreme environments.

Heavy-metal poisoning is treated by administering chelating agents—substances that combine with the metal ions and hold them very tightly. One effective antidote is the chelating agent ethylenediaminetetraacetic acid, or EDTA.

The calcium salt of EDTA is administered intravenously. In the body, calcium ions of the salt are displaced by heavy-metal ions, such as lead, that bind to the chelate more tightly. The lead–EDTA complex is soluble in body fluids and is excreted in the urine.

$$CaEDTA^{2-} + Pb^{2+} \rightarrow PbEDTA^{2-} + Ca^{2+} \qquad (10.11)$$

Not all enzyme inhibitors act as poisons toward the body; some, in fact, are useful therapeutic agents. Sulfa drugs and the group of compounds known as

antibiotic
A substance produced by one microorganism that kills or inhibits the growth of other microorganisms.

penicillins are two well-known families of **antibiotics** that inhibit specific enzymes essential to the life processes of bacteria. The first penicillin was discovered in 1928 by Alexander Fleming. He noticed that the bacteria on a culture plate did not grow in the vicinity of a mold that had contaminated the culture; the mold was producing penicillin. The general penicillin structure is shown in ■ Table 10.5, along with four widely used forms. Penicillins interfere with transpeptidase, an enzyme that is important in bacterial cell wall construction. The inability to form strong cell walls prevents the bacteria from surviving.

Reversible Inhibition

A reversible inhibitor (in contrast to one that is irreversible) reversibly binds to an enzyme. An equilibrium is established; therefore, the inhibitor can be removed from the enzyme by shifting the equilibrium:

$$E + I \rightleftharpoons EI$$

competitive inhibitor
An inhibitor that competes with substrate for binding at the active site of the enzyme.

ThomsonNOW™ Go to Coached Problems to explore **competitive and noncompetitive enzyme inhibition.**

There are two types of reversible inhibitors: competitive and noncompetitive.

A **competitive inhibitor** binds to the active site of an enzyme and thus "competes" with substrate molecules for the active site. Competitive inhibitors often have molecular structures that are similar to the normal substrate of the enzyme. The competitive inhibition of succinate dehydrogenase by malonate is a classic example. Succinate dehydrogenase catalyzes the oxidation of the substrate succinate to form fumarate by transferring two hydrogens to the coenzyme FAD:

CHEMISTRY AROUND US 10.2 Mercury in Fish

Most healthful diets extol the wisdom of eating fish. After all, fish contain high-quality protein and other essential nutrients, are low in saturated fat, and contain omega-3 fatty acids. However, a growing concern in the United States is that the health benefits of eating fish might be undermined by the level of mercury contained in some types. Mercury, especially in the form of compounds, is recognized as a powerful neural toxin. Women who are pregnant, women who may become pregnant, women who are nursing infants, and young children are especially susceptible to the negative side effects of dietary mercury. Unfortunately, this is the group of people who would benefit most from consuming fish as part of a well-balanced diet.

Women who eat a large amount of fish during pregnancy, or even as little as a single serving of highly contaminated fish, run the risk of exposing their developing child to excessive levels of mercury. Compounds of the toxic metal can cross the placenta and harm the brain and nervous system of the rapidly developing child. Exposure to mercury while in the womb has been associated with learning deficiencies, delayed mental development, and other neurological problems in children.

Mercury occurs naturally in the environment and is emitted from sources such as coal-fired power plants. Mercury that falls to earth can accumulate in mud and sediments of streams and oceans where it can be aerobically converted to compounds such as methylmercury. These compounds find their way into fish and are amplified as larger fish eat smaller fish in the food chain. Thus, the greatest accumulation is found in fish that are highest in the food chain. Fish like tuna, sea bass, marlin, and halibut show some of the worst contamination as a result of this process.

One interpretation of a survey conducted by the U.S. Environmental Protection Agency found that more than half the fish in the nation's lakes and reservoirs have levels of mercury compounds that exceed government standards for children and for women of child-bearing age.

The following guidelines are helpful in deciding what and how much fish to eat weekly. This allows individuals to reap the benefits of eating fish, while avoiding excessive exposure to the harmful effects of mercury.

- Avoid shark, swordfish, king mackerel, or tilefish.
- Eat up to 12 ounces (two average meals) a week of the following fish and shellfish that are lower in mercury: Shrimp, canned light tuna, salmon, polloc, and catfish. Albacore ("white") tuna has more mercury than canned light tuna, so no more than 6 ounces (one average meal) of albacore tuna should be eaten per week.
- Check local advisories about the safety of fish caught in local lakes, rivers, and coastal areas. If no advice is available, eat up to 6 ounces (one average meal) per week of fish caught in these waters. Don't consume any other fish during that week.

TABLE 10.5 Four widely used penicillins

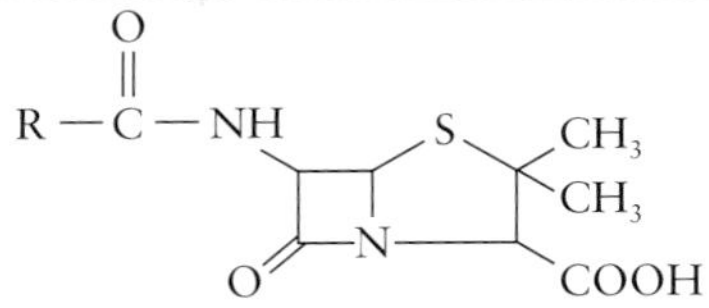

General penicillin structure: R—C(=O)—NH—(β-lactam/thiazolidine ring with S, N, O, CH_3, CH_3, COOH)

Name	Year marketed	Side chain (R—)	Comments
Penicillin G	1943	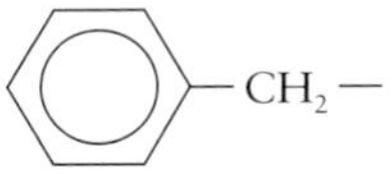 $C_6H_5-CH_2-$	Usually given by injection
Penicillin V	1953	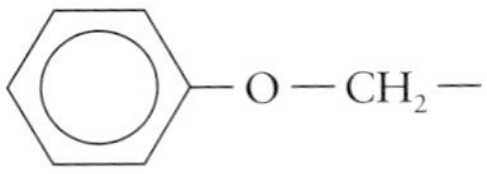 $C_6H_5-O-CH_2-$	An oral penicillin; resistant to stomach hydrolysis
Methicillin	1960	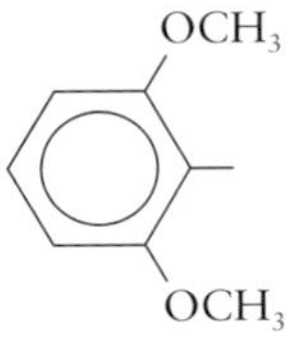 2,6-(OCH_3)$_2C_6H_3-$	Given by injection
Ampicillin	1961	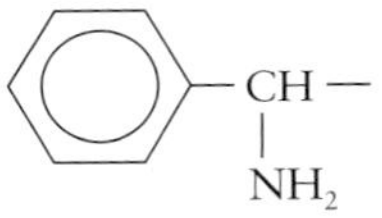 $C_6H_5-CH(NH_2)-$	Given by injection or taken orally; effective against a broad spectrum of organisms

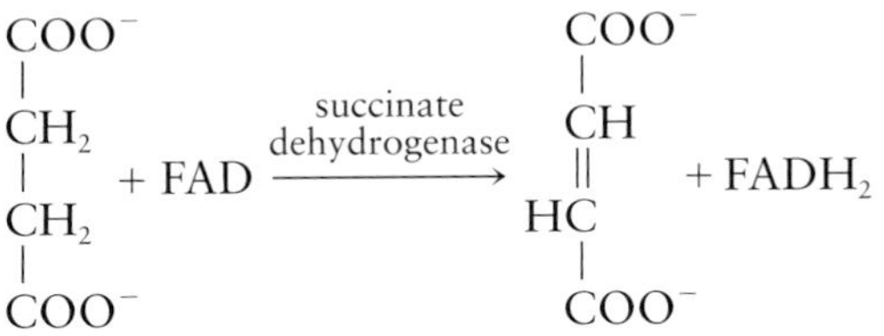

$$\underset{\text{succinate}}{\begin{array}{c}COO^-\\|\\CH_2\\|\\CH_2\\|\\COO^-\end{array}} + FAD \xrightarrow{\text{succinate dehydrogenase}} \underset{\text{fumarate}}{\begin{array}{c}COO^-\\|\\CH\\||\\HC\\|\\COO^-\end{array}} + FADH_2 \qquad (10.12)$$

Malonate, having a structure similar to succinate, competes for the active site of succinate dehydrogenase and thus inhibits the enzyme:

$$\underset{\text{succinate}}{\begin{array}{c}COO^-\\|\\CH_2\\|\\CH_2\\|\\COO^-\end{array}} \qquad \underset{\text{malonate}}{\begin{array}{c}COO^-\\|\\CH_2\\|\\COO^-\end{array}}$$

The action of sulfa drugs on bacteria is another example of competitive enzyme inhibition. Folic acid, a substance needed for growth by some disease-causing bacteria, is normally synthesized within the bacteria by a chemical process that requires *p*-aminobenzoic acid. Because sulfanilamide, the first sulfa drug, resembles *p*-aminobenzoic acid and competes with it for the active site of the bacterial enzyme involved, it can prevent bacterial growth (see ■ Figure 10.10). This

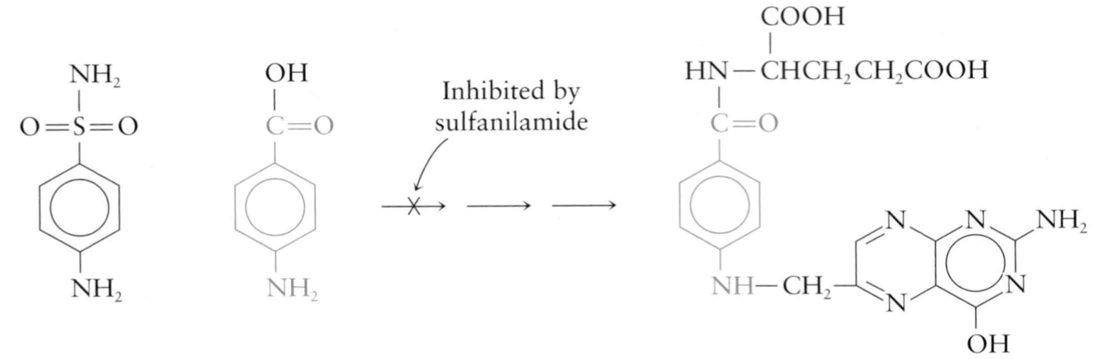

sulfanilamide

p-aminobenzoic acid

Inhibited by sulfanilamide

folic acid

■ **FIGURE 10.10** Structural relationships of sulfanilamide, *p*-aminobenzoic acid, and folic acid.

is possible because the enzyme binds readily to either of these molecules. Therefore, the introduction of large quantities of sulfanilamide into a patient's body causes most of the active sites to be bound to the wrong (from the bacterial viewpoint) substrate. Thus, the synthesis of folic acid is stopped or slowed, and the bacteria are prevented from multiplying. Meanwhile, the normal body defenses work to destroy them.

Human beings also require folic acid, but they get it from their diet. Consequently, sulfa drugs exert no toxic effect on humans. However, one danger of sulfa drugs, or of any antibiotic, is that excessive amounts can destroy important intestinal bacteria that perform many symbiotic, life-sustaining functions. Although sulfa drugs have largely been replaced by antibiotics such as the penicillins and tetracyclines, some are still used in treating certain bacterial infections, such as those occurring in the urinary tract.

The nature of competitive inhibition is represented in ■ Figure 10.11. There is competition between the substrate and the inhibitor for the active site. Once the inhibitor combines with the enzyme, the active site is blocked, preventing further catalytic action.

Competitive inhibition can be reversed by increasing the concentration of substrate and letting Le Châtelier's principle operate. This is illustrated by the following equilibria that would exist in a solution containing enzyme (E), substrate (S), and competitive inhibitor (I):

$$\text{equilibrium 1:}\quad E + S \underset{}{\overset{\text{Increasing [S]}}{\rightleftharpoons}} ES$$

$$\text{equilibrium 2:}\quad E + I \underset{\text{Decreasing [E]}}{\rightleftharpoons} EI$$

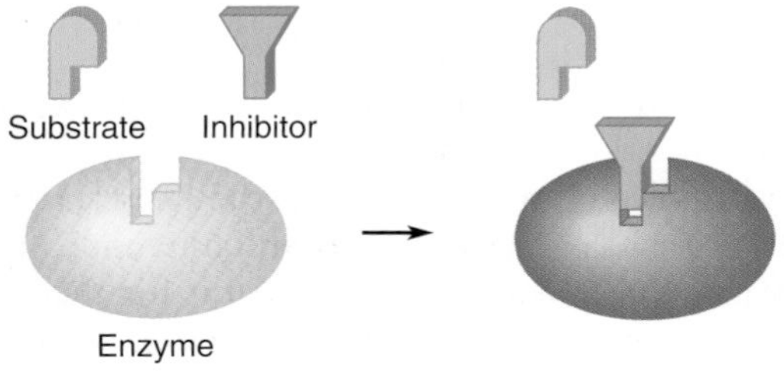

■ **FIGURE 10.11** The behavior of competitive inhibitors. The active site is occupied by the inhibitor, and the substrate is prevented from bonding.

Substrate

Inhibitor

Enzyme

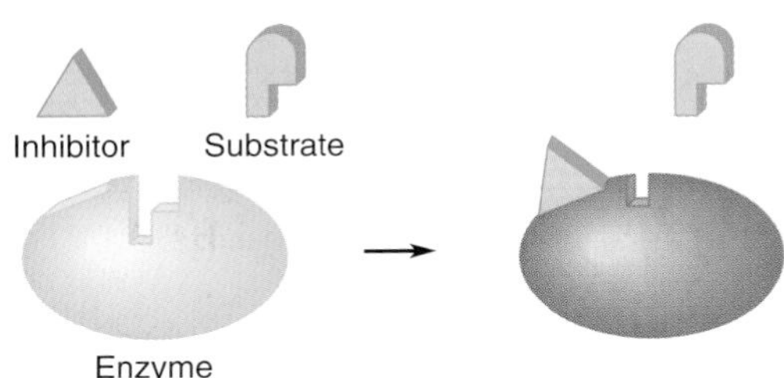

■ **FIGURE 10.12** The behavior of noncompetitive inhibitors. The inhibitor bonds to the enzyme at a site other than the active site and changes the shape of the active site.

If the concentration of substrate is increased, then according to Le Châtelier's principle, the first equilibrium should shift to the right. As it does, the concentration of enzyme decreases, which causes equilibrium 2 to shift to the left. The overall effect is that the concentration of ES has increased while the concentration of EI has decreased. Thus, the competition between inhibitor and substrate is won by whichever molecular species is in greater concentration.

A **noncompetitive inhibitor** bears no resemblance to the normal enzyme substrate and binds reversibly to the surface of an enzyme at a site other than the catalytically active site (see ■ Figure 10.12). The interaction between the enzyme and the noncompetitive inhibitor causes the three-dimensional shape of the enzyme and its active site to change. The enzyme no longer binds the normal substrate, or the substrate is improperly bound in a way that prevents the catalytic groups of the active site from participating in catalyzing the reaction.

noncompetitive inhibitor An inhibitor that binds to the enzyme at a location other than the active site.

Unlike competitive inhibition, noncompetitive inhibition cannot be reversed by the addition of more substrate because additional substrate has no effect on the enzyme-bound inhibitor (it can't displace the inhibitor because it can't bond to the site occupied by the inhibitor).

■ **Learning Check 10.4** Compare a noncompetitive and a competitive inhibitor with regard to the following:

a. Resemblance to substrate
b. Binding site on the enzyme
c. The effect of increasing substrate concentration

10.8 The Regulation of Enzyme Activity

LEARNING OBJECTIVE

8. Describe the three methods of cellular control over enzyme activity.

Enzymes work together in an organized yet complicated way to facilitate all the biochemical reactions needed by a living organism. For an organism to respond to changing conditions and cellular needs, very sensitive controls over enzyme activity are required. We will discuss three mechanisms by which this is accomplished: activation of zymogens, allosteric regulation, and genetic control.

ThomsonNOW™ Go to Coached Problems to explore the **control of enzymatic reactions.**

The Activation of Zymogens

A common mechanism for regulating enzyme activity is the synthesis of enzymes in the form of inactive precursors called **zymogens** or **proenzymes.** Some enzymes in their active form would degrade the internal structures of the cells and, ultimately, the cells themselves. Such enzymes are synthesized and stored as zymogens. When the active enzyme is needed, the stored zymogen is released from storage and activated at the location of the reaction. Activation usually involves the cleavage of one or more peptide bonds of the zymogen. The digestive enzymes

zymogen or proenzyme The inactive precursor of an enzyme.

TABLE 10.6 Examples of zymogens

Zymogen	Active enzyme	Function
chymotrypsinogen	chymotrypsin	Digestion of proteins
pepsinogen	pepsin	Digestion of proteins
procarboxypeptidase	carboxypeptidase	Digestion of proteins
proelastase	elastase	Digestion of proteins
prothrombin	thrombin	Blood clotting
trypsinogen	trypsin	Digestion of proteins

pepsin, trypsin, and chymotrypsin and several of the enzymes involved in blood clotting are produced and activated this way. Trypsinogen, the inactive form of trypsin, is synthesized in the pancreas and activated in the small intestine by the hydrolysis of a single peptide bond. The products are the active enzyme (trypsin) and a hexapeptide:

$$\text{trypsinogen} + H_2O \xrightarrow{\text{enteropeptidase}} \text{trypsin} + \text{hexapeptide} \qquad (10.13)$$

Several examples of zymogens are listed in Table 10.6.

Allosteric Regulation

modulator
A substance that binds to an enzyme at a location other than the active site and alters the catalytic activity.

allosteric enzyme
An enzyme with quaternary structure whose activity is changed by the binding of modulators.

activator
A substance that binds to an allosteric enzyme and increases its activity.

A second method of enzyme regulation involves the combination of the enzyme with some other compound such that the three-dimensional conformation of the enzyme is altered and its catalytic activity is changed. Compounds that alter enzymes this way are called **modulators** of the activity; they may increase the activity (activators) or decrease the activity (inhibitors). The noncompetitive inhibitors we studied in Section 10.7 are examples of modulators. Enzymes that have a quaternary protein structure with distinctive binding sites for modulators are referred to as **allosteric enzymes.** These variable-rate enzymes are often located at key control points in cellular processes. In general, **activators** act as signals for increased production, whereas inhibitors are used to stop the formation of excess products.

An excellent example of allosteric regulation—the control of an allosteric enzyme—is the five-step synthesis of the amino acid isoleucine (see Active Figure 10.13). Threonine deaminase, the enzyme that catalyzes the first step in the conversion of threonine to isoleucine, is subject to inhibition by the final prod-

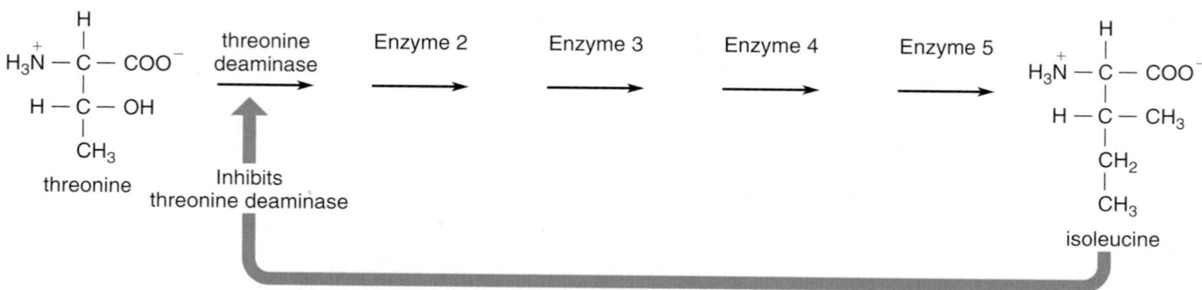

ACTIVE FIGURE 10.13 The allosteric regulation of threonine deaminase by isoleucine, an example of feedback inhibition. **Sign in at www.thomsonedu.com to explore an interactive version of this figure.**

STUDY SKILLS 10.1 A Summary Chart of Enzyme Inhibitors

At this point in your study of enzymes, a summary chart is a useful tool for representing the various types of enzyme inhibitors in a compact, easy-to-review way.

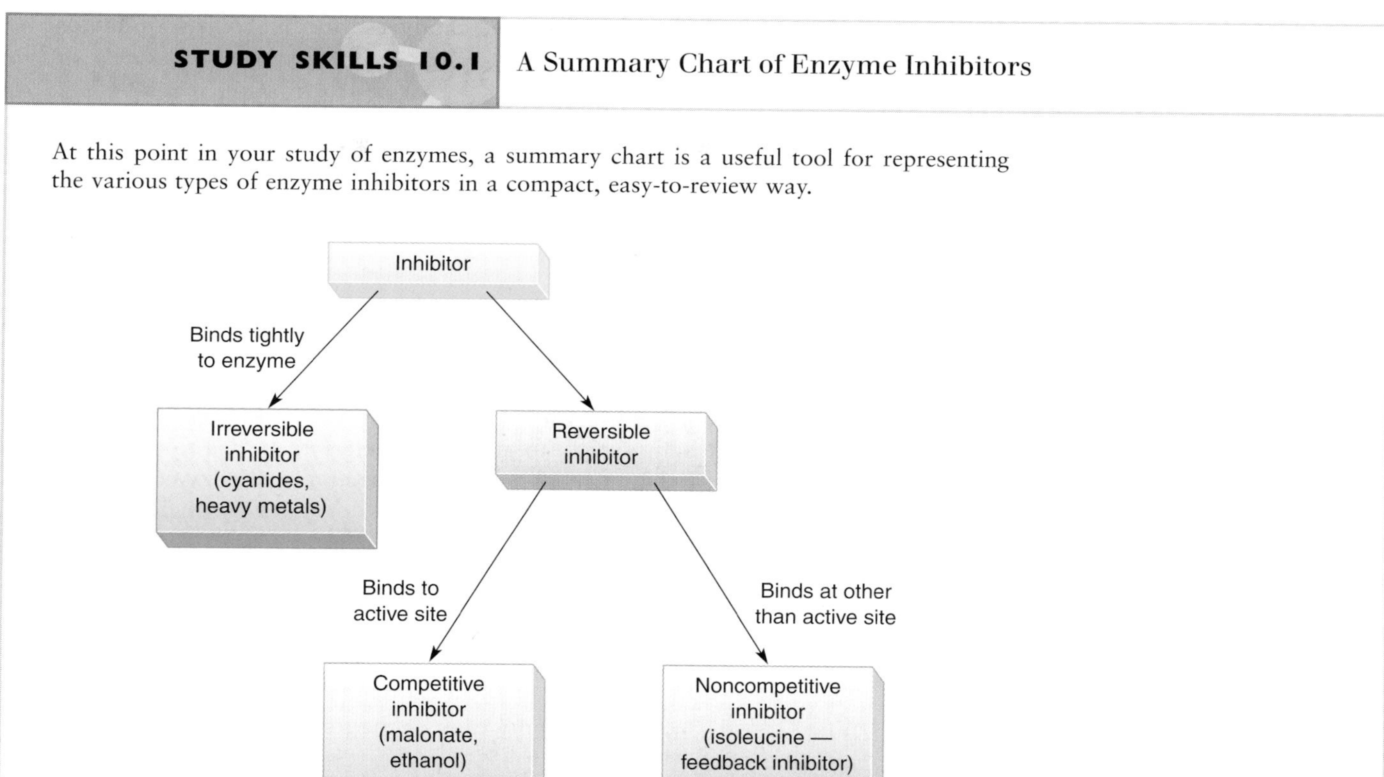

uct, isoleucine. The structures of isoleucine and threonine are quite different, so isoleucine is not a competitive inhibitor. Also, the site to which isoleucine binds to the enzyme is different from the enzyme active site that binds to threonine. This second site, called the allosteric site, specifically recognizes isoleucine, whose presence there induces a change in the conformation of the enzyme such that threonine binds poorly to the active site. Thus, isoleucine exerts an inhibiting effect on the enzyme activity. As a result, the reaction slows as the concentration of isoleucine increases, and no excess isoleucine is produced. When the concentration of isoleucine falls to a low enough level, the enzyme becomes more active, and more isoleucine is synthesized. This type of allosteric regulation in which the enzyme that catalyzes the first step of a series of reactions is inhibited by the final product is called **feedback inhibition.**

feedback inhibition
A process in which the end product of a sequence of enzyme-catalyzed reactions inhibits an earlier step in the process.

The control of enzyme activity by allosteric regulation of key enzymes is of immense benefit to an organism because it allows the concentration of cellular products to be maintained within very narrow limits.

Genetic Control

One way to increase production from an enzyme-catalyzed reaction, given a sufficient supply of substrate, is for a cell to increase the number of enzyme molecules present. The synthesis of all proteins, including enzymes, is under genetic control by nucleic acids (Chapter 11). An example of the genetic control of enzyme activity involves **enzyme induction,** the synthesis of enzymes in response to a temporary need of the cell.

enzyme induction
The synthesis of an enzyme in response to a cellular need.

The first demonstrated example of enzyme induction came from studies conducted in the 1950s on the bacterium *Escherichia coli.* β-galactosidase, an enzyme required for the utilization of lactose, catalyzes the hydrolysis of lactose to D-galactose and D-glucose.

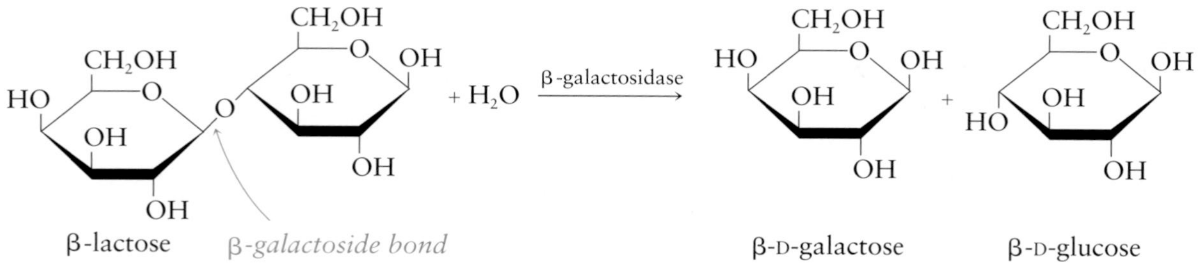

$$\beta\text{-lactose} + H_2O \xrightarrow{\beta\text{-galactosidase}} \beta\text{-D-galactose} + \beta\text{-D-glucose} \quad (10.14)$$

In the absence of lactose in the growth medium, there were fewer than ten β-galactosidase molecules per cell of *E. coli*. However, if *E. coli* are added to a lactose-containing medium, within minutes the bacterium begins to produce thousands of molecules of β-galactosidase. This is an example of enzyme induction. If lactose is removed from the growth medium, the production of β-galactosidase again falls back to a low level.

ThomsonNOW™ Go to Coached Problems to explore the **environmental control of enzymatic reactions.**

The biological significance of the genetic control of enzyme activity is that it allows an organism to adapt to environmental changes. The coupling of genetic control and allosteric regulation of enzyme activity puts the cellular processes in the body under constant control.

Learning Check 10.5 Explain how each of the following processes is involved in the regulation of enzyme activity:

a. Activation of zymogens
b. Allosteric regulation
c. Genetic control

10.9 Medical Application of Enzymes

LEARNING OBJECTIVE

9. Discuss the importance of measuring enzyme levels in the diagnosis of disease.

Enzymes in Clinical Diagnosis

Certain enzymes are normally found almost exclusively inside tissue cells and are released into the blood and other biological fluids only when these cells are damaged or destroyed. Because of the normal breakdown and replacement of tissue cells that go on constantly, the blood serum contains these enzymes but at very low concentrations—often a million times lower than the concentration inside cells. However, blood serum levels of cellular enzymes increase significantly when excessive cell injury or destruction occurs, or when cells grow rapidly as a result of cancer. Changes in blood serum concentrations of specific enzymes can be used clinically to detect cell damage or uncontrolled growth, and even to suggest the site of the damage or cancer. Also, the extent of cell damage can often be estimated by the magnitude of the serum concentration increase above normal levels. For these reasons, the measurement of enzyme concentrations in blood serum and other biological fluids has become a major diagnostic tool, particularly in diagnosing diseases of the heart, liver, pancreas, prostate, and bones. In fact, certain enzyme determinations are performed so often that they have become routine procedures in the clinical chemistry laboratory. ■ Table 10.7 lists some enzymes used in medical diagnosis.

TABLE 10.7 Diagnostically useful assays of blood serum enzymes

Enzyme	Pathological condition
acid phosphatase	Prostate cancer
alkaline phosphatase (ALP)	Liver or bone disease
amylase	Diseases of the pancreas
creatine kinase (CK)	Heart attack
aspartate transaminase (AST)	Heart attack or hepatitis
alanine transaminase (ALT)	Hepatitis
lactate dehydrogenase (LDH)	Heart attack, liver damage
lipase	Acute pancreatitis
lysozyme	Monocytic leukemia

Isoenzymes

Lactate dehydrogenase (LDH) and creatine kinase (CK), two of the enzymes listed in Table 10.7, are particularly useful in clinical diagnosis because each occurs in multiple forms called **isoenzymes.** Although all forms of a particular isoenzyme catalyze the same reaction, their molecular structures are slightly different and their locations within the body tissues may vary.

isoenzyme
A slightly different form of the same enzyme produced by different tissues.

Probably the most fully studied group of isoenzymes is that of lactate dehydrogenase (LDH). The quaternary structure of the enzyme consists of four subunits that are of two different types. One is labeled H because it is the predominant subunit present in the LDH enzyme found in heart muscle cells. The other, labeled M, predominates in other muscle cells. There are five possible ways to combine these four subunits to form the enzyme (see ■ Figure 10.14). Each combination has slightly different properties, which allows them to be separated and identified by electrophoresis.

■ Table 10.8 shows the distribution of the five LDH isoenzymes in several human tissues. Notice that each type of tissue has a distinct pattern of isoenzyme percentages. Liver and skeletal muscle, for example, contain particularly high percentages of LDH_5. Also notice the differences in LDH_1 and LDH_2 levels in heart and skeletal muscles.

Because of the difference in tissue distribution of LDH isoenzymes, serum levels of LDH are used in the diagnosis of a wide range of diseases: anemias involving the rupture of red blood cells, acute liver diseases, congestive heart failure, and

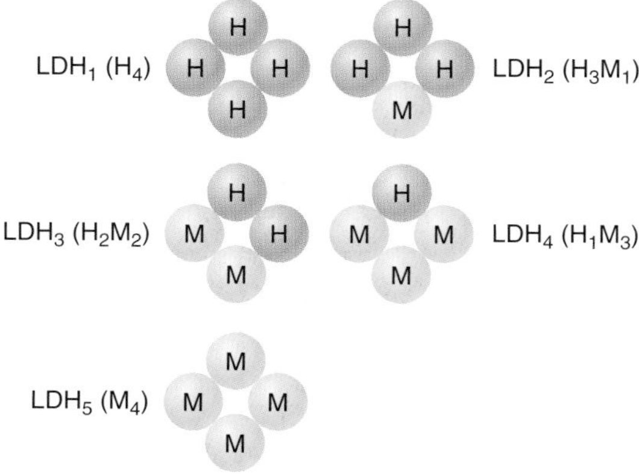

■ **FIGURE 10.14** Isomeric forms (isoenzymes) of lactate dehydrogenase.

TABLE 10.8 Tissue distribution of LDH isoenzymes

Tissue	LDH_1 (%)	LDH_2 (%)	LDH_3 (%)	LDH_4 (%)	LDH_5 (%)
Brain	23	34	30	10	3
Heart	50	36	9	3	2
Kidney	28	34	21	11	6
Liver	4	6	17	16	57
Lung	10	20	30	25	15
Serum	28	41	19	7	5
Skeletal muscle	5	5	10	22	58

muscular diseases such as muscular dystrophy. This wide range of diseases makes an LDH assay a good initial diagnostic test. Elevated levels of LDH_1 and LDH_2, for example, indicate myocardial infarction, whereas elevated LDH_5 indicates possible liver damage. If LDH activity is elevated, then this and other isoenzyme assays can be used to pinpoint the location and type of disease more accurately.

Concept Summary

General Characteristics of Enzymes. Enzymes are highly efficient protein catalysts involved in almost every biological reaction. They are often quite specific in terms of the substance acted on and the type of reaction catalyzed. **OBJECTIVE 1 (Section 10.1), Exercise 10.2**

Enzyme Nomenclature and Classification. Enzymes are grouped into six major classes on the basis of the type of reaction catalyzed. Common names for enzymes often end in *-ase* and are based on the substrate and/or the type of reaction catalyzed. **OBJECTIVE 2 (Section 10.2), Exercise 10.12**

Enzyme Cofactors. Cofactors are nonprotein molecules required for an enzyme to be active. Cofactors are either organic (coenzymes) or inorganic ions. **OBJECTIVE 3 (Section 10.3), Exercise 10.14**

The Mechanism of Enzyme Action. The behavior of enzymes is explained by a theory in which the formation of an enzyme–substrate complex is assumed to occur. The specificity of enzymes is explained by the lock-and-key theory and the induced-fit theory. **OBJECTIVE 4 (Section 10.4), Exercise 10.20**

Enzyme Activity. The catalytic ability of enzymes is described by turnover number and enzyme international units. Experiments that measure enzyme activity are referred to as enzyme assays. **OBJECTIVE 5 (Section 10.5), Exercise 10.26**

Factors Affecting Enzyme Activity. The catalytic activity of enzymes is influenced by numerous factors. The most important are substrate concentration, enzyme concentration, temperature, and pH. **OBJECTIVE 6 (Section 10.6), Exercise 10.28**

Enzyme Inhibition. Chemical substances called inhibitors decrease the rates of enzyme-catalyzed reactions. Irreversible inhibitors render enzymes permanently inactive and include several very toxic substances such as the cyanide ion and heavy-metal ions. Reversible inhibitors are of two types: competitive and noncompetitive. **OBJECTIVE 7 (Section 10.7), Exercise 10.34**

The Regulation of Enzyme Activity. Three mechanisms of cellular control over enzyme activity exist. One method involves the synthesis of enzyme precursors called zymogens, which are activated when needed by the cell. The second mechanism relies on the binding of small molecules (modulators), which increase or decrease enzyme activity. Genetic control of enzyme synthesis, the third method, regulates the amount of enzyme available. **OBJECTIVE 8 (Section 10.8), Exercise 10.38**

Medical Applications of Enzymes. Numerous enzymes have become useful as aids in diagnostic medicine. The presence of specific enzymes in body fluids such as blood has been related to certain pathological conditions. **OBJECTIVE 9 (Section 10.9), Exercise 10.46**

Key Terms and Concepts

Absolute specificity (10.1)
Activator (10.8)
Active site (10.4)
Allosteric enzyme (10.8)
Antibiotic (10.7)
Apoenzyme (10.3)
Coenzyme (10.3)
Cofactor (10.3)
Competitive inhibitor (10.7)
Enzyme (10.1)
Enzyme activity (10.5)
Enzyme induction (10.8)
Enzyme inhibitor (10.7)
Enzyme international unit (IU) (10.5)
Extremozyme (10.7)
Feedback inhibition (10.8)
Inborn error of metabolism (10.3)
Induced-fit theory (10.4)
Isoenzymes (10.9)
Lock-and-key theory (10.4)
Modulator (10.8)
Noncompetitive inhibitor (10.7)
Optimum pH (10.6)
Optimum temperature (10.6)
Relative specificity (10.1)
Stereochemical specificity (10.1)
Substrate (10.2)
Turnover number (10.5)
Zymogen or proenzyme (10.8)

Key Reactions

1. Formation of an active enzyme (Section 10.3):

apoenzyme + cofactor (coenzyme or inorganic ion) → active enzyme — Reaction 10.4

2. Mechanism of enzyme action (Section 10.4):

$$E + S \rightleftarrows ES \rightarrow E + P$$ Reaction 10.6

Exercises

SYMBOL KEY
Even-numbered exercises are answered in Appendix B.
Blue-numbered exercises are more challenging.
■ denotes exercises available in ThomsonNow and assignable in OWL.
ThomsonNOW™ To assess your understanding of this chapter's topics with sample tests and other resources, sign in at www.thomsonedu.com.

GENERAL CHARACTERISTICS OF ENZYMES (SECTION 10.1)

10.1 What is the role of enzymes in the body?

10.2 List two ways that enzyme catalysis of a reaction is superior to normal laboratory conditions.

10.3 What is the relationship between an enzyme and the energy of activation for a reaction?

10.4 Why are so many different enzymes needed?

10.5 Define what is meant by the term *enzyme specificity.*

10.6 Define what is meant by the term *absolute specificity.*

10.7 List three ways in which enzymes are particularly well suited for their essential roles in living organisms.

ENZYME NOMENCLATURE AND CLASSIFICATION (SECTION 10.2)

10.8 What is the ending of EC names and most common names of enzymes?

10.9 ■ What is the relationship between urea and urease? Between maltose and maltase?

10.10 ■ Match the following enzymes and substrates:

ENZYME	SUBSTRATE
a. sucrase	maltose
b. amylase	amylose
c. lactase	sucrose
d. maltase	arginine
e. arginase	lactose

10.11 ■ Match the following general enzyme names and reactions catalyzed:

ENZYME	REACTION CATALYZED
a. decarboxylase	formation of ester linkages
b. phosphatase	removal of carboxyl groups from compounds
c. peptidase	hydrolysis of peptide linkages
d. esterase	hydrolysis of phosphate ester linkages

10.12 ■ Because one substrate may undergo a number of reactions, it is often convenient to use an enzyme nomenclature system that includes both the substrate name (or general type) and the type of reaction catalyzed. Identify the substrate and type of reaction for the following enzyme names:

a. succinate dehydrogenase
b. L-amino acid reductase
c. cytochrome oxidase
d. glucose-6-phosphate isomerase

ENZYME COFACTORS (SECTION 10.3)

10.13 ■ Some enzymes consist of protein plus another component. Which of the terms *cofactor* or *coenzyme* correctly describes each of the following nonprotein components?

a. an inorganic ion

b. a nonspecific component

c. an organic material

d. nicotinic acid,

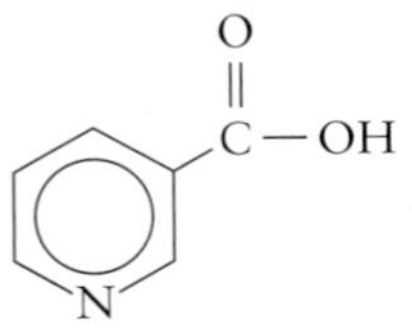

10.14 What are the relationships among the terms *cofactor, active enzyme,* and *apoenzyme?*

10.15 State the relationship between vitamins and enzyme activity.

10.16 List some typical inorganic ions that serve as cofactors.

10.17 What vitamins are required for FAD and NAD^+ formation?

THE MECHANISM OF ENZYME ACTION (SECTION 10.4)

10.18 Explain what is meant by the following equation:

$$E + S \rightleftarrows ES \rightarrow E + P$$

10.19 ■ In what way are the substrate and active site of an enzyme related?

10.20 How is enzyme specificity explained by the lock-and-key theory?

10.21 ■ Compare the lock-and-key theory with the induced-fit theory.

10.22 An enzyme can catalyze reactions involving propanoic acid, butanoic acid, and pentanoic acid. Would the lock-and-key theory or the induced-fit theory best explain this enzyme's mechanism of action? Why?

ENZYME ACTIVITY (SECTION 10.5)

10.23 List two ways of describing enzyme activity.

10.24 What observations may be used in experiments to determine enzyme activity?

10.25 Define the term *turnover number.*

10.26 What is an enzyme international unit? Why is the international unit a useful method of expressing enzyme activity in medical diagnoses?

FACTORS AFFECTING ENZYME ACTIVITY (SECTION 10.6)

10.27 ■ Use graphs to illustrate enzyme activity as a function of the following:

a. substrate concentration

b. enzyme concentration

c. pH

d. temperature

10.28 Write a single sentence to summarize the information of each graph in Exercise 10.27.

10.29 ■ What happens to the rate of an enzyme-catalyzed reaction as substrate concentration is raised beyond the saturation point?

10.30 How might V_{max} for an enzyme be determined?

10.31 ■ How would you expect hypothermia to affect enzyme activity in the body?

10.32 When handling or storing solutions of enzymes, the pH is usually kept near 7.0. Explain why.

ENZYME INHIBITION (SECTION 10.7)

10.33 Distinguish between irreversible and reversible enzyme inhibition.

10.34 Distinguish between competitive and noncompetitive enzyme inhibition.

10.35 ■ How does each of the following irreversible inhibitors interact with enzymes?

a. cyanide

b. heavy-metal ions

10.36 List an antidote for each of the two poisons in Exercise 10.35 and describe how each functions.

THE REGULATION OF ENZYME ACTIVITY (SECTION 10.8)

10.37 Why is the regulation of enzyme activity necessary?

10.38 List three mechanisms for the control of enzyme activity.

10.39 Describe the importance of zymogens in the body. Give an example of an enzyme that has a zymogen.

10.40 The clotting of blood occurs by a series of zymogen activations. Why are the enzymes that catalyze blood clotting produced as zymogens?

10.41 ■ What type of enzyme possesses a binding site for modulators?

10.42 Name and contrast the two types of modulators.

10.43 Explain how feedback enzyme inhibition works and why it is advantageous for the cell.

10.44 What is meant by genetic control of enzyme activity and enzyme induction?

MEDICAL APPLICATION OF ENZYMES (SECTION 10.9)

10.45 Why are enzyme assays of blood serum useful in a clinical diagnosis?

10.46 How is each of the following enzyme assays useful in diagnostic medicine?

a. CK

b. ALP

c. amylase

10.47 What are isoenzymes? List two examples of isoenzymes.

10.48 Why is an LDH assay a good initial diagnostic test?

10.49 What specific enzyme assays are particularly useful in diagnosing liver damage or disease?

ADDITIONAL EXERCISES

10.50 Carboxypeptidase is a proteolytic enzyme that cleaves the C-terminal amino acid residue from a polypeptide. The amino acid arginine is found in the active site of carboxypeptidase and is responsible for holding the C-terminal end of the polypeptide in place so the cleavage of the peptide bond can occur. What type of interaction occurs between the C-terminal end of the polypeptide and the arginine side chain?

10.51 An enzyme preparation that decarboxylates butanoic acid has an activity of 50 IU. How many grams of butanoic acid will this enzyme preparation decarboxylate in 10 minutes? Assume an unlimited supply of substrate.

10.52 Propanoic acid chloride was found to react with the side chain of a serine residue on an enzyme to cause irreversible inhibition of that enzyme. Suggest a mechanism of how this occurs.

$$CH_3CH_2\overset{\overset{\displaystyle O}{\|}}{C}-Cl$$

10.53 1.0×10^{12} enzyme molecules were added to 1.0 L of substrate solution with an initial pH of 7.0. After 1.0 minutes, the solution had a pH of 6.0. Assume an unlimited supply of substrate and any change in hydrogen ion concentration was due to the following reaction.

$$\text{Substrate} \xrightarrow{\text{enzyme}} \text{Product} + H^+$$

a. What was the change in H^+ concentration?

b. How many molecules of substrate were reacted to cause this change in H^+ concentration?

c. What is the turnover number of this enzyme?

10.54 The following reaction has come to equilibrium and it is found that there are 100 times as many molecules of A as there are of B in the solution.

$$A \rightleftharpoons B$$

An enzyme that catalyzes this reaction is then added to the solution. In which direction will the equilibrium shift?

ALLIED HEALTH EXAM CONNECTION

Reprinted with permission from Nursing School and Allied Health Entrance Exams, COPYRIGHT 2005 Petersons.

10.55 Which of the following is not a characteristic of enzymes?

a. They are macromolecules.

b. They act on substances.

c. They are phospholipids.

d. They initiate and decelerate chemical reactions.

e. They act as catalysts.

10.56 The human body has an average pH of about 7 and a temperature of about 37°C. Use graphs to illustrate enzyme activity in the human body as a function of the following:

a. Substrate concentration

b. Enzyme concentration

c. pH (include pH optimum value)

d. Temperature (include temperature optimum value)

10.57 Look at Table 10.4 and identify the group of proteolytic enzymes that begin protein digestion in the stomach. Does this name use an EC name ending?

10.58 Saliva contains mucus, water, and ______, which partially digests polysaccharides. Fill in the blank with the correct response from below:

a. lipase

b. amylase

c. trypsin

d. insulin

CHEMISTRY FOR THOUGHT

10.59 ■ Explain how the pasteurization of milk utilizes one of the factors that influence enzyme activity.

10.60 Explain how food preservation by freezing utilizes one of the factors that influence enzyme activity.

10.61 Some pottery glazes contain lead compounds. Why would it be unwise to use such pottery to store or serve food?

10.62 ■ Urease can catalyze the hydrolysis of urea, but not the hydrolysis of methyl urea. Why?

10.63 The active ingredient in meat tenderizer is the enzyme papain, a protease. Explain how treating meat with this material before cooking would make it more tender.

10.64 Why are enzymes that are used for laboratory or clinical work stored in refrigerators?

10.65 Describe the differences between graphs showing temperature versus reaction rate for an enzyme-catalyzed reaction and an uncatalyzed reaction.

10.66 Answer the question associated with Figure 10.3. How else might ammonia be detected?

10.67 In Figure 10.9, it is noted that pickles resist spoilage. Why is that true for this acidic food?

10.68 One of the steps in preparing frozen corn for the grocery market involves "blanching" (placing the corn in boiling water). Why is blanching necessary?

10.69 ■ What experiment would you conduct to determine if an inhibitor is competitive or noncompetitive?

10.70 Why might an enzyme be selected over an inorganic catalyst for a certain industrial process?

10.71 Enzymes are enormous molecules, and yet the active site is relatively small. Why is a huge molecule necessary for enzymes?

CHAPTER 11

Nucleic Acids and Protein Synthesis

LEARNING OBJECTIVES

When you have completed your study of this chapter, you should be able to:

1. Identify the components of nucleotides and correctly classify the sugars and bases. (Section 11.1)
2. Describe the structure of DNA. (Section 11.2)
3. Outline the process of DNA replication. (Section 11.3)
4. Contrast the structures of DNA and RNA and list the function of the three types of cellular RNA. (Section 11.4)
5. Describe what is meant by the terms *transcription* and *translation*. (Section 11.5)
6. Describe the process by which RNA is synthesized in cells. (Section 11.6)
7. Explain how the genetic code functions in the flow of genetic information. (Section 11.7)
8. Outline the process by which proteins are synthesized in cells. (Section 11.8)
9. Describe how genetic mutations occur and how they influence organisms. (Section 11.9)
10. Describe the technology used to produce recombinant DNA. (Section 11.10)

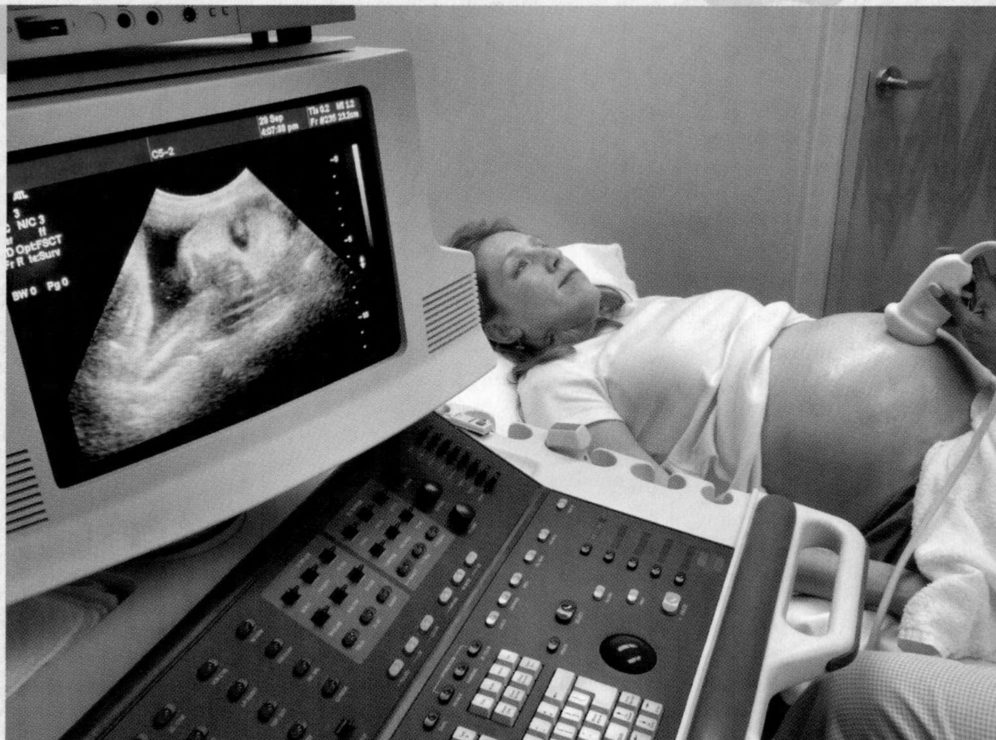

Numerous technologies are available to visualize soft tissues, and organs such as the heart, by using various forms of radiation. However, sonography technologists (sonographers) use sound waves and ultrasound scanning equipment rather than radiation. Sonography is often used to view a fetus as it develops prior to birth. Although ultrasound may reveal a developing fetus, the blueprint for its growth has already been determined through nucleic acids, the topic of this chapter.

One of the most remarkable properties of living cells is their ability to produce nearly exact replicas of themselves through hundreds of generations. Such a process requires that certain types of information be passed unchanged from one generation to the next. The transfer of necessary genetic information to new cells is accomplished by means of biomolecules called **nucleic acids.** The high molecular weight compounds represent coded information, much as words represent information in a book. It is the nearly infinite variety of possible structures that enables nucleic acids to represent the huge amount of information that must be transmitted sexually or asexually to reproduce a living organism. This information controls the inherited characteristics of the individuals in the new generation and determines the life processes as well (see ■ Figure 11.1).

ThomsonNOW™ Throughout the chapter this icon introduces resources on the ThomsonNOW website for this text. Sign in at **www.thomsonedu.com** to:

- Evaluate your knowledge of the material
- Take an exam prep quiz
- Identify areas you need to study with a **Personalized Learning Plan**.

nucleic acid
A biomolecule involved in the transfer of genetic information from existing cells to new cells.

11.1 Components of Nucleic Acids

LEARNING OBJECTIVE

1. Identify the components of nucleotides and correctly classify the sugars and bases.

Nucleic acids are classified into two categories: **ribonucleic acid (RNA),** found mainly in the cytoplasm of living cells, and **deoxyribonucleic acid (DNA),** found primarily in the nuclei of cells. Both DNA and RNA are polymers, consisting of long, linear molecules. The repeating structural units, or monomers, of the nucleic acids are called **nucleotides.** Nucleotides, however, are composed of three simpler components: a heterocyclic base, a sugar, and a phosphate (see ■ Figure 11.2). These components will be discussed individually.

ribonucleic acid (RNA)
A nucleic acid found mainly in the cytoplasm of cells.

deoxyribonucleic acid (DNA)
A nucleic acid found primarily in the nuclei of cells.

nucleotide
The repeating structural unit or monomer of polymeric nucleic acids.

Each of the five bases commonly found in nucleic acids are heterocyclic compounds that can be classified as either a pyrimidine or a purine, the parent compounds from which the bases are derived:

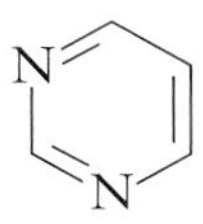

pyrimidine purine

The three pyrimidine bases are uracil, thymine, and cytosine, usually abbreviated U, T, and C. Adenine (A) and guanine (G) are the two purine bases. Adenine, guanine, and cytosine are found in both DNA and RNA, but uracil is ordinarily

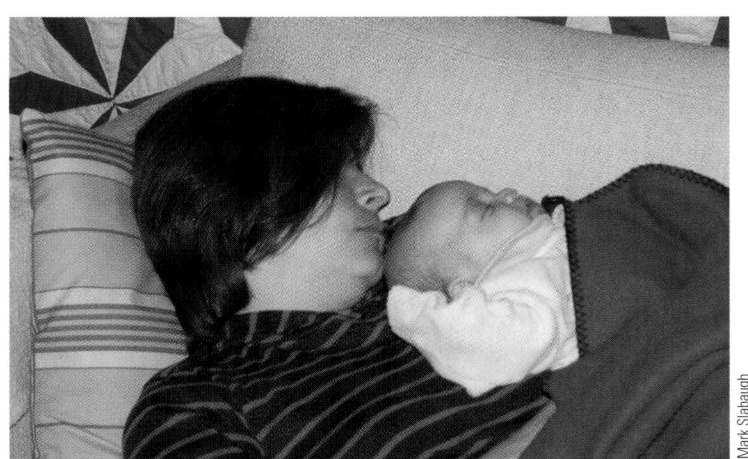

■ **FIGURE 11.1** Nucleic acids passed from the parents to offspring determine the inherited characteristics.

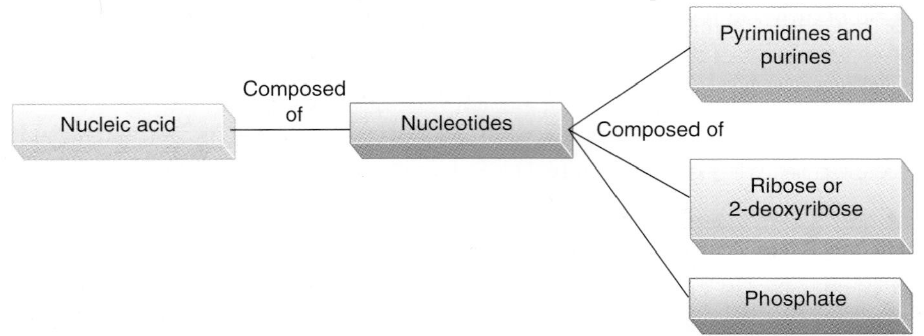

■ **FIGURE 11.2** The composition of nucleic acids.

found only in RNA, and thymine only in DNA. Structural formulas of the five bases are given in ■ Figure 11.3.

ThomsonNOW™ Go to Coached Problems to explore the **nucleotide structure.**

The sugar component of RNA is D-ribose, as the name ribonucleic acid implies. In deoxyribonucleic acid (DNA), the sugar is D-deoxyribose. Both sugars occur in the β-configuration.

HO—CH_2 OH OH OH — D-ribose

HO—CH_2 OH ← *β-configuration* OH — D-deoxyribose; *no —OH group*

Phosphate, the third component of nucleotides, is derived from phosphoric acid (H_3PO_4), which under cellular pH conditions exists in ionic form:

$$^{-}O-\overset{\overset{\displaystyle O}{\|}}{\underset{\underset{\displaystyle O^{-}}{|}}{P}}-OH$$

■ **FIGURE 11.3** The major bases of nucleic acids.

Pyrimidines

uracil (only in RNA) thymine (only in DNA) cytosine

Purines

adenine guanine

The formation of a nucleotide from these three components is represented in Reaction 11.1:

adenine

phosphate

ribose

$$^{-}O-P(=O)(O^{-})-OH + HO-CH_2\text{-ribose-adenine} \longrightarrow {}^{-}O-P(=O)(O^{-})-O-CH_2\text{-ribose-adenine} + 2H_2O \qquad (11.1)$$

adenosine 5′-monophosphate (AMP)

In a nucleotide, the base is always attached to the 1′ position of the sugar, and the phosphate is generally located at the 5′ position. The carbon atoms in the sugar are designated with a number followed by a prime to distinguish them from the atoms in the bases. Because the nucleotide of Reaction 11.1 contains ribose as the sugar, the nucleotide must be one found in RNA. The general structure of a nucleotide can be represented as

$$^{-}O-P(=O)(O^{-})-O-CH_2\text{-sugar-base}$$

OH present in RNA

or

phosphate — sugar — base

11.2 The Structure of DNA

LEARNING OBJECTIVE

2. Describe the structure of DNA.

DNA molecules, among the largest molecules known, contain between 1 and 100 million nucleotide units. The nucleotides are joined together in nucleic acids by phosphate groups that connect the 5′ carbon of one nucleotide to the 3′ carbon of the next. These linkages are referred to as *phosphodiester bonds*. The result is a chain of alternating phosphate and sugar units to which the bases are attached. The sugar–phosphate chain is referred to as the **nucleic acid backbone**, and it is constant throughout the entire DNA molecule (see ■ Figure 11.4). Thus, one DNA molecule differs from another only in the sequence, or order of attachment, of the bases along the backbone. Just as the amino acid sequence of a protein determines the primary structure of the protein, the order of the bases provides the primary structure of DNA.

ThomsonNOW™ Go to Coached Problems to explore the **DNA nucleotides** and **DNA and RNA primary structure.**

nucleic acid backbone
The sugar–phosphate chain that is common to all nucleic acids.

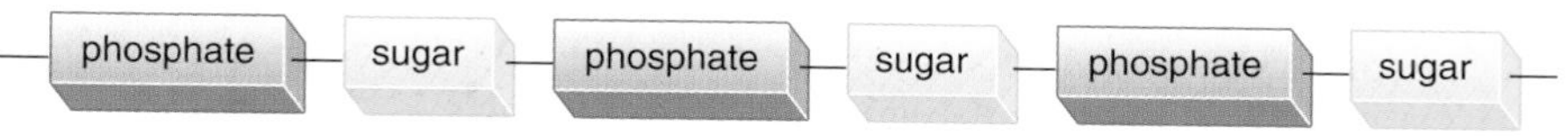

■ **FIGURE 11.4** The nucleic acid backbone.

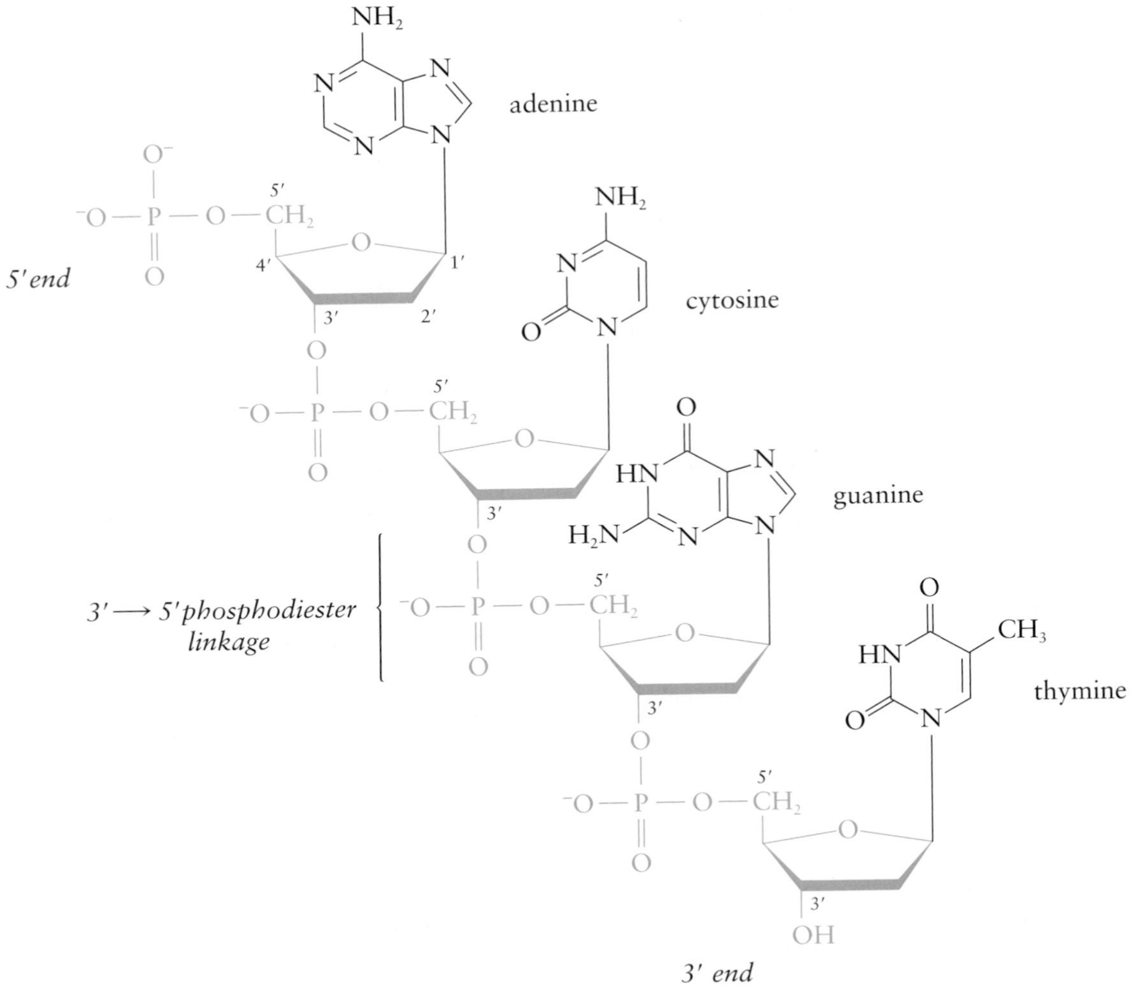

■ **FIGURE 11.5** The structure of ACGT, a tetranucleotide segment of DNA. The nucleic acid backbone is in color.

■ Figure 11.5 shows the structure of a tetranucleotide that represents a partial structural formula of a DNA molecule. The end of the polynucleotide segment of DNA in Figure 11.5 that has no nucleotide attached to the 5′ OH is called the 5′ end of the segment. The other end of the chain is the 3′ end. By convention, the sequence of bases along the backbone is read from the 5′ end to the 3′ end. Because the backbone structure along the chain does not vary, a polynucleotide structure may conveniently be abbreviated by giving only the sequence of the bases along the backbone. For example, ACGT represents the tetranucleotide in Figure 11.5.

■ **Learning Check 11.1** Draw the structural formula for a trinucleotide portion of DNA with the sequence CTG. Point out the 5′ and 3′ ends of the molecule. Indicate with arrows the phosphodiester linkages. Enclose the nucleotide corresponding to T in a box with dotted lines.

The Secondary Structure of DNA

The secondary structure of DNA was proposed in 1953 by American molecular biologist James D. Watson and English biologist Francis H. C. Crick (see ■ Fig-

ThomsonNOW™ Go to Coached Problems to explore **DNA and RNA secondary structure.**

ure 11.6). This was perhaps the greatest discovery of modern biology, and it earned its discoverers the Nobel Prize in Physiology or Medicine in 1962.

The model of DNA established by Watson and Crick was based on key contributions of other researchers, including the analysis of the base composition of DNA. The analysis of DNA from many different forms of life revealed an interesting pattern. The relative amounts of each base often varied from one organism to another, but in all DNA the percentages of adenine and thymine were always equal to each other as were the percentages of guanine and cytosine. For example, human DNA contains 20% guanine, 20% cytosine, 30% adenine, and 30% thymine. Watson and Crick concluded from these and other types of data that DNA is composed of two strands entwined around each other in a double helix, as shown in ■ Figure 11.7.

The two intertwined polynucleotide chains of the DNA double helix run in opposite (antiparallel) directions. Thus, each end of the double helix contains the 5′ end of one chain and the 3′ end of the other. The sugar–phosphate backbone is on the outside of the helix, and the bases point inward. The unique feature of the Watson and Crick structure is the way in which the chains are held together to form the double helix. They theorized that the DNA structure is stabilized by

■ **FIGURE 11.6** J. D. Watson (left) and F. H. C. Crick (right), working in the Cavendish Laboratory at Cambridge, England, built scale models of the DNA double helix based on the X-ray crystallography data of M. H. F. Wilkins and Rosalind Franklin.

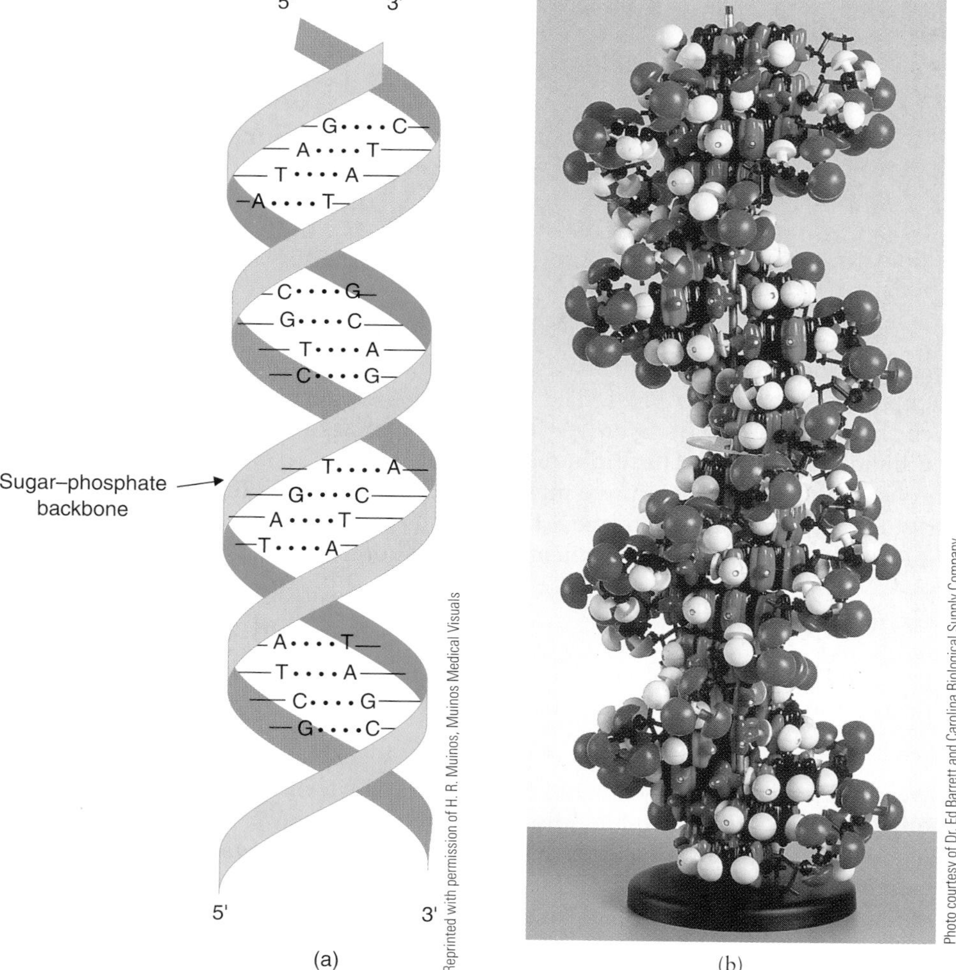

■ **FIGURE 11.7** The double helix of DNA: (a) a schematic drawing and (b) a three-dimensional molecular model. Both representations show the bases pointing toward the center of the helix away from the sugar–phosphate backbone.

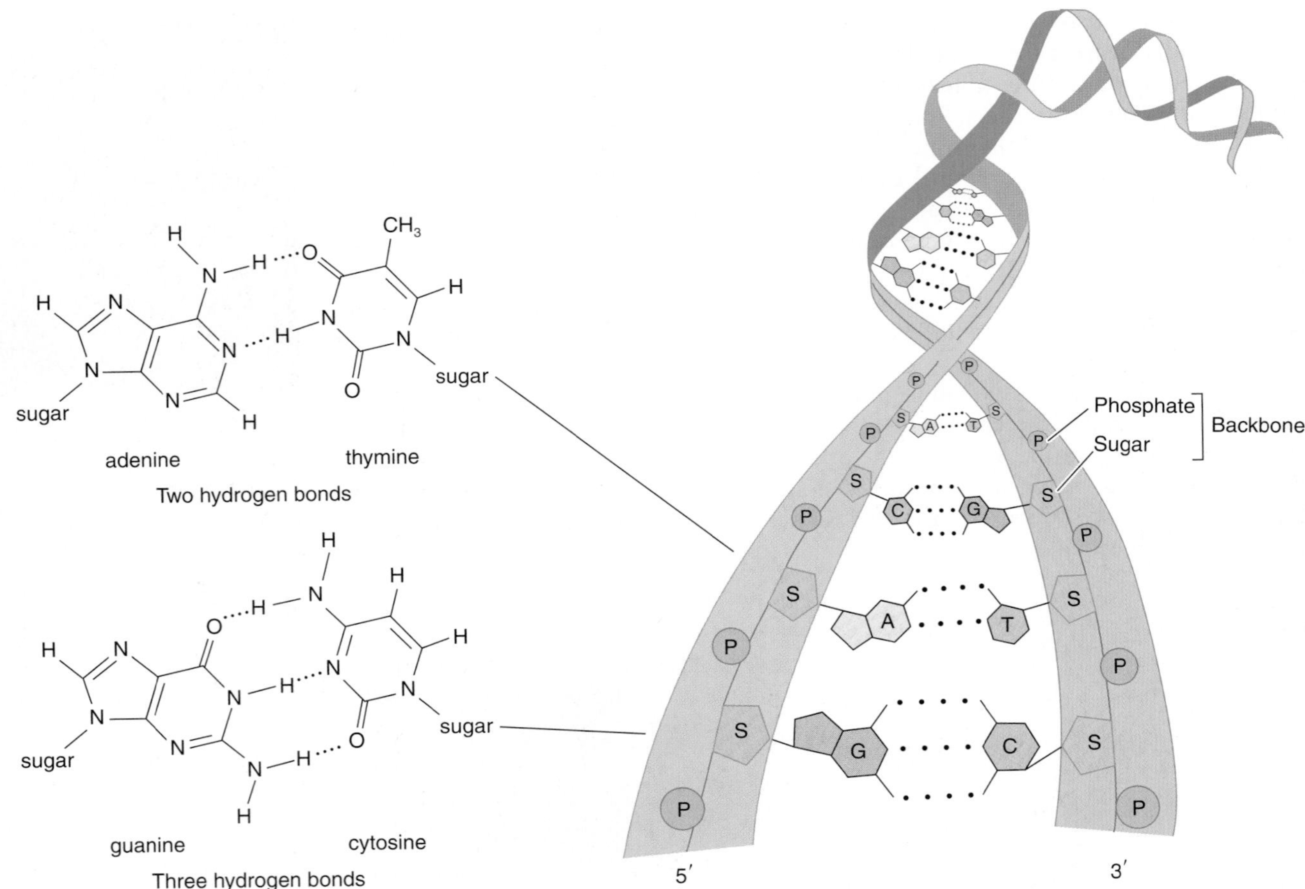

■ **FIGURE 11.8** Base pairing. Hydrogen bonding between complementary base pairs holds the two strands of a DNA molecule together.

hydrogen bonding between the bases that extend inward from the sugar–phosphate backbone (see ■ Figure 11.8). The spacing in the interior of the double helix is such that adenine always hydrogen bonds to thymine, and guanine always hydrogen bonds to cytosine. Thus, in double-helical DNA, wherever there is an adenine on one strand of the helix, there must be a thymine on the other strand, and similarly for guanine and cytosine. The two DNA strands with these matched sequences are said to be **complementary** to each other.

complementary DNA strands Two strands of DNA in a double-helical form such that adenine and guanine of one strand are matched and hydrogen bonded to thymine and cytosine, respectively, of the second strand.

EXAMPLE 11.1

One strand of a DNA molecule has the base sequence CCATTG. What is the base sequence for the complementary strand?

Solution

Three things must be remembered and used to solve this problem: (1) The base sequence of a DNA strand is always written from the 5′ end to the 3′ end; (2) adenine (A) is always paired with its complementary base thymine (T), and guanine (G) is always paired with its complement cytosine (C); and (3) in double-stranded DNA, the two strands run in opposite directions, so that the 5′ end of one strand is associated with the 3′ end of the other strand. These three ideas lead to the following:

Original strand:	5′ end C–C–A–T–T–G 3′ end
Complementary stand:	3′ end G–G–T–A–A–C 5′ end

When we follow convention and write the sequence of the complementary strand in the 5′ to 3′ direction, it becomes

$$5' \text{ C–A–A–T–G–G } 3'$$

■ **Learning Check 11.2** Write the complementary base sequence for the DNA strand TTACG.

11.3 DNA Replication

LEARNING OBJECTIVE

3. Outline the process of DNA replication.

DNA is responsible for one of the most essential functions of living organisms, the storage and transmission of hereditary information. A human cell normally contains 46 structural units called **chromosomes** (see ■ Figure 11.9). Each chromosome contains one molecule of DNA coiled tightly about a group of small, basic proteins called histones. Individual sections of DNA molecules make up **genes**, the fundamental units of heredity. In Section 11.5, you'll learn that each gene directs the synthesis of a specific protein. The number of genes contained in the structural unit of an organism varies with the type of organism. For example, a virus, the

chromosome
A tightly packed bundle of DNA and protein that is involved in cell division.

gene
An individual section of a chromosomal DNA molecule that is the fundamental unit of heredity.

OVER THE COUNTER 11.1 Nucleic Acid Supplements

Because of the fundamental roles nucleic acids play in life itself, it is not surprising that many claims of health benefits related to these substances have been made. As a result of these claims, DNA, RNA, and their various derivatives are well represented on the dietary supplements shelves of retailers. In addition, numerous shampoos and other cosmetics that contain DNA and RNA are also available. The idea behind the use of such products is that as we age, our DNA/RNA becomes depleted or defective, and must be replaced if good health is to be maintained or restored. It has been suggested that injections or oral supplements could be used to accomplish the replacement.

The health benefits claimed for such therapy include slowing the aging process, improving memory and mental processes, stimulating the immune system, and fighting cancer. There has been almost universal doubt about such health benefits, especially if the nucleic acids were taken orally. It has been argued that these substances would be destroyed during digestion and therefore could not reach target organs and tissues of the body.

Only very recently have reports been made of scientific evidence that nucleic acid supplementation might provide some benefits in terms of positively influencing the immune response of the body. Evidence to support the claims of slowing the aging process and improving memory and mental processes is very inconclusive and not promising at this time. There has been very little investigation of the anti-cancer properties of RNA and DNA in their native forms. However, some synthetic forms of the nucleic acid have been studied in this regard and some promising results have been obtained in the treatment of specific types of cancer.

The only negative claim about using nucleic acid supplements is that they may produce excessive uric acid and promote or aggravate gout. Because of this, anyone with gout or a tendency to develop it should not use nucleic acid supplements without first obtaining a physician's approval.

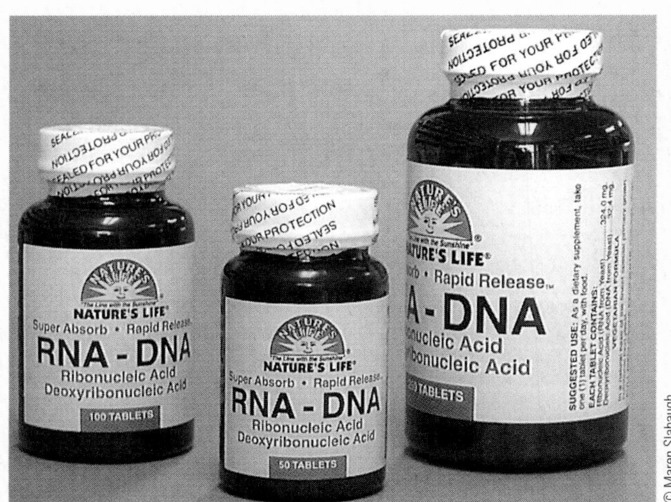

Dietary supplements containing nucleic acids.

■ **FIGURE 11.9** Chromosomes. (a) The 46 chromosomes of a human cell. (b) Each chromosome is a protein-coated strand of multicoiled DNA.

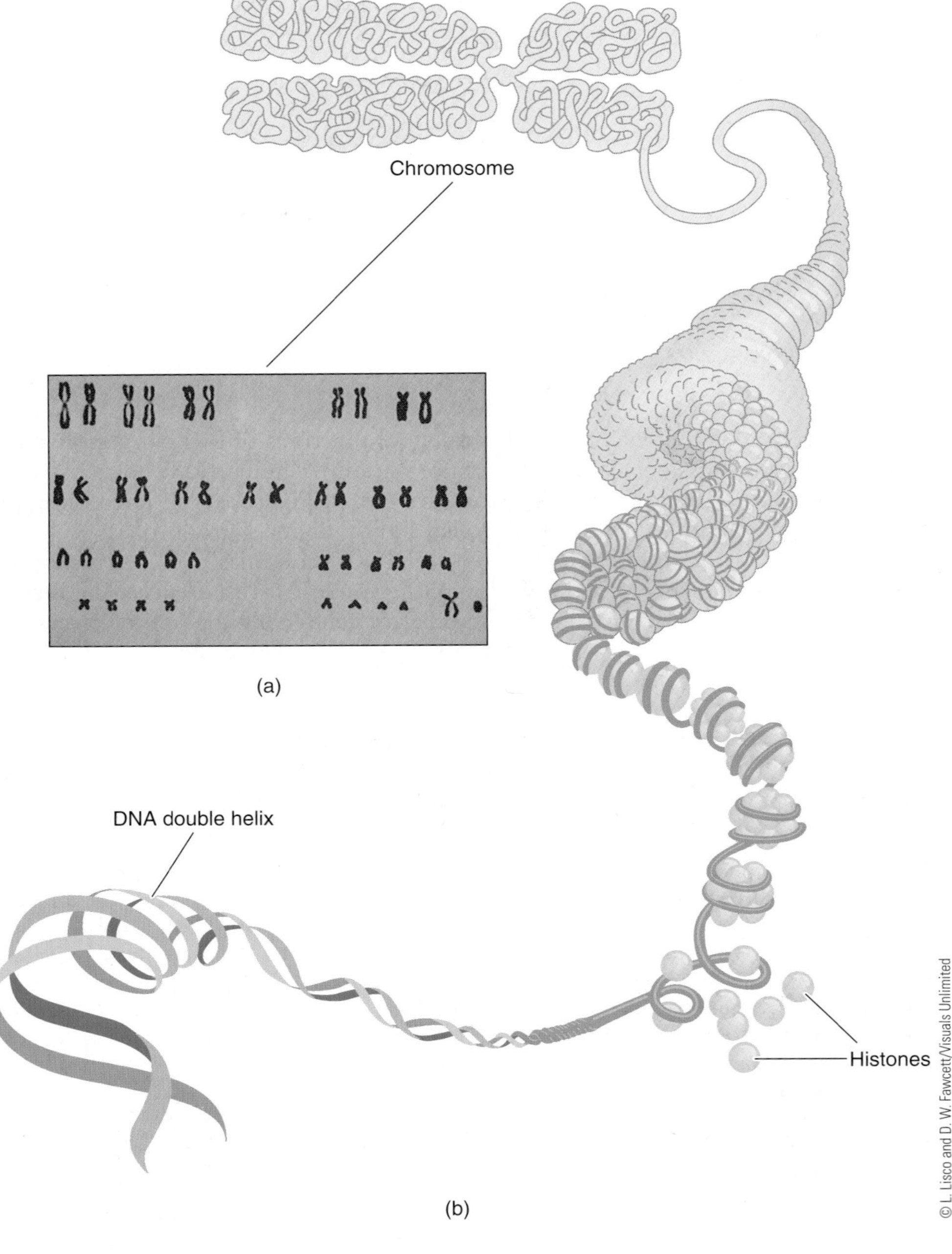

ThomsonNOW™ Go to Chemistry Interactive to explore the **DNA replication.**

smallest structure known to carry genetic information, is thought to contain from a few to several hundred genes. A bacterial cell, such as *Escherichia coli (E. coli),* contains about 1000 genes, whereas a human cell contains approximately 25,000.

One of the elegant features of the Watson–Crick three-dimensional structure for DNA was that it provided the first satisfactory explanation for the transmission of hereditary information—it explained how DNA is duplicated for the next generation.

replication
The process by which an exact copy of a DNA molecule is produced.

The process by which an exact copy of DNA is produced is called **replication.** It occurs when two strands of DNA separate and each serves as a template (pattern) for the construction of its own complement. The process generates new DNA double-stranded molecules that are exact replicas of the original DNA molecule. A schematic diagram of the replication process is presented in ■ Active Figure 11.10. Notice that the two daughter DNA molecules have exactly the same base

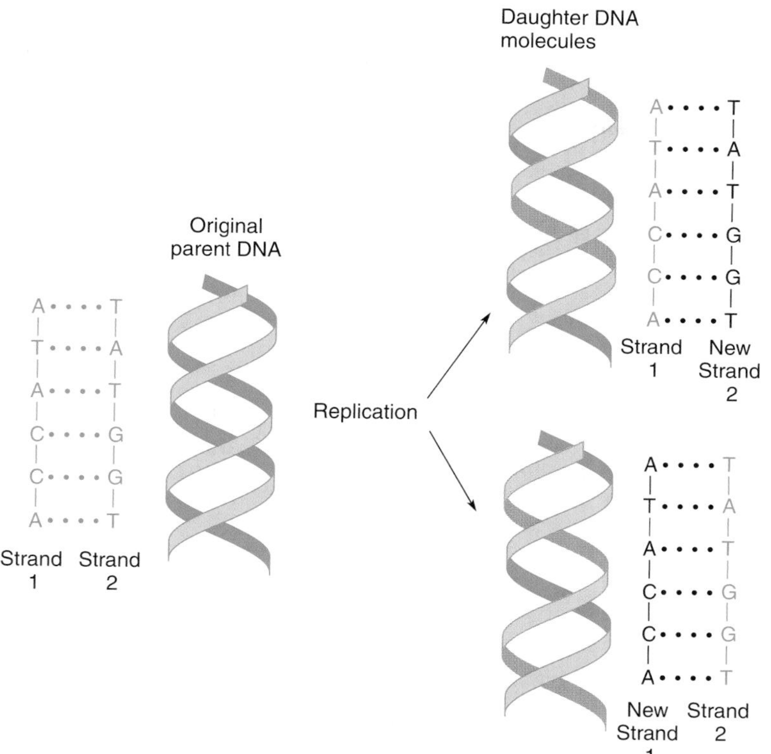

ACTIVE FIGURE 11.10 A schematic diagram of the replication of DNA. **Sign in at www.thomsonedu.com to explore an interactive version of this figure.**

sequences as the original parent DNA. Notice, too, that each daughter contains one strand of the parent and one new strand that is complementary to the parent strand. This type of replication is called **semiconservative replication.**

semiconservative replication A replication process that produces DNA molecules containing one strand from the parent and a new strand that is complementary to the strand from the parent.

The process of replication can be viewed as occurring in three steps.

Step 1. Unwinding of the double helix: Replication begins when the enzyme helicase catalyzes the separation and unwinding of the nucleic acid strands at a specific point along the DNA helix. In this process, hydrogen bonds between complementary base pairs are broken, and the bases that were formerly in the center of the helix are exposed. The point where this unwinding takes place is called a **replication fork** (see Figure 11.11).

replication fork A point where the double helix of a DNA molecule unwinds during replication.

Step 2. Synthesis of DNA segments: DNA replication takes place along both nucleic acid strands separated in Step 1. The process proceeds from the 3′ end toward the 5′ end of the exposed strand (the template). Because the two strands are antiparallel, the synthesis of new nucleic acid strands proceeds toward the replication fork on one exposed strand and away from the replication fork on the other strand (Figure 11.11). New (daughter) DNA strands form as nucleotides, complementary to those on the exposed strands, are linked together under the influence of the enzyme DNA polymerase. Notice that the daughter chain grows from the 5′ end toward the 3′ end as the process moves from the 3′ end of the exposed (template) strand toward the 5′ end. As the synthesis proceeds, a second replication fork is created when a new section of DNA unwinds. The daughter strand that was growing toward the first replication fork continues growing smoothly toward the new fork. However, the other daughter strand was growing away from the first fork. A new segment of this strand begins growing from the new fork but is not initially bound to the segment that grew from the first fork. Thus, as the parent DNA progressively unwinds, this daughter strand is synthesized as a series of fragments that are bound together in Step 3. The gaps or

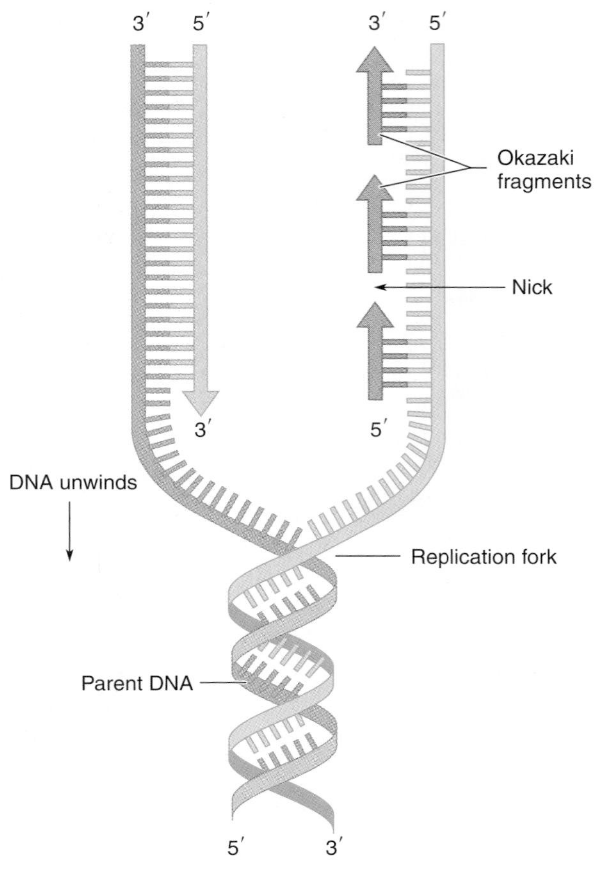

■ **FIGURE 11.11** The replication of DNA. Both new strands are growing in the 5′ to 3′ direction.

Okazaki fragment
A DNA fragment produced during replication as a result of strand growth in a direction away from the replication fork.

breaks between segments in this daughter strand are called nicks, and the DNA fragments separated by the nicks are called **Okazaki fragments** after their discoverer, Reiji Okazaki (Figure 11.11).

Step 3. Closing the nicks: One daughter DNA strand is synthesized without any nicks, but the Okazaki fragments of the other strand must be joined together. An enzyme called DNA ligase catalyzes this final step of DNA replication. The result is two DNA double-helical molecules that are identical to the original molecule.

Observations with an electron microscope show that the replication of DNA molecules in eukaryotic cells occurs simultaneously at many points along the original DNA molecule. These replication zones blend together as the process of replication continues. This is necessary if long molecules are to be replicated rapidly. For example, it is estimated that the replication of the largest chromosome in *Drosophila* (the fruit fly) would take more than 16 days if there were only one origin for replication. Research results indicate that the actual replication is accomplished in less than 3 minutes because it takes place simultaneously at more than 6000 replication forks per DNA molecule. ■ Active Figure 11.12 shows a schematic diagram in which replication is proceeding at several points in the DNA chain.

A knowledge of the DNA replication process led scientist Kary Mullis in 1983 to a revolutionary laboratory technique called the *polymerase chain reaction (PCR)*. The PCR technique mimics the natural process of replication, in which the DNA double helix unwinds. As the two strands separate, DNA polymerase makes a copy using each strand as a template. To perform PCR, a small quantity of the target DNA is added to a test tube along with a buffered solution containing DNA polymerase, the cofactor $MgCl_2$, the four nucleotide building blocks, and primers.

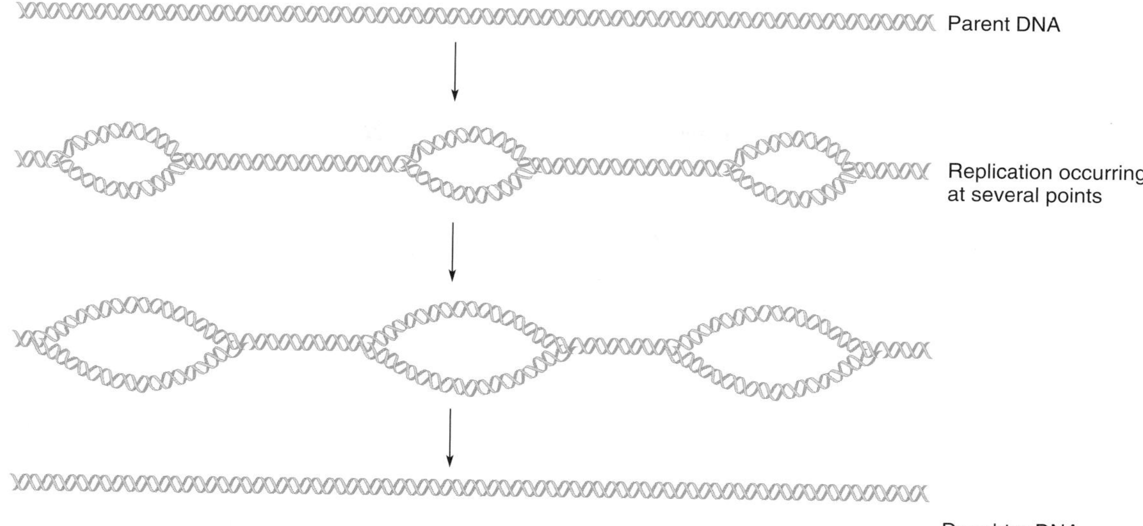

ACTIVE FIGURE 11.12 A schematic representation of eukaryotic chromosome replication. The new DNA is shown in purple. **Sign in at www.thomsonedu.com to explore an interactive version of this figure.**

The primers are short polynucleotide segments that will bind to the separated DNA strands and serve as starting points for new chain growth. The PCR mixture is taken through three-step replication cycles:

1. Heat (94°C–96°C) is used for one to several minutes to unravel (denature) the DNA into single strands.
2. The tube is cooled to 50°C–65°C for one to several minutes, during which the primers hydrogen-bond to the separated strands of target DNA.
3. The tube is heated to 72°C for one to several minutes, during which time the DNA polymerase synthesizes new strands.

Each cycle doubles the amount of DNA. Following 30 such cycles, a theoretical amplification factor of 1 billion is attained.

Almost overnight, PCR became a standard research technique for detecting all manner of mutations associated with genetic disease (Section 11.9). PCR can also be used to detect the presence of unwanted DNA, as in the case of a bacterial or viral infection. Conventional tests that involve the culture of microorganisms or the use of antibodies can take weeks to perform. PCR offers a fast and simple alternative. The ability of PCR to utilize degraded DNA samples and sometimes the DNA from single cells is of great interest to forensic scientists. PCR has also permitted DNA to be amplified from some unusual sources, such as extinct mammals, Egyptian mummies, and ancient insects trapped in amber.

11.4 Ribonucleic Acid (RNA)

LEARNING OBJECTIVE

4. Contrast the structures of DNA and RNA and list the function of the three types of cellular RNA.

RNA, like DNA, is a long, unbranched polymer consisting of nucleotides joined by $3' \rightarrow 5'$ phosphodiester bonds. The number of nucleotides in an RNA molecule ranges from as few as 73 to many thousands. The primary structure of RNA

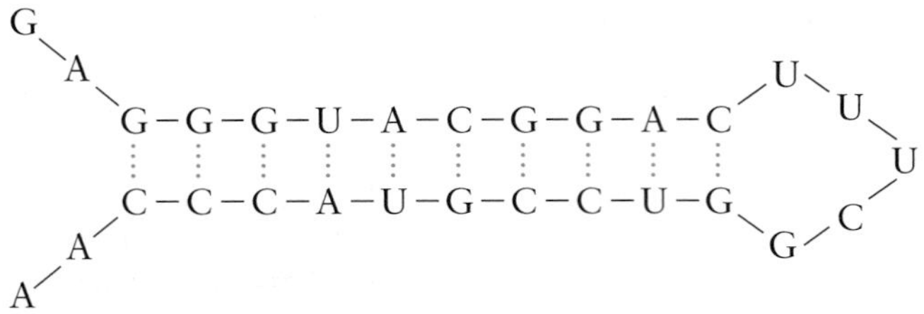

FIGURE 11.13 A portion of RNA that has folded back on itself and formed a double-helical region.

```
G
 \
  A                                  U
   \                               /   \
    G-G-G-U-A-C-G-G-A-C             U
    : : : : : : : : : :              \
    C-C-C-A-U-G-C-C-U-G               U
   /                   \             /
  A                     G         C
 /                        \     /
A                           G
```

differs from that of DNA in two ways. As we learned in Section 11.1, the sugar unit in RNA is ribose rather than deoxyribose. The other difference is that RNA contains the base uracil (U) instead of thymine (T).

The secondary structure of RNA is also different from that of DNA. RNA molecules are single-stranded, except in some viruses. Consequently, an RNA molecule need not contain complementary base ratios of 1:1. However, RNA molecules do contain regions of double-helical structure where they form loops (see Figure 11.13). In these regions, adenine (A) pairs with uracil (U), and guanine

CHEMISTRY AROUND US 11.1 The Clone Wars

Any gardener who has used a cutting from one plant to grow an identical plant has created a clone of the original plant, since clones are organisms that contain identical DNA. Based on that definition, identical twins are essentially clones of each other. The first successful laboratory cloning of an animal involved a tadpole, and occurred in 1952. It took more than 40 years of research before a mammal, Dolly the sheep, was cloned from an adult cell rather than an embryonic cell. Since that success in 1996, numerous other successful mammal clonings have been achieved in the laboratory.

Research into cloning methods and procedures continues today and has proved to be a sensitive ethical and societal issue. A repeatedly asked question is: Why do it? Supporters of the practice give numerous answers to the question, including the need to save rare or endangered animal species, and the need to produce more quality food for an ever-increasing world population.

The cloning procedure involves three basic steps. First, a donor cell is obtained from an animal that is to be cloned. Next, an egg cell is obtained from an animal that will act as the surrogate mother for the cloned animal. The DNA-containing nucleus of the egg cell is removed and replaced with the nucleus (and the DNA) from the donor cell of the animal being cloned. Finally, the modified egg cell is re-implanted in the surrogate mother, and the embryo, which now contains DNA identical to that of the donor animal, develops. The resulting cloned animal is an exact genetic copy of the donor animal and has all the properties, characteristics, etc., of the donor animal.

This cloning technique has already been used to produce cloned copies of two endangered species of cows: the guar, an ox-like animal from India, and the banteng, a cow from southeast Asia. The San Diego Zoo maintains a "frozen zoo" that contains preserved frozen cells of animals that can be used as future sources of DNA to reproduce or expand the numbers of selected animals.

The high cost of cloning has not prevented some enterprising individuals from commercializing the process. For example, a California company will clone your favorite pet for a nominal fee of $50,000. Surprisingly (or maybe not for pet lovers), the company has had some customers interested in paying for their services.

The obvious extension of this technique to the cloning of humans is the basis for strong opposition from some individuals and organizations. They take the position that no one should play God and create animals, especially humans, even if it is scientifically possible. Cloning supporters point out that rather than creating humans, cloning could help humans have better-quality lives. They say that cloning is not an end in itself but can lead to scientific advances that will benefit human beings. One potential use in this category is to genetically engineer and then mass produce animals that can produce and secrete specific human proteins in their blood or milk. Potentially, such transgenic animals could produce therapeutic proteins for treating or studying human diseases. Another potential application is the use of cloning techniques to produce animal cells and organs that would not trigger the immune system of humans to reject such cells and organs used as transplants. Thus, an animal pancreas with such characteristics could be transplanted into a human suffering from diabetes and cure the disease.

It appears likely that disagreements about the appropriateness of conducting cloning research will continue. If cool heads prevail, a reasonable compromise will be found, and a level of research that provides minimal upset to opponents and maximum benefits to humanity in general will continue.

(G) pairs with cytosine (C). The proportion of helical regions varies over a wide range depending on the kind of RNA studied, but a value of 50% is typical.

RNA is distributed throughout cells; it is present in the nucleus, the cytoplasm, and in mitochondria (Table 8.3; Section 12.7). Cells contain three kinds of RNA: messenger RNA (mRNA), ribosomal RNA (rRNA), and transfer RNA (tRNA). Each of these kinds of RNA performs an important function in protein synthesis.

Messenger RNA

Messenger RNA (mRNA) functions as a carrier of genetic information from the DNA of the cell nucleus directly to the cytoplasm, where protein synthesis takes place. The bases of mRNA are in a sequence that is complementary to the base sequence of one of the strands of the nuclear DNA. In contrast to DNA, which remains intact and unchanged throughout the life of the cell, mRNA has a short lifetime—usually less than an hour. It is synthesized as needed and then rapidly degraded to the constituent nucleotides.

messenger RNA (mRNA)
RNA that carries genetic information from the DNA in the cell nucleus to the site of protein synthesis in the cytoplasm.

Ribosomal RNA

Ribosomal RNA (rRNA) constitutes 80–85% of the total RNA of the cell. It is located in the cytoplasm in organelles called **ribosomes,** which are about 65% rRNA and 35% protein. The function of ribosomes as the sites of protein synthesis is discussed in Section 11.8.

ribosomal RNA (rRNA)
RNA that constitutes about 65% of the material in ribosomes, the sites of protein synthesis.

ribosome
A subcellular particle that serves as the site of protein synthesis in all organisms.

Transfer RNA

Transfer RNA (tRNA) molecules deliver amino acids, the building blocks of proteins, to the site of protein synthesis. Cells contain at least one specific type of tRNA for each of the 20 common amino acids found in proteins. These tRNA molecules are the smallest of all the nucleic acids, containing 73–93 nucleotides per chain.

transfer RNA (tRNA)
RNA that delivers individual amino acid molecules to the site of protein synthesis.

A comparison of RNA characteristics for the three different forms is summarized in ■ Table 11.1.

Because of their small size, a number of tRNAs have been studied extensively. ■ Figure 11.14 shows a representation of the secondary structure of a typical tRNA. This tRNA molecule, like all others, has regions where there is hydrogen bonding between complementary bases, and regions (loops) where there is no hydrogen bonding.

Two regions of tRNA molecules have important functions during protein synthesis. One of these regions, designated the **anticodon,** enables the tRNA to bind to mRNA during protein synthesis. The second important site is the 3′ end

anticodon
A three-base sequence in tRNA that is complementary to one of the codons in mRNA.

TABLE 11.1 Different forms of RNA molecules in *Escherichia coli* (*E. coli*)

Class of RNA	% in cells	Number of RNA subtypes	Number of nucleotides
Ribosomal RNA (rRNA)	80	3	120 1700 3700
Transfer RNA (tRNA)	15	46	73–93
Messenger RNA (mRNA)	5	Many	75–3000

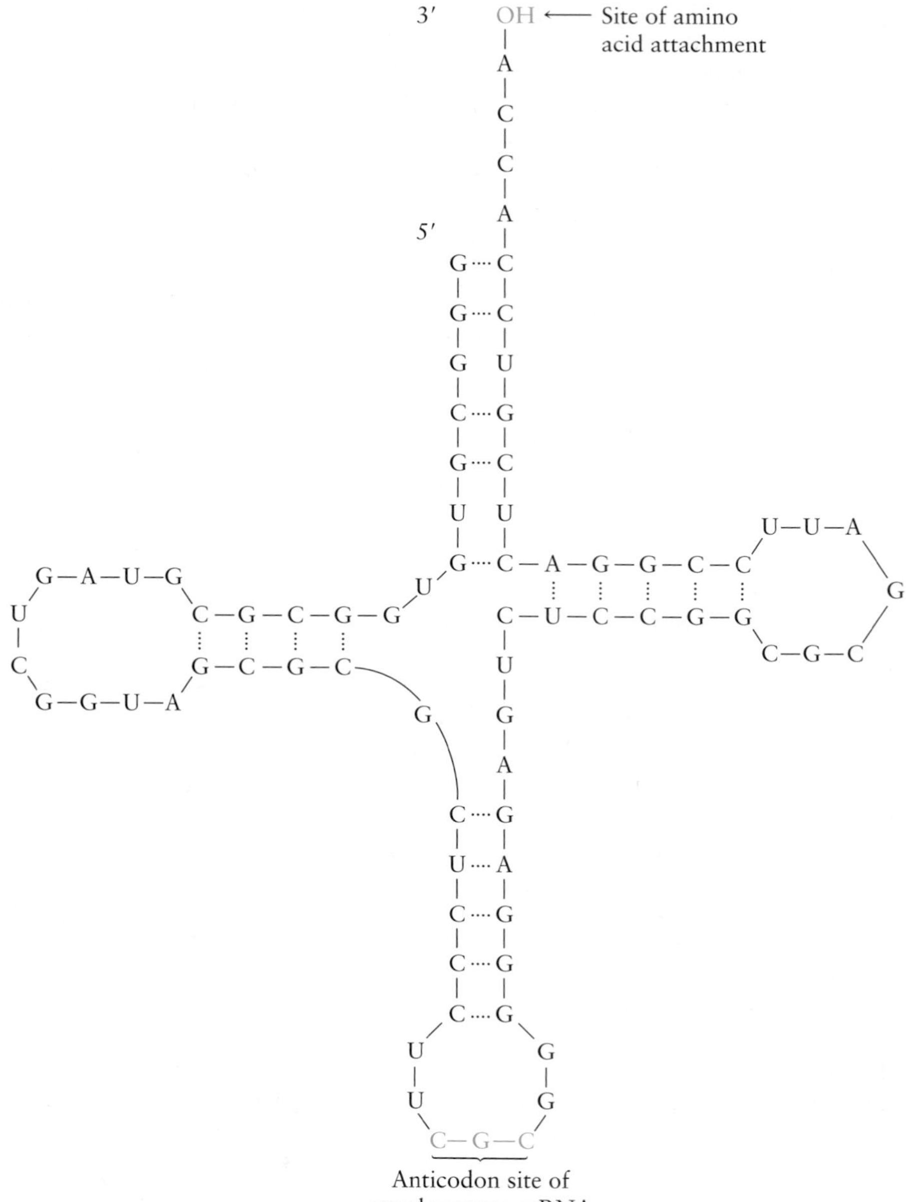

■ **FIGURE 11.14** The typical tRNA cloverleaf structure.

of the molecule, which binds to an amino acid and transports it to the site of protein synthesis. Each amino acid is joined to the 3′ end of its specific tRNA by an ester bond that forms between the carboxyl group of the amino acid and the 3′ hydroxy group of ribose. The reaction is catalyzed by an enzyme that matches tRNA molecules to their proper amino acids. These enzymes, which are very specific for both the structure of the amino acid and the tRNA, rarely cause a bond to form between an amino acid and the wrong tRNA. When a tRNA molecule is attached to its specific amino acid, it is said to be "activated" because it is ready to participate in protein synthesis. ■ Figure 11.15 shows the general structure of an activated tRNA and a schematic representation that will be used in Section 11.8 to describe protein synthesis.

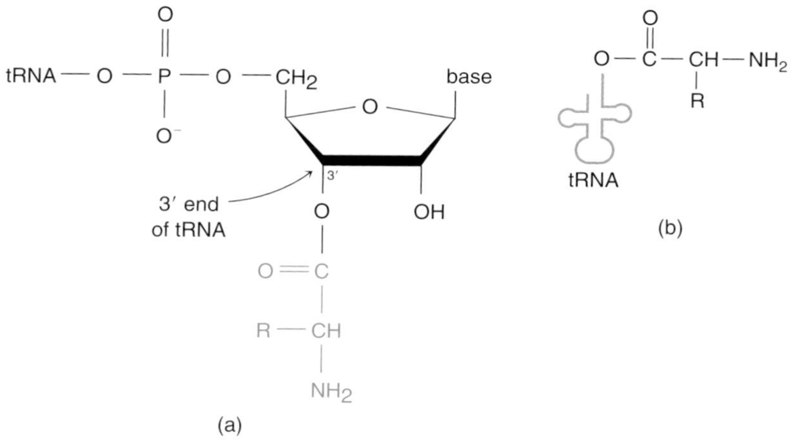

■ **FIGURE 11.15** An activated tRNA: (a) general structure and (b) a schematic representation.

11.5 The Flow of Genetic Information

LEARNING OBJECTIVE

5. Describe what is meant by the terms *transcription* and *translation*.

You learned in Section 11.3 that DNA is the storehouse of genetic information in the cell and that the stored information can be passed on to new cells as DNA undergoes replication. In this section, you will discover how the information stored in DNA is expressed within the cell.

This process of expression is so well established that it is called the **central dogma of molecular biology.** The accepted dogma, or principle, says that genetic information contained in DNA molecules is transferred to RNA molecules. The transferred information of RNA molecules is then expressed in the structure of synthesized proteins. In other words, the genetic information in DNA (genes) directs the synthesis of certain proteins. In fact, there is a specific DNA gene for every protein in the body. DNA does not direct the synthesis of carbohydrates, lipids, or the other nonprotein substances essential for life. However, these other materials are manufactured by the cell through reactions made possible by enzymes (proteins) produced under the direction of DNA. Thus, in this respect, the information stored in DNA really does determine every characteristic of the living organism.

central dogma of molecular biology The well-established process by which genetic information stored in DNA molecules is expressed in the structure of synthesized proteins.

Two steps are involved in the flow of genetic information: transcription and translation. In higher organisms (eukaryotes), the DNA containing the stored information is located in the nucleus of the cell, and protein synthesis occurs in the cytoplasm. Thus, the stored information must first be carried out of the nucleus. This is accomplished by **transcription,** or transferring the necessary information from a DNA molecule onto a molecule of messenger RNA. The appropriately named messenger RNA carries the information (the message) from the nucleus to the site of protein synthesis in the cellular cytoplasm.

transcription The transfer of genetic information from a DNA molecule to a molecule of messenger RNA.

In the second step, the mRNA serves as a template on which amino acids are assembled in the proper sequence necessary to produce the specified protein. This takes place when the code or message carried by mRNA is **translated** into an amino acid sequence by tRNA. There is an exact word-to-word translation from mRNA to tRNA. Thus, each word in the mRNA language has a corresponding word in the amino acid language of tRNA. This communicative relationship between mRNA nucleotides and amino acids is called the *genetic code* (Section 11.7). ■ Figure 11.16 summarizes the mechanisms for the flow of genetic information in the cell.

translation The conversion of the code carried by messenger RNA into an amino acid sequence of a protein.

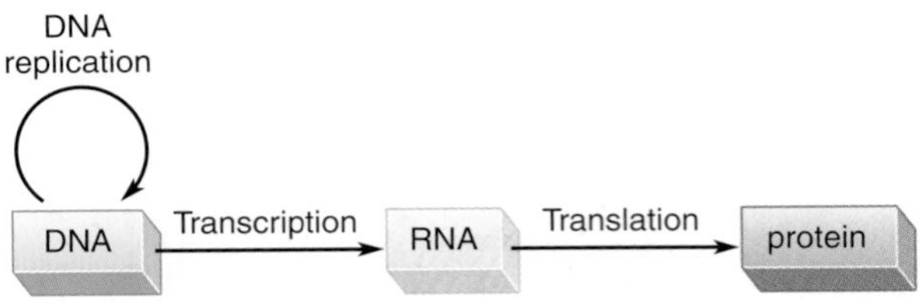

■ **FIGURE 11.16** The flow of genetic information in the cell.

11.6 Transcription: RNA Synthesis

LEARNING OBJECTIVE

6. Describe the process by which RNA is synthesized in cells.

ThomsonNOW™ Go to Chemistry Interactive to explore **DNA transcription.**

An enzyme called RNA polymerase catalyzes the synthesis of RNA. During the first step of the process, the DNA double helix begins to unwind at a point near the gene that is to be transcribed. Only one strand of the DNA molecule is transcribed. Ribonucleotides are linked together along the unwound DNA strand in a sequence determined by complementary base pairing of the DNA strand bases and ribonucleotide bases.

The DNA strand always has one sequence of bases recognized by RNA polymerase as the initiation or starting point. Starting at this point, the enzyme catalyzes the synthesis of mRNA in the 5′ to 3′ direction until it reaches another sequence of bases that designates the termination point. Because the complementary chains of RNA and DNA run in opposite directions, the enzyme must move along the DNA template in the 3′ to 5′ direction (see ■ Figure 11.17). Once the mRNA molecule has been synthesized, it moves away from the DNA template, which then rewinds to form the original double-helical structure. Transfer RNA and ribosomal RNA are synthesized in the same way, with DNA serving as a template.

EXAMPLE 11.2

Write the sequence for the mRNA that could be synthesized using the following DNA base sequence as a template:

5′ G–C–A–A–C–T–T–G 3′

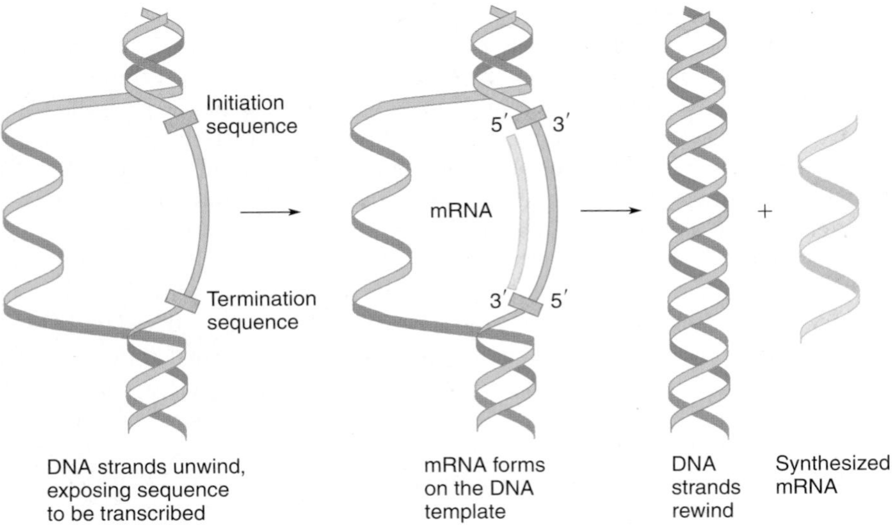

■ **FIGURE 11.17** The synthesis of mRNA.

Solution

RNA synthesis begins at the 3′ end of the DNA template and proceeds toward the 5′ end. The complementary RNA strand is formed from the bases C, G, A, and U. Uracil (U) is the complement of adenine (A) on the DNA template.

		direction of strand →	
DNA template:	5′	G–C–A–A–C–T–T–G	3′
New mRNA:	3′	C–G–U–U–G–A–A–C	5′
		← direction of strand	

Writing the sequence of the new mRNA in the 5′ to 3′ direction, it becomes

5′ C–A–A–G–U–U–G–C 3′

Learning Check 11.3 Write the sequence for the mRNA that could be synthesized on the following DNA template:

5′ A–T–T–A–G–C–C–G 3′

Although the general process of transcription operates universally, there are differences in detail between the processes in prokaryotes and eukaryotes. Each gene of prokaryotic cells exists as a continuous segment of a DNA molecule. Transcription of this gene segment produces mRNA that undergoes translation into a

CHEMISTRY AROUND US 11.2 The Race Against Avian Flu

Viruses are infectious particles with a simple structure consisting of DNA or RNA enclosed in a protein coat. Their simple structure prevents them from independently reproducing themselves. In order to reproduce, they must invade host cells and use the host cells' reproductive processes to synthesize virus components. The host cells then rupture, releasing the new virus particles, which then invade other host cells to repeat the process.

A serious concern today is focused on the avian flu or bird flu virus that is designated as H5N1. It has been found primarily in Asian countries and accounted for 140 million bird deaths between 2004 and 2006. During that same time, 116 humans were also infected, and 50% of them died. The concern today is based on the characteristic of viruses to mutate and the agreement of authorities that the Avian Flu virus will eventually mutate into a form that can be passed from human to human. When this occurs, it will cause a worldwide epidemic (pandemic). Two important questions remain unanswered: When will this occur? and How severe will it be?

Bird flu in humans is difficult to treat, so it is wise to take precautions. The following recommendations are designed to provide at least a minimal level of protection against contracting bird flu because of carelessness.

1. Be knowledgeable. Refer to reliable news sources and current reports.
2. Protect yourself by getting the standard influenza vaccination. This yearly shot won't protect against a bird flu pandemic, but it does protect against common flu viruses.
3. See your doctor within two days of showing flu symptoms, such as cough, fever, sore throat, and aching muscles.
4. Wash your hands frequently with soap and water, or use alcohol-based hand sanitizers. Hand washing is especially important before handling food or touching your nose, mouth, or eyelids. Lather your hands for at least 15 seconds during each washing.
5. Stay healthy by eating a healthy diet, getting adequate sleep, and exercising regularly.
6. Think carefully about travel. Influenza viruses spread easily when people are confined to small spaces such as airplanes, trains, or buses.

Avian flu accounted for 140 million bird deaths in recent years.

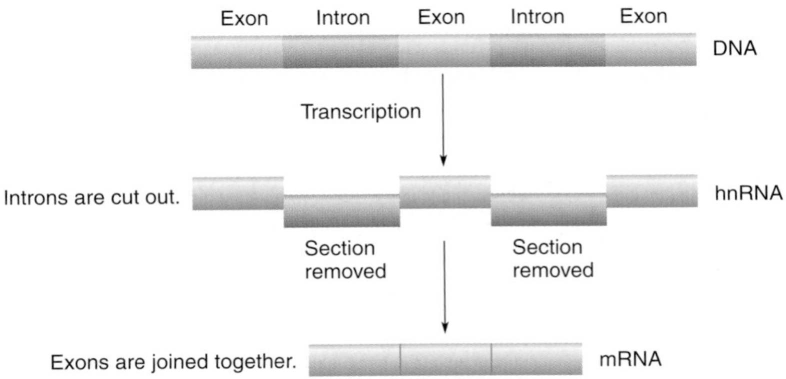

■ **FIGURE 11.18** Segments of hnRNA formed from introns of eukaryotic cells are removed by special enzymes to produce mRNA.

protein almost immediately because there is no nuclear membrane in a prokaryote separating DNA from the cytoplasm. In 1977, however, it was discovered that the genes of eukaryotic cells are segments of DNA that are "interrupted" by segments that do not code for amino acids. These DNA segments that carry no amino acid code are called **introns,** and the coded DNA segments are called **exons.**

intron
A segment of a eukaryotic DNA molecule that carries no codes for amino acids.

exon
A segment of a eukaryotic DNA molecule that is coded for amino acids.

heterogeneous nuclear RNA (hnRNA)
RNA produced when both introns and exons of eukaryotic cellular DNA are transcribed.

When transcription occurs in the nuclei of eukaryotic cells, both introns and exons are transcribed. This produces what is called **heterogeneous nuclear RNA (hnRNA).** This long molecule of hnRNA then undergoes a series of enzyme-catalyzed reactions that cut and splice the hnRNA to produce mRNA (see ■ Figure 11.18). The mRNA resulting from this process contains only the sequence of bases that actually codes for protein synthesis. Although the function of introns in eukaryotic DNA is not yet understood, it is an area of active research.

11.7 The Genetic Code

LEARNING OBJECTIVE

7. Explain how the genetic code functions in the flow of genetic information.

Discovery of the Genetic Code

By 1961, it was clear that the sequence of bases in DNA serves to direct the synthesis of mRNA, and that the sequence of bases in mRNA corresponds to the order of amino acids in a particular protein. However, the genetic code, the exact relationship between mRNA sequences and amino acids, was unknown. At least 20 mRNA "words" are needed to represent uniquely each of the 20 amino acids found in proteins. If the mRNA words consisted of a single letter represented by a base (A, C, G, or U), only four amino acids could be uniquely represented. Thus, it was proposed that it is not one mRNA base but a combination of bases that codes for each amino acid. For example, if a code word consists of a sequence of two mRNA bases, there are $4^2 = 16$ possible combinations, so 16 amino acids could be represented uniquely. This is a more extensive code, but it still contains too few words to do the job for 20 amino acids. If the code consists of a sequence of three bases, there are $4^3 = 64$ possible combinations, more than enough to specify uniquely each amino acid in the primary sequence of a protein. Research has confirmed that nature does indeed use three-letter code words (a triplet code) to store and express genetic information. Each sequence of three nucleotide bases that represents code words on mRNA molecules is called a **codon.**

codon
A sequence of three nucleotide bases that represents a code word on mRNA molecules.

After the discovery of three-letter codons, researchers were anxious to answer the next question: Which triplets of bases (codons) code for which amino acids? In 1961, Marshall Nirenberg and his coworkers attempted to break the code in a

TABLE 11.2 The genetic code: mRNA codons for each of the 20 amino acids

Amino acid	Codons	Number of codons
alanine	GCA, GCC, GCG, GCU	4
arginine	AGA, AGG, CGA, CGC, CGG, CGU	6
asparagine	AAC, AAU	2
aspartic acid	GAC, GAU	2
cysteine	UGC, UGU	2
glutamic acid	GAA, GAG	2
glutamine	CAA, CAG	2
glycine	GGA, GGC, GGG, GGU	4
histidine	CAC, CAU	2
isoleucine	AUA, AUC, AUU	3
leucine	CUA, CUC, CUG, CUU, UUA, UUG	6
lysine	AAA, AAG	2
methionine, initiation	AUG	1
phenylalanine	UUC, UUU	2
proline	CCA, CCC, CCG, CCU	4
serine	UCA, UCC, UCG, UCU, AGC, AGU	6
threonine	ACA, ACC, ACG, ACU	4
tryptophan	UGG	1
tyrosine	UAC, UAU	2
valine	GUA, GUC, GUG, GUU	4
Stop signals	UAG, UAA, UGA	3
Total number of codons		64

ThomsonNOW™ Go to Coached Problems to practice **identifying codons.**

very ingenious way. They made a synthetic molecule of mRNA consisting of uracil bases only. Thus, this mRNA contained only one codon, the triplet UUU. They incubated this synthetic mRNA with ribosomes, amino acids, tRNAs, and the appropriate enzymes for protein synthesis. The exciting result of this experiment was that a polypeptide that contained only phenylalanine was synthesized. Thus, the first word of the genetic code had been deciphered; UUU = phenylalanine.

A series of similar experiments by Nirenberg and other researchers followed, and by 1967 the entire genetic code had been broken. The complete code is shown in ■ Table 11.2.

STUDY SKILLS 11.1 Remembering Key Words

Three very important processes discussed in this chapter are replication, transcription, and translation. New terminology is always easier to learn if we can make associations with something we already know. See if these ideas help. To replicate means to copy or perhaps to repeat. Associate the word *replicate* with *repeat.*

For the processes of transcription and translation, think of an analogy with the English language. If you were to transcribe your lecture notes, you would rewrite them in the same language. In the same way, transcription of DNA to mRNA is a rewriting of the information using the same nucleic acid language. If you translated your lecture notes, you would convert them to another language, such as German. Translation of mRNA to an amino acid sequence is converting information from a nucleic acid language to a language of proteins.

TABLE 11.3 General characteristics of the genetic code

Characteristic	Example
Codons are three-letter words	GCA = alanine
The code is degenerate	GCA, GCC, GCG, GCU all represent alanine
The code is precise	GCC represents only alanine
Chain initiation is coded	AUG
Chain termination is coded	UAA, UAG, and UGA
The code is almost universal	GCA = alanine, perhaps, in all organisms

Characteristics of the Genetic Code

The first important characteristic of the genetic code is that it applies almost universally. With very minor exceptions, the same amino acid is represented by the same three-base codon (or codons) in every organism.

Another interesting feature of the genetic code is apparent from Table 11.2. Most of the amino acids are represented by more than one codon, a condition known as *degeneracy.* Only methionine and tryptophan are represented by single codons. Leucine, serine, and arginine are the most degenerate, with each one represented by six codons. The remaining 15 amino acids are each coded for by at least two codons. Even though most amino acids are represented by more than one codon, it is significant that the reverse is not true; no single codon represents more than one amino acid. Each three-base codon represents one and only one amino acid.

Only 61 of the possible 64 base triplets represent amino acids. The remaining three (UAA, UAG, and UGA) are signals for chain terminations. They tell the protein-synthesizing process when the primary structure of the synthesized protein is complete and it is time to stop adding amino acids to the chain. These three codons are indicated in Table 11.2 by stop signals.

The presence of stop signals in the code implies that start signals must also exist. There is only one initiation (start) codon, and it is AUG, the codon for the amino acid methionine. AUG functions as an initiation codon only when it occurs as the first codon of a sequence. When this happens, protein synthesis begins at that point (Section 11.8).

These general features of the genetic code are summarized in ■ Table 11.3.

11.8 Translation and Protein Synthesis

LEARNING OBJECTIVE

8. Outline the process by which proteins are synthesized in cells.

ThomsonNOW™ Go to Chemistry Interactive to explore **DNA translation.**

To this point, you have become familiar with the molecules that participate in protein synthesis and the genetic code, the language that directs the synthesis. We now investigate the actual process by which polypeptide chains are assembled. There are three major stages in protein synthesis: initiation of the polypeptide chain, elongation of the chain, and termination of the completed polypeptide chain.

Initiation of the Polypeptide Chain

The first amino acid to be involved in protein synthesis in prokaryotic (bacterial) cells is a derivative of methionine. This compound, *N*-formylmethionine, initiates the growing polypeptide chain as the N-terminal amino acid. The fact that most

CHEMISTRY AROUND US 11.3 Stem Cell Research

"**Stem cell** research" is a phrase many individuals are familiar with but only understand in a superficial way. This ethically charged and very controversial research has great potential to provide treatments and possibly cures for serious conditions such as diabetes, heart disease, and Parkinson's disease. Despite this potential to benefit the human society, numerous countries have limited and even banned specific practices associated with the research on the basis of ethical concerns.

Stem cells possess the unique capability to develop into many different types of cells found in the body. When a stem cell divides, each new cell has the ability to either remain as a stem cell or transform into another type of specialized cell such as a muscle cell, a red blood cell, or a brain cell. Because of this ability, stem cells can serve as a repair system for the body and theoretically divide without limit to replenish other cells that are lost through injury, disease, or normal wear and tear.

There are two types of stem cells: embryonic stem cells and adult stem cells. Human embryonic stem cells are thought to have much greater developmental potential than adult stem cells because they are *pluripotent,* which means they are able to develop into any type of cell. Adult stem cells are *multipotent,* meaning they can become cells corresponding to the type of tissue in which they reside.

After many years of research, it is now possible to isolate stem cells from human embryos, and get them to multiply in the laboratory. These are called human embryonic stem cells. The embryos used in these studies were created by in vitro fertilization processes in order to treat infertility. When the embryos were no longer needed, they were donated for research with the informed consent of the donor.

The existence of two types of stem cells has prompted some critics of the embryonic cell research to ask why adult stem cells are not used instead of embryonic stem cells. Supporters of the research point out that embryonic stem cells have a much greater versatility because they can become any type of cell, not just the same type of cell as the cells of the tissue in which they are found. Embryonic stem cells could be used to treat a damaged organ, a broken bone, or an injured spinal cord. In some research, embryonic stem cells have been grown into heart muscle cells that clump together in a laboratory dish and pulse in unison. And when those heart cells were injected into mice and pigs with heart disease, they filled in for the injured or dead cells and speeded the animal's recovery. Similar studies have suggested embryonic stem cells have the potential to provide similar cures for conditions such as diabetes and spinal cord injury. In our society where waiting lists for donated organs are too long to help many patients, stem cell research could hold the answer.

stem cell
An unspecialized cell that has the ability to replicate and differentiate, giving rise to a specialized cell.

proteins do not have *N*-formylmethionine as the N-terminal amino acid indicates that when protein synthesis is completed, the *N*-formylmethionine is cleaved from the finished protein.

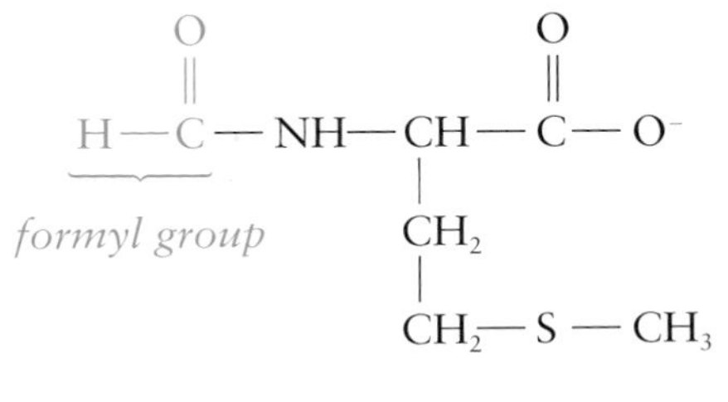

$$\underbrace{\mathrm{H-\overset{\overset{\displaystyle O}{\|}}{C}}}_{\textit{formyl group}}\mathrm{-NH-\underset{\underset{\displaystyle CH_2-S-CH_3}{|}}{\underset{\displaystyle CH_2}{\underset{|}{CH}}}-\overset{\overset{\displaystyle O}{\|}}{C}-O^-}$$

N-formylmethionine
(fMet)

A ribosome comprises two subunits, a large subunit and a small subunit. The initiation process begins when mRNA is aligned on the surface of a small ribosomal subunit in such a way that the initiating codon, AUG, occupies a specific site on the ribosome called the P site (peptidyl site). When the AUG codon is used this way to initiate synthesis of a polypeptide chain in prokaryotic cells, it represents *N*-formylmethionine (fMet) instead of methionine. When AUG is located anywhere else on the mRNA, it simply represents methionine. For the eukaryotic cells of humans, AUG always specifies methionine even when it is the initiating codon. Next, a tRNA molecule with its attached fMet binds to the codon through hydrogen bonds. The resulting complex binds to the large ribosomal subunit to form a unit called an *initiation complex* (see ■ Active Figure 11.19).

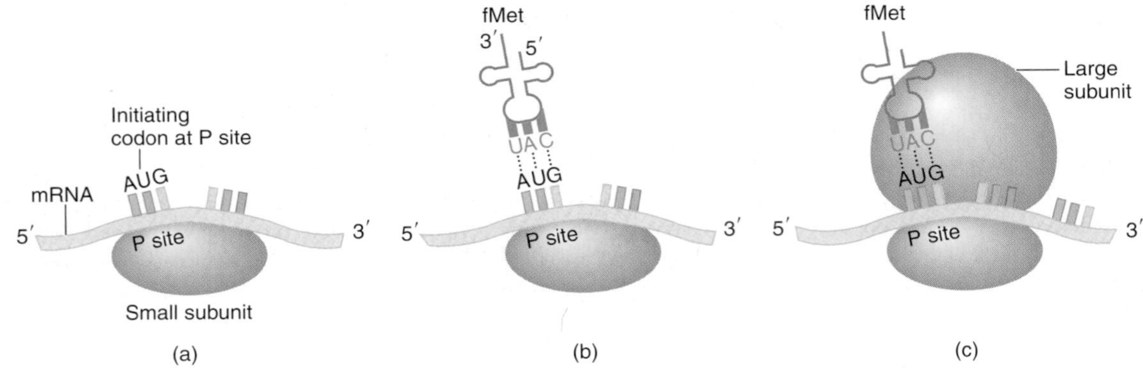

■ **ACTIVE FIGURE 11.19** Initiation complex formation. (a) mRNA aligns on a small ribosomal subunit so that AUG, the initiating codon, is at the ribosomal P site. (b) A tRNA with an attached *N*-formylmethionine forms hydrogen bonds with the codon. (c) A large ribosomal subunit completes the initiation complex. **Sign in at www.thomsonedu.com to explore an interactive version of this figure.**

Elongation of the Chain

A second site, called the A site (aminoacyl site), is located on the mRNA–ribosome complex next to the P site. The A site is where an incoming tRNA carrying the next amino acid will bond. Each of the tRNA molecules representing the 20 amino acids can try to fit the A site, but only the one with the correct anticodon that is complementary to the next codon on the mRNA will fit properly.

Once at the A site, the second amino acid (in this case, phenylalaline) is linked to *N*-formylmethionine by a peptide bond whose formation is catalyzed by the enzyme peptidyl transferase (see ■ Figure 11.20). After the peptide bond forms, the tRNA bound to the P site is "empty," and the growing polypeptide chain is now attached to the tRNA bound to the A site.

In the next phase of elongation, the whole ribosome moves one codon along the mRNA toward the 3′ end. As the ribosome moves, the empty tRNA is released from the P site, and tRNA attached to peptide chain moves from the A site and takes its place on the P site. This movement of the ribosome along the mRNA is called *translocation* and makes the A site available to receive the next tRNA with the proper anticodon. The amino acid carried by this tRNA bonds to the peptide chain, and the elongation process is repeated. This occurs over and over until the entire polypeptide chain is synthesized. The elongation process is represented in ■ Active Figure 11.21 for the synthesis of the tripeptide fMet—Phe—Val.

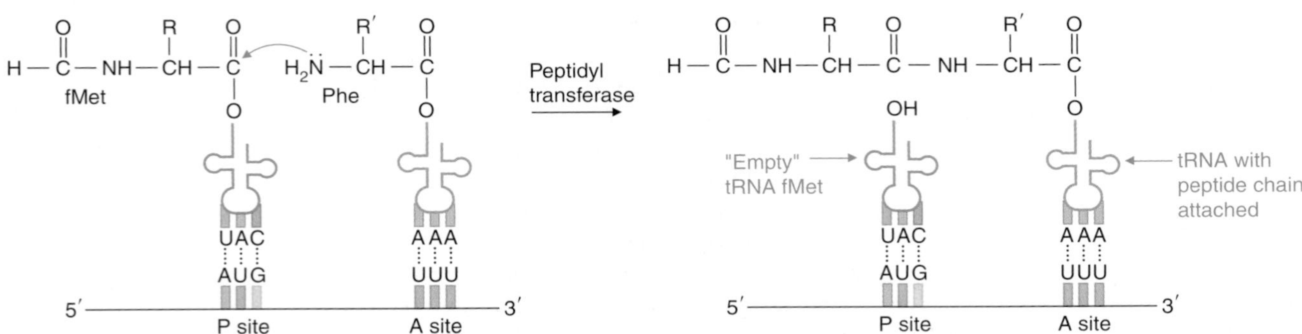

■ **FIGURE 11.20** The amino acid at the P site bonds through a peptide bond to the amino acid at the A site.

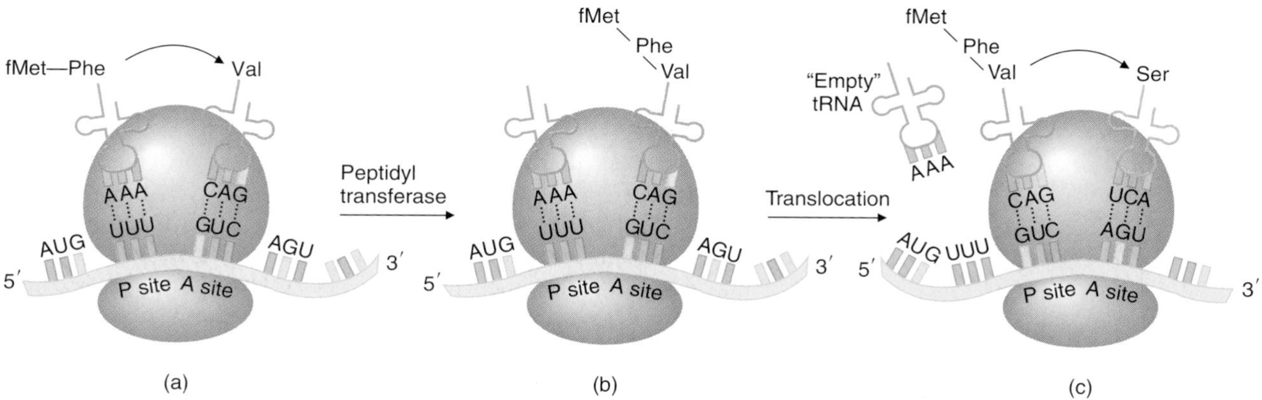

■ **ACTIVE FIGURE 11.21** Polypeptide chain elongation. (a) The P site is occupied by the tRNA with the growing peptide chain, and Val—tRNA is located at the A site. (b) The formation of a peptide bond between Val and the dipeptide fMet—Phe takes place under the influence of peptidyl transferase. (c) During translocation when the ribosome shifts to the right, the empty tRNA leaves, the polypeptide—tRNA moves to the P site, and the next tRNA carrying Ser arrives at the A site. **Sign in at www.thomsonedu.com to explore an interactive version of this figure.**

The Termination of Polypeptide Synthesis

The chain elongation process continues and polypeptide synthesis continues until the ribosome complex reaches a stop codon (UAA, UAG, or UGA) on mRNA. At that point, a specific protein known as a termination factor binds to the stop codon and catalyzes the hydrolysis of the completed polypeptide chain from the final tRNA. The "empty" ribosome dissociates and can then bind to another strand of mRNA to once again begin the process of protein synthesis.

Several ribosomes can move along a single strand of mRNA one after another (see ■ Figure 11.22). Thus, several identical polypeptide chains can be synthesized almost simultaneously from a single mRNA molecule. This markedly increases the efficiency of utilization of the mRNA. Such complexes of several ribosomes and mRNA are called **polyribosomes** or **polysomes.** Growing polypeptide chains extend from the ribosomes into the cellular cytoplasm and spontaneously fold to assume characteristic three-dimensional secondary and tertiary configurations.

polyribosome or polysome A complex of mRNA and several ribosomes.

■ **Learning Check 11.4** Write the primary structure of the polypeptide produced during translation of the following mRNA sequence:

5′ AUG–CAC–CAU–GUA–UUG–UGU–UAG 3′

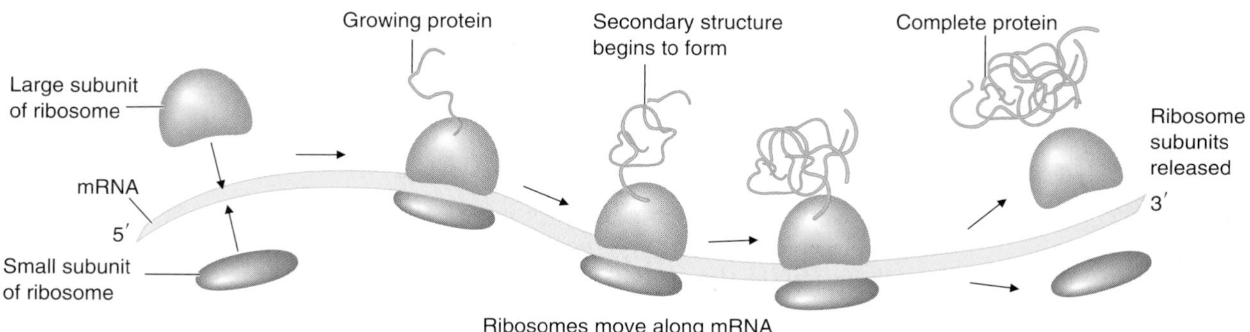

■ **FIGURE 11.22** A polyribosome, several ribosomes proceeding simultaneously along mRNA.

11.9 Mutations

LEARNING OBJECTIVE

9. Describe how genetic mutations occur and how they influence organisms.

In Section 11.3, we pointed out that the base-pairing mechanism provides a nearly perfect way to copy a DNA molecule during replication. However, not even the copying mechanism involved in DNA replication is totally error free. It is estimated that, on average, one in every 10^{10} bases of DNA (i.e., 1 in 10 billion) is incorrect. Any change resulting in an incorrect sequence of bases on DNA is called a **mutation.** The faithful transcription of mutated DNA leads to an incorrect base sequence in mRNA. This can lead to an incorrect amino acid sequence for a protein, or possibly the failure of a protein to be synthesized at all.

mutation
Any change resulting in an incorrect base sequence on DNA.

Some mutations occur naturally during DNA replication; others can be induced by environmental factors such as ionizing radiation (X rays, ultraviolet light, gamma rays, etc.). Repeated exposure to X rays greatly increases the rate of mutation. Thus, patients are given X rays only when necessary, and technicians who administer X rays remain behind protective barriers. A large number of chemicals (e.g., nitrous acid and dimethyl sulfate) can also induce mutations by reacting with DNA. Such chemicals are called **mutagens.**

mutagen
A chemical that induces mutations by reacting with DNA.

Some mutations might be beneficial to an organism by making it more capable of surviving in its environment. Mutations may also occur without affecting an organism. Conversely, mutations may be lethal or produce genetic diseases whenever an important protein (or enzyme) is not correctly synthesized. Sickle-cell disease, phenylketonuria (PKU), hemophilia, muscular dystrophy, and many other disorders are results of such mutations that have become permanently incorporated into the genetic makeup of certain individuals.

11.10 Recombinant DNA

LEARNING OBJECTIVE

10. Describe the technology used to produce recombinant DNA.

Remarkable technology is available that allows segments of DNA from one organism to be introduced into the genetic material of another organism. The resulting new DNA (containing the foreign segment) is referred to as **recombinant DNA.** The application of this technology, commonly called *genetic engineering*, has produced major advances in human health care and holds the promise for a future of exciting advances in biology, agriculture (see ■ Figure 11.23), and many other areas of study.

recombinant DNA
DNA of an organism that contains genetic material from another organism.

An early success of genetic engineering was the introduction of the gene for human insulin into the DNA of the common bacterium *E. coli.* The bacterium then transcribed and translated the information carried by the gene and produced the protein hormone. By culturing such *E. coli*, it has become possible to produce and market large quantities of human insulin. The availability of human insulin is very important for diabetics allergic to the insulin traditionally used, which is isolated from the pancreatic tissue of slaughtered pigs or cattle. ■ Table 11.4 lists several other medically important materials that have been produced through genetic engineering.

PhotoDisc

■ **FIGURE 11.23** This tomato, a beneficiary of genetic engineering, has an increased shelf life.

Enzymes Used in Genetic Engineering

The discovery of restriction enzymes in the 1960s and 1970s made genetic engineering possible. Restriction enzymes, which are found in a wide variety of bacterial cells, catalyze the cleaving of DNA molecules. These enzymes are normally part of a mechanism that protects certain bacteria from invasion by foreign DNA such as that of a virus. In these bacteria, some of the bases in their DNA have methyl groups attached. For example, the following compounds are found:

TABLE 11.4 Some substances produced by genetic engineering

Substance	Use
Human insulin	Treatment of diabetes mellitus
Human growth hormone	Treatment of dwarfism
Interferon	Fights viral infections
Hepatitis B vaccine	Protection against hepatitis
Malaria vaccine	Protection against malaria
Tissue plasminogen activator	Promotes the dissolving of blood clots

1-methylguanine 5-methylcytosine

The methylated DNA of these bacteria is left untouched by the restriction enzymes, but foreign DNA that lacks the methylated bases undergoes rapid cleavage of both strands and thus becomes nonfunctional. Because there is a "restriction" on the type of DNA allowed in the bacterial cell, the protective enzymes are called **restriction enzymes.**

restriction enzyme
A protective enzyme found in some bacteria that catalyzes the cleaving of all but a few specific types of DNA.

Restriction enzymes act at sites on DNA called palindromes. In language, a palindrome is any word or phrase that reads the same in either direction, such as "radar" or "Madam, I'm Adam." For double-stranded DNA, a palindrome is a section in which the two strands have the same sequence but run in opposite directions. Examples of DNA palindromes follow; the arrows indicate the points of attack by restriction enzymes:

```
           ↓                      ↓
5′ C–C–G–C–G–G 3′      5′ G–G–A–T–C–C 3′
   ⋮ ⋮ ⋮ ⋮ ⋮ ⋮            ⋮ ⋮ ⋮ ⋮ ⋮ ⋮
3′ G–G–C–G–C–C 5′      3′ C–C–T–A–G–G 5′
       ↑                            ↑
```

At least 100 restriction enzymes are known, and each catalyzes DNA cleavage in a specific and predictable way. These enzymes are the tools used to take DNA apart and reduce it to fragments of known size and nucleotide sequence.

Another set of enzymes important in genetic engineering, called DNA ligases, has been known since 1967. These enzymes normally function to connect DNA fragments during replication, and they are used in genetic engineering to put together pieces of DNA produced by restriction enzymes.

Plasmids

The introduction of a new DNA segment (gene) into a bacterial cell requires the assistance of a DNA carrier called a **vector.** The vector is often a circular piece of double-stranded DNA called a **plasmid.** Plasmids range in size from 2000 to several hundred thousand nucleotides and are found in the cytoplasm of bacterial cells. Plasmids function as accessories to chromosomes by carrying genes for the inactivation of antibiotics and the production of toxins. They have the unusual

vector
A carrier of foreign DNA into a cell.

plasmid
Circular, double-stranded DNA found in the cytoplasm of bacterial cells.

CHEMISTRY AROUND US 11.4 DNA and the Crime Scene

DNA, commonly referred to as the blueprint of life, passes along hereditary information to offspring and directs all life processes. DNA also plays a key role in forensic science, made popular by TV shows such as *CSI: Crime Scene Investigation*. However, DNA fingerprinting isn't just a gimmick used in Hollywood scripts; it is a valuable tool in solving real-world crimes.

DNA profiling—also called DNA typing or DNA fingerprinting—is possible because most of the DNA molecule is common to all human, but small parts of it differ from person to person. In fact, 99.9% of human DNA is identical, but 0.1% differs from person to person. Small as this difference seems, it is equivalent to 3 million bases in an individual DNA. Therefore, with the exception of identical twins, the DNA of every human is unique. Unlike a conventional fingerprint, which occurs only on the fingertips, a DNA fingerprint is identical for every cell of each tissue and organ of the person's body. Consequently, DNA fingerprinting has become the method of choice for identifying and distinguishing among individual human beings.

Law enforcement laboratories around the world routinely use DNA fingerprints to link suspects to crime-scene biological evidence such as blood, semen, or hair. DNA fingerprints used to establish paternity in custody and child-support litigations provide nearly perfect accuracy in parental identification. The U.S. armed services have switched from dog tags and dental records to DNA fingerprint records as a means to identify all personnel, including those killed in combat.

The use of DNA by the U.S. armed services has even made it possible to identify soldiers from past wars. In 1997, DNA collected from bones of American servicemen killed during the Vietnam War helped the Army identify the bodies. It wasn't the only method used in the identification process, but it provided conclusive evidence.

Permission given by the United States military has made it possible to run DNA tests on the remains of the last Unknown Soldier at Arlington National Cemetery. A controversy has developed about whether the man was Lt. Michael J. Blassie or Capt. Rodney L. Strobridge. Blassie's relatives claim he is the Unknown Soldier on the basis of an ID tag found near the body, but others believe it is Strobridge, whose helicopter crashed near the body's location. DNA tests should be able to clear up the mystery soon. When the remains are identified, should they still be classified as "unknown"?

The technique of DNA fingerprinting involves cleaving DNA molecules at specific points using restriction enzymes. The mixture of DNA fragments is then separated into specific patterns by a process called gel electrophoresis. In this process, the cleaved DNA is put into a gel through which electricity is passed. The resulting patterns are made visible by attaching radioactive reagents to the DNA fragments and exposing X-ray film to the results of the electrophoresis. The film shows a specific pattern of dark bands when it is developed.

The ability to obtain DNA fingerprints from exceedingly small tissue samples has become possible by using the polymerase chain reaction technique (Section 11.3).

The PCR technique allows billions of copies of a DNA sample to be prepared in a few hours. PCR enables researchers to study extremely faint and tiny traces of DNA obtained from biological samples, such as specks of dried blood, strands of hair, chips of bone, or even from fingerprints. PCR amplification of the DNA in a single sperm cell has been used to link suspects to rape victims. A single cell from saliva on the stamp of a letter bomb would provide enough DNA to help identify the person who licked the stamp.

These few examples illustrate the tremendous identification potential of the PCR technique coupled with the uniqueness of each individual's DNA. It is getting more and more difficult to hide.

ability to replicate independently of chromosomal DNA. A typical bacterial cell contains about 20 plasmids and one or two chromosomes.

The Formation of Recombinant DNA

The recombinant DNA technique begins with the isolation of a plasmid from a bacterium. A restriction enzyme is added to the plasmid, which is cleaved at a specific site:

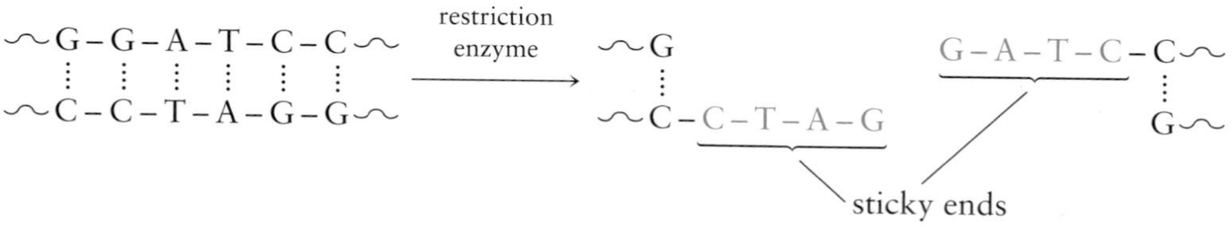

```
~G–G–A–T–C–C~    restriction     ~G              G–A–T–C–C~
 ⋮  ⋮  ⋮  ⋮  ⋮  ⋮     enzyme         ⋮                      ⋮
~C–C–T–A–G–G~   ──────────►   ~C–C–T–A–G               G~
                                        \_______ sticky ends ___/
```

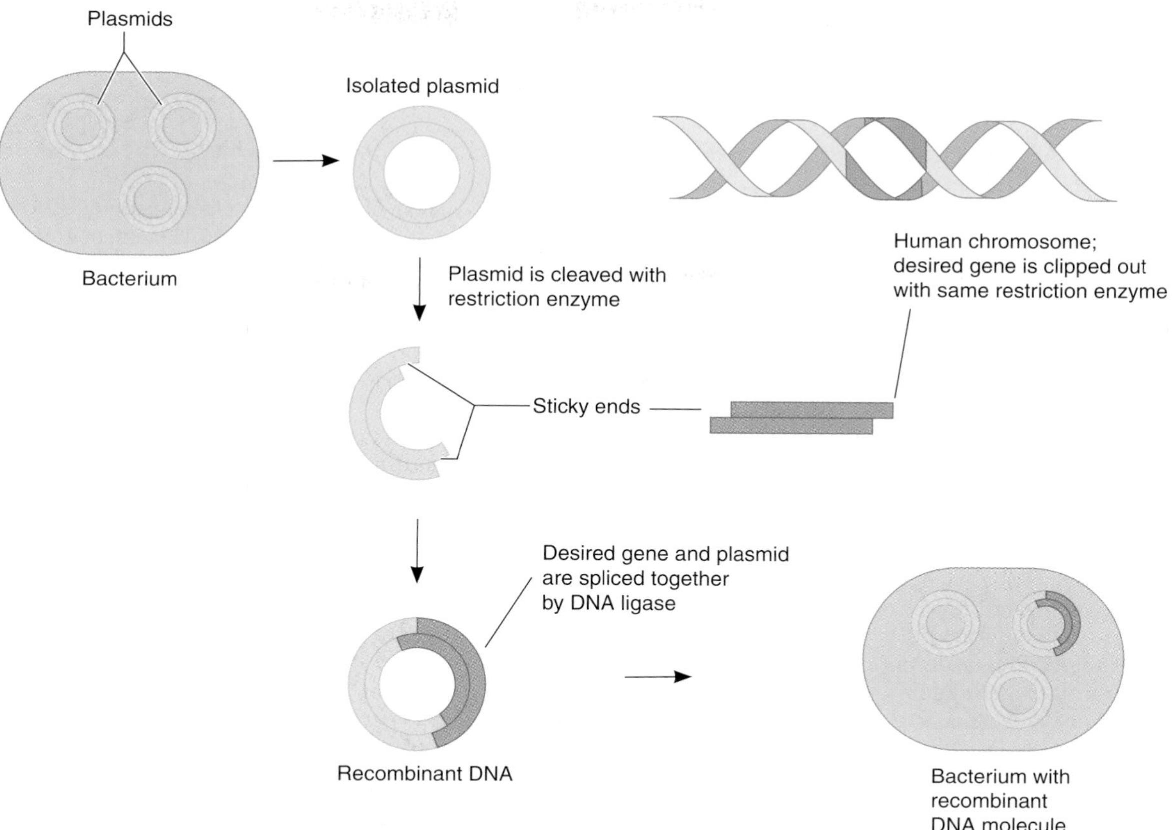

■ **FIGURE 11.24** The formation of recombinant DNA.

Because a plasmid is circular, cleaving it this way produces a double-stranded chain with two ends (see ■ Figure 11.24). These are called "sticky ends" because each has one strand in which several bases are unpaired and ready to pair up with a complementary section if available.

The next step is to provide the sticky ends with complementary sections for pairing. A human chromosome is cleaved into several fragments using the same restriction enzyme. Because the same enzyme is used, the human DNA is cleaved such that the same sticky ends result:

G–A–T–C–C∼∼∼∼∼∼∼∼∼∼G
⋮ human DNA segment ⋮
G∼∼∼∼∼∼∼∼∼∼C–C–T–A–G

To splice the human gene into the plasmid, the two are brought together under conditions suitable for the formation of hydrogen bonds between the complementary bases of the sticky ends. The breaks in the strands are then joined by using DNA ligase, and the plasmid once again becomes a circular piece of double-stranded DNA (Figure 11.24); recombinant DNA has been formed.

Bacterial cells are bathed in a solution containing recombinant DNA plasmids, which diffuse into the cells. When the bacteria reproduce, they replicate all the genes, including the recombinant DNA plasmid. Bacteria multiply quickly, and soon a large number of bacteria, all containing the modified plasmid, are manufacturing the new protein as directed by the recombinant DNA.

The use of this technology makes possible (in principle) the large-scale production of virtually any polypeptide or protein. By remembering the many essential roles these substances play in the body, it is easy to predict that genetic engineering will make tremendous contributions to improving health care in the future.

CHEMISTRY AND YOUR HEALTH 11.1 Genetically Modified Foods

Many people think that the practice of genetically altering food is the result of recent discoveries in the science of genetics. The reality is that such bioengineering of food has been going on for centuries in the practices of cross-pollinating plants and selective crossbreeding of animals. These activities were done in successful attempts to produce plant and animal food sources with desirable characteristics.

The goals of bioengineering foods today are essentially the same as those earlier goals, but the goals are more sophisticated. Instead of focusing on such things as increased animal size or drought-resistant plants, today's bioengineering has goals such as producing plants that repel or kill insects that attack them or producing animals that yield meat low in artery-clogging fats.

The following successes will serve to illustrate the tremendous potential of bioengineering in helping to alleviate an increasing problem of supplying the world's inhabitants with quality food. Corn is commonly modified by adding a gene that produces plants that kill corn-devouring caterpillars. A recently developed strain of rice contains genes that produce beta-carotene, the precursor of vitamin A, in the rice grains. When this so-called golden rice is substituted for regular rice as the major source of food in the diet, it can help prevent the onset in children of blindness that results from vitamin A deficiency. Scientists have found a way to produce tomato plants that contain three times the normal level of lycopene, a known cancer-preventing compound.

In spite of this great potential, the bioengineering of food is not accepted as desirable by everyone. The extent of concern is not uniform. For example, there are few labeling guidelines for genetically modified food in the United States. The U.S. Food and Drug Administration only requires that food be labeled as genetically modified if the nutritional content of the food has been changed, or if a potential allergen (such as a peanut gene) has been added. By contrast, the European Union was concerned enough to place a moratorium on the imports of modified sweet corn from the United States. The import of other bioengineered crops has since been allowed.

The greatest concerns of opponents to bioengineered foods are related to human health. It is feared that genes introduced into plants might accidentally make the plants into sources of protein allergens dangerous to humans. The well-known peanut proteins that are serious allergens to some people are sometimes used as an example.

It appears that bioengineering of foods holds great promise of providing significant benefits to humans. However, more education and research are needed to minimize concerns and demonstrate the safety of the procedure to those who are in doubt.

Concept Summary

ThomsonNOW™ Sign in at **www.thomsonedu.com** to:

- Assess your understanding with Exercises keyed to each learning objective.
- Check your readiness for an exam by taking the **Pre-test** and exploring the modules recommended in your **Personalized Learning Plan.**

Components of Nucleic Acids. Nucleic acids are classified into two categories: ribonucleic acids (RNA) and deoxyribonucleic acids (DNA). Both types are polymers made up of monomers called nucleotides. All nucleotides are composed of a pyrimidine or purine base, a sugar, and phosphate. The sugar component of RNA is ribose, and that of DNA is deoxyribose. The bases adenine, guanine, and cytosine are found in all nucleic acids. Uracil is found only in RNA and thymine only in DNA. **OBJECTIVE 1 (Section 11.1), Exercises 11.2 and 11.4**

The Structure of DNA. The nucleotides of DNA are joined by linkages between phosphate groups and sugars. The resulting sugar–phosphate backbone is the same for all DNA molecules, but the order of attached bases along the backbone varies. This order of nucleotides with attached bases is the primary structure of nucleic acids. The secondary structure is a double-stranded helix held together by hydrogen bonds between complementary base pairs on the strands. **OBJECTIVE 2 (Section 11.2), Exercise 11.10**

DNA Replication. The replication of DNA occurs when a double strand of DNA unwinds at specific points. The exposed bases match up with complementary bases of nucleotides. The nucleotides bind together to form two new strands that are complementary to the strands that separated. Thus, the two new DNA molecules each contain one old strand and one new strand. **OBJECTIVE 3 (Section 11.3), Exercise 11.20**

Ribonucleic Acid (RNA). Three forms of ribonucleic acid are found in cells: messenger RNA, ribosomal RNA, and transfer RNA. Each form serves an important function during protein synthesis. All RNA molecules are single stranded, but some contain loops or folds. **OBJECTIVE 4 (Section 11.4), Exercises 11.26 and 11.28**

The Flow of Genetic Information. The flow of genetic information occurs in two steps called transcription and translation. In transcription, information stored in DNA molecules is passed to molecules of messenger RNA. In translation, the messenger RNA serves as a template that directs the assembly of amino acids into proteins. **OBJECTIVE 5 (Section 11.5), Exercise 11.32**

Transcription: RNA Synthesis. The various RNAs are synthesized in much the same way as DNA is replicated. Nucleotides with complementary bases align themselves against one strand of a partially unwound DNA segment that contains the genetic information that is to be transcribed. The aligned nucleotides bond together to form the RNA. In eukaryotic cells, the produced RNA is heterogeneous and is cut and spliced after being synthesized to produce the functional RNA. **OBJECTIVE 6 (Section 11.6), Exercise 11.34**

The Genetic Code. The genetic code is a series of three-letter words that represent the amino acids of proteins as well as start and stop signals for protein synthesis. The letters of the words are the bases found on mRNA, and the words on mRNA are called codons. The genetic code is the same for all organisms and is degenerate for most amino acids. **OBJECTIVE 7 (Section 11.7), Exercise 11.38**

Translation and Protein Synthesis. The translation step in the flow of genetic information results in the synthesis of proteins. The synthesis takes place when properly coded mRNA forms a complex with the component of a ribosome. Transfer RNA molecules carrying amino acids align themselves along the mRNA in an order representing the correct primary structure of the protein. The order is determined by the matching of complementary codons on the mRNA to anticodons on the tRNA. The amino acids sequentially bond together to form the protein, which then spontaneously forms characteristic secondary and tertiary structures. **OBJECTIVE 8 (Section 11.8), Exercise 11.44**

Mutations. Any change that results in an incorrect sequence of bases on DNA is called a mutation. Some mutations occur naturally during DNA replication, whereas others are induced by environmental factors. Some mutations are beneficial to organisms; others may be lethal or result in genetic diseases. **OBJECTIVE 9 (Section 11.9), Exercise 11.48**

Recombinant DNA. The discovery and application of restriction enzymes and DNA ligases have resulted in a technology called genetic engineering. The primary activities of genetic engineers are the isolation of genes (DNA) that code for specific useful proteins and the introduction of these genes into the DNA of bacteria. The new (recombinant) DNA in the rapidly reproducing bacteria mediates production of the useful protein, which is then isolated for use. **OBJECTIVE 10 (Section 11.10), Exercise 11.52**

Key Terms and Concepts

Anticodon (11.4)
Central dogma of molecular biology (11.5)
Chromosome (11.3)
Codon (11.7)
Complementary DNA strands (11.2)
Deoxyribonucleic acid (DNA) (11.1)
Exon (11.6)
Gene (11.3)
Heterogeneous nuclear RNA (hnRNA) (11.6)
Intron (11.6)
Messenger RNA (mRNA) (11.4)
Mutagen (11.9)
Mutation (11.9)
Nucleic acid (Introduction)
Nucleic acid backbone (11.2)
Nucleotide (11.1)
Okazaki fragment (11.3)
Plasmid (11.10)
Polyribosome (polysome) (11.8)
Recombinant DNA (11.10)
Replication (11.3)
Replication fork (11.3)
Restriction enzyme (11.10)
Ribonucleic acid (RNA) (11.1)
Ribosomal RNA (rRNA) (11.4)
Ribosome (11.4)
Semiconservative replication (11.3)
Stem cell (11.8)
Transcription (11.5)
Transfer RNA (tRNA) (11.4)
Translation (11.5)
Vector (11.10)

Exercises

SYMBOL KEY

Even-numbered exercises are answered in Appendix B.

Blue-numbered exercises are more challenging.

■ denotes exercises available in ThomsonNow and assignable in OWL.

ThomsonNOW™ To assess your understanding of this chapter's topics with sample tests and other resources, sign in at www.thomsonedu.com.

COMPONENTS OF NUCLEIC ACIDS (SECTION 11.1)

11.1 ■ What is the principal location of DNA within the eukaryotic cell?

11.2 Which pentose sugar is present in DNA? In RNA?

11.3 Name the three components of nucleotides.

11.4 ■ Indicate whether each of the following is a pyrimidine or a purine:

a. guanine
b. thymine
c. uracil
d. cytosine
e. adenine

11.5 Which bases are found in DNA? In RNA?

11.6 Write the structural formula for the nucleotide thymidine 5′-monophosphate. The base component is thymine.

THE STRUCTURE OF DNA (SECTION 11.2)

11.7 In what way might two DNA molecules that contain the same number of nucleotides differ?

11.8 ■ Identify the 3′ and 5′ ends of the DNA segment AGTCAT.

11.9 What data obtained from the chemical analysis of DNA supported the idea of complementary base pairing in DNA proposed by Watson and Crick?

11.10 Describe the secondary structure of DNA as proposed by Watson and Crick.

11.11 Describe the role of hydrogen bonding in the secondary structure of DNA.

11.12 ■ How many total hydrogen bonds would exist between the following strands of DNA and their complementary strands?

a. CAGTAG

b. TTGACA

11.13 How many total hydrogen bonds would exist between the following strands of DNA and their complementary strands?

a. CACGGT b. TTTAAA

11.14 A strand of DNA has the base sequence ATGCATC. Write the base sequence for the complementary strand.

11.15 ■ A strand of DNA has the base sequence GATTCA. Write the base sequence for the complementary strand.

DNA REPLICATION (SECTION 11.3)

11.16 What is a chromosome? How many chromosomes are in a human cell? What is the approximate number of genes in a human cell?

11.17 What is meant by the term *semiconservative replication*?

11.18 What is a replication fork?

11.19 Describe the function of the enzyme helicase in the replication of DNA.

11.20 ■ List the steps involved in DNA replication.

11.21 What enzymes are involved in DNA replication?

11.22 In what direction is a new DNA strand formed?

11.23 Explain how the synthesis of a DNA daughter strand growing toward a replication fork differs from the synthesis of a daughter strand growing away from a replication fork.

11.24 The base sequence ACGTCT represents a portion of a single strand of DNA. Represent the complete double-stranded molecule for this portion of the strand and use the representation to illustrate the replication of the DNA strand. Be sure to clearly identify the nucleotide bases involved, the new strands formed, and the daughter DNA molecules.

11.25 Explain the origin of Okazaki fragments.

RIBONUCLEIC ACID (RNA) (SECTION 11.4)

11.26 How does the sugar–phosphate backbone of RNA differ from the backbone of DNA?

11.27 Compare the secondary structures of RNA and DNA.

11.28 ■ Briefly describe the characteristics and functions of the three types of cellular RNA.

11.29 Must the ratio of guanine to cytosine be 1:1 in RNA? Explain.

11.30 ■ What are the two important regions of a tRNA molecule?

THE FLOW OF GENETIC INFORMATION (SECTION 11.5)

11.31 What is the central dogma of molecular biology?

11.32 ■ In the flow of genetic information, what is meant by the terms *transcription* and *translation?*

TRANSCRIPTION: RNA SYNTHESIS (SECTION 11.6)

11.33 Briefly describe the synthesis of mRNA in the transcription process.

11.34 ■ Write the base sequence for the mRNA that would be formed during transcription from the DNA strand with the base sequence CTAAGATCG.

11.35 Describe the differences in the transcription and translation processes as they occur in prokaryotic cells and eukaryotic cells.

11.36 What is the relationship among exons, introns, and hnRNA?

THE GENETIC CODE (SECTION 11.7)

11.37 What is a codon?

11.38 ■ For each of the following mRNA codons, give the tRNA anticodon and use Table 11.2 to determine the amino acid being coded for by the codon.

a. UAU

b. CAU

c. UCA

d. UCU

11.39 Describe the experiment that allowed researchers to first identify the codon for a specific amino acid.

11.40 ■ Which of the following statements about the genetic code are true and which are false? Correct each false statement.

a. Each codon is composed of four bases.

b. Some amino acids are represented by more than one codon.

c. All codons represent an amino acid.

d. Each living species is thought to have its own unique genetic code.

e. The codon AUG at the beginning of a sequence is a signal for protein synthesis to begin at that codon.

f. It is not known whether or not the code contains stop signals for protein synthesis.

TRANSLATION AND PROTEIN SYNTHESIS (SECTION 11.8)

11.41 ■ The β-chain of hemoglobin is a protein that contains 146 amino acid residues. What minimum number of nucleotides must be present in a strand of mRNA that is coded for this protein?

11.42 What is a polysome?

11.43 Beginning with DNA, describe in simple terms (no specific codons, etc.) how proteins are coded and synthesized.

11.44 ■ List the three major stages in protein synthesis.

11.45 Does protein synthesis begin with the N-terminal or with the C-terminal amino acid?

11.46 What is meant by the terms *A site* and *P site*?

11.47 Beginning with DNA, describe specifically the coding and synthesis of the following tetrapeptide that represents the first four amino acid residues of the hormone oxytocin: Gly–Leu–Pro–Cys. Be sure to include processes such as formation of mRNA (use correct codons, etc.), attachment of mRNA to a ribosome, attachment of tRNA to mRNA-ribosome complex, and so on.

MUTATIONS (SECTION 11.9)

11.48 What is a genetic mutation?

11.49 Briefly explain how genetic mutations can do the following:

a. Harm an organism

b. Help an organism

11.50 What is the result of a genetic mutation that causes the mRNA sequence GCC to be replaced by CCC?

RECOMBINANT DNA (SECTION 11.10)

11.51 Explain the function and importance of restriction enzymes and DNA ligase in the formation of recombinant DNA.

11.52 Explain how recombinant DNA differs from normal DNA.

11.53 Explain what plasmids are and how they are used to get bacteria to synthesize a new protein that they normally do not synthesize.

11.54 List three substances likely to be produced on a large scale by genetic engineering, and give an important use for each.

ADDITIONAL EXERCISES

11.55 Explain how heating a double-stranded DNA fragment to 94°C–96°C when performing PCR causes the DNA to unravel into single strands.

11.56 Review Figure 11.5 and suggest a reaction that would be used to excise introns from hnRNA. Also suggest a reaction that would be used to join the exon segments together.

11.57 A segment of DNA has an original code of –ACA–. This segment was mutated to –AAA–. After the mutation occurred, the protein that was coded for was no longer able to maintain its tertiary structure. Explain why.

11.58 A species of bacteria can use sucrose as its sole source of carbon energy. The bacterial enzyme Q is used to hydrolyze sucrose into its constituent monosaccharides. In the absence of sucrose, there are fewer than 10 enzyme Q molecules in a bacterial cell. In the presence of sucrose there are more than several thousand enzyme molecules in a single cell. Enzyme Q does not exist as a zymogen.

a. What are the constituent monosaccharides of sucrose?

b. What is the specific type of enzyme regulation that controls enzyme Q?

c. Suggest a mechanism of how this type of enzyme regulation occurs.

11.59 During DNA replication, about 1000 nucleotides are added per minute per molecule of DNA polymerase. The genetic material in a single human cell consists of about 3 billion nucleotides.

a. How many years would it take one DNA polymerase enzyme to replicate the genetic material in one cell?

b. How many DNA polymerase molecules would be needed to replicate a cell's genetic material in 10 minutes (assuming equidistance between replication forks)?

ALLIED HEALTH EXAM CONNECTION

Reprinted with permission from Nursing School and Allied Health Entrance Exams, COPYRIGHT 2005 Petersons.

11.60 Put the following terms in the correct order of information transfer according to the central dogma of molecular biology:

RNA, DNA, protein

11.61 ■ Which of the following are components of a nucleotide in a DNA molecule?

a. Sugar

b. A phospholipid

c. A nitrogen base

11.62 Explain the function and importance of each of the following in the formation of recombinant DNA:

a. Plasmid

b. Restriction enzyme

c. DNA ligase

CHEMISTRY FOR THOUGHT

11.63 Genetic engineering shows great promise for the future but has been controversial at times. Discuss with some classmates the pros and cons of genetic engineering. List two benefits and two concerns that come from your discussion.

11.64 Two samples of DNA are compared, and one has a greater percentage of guanine-cytosine base pairs. How should that greater percentage affect the attractive forces holding the double helix together?

11.65 If DNA replication were not semiconservative, what might be another possible structure for the daughter DNA molecules?

11.66 The genetic code contains three stop signals and 61 codons that code for 20 amino acids. From the standpoint of mutations, why are we fortunate that the genetic code does not have only the required 20 amino acid codons and 44 stop signals?

11.67 Azidothymidine (AZT), a drug used to fight HIV, is believed to act as an enzyme inhibitor. What type of enzyme inhibition is most likely caused by this drug?

11.68 If DNA specifies only the primary structure of a protein, how does the correct three-dimensional protein structure develop?

11.69 When were the first experiments carried out that produced genetically modified plants?

11.70 How does the DNA content determine what reactions occur within an organism?

11.71 What would be the ramifications if DNA were single stranded?

CHAPTER 12

Nutrition and Energy for Life

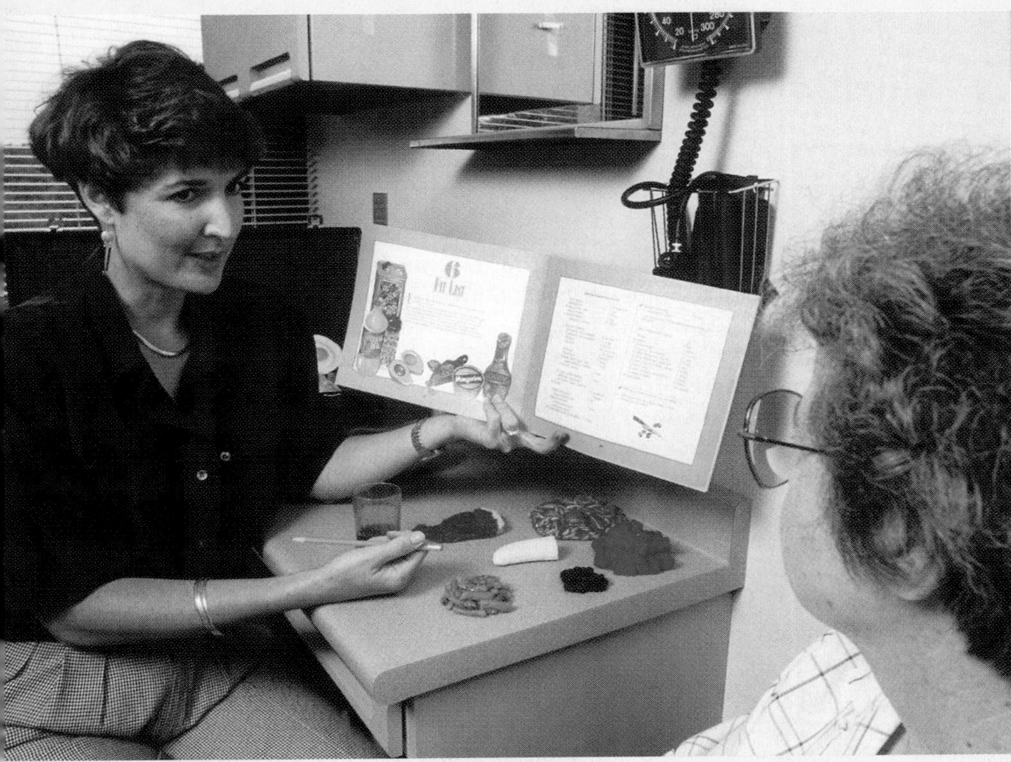

In addition to working in health care facilities, dietitians are employed in other capacities, providing nutritional counseling to individuals and groups. Large institutions such as schools and businesses often rely on dietitians to set up and supervise food services. This chapter will familiarize you with the nutritional requirements for carbohydrates, lipids, and proteins and provide an overview of energy production from food.

LEARNING OBJECTIVES

When you have completed your study of this chapter, you should be able to:

1. Describe the difference between macronutrients and micronutrients in terms of amounts required and their functions in the body. (Section 12.1)
2. Describe the primary functions in the body of each macronutrient. (Section 12.2)
3. Distinguish between and classify vitamins as water-soluble or fat-soluble on the basis of name and behavior in the body. (Section 12.3)
4. List a primary function in the body for each major mineral. (Section 12.4)
5. Describe the major steps in the flow of energy in the biosphere. (Section 12.5)
6. Differentiate among metabolism, anabolism, and catabolism. (Section 12.6)
7. Outline the three stages in the extraction of energy from food. (Section 12.6)
8. Explain how ATP plays a central role in the production and use of cellular energy. (Section 12.7)
9. Explain the role of coenzymes in the common catabolic pathway. (Section 12.8)

ThomsonNOW™ Throughout the chapter this icon introduces resources on the ThomsonNOW website for this text. Sign in at **www.thomsonedu.com** to:

- Evaluate your knowledge of the material
- Take an exam prep quiz
- Identify areas you need to study with a **Personalized Learning Plan**.

nutrition
An applied science that studies food, water, and other nutrients and the ways living organisms use them.

The science of **nutrition** is an applied field that focuses on the study of food, water, and other nutrients and the ways in which living organisms utilize them. Nutrition scientists use the principles of nutrition to obtain answers to questions that have a great deal of practical significance. For example, they might try to determine the proper components of a sound diet, the best way to maintain proper body weight, or the foods and other nutrients needed by people with specific illnesses or injuries.

The fuels of the human body are the sugars, lipids, and proteins derived from food (see ■ Figure 12.1). The reactions that release energy from these substances are among the body's most important biochemical processes. In this chapter, we will study an introduction to nutrition and how cells extract energy from food.

12.1 Nutritional Requirements

LEARNING OBJECTIVE

1. Describe the difference between macronutrients and micronutrients in terms of amounts required and their functions in the body.

macronutrient
A substance needed by the body in relatively large amounts.

micronutrient
A substance needed by the body only in small amounts.

A human body must be supplied with appropriate nutrients if it is to function properly and remain healthy. Some nutrients are required in relatively large amounts (gram quantities daily) because they provide energy and materials required to repair damaged tissue or build new tissue. These **macronutrients** are the carbohydrates, lipids, and proteins contained in food. **Micronutrients** are required by the body in only small amounts (milligrams or micrograms daily). The small quantities needed suggest correctly that at least some micronutrients are utilized in enzymes. Micronutrients are classified as either vitamins or minerals.

In addition to macro- and micronutrients, the human body must receive appropriate amounts of water and fiber. The importance of water becomes obvious when we learn that 45–75% of the human body mass is water. Fiber is an indigestible plant material composed primarily of cellulose. Fiber makes no contribution to the body in the form of macro- or micronutrients, but it prevents or relieves constipation by absorbing water and softening the stool for easier elimination.

reference daily intakes (RDI)
A set of standards for protein, vitamins, and minerals used on food labels as part of the Daily Values.

A number of countries of the world have established nutritional guidelines for their citizens in an attempt to improve and maintain good national health. In the United States, the Nutrition Labeling and Education Act of 1990 brought sweeping changes to the regulations that define what is required on a food label. The official guidelines are called **Reference Daily Intakes (RDI)** for proteins and 19 vitamins and

© Michael C. Slabaugh

■ **FIGURE 12.1** Carbohydrates, lipids, and proteins from food supply the energy for all of our activities.

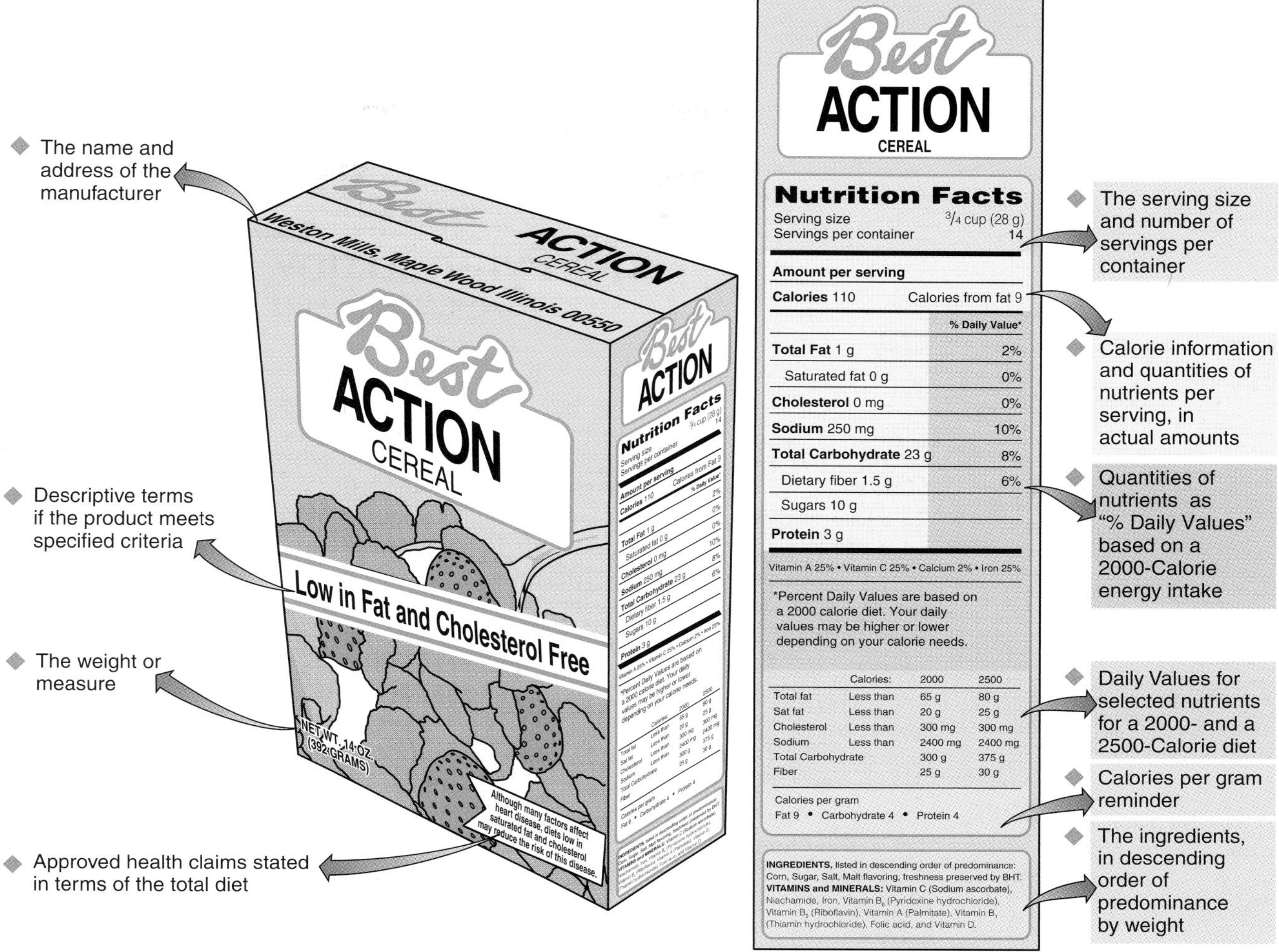

■ **FIGURE 12.2** An example of a food label.

minerals, and **Daily Reference Values (DRV)** for other nutrients of public health importance. For simplicity, all reference values on food labels are referred to as **Daily Values (DV)** (see ■ Figure 12.2). The Food and Drug Administration (FDA) decided to use 2000 Calories as a standard for energy intake in calculating the DRVs. A 2000-Calorie diet is considered about right for many adults. Recall from Chapter 1 that the nutritional Calorie (written with a capital C) is equal to 1 kilocalorie (kcal) of energy. The guidelines are reviewed and revised about every five years. The U.S. Department of Agriculture (USDA) has issued the Food Guide Pyramid, designed to replace the old Four Basic Food Groups posters with new recommendations (see ■ Figure 12.3).

daily reference values (DRV) A set of standards for nutrients and food components (such as fat and fiber) that have important relationships with health; used on food labels as part of the Daily Values.

daily values (DV) Reference values developed by the FDA specifically for use on food labels. The Daily Values represent two sets of standards: Reference Daily Intakes (RDI) and Daily Reference Values (DRV).

ThomsonNOW™ Go to Chemistry Interactive to explore the **Food Guide Pyramid.**

12.2 The Macronutrients

LEARNING OBJECTIVE

2. Describe the primary functions in the body of each macronutrient.

Carbohydrates

Carbohydrates are ideal energy sources for most body functions and also provide useful materials for the synthesis of cell and tissue components. These facts, plus

■ **FIGURE 12.3** The Food Guide Pyramid developed by the U.S. Department of Agriculture. Mypyramid.gov will help you choose foods and amounts that are right for you.

MyPyramid.gov
STEPS TO A HEALTHIER YOU

the relatively low cost and ready availability of carbohydrates, have led to their worldwide use as the main dietary source of energy.

Despite their importance as a dominant source of energy, many people consider foods rich in carbohydrates to be inferior, at least in part because of their reputation of being fattening. It is now recognized that this reputation is generally not deserved. Most of the excess calories associated with eating carbohydrates are actually due to the high-calorie foods eaten *with* the carbohydrates—for example, potatoes and bread are often eaten with butter, a high-energy lipid.

simple carbohydrates Monosaccharides and disaccharides, commonly called sugars.

complex carbohydrates The polysaccharides amylose and amylopectin, collectively called starch.

Dietary carbohydrates are often classified as simple or complex. **Simple carbohydrates** are the sugars we classified earlier as monosaccharides and disaccharides (Chapter 7). **Complex carbohydrates** consist essentially of the polysaccharides amylose and amylopectin, which are collectively called starch. Cellulose, another polysaccharide, is also a complex carbohydrate; however, because it cannot be digested by humans, it serves a nonnutritive role as fiber.

Numerous nutritional studies include recommendations about dietary carbohydrates in their reports. One conclusion common to most of these studies is that a typical American diet does not include enough complex carbohydrates. These studies recommend that about 58% of daily calories should come from carbohydrate food. Currently, only about 46% of the typical diet comes from carbohydrates, and too much of that total comes from simple carbohydrates (see ■ Figure 12.4).

Lipids

Lipids, or fats, like carbohydrates, have a somewhat negative dietary image. In part, this is because the word *fat* has several meanings, and its relationship to health is sometimes misunderstood. In our modern society, some people think of fat as something to be avoided at all costs, regardless of body size.

As we saw in Chapter 8, a number of different substances are classified as lipids. However, about 95% of the lipids in foods and in our bodies are triglycerides, in which three fatty acid molecules are bonded to a molecule of glycerol by ester linkages. The fatty acids in triglycerides can be either saturated or unsaturated. Generally, triglycerides containing a high percentage of unsaturated fatty acids are liquids at room temperature and are called oils (see ■ Figure 12.5). A higher concentration of saturated fatty acids causes triglyceride melting points to increase, and they are solids (fats) at room temperature.

Lipids are important dietary constituents for a number of reasons: They are a concentrated source of energy and provide more than twice the energy of an equal mass of carbohydrate; they contain some fat-soluble vitamins and help carry them through the body; and some fatty acids needed by the body cannot be synthesized and must come from the diet—they are the essential fatty acids.

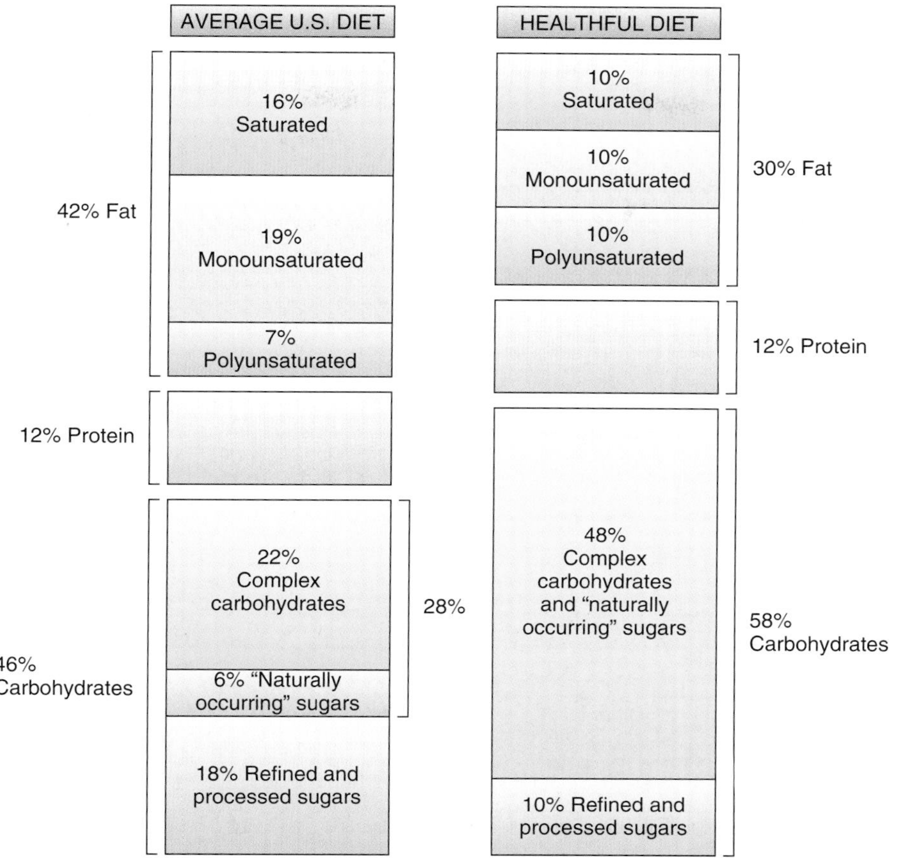

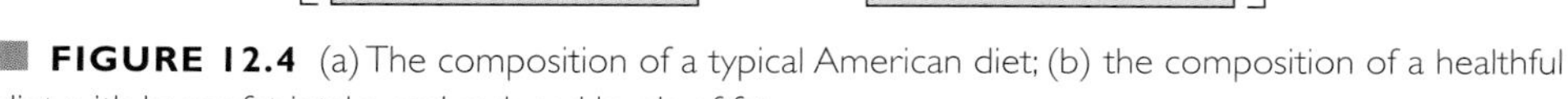
FIGURE 12.4 (a) The composition of a typical American diet; (b) the composition of a healthful diet with lower fat intake and reduced levels of fat.

© Clark Taylor

FIGURE 12.5 Oil-fried foods are rich in triglycerides.

In addition to metabolic needs, dietary lipids are desirable for other reasons. Lipids improve the texture of foods, and absorb and retain flavors, thus making foods more palatable. Lipids are digested more slowly than other foods and therefore prolong **satiety,** a feeling of satisfaction and fullness after a meal.

satiety
A state of satisfaction or fullness.

The essential fatty acids are the polyunsaturated linoleic and linolenic acids. Generally, vegetable oils are good sources of unsaturated fatty acids (see Figure 12.6). Some oils that are especially rich in linoleic acid come from corn, cottonseed, soybean, and wheat germ. Infants especially need linoleic acid for good health and growth; human breast milk contains a higher percentage of it than does cow's milk.

We still do not have a complete understanding of the relationship between dietary lipids and health. A moderate amount of fat is needed in everyone's diet, but many people consume much more than required. Research results indicate a correlation between the consumption of too much fat, and fat of the wrong type (saturated fatty acids), and two of the greatest health problems: obesity and cardiovascular disease. Because of these results, there is a recommended reduction in the percentage of Calories obtained from fats—from the national average of almost 42% to no more than 30%.

TABLE 12.1 The RDI for protein

Group	RDI (g)
Pregnant women	60
Nursing mothers	65
Infants under age 1	14
Children age 1 to 4	16
Adults	50

Proteins

Proteins are the only macronutrients for which an RDI has been established. The RDI varies for different groups of people, as shown in Table 12.1, which shows

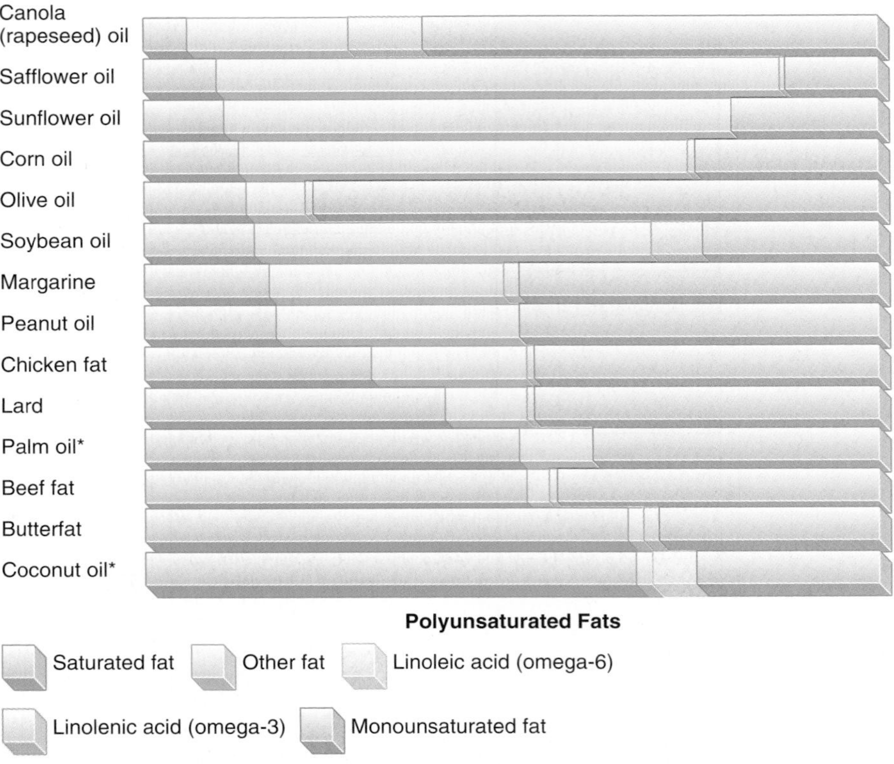

■ **FIGURE 12.6** The fatty acid composition of common food fats.

that children have a greater need for protein than do adults. This reflects one of the important uses of protein in the body, the production of new tissue as the body grows. In addition, proteins are needed for the maintenance and repair of cells and for the production of enzymes, hormones, and other important nitrogen-containing compounds of the body. Proteins can also be used to supply energy; they provide about 4 Calories/gram, the same as carbohydrates. It is recommended that 12% of the Calories obtained from food be obtained from dietary proteins (Figure 12.4).

As we learned in Chapter 9, proteins are natural polymers of amino acids joined by amide (or peptide) linkages. On digestion, proteins are broken down to individual amino acids that are absorbed into the body's amino acid pool and used for the purposes mentioned earlier. The **essential amino acids** listed in ■ Table 12.2 must be obtained from the diet because they cannot be synthesized in the body in sufficient amounts to satisfy the body's needs.

essential amino acid
An amino acid that cannot be synthesized within the body at a rate adequate to meet metabolic needs.

The minimum quantity of each essential amino acid needed per day by an individual will vary, depending on the use in the body. For example, results of one study showed that young men need a daily average of only 7 g (0.034 mol) of tryptophan but 31 g (0.21 mol) of methionine. According to this study, the body of an average young man requires about seven times as many moles of methionine as it does tryptophan to carry out the processes involving amino acids.

Proteins in foods are classified as **complete proteins** if they contain all the essential amino acids in the proper proportions needed by the body. Protein foods that don't meet this requirement are classified as incomplete. Several protein-containing foods are shown in ■ Figure 12.7.

complete protein
Protein in food that contains all essential amino acids in the proportions needed by the body.

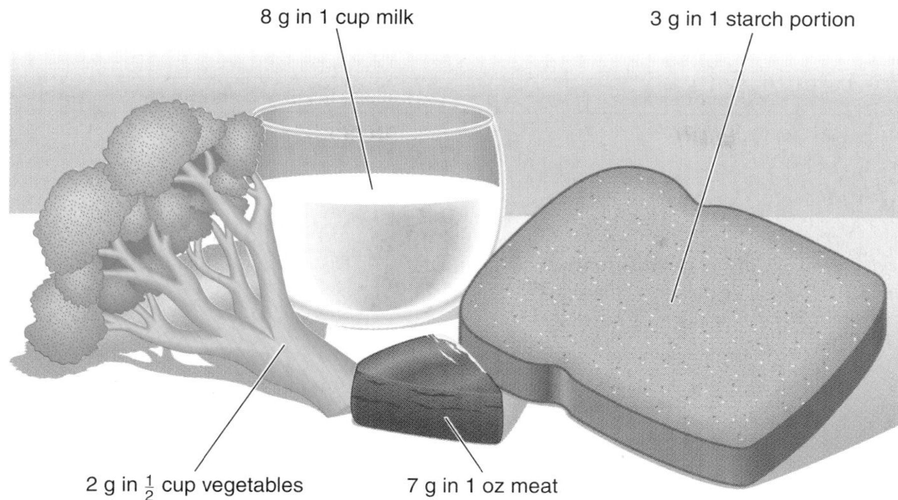

FIGURE 12.7 The protein content of several foods.

TABLE 12.2 The essential amino acids

Histidine
Isoleucine
Leucine
Lysine
Methionine
Phenylalanine
Threonine
Tryptophan
Valine

12.3 Micronutrients I: Vitamins

LEARNING OBJECTIVE

3. Distinguish between and classify vitamins as water-soluble or fat-soluble on the basis of name and behavior in the body.

Vitamins are organic micronutrients that the body cannot produce in amounts needed for good health. The highly polar nature of some vitamins renders them

CHEMISTRY AROUND US 12.1 The Ten Most Dangerous Foods to Eat While Driving

We don't often think of it, but every time we drive a car, we are maneuvering a two-ton piece of machinery at speeds up to 80 mph! And yet, as if that weren't enough, some drivers eat and drink as they do it! Eating ranks as the second driving distraction for Americans, right behind tuning the radios, which tops the list. Exxon conducted a survey in 2001 and found that 70% of auto drivers eat while they are driving; 85% drink coffee, juice, or soda; and a few confessed they would use a microwave in the car if they could! The National Highway Traffic Safety Administration tracked 6.3 million auto crashes in one year, and attributed 25% to food and drink distractions.

With the hurried pace of society, it is no wonder many of us try to eat and drive at the same time. However, some foods and drinks have disaster written all over them when combined with driving. Liquids are especially problematic as they spill easily. Most drivers do not take time to pull over and clean up a mess, but instead continue to drive and simultaneously clean the car and/or themselves. According to studies, the following are the top ten food and drink culprits that contribute to causing accidents and collisions:

1. Coffee: It is usually way too hot and burns the mouth. The coffee also finds a way out of the cup and onto a clean white blouse.
2. Hot Soup: It has the same problem as coffee because most people drink it.
3. Tacos: Notorious for coming apart at the first bite, most tacos end up scattered all over the seat, creating a salad bar effect.
4. Chili: The spoon rarely makes it to the mouth without losing some of the food.
5. Hamburgers: Our hands get greasy, and the mustard and ketchup squeeze out the other end, making steering almost impossible!
6. Barbecued food: It tastes good, but that slippery, messy sauce will end up everywhere!
7. Fried chicken: The grease ends up on your hands, face, seat, and clothes. The steering wheel usually becomes greasy, which makes it nearly impossible to steer.
8. Jelly or cream-filled doughnuts: No one has ever eaten these types of doughnuts without creating some sort of mess. Why would you want to try to eat them while driving a car?
9. Soft drinks: They are subject to spills, and the carbonation can also trigger that moment when the nose gets all tingly and distracts you from your driving.
10. Chocolate: It tastes yummy but ends up all over the hands and everything you touch! If the car is warm, good luck trying to keep the chocolate from melting!

TABLE 12.3 Vitamin sources, functions, and deficiency conditions

Vitamin	Dietary sources	Functions	Deficiency conditions
Water-soluble			
B_1 (thiamin)	Bread, beans, nuts, milk, peas, pork	Coenzyme in decarboxylation reactions	Beriberi: nausea, severe exhaustion, paralysis
B_2 (riboflavin)	Milk, meat, eggs, dark green vegetables, bread, beans, peas	Forms the coenzymes FMN and FAD, which are hydrogen transporters	Dermatitis (skin problems)
Niacin	Meat, whole grains, poultry, fish	Forms the coenzyme NAD^+, which is a hydride transporter	Pellagra: weak muscles, no appetite, diarrhea, dermatitis
B_6 (pyridoxine)	Meats, whole grains, poultry, fish	Coenzyme form carries amino and carboxyl groups	Dermatitis, nervous disorders
B_{12} (cobalamin)	Meat, fish, eggs, milk	Coenzyme in amino acid metabolism	Rare except in vegetarians; pernicious anemia
Folic acid	Leafy green vegetables, peas, beans	Coenzyme in methyl group transfers	Anemia
Pantothenic acid	All plants and animals, nuts; whole-grain cereals	Part of coenzyme A, acyl group carrier	Anemia
Biotin	Found widely; egg yolk, liver, yeast, nuts	Coenzyme form used in fatty acid synthesis	Dermatitis, muscle weakness
C (ascorbic acid)	Citrus fruits, tomatoes, green pepper, strawberries, leafy green vegetables	Synthesis of collagen for connective tissue	Scurvy: tender tissues; weak, bleeding gums; swollen joints
Fat-soluble			
A (retinol)	Eggs, butter, cheese, dark green and deep orange vegetables	Synthesis of visual pigments	Inflamed eye membranes, night blindness, scaliness of skin
D (calciferol)	Fish-liver oils, fortified milk	Regulation of calcium and phosphorus metabolism	Rickets (malformation of the bones)
E (tocopherol)	Whole-grain cereals, margarine, vegetable oil	Prevention of oxidation of vitamin A and fatty acids	Breakage of red blood cells
K	Cabbage, potatoes, peas, leafy green vegetables	Synthesis of blood-clotting substances	Blood-clotting disorders

water-soluble. Nine water-soluble vitamins have been identified. Some are designated by names, some by letters, and some by letters and numbers. ■ Table 12.3 presents the names nutritionists use (the "correct" names), along with other common names you will likely encounter in books and on product labels. With the exception of vitamin C, all water-soluble vitamins have been shown to function as coenzymes.

Some vitamins (A, D, E, and K) have very nonpolar molecular structures and therefore dissolve only in nonpolar solvents. In the body, the nonpolar solvents are the lipids we have classified as fats, so these vitamins are called fat-soluble. The fat-soluble vitamins have diverse functions in the body (Table 12.3). Care must be taken to avoid overdoses of the fat-soluble vitamins. Toxic effects are known to occur, especially with vitamin A, when excess amounts of these vitamins accumulate in body tissue. Excesses of water-soluble vitamins are excreted readily through the kidneys and are not normally a problem.

CHEMISTRY AND YOUR HEALTH 12.1 The Health Gauge

Your BMI (Body Mass Index) is an objective scientific measure that uses your height and weight to decide if you are overweight or obese. The calculation is done by dividing your weight in kilograms by your height in meters squared.

$$\text{BMI} = \frac{\text{Body weight (kg)}}{\text{Body height (m)}^2}$$

However, BMI classification can be in error for healthy muscular individuals with very low body fat. A couch potato and weight lifter might have the same height and weight, and thus the same score on the BMI scale. The couch potato could rightly be classified as "obese," but the weight lifter hardly deserves the same classification.

In 1998, the U.S. federal government announced BMI values that defined healthy weight. A BMI of 24 or less was defined as healthy weight. Those with a BMI of 25 to 29.9 are defined as overweight. Individuals who fall into the BMI range of 25 to 34.9, and have a waist size of over 40 inches for men and 35 inches for women, are considered to be especially at high risk for health problems.

These broad definitions need some refining to increase their usefulness. A BMI less than 20 is a lean BMI, which indicates a low amount of body fat. This can be desirable for athletes, but for nonathletes it might indicate an excessively low body weight, which can lower immunity responses. Such individuals should consider increasing muscle mass through good diet and exercise.

A BMI between 20 and 22 indicates the ideal, healthy amount of body fat, which is associated with a long, healthy life. Coincidentally, this ratio is perceived by many individuals to be the most physically attractive.

A BMI between 22 and 25 is considered to be quite good and is associated with good health.

"Hefty" individuals have a BMI between 25 and 30 and should try to lower their weight through diet and exercise. Such people are at increased risk for a variety of illnesses.

A BMI greater than 30 indicates a definitely unhealthy condition with an increased risk for heart disease, diabetes, high blood pressure, gallbladder disease, and some cancers. Individuals in this category should make a serious attempt to lose weight by changing their diet and exercising more.

BMI values can provide useful guidelines, especially when body type is also considered: athlete or couch potato, tall or short, big bones or petite.

12.4 Micronutrients II: Minerals

LEARNING OBJECTIVE

4. List a primary function in the body for each major mineral.

The word **mineral** is generally used to describe inorganic substances, often as if they were in the elemental form. Thus, we might hear that a food is a good source of phosphorus, when in fact the food does not contain elemental phosphorus, a caustic nonmetal, but probably contains phosphate ions such as HPO_4^{2-} and $H_2PO_4^-$ or, more likely, organic substances that contain phosphorus, such as nucleic acids and phospholipids.

mineral A metal or nonmetal used in the body in the form of ions or compounds.

In the body, the elements classified as minerals are never used in elemental form, but rather in the form of ions or compounds. They are classified as **major minerals** and **trace minerals** on the basis of the amount present in the body, as shown in ■ Figure 12.8. In general, the major minerals are required in larger daily amounts than are the trace minerals. For example, the RDIs for calcium and phosphorus, the major minerals found in the body in largest amounts, are 800 mg. The RDIs for iron and iodine, two of the trace minerals, are 10–18 mg and 0.12 mg, respectively. By way of comparison, the RDIs for vitamin C and niacin are, respectively, 55–60 mg and 18 mg.

major mineral A mineral found in the body in quantities greater than 5 g.

trace mineral A mineral found in the body in quantities smaller than 5 g.

As shown in ■ Table 12.4, the functions of minerals are consistent with their classification as major or trace and with the amount required daily in the diet. For example, compounds of some major minerals (Ca and P) are the primary inorganic structural components of bones and teeth. Other major minerals (Na, K, Cl, and Mg) form principal ions that are distributed throughout the body's various fluids. Some trace minerals are components of vitamins (Co), enzymes (Zn and Se), hormones (I), or specialized proteins (Fe and Cu). Thus, we see that even

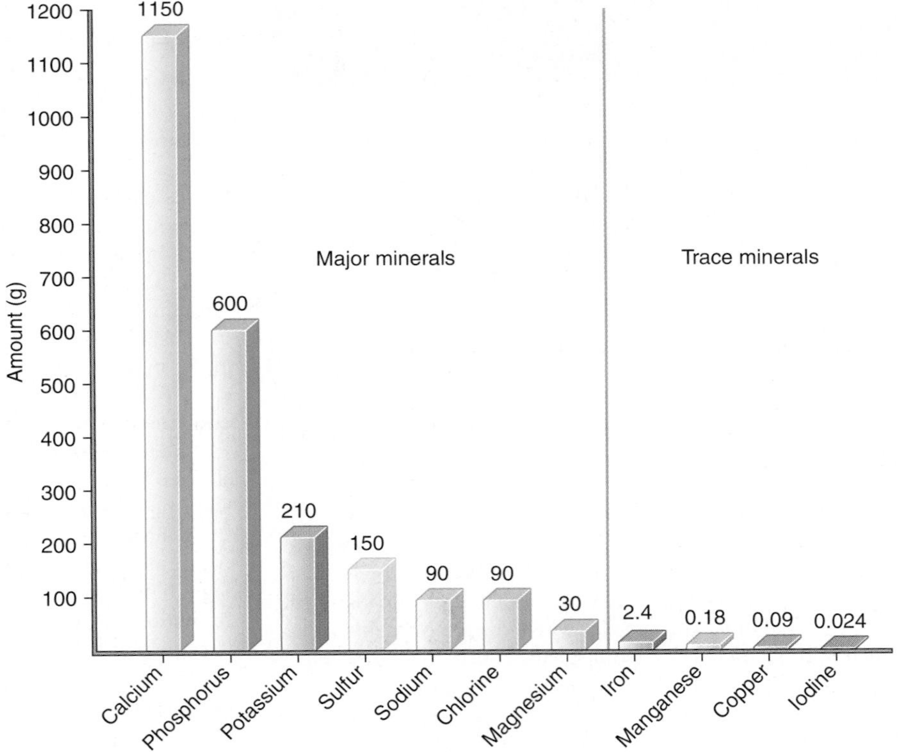

FIGURE 12.8 Minerals in a 60-kg person. The major minerals are those present in amounts larger than 5 g (a teaspoon); only four of the numerous trace minerals are shown. Only calcium and phosphorus appear in amounts larger than a pound (454 g).

though trace minerals are required in small quantities, their involvement in critical enzymes, hormones, and the like makes them equally as important for good health as the major minerals.

12.5 The Flow of Energy in the Biosphere

LEARNING OBJECTIVE

5. Describe the major steps in the flow of energy in the biosphere.

The sun is the ultimate source of energy used in all biological processes. The sun's enormous energy output is the result of the fusion of hydrogen atoms into helium atoms. The net reaction for this process is

$$4\text{H} \xrightarrow{\text{nuclear fusion}} \text{He} + \text{energy} \tag{12.1}$$

A portion of the liberated energy reaches Earth as sunlight and is absorbed by chlorophyll pigments in plants. In the plants, the energy drives the reactions of photosynthesis that convert carbon dioxide and water first into glucose, and then into starch, triglycerides, and other storage forms of energy. The net reaction for glucose production is

$$6\text{CO}_2 + 6\text{H}_2\text{O} + \text{energy} \xrightarrow{\text{photosynthesis}} \underset{\text{glucose}}{\text{C}_6\text{H}_{12}\text{O}_6} + 6\text{O}_2 \tag{12.2}$$

All animals obtain energy either directly or indirectly from these energy stores in plants.

TABLE 12.4 Major and trace mineral sources, functions, and deficiency conditions

Mineral	Dietary sources	Functions	Deficiency conditions
Major minerals			
Calcium	Dairy foods, dark green vegetables	Bone and teeth formation, blood clotting, nerve impulse transmission	Stunted growth, rickets, weak and brittle bones
Chlorine	Table salt, seafood, meat	HCl in gastric juice, acid–base balance	Muscle cramps, apathy, reduced appetite
Magnesium	Whole-grain cereals, meat, nuts, milk, legumes	Activation of enzymes, protein synthesis	Inhibited growth, weakness, spasms
Phosphorus	Milk, cheese, meat, fish, grains, legumes, nuts	Enzyme component, acid–base balance, bone and tooth formation	Weakness, calcium loss, weak bones
Potassium	Meat, milk, many fruits, cereals, legumes	Acid–base and water balance, nerve function	Muscle weakness, paralysis
Sodium	Most foods except fruit	Acid–base and water balance, nerve function	Muscle cramps, apathy, reduced appetite
Sulfur	Protein foods	Component of proteins	Deficiencies are very rare
Trace minerals			
Arsenic[a]	Many foods	Growth and reproduction	Poor growth and reproduction
Cobalt	Meat, liver, dairy foods	Component of vitamin B_{12}	Pernicious anemia (vitamin-deficiency symptom)
Copper	Drinking water, liver, grains, legumes, nuts	Component of numerous enzymes, hemoglobin formation	Anemia, fragility of arteries
Chromium	Fats, vegetable oils, grains, meat	Enhances insulin action	Reduced ability to metabolize glucose
Fluorine	Drinking water, seafood, onions, spinach	Maintenance of bones and teeth	Higher frequency of tooth decay
Iodine	Iodized salt, fish, dairy products	Component of thyroid hormones	Hypothyroidism, goiter
Iron	Liver, lean meat, whole grains, dark green vegetables	Component of enzymes and hemoglobin	Anemia
Manganese	Grains, beet greens, legumes, fruit	Component of enzymes	Deficiencies are rare
Molybdenum	Legumes, cereals, organ meats, dark green vegetables	Component of enzymes	Deficiencies are rare
Nickel	Many foods	Needed for health of numerous tissues	Organ damage, deficiencies are rare
Selenium	Grains, meats, poultry, milk	Component of enzymes	Deficiencies are rare
Silicon[a]	Many foods	Bone calcification	Poor bone development, deficiencies are rare
Tin[a]	Many foods	Needed for growth	Poor growth
Vanadium[a]	Many foods	Growth, bone development, reproduction	Poor growth, bone development, and reproduction
Zinc	Milk, liver, shellfish, wheat bran	Component of numerous enzymes	Poor growth, lack of sexual maturation, loss of appetite, abnormal glucose tolerance

[a]Need and deficiency symptoms determined for animals, probable in humans, but not yet recognized.

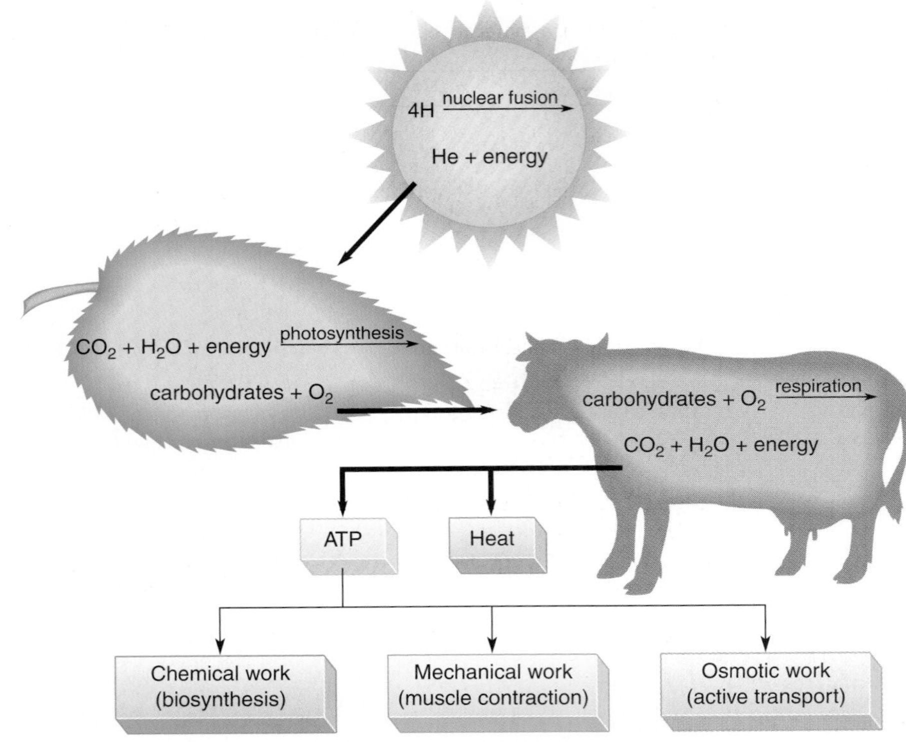

■ **FIGURE 12.9** Energy flow in the biosphere.

cellular respiration The entire process involved in the use of oxygen by cells.

During **cellular respiration,** both plants and animals combine these energy-rich compounds with oxygen from the air, producing carbon dioxide and water, and releasing energy. Cellular respiration, which should not be confused with pulmonary respiration (breathing), is represented as

$$\text{glucose and other storage forms of energy} + O_2 \xrightarrow{\text{respiration}} CO_2 + H_2O + \text{energy released} \qquad (12.3)$$

A portion of the energy released during respiration is captured within the cells in the form of adenosine triphosphate (ATP), a chemical carrier of energy, from which energy can be obtained and used directly for the performance of biological work. The remainder of the energy from respiration is liberated as heat. Steps in the flow of energy in the biosphere are summarized in ■ Figure 12.9.

Notice in Figure 12.9 that atmospheric carbon dioxide is the initial source of carbon for carbohydrates. Another important feature of Figure 12.9 is that overall the respiration process is the reverse of the photosynthetic process. Thus, some of Earth's carbon compounds are repeatedly recycled by living organisms (see ■ Figure 12.10).

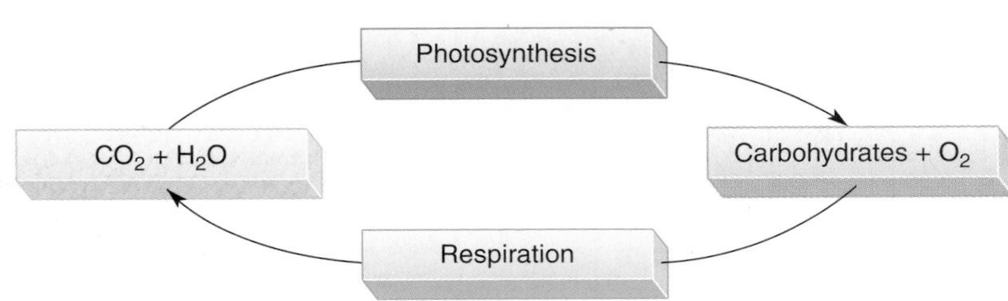

■ **FIGURE 12.10** A simplified carbon cycle.

12.6 Metabolism and an Overview of Energy Production

LEARNING OBJECTIVES

6. Differentiate among metabolism, anabolism, and catabolism.
7. Outline the three stages in the extraction of energy from food.

Metabolism

Living cells are very active chemically, with thousands of different reactions occurring at the same time. The sum total of all the chemical reactions involved in maintaining a living cell is called **metabolism.** The reactions of metabolism are divided into two categories, catabolism and anabolism. **Catabolism** consists of all reactions that lead to the breakdown of biomolecules. **Anabolism** includes all reactions that lead to the synthesis of biomolecules. In general, energy is released during catabolism and required during anabolism.

A characteristic of all living organisms is a high degree of order in their structures and chemical processes. Even the simplest organisms exhibit an enormous array of chemical reactions that are organized into orderly, well-regulated sequences known as **metabolic pathways.** Each metabolic pathway consists of a series of consecutive chemical reactions or steps that convert a starting material into an end product. For example, two pathways that will be covered in Chapter 13 are the citric acid cycle and the electron transport chain.

Fortunately for those who study the biochemistry of living systems, a great many similarities exist among the major metabolic pathways found in all life forms. These similarities enable scientists to study the metabolism of simple organisms and use the results to help explain the corresponding metabolic pathways in more complex organisms, including humans. This text concentrates on human biochemistry, but you should realize that much of what is said about human metabolism can be applied equally well to the metabolism of most other organisms.

metabolism
The sum of all reactions occurring in an organism.

catabolism
All reactions involved in the breakdown of biomolecules.

anabolism
All reactions involved in the synthesis of biomolecules.

metabolic pathway
A sequence of reactions used to produce one product or accomplish one process.

The Catabolism of Food

The three stages involved in the catabolic extraction of energy from food are illustrated in ■ Figure 12.11.

Stage I is digestion, in which large, complex molecules are chemically broken into relatively small, simple ones (see ■ Figure 12.12). The most common reaction in digestion is hydrolysis, in which proteins are converted to amino acids, carbohydrates are converted to monosaccharides (primarily glucose, galactose, and fructose), and fats are converted to fatty acids and glycerol. The smaller molecules are then absorbed into the body through the lining of the small intestine.

In Stage II, the small molecules from digestion are degraded to even simpler units, primarily the two-carbon acetyl portion of acetyl coenzyme A (acetyl CoA):

$$\underbrace{CH_3-\overset{\overset{\displaystyle O}{\|}}{C}}_{\text{acetyl}}\underbrace{-S-CoA}_{\text{coenzyme A}}$$

Some energy is released in the second stage, but much more is produced during the oxidation of the acetyl units of acetyl CoA in the third stage. Stage III consists of the citric acid cycle followed by electron transport and oxidative phosphorylation. Because the reactions of Stage III are the same regardless of the type of food being degraded, Stage III is sometimes referred to as the **common catabolic pathway.** Energy released during Stage III appears in the form of energy-rich molecules

ThomsonNOW Go to Chemistry Interactive to explore the **carbon balance in the citric acid cycle.**

common catabolic pathway
The reactions of the citric acid cycle plus those of the electron transport chain and oxidative phosphorylation.

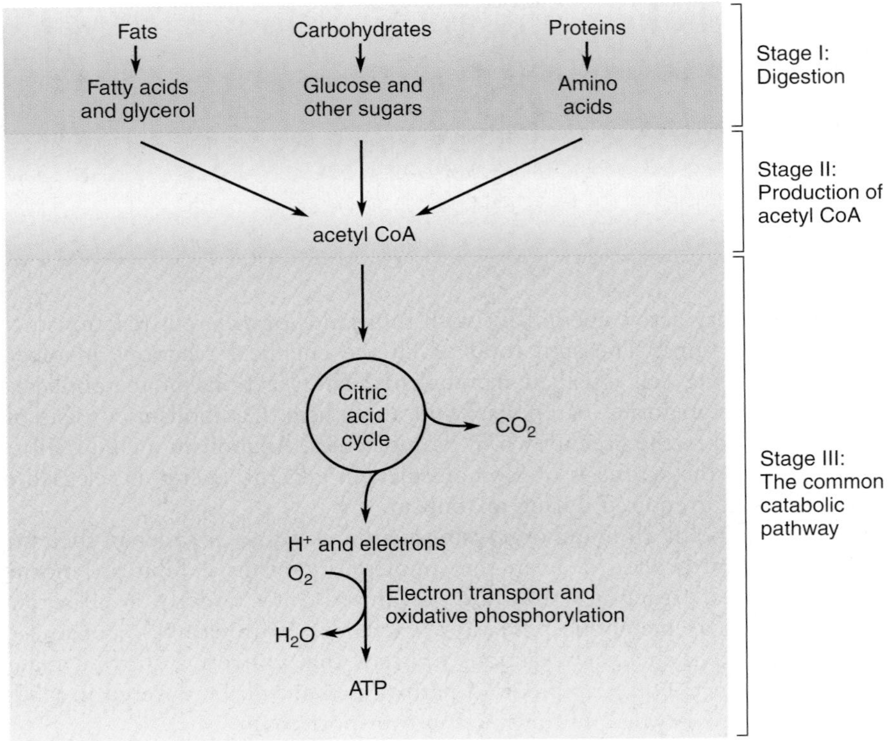

FIGURE 12.11 The three stages in the extraction of energy from food.

of ATP. The whole purpose of the catabolic pathway is to convert the chemical energy in foods to molecules of ATP. The common catabolic pathway will be covered in Chapter 13. Chapters 13–15 discuss the ways in which carbohydrates, lipids, and proteins provide molecules that are degraded by reactions of the common catabolic pathway.

12.7 ATP: The Primary Energy Carrier

LEARNING OBJECTIVE

8. Explain how ATP plays a central role in the production and use of cellular energy.

Certain bonds in ATP store the energy released during the oxidation of carbohydrates, lipids, and proteins. ATP molecules act as energy carriers and deliver the energy to the parts of the cell where energy is needed to power muscle contraction, biosynthesis, and other cellular work.

The Structure and Hydrolysis of ATP

The structure of the ATP molecules is relatively simple (see Figure 12.13). It consists of a heterocyclic base, adenine, bonded to the sugar ribose. Taken together, this portion of the molecule is called adenosine. The triphosphate portion

$$\text{proteins} \xrightarrow{\text{hydrolysis}} \text{amino acids}$$

$$\text{disaccharides and polysaccharides} \xrightarrow{\text{hydrolysis}} \text{monosaccharides}$$

$$\text{fats and oils} \xrightarrow{\text{hydrolysis}} \text{fatty acids + glycerol}$$

FIGURE 12.12 The digestion of food involves enzyme-catalyzed hydrolysis reactions.

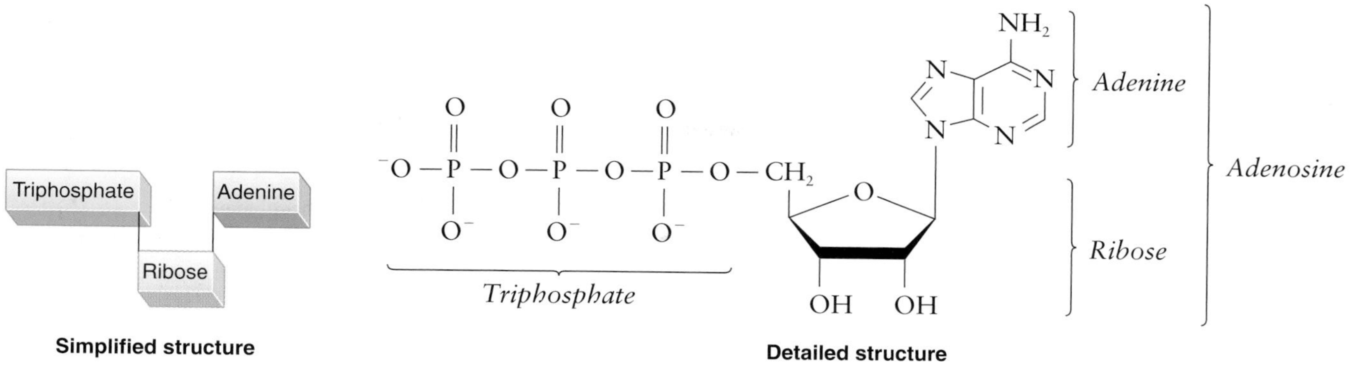

FIGURE 12.13 The structure of ATP.

is bonded to the ribose to give adenosine triphosphate, or ATP. At the cell pH of 7.4, all the protons of the triphosphate group are ionized, giving the ATP molecule a charge of -4. In the cell, ATP is complexed with Mg^{2+} in a 1:1 ratio. Thus, the 1:1 complex still exhibits a net charge of -2. Two other triphosphates—GTP, containing the base guanine, and UTP, containing the base uracil—are important in carbohydrate metabolism (Chapter 13).

The triphosphate end of ATP is the part of the molecule that is important in the transfer of biochemical energy. The key reaction in this energy delivery system is the transfer of a phosphoryl group, $—PO_3^{2-}$, from ATP to another molecule. For example, during the hydrolysis of ATP in water, a phosphoryl group is transferred from ATP to water. The products of this hydrolysis are adenosine diphosphate (ADP) and a phosphate ion, often referred to as an inorganic phosphate, P_i, or simply as *phosphate*.

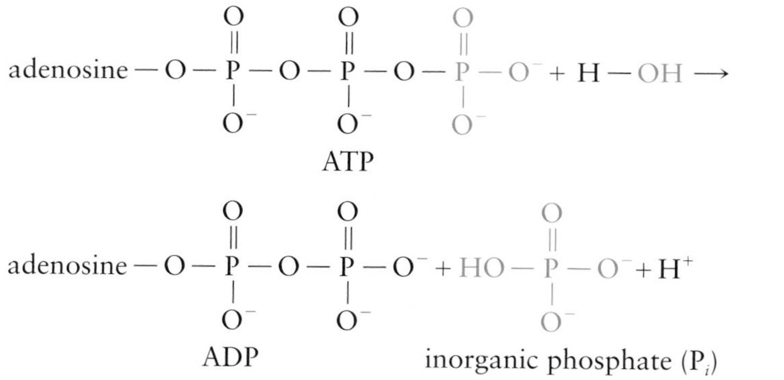

$$\underbrace{\text{adenosine}-O-\underset{O^-}{\overset{O}{\overset{\|}{\underset{|}{P}}}}-O-\underset{O^-}{\overset{O}{\overset{\|}{\underset{|}{P}}}}-O-\underset{O^-}{\overset{O}{\overset{\|}{\underset{|}{P}}}}-O^-}_{\text{ATP}} + H-OH \longrightarrow$$

$$\underbrace{\text{adenosine}-O-\underset{O^-}{\overset{O}{\overset{\|}{\underset{|}{P}}}}-O-\underset{O^-}{\overset{O}{\overset{\|}{\underset{|}{P}}}}-O^-}_{\text{ADP}} + \underbrace{HO-\underset{O^-}{\overset{O}{\overset{\|}{\underset{|}{P}}}}-O^-}_{\text{inorganic phosphate }(P_i)} + H^+ \qquad (12.4)$$

The transfer of a phosphoryl group from ATP to water is accompanied by a release of energy. This energy is called free energy and is represented by the symbol ΔG. When energy is released by a reaction or process, ΔG will have a negative value. When energy is absorbed, ΔG will have a positive value. When ΔG is measured under standard conditions (temperature, concentration, etc.), it is represented by $\Delta G°$, and a superscript prime ($\Delta G°'$) is added if $\Delta G°$ is measured at body conditions. The liberated free energy is available for use by the cell to carry out processes requiring an input of energy. This energy-releasing hydrolysis reaction is represented by the following equation:

$$ATP + H_2O \rightarrow ADP + P_i + H^+ \qquad \Delta G°' = -7.3 \text{ kcal/mol} \qquad (12.5)$$

Many other phosphate-containing compounds found in biological systems also liberate energy on hydrolysis. Several are listed in Table 12.5, along with values for their free standard energy of hydrolysis.

TABLE 12.5 Free energy changes for the hydrolysis of some phosphate compounds

Reaction			$\Delta G^{\circ\prime}$ (kcal/mol)
Phosphoenolpyruvate + H_2O	→	pyruvate + phosphate	−14.8
Carbamoyl phosphate + H_2O	→	ammonia + CO_2 + phosphate	−12.3
1,3-bisphosphoglycerate + H_2O	→	3-phosphoglycerate + phosphate	−11.8
Phosphocreatine + H_2O	→	creatine + phosphate	−10.3
ATP + H_2O	→	ADP + phosphate	−7.3
ADP + H_2O	→	AMP + phosphate	−7.3
Glucose 1-phosphate + H_2O	→	glucose + phosphate	−5.0
AMP + H_2O	→	adenosine + phosphate	−3.4
Glucose 6-phosphate + H_2O	→	glucose + phosphate	−3.3
Glycerol 3-phosphate + H_2O	→	glycerol + phosphate	−2.2

high-energy compound
A substance that on hydrolysis liberates a great amount of free energy.

Compounds that liberate a great amount of free energy on hydrolysis are called **high-energy compounds.** Remember, a negative sign for $\Delta G^{\circ\prime}$ means the energy is liberated when the reaction takes place. Thus, ATP ($\Delta G^{\circ\prime} = -7.3$ kcal/mol) is considered a high-energy compound, but not the only one found in cells, as you can see from Table 12.5.

The hydrolysis of ATP to ADP is the principal energy-releasing reaction for ATP. However, some other hydrolysis reactions also play important roles in energy transfer. An example is the hydrolysis of ATP to adenosine monophosphate (AMP) and pyrophosphate (PP_i):

$$\text{adenosine}-O-\underset{O^-}{\overset{O}{\overset{\|}{\underset{|}{P}}}}-O-\underset{O^-}{\overset{O}{\overset{\|}{\underset{|}{P}}}}-O-\underset{O^-}{\overset{O}{\overset{\|}{\underset{|}{P}}}}-O^- + H_2O \longrightarrow \text{adenosine}-O-\underset{O^-}{\overset{O}{\overset{\|}{\underset{|}{P}}}}-O^- + {}^-O-\underset{O^-}{\overset{O}{\overset{\|}{\underset{|}{P}}}}-O-\underset{O^-}{\overset{O}{\overset{\|}{\underset{|}{P}}}}-O^- + 2H^+ \quad (12.6)$$

ATP; AMP adenosine monophosphate; PP_i pyrophosphate

The reaction may also be written as

$$ATP + H_2O \rightarrow AMP + PP_i + 2H^+ \qquad \Delta G^{\circ\prime} = -8.0 \text{ kcal/mol} \quad (12.7)$$

STUDY SKILLS 12.1 Bioprocesses

The last four chapters of the text complete our study of the molecular basis of life. This chapter and the next three build on your knowledge of the biomolecules (carbohydrates, lipids, proteins, and nucleic acids) and examine the processes that support living organisms. Your attention should now be directed toward gaining an understanding of these processes, where they occur in the cell, their purpose, and, in some cases, key summary reactions. A number of detailed pathways such as glycolysis, the fatty acid spiral, and the urea cycle will be presented. Don't let the numerous reactions in these pathways intimidate you. Although each reaction could be looked at individually, we have tried to help you focus your attention by emphasizing the overall importance of each pathway by identifying its purpose, cellular location, starting material, products, energy requirements, and regulation.

CHEMISTRY AROUND US 12.2 Eating Disorders

Food is there for us to enjoy, but sometimes it becomes part of a dangerous and even deadly disorder. Bulimia and anorexia are eating disorders, affecting 10 million females and 1 million males in the United States alone. Another 25 million more struggle with binge eating disorder. The great difficulty with diagnosing and treating this disorder is that it has many negative associations in society, and individuals go to great lengths to cover up the signs. However, families and friends would do well to look for some oft-repeated symptoms. Although individuals may exhibit these signs and not have an eating disorder, these indicate a predisposition that could easily transform to a full disorder.

1. Preoccupation with weight: People become concerned with their weight and image for a number of reasons, including establishing social status, earning a spot on a sports team, or simply following media trends. Individuals may be considered healthy by their peers but view themselves as fat. This negative self image soon becomes the only view these individuals recognize and creates the drive to eat very little or to purge after consuming large quantities of food.
2. Eating alone and hiding food: Many times, roommates, friends, or families will notice food missing from their cupboards. An individual might be eating in secret, ashamed of eating in public. Watch for wrappers in the car or food items hidden in nontraditional places. Although they may be harmless eating habits, they often denote a serious problem.
3. Mood changes: When an individual does not get enough food to sustain his or her daily routine, mood changes become more extreme. Extreme depression, irritability, isolation, perfectionism, and impulsive behavior are potential signs that an individual is experiencing an eating disorder.
4. Overexercising: Exercise is a wonderful habit, but it should always be practiced in moderation. A standard rule is that a person should exercise 30–60 minutes a day, five or six times a week. If an individual's exercise routine becomes extreme and he or she is not refueling with good foods, the result is the same as using vomiting, laxatives, or other diuretics to get rid of unwanted food.

As a friend or family member, be aware that confronting an individual about an eating disorder might bring a lot of emotion to a situation. However, an eating disorder is not something to treat lightly and should never be passed over as a "phase" that will soon pass. An eating disorder is a serious condition that many times requires medical attention in order to help an individual overcome the physical and mental addiction.

When the hydrolysis of ATP to AMP occurs during metabolism, it is usually followed by the immediate hydrolysis of the pyrophosphate, which releases even more free energy:

$$\underset{PP_i}{{}^{-}O-\overset{\overset{\large O}{\|}}{\underset{\underset{\large O^-}{|}}{P}}-O-\overset{\overset{\large O}{\|}}{\underset{\underset{\large O^-}{|}}{P}}-O^-} + H_2O \longrightarrow 2\underset{P_i}{HO-\overset{\overset{\large O}{\|}}{\underset{\underset{\large O^-}{|}}{P}}-O^-} \quad \Delta G^{\circ\prime} = -6.5 \text{ kcal/mol} \qquad (12.8)$$

The hydrolyses of ATP and related compounds are summarized in Table 12.6.

The ATP–ADP Cycle

In biological systems, ATP functions as an immediate donor of free energy rather than as a storage form of free energy. In a typical cell, an ATP molecule is

TABLE 12.6 Hydrolyses of some ATP-related compounds

Reaction			$\Delta G^{\circ\prime}$ (kcal/mol)
ATP + H_2O	→	ADP + P_i + H^+	−7.3
ATP + H_2O	→	AMP + PP_i + $2H^+$	−8.0
PP_i + H_2O	→	$2P_i$	−6.5
ADP + H_2O	→	AMP + P_i + H^+	−7.3

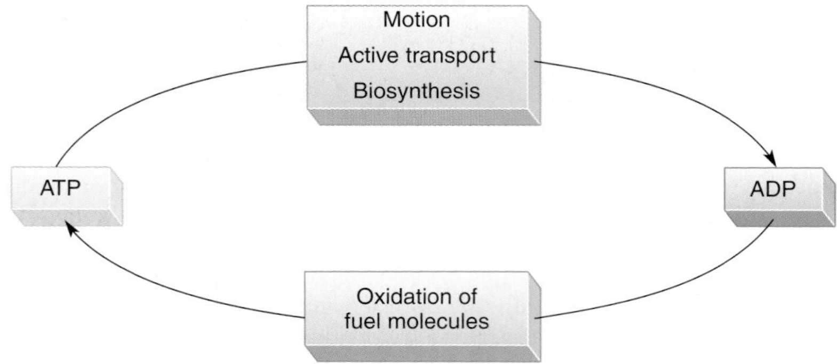

■ **FIGURE 12.14** The ATP–ADP cycle. ATP, which supplies the energy for cellular work, is continuously regenerated from ADP during the oxidation of fuel molecules.

hydrolyzed within 1 minute following its formation. Thus, the turnover rate of ATP is very high. For example, a resting human body hydrolyzes ATP at the rate of about 40 kg every 24 hours. During strenuous exertion, this rate of ATP utilization may be as high as 0.5 kg per minute. Motion, biosynthesis, and other forms of cellular work can occur only if ATP is continuously regenerated from ADP. ■ Figure 12.14 depicts the central role served by the ATP–ADP cycle in linking energy production with energy utilization.

Mitochondria

mitochondrion
A cellular organelle where reactions of the common catabolic pathway occur.

Enzymes that catalyze the formation of ATP, as well as the other processes in the common catabolic pathway, are all located within the cell in organelles called **mitochondria** (see ■ Figure 12.15). Mitochondria are often called cellular "power stations" because they are the sites for most ATP synthesis in the cells.

A mitochondrion contains both inner and outer membranes. As a result of extensive folding, the inner membrane has a surface area that is many times that of the outer membrane. The folds of the inner membrane are called *cristae*, and the space that surrounds them is called the *matrix*. The enzymes for ATP synthesis (electron transport and oxidative phosphorylation) are located within the inner membrane. The enzymes for the citric acid cycle are found within the matrix, attached or near to the surface of the inner membrane. Thus, all the enzymes involved in the common catabolic pathway are in close proximity.

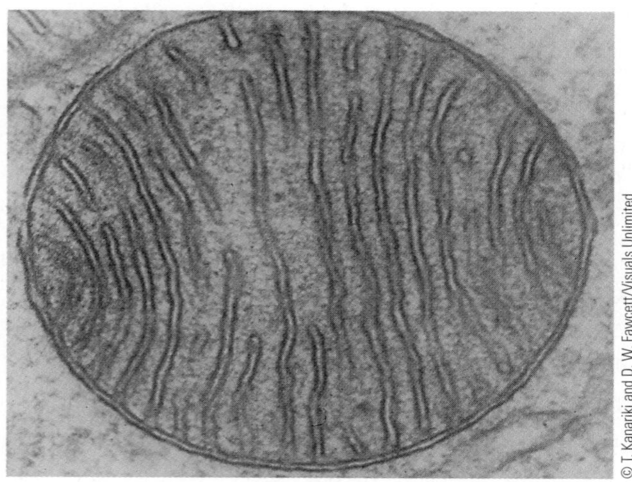

(a)

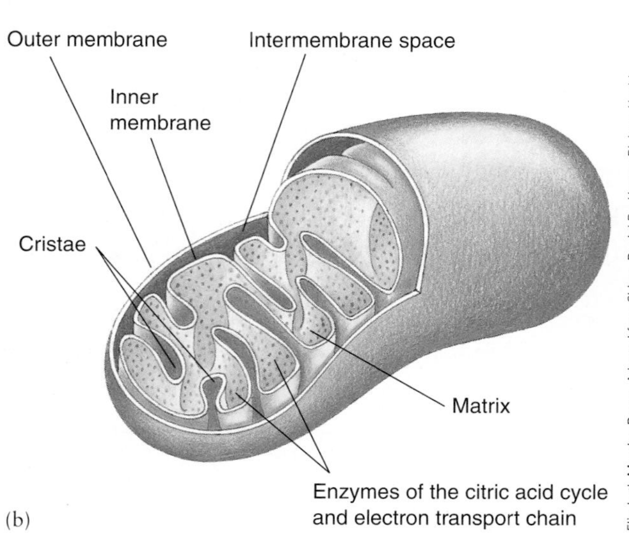

(b)

■ **FIGURE 12.15** The mitochondrion: (a) a photomicrograph; (b) a schematic drawing, cut away to reveal the internal organization.

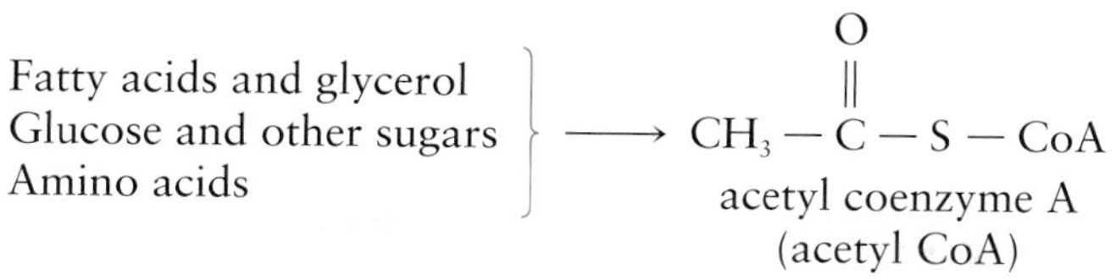

$$\left.\begin{array}{l}\text{Fatty acids and glycerol}\\ \text{Glucose and other sugars}\\ \text{Amino acids}\end{array}\right\} \longrightarrow \underset{\substack{\text{acetyl coenzyme A}\\ \text{(acetyl CoA)}}}{CH_3-\overset{\overset{\displaystyle O}{\|}}{C}-S-CoA}$$

FIGURE 12.16 Stage II of the oxidation of foods. Acetyl CoA is formed as the pathways of catabolism converge.

12.8 Important Coenzymes in the Common Catabolic Pathway

LEARNING OBJECTIVE

9. Explain the role of coenzymes in the common catabolic pathway.

As discussed in Section 10.3, coenzymes often participate in enzyme-catalyzed reactions by acting as shuttle systems in the transfer of chemical groups from one compound to another. For example, a coenzyme may accept hydrogen from one compound and later donate hydrogen to another compound. We also pointed out that many important coenzymes are formed from vitamins. The three substances discussed in this section exhibit both of these coenzyme characteristics.

Coenzyme A

Coenzyme A is a central compound in metabolism. It is a part of acetyl coenzyme A (acetyl CoA), the substance formed from all foods as they pass through Stage II of catabolism (see Figure 12.16).

Coenzyme A is derived from the B vitamin pantothenic acid (see Figure 12.17). It also contains two other components, a phosphate derivative of ADP and β-mercaptoethylamine. A key feature in the structure of coenzyme A is the presence of the sulfhydryl group (—SH), which is the reactive portion of the molecule. For this reason, coenzyme A is often abbreviated as CoA—SH.

The letter *A* was included in the coenzyme name to signify its participation in the transfer of acetyl groups. It is now known that coenzyme A transfers acyl groups in general rather than just the acetyl group. This is demonstrated in fatty acid oxidations (Section 14.4), fatty acid synthesis (Section 14.7), and other biological

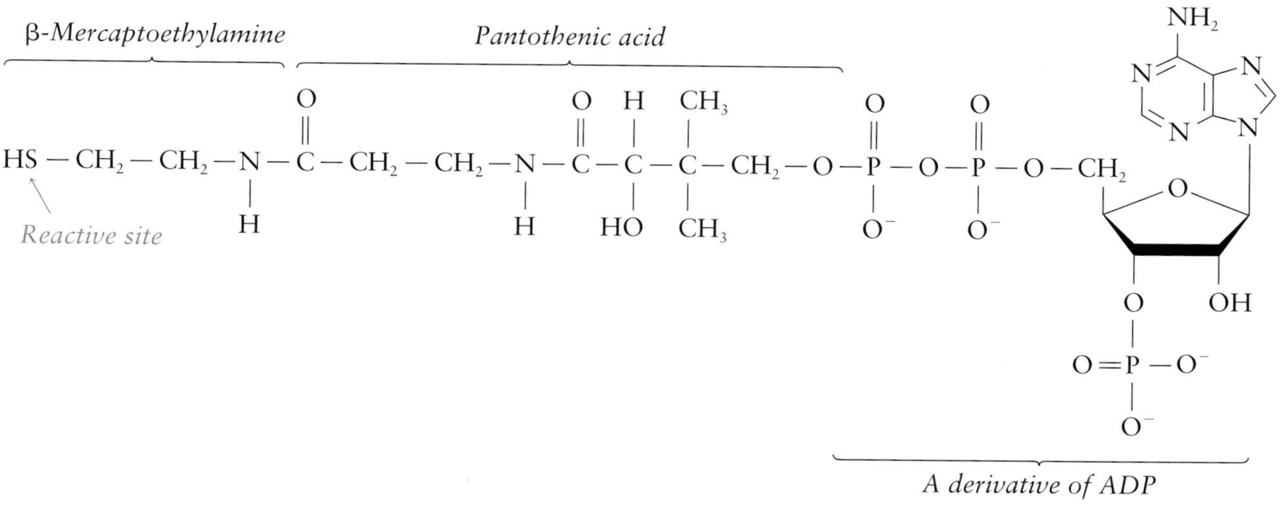

β-Mercaptoethylamine | *Pantothenic acid* | *A derivative of ADP*

$$HS-CH_2-CH_2-\underset{\underset{\displaystyle H}{|}}{N}-\overset{\overset{\displaystyle O}{\|}}{C}-CH_2-CH_2-\underset{\underset{\displaystyle H}{|}}{N}-\overset{\overset{\displaystyle O}{\|}}{C}-\overset{\overset{\displaystyle H}{|}}{\underset{\underset{\displaystyle HO}{|}}{C}}-\overset{\overset{\displaystyle CH_3}{|}}{\underset{\underset{\displaystyle CH_3}{|}}{C}}-CH_2-O-\overset{\overset{\displaystyle O}{\|}}{\underset{\underset{\displaystyle O^-}{|}}{P}}-O-\overset{\overset{\displaystyle O}{\|}}{\underset{\underset{\displaystyle O^-}{|}}{P}}-O-CH_2-\text{(ribose-3'-phosphate)-adenine}$$

Reactive site (HS); NH_2; $O=P-O^-$, O^-; OH

FIGURE 12.17 The structure of coenzyme A. The sulfhydryl group (—SH) is the reactive site.

thioester bond
The carbon–sulfur bond of the $-\overset{O}{\overset{\|}{C}}-S-$ functional group.

reactions. Acyl groups are linked to coenzyme A through the sulfur atom in a **thioester bond.**

$$\underset{\text{acyl CoA}}{R-\overset{O}{\overset{\|}{C}}-S-CoA} \qquad \underset{\text{acetyl CoA}}{CH_3-\overset{O}{\overset{\|}{C}}-S-CoA}$$

The Coenzymes NAD^+ and FAD

The majority of energy for ATP synthesis is released when the oxygen we breathe is reduced. During that process, oxygen accepts electrons and hydrogen ions, and water is produced. The electrons come from the oxidation of fuel molecules, such as glucose, but the electrons are not transferred directly to the oxygen from the fuel molecules or their breakdown products. Instead, these substrates first transfer the electrons to special coenzyme carriers. The reduced forms of these coenzymes then transfer the electrons to oxygen through the reactions of the electron transport chain. ATP is formed from ADP and P_i as a result of this flow of electrons.

The first of these coenzymes, nicotinamide adenine dinucleotide (NAD^+), is a derivative of ADP and the vitamin nicotinamide (see ■ Figure 12.18). The reactive site of NAD^+ is located in the nicotinamide portion of the molecule. During the oxidation of a substrate, the nicotinamide ring of NAD^+ accepts two electrons and one proton, which forms the reduced coenzyme NADH.

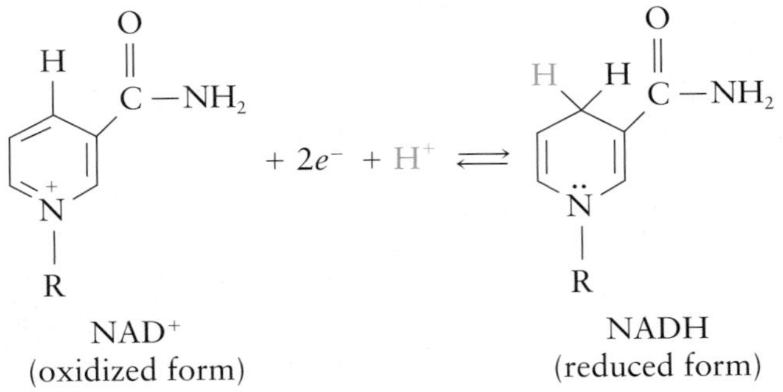

$$\underset{\substack{NAD^+ \\ \text{(oxidized form)}}}{\text{[nicotinamide ring: H, C(=O)–NH}_2\text{, N}^+\text{–R]}} + 2e^- + H^+ \rightleftharpoons \underset{\substack{NADH \\ \text{(reduced form)}}}{\text{[ring: H, H, C(=O)–NH}_2\text{, }\ddot{N}\text{–R]}} \qquad (12.9)$$

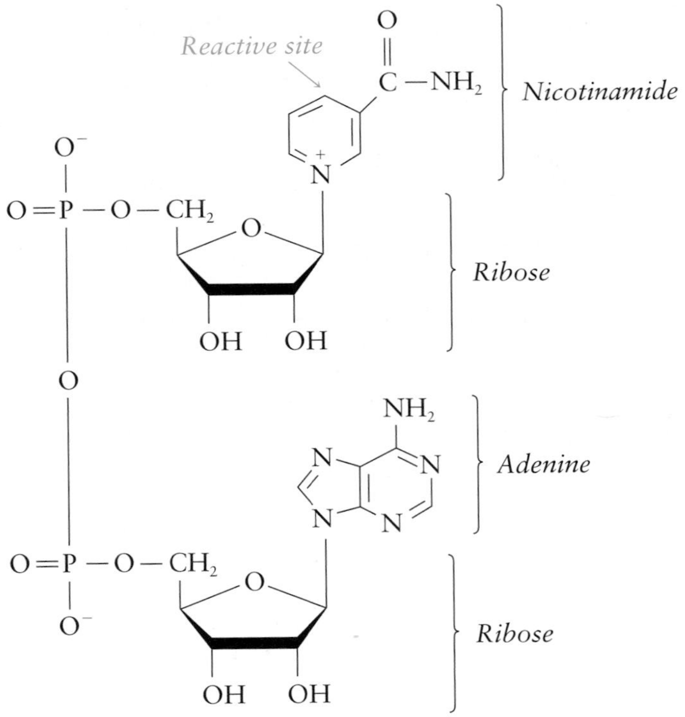

■ **FIGURE 12.18** The structure of nicotinamide adenine dinucleotide (NAD^+). The reactive site is in the nicotinamide part of the molecule.

OVER THE COUNTER 12.1 Creatine Supplements: The Jury Is Still Out

Creatine has become a popular dietary supplement among professional athletes and other individuals interested in building muscles. Users find that supplemental creatine in the diet helps them retain strength and size, and it increases their energy levels. It also seems to help some athletes recover from injury more quickly. Preliminary scientific research results support the claim that muscle size and strength increase significantly faster in creatine users than in nonusers when both groups follow the same training and nutritional programs.

However, the advantages are accompanied by negative side effects. Supplemental creatine in the diet may influence some electrolyte levels in body tissues. In muscle tissue, the ratio of calcium ions to phosphate ions may be reversed, causing a disruption of the muscle's normal contraction-relaxation processes. As a result, a muscle might contract and shorten when it should be relaxing and lengthening. This can cause muscle cramping and in some cases serious muscle strains or even muscle tears.

Electrolyte imbalance can also contribute to dehydration and other heat-related illnesses. Electrolyte levels influence the fluid balance in the body, and when muscle cells retain increased levels of fluids, the blood plasma volume of the body is reduced. This has a negative effect on the body's ability to dissipate heat, a potentially serious condition in hot, humid environments.

Research results concerning the use of dietary creatine supplements are preliminary and are generally based on short-term studies. Some factors have not been studied, such as the effects of combining creatine supplementation with strength training and a conditioning program. Some research seems to indicate that creatine may be more useful in developing "ornamental" muscle, important to bodybuilding, rather than "functional" muscle, which adds strength and is important in athletics.

Three of the creatine supplements being marketed.

A typical cellular reaction in which NAD^+ serves as an electron acceptor is the oxidation of an alcohol:

General reaction:

$$\underset{\text{alcohol}}{\mathrm{R-\overset{\overset{\displaystyle OH}{|}}{\underset{\underset{\displaystyle H}{|}}{C}}-H}} + \mathrm{NAD^+} \longrightarrow \underset{\text{aldehyde}}{\mathrm{R-\overset{\overset{\displaystyle O}{\|}}{C}-H}} + \mathrm{NADH + H^+} \qquad (12.10)$$

Specific example:

$$\underset{\text{ethanol}}{\mathrm{CH_3CH_2-OH}} + \mathrm{NAD^+} \xrightarrow[\text{dehydrogenase}]{\text{alcohol}} \underset{\text{acetaldehyde}}{\mathrm{CH_3-\overset{\overset{\displaystyle O}{\|}}{C}-H}} + \mathrm{NADH + H^+} \qquad (12.11)$$

In this reaction, one hydrogen atom of the alcohol substrate is directly transferred to NAD^+, whereas the other appears in solution as H^+. Both electrons lost by the alcohol have been transferred to the nicotinamide ring in NADH.

Biochemical reactions in which coenzymes participate are often written in a more concise way. For example, Reaction 12.10 may be written

$$\mathrm{R-\overset{\overset{\displaystyle OH}{|}}{\underset{\underset{\displaystyle H}{|}}{C}}-H} \xrightarrow[\mathrm{NAD^+ \;\curvearrowright\; NADH + H^+}]{} \mathrm{R-\overset{\overset{\displaystyle O}{\|}}{C}-H} \qquad (12.12)$$

This format allows for the separate emphasis of the oxidation reaction and the involvement of a coenzyme.

Learning Check 12.1 Write the following reaction in the normal form of a chemical equation:

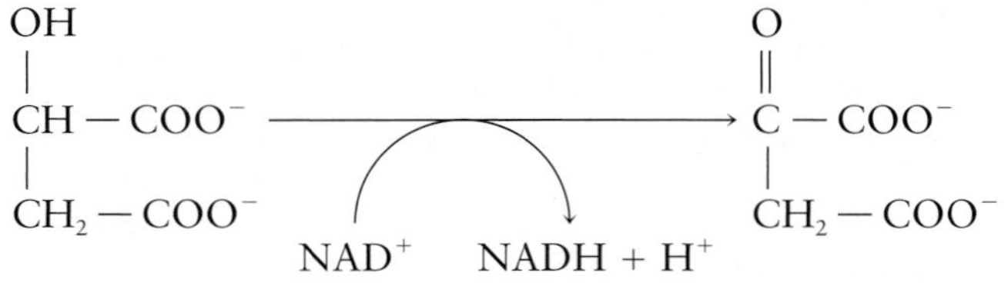

$$\begin{array}{c}OH\\|\\CH-COO^-\\|\\CH_2-COO^-\end{array} \xrightarrow[NAD^+ \quad NADH + H^+]{} \begin{array}{c}O\\\|\\C-COO^-\\|\\CH_2-COO^-\end{array}$$

Flavin adenine dinucleotide (FAD) is the other major electron carrier in the oxidation of fuel molecules. This coenzyme is a derivative of ADP and the vitamin riboflavin. The reactive site is located within the riboflavin ring system (see Figure 12.19).

As in the reduction of NAD^+, substrates of enzymes that use FAD as the coenzyme give up two electrons to the coenzyme. However, unlike NAD^+, the FAD coenzyme accepts both of the hydrogen atoms lost by the substrate. Thus, the abbreviation for the reduced form of FAD is $FADH_2$:

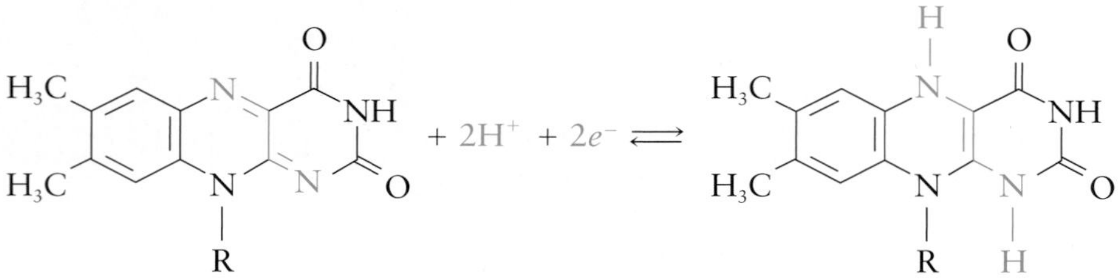

$$\text{FAD (oxidized form)} + 2H^+ + 2e^- \rightleftharpoons \text{FADH}_2 \text{ (reduced form)} \quad (12.13)$$

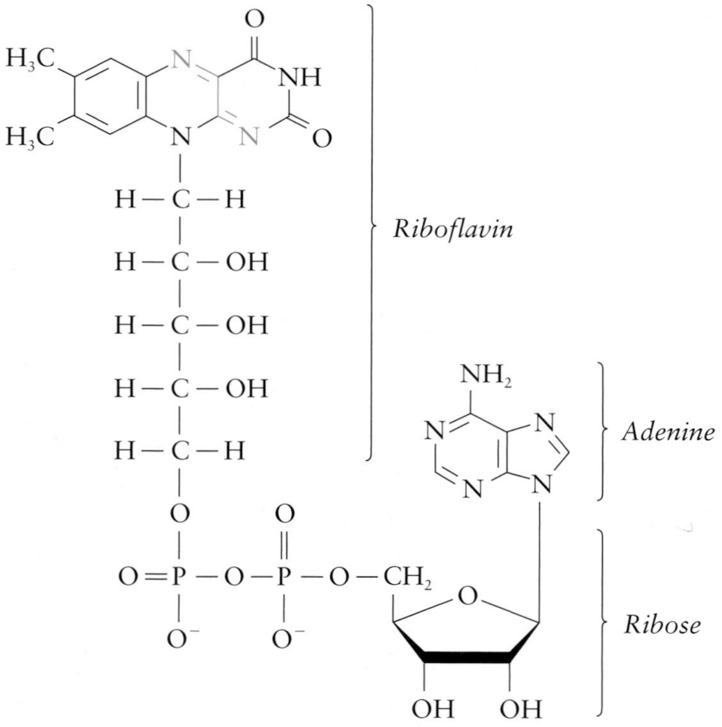

FAD
(oxidized form)

$FADH_2$
(reduced form)

FIGURE 12.19 The structure of flavin adenine dinucleotide (FAD). The reactive portion (shown in blue) is in the riboflavin ring system.

Riboflavin

Adenine

Ribose

The substrates for oxidation reactions in which FAD is the coenzyme differ from those involving NAD^+. FAD is often involved in oxidation reactions in which a $-CH_2-CH_2-$ portion of the substrate is oxidized to a double bond:

General reaction:

$$\underset{\substack{\text{saturated}\\\text{(reduced)}}}{R-\overset{\overset{\large H}{|}}{\underset{\underset{\large H}{|}}{C}}-\overset{\overset{\large H}{|}}{\underset{\underset{\large H}{|}}{C}}-R} + FAD \longrightarrow \underset{\substack{\text{unsaturated}\\\text{(oxidized)}}}{R-CH=CH-R} + FADH_2 \qquad (12.14)$$

Specific example:

$$\underset{\text{butyryl CoA}}{CH_3CH_2CH_2-\overset{\overset{\large O}{\|}}{C}-S-CoA} \xrightarrow[\text{FAD} \quad \text{FADH}_2]{} CH_3CH=CH-\overset{\overset{\large O}{\|}}{C}-SCoA \qquad (12.15)$$

Learning Check 12.2 Complete the following reactions:

a. $^{-}OOC-CH_2-CH_2-COO^- + FAD \longrightarrow$

b. $CH_3-\overset{\overset{\large OH}{|}}{CH}-COO^- + NAD^+ \longrightarrow$

Concept Summary

ThomsonNOW™ Sign in at **www.thomsonedu.com** to:

- Assess your understanding with Exercises keyed to each learning objective.
- Check your readiness for an exam by taking the **Pre-test** and exploring the modules recommended in your **Personalized Learning Plan.**

Nutritional Requirements. Every human body requires certain amounts of various macro- and micronutrients, water, and fiber to function properly. **OBJECTIVE 1 (Section 12.1), Exercise 12.2.** The amounts needed vary with a number of factors such as body size, age, and sex. Various countries have established nutritional guidelines in attempts to maintain good health for their citizens. In the United States, the Reference Daily Intake (RDI) and Daily Reference Values (DRV) are designed to represent appropriate nutrient intake for 95% of the population.

The Macronutrients. Macronutrients are substances required by the body in relatively large amounts. They are used by the body to provide energy and the materials necessary to form new or replacement tissue. The macronutrients are carbohydrates, lipids, and proteins. **OBJECTIVE 2 (Section 12.2), Exercise 12.4.** An RDI has been established only for proteins, but various groups have recommended the amounts of carbohydrates, lipids, and proteins that should be included in the diet.

Micronutrients I: Vitamins. Vitamins are organic micronutrients that the body cannot produce in the amounts needed for good health. A number of vitamins have a high water solubility resulting from the highly polar nature of their molecules. All but one of the water-soluble vitamins are known to function as coenzymes in the body and are involved in many important metabolic processes. Fat-soluble vitamins have nonpolar molecular structures. As a result, they are insoluble in water but soluble in fat or other nonpolar solvents. Fat-soluble vitamins act somewhat like hormones in the body. Fat-soluble vitamins do not dissolve in water-based body fluids, so they are not excreted through the kidneys. Amounts in excess of bodily requirements are stored in body fat. Thus, it is much easier to produce toxic effects by overdosing with fat-soluble vitamins than with water-soluble vitamins. **OBJECTIVE 3 (Section 12.3), Exercise 12.10**

Micronutrients II: Minerals. Minerals are inorganic substances needed by the body. Those present in the body in amounts equal to or greater than 5 g are called major minerals. Those present in smaller amounts are called trace minerals. Minerals perform many useful functions in the body. **OBJECTIVE 4 (Section 12.4), Exercise 12.16**

The Flow of Energy in the Biosphere. Energy for all living processes comes from the sun. A portion of the sun's energy is trapped in the process of photosynthesis, which produces carbohydrates. We derive energy from the foods we eat: carbohydrates, lipids, and proteins. As these substances are oxidized, molecules of ATP that supply the energy needed for cellular processes are formed. **OBJECTIVE 5 (Section 12.5), Exercise 12.24**

Metabolism and an Overview of Energy Production. Metabolism, the sum of all cellular reactions, involves the breakdown (catabolism) and synthesis (anabolism) of molecules. **OBJECTIVE 6 (Section 12.6), Exercise 12.26.** There are three stages in the catabolism of foods to provide energy. In Stage I, digestion converts foods into smaller molecules. In Stage II, these smaller molecules are converted into two-carbon acetyl units that combine with coenzyme A, forming acetyl CoA. Stage III, which is called the common catabolic pathway, consists of the citric acid cycle followed by the electron transport chain and oxidative phosphorylation. The main function of catabolism is to produce ATP molecules. **OBJECTIVE 7 (Section 12.6), Exercise 12.28**

ATP: The Primary Energy Carrier. Molecular ATP is the link between energy production and energy use in cells. ATP is called a high-energy compound because of the large amount of free energy liberated on hydrolysis. Cells are able to harness the free energy liberated by ATP to carry out cellular work. ATP is consumed immediately following its formation and is regenerated from ADP as fuel molecules are oxidized. Mitochondria play a key role in energy production because they house the enzymes for both the citric acid cycle and the electron transport chain. **OBJECTIVE 8 (Section 12.7), Exercise 12.34**

Important Coenzymes in the Common Catabolic Pathway. Three very important coenzymes are involved in catabolism: coenzyme A, NAD^+, and FAD. Coenzyme A binds to the two-carbon fragments produced in Stage II to form acetyl CoA, the direct fuel of the citric acid cycle. NAD^+ and FAD are the oxidizing agents that participate in the oxidation reactions of the citric acid cycle. They transport hydrogen atoms and electrons from the citric acid cycle to the electron transport chain. **OBJECTIVE 9 (Section 12.8), Exercise 12.46**

Key Terms and Concepts

Anabolism (12.6)
Catabolism (12.6)
Cellular respiration (12.5)
Common catabolic pathway (12.6)
Complete protein (12.2)
Complex carbohydrate (12.2)
Daily Reference Values (DRV) (12.1)
Daily Values (DV) (12.1)
Essential amino acid (12.2)
High-energy compound (12.7)
Macronutrient (12.1)
Major mineral (12.4)
Metabolic pathway (12.6)
Metabolism (12.6)
Micronutrient (12.1)
Mineral (12.4)
Mitochondrion (12.7)
Nutrition (Introduction)
Reference Daily Intakes (RDI) (12.1)
Satiety (12.2)
Simple carbohydrates (12.2)
Thioester bond (12.8)
Trace mineral (12.4)

Key Reactions

1. Photosynthesis (Section 12.5):

$$6CO_2 + 6H_2O + \text{energy} \rightarrow C_6H_{12}O_6 + 6O_2 \qquad \text{Reaction 12.2}$$

2. Hydrolysis of ATP (Section 12.7):

$$\text{ATP} + H_2O \rightarrow \text{ADP} + P_i + H^+ \qquad \text{Reaction 12.5}$$

3. Oxidation by NAD^+ (Section 12.8):

$$R-\underset{H}{\overset{OH}{\underset{|}{\overset{|}{C}}}}-H + NAD^+ \longrightarrow R-\overset{O}{\overset{\|}{C}}-H + NADH + H^+ \qquad \text{Reaction 12.10}$$

4. Oxidation by FAD (Section 12.8):

$$R-\underset{H}{\overset{H}{\underset{|}{\overset{|}{C}}}}-\underset{H}{\overset{H}{\underset{|}{\overset{|}{C}}}}-R + FAD \longrightarrow R-CH=CH-R + FADH_2 \qquad \text{Reaction 12.14}$$

Exercises

SYMBOL KEY

Even-numbered exercises are answered in Appendix B.

Blue-numbered exercises are more challenging.

■ denotes exercises available in ThomsonNow and assignable in OWL.

ThomsonNOW To assess your understanding of this chapter's topics with sample tests and other resources, sign in at www.thomsonedu.com.

NUTRITIONAL REQUIREMENTS (SECTION 12.1)

12.1 ■ What is the principal component of dietary fiber?

12.2 ■ What is the primary difference between a macronutrient and a micronutrient?

12.3 Explain the importance of sufficient fiber in the diet.

THE MACRONUTRIENTS (SECTION 12.2)

12.4 List two general functions in the body for each of the macronutrients.

12.5 For each of the following macronutrients, list a single food item that would be a good source:

a. carbohydrate (simple)

b. carbohydrate (complex)

c. lipid

d. protein

12.6 ■ List the types of macronutrients found in each of the following food items. List the nutrients in approximate decreasing order of abundance (most abundant first, etc.).

a. Potato chips

b. Buttered toast

c. Plain toast with jam

d. Cheese sandwich

e. A lean steak

f. A fried egg

12.7 What makes a fatty acid essential? What are the essential fatty acids?

12.8 Use Figure 12.2 and list the Daily Values (DV) for fat, saturated fat, total carbohydrate, and fiber based on a 2000-Calorie diet.

12.9 Use the Daily Values (DV) and the Calories per gram conversions given in Figure 12.2 to calculate the number of Calories in a 2000-Calorie diet that should be from fat, saturated fat, and total carbohydrate.

MICRONUTRIENTS I: VITAMINS (SECTION 12.3)

12.10 ■ Identify each of the following vitamins as water-soluble or fat-soluble.

a. tocopherol

b. niacin

c. folic acid

d. retinol

12.11 What general function is served by eight of the water-soluble vitamins?

12.12 Why is there more concern about large doses of fat-soluble vitamins than of water-soluble vitamins?

12.13 What fat-soluble vitamin or deficiency is associated with the following?

a. Night blindness

b. Blood clotting

c. Calcium and phosphorus use in forming bones and teeth

d. Preventing oxidation of fatty acids

12.14 ■ What water-soluble vitamin deficiency is associated with the following?

a. Scurvy

b. Beriberi

c. Pernicious anemia

d. Pellagra

MICRONUTRIENTS II: MINERALS (SECTION 12.4)

12.15 What is the difference between a vitamin and a mineral?

12.16 ■ List at least one specific function for each major mineral of the body.

12.17 What determines whether a mineral is classified as a major or trace mineral?

12.18 What are the general functions of trace minerals in the body?

12.19 How can the ingestion of a small lump of elemental sodium metal be toxic when sodium is a required major mineral?

THE FLOW OF ENERGY IN THE BIOSPHERE (SECTION 12.5)

12.20 What element serves as a fuel for the fusion process occurring on the sun?

12.21 What is the source of energy for life on Earth?

12.22 Write a net equation for the photosynthetic process.

12.23 Explain how Earth's supply of carbon compounds may be recycled by the processes of photosynthesis and respiration.

12.24 Describe in general terms the steps in the flow of energy from the sun to molecules of ATP.

METABOLISM AND AN OVERVIEW OF ENERGY PRODUCTION (SECTION 12.6)

12.25 What is a metabolic pathway?

12.26 ■ Classify catabolism and anabolism as synthetic or breakdown processes.

12.27 Outline the three stages in the extraction of energy from food.

12.28 ■ Which stage of energy production is concerned primarily with the following?

a. Formation of ATP

b. Digestion of fuel molecules

c. Consumption of O_2

d. Generation of acetyl CoA

12.29 Explain why Stage III of energy production is referred to as the common catabolic pathway.

12.30 In terms of energy production, what is the main purpose of the catabolic pathway?

12.31 Would the digestion of protein to amino acids that are then used to build muscle in the body be considered a catabolic pathway? Why or why not?

ATP: THE PRIMARY ENERGY CARRIER (SECTION 12.7)

12.32 ■ Is ATP involved in anabolic processes or catabolic processes?

12.33 ■ What do the symbols ATP, ADP, and AMP represent?

12.34 Explain why ATP is referred to as the primary energy carrier.

12.35 Using the partial structure shown below, write the structure of ATP, ADP, and AMP:

$$-\overset{\overset{\displaystyle O}{\|}}{\underset{\underset{\displaystyle O^-}{|}}{P}}-O-\text{adenosine}$$

12.36 What do the symbols P_i and PP_i represent?

12.37 Using symbolic formulas such as ADP and PP_i, write equations for the hydrolysis of ATP to ADP and the hydrolysis of ATP to AMP.

12.38 Using Table 12.5, explain why phosphoenolpyruvate is called a high-energy compound but glycerol 3-phosphate is not.

12.39 What does the symbol $\Delta G°'$ represent? How does the sign for $\Delta G°'$ indicate whether a reaction is exergonic or endergonic?

12.40 Which portion of the ATP molecule is particularly responsible for its being described as a high-energy compound?

12.41 Why is it more accurate to refer to ATP as a carrier or donor of free energy rather than as a storage form of free energy?

12.42 What is the role of mitochondria in the use of energy by living organisms?

12.43 Describe the importance and function of the ATP–ADP cycle.

12.44 Describe the structure of a mitochondrion and identify the location of enzymes important in energy production.

IMPORTANT COENZYMES IN THE COMMON CATABOLIC PATHWAY (SECTION 12.8)

12.45 ■ What coenzymes are produced by the body from the following vitamins?

a. riboflavin

b. pantothenic acid

c. nicotinamide

12.46 ■ What roles do the following coenzymes serve in the catabolic oxidation of foods?

a. coenzyme A b. FAD c. NAD^+

12.47 Write the abbreviations for the oxidized and reduced forms of

a. flavin adenine dinucleotide

b. nicotinamide adenine dinucleotide

12.48 ■ Which compound is oxidized in the following reaction? Which compound is reduced?

$$^-OOC-CH_2-CH_2-COO^- + FAD \longrightarrow$$
$$^-OOC-CH{=}CH-COO^- + FADH_2$$

12.49 Which compound is oxidized in the following reaction? Which compound is reduced?

$$CH_3CH_2-OH + NAD^+ \longrightarrow CH_3\overset{\overset{\displaystyle O}{\|}}{C}-H + NADH + H^+$$

12.50 Write balanced reactions for the following oxidations using NAD^+ or FAD:

a. $HO-CH_2-COOH \rightarrow O{=}CH-COOH$

b. $HO-CH_2CH_2-OH \rightarrow HO-CH{=}CH-OH$

12.51 ■ Write balanced reactions for the following oxidations using NAD^+:

a. $CH_3-\overset{\overset{\displaystyle OH}{|}}{CH}-COO^- \longrightarrow CH_3-\overset{\overset{\displaystyle O}{\|}}{C}-COO^-$

b. $CH_3-OH \longrightarrow H-\overset{\overset{\displaystyle O}{\|}}{C}-H$

12.52 Draw abbreviated structural formulas to show the reactive portion of FAD and the reduced compound $FADH_2$.

ADDITIONAL EXERCISES

12.53 Based on the solubility characteristics of vitamin B_1, would you expect to find it in the interior of an enzyme or on the exterior surface of the enzyme when it acts as a coenzyme?

12.54 Balance the following equation. How many electrons will be transferred to NADH?

$$CH_4 + O_2 + \_\_\_\_ NAD^+ \rightarrow CO_2 + \_\_\_\_ NADH + \_\_\_\_ H^+$$

12.55 Draw an energy diagram that depicts the two hydrolysis reactions of ATP below.

$$ATP \rightarrow ADP + P_i \rightarrow AMP + 2P_i$$

12.56 What would be the main process in the human body that would be interrupted by a deficiency of an essential amino acid? What specific part of that process would be affected?

12.57 Average 18 year olds should get 1300 mg of calcium in their diet daily. Use Figure 12.8 and calculate the percentage that this daily intake is of the total amount of calcium found in a 60-kg person represents.

ALLIED HEALTH EXAM CONNECTION

Reprinted with permission from Nursing School and Allied Health Entrance Exams, COPYRIGHT 2005 Petersons.

12.58 Identify which vitamin(s) is indicated by the following:

a. The fat-soluble vitamins

b. Thiamin

c. Vitamin that helps prevent rickets

d. Vitamin that helps clot blood

e. Vitamin useful in combating pernicious anemia

12.59 ■ Answer the following:

a. Which mineral is the most abundant in the human body?

b. Which mineral is necessary for the proper functioning of the thyroid gland?

c. Which mineral is necessary for the proper formation of hemoglobin?

d. What is the difference between a vitamin and a mineral?

12.60 Which of the following are not a requirement for photosynthesis?

a. oxygen

b. carbon dioxide

c. sunlight

d. chlorophyll

12.61 With which organelle is the synthesis of ATP associated?

12.62 Use Figure 12.2 to determine which macronutrient would produce the greatest amount of energy from 1 gram of the macronutrient.

12.63 Outline the three stages in the extraction of energy from food. What type of chemical reaction occurs most often during digestion?

CHEMISTRY FOR THOUGHT

12.64 At one time a system of MDR (minimum daily requirement) values for nutrients was used. Why do you think Daily Values (DV) might be more informative and useful than MDRs?

12.65 Briefly explain why you think it would be undesirable or beneficial (decide which position to defend in each case) to follow a diet that is (a) extremely low in carbohydrates, (b) extremely low in lipids, or (c) extremely low in proteins.

12.66 Doughnuts, even those without frosting, are high in triglycerides. Explain.

12.67 One of the oils listed in Figure 12.6 is palm oil, which for several years was used in fast-food restaurants to fry potatoes. Why was this use considered an unhealthy practice?

12.68 In Figure 12.7, it is observed that vegetables are a source of protein. We most often think of meats, grains, and dairy products as sources of protein, but not vegetables. Use Table 9.4 to list the types of protein that might be found in vegetables.

12.69 Vitamin tablets and oranges are both plentiful sources of vitamin C. Why might an orange be better for you than taking a vitamin C tablet?

12.70 A cold cereal manufactured from corn is deficient in the amino acid lysine. What other foods in a breakfast meal might supply lysine?

12.71 The body has a storage capacity for excess carbohydrates and fats but not for excess proteins. What consequences does this fact have for a daily diet?

12.72 Which macronutrient is a major source of dietary sulfur?

CHAPTER 13

Carbohydrate Metabolism

LEARNING OBJECTIVES

When you have completed your study of this chapter, you should be able to:

1. Identify the products of carbohydrate digestion. (Section 13.1)
2. Explain the importance to the body of maintaining proper blood sugar levels. (Section 13.2)
3. List the starting material and products of the glycolysis pathway. (Section 13.3)
4. Describe how the glycolysis pathway is regulated in response to cellular needs. (Section 13.3)
5. Name the three fates of pyruvate. (Section 13.4)
6. Identify the two major functions of the citric acid cycle. (Section 13.5)
7. Describe how the citric acid cycle is regulated in response to cellular energy needs. (Section 13.5)
8. Explain the function of the electron transport chain and describe how electrons move down the chain. (Section 13.6)
9. List the major features of the chemiosmotic hypothesis. (Section 13.7)
10. Calculate the amount of ATP produced by the complete oxidation of a mole of glucose. (Section 13.8)
11. Explain the importance of the processes of glycogenesis and glycogenolysis. (Section 13.9)
12. Describe gluconeogenesis and the operation of the Cori cycle. (Section 13.10)
13. Describe how hormones regulate carbohydrate metabolism. (Section 13.11)

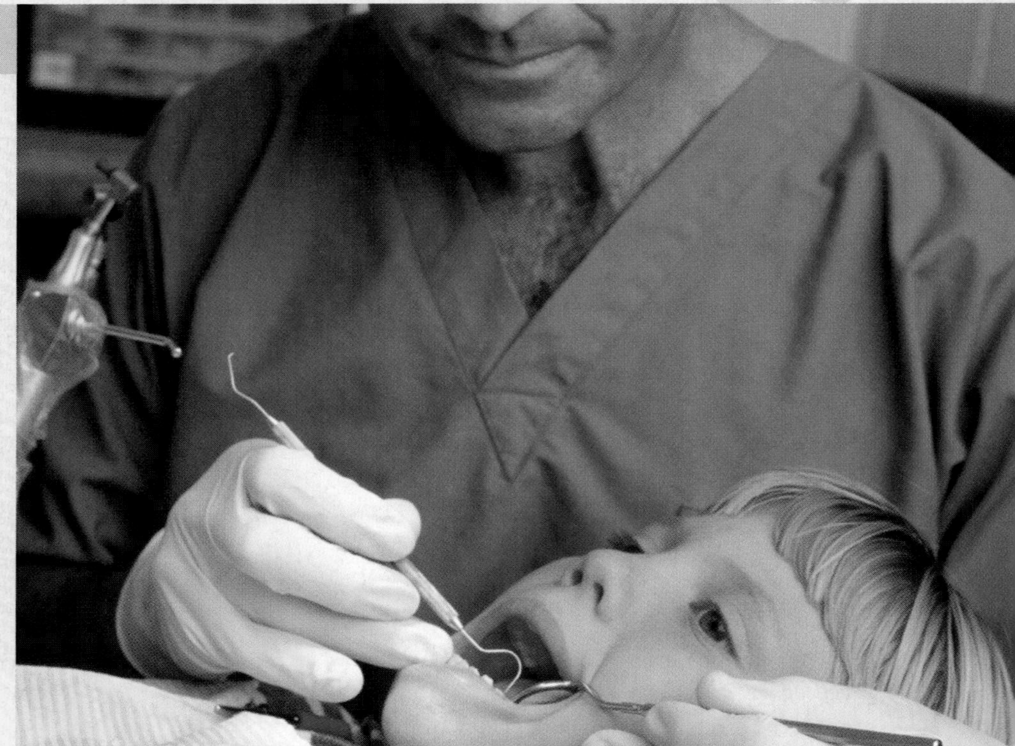

Dental hygienists provide a variety of health care services. A well-known example is the cleaning of teeth (prophylaxis). In addition, they help patients understand how nutrition affects dental health. In this chapter, you will be introduced to several nutritionally important topics, including the production and regulation of energy from carbohydrates.

Carbohydrates play major roles in cell metabolism. The central substance in carbohydrate metabolism is glucose; it is the key food molecule for most organisms. As the major carbohydrate metabolic pathways are discussed, you will see that each one involves either the use or the synthesis of glucose. You will also see how the chemical degradation of glucose provides energy in the cell, how glucose is stored, how the blood glucose level is regulated, and how diabetes results from changes in blood glucose regulation mechanisms.

ThomsonNOW™ Throughout the chapter this icon introduces resources on the ThomsonNOW website for this text. Sign in at **www.thomsonedu.com** to:

- Evaluate your knowledge of the material
- Take an exam prep quiz
- Identify areas you need to study with a **Personalized Learning Plan.**

13.1 The Digestion of Carbohydrates

LEARNING OBJECTIVE

1. Identify the products of carbohydrate digestion.

The major function of dietary carbohydrates is to serve as an energy source. In a typical American diet, carbohydrates meet about 45–55% of the daily energy needs. Fats and proteins furnish the remaining energy. During carbohydrate digestion, di- and polysaccharides are hydrolyzed to monosaccharides, primarily glucose, fructose, and galactose:

$$\text{polysaccharides} + H_2O \xrightarrow{\text{digestion}} \text{glucose} \qquad (13.1)$$

$$\text{sucrose} + H_2O \xrightarrow{\text{digestion}} \text{glucose} + \text{fructose} \qquad (13.2)$$

$$\text{lactose} + H_2O \xrightarrow{\text{digestion}} \text{glucose} + \text{galactose} \qquad (13.3)$$

$$\text{maltose} + H_2O \xrightarrow{\text{digestion}} \text{glucose} \qquad (13.4)$$

After digestion is completed, glucose, fructose, and galactose are absorbed into the bloodstream through the lining of the small intestine and transported to the liver. In the liver, fructose and galactose are rapidly converted to glucose or to compounds that are metabolized by the same pathway as glucose.

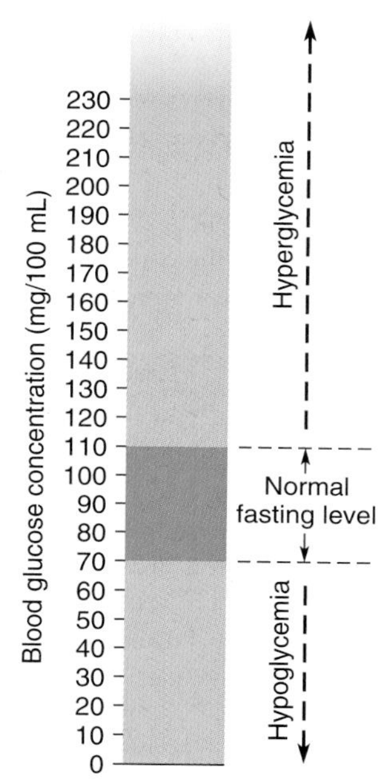

FIGURE 13.1 Blood glucose levels.

13.2 Blood Glucose

LEARNING OBJECTIVE

2. Explain the importance to the body of maintaining proper blood sugar levels.

Glucose is by far the most plentiful monosaccharide found in blood, and the term *blood sugar* usually refers to glucose. In adults, the normal **blood sugar level** measured after a fast of 8–12 hours is in the range of 70–110 mg/100 mL. (In clinical reports, such levels are often expressed in units of milligrams per deciliter, mg/dL.) The blood sugar level reaches a maximum of approximately 140–160 mg/100 mL about 1 hour after a carbohydrate-containing meal. It returns to normal after 2–2½ hours.

If a blood sugar level is below the normal fasting level, a condition called **hypoglycemia** exists. Because glucose is the only nutrient normally used by the brain for energy, mild hypoglycemia leads to dizziness and fainting as brain cells are deprived of energy. Severe hypoglycemia can cause convulsions and shock. When the blood glucose concentration is above normal, the condition is referred to as **hyperglycemia.** If blood glucose levels exceed approximately 180 mg/100 mL, the sugar is not completely reabsorbed by the kidneys, and glucose is excreted in the urine (see ■ Figure 13.1). The blood glucose level at which this occurs is called the **renal threshold,** and the condition in which glucose appears in the urine is called **glucosuria.** Prolonged hyperglycemia at a glucosuric level must be considered serious because it indicates that something is wrong with the body's normal ability to control the blood sugar level.

blood sugar level
The amount of glucose present in blood, normally expressed as milligrams per 100 mL of blood.

hypoglycemia
A lower-than-normal blood sugar level.

hyperglycemia
A higher-than-normal blood sugar level.

renal threshold
The blood glucose level at which glucose begins to be excreted in the urine.

glucosuria
A condition in which elevated blood sugar levels result in the excretion of glucose in the urine.

The liver is the key organ involved in regulating the blood glucose level. The liver responds to the increase in blood glucose that follows a meal by removing glucose from the bloodstream. The removed glucose is primarily converted to glycogen or triglycerides for storage. Similarly, when blood glucose levels are low, the liver responds by converting stored glycogen to glucose and by synthesizing new glucose from noncarbohydrate substances.

13.3 Glycolysis

LEARNING OBJECTIVES

3. List the starting material and products of the glycolysis pathway.
4. Describe how the glycolysis pathway is regulated in response to cellular needs.

glycolysis
A series of reactions by which glucose is oxidized to pyruvate.

ThomsonNOW™ Go to Chemistry Interactive to explore **glycolysis.**

lactose intolerance
The inability to digest milk and other products containing lactose.

Glycolysis is an important part of the catabolic process by which the body gets energy from carbohydrates. It consists of a series of ten reactions, with the net result being the conversion of a glucose molecule into two molecules of pyruvate. All enzymes for glycolysis are present in cellular cytoplasm. The ten reactions of

OVER THE COUNTER 13.1 Lactose Intolerance

Lactose makes up about 40% of the nutrients in an infant's diet during the first year of life. The small intestines of infants and small children contain a particular form of lactase, the enzyme that catalyzes the digestion of lactose. This form of the enzyme is very active and helps them digest the lactose in their diet easily. The digestive system of adults usually contains a less active form of the enzyme, and in cultures where milk is not used much as an adult food, lactase deficiency is prevalent. Even in the United States, about 20% of the adult population suffers from **lactose intolerance** to some degree.

When lactose is not digested in the small intestine, it passes into the colon (large intestine), where it causes water to be drawn by osmosis from the interstitial fluid into the colon itself. Bacteria in the colon act on the lactose and produce gases and organic acids. The presence in the colon of excess water, gases, and organic acids leads to the symptoms of lactose intolerance—abdominal distention (a bloated feeling), cramps, and diarrhea.

The main treatment for lactose intolerance is dietary management; patients must be aware of the amount of lactose in all products they consume. If lactose intolerance is suspected, the first step is total abstention from lactose, to determine whether the symptoms really are caused by the enzyme deficiency. Dietary restrictions can be modified once the cause has been identified, and a person's tolerance level can be determined by slowly adding lactose-containing foods to the diet. If total abstention from lactose is required, it may be necessary to add calcium supplements to the diet to replace the calcium normally contained in dairy foods.

Hard cheeses like cheddar and Swiss are quite low in lactose and can often be tolerated, except in extreme cases. Even though yogurt has a high lactose content, it is better tolerated than milk. This is probably because the live bacteria in yogurt release lactase when the bacteria are digested by proteolytic enzymes and stomach acid. The released lactase then helps digest the lactose. In addition to monitoring their diets, some people with lactose intolerance use commercially available products that contain lactase. Available in liquid or tablet form, these products are taken just before eating or along with food that contains lactose. Some of the lactase products are Lactaid®, DairyEase®, and Lactrase®.

Many prescription and OTC drugs contain lactose as a carrier or filler. The FDA provides a list identifying such drugs, and people with lactose intolerance should consult the list to avoid inadvertently using a medication that contains lactose.

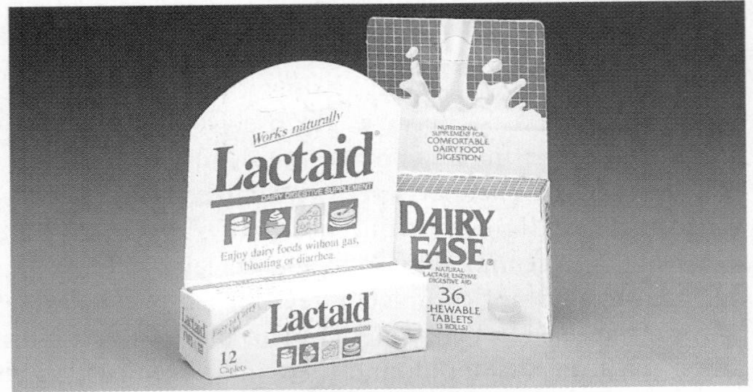

© West

Two commercial products containing lactase.

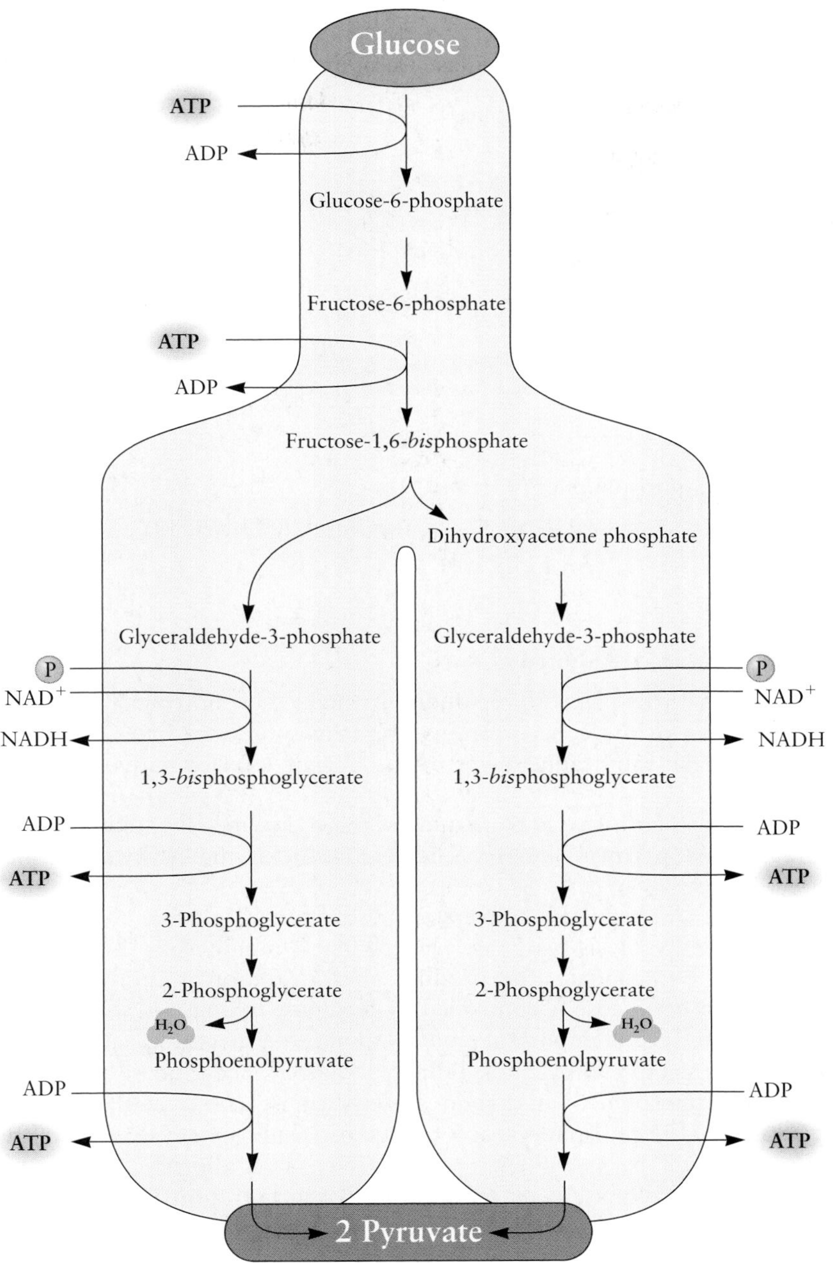

■ **FIGURE 13.2** A summary of glycolysis. The Ⓟ stands for

$$-\overset{\displaystyle O^-}{\underset{\displaystyle O^-}{\overset{|}{\underset{|}{P}}}}=O\ ,$$

and the prefix "bis-" is used to denote two of these groups.

glycolysis catalyzed by these enzymes are represented in ■ Figure 13.2. The addition of all these reactions gives a net reaction for glycolysis:

$$\text{glucose} + 2P_i + 2ADP + 2NAD^+ \rightarrow$$
$$2\ \text{pyruvate} + 2ATP + 2NADH + 4H^+ + 2H_2O \qquad (13.5)$$

This equation indicates a net gain of 2 mol of ATP for every mole of glucose converted to pyruvate. In addition, 2 mol of the coenzyme NADH is formed when NAD^+ serves as an oxidizing agent in glycolysis. Fructose and galactose from digested food are both converted into intermediates that enter into the glycolysis pathway. Fructose enters glycolysis as dihydroxyacetone phosphate and glyceraldehyde 3-phosphate, whereas galactose enters in the form of glucose 6-phosphate.

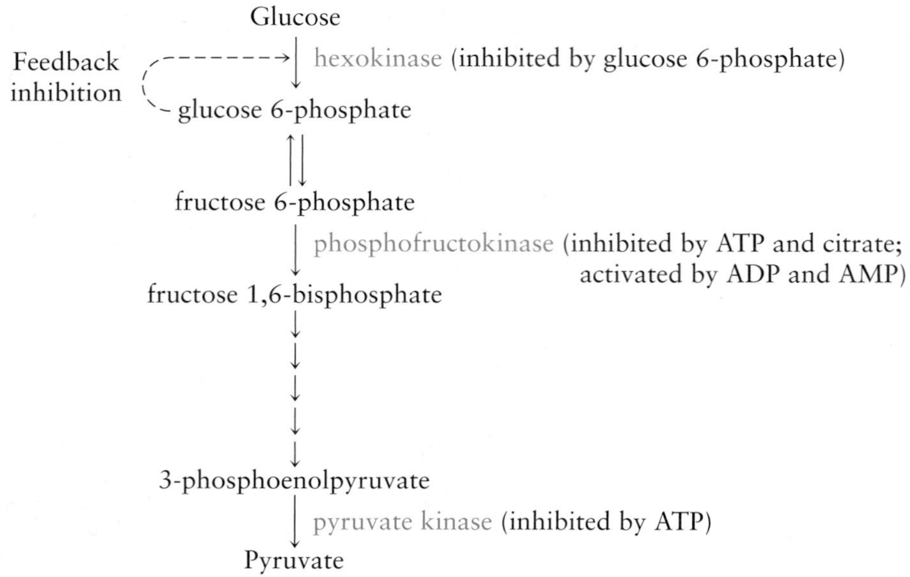

FIGURE 13.3 The regulation of the glycolysis pathway. Glucose 6-phosphate is a noncompetitive inhibitor of hexokinase. Phosphofructokinase and pyruvate kinase are also allosteric enzymes subject to control.

The Regulation of Glycolysis

The glycolysis pathway, like all metabolic pathways, is controlled constantly. The precise regulation of the concentrations of pathway intermediates is efficient and makes possible the rapid adjustment of the output level of pathway products as the need arises.

The glycolysis pathway is regulated by three enzymes: hexokinase, phosphofructokinase, and pyruvate kinase. The regulation of the pathway is shown in ■ Figure 13.3.

Hexokinase catalyzes the conversion of glucose to glucose 6-phosphate and thus initiates the glycolysis pathway. The enzyme is inhibited by a high concentration of glucose 6-phosphate, the product of the reaction it catalyzes. Thus, the phosphorylation of glucose is controlled by feedback inhibition (Section 10.8).

A second and very important control point for glycolysis shown in Figure 13.3 is the step catalyzed by phosphofructokinase. Once glucose is converted to fructose 1,6-bisphosphate, its carbon skeleton must continue through the glycolytic pathway. The phosphorylation of fructose 6-phosphate is a committed step that will occur as long as the enzyme is active. As an allosteric enzyme (Section 10.8), phosphofructokinase is inhibited by high concentrations of ATP and citrate, and it is activated by high concentrations of ADP and AMP.

The third control point is the last step of glycolysis, the conversion of phosphoenolpyruvate to pyruvate catalyzed by pyruvate kinase. This enzyme, too, is an allosteric enzyme that is inhibited by higher concentrations of ATP.

To understand how these controls interact, remember that when the glycolysis pathway is operating at a maximum rate, so also are the citric acid cycle and the electron transport chain. As a result of their activity, large quantities of ATP are produced. The entire pathway continues to operate at a high rate, as long as the produced ATP is used up by processes within the cell. However, if ATP use decreases, the concentration of ATP increases, and more diffuses to the binding sites of the allosteric enzymes sensitive to ATP. Thus, the activity of these enzymes decreases. When phosphofructokinase is inhibited in this way, the entire glycolysis pathway slows or stops. As a result, the concentration of glucose 6-phosphate increases, thus inhibiting hexokinase and the initial step of the glycolysis pathway.

Conversely, when ATP concentrations are low, the concentrations of AMP and ADP are high. The AMP and ADP function as activators (Section 10.8) that influence phosphofructokinase and thereby speed up the entire glycolysis pathway.

Learning Check 13.1 Explain how each of the following enzymes helps regulate the glycolysis pathway:

a. hexokinase
b. phosphofructokinase
c. pyruvate kinase

13.4 The Fates of Pyruvate

LEARNING OBJECTIVE

5. Name the three fates of pyruvate.

The sequence of reactions that converts glucose to pyruvate is very similar in all organisms and in all kinds of cells. However, the fate of the pyruvate as it is used to generate energy is variable. Three different fates of pyruvate are considered here.

As glucose is oxidized to pyruvate in glycolysis, NAD^+ is reduced to NADH. The need for a continuous supply of NAD^+ for glycolysis is a key to understanding the fates of pyruvate.

$$\text{glucose} + 2\text{NAD}^+ \xrightarrow[2\text{ADP} + 2\text{P}_i \;\to\; 2\text{ATP}]{} 2\text{ pyruvate} + 2\text{NADH} + 2\text{H}^+ \qquad (13.6)$$

In each case, pyruvate is metabolized in ways that regenerate NAD^+ so that glycolysis can continue.

Oxidation to Acetyl CoA

The mitochondrial membrane is permeable to pyruvate formed by glycolysis in the cytoplasm, and under **aerobic** conditions (a plentiful oxygen supply), pyruvate is converted inside the mitochondria to acetyl CoA.

aerobic
In the presence of oxygen.

$$\underset{\text{pyruvate}}{CH_3-\overset{\overset{\displaystyle O}{\|}}{C}-COO^-} + CoA-S-H + NAD^+ \xrightarrow{\text{pyruvate dehydrogenase complex}} \underset{\text{acetyl CoA}}{CH_3-\overset{\overset{\displaystyle O}{\|}}{C}-S-CoA} + NADH + CO_2 \qquad (13.7)$$

Acetyl CoA formed in this reaction can enter the citric acid cycle on its way to complete oxidation to CO_2. The NAD^+ needed for the conversion of pyruvate to acetyl CoA (Reaction 13.7) and for the operation of the glycolysis pathway (Step 6) is regenerated when NADH transfers its electrons to O_2 in the electron transport chain.

Not all acetyl CoA generated metabolically enters the citric acid cycle. Some also serves as a starting material for biosynthesis (Section 14.7). The fatty acid components of triglycerides, for example, are synthesized using acetyl CoA. Thus, as most people are aware, the intake of excess dietary carbohydrates can lead to an increase in body fat storage.

Reduction to Lactate

Under **anaerobic** conditions, such as those that accompany strenuous or long-term muscle activity, the cellular supply of oxygen is not adequate for the reoxidation of NADH to NAD^+. Under such conditions, the cells begin reducing pyruvate to lactate as a means of regenerating NAD^+:

anaerobic
In the absence of oxygen.

$$\underset{\text{pyruvate}}{CH_3-\overset{\overset{\displaystyle O}{\|}}{C}-COO^-} + NADH + H^+ \underset{}{\overset{\text{lactate dehydrogenase}}{\rightleftarrows}} \underset{\text{lactate}}{CH_3-\overset{\overset{\displaystyle OH}{|}}{CH}-COO^-} + NAD^+ \qquad (13.8)$$

lactate fermentation The production of lactate from glucose.

When Reaction 13.8 is added to the net reaction of glycolysis, an overall reaction for **lactate fermentation** results:

$$\underset{\text{glucose}}{C_6H_{12}O_6} + 2ADP + 2P_i \xrightarrow{\text{lactate fermentation}} 2\underset{\text{lactate}}{CH_3-\overset{\overset{\displaystyle OH}{|}}{CH}-COO^-} + 2ATP \qquad (13.9)$$

Notice that the NAD^+ produced when pyruvate is reduced (Reaction 13.8) is used up and does not appear in lactate fermentation (Reaction 13.9). Although much more energy is available from the complete oxidation of pyruvate under aerobic conditions, the two ATPs produced from lactate fermentation are sufficient to sustain the life of anaerobic microorganisms. In human metabolism, those two molecules of ATP also play a critical role by furnishing energy when cellular supplies of oxygen are insufficient for ATP production through the complete oxidation of pyruvate.

Reduction to Ethanol

alcoholic fermentation The conversion of glucose to ethanol.

Several organisms, including yeast, have developed an alternative pathway for the regeneration of NAD^+ under anaerobic conditions. The process, called **alcoholic fermentation,** involves the decarboxylation of pyruvate to form acetaldehyde:

$$\underset{\text{pyruvate}}{CH_3-\overset{\overset{\displaystyle O}{\|}}{C}-COO^-} + H^+ \xrightarrow[\text{decarboxylase}]{\text{pyruvate}} \underset{\text{acetaldehyde}}{CH_3-\overset{\overset{\displaystyle O}{\|}}{C}-H} + CO_2 \qquad (13.10)$$

FIGURE 13.4 Bread, wine, and cheese are all produced through fermentation processes.

The carbon dioxide produced in this reaction causes beer to foam and provides carbonation for naturally fermented wines and champagnes (see Figure 13.4).

The second step in alcoholic fermentation is the reduction of acetaldehyde to ethanol by NADH:

$$\underset{\text{acetaldehyde}}{CH_3-\overset{\overset{\displaystyle O}{\|}}{C}-H} + NADH + H^+ \underset{\text{dehydrogenase}}{\overset{\text{alcohol}}{\rightleftharpoons}} \underset{\text{ethanol}}{CH_3-\overset{\overset{\displaystyle OH}{|}}{\underset{\underset{\displaystyle H}{|}}{C}}-H} + NAD^+ \qquad (13.11)$$

The NAD^+ regenerated by this reaction may then be used in the glycolysis pathway.

An overall reaction for alcoholic fermentation is obtained by combining the reaction for the conversion of pyruvate to ethanol and the reactions of glycolysis. Notice that because NAD^+ is used up in glycolysis but regenerated in ethanol formation, it does not appear in the overall equation:

$$\underset{\text{glucose}}{C_6H_{12}O_6} + 2ADP + 2P_i \xrightarrow{\text{alcoholic fermentation}} 2\underset{\text{ethanol}}{CH_3CH_2-OH} + 2CO_2 + 2ATP \qquad (13.12)$$

Figure 13.5 summarizes the various pathways pyruvate can follow as it is used to regenerate NAD^+.

Learning Check 13.2 Why is it important that each biochemical fate of pyruvate involve the production of NAD^+?

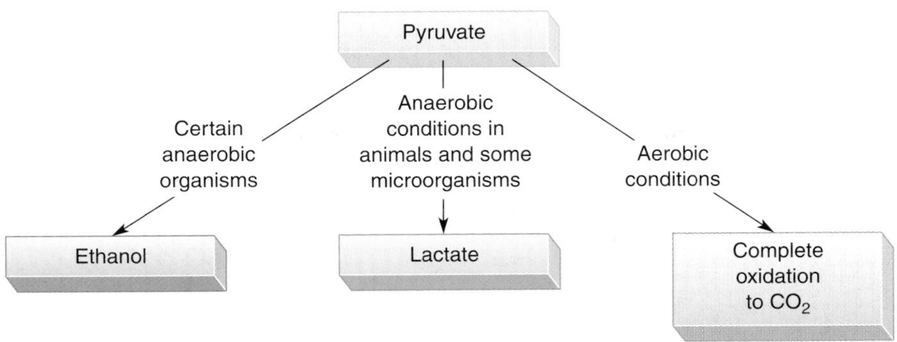

■ **FIGURE 13.5** The fates of pyruvate.

13.5 The Citric Acid Cycle

LEARNING OBJECTIVES

6. Identify the two major functions of the citric acid cycle.
7. Describe how the citric acid cycle is regulated in response to cellular energy needs.

Stage III in the oxidation of fuel molecules (Section 12.6) begins when the two-carbon acetyl units (of acetyl CoA) enter the **citric acid cycle.** This process is called the citric acid cycle because one of the key intermediates is citric acid. However, it is also called the tricarboxylic acid cycle in reference to the three carboxylic acid groups in citric acid, and the Krebs cycle in honor of Sir Hans A. Krebs, who deduced its reaction sequence in 1937 (see ■ Figure 13.6).

citric acid cycle
A series of reactions in which acetyl CoA is oxidized to carbon dioxide and reduced forms of coenzymes FAD and NAD^+ are produced.

The citric acid cycle is the principal process for generating the reduced coenzymes NADH and $FADH_2$, which are necessary for the reduction of oxygen and ATP synthesis in the electron transport chain. Another important function of the citric acid cycle as a source of intermediates for biosynthesis is discussed in Section 14.11.

■ **FIGURE 13.6** Sir Hans Adolf Krebs (1900–1981), German-born British biochemist, received the 1953 Nobel Prize in physiology or medicine.

Reactions of the Citric Acid Cycle

The reactions of the citric acid cycle occur within the matrix of the mitochondrion. ■ Figure 13.7 summarizes the details of the eight reactions that make up the citric acid cycle. Of particular importance are those reactions in which molecules of NADH and $FADH_2$ are produced. Notice that a two-carbon unit (the acetyl group) enters the cycle and that two carbon atoms are liberated in the form of two CO_2 molecules, one each in Reactions 3 and 4. Thus, this series of reactions both begins and ends with a C_4 compound, oxaloacetate. This is why the pathway is called a cycle. In each trip around the cycle, the starting material is regenerated and the reactions can proceed again as long as more acetyl CoA is available as fuel.

A Summary of the Citric Acid Cycle

The individual reactions of the cycle can be added to give an overall net equation:

$$\text{acetyl CoA} + 3\text{NAD}^+ + \text{FAD} + \text{GDP} + \text{P}_i + 2\text{H}_2\text{O} \rightarrow$$
$$2\text{CO}_2 + \text{CoA—S—H} + 3\text{NADH} + 2\text{H}^+ + \text{FADH}_2 + \text{GTP} \quad (13.13)$$

ThomsonNOW™ Go to Chemistry Interactive to explore the **carbon balance in the citric acid cycle.**

Some important features of the citric acid cycle are the following:

1. Acetyl CoA, available from the breakdown of carbohydrates, lipids, and amino acids, is the fuel of the citric acid cycle.
2. The operation of the cycle requires a supply of the oxidizing agents NAD^+ and FAD. The cycle is dependent on reactions of the electron transport chain to supply the necessary NAD^+ and FAD (Section 13.6). Because oxygen is the final

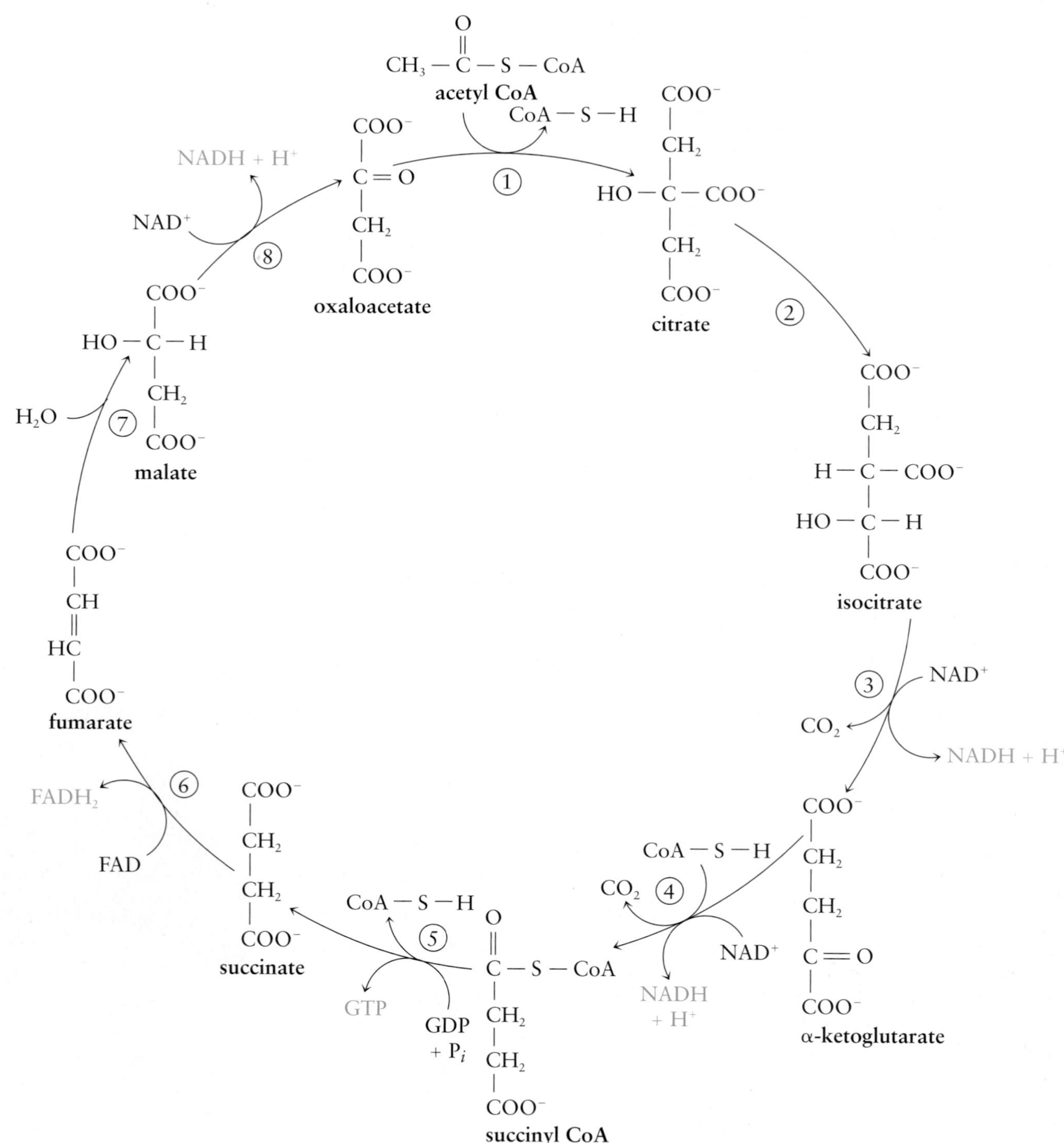

FIGURE 13.7 The citric acid cycle.

acceptor of electrons in the electron transport chain, the continued operation of the citric acid cycle depends ultimately on an adequate supply of oxygen.

3. Two carbon atoms enter the cycle as an acetyl unit, and two carbon atoms leave the cycle as two molecules of CO_2. However, the carbon atoms leaving the cycle correspond to carbon atoms that entered in the previous cycle; there is a one-cycle delay between the entry of two carbon atoms as an acetyl unit and their release as CO_2.

4. In each complete cycle, four oxidation–reduction reactions produce three molecules of NADH and one molecule of $FADH_2$.
5. One molecule of the high-energy phosphate compound guanosine triphosphate (GTP, Section 12.7) is generated.

Regulation of the Citric Acid Cycle

The rate at which the citric acid cycle operates is precisely adjusted to meet cellular needs for ATP. One important control point is the very first step in the cycle, the synthesis of citrate. The catalyst for this reaction, citrate synthetase, is an allosteric enzyme that is inhibited by ATP and NADH and activated by ADP. When cellular needs for ATP are met, ATP interacts with citrate synthetase and reduces its affinity for acetyl CoA. Thus, ATP acts as a modulator (inhibitor) of citrate synthetase and inhibits the entry of acetyl CoA into the citric acid cycle. Similarly, NADH acts as an inhibitor and signals the citric acid cycle to slow the production of NADH and $FADH_2$. On the other hand, when ATP levels are low, ADP levels are usually high. ADP, an activator of citrate synthetase, stimulates the entry of acetyl CoA into the citric acid cycle and the production of NADH and $FADH_2$.

Isocitrate dehydrogenase (Reaction 3), a second controlling enzyme, is an allosteric enzyme that is activated by ADP and inhibited by NADH in very much the same way as is citrate synthetase. The third control point (Reaction 4) is catalyzed by the α-ketoglutarate dehydrogenase complex. This group of enzymes is inhibited by succinyl CoA and NADH, products of the reaction it catalyzes, and also by ATP. These control points are indicated in ■ Figure 13.8. In short, the entry of acetyl CoA into the citric acid cycle and the rate at which the cycle operates are

CHEMISTRY AROUND US 13.1 Lactate Accumulation

During vigorous exercise, the oxygen supply to muscles can become too low to support increased aerobic conversion of pyruvate to acetyl CoA. Under such conditions, a shift occurs from aerobic pyruvate oxidation to anaerobic production of lactate as the means for producing ATP. The resulting accumulation of lactate causes the muscle pain and cramps that accompany physical exhaustion. The buildup of lactate also causes a slight decrease in blood pH, which triggers an increase in the rate and depth of breathing, a response that provides more oxygen to the cells. The combination of pain from lactate accumulation and the maximum rate of breathing represents a limit to physical activities. Regular training and exercise increase this limit by improving the efficiency with which oxygen is delivered to the cells. The muscle cells of a well-conditioned athlete can maintain the aerobic oxidation of pyruvate for a much longer period of activity than can those of a nonathlete.

The shift to anaerobic lactate formation also occurs in the heart muscle during an episode of cardiac arrest, in which the oxygen-carrying blood supply to the heart is cut off by the blockage of an artery leading to the heart muscles. Aerobic oxidation of pyruvate and its ATP production are stopped. Glycolysis proceeds at an accelerated rate to meet the energy needs, and lactate accumulates. The heart muscle contracts, producing a cramp, and stops beating. Massaging the heart muscle relieves such cramps, just as it does for skeletal muscles. Thus, it is sometimes possible to start the heart beating by massaging it.

Premature infants, whose lungs are not fully developed, are given pure oxygen to breathe in order to prevent a shift to anaerobic lactate production. In addition, they are given bicarbonate solutions intravenously to combat the increasing acidity of the blood due to lactate buildup.

Vigorous exercise causes a buildup of lactate in tissues.

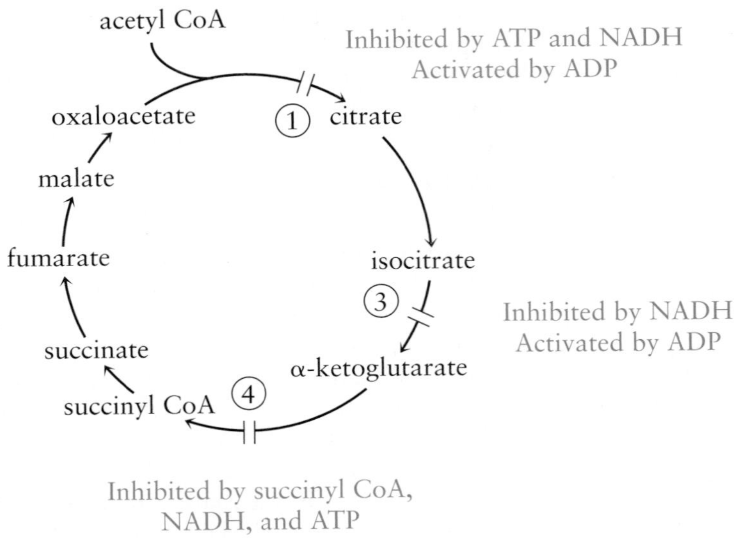

FIGURE 13.8 Control points in the citric acid cycle.

reduced when cellular ATP levels are high. When ATP supplies are low (and ADP levels are high), the cycle is stimulated.

13.6 The Electron Transport Chain

LEARNING OBJECTIVE

8. Explain the function of the electron transport chain and describe how electrons move down the chain.

ThomsonNOW Go to Chemistry Interactive and/or Coached Problems to explore **electron and proton transport.**

The reduced coenzymes NADH and $FADH_2$ are end products of the citric acid cycle. In the final stage of food oxidation, the hydrogen ions and electrons carried by these coenzymes combine with oxygen and form water.

$$4H^+ + 4e^- + O_2 \rightarrow 2H_2O \qquad (13.14)$$

electron transport chain
A series of reactions in which protons and electrons from the oxidation of foods are used to reduce molecular oxygen to water.

This simple energy-releasing process does not take place in a single step. Instead, it occurs as a series of steps that involve a number of enzymes and cofactors located within the inner membrane of the mitochondria. Electrons from the reduced coenzymes are passed from one electron carrier to another within the membrane in assembly-line fashion and are finally combined with molecular oxygen, the final electron acceptor. The series of reactions involved in the process is called the **electron transport chain.** As illustrated in ■ Figure 13.9, there are several intermediate electron carriers between NADH and molecular oxygen. The electron carriers are lined up in order of increasing affinity for electrons.

The first electron carrier in the electron transport chain is an enzyme that contains a tightly bound coenzyme. The coenzyme has a structure similar to FAD. The enzyme formed by the combination of this coenzyme with a protein is called flavin mononucleotide (FMN). Two electrons and one H^+ ion from NADH plus another H^+ ion from a mitochondrion pass to FMN, then to an iron–sulfur (Fe—S) protein, and then to coenzyme Q (CoQ). CoQ is also the entry point into the electron transport chain for the two electrons and two H^+ ions from $FADH_2$. As NADH and $FADH_2$ release their hydrogen atoms and electrons, NAD^+ and FAD are regenerated for reuse in the citric acid cycle.

cytochrome
An iron-containing enzyme located in the electron transport chain.

Four of the five remaining electron carriers are **cytochromes** (abbreviated cyt), which are structurally related proteins that contain an iron group. CoQ passes along the two electrons to two molecules of cytochrome *b*, and the two H^+ ions

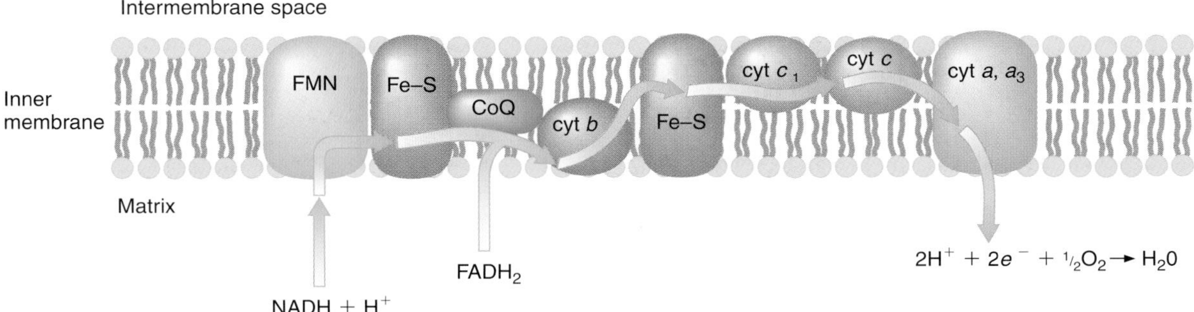

■ **FIGURE 13.9** The electron transport chain. The heavy arrows show the path of the electrons as they pass from one carrier to the next.

are given up to the mitochondrion. The electrons are passed along the chain, and in the final step, an oxygen atom accepts the electrons and combines with two H^+ ions to form water.

13.7 Oxidative Phosphorylation

LEARNING OBJECTIVE

9. List the major features of the chemiosmotic hypothesis.

As electrons are transported along the electron transport chain, a significant amount of free energy is released (52.6 kcal/mol). Some of this energy is conserved by the synthesis of ATP from ADP and P_i. Because this synthesis of ATP (a phosphorylation reaction) is linked to the oxidation of NADH and $FADH_2$, it is termed **oxidative phosphorylation.** It takes place at three different locations of the electron transport chain (see ■ Figure 13.10).

oxidative phosphorylation A process coupled with the electron transport chain whereby ADP is converted to ATP.

For each mole of NADH oxidized in the electron transport chain, 2.5 mol of ATP is formed. The free energy required for the phosphorylation of ADP is 7.3 kcal/mol, or 18 kcal for 2.5 mol, of ATP. Because the oxidation of NADH liberates 52.6 kcal/mol, coupling the oxidation and phosphorylation reactions conserves 18 kcal/52.6 kcal, or approximately 34% of the energy released.

The electron donor $FADH_2$ enters the electron transport chain later than does NADH. Thus, for every $FADH_2$ oxidized to FAD, only 1.5 ATP molecules are formed during oxidative phosphorylation, and approximately 25% of the released free energy is conserved as ATP.

Now we can calculate the energy yield for the entire catabolic pathway (citric acid cycle, electron transport chain, and oxidative phosphorylation combined). According to Section 13.5, every acetyl CoA entering the citric acid cycle produces 3 NADH and 1 $FADH_2$ plus 1 GTP, which is equivalent in energy to 1 ATP. Thus, 10 ATP molecules are formed per molecule of acetyl CoA catabolized.

3 NADH ultimately produce	7.5 ATP
1 $FADH_2$ ultimately produces	1.5 ATP
1 GTP is equivalent to	1 ATP
	10 ATP total produced

The complex mechanism by which the cell couples the oxidations of the electron transport chain and the synthesis of ATP is believed to involve a flow of protons. This explanation, called the **chemiosmotic hypothesis,** proposes that the flow of electrons through the electron transport chain causes protons to be "pumped" from the matrix across the inner membrane and into the space between the inner

chemiosmotic hypothesis The postulate that a proton flow across the inner mitochondrial membrane during operation of the electron transport chain provides energy for ATP synthesis.

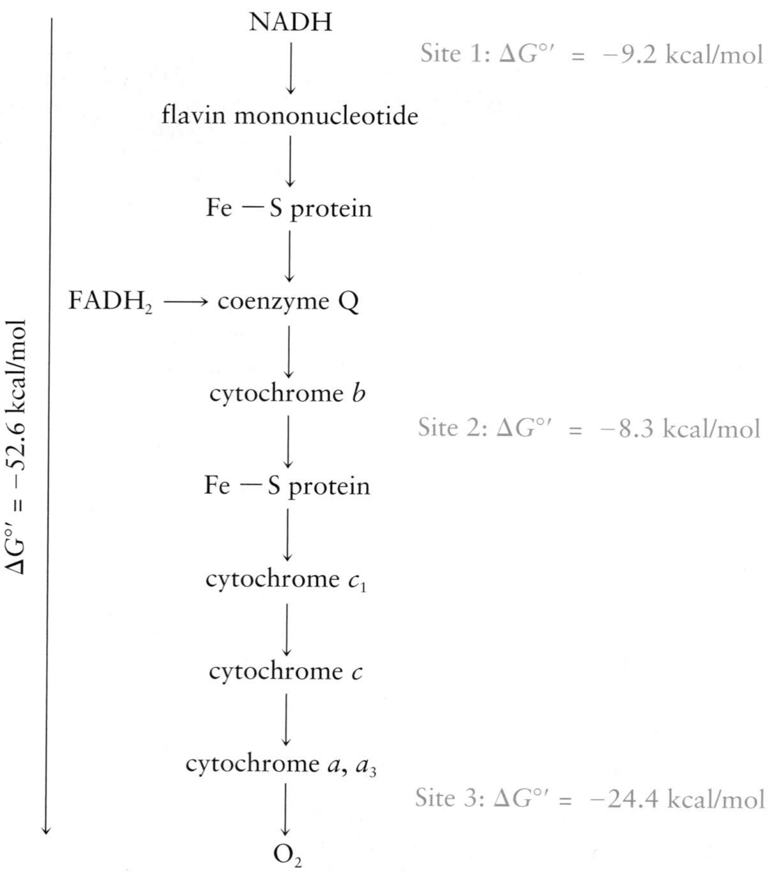

■ **FIGURE 13.10** ATP is synthesized at three sites within the electron transport chain. Energy decreases at other locations result in the release of heat energy.

and outer mitochondrial membranes (see ■ Figure 13.11). This creates a difference in proton concentration across the inner mitochondrial membrane, along with an electrical potential difference. As a result of the concentration and potential differences, protons flow back through the membrane through a channel formed by the enzyme F_1-ATPase. The flow of protons through this enzyme is believed to drive the phosphorylation reaction.

$$\text{ADP} + \text{P}_i \xrightarrow[\text{F}_1\text{-ATPase}]{\text{H}^+ \text{ flow through}} \text{ATP} \qquad (13.15)$$

■ **Learning Check 13.3** How many molecules of ATP are produced from the complete catabolism of the following?

a. Two molecules of $FADH_2$
b. Two molecules of NADH
c. Two molecules of acetyl CoA

13.8 The Complete Oxidation of Glucose

LEARNING OBJECTIVE

10. Calculate the amount of ATP produced by the complete oxidation of a mole of glucose.

The information given to this point about glycolysis, the citric acid cycle, and the electron transport chain almost makes it possible to calculate the ATP yield from the

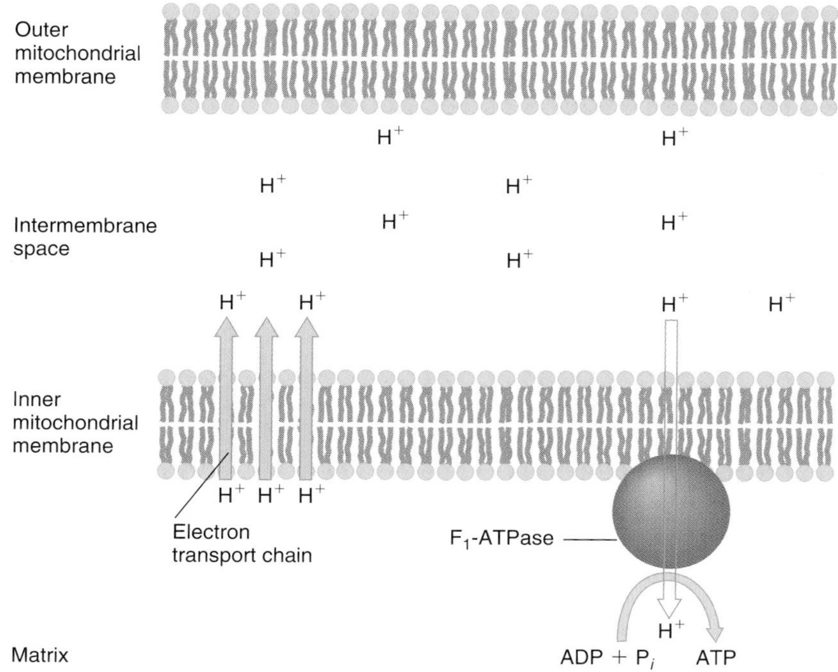

■ **FIGURE 13.11** The chemiosmotic hypothesis. Electron transport pumps protons (H^+) out of the matrix. The formation of ATP accompanies the flow of protons through F_1-ATPase back into the matrix.

total oxidation of 1 mol of glucose. However, one more factor must be considered. NADH produced in the cytoplasm during glycolysis does not pass through the mitochondrial membrane to the site of the electron transport chain. Brain and muscle cells employ a transport mechanism that passes electrons from the cytoplasmic NADH through the membrane to FAD molecules inside the mitochondria, forming $FADH_2$. In Section 13.7, we pointed out that one molecule of $FADH_2$ gives 1.5 molecules of ATP. Thus, with this transport mechanism, one molecule of cytoplasmic NADH generates only 1.5 molecules of ATP, whereas 2.5 molecules of ATP come from each molecule of mitochondrial NADH. A more efficient shuttle mechanism is found in liver, heart, and kidney cells, where one cytoplasmic NADH results in one mitochondrial NADH. Thus, 2.5 molecules of ATP are formed in this case.

This production of ATP from the complete aerobic catabolism of 1 mol of glucose in the liver is summarized in ■ Table 13.1. Notice from the table that of the 32 mol of ATP generated, the great majority (28 mol) is formed as a result of oxidative phosphorylation. In Section 13.4, we learned that only 2 mol of ATP is produced per mole of glucose by lactate fermentation and by alcoholic fermentation. Thus, in terms of ATP production from fuels, the complete aerobic oxidation of glucose is 16 times more efficient than either lactate or alcoholic fermentation.

Another useful way to look at the efficiency of the complete oxidation of glucose is to compare the amount of free energy available in glucose and the amount of free energy stored in synthesized ATP.

Glucose oxidation:

$$C_6H_{12}O_6 + 6O_2 \rightarrow 6CO_2 + 6H_2O \qquad \Delta G^{\circ\prime} = -686 \text{ kcal/mol} \qquad (13.16)$$

ATP synthesis:

$$32ADP + 32P_i \rightarrow 32ATP + 32H_2O \qquad \Delta G^{\circ\prime} = +234 \text{ kcal/mol} \qquad (13.17)$$

Overall reaction:

$$C_6H_{12}O_6 + 6O_2 + 32ADP + 32P_i \rightarrow 6CO_2 + 32ATP + 38H_2O \qquad (13.18)$$

$$\text{Total: } \Delta G^{\circ\prime} = -452 \text{ kcal/mol}$$

TABLE 13.1 The ATP yield from the complete oxidation of one mol of glucose

Reaction	Comments	Yield of ATP (moles)
Cytoplasmic reactions		
Glycolysis		
glucose → glucose 6-phosphate	1 mol consumes 1 ATP	−1
glucose 6-phosphate → fructose 1,6-bisphosphate	1 mol consumes 1 ATP	−1
glyceraldehyde 3-phosphate → 1,3-bisphosphoglycerate	2 mol produces 2 mol of cytoplasmic NADH	
1,3-bisphosphoglycerate → 3-phosphoglycerate	2 mol produces 2 ATP	+2
phosphoenolpyruvate → pyruvate	2 mol produces 2 ATP	+2
Mitochondrial reactions		
Oxidation of pyruvate		
pyruvate → acetyl CoA + CO_2	2 mol produces 2 NADH	
Citric acid cycle		
isocitrate → α-ketoglutarate + CO_2	2 mol produces 2 NADH	
α-ketoglutarate → succinyl CoA + CO_2	2 mol produces 2 NADH	
succinyl CoA → succinate	2 mol produces 2 GTP	+2
succinate → fumarate	2 mol produces 2 $FADH_2$	
malate → oxaloacetate	2 mol produces 2 NADH	
Electron transport chain and oxidative phosphorylation		
2 cytoplasmic NADH formed in glycolysis	each yields 2.5 ATP	+5[a]
2 NADH formed in the oxidation of pyruvate	each yields 2.5 ATP	+5
2 $FADH_2$ formed in the citric acid cycle	each yields 1.5 ATP	+3
6 NADH formed in the citric acid cycle	each yields 2.5 ATP	+15
	Net yield of ATP	+32

[a]In muscle and brain cells, 3 ATP are produced in this step and only 30 ATP overall.

Thus, glucose oxidation liberates 686 kcal/mol, whereas the synthesis of 32 mol of ATP stores 234 kcal/mol. The efficiency of energy stored is

$$\frac{\text{energy stored}}{\text{energy available}} \times 100 = \frac{234 \text{ kcal/mol}}{686 \text{ kcal/mol}} \times 100 = 34.1\%$$

It is remarkable that living cells can capture 34% of the released free energy and make it available to do biochemical work. By contrast, an automobile engine makes available only about 20–30% of the energy actually released by burning gasoline.

13.9 Glycogen Metabolism

LEARNING OBJECTIVE

11. Explain the importance of the processes of glycogenesis and glycogenolysis.

Glycogen Synthesis

glycogenesis
The synthesis of glycogen from glucose.

Glucose consumed in excess of immediate body requirements is converted to glycogen (Section 7.8), which is stored primarily in the liver and muscle tissue. The liver of an average adult can store about 110 g of glycogen, and the muscles can store about 245 g. The process by which glucose is converted to glycogen is called **glycogenesis.** Glycogenesis can occur in all cells, but it is an especially important function of liver and muscle cells. The net result of this anabolic process is the bonding of glucose units to a growing glycogen chain, with the hydrolysis of a high-energy nucleotide, UTP (uridine triphosphate, Section 12.7), providing the energy. Each glucose unit lengthening the chain forms a new $\alpha(1 \rightarrow 4)$ linkage. The $\alpha(1 \rightarrow 6)$ branch points are inserted by a "branching enzyme."

The overall process of glycogenesis is summarized by Reaction 13.19:

$$n \text{ glucose} \xrightarrow[\text{UTP} \quad \text{UDP} + P_i]{} \underset{\text{glycogen}}{(\text{glucose})_n} \qquad (13.19)$$

Glycogen Breakdown

The breakdown of glycogen back to glucose is called **glycogenolysis.** Although major amounts of glycogen are stored in both muscle tissue and the liver, glycogenolysis can occur in the liver (and kidney and intestinal cells) but not in muscle tissue because one essential enzyme (glucose 6-phosphatase) is lacking. The first step in glycogen breakdown is the cleavage of $\alpha(1 \rightarrow 4)$ linkages, catalyzed by glycogen phosphorylase. In this reaction, glucose units are released from the glycogen chain as glucose 1-phosphate:

glycogenolysis The breakdown of glycogen to glucose.

$$\underset{\text{glycogen}}{(\text{glucose})_n} + P_i \rightarrow \underset{\text{glycogen with one fewer glucose unit}}{(\text{glucose})_{n-1}} + \text{glucose 1-phosphate} \qquad (13.20)$$

Also associated with the phosphorylase enzyme is a second enzyme, called the debranching enzyme. This debranching enzyme catalyzes the hydrolysis of the $\alpha(1 \rightarrow 6)$ linkages, eliminating the branch points. The phosphorylase can then continue to act on the rest of the chain.

In the second reaction, phosphoglucomutase catalyzes the isomerization of glucose 1-phosphate to glucose 6-phosphate:

$$\text{glucose 1-phosphate} \rightleftarrows \text{glucose 6-phosphate} \qquad (13.21)$$

STUDY SKILLS 13.1 Key Numbers for ATP Calculations

It is important to be able to determine the energy yield in terms of the number of ATP molecules produced by the stepwise catabolism of glucose. Some of the relationships used in this determination will be used in the next chapter to calculate the energy yield from fatty acid catabolism. The three key substances to focus on are acetyl CoA, NADH, and $FADH_2$. Every mitochondrial NADH produces 2.5 ATP molecules, and every $FADH_2$ produces 1.5 ATP molecules. Every acetyl CoA that passes through the citric acid cycle gives 3 NADH, 1 $FADH_2$, and 1 GTP, which in turn produce a total of 10 ATP molecules. These reactions are outlined here.

Catabolic step		Energy production	
Glycolysis (1 glucose ⟶ 2 pyruvate)	produces	2NADH (cytoplasmic) +	2 ATP
		liver, heart, kidneys →	5 ATP
		brain, muscle →	(3 ATP)
Oxidation of 2 pyruvates to 2 acetyl CoA	produces	2 NADH (mitochondrial) →	5 ATP
Citric acid cycle, one turn	produces	3 NADH →	7.5 ATP
		1 $FADH_2$ →	1.5 ATP
		1 GTP →	1 ATP
			10 ATP
Citric acid cycle, two turns	produce	2 × 10 =	20 ATP
		Total/glucose molecule =	32 ATP (30 ATP)

The final step in glycogen breakdown is the hydrolysis of glucose 6-phosphate to produce free glucose. The enzyme for this reaction, glucose 6-phosphatase, is the one found only in liver, kidney, and intestinal cells:

$$\text{glucose 6-phosphate} + H_2O \rightarrow \text{glucose} + P_i \qquad (13.22)$$

Without glucose 6-phosphatase, muscle cells cannot form free glucose from glycogen. However, they can carry out the first two steps of glycogenolysis to produce glucose 6-phosphate. This form of glucose is the first intermediate in the glycolysis pathway and can be used to produce energy. In this way, muscle cells utilize glycogen only for energy production. On the other hand, an important function of the liver is to maintain a relatively constant level of blood glucose. Thus, it has the capacity to degrade glycogen all the way to glucose, which is released into the blood during muscular activity and between meals.

■ **Learning Check 13.10** What enzyme enables the liver to form glucose from glycogen? Why can't muscles carry out the same conversion?

13.10 Gluconeogenesis

LEARNING OBJECTIVE

12. Describe gluconeogenesis and the operation of the Cori cycle.

The supply of glucose in the form of liver and muscle glycogen can be depleted by about 12–18 hours of fasting. These stores of glucose are also depleted in a short time as a result of heavy work or strenuous exercise (see ■ Figure 13.12). Nerve tissue, including the brain, would be deprived of glucose under such conditions if the only source was glycogen. Fortunately, a metabolic pathway exists that overcomes this problem.

ThomsonNOW™ Go to Chemistry Interactive to explore **gluconeogenesis.**

gluconeogenesis The synthesis of glucose from noncarbohydrate molecules.

Glucose can be synthesized from noncarbohydrate materials in a process called **gluconeogenesis.** When carbohydrate intake is low and when glycogen stores are depleted, the carbon skeletons of lactate, glycerol (derived from the hydrolysis of fats), and certain amino acids are used to synthesize pyruvate, which is then converted to glucose:

$$\text{(lactate, certain amino acids, glycerol)} \rightarrow \text{pyruvate} \rightarrow \text{glucose} \qquad (13.23)$$

About 90% of gluconeogenesis takes place in the liver. Very little occurs in the brain, skeletal muscle, or heart, even though these organs have a high demand for glucose. Thus, gluconeogenesis taking place in the liver helps maintain the blood glucose level so that tissues needing glucose can extract it from the blood.

Gluconeogenesis involving lactate is especially important. As might be expected, this process follows a separate and different pathway from glycolysis. During active exercise, lactate levels increase in muscle tissue, and the compound diffuses out of the tissue into the blood. It is transported to the liver, where lactate dehydrogenase, the enzyme that catalyzes lactate formation in muscle, converts it back to pyruvate:

$$\underset{\text{lactate}}{CH_3-\overset{\overset{\displaystyle OH}{|}}{C}H-COO^-} + NAD^+ \rightleftharpoons \underset{\text{pyruvate}}{CH_3-\overset{\overset{\displaystyle O}{\|}}{C}-COO^-} + NADH + H^+ \qquad (13.24)$$

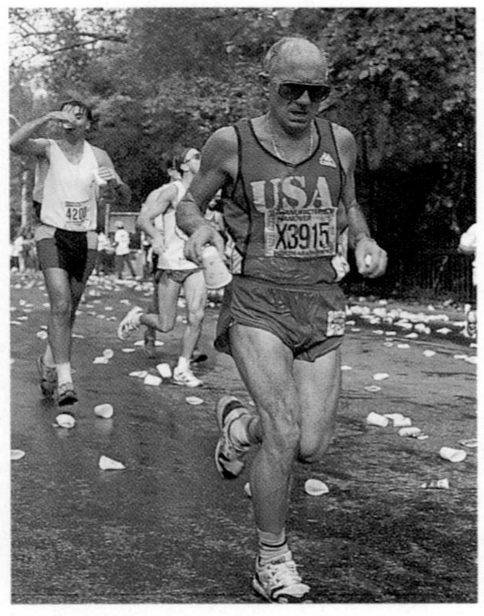

© Robert Brenner/PhotoEdit

■ **FIGURE 13.12** Gluconeogenesis provides a source of glucose when glycogen supplies are exhausted after vigorous exercise.

The pyruvate is then converted to glucose by the gluconeogenesis pathway, and the glucose enters the blood. In this way, the liver increases a low blood glucose level and makes glucose available to muscle. This cyclic process, involving the transport of lactate from muscle to the liver, the resynthesis of glucose by gluco-

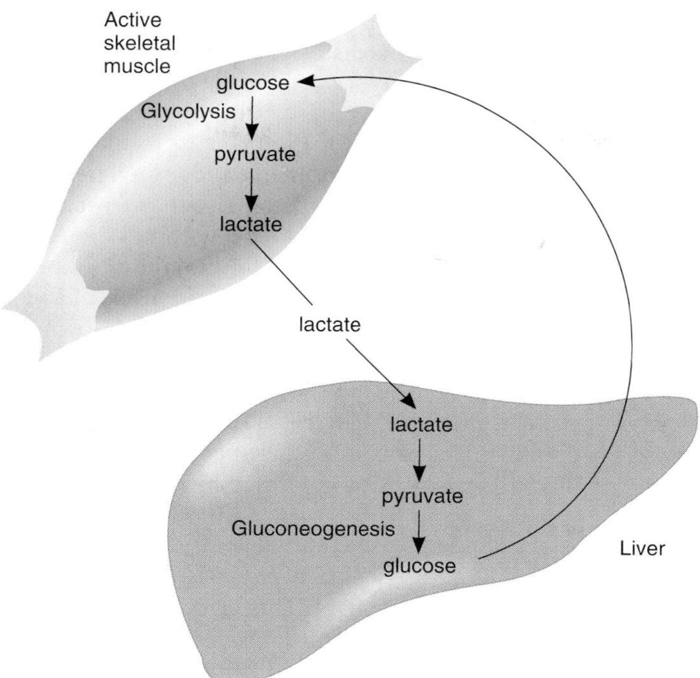

■ **FIGURE 13.13** The Cori cycle, showing the relationship between glycolysis (in the muscle) and gluconeogenesis (in the liver).

neogenesis, and the return of glucose to muscle tissue, is called the **Cori cycle** (see ■ Figure 13.13). ■ Figure 13.14 shows the relationship of gluconeogenesis to the other major pathways of glucose metabolism.

Cori cycle
The process in which glucose is converted to lactate in muscle tissue, lactate is reconverted to glucose in the liver, and glucose is returned to the muscle.

13.11 The Hormonal Control of Carbohydrate Metabolism

LEARNING OBJECTIVE

13. Describe how hormones regulate carbohydrate metabolism.

It is important that metabolic pathways be responsive to cellular conditions so that energy is not wasted in producing unneeded materials. Besides the regulation of enzymes at key control points (Section 13.3), the body also uses three important regulatory hormones: epinephrine, glucagon, and insulin.

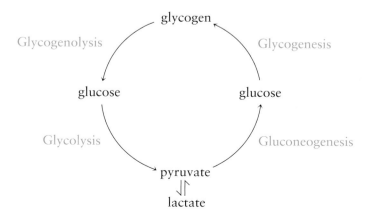

glycogen
Glycogenolysis
Glycogenesis
glucose
glucose
Glycolysis
Gluconeogenesis
pyruvate
lactate

■ **FIGURE 13.14** A summary of the major pathways in glucose metabolism. Note that the breakdown processes end in *-lysis,* whereas the synthesis pathways end in *-genesis.*

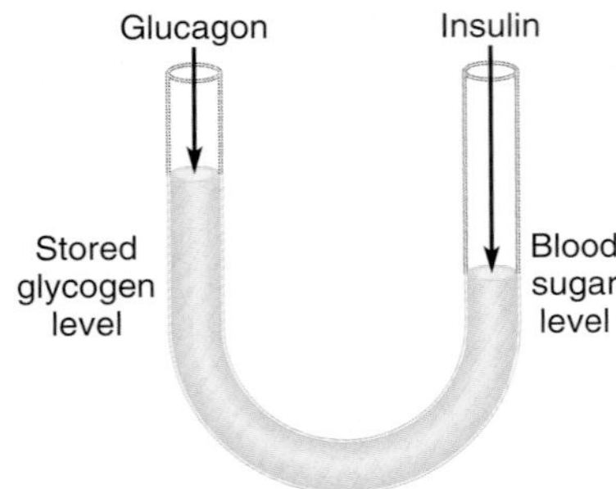

■ **FIGURE 13.15** The biochemical balance between two opposing hormones, insulin and glucagon.

In Section 13.2, we pointed out that when carbohydrates are consumed, the blood glucose level increases. This increase signals the β-cells of the pancreas to release a small amount of the hormone insulin into the bloodstream. The hormone enhances the absorption of glucose from the blood into cells of active tissue such as skeletal and heart muscles (Section 9.6). In addition, insulin increases the rate of synthesis of glycogen, fatty acids, and proteins and stimulates glycolysis. Each of these activities uses up glucose. As a result of insulin released into the bloodstream, the blood glucose level begins to decrease within 1 hour after carbohydrates have been ingested and returns to normal within about 3 hours.

Glucagon, a second polypeptide hormone, is synthesized and secreted by the α-cells of the pancreas. Glucagon activates the breakdown of glycogen (glycogenolysis) in the liver and thus counteracts the effect of insulin by raising blood glucose levels. ■ Figure 13.15 emphasizes the fact that insulin and glucagon work in opposition to each other and that the blood sugar level at any given time depends, in part, on the biochemical balance between these two hormones. Blood levels of glucose are also influenced to some degree by growth hormone and adrenal cortex steroids.

The third hormone affecting glucose metabolism is epinephrine, synthesized by the adrenal gland (Section 6.6). Epinephrine stimulates glycogen breakdown in muscles and, to a smaller extent, in the liver. This glycogenolysis reaction—usually a response to pain, anger, or fear—provides energy for a sudden burst of muscular activity. ■ Table 13.2 summarizes the hormonal control of glycogen.

CHEMISTRY AROUND US 13.2 Carbohydrate Loading

Among the meanings of the word *endurance* is "to continue without impairment or yielding." Endurance athletes such as long-distance runners, cyclists, or swimmers certainly fit that description, but no one can perform such feats without a steady source of energy. Food is the source of energy for endurance athletes just as it is for other people. Numerous ideas have been developed about which foods are best at enhancing specific athletic characteristics. Certainly, it seems to make sense that if an athlete wants to build up protein-containing muscle tissue, he or she would include an ample amount of protein in their diet.

The body, through various metabolic processes, can extract energy from any type of food, be it carbohydrates, lipids, or proteins. But, which type provides the readily available energy over the long durations needed by endurance athletes? The answer, according to many successful endurance athletes, is carbohydrates in the form of glycogen stored in the muscles. This answer is the basis for the common practice among endurance athletes of carbohydrate-loading or carb-loading.

Contrary to what some believe, carb-loading involves more than simply eating a carbohydrate-laden meal just before a big race. To be most effective, the muscles of the body have to undergo significant training in order to maximize their ability to store greater-than-normal amounts of glycogen. Research results from the Human Performance Laboratory at Ball State University suggest that the muscles of well-trained athletes are able to store up to 25% more glycogen than the muscles of untrained individuals.

After training, the next part of the carb-loading process is the actual loading up of the trained muscle with glycogen. This is accomplished by increasing the amount of carbohydrates in the diet to the extent that they constitute 60–70% of the daily caloric intake. When the practice of carb-loading first began to be followed, it was thought to be most effective by starting with a glycogen depletion phase in which muscles were drained of glycogen. This was accomplished by hard workouts for about a week followed by a few days of a low-carbohydrate diet. Then, during the final days before a race, the athletes would switch to a very high-carbohydrate diet in order to saturate the muscles with glycogen.

This approach seemed to work well, but a study published in the *International Journal of Sports Medicine* in 1981 indicated that equally good results were obtained by eliminating the depletion phase. This new approach also prevented the athletes from feeling sluggish and irritable the week before their race.

Today, the best plan seems to be to taper from intensive training as the race day approaches, eat well and increase the percentage of carbohydrates in the diet to the 60–70% range a few days before the race. Each person's body responds to carbohydrates a little differently, so the prerace training is a good time to figure out what foods your own body digests and utilizes the best. Carb-loading should not be used as an excuse to ignore other nutritional needs. It is still important to include foods such as fruits, vegetables, and legumes that are rich in vitamins and minerals.

TABLE 13.2 The hormonal control of glycogen

Hormone	Source	Effect on glycogen	Impact on blood glucose
Insulin	β-cells of pancreas	Increases formation	Lowers blood glucose levels
Glucagon	α-cells of pancreas	Activates breakdown of liver glycogen	Raises blood glucose levels
Epinephrine	Adrenal medulla	Stimulates breakdown	Raises blood glucose levels

CHEMISTRY AND YOUR HEALTH 13.1 Prediabetic . . . or Already There?

Many people show signs of becoming diabetic, but brush them aside considering "prediabetes" a long-term condition. This is a big mistake that could speed up the onset of Type II diabetes, which is rapidly becoming a major health problem in American society.

The following dietary changes can postpone the final onset of diabetes for years if the changes are consistent and become a way of life.

1. Lose weight. It has been shown that dropping 5–10% of a person's body weight can bring blood sugar back to the normal range.
2. Get 30–60 minutes of moderate to brisk exercise each day. Exercise is an absolutely necessary ingredient in maintaining good health. It helps maintain the normal levels of blood sugar and also burns calories, which help in the achievement of weight loss.
3. Follow a healthy eating plan. Be certain to emphasize the inclusion of whole grains, fruits and vegetables, lean protein, and unsaturated fats in the diet. Simple carbohydrates, such as white bread, cakes, cookies, and other processed foods made with saturated fats should largely be avoided.

Individuals who are known to be prediabetic and who experience two or more of the following symptoms should check with their doctor to determine if they have developed diabetes:

- Frequent urination
- Excessive thirst
- Extreme hunger
- Unusual weight loss
- Increased fatigue
- Irritability
- Blurry vision

Concept Summary

ThomsonNOW™ Sign in at **www.thomsonedu.com** to:

- Assess your understanding with Exercises keyed to each learning objective.
- Check your readiness for an exam by taking the **Pre-test** and exploring the modules recommended in your **Personalized Learning Plan.**

The Digestion of Carbohydrates. Glucose is the key food molecule for most organisms, and it is the central substance in carbohydrate metabolism. During digestion, carbohydrates are hydrolyzed to the monosaccharides glucose, fructose, and galactose, which are absorbed into the bloodstream through the lining of the small intestine. **OBJECTIVE 1 (Section 13.1), Exercise 13.2**

Blood Glucose. Glucose is the most abundant carbohydrate in the blood. The liver regulates blood glucose levels so that a sufficient concentration is always available to meet the body's energy needs. **OBJECTIVE 2 (Section 13.2), Exercise 13.6**. A lower-than-normal blood glucose level is referred to as hypoglycemia; a higher-than-normal blood glucose level is termed hyperglycemia.

Glycolysis. Glycolysis, a series of ten reactions that occurs in the cytoplasm, is a process in which one glucose molecule is converted into two molecules of pyruvate. Two molecules of ATP and two molecules of NADH are produced in the process. **OBJECTIVE 3 (Section 13.3), Exercise 13.10**. Both galactose and fructose are converted into intermediates that enter the glycolysis pathway. The glycolysis pathway is controlled by the regulation of three key enzymes in the reaction sequence. **OBJECTIVE 4 (Section 13.3), Exercise 13.14**

The Fates of Pyruvate. Pyruvate, the product of glycolysis, has three different fates. Under aerobic conditions, pyruvate enters mitochondria and is converted to acetyl CoA. The acetyl CoA enters the citric acid cycle. Products of the citric acid cycle enter the electron transport chain, where oxidation is completed and ATP is synthesized. A second fate of pyruvate is conversion to lactate. This takes place in anaerobic microorganisms and in humans when glycolysis occurs faster than the oxygen-dependent citric acid cycle and electron transport chain can operate. Some microorganisms convert pyruvate to ethanol, the third fate of pyruvate. In each of these three processes, NAD^+ is regenerated so that glycolysis can continue. **OBJECTIVE 5 (Section 13.4), Exercise 13.20**

The Citric Acid Cycle. The citric acid cycle generates NADH and $FADH_2$, which are necessary for ATP synthesis. It also produces intermediates used for biosynthesis. **OBJECTIVE 6 (Section 13.5), Exercise 13.24**. This pathway consists of eight reactions that process incoming molecules of acetyl CoA. The carbon atoms leave the cycle in the form of molecules of CO_2. The hydrogen atoms and electrons leave the cycle in the form of reduced coenzymes NADH and $FADH_2$. The cycle is regulated by three allosteric enzymes in response to cellular levels of ATP. **OBJECTIVE 7 (Section 13.5), Exercise 13.32**. One acetyl CoA molecule entering the citric acid cycle produces three molecules of NADH, one of $FADH_2$, and one of GTP.

The Electron Transport Chain. This pathway involves a series of reactions that pass electrons from NADH and $FADH_2$ to molecular oxygen. Each carrier in the series has an increasing affinity for electrons. Four of the carriers, which are referred to as cytochromes, contain iron, which accepts and then transfers the electrons. As NADH and $FADH_2$ release their hydrogen atoms and electrons, NAD^+ and FAD are regenerated for return to the citric acid cycle. **OBJECTIVE 8 (Section 13.6), Exercises 13.34 and 13.36**

Oxidative Phosphorylation. At three sites within the electron transport chain, the decrease in free energy is sufficient to convert ADP to ATP. One molecule of NADH produces 2.5 molecules of ATP. $FADH_2$, which enters the chain one step later, produces 1.5 molecules of ATP. According to the chemiosmotic hypothesis, the synthesis of ATP takes place because of a flow of protons across the inner mitochondrial membrane. **OBJECTIVE 9 (Section 13.7), Exercise 13.46**

The Complete Oxidation of Glucose. The complete oxidation of 1 mol of glucose produces 32 (or 30) mol of ATP. **OBJECTIVE 10 (Section 13.8), Exercise 13.48**. In contrast, lactate and alcoholic fermentation produce only 2 mol of ATP.

Glycogen Metabolism. When blood glucose levels are high, excess glucose is converted into glycogen by the process of glycogenesis. The glycogen is stored within the liver and muscle tissue. Glycogenolysis, the breakdown of glycogen, occurs when muscles need energy and when the liver is restoring a low blood sugar level to normal. **OBJECTIVE 11 (Section 13.9), Exercise 13.50**

Gluconeogenesis. Lactate, certain amino acids, and glycerol can be converted into glucose. This process, called gluconeogenesis, takes place in the liver when glycogen supplies are being depleted and when carbohydrate intake is low. Gluconeogenesis from lactate is especially important during periods of high muscle activity; the liver converts excess lactate from the muscles into glucose, which is then cycled back to the muscles. This process is called the Cori cycle. **OBJECTIVE 12 (Section 13.10), Exercise 13.58**

The Hormonal Control of Carbohydrate Metabolism. Three hormones exert major interactive effects on carbohydrate metabolism: insulin, glucagon, and epinephrine. Insulin promotes the uptake and utilization of glucose by the cells. Thus, insulin lowers blood sugar levels. Glucagon stimulates the conversion of glycogen to glucose and thus raises blood glucose levels. Epinephrine is released in response to anger, fear, and excitement. It also stimulates the release of glucose from glycogen and raises blood glucose levels. **OBJECTIVE 13 (Section 13.11), Exercise 13.62**

Key Terms and Concepts

Aerobic (13.4)
Alcoholic fermentation (13.4)
Anaerobic (13.4)
Blood sugar level (13.2)
Chemiosmotic hypothesis (13.7)
Citric acid cycle (13.5)
Cori cycle (13.10)
Cytochrome (13.6)
Electron transport chain (13.6)
Gluconeogenesis (13.10)
Glucosuria (13.2)
Glycogenesis (13.9)
Glycogenolysis (13.9)
Glycolysis (13.3)
Hyperglycemia (13.2)
Hypoglycemia (13.2)
Lactate fermentation (13.4)
Lactose intolerance (13.3)
Oxidative phosphorylation (13.7)
Renal threshold (13.2)

Key Reactions

1. Glycolysis (Section 13.3):

$$\text{glucose} + 2P_i + 2ADP + 2NAD^+ \longrightarrow 2\ \text{pyruvate} + 2ATP + 2NADH + 4H^+ + 2H_2O \qquad \text{Reaction 13.5}$$

2. Oxidation of pyruvate to acetyl CoA (Section 13.4):

$$CH_3-\overset{\overset{\displaystyle O}{\|}}{C}-COO^- + CoA-S-H + NAD^+ \longrightarrow CH_3-\overset{\overset{\displaystyle O}{\|}}{C}-S-CoA + NADH + CO_2 \qquad \text{Reaction 13.7}$$

3. Reduction of pyruvate to lactate (Section 13.4):

$$CH_3-\overset{\overset{\large O}{\|}}{C}-COO^- + NADH + H^+ \rightleftharpoons CH_3-\overset{\overset{\large OH}{|}}{CH}-COO^- + NAD^+$$ Reaction 13.8

4. Net reaction of lactate fermentation (Section 13.4):

$$C_6H_{12}O_6 + 2ADP + 2P_i \longrightarrow 2CH_3-\overset{\overset{\large OH}{|}}{CH}-COO^- + 2ATP$$ Reaction 13.9

5. Reduction of pyruvate to ethanol (Section 13.4):

$$CH_3-\overset{\overset{\large O}{\|}}{C}-COO^- + 2H^+ + NADH \longrightarrow CH_3CH_2-OH + NAD^+ + CO_2$$ Reactions 13.10 and 13.11

6. Net reaction of alcoholic fermentation (Section 13.4):

$$C_6H_{12}O_6 + 2ADP + 2P_i \longrightarrow 2CH_3CH_2-OH + 2CO_2 + 2ATP$$ Reaction 13.12

7. The citric acid cycle (Section 13.5):

$$\text{acetyl CoA} + 3NAD^+ + FAD + GDP + P_i + 2H_2O \longrightarrow 2CO_2 + CoA-S-H + 3NADH + 2H^+ + FADH_2 + GTP$$ Reaction 13.13

8. The electron transport chain (Section 13.6):

$$4H^+ + 4e^- + O_2 \longrightarrow 2H_2O$$ Reaction 13.14

9. Oxidative phosphorylation (Section 13.7):

$$ADP + P_i \longrightarrow ATP$$ Reaction 13.15

10. Complete oxidation of glucose (Section 13.8):

$$C_6H_{12}O_6 + 6O_2 + 32ADP + 32P_i \longrightarrow 6CO_2 + 32ATP + 38H_2O$$ Reaction 13.18

11. Glycogenesis (Section 13.9):

$$\text{glucose} \longrightarrow \text{glycogen}$$ Reaction 13.1

12. Glycogenolysis (Section 13.9):

$$\text{glycogen} \longrightarrow \text{glucose}$$ Reactions 13.20, 13.21, and 13.23

13. Gluconeogenesis (Section 13.10):

$$\left\{\begin{matrix}\text{lactate,}\\ \text{certain amino acids,}\\ \text{glycerol}\end{matrix}\right\} \longrightarrow \text{pyruvate} \longrightarrow \text{glucose}$$ Reaction 13.23

Exercises

SYMBOL KEY

Even-numbered exercises are answered in Appendix B.

Blue-numbered exercises are more challenging.

■ denotes exercises available in ThomsonNow and assignable in OWL.

ThomsonNOW™ To assess your understanding of this chapter's topics with sample tests and other resources, sign in at **www.thomsonedu.com.**

THE DIGESTION OF CARBOHYDRATES (SECTION 13.1)

13.1 Why is glucose considered the pivotal compound in carbohydrate metabolism?

13.2 ■ Name the products produced by the digestion of
a. starch
b. lactose
c. sucrose
d. maltose

13.3 What type of reaction is involved as carbohydrates undergo digestion?

BLOOD GLUCOSE (SECTION 13.2)

13.4 Describe what is meant by the terms *blood sugar level* and *normal fasting level.*

13.5 What range of concentrations for glucose in blood is considered a normal fasting level?

13.6 ■ How do each of the following terms relate to blood sugar level?

a. Hypoglycemia

b. Hyperglycemia

c. Renal threshold

d. Glucosuria

13.7 Explain how the synthesis and degradation of glycogen in the liver help regulate blood sugar levels.

13.8 Explain why severe hypoglycemia can be very serious.

GLYCOLYSIS (SECTION 13.3)

13.9 ■ What is the main purpose of glycolysis?

13.10 ■ What is the starting material for glycolysis? The product?

13.11 ■ In what part of the cell does glycolysis take place?

13.12 How many steps in glycolysis require ATP? How many steps produce ATP? What is the net production of ATP from one glucose molecule undergoing glycolysis?

13.13 Which coenzyme serves as an oxidizing agent in glycolysis?

13.14 What is the third control point in glycolysis? How is it responsive to cellular needs?

13.15 Number the carbon atoms of glucose 1 through 6, and show the location of each carbon in the two molecules of pyruvate produced by glycolysis.

13.16 Explain how a high concentration of glucose 6-phosphate causes a decrease in the rate of glycolysis.

13.17 Why is the conversion of fructose 6-phosphate to fructose 1,6-bisphosphate sometimes called the "committed step" of glycolysis?

THE FATES OF PYRUVATE (SECTION 13.4)

13.18 ■ Distinguish between the terms *aerobic* and *anaerobic*.

13.19 Explain how lactate formation allows glycolysis to continue under anaerobic conditions.

13.20 ■ Compare the fate of pyruvate in the body under (a) aerobic conditions and (b) anaerobic conditions.

13.21 Write a summary equation for the conversion of glucose to

a. lactate

b. pyruvate

c. ethanol

13.22 What is the fate of the pyruvate molecule's carboxylate group during alcoholic fermentation? Explain how ethanol formation from pyruvate allows yeast to survive with limited oxygen supplies.

THE CITRIC ACID CYCLE (SECTION 13.5)

13.23 What are two other names for the citric acid cycle?

13.24 What is the primary function of the citric acid cycle in ATP production? What other vital role is served by the citric acid cycle?

13.25 Why is it appropriate to refer to the sequence of reactions in the citric acid cycle as a cycle?

13.26 ■ Describe the citric acid cycle by identifying the following:

a. The fuel needed by the cycle

b. The form in which carbon atoms leave the cycle

c. The form in which hydrogen atoms and electrons leave the cycle

13.27 The citric acid cycle is located in close cellular proximity to the electron transport chain. Why?

13.28 ■ How many molecules of each of the following are produced by one run through the citric acid cycle?

a. NADH b. $FADH_2$ c. GTP d. CO_2

13.29 Write reactions for the steps in the citric acid cycle that involve oxidation by NAD^+.

13.30 Write a reaction for the step in the citric acid cycle that involves oxidation by FAD.

13.31 Write a reaction for the step in the citric acid that involves the phosphorylation of GDP. What is the structural difference between ATP and GTP? Explain why GTP can function as a high-energy compound.

13.32 The reactions in the citric acid cycle involve enzymes.

a. Identify the enzymes at the three control points.

b. Each control-point enzyme is inhibited by NADH and ATP. Explain how this enables the cell to be responsive to energy needs.

c. Two of the control-point enzymes are activated by ADP. Explain the benefit of this regulation to the cell.

THE ELECTRON TRANSPORT CHAIN (SECTION 13.6)

13.33 What is the primary function of the electron transport chain?

13.34 ■ Which coenzymes bring electrons to the electron transport chain?

13.35 List the participants of the electron transport chain in the order in which they accept electrons.

13.36 What is the role of the cytochromes in the electron transport chain?

13.37 What is the role of oxygen in the functioning of the electron transport chain? Write a reaction that summarizes the fate of the oxygen.

OXIDATIVE PHOSPHORYLATION (SECTION 13.7)

13.38 ■ The third stage in the oxidation of foods to provide energy involves oxidative phosphorylation. What is oxidized? What is phosphorylated?

13.39 How many ATP molecules are synthesized when NADH passes electrons to molecular oxygen?

13.40 ■ How many ATP molecules are synthesized when $FADH_2$ passes electrons to molecular oxygen?

13.41 Explain what is meant by the statement: "Oxidation of NADH in the electron transport chain results in 34% energy conservation."

13.42 How many ATP molecules can be formed by the oxidation of ten acetyl CoA molecules in the common catabolic pathway?

13.43 Identify the steps in the electron transport chain where ATP is synthesized.

13.44 The change in free energy for the complete oxidation of glucose to CO_2 and H_2O is −686 kcal per mole of glucose. If all this free energy could be used to drive the synthesis of ATP, how many moles of ATP could be formed from 1 mol of glucose?

13.45 What is the function of the protein F_1-ATPase?

13.46 ■ According to the chemiosmotic theory, ATP synthesis results from the flow of what particles? Across what membrane does this flow occur?

THE COMPLETE OXIDATION OF GLUCOSE (SECTION 13.8)

13.47 Explain why NADH produced in muscles during glycolysis ultimately gives rise to just 1.5 molecules of ATP.

13.48 A total of 32 mol of ATP can be produced by the complete oxidation of 1 mol of glucose in the liver. Indicate the number of these moles of ATP produced by

a. glycolysis

b. the citric acid cycle

c. the electron transport chain and oxidative phosphorylation

13.49 The complete biochemical oxidation of glucose to ATP is only 34% efficient in capturing the total amount of energy released. The remaining 66% is released as what form of energy?

GLYCOGEN METABOLISM (SECTION 13.9)

13.50 Compare the terms *glycogenesis* and *glycogenolysis*.

13.51 Fructose and galactose from digested food enter the glycolysis pathway. Name the intermediates of the pathway which serve as entry points for these substances.

13.52 What high-energy compound is involved in the conversion of glucose to glycogen?

13.53 Why is it important that the liver store glycogen? How does this aid the controlling of blood sugar levels?

13.54 Why can liver glycogen, but not muscle glycogen, be used to raise blood sugar levels?

13.55 Summarize the reactions for glycogenolysis.

GLUCONEOGENESIS (SECTION 13.10)

13.56 What organ serves as the principal site for gluconeogenesis?

13.57 ■ A friend has been fasting for three days. Where does the glucose come from to keep his brain functioning?

13.58 What are the sources of compounds that undergo gluconeogenesis?

13.59 Explain the operation of the Cori cycle and its importance to muscles.

13.60 Would the Cori cycle operate if the body's muscles were completely oxidizing the glucose supplied? Why or why not?

THE HORMONAL CONTROL OF CARBOHYDRATE METABOLISM (SECTION 13.11)

13.61 ■ What cells produce insulin and glucagon in the body?

13.62 Describe the influence of glucagon and insulin on the following:

a. Blood sugar level

b. Glycogen formation

13.63 Why is it more important that epinephrine stimulates the breakdown of glycogen in the muscles than that it elevates blood sugar levels?

ADDITIONAL EXERCISES

13.64 You eat a dextrose tablet that contains 10.0 g of glucose for an energy boost. How many molecules of ATP can be produced with the complete oxidation of this glucose by muscle cells?

13.65 As an athlete's cells begin to go anaerobic, what happens to the rate of pyruvate oxidation by the citric acid cycle? What happens to the rate of glycolysis? Explain how each of these rate changes occurs and which enzymes are affected.

13.66 Lactate dehydrogenase catalyzes the following equilibrium reaction in both liver and muscle cells:

$$\text{pyruvate} + \text{NADH} + \text{H}^+ \leftrightarrows \text{lactate} + \text{NAD}^+$$

In terms of equilibrium, explain why lactate must be removed from muscle cells under anaerobic conditions and why lactate does not accumulate in the liver under these conditions. (HINT: Think of the Cori cycle.)

13.67 Identify all chiral carbon atoms in citrate and isocitrate.

```
     COO⁻                 COO⁻
      |                    |
     CH₂                  CH₂
      |                    |
  H—C—COO⁻             H—C—COO⁻
      |                    |
     CH₂               HO—C—H
      |                    |
     COO⁻                 COO⁻
   citrate             isocitrate
```

13.68 The pyruvate dehydrogenase enzyme complex becomes inactive when a specific serine residue in the enzyme complex is phosphorylated by ATP. Phosphorylation is the covalent addition of a phosphate group. Show how this reaction occurs and name the type of bond formed.

ALLIED HEALTH EXAM CONNECTION

Reprinted with permission from Nursing School and Allied Health Entrance Exams, COPYRIGHT 2005 Petersons.

13.69 In order to be absorbed into the blood, complex carbohydrates must first be hydrolyzed into _____ units, primarily _____.

13.70 Which of the following are true statements about glycolysis?

a. It requires 1 ATP molecule to activate a molecule of glucose.

b. It occurs in the mitochondria.

c. It does not require oxygen.

13.71 Compare the fate of pyruvate (a) in the body under aerobic conditions, (b) in the body under anaerobic conditions, and (c) in alcoholic fermentative microbes under anaerobic conditions.

13.72 Which of the following are true statements about the citric acid cycle?

a. It produces H_2O, CO_2, and NADH.

b. It occurs in the mitochondria.

c. It is the first stage of aerobic cellular respiration.

d. It is also called the Krebs cycle.

13.73 What is the final electron acceptor of the electron transport system? Write the reaction for the last step of the electron transport chain.

13.74 Glycogenesis and glycogen storage are important functions of which two cell types?

CHEMISTRY FOR THOUGHT

13.75 The citric acid cycle does not utilize oxygen, yet it cannot function without oxygen. Why?

13.76 Malonate ($^-OOC—CH_2—COO^-$) is an extremely effective competitive inhibitor of the citric acid cycle. Which enzyme do you think malonate inhibits? Explain.

13.77 A friend started to make wine by adding yeast to fresh grape juice and placing the mixture in a sealed bottle. Several days later the bottle exploded. Explain.

13.78 Two students are debating whether 30 or 32 molecules of ATP can be formed by the catabolism of one molecule of glucose. Who is correct?

13.79 Explain why monitoring blood lactate levels might be a useful technique in following the conditioning level of an Olympic runner.

13.80 You may have experienced a rush of epinephrine during a recent exam or athletic event. Are there other physiological responses to epinephrine in addition to higher blood sugar level?

13.81 In Figure 13.4, it is noted that making bread involves a fermentation process. Why, then, is the alcohol content of bread zero?

13.82 A friend has the habit of eating four or five candy bars per day as snacks. What might be the metabolic consequences of such a practice?

13.83 List some pros and cons of a low-carbohydrate diet.

13.84 Cyanide is believed to act as an inhibitor of the electron transport chain. Why is such an inhibitor so toxic?

13.85 Drinking milk always seems to make you slightly nauseous. What might be an explanation for this?

CHAPTER 14

Lipid and Amino Acid Metabolism

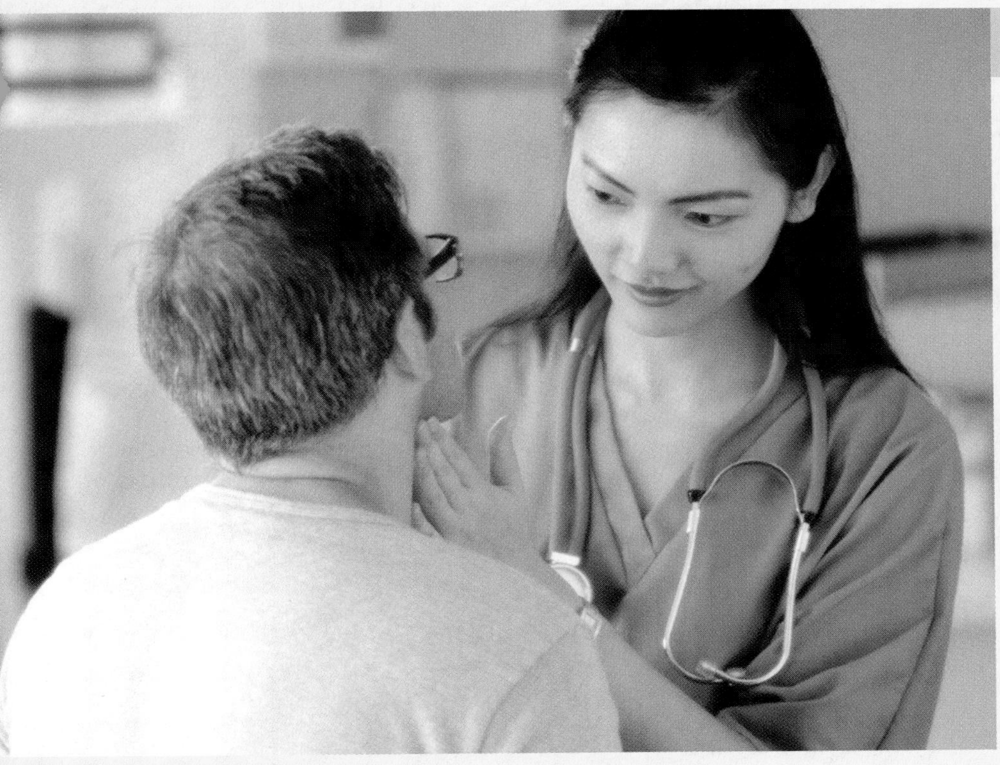

Besides the familiar work nurses do in hospitals, they also practice as community or public health nurses, school nurses, and occupational nurses. Nurse practitioners are trained to perform a number of procedures that are generally done only by physicians. Nurse practitioners who set up an independent practice in rural areas sometimes provide the only source of health information and immediate medical treatment for the people living there. In this chapter, you will encounter a number of special topics that deal with health issues such as blood lipid levels and the disease PKU.

LEARNING OBJECTIVES

When you have completed your study of this chapter, you should be able to:

1. Describe the digestion, absorption, and distribution of lipids in the body. (Section 14.1)
2. Explain what happens during fat mobilization. (Section 14.2)
3. Identify the metabolic pathway by which glycerol is catabolized. (Section 14.3)
4. Outline the steps of β-oxidation for fatty acids. (Section 14.4)
5. Determine the amount of ATP produced by the complete catabolism of a fatty acid. (Section 14.5)
6. Name the three ketone bodies and list the conditions that cause their overproduction. (Section 14.6)
7. Describe the pathway for fatty acid synthesis. (Section 14.7)
8. Describe the source and function of the body's amino acid pool. (Section 14.8)
9. Write equations for transamination and deamination reactions. (Section 14.9)
10. Explain the overall results of the urea cycle. (Section 14.9)
11. Describe how amino acids can be used for energy production, the synthesis of triglycerides, and gluconeogenesis. (Section 14.10)
12. State the relationship between intermediates of carbohydrate metabolism and the synthesis of nonessential amino acids. (Section 14.11)

ThomsonNOW Throughout the chapter this icon introduces resources on the ThomsonNOW website for this text. Sign in at **www.thomsonedu.com** to:

- Evaluate your knowledge of the material
- Take an exam prep quiz
- Identify areas you need to study with a **Personalized Learning Plan**.

Triglycerides (fats and oils) are important dietary sources of energy. Fat also functions as the major form of energy storage in humans. Fat, with a caloric value of 9 Calories/g, contains more than twice the energy per gram of glycogen or starch. Thus, fat is a more concentrated and efficient storage form of energy than carbohydrates. Because it is water insoluble, fat can be stored in much larger quantities in our bodies. For example, the limited carbohydrate reserves in the body are depleted after about 1 day without food, but stored fat can provide needed calories (but not other nutrients) in an individual for 30–40 days.

Amino acids perform three vital functions in the human body: They are the building blocks for the synthesis of proteins, they provide carbon and nitrogen atoms for the synthesis of other biomolecules, and they can be sources of energy.

In this chapter, we discuss the storage, degradation, and synthesis of lipids. We also study the degradation and synthesis of amino acids and the relationships between amino acid metabolism and the metabolism of carbohydrates and lipids.

14.1 Blood Lipids

LEARNING OBJECTIVE

1. Describe the digestion, absorption, and distribution of lipids in the body.

During digestion, triglycerides are hydrolyzed to glycerol, fatty acids, and monoglycerides (one fatty acid attached to glycerol). Phosphoglycerides are also hydrolyzed to their component substances. These smaller molecules, along with cholesterol, are absorbed into cells of the intestinal mucosa, where resynthesis of

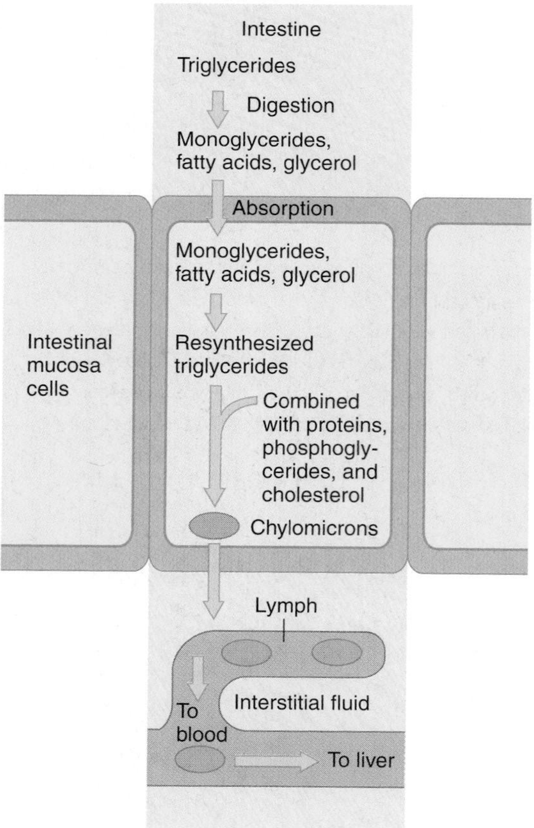

FIGURE 14.1 The digestion and absorption of triglycerides.

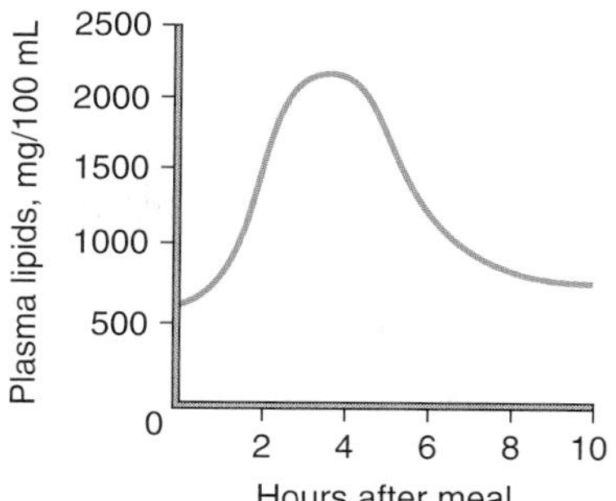

■ **FIGURE 14.2** The rise in blood lipids following a meal containing fat.

the triglycerides and phosphoglycerides occurs. For transport within the aqueous environment of the lymph and blood, these insoluble lipids are complexed with proteins to form lipoprotein aggregates called **chylomicrons**. The chylomicrons (which contain triglycerides, phosphoglycerides, and cholesterol) pass into the lymph system and then into the bloodstream. Chylomicrons are modified by the liver into smaller lipoprotein particles, the form in which most lipids are transported to various parts of the body by the bloodstream (see ■ Figure 14.1).

chylomicron
A lipoprotein aggregate found in the lymph and the bloodstream.

To some extent, the behavior of blood lipids parallels that of blood sugar. The concentration of both substances increases following a meal and returns to normal as a result of storage in fat depots and of oxidation to provide energy. As shown in ■ Figure 14.2, the concentration of plasma lipids usually begins to rise within 2 hours after a meal, reaches a peak in 4–6 hours, and then drops rather rapidly to a normal level.

One general method of classifying lipoproteins is by density. Because lipids are less dense than proteins, lipoprotein particles with a higher proportion of lipids are less dense than those with a lower proportion. Chylomicrons are the least dense because they have the highest ratio of lipid to protein. ■ Figure 14.3 is a schematic model of a low-density lipoprotein (LDL). The compositions of the four classes of lipoprotein are compared in ■ Figure 14.4. We see

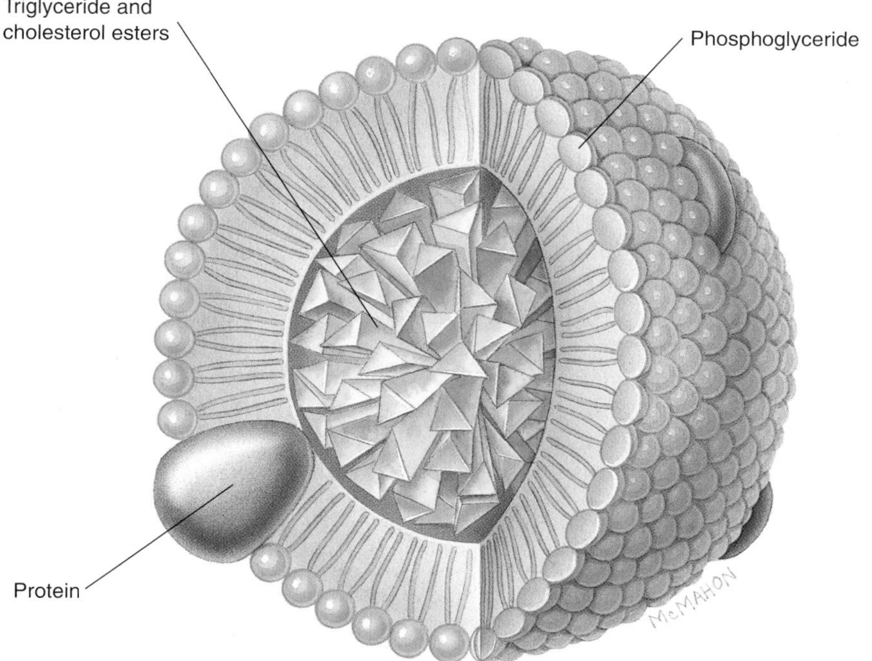

■ **FIGURE 14.3** A schematic model of low-density lipoprotein (LDL).

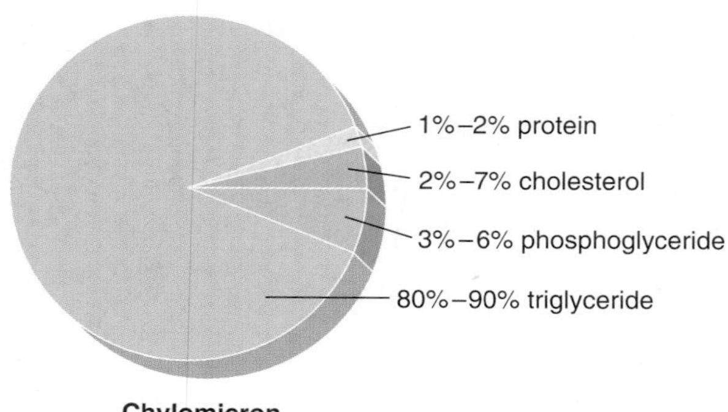

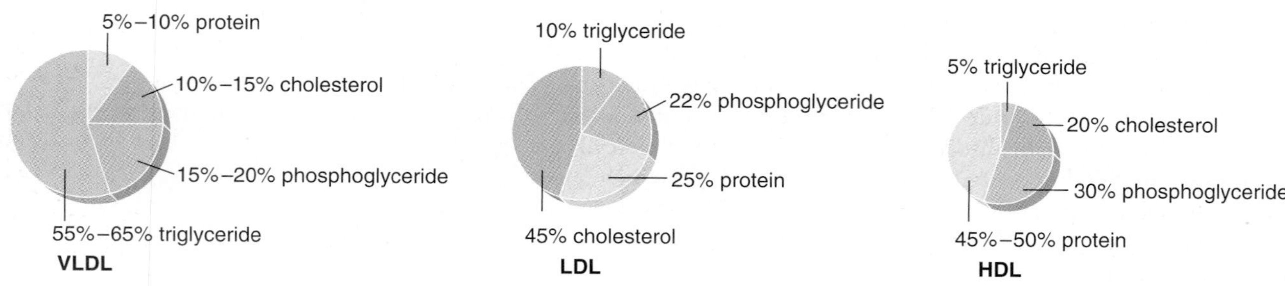

FIGURE 14.4 Relative amounts of cholesterol, protein, triglyceride, and phosphoglyceride in the four classes of lipoproteins.

from Figure 14.4 that the protein content of chylomicrons is significantly lower than that of other plasma lipoproteins, and that this low percentage results in a density that is less than 0.95 g/mL. Notice that the so-called high-density lipoproteins (HDL) have a protein mass that is 45–50% of the total mass of the particles. The densities of HDL particles have correspondingly high values, ranging from 1.06–1.21 g/mL.

14.2 Fat Mobilization

LEARNING OBJECTIVE

2. Explain what happens during fat mobilization.

Carbohydrates from dietary sources and glycogen catabolism are the compounds utilized preferentially for energy production by some tissues. The brain and active skeletal muscles are examples of such tissues. But as we have seen, the body's stores of glycogen are depleted after only a few hours of fasting. When this happens, fatty acids stored in triglycerides are called on as energy sources. Even when glycogen supplies are adequate, many body cells, such as those of resting muscle and liver, utilize energy derived from the breakdown of fatty acids. This helps conserve the body's glycogen stores and glucose for use by brain cells and red blood cells. Red blood cells cannot oxidize fatty acids because red blood cells have no mitochondria, the site of fatty acid oxidation. Brain cells are bathed in cerebrospinal fluid and do not obtain nutrients directly from the blood (the blood-brain barrier). Glucose and many other substances can cross the barrier into the cerebrospinal fluid, but fatty acids cannot.

adipose tissue
A kind of connective tissue where triglycerides are stored.

When body cells need fatty acids for energy, the endocrine system is stimulated to produce several hormones, including epinephrine, which interact with **adipose tissue.** Earlier we saw that epinephrine stimulates the conversion of glycogen to glucose. In the fat cells of adipose tissue, epinephrine stimulates the hydrolysis

OVER THE COUNTER 14.1 Cholesterol-Lowering Drugs

One of many kinds of lipids found in the blood, cholesterol is essential for several of the body's processes. Cholesterol is involved in the formation of cell membranes, the insulation of nerves, the synthesis of several hormones, and the digestion of food.

Cholesterol and other lipids are insoluble in water. In order for them to be transported in the water-based blood, they are combined with proteins to form lipoproteins, which have a higher water solubility than the lipids. Lipoproteins are classified according to their densities, and two that involve cholesterol are called high-density lipoprotein (HDL) and low-density lipoprotein (LDL). Low-density lipoproteins carry more cholesterol in the bloodstream than any other type of lipoprotein. It has been determined that a high level of LDL in the blood increases the risk of coronary heart disease. The level of LDL in the blood can often be reduced by changes in lifestyle, such as increasing the amount of exercise, reducing the intake of dietary fat, stopping smoking, and losing weight.

When lifestyle changes fail to reduce the level of blood LDL enough, several types of cholesterol-lowering drugs are available. Niacin is the oldest medication and one of the few OTC treatments for high cholesterol levels. It is a useful lipid-modifying drug because it (1) decreases LDL cholesterol, total cholesterol, and triglycerides, and (2) raises HDL cholesterol. The chief constraints against niacin use have been flushing, gastrointestinal discomfort, and metabolic effects including liver toxicity.

The resin drugs have been in use for about 20 years and include cholestyramine (Questran®) and colestipol (Colestid®). These drugs bind with bile acids in the digestive tract. Bile acids are digestive aids that are synthesized from cholesterol in the liver. When they are tied up and removed from the digestive processes, the liver synthesizes more. Because cholesterol is used in the synthesis, less is available to be released into the blood to form LDL.

A second type of cholesterol-lowering drug acts by reducing the production of triglycerides, which are involved in the formation of LDL. With reduced triglyceride levels, less LDL is formed and less cholesterol circulates in the blood. The drug gemfibrozil (Lopid®), or large doses of niacin, a vitamin, function this way to lower blood cholesterol levels.

A third type of cholesterol-lowering drug is called statins. Statins function in the liver to block the synthesis of cholesterol. As a result, the cholesterol level in liver cells is reduced, thereby causing the cells to remove cholesterol from the circulating blood, thus lowering the blood cholesterol level. These drugs can also help the body reabsorb cholesterol from plaque that has formed in blood vessels. This helps to slowly break down the plaque and unplug the vessels.

Statin drugs can reduce LDL cholesterol by up to 40% depending on the dosage, and they are the only type of lipid-lowering drug that has proven to reduce the risk of death from cardiovascular disease. They have also been shown to reduce the risk of a second heart attack. Examples of the statin drugs are fluvastatin (Lescol®), lovastatin (Mevacor®), simvastatin (Zocor®), pravastatin (Pravachol®), and atorvastatin (Lipitor®).

lovastatin

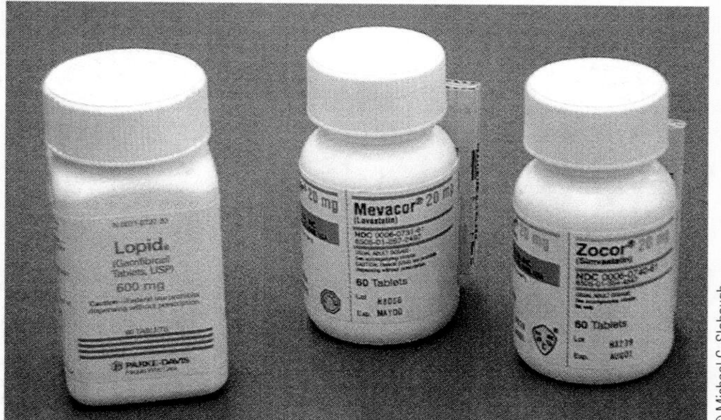

A few of the cholesterol-lowering drugs available by prescription.

of triglycerides to fatty acids and glycerol, which enter the bloodstream. This process is called **fat mobilization**. In the blood, mobilized fatty acids form a lipoprotein with the plasma protein called serum albumin. In this form, the fatty acids are transported to the tissue cells that need them. The glycerol produced by the hydrolysis of the triglyceride is water soluble, so it dissolves in the blood and is also transported to cells that need it.

fat mobilization
The hydrolysis of stored triglycerides, followed by the entry of fatty acids and glycerol into the bloodstream.

14.3 Glycerol Metabolism

LEARNING OBJECTIVE

3. Identify the metabolic pathway by which glycerol is catabolized.

The glycerol hydrolyzed from fats can provide energy to cells. Glycerol is converted by cells to dihydroxyacetone phosphate in two steps:

$$\underset{\text{glycerol}}{\begin{array}{c} H_2C-OH \\ | \\ HC-OH \\ | \\ H_2C-OH \end{array}} \xrightarrow[\text{ATP} \rightarrow \text{ADP}]{} \underset{\text{glycerol 3-phosphate}}{\begin{array}{c} H_2C-OH \\ | \\ HC-OH \\ | \\ H_2C-OPO_3^{2-} \end{array}} \xrightarrow[\text{NAD}^+ \rightarrow \text{NADH} + \text{H}^+]{} \underset{\text{dihydroxyacetone phosphate}}{\begin{array}{c} H_2C-OH \\ | \\ C=O \\ | \\ H_2C-OPO_3^{2-} \end{array}} \longrightarrow \text{Glycolysis} \quad (14.1)$$

Dihydroxyacetone phosphate is one of the chemical intermediates of glycolysis (Section 13.3). Thus, glycerol, by entering glycolysis, can be converted to pyruvate and contribute to cellular energy production. The pyruvate formed from glycerol can also be converted to glucose through gluconeogenesis (Section 13.10).

14.4 The Oxidation of Fatty Acids

LEARNING OBJECTIVE

4. Outline the steps of β-oxidation for fatty acids.

ThomsonNOW™ Go to Chemistry Interactive to explore **beta-oxidation of fatty acids.**

Fatty acids that enter tissue cells cannot be oxidized to produce energy until they pass through the mitochondrial membrane. This cannot occur until the fatty acid is converted into fatty acyl CoA by reaction with coenzyme A. Fatty acyl refers to the $R-\overset{\overset{\large O}{\|}}{C}-$ portion of the molecule.

$$\underset{\text{fatty acid}}{R-\overset{\overset{\large O}{\|}}{C}-OH} + HS-CoA \xrightarrow[\text{ATP} \rightarrow \text{AMP} + \text{PP}_i]{} \underset{\text{fatty acyl CoA}}{R-\overset{\overset{\large O}{\|}}{C}-S-CoA} + H_2O \quad (14.2)$$

This reaction, which is catalyzed by acyl CoA synthetase, is referred to as activation of the fatty acid because the fatty acyl CoA is a high-energy compound. The energy needed for its synthesis is provided by the hydrolysis of ATP to AMP and PP_i and the subsequent hydrolysis of PP_i to $2P_i$. The process of forming fatty acyl CoA molecules demonstrates how the energy stored in the phosphate bonds of ATP can be used to drive chemical processes that normally could not occur. The cost to the cell of activating one molecule of fatty acid is the loss of two high-energy bonds in an ATP molecule.

β-oxidation process
A pathway in which fatty acids are broken down into molecules of acetyl CoA.

The fatty acyl CoA molecules that enter mitochondria are then degraded in a catabolic process called **β-oxidation**. During β-oxidation, the second (or beta) carbon down the chain from the carbonyl of the fatty acyl CoA molecule is oxidized to a ketone:

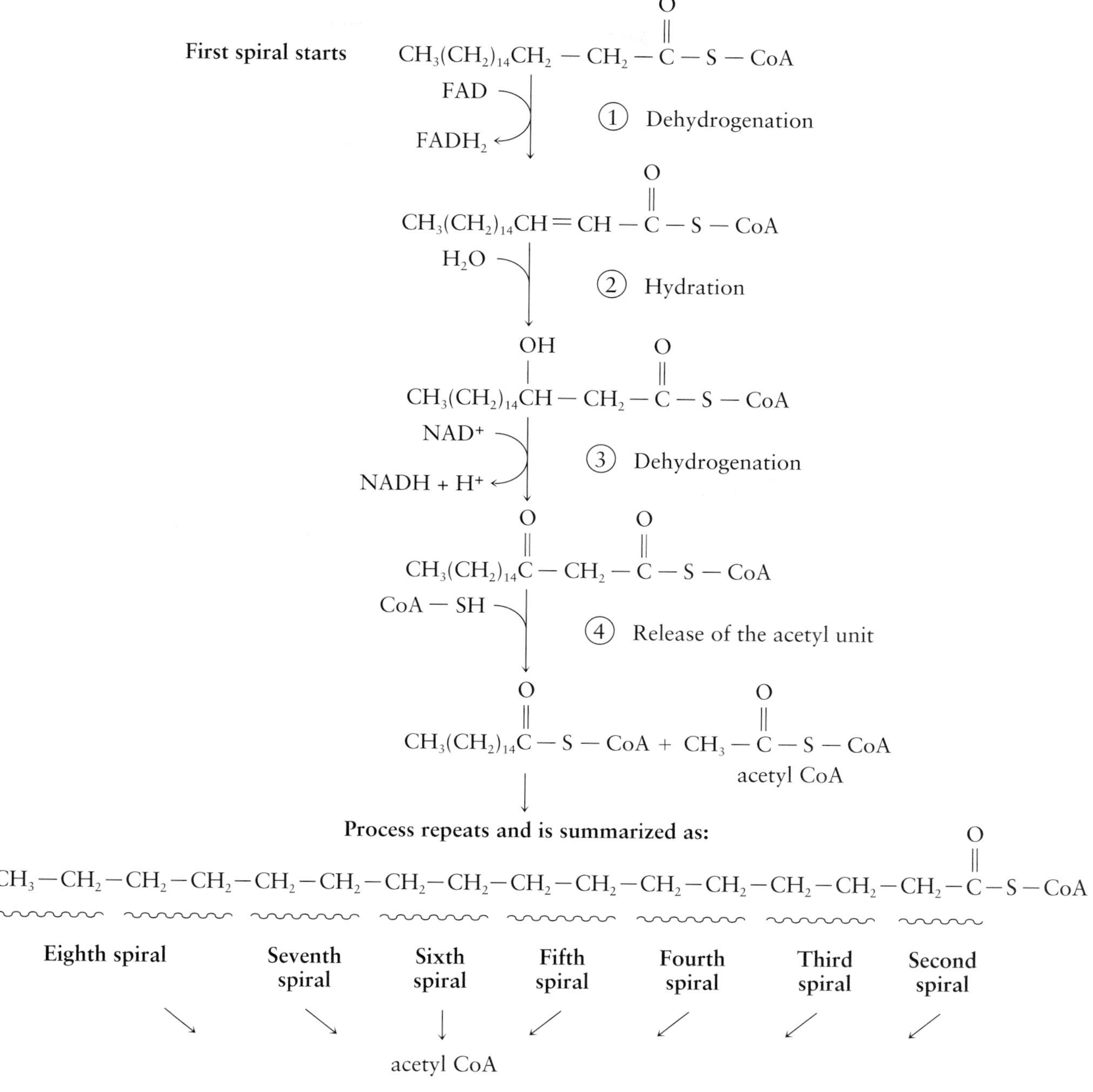

FIGURE 14.5 The fatty acid spiral: β-oxidation of stearic acid.

$$\underset{\text{fatty acyl CoA}}{R-\overset{\beta}{C}H_2-\overset{\alpha}{C}H_2-\overset{O}{\overset{\|}{C}}-S-CoA} \xrightarrow{\text{Oxidation}} R-\overset{O}{\overset{\|}{\underset{\beta}{C}}}-CH_2-\overset{O}{\overset{\|}{C}}-S-CoA \qquad (14.3)$$

Four reactions are involved in oxidizing the β-carbon to a ketone. We will use the fatty acyl CoA formed from one 18-carbon acid, stearic acid, as an example of the reactions. We will also keep track of any NADH or $FADH_2$ produced in the reactions because these molecules can enter the respiration process (electron transport chain). The reactions involved in β-oxidation are shown in ■ Figure 14.5.

In the final step of β-oxidation (Step 4), the chain is broken between the α- and β-carbons by reaction with coenzyme A. As a result, a new fatty acyl CoA is formed with a chain shortened by two carbon atoms, and a unit of acetyl CoA is released:

$$\underset{\substack{\uparrow\\ \beta\text{-carbon}\\ 18\text{ carbons}}}{CH_3(CH_2)_{14}-\overset{\overset{\text{O}}{\|}}{C}}-\overset{\alpha\text{-carbon}\downarrow}{CH_2}-\overset{\overset{\text{O}}{\|}}{C}-S-CoA + H-S-CoA \longrightarrow \underset{\substack{\text{new fatty acyl CoA}\\ 16\text{ carbons}}}{CH_3(CH_2)_{14}-\overset{\overset{\text{O}}{\|}}{C}-S-CoA} + \underset{\substack{\text{acetyl CoA}\\ 2\text{ carbons}}}{CH_3-\overset{\overset{\text{O}}{\|}}{C}-S-CoA} \tag{14.4}$$

The new fatty acyl compound enters the β-oxidation process at Step 1, and the sequence is repeated until the fatty acyl CoA is completely degraded to acetyl CoA (Figure 14.5). The fatty acyl CoA molecule that starts each run through the β-oxidation process is two carbons shorter than the one going through the previous run. For this reason, the β-oxidation pathway for fatty acid degradation to acetyl CoA is often called the fatty acid spiral (rather than cycle).

Every run through the spiral produces one molecule each of acetyl CoA, NADH, and $FADH_2$ until the fatty acyl CoA is only four carbons long:

$$\underset{\substack{\text{fatty acyl CoA}\\ \text{entering the spiral}}}{R-CH_2CH_2-\overset{\overset{\text{O}}{\|}}{C}-S-CoA} + NAD^+ + FAD + H_2O + CoA-SH \longrightarrow$$

$$\underset{\substack{\text{fatty acyl CoA}\\ 2\text{ carbons shorter}}}{R-\overset{\overset{\text{O}}{\|}}{C}-S-CoA} + \underset{\text{acetyl CoA}}{CH_3\overset{\overset{\text{O}}{\|}}{C}-S-CoA} + NADH + H^+ + FADH_2 \tag{14.5}$$

In the last spiral, the four-carbon chain of butyryl CoA passes through the β-oxidation sequence, and it produces one molecule of $FADH_2$, one molecule of NADH, and two molecules of acetyl CoA:

$$\underset{\text{butyryl CoA}}{CH_3-CH_2-CH_2-\overset{\overset{\text{O}}{\|}}{C}-S-CoA} + NAD^+ + H_2O + FAD + CoA-SH \longrightarrow$$

$$\underset{\text{2 acetyl CoA molecules}}{2CH_3-\overset{\overset{\text{O}}{\|}}{C}-S-CoA} + NADH + H^+ + FADH_2 \tag{14.6}$$

In other words, the complete conversion of a fatty acyl CoA to two-carbon fragments of acetyl CoA always produces one more molecule of acetyl CoA than of $FADH_2$ or NADH. Thus, the breakdown of stearic acid requires eight passes through the β-oxidation sequence and produces nine molecules of acetyl CoA, but only eight molecules of $FADH_2$ and eight molecules of NADH.

Net reaction:

$$CH_3(CH_2)_{16}\overset{\overset{\text{O}}{\|}}{C}-S-CoA + 8FAD + 8NAD^+ + 8H_2O + 8CoA-SH \longrightarrow$$

$$9CH_3\overset{\overset{\text{O}}{\|}}{C}-S-CoA + 8FADH_2 + 8NADH + 8H^+ \tag{14.7}$$

Learning Check 14.1 Capric acid (ten carbon atoms) is converted to acetyl CoA through β-oxidation. What are the yields of the following?

a. acetyl CoA b. $FADH_2$ c. NADH

14.5 The Energy from Fatty Acids

LEARNING OBJECTIVE

5. Determine the amount of ATP produced by the complete catabolism of a fatty acid.

To compare the energy yield from fatty acids with that from glucose, let's continue our discussion with stearic acid, a typical and abundant fatty acid.

We start with the activation of stearic acid by reacting it with coenzyme A to form stearoyl CoA. The energy needed to cause this reaction to occur comes from the hydrolysis of ATP to AMP and PP_i and the subsequent hydrolysis of PP_i to $2P_i$. This is equivalent to hydrolyzing two molecules of ATP to ADP.

As a stearoyl CoA molecule (18 carbons) passes through the β-oxidation spiral, 9 acetyl CoA, 8 $FADH_2$, and 8 NADH molecules are produced. Acetyl CoA produced in the fatty acid spiral can enter the citric acid cycle (followed by the electron transport chain), where each molecule of acetyl CoA results in the production of 10 ATP molecules. In addition, when the $FADH_2$ and NADH molecules enter the electron transport chain, each $FADH_2$ yields 1.5 ATP molecules, and each NADH yields 2.5 molecules. The calculations are summarized in Table 14.1, which shows a total of 120 molecules of ATP formed from the 18-carbon fatty acid.

It is instructive to compare the energy yield from the complete oxidation of fatty acids with that obtained from an equivalent amount of glucose because both are important constituents of the diet. In Section 13.8, we saw that the oxidation of a single glucose molecule produces 32 ATP molecules. The complete oxidation of three 6-carbon glucose molecules (18 carbons) yields 3×32, or 96, ATP molecules. Thus, on the basis of equal numbers of carbons, lipids are nearly 25% more efficient than carbohydrates as energy-storage systems.

On a mass basis, the difference is even more striking. One mole of stearic acid weighs 284 g, and, as we have seen, it yields 120 mol of ATP on complete degradation. Three moles of glucose weigh 540 g and yield 96 mol of ATP. On this basis, 284 g of glucose would generate 50 mol of ATP. On an equal-mass basis, lipids contain more than twice the energy of carbohydrates. The reason for this difference is that fatty acids are a more reduced form of fuel, with a number of $—CH_2—$ groups as plentiful sources of H atoms for the energy-yielding oxidation process. Glucose, in contrast, is already partially oxidized, with an oxygen atom located on each carbon atom.

TABLE 14.1 Energy produced by the oxidation of stearic acid

Oxidation product or step	ATP/unit	Total ATP
Activation step		−2
9 acetyl CoA	10	90
8 $FADH_2$	1.5	12
8 $(NADH + H^+)$	2.5	20
Total ATP from the 18-carbon fatty acid		120

Learning Check 14.2 Calculate the number of molecules of ATP produced by the complete oxidation of palmitic acid (16 carbon atoms) to carbon dioxide and water.

14.6 Ketone Bodies

LEARNING OBJECTIVE

6. Name the three ketone bodies and list the conditions that cause their overproduction.

It is generally agreed by nutrition scientists that a diet balanced in carbohydrates and fats as energy sources is best for good health. When such a balanced diet is followed, most of the acetyl CoA produced by fatty acid metabolism is processed through the citric acid cycle.

Under certain conditions, the balance between carbohydrate and fatty acid metabolism is lost. During a fast, for example, fatty acids from stored fats become the body's primary energy source. With a minimum amount of cellular glucose available, the level of glycolysis decreases, and a reduced amount of oxaloacetate is synthesized. In addition, oxaloacetate is used for gluconeogenesis to a greater-than-normal extent as the cells react to make their own glucose. The lack of oxaloacetate reduces the activity of the citric acid cycle. As a result of these events, more acetyl CoA is produced by fatty acid oxidation than can be processed through the citric acid cycle. As the concentration of acetyl CoA builds up, the excess is converted within the liver to three substances called **ketone bodies**—acetoacetate, β-hydroxybutyrate, and acetone.

ketone bodies
Three compounds—acetoacetate, β-hydroxybutyrate, and acetone—formed from acetyl CoA.

$$CH_3-\overset{\overset{O}{\|}}{C}-CH_2-\overset{\overset{O}{\|}}{C}-O^- \qquad CH_3-\overset{\overset{OH}{|}}{CH}-CH_2-\overset{\overset{O}{\|}}{C}-O^- \qquad CH_3-\overset{\overset{O}{\|}}{C}-CH_3$$

acetoacetate — β-hydroxybutyrate (not a ketone) — acetone

The ketone bodies are carried by the blood to body tissues, mainly the brain, heart, and skeletal muscle, where they may be oxidized to meet energy needs (see ■ Figure 14.6). You can see from their structural formulas that β-hydroxybutyrate does

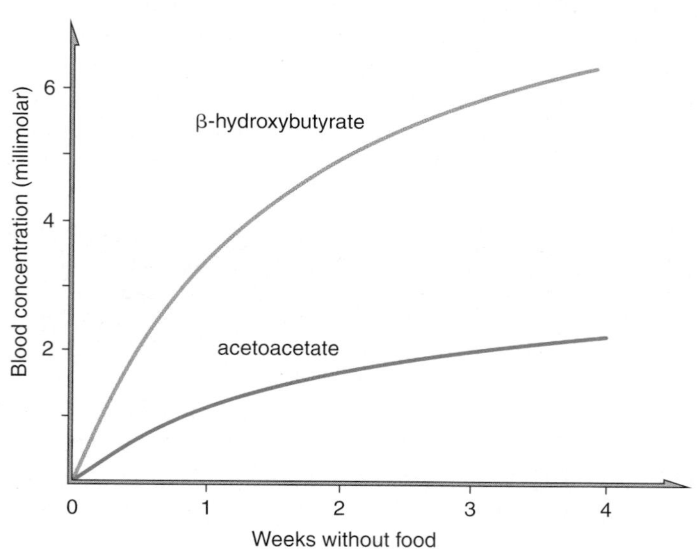

■ **FIGURE 14.6** The effect of starvation on blood levels of β-hydroxybutyrate and acetoacetate.

STUDY SKILLS 14.1 Key Numbers for ATP Calculations

The calculation of the energy yield in terms of ATP molecules from the metabolism of a fatty acid is a useful exercise because it may help you better understand this metabolic process. The three key substances to account for in such calculations are acetyl CoA, NADH, and $FADH_2$. Because acetyl CoA consists of two carbon atoms transported by coenzyme A, the number of acetyl CoA molecules formed is determined by the following equation:

$$\text{acetyl CoA} = \frac{\text{number of fatty acid carbons}}{2}$$

For example, a 10-carbon fatty acid would produce 10/2, or 5, acetyl CoA molecules.

Every trip through the fatty acid spiral produces one $NADH + H^+$ molecule and one $FADH_2$ molecule. Thus,

$$\text{number of trips} = \text{molecules NADH} = \text{molecules } FADH_2$$

Furthermore, the number of trips equals one less than the number of acetyl CoA molecules produced (because the last trip of the spiral produces two molecules of acetyl CoA). Thus, for the 10-carbon fatty acid above, 5 acetyl CoA molecules are produced during four trips through the spiral. Also, the four trips produce four $NADH + H^+$ molecules and four $FADH_2$ molecules.

Finally, remember the multipliers: Every acetyl CoA gives rise to 10 ATPs, every NADH produces 2.5 ATPs, and every $FADH_2$ gives 1.5 ATPs. The summary for the 10-carbon fatty acid is

$$\begin{aligned}
(5 \text{ acetyl CoA}) \times 10 &= 50 \text{ ATP}\\
(4 \text{ NADH} + H^+) \times 2.5 &= 10 \text{ ATP}\\
(4\ FADH_2) \times 1.5 &= 6 \text{ ATP}\\
\text{Activation step} &= -2 \text{ ATP}\\
\text{Total ATP} &= 64 \text{ ATP}
\end{aligned}$$

not actually contain a ketone group, but it is nevertheless included in the designation because it is always produced along with the other two. Under normal conditions, the concentration of ketone bodies in the blood averages a very low 0.5 mg/100 mL.

Diabetes mellitus, like a long fast or starvation, also produces an imbalance in carbohydrate and lipid metabolism. Even though blood glucose reaches hyperglycemic levels, a deficiency of insulin prevents the glucose from entering tissue cells in sufficient amounts to meet cellular energy needs. The resulting increase in fatty acid metabolism leads to excessive production of acetyl CoA and a substantial increase in the level of ketone bodies in the blood of untreated diabetics. A concentration higher than about 20 mg/100 mL of blood is called **ketonemia** ("ketones in the blood"). At a level of about 70 mg/100 mL of blood, the renal threshold for ketone bodies is exceeded, and ketone bodies are excreted in the urine. This condition is called **ketonuria** ("ketones in the urine"). A routine check of urine for ketone bodies is useful in the diagnosis of diabetes. Occasionally, the concentration of acetone in the blood reaches levels that cause it to be expelled through the lungs, and the odor of acetone can be detected on the breath. When ketonemia, ketonuria, and **acetone breath** exist simultaneously, the condition is called **ketosis.**

Two of the ketone bodies are acids, and their accumulation in the blood as ketosis worsens results in a condition of **acidosis** called **ketoacidosis** because it is caused by ketone bodies in the blood. If ketoacidosis is not controlled, the person becomes severely dehydrated because the kidneys excrete excessive amounts of water in response to low blood pH. Prolonged ketoacidosis leads to general debilitation, coma, and even death.

Patients suffering from diabetes-related ketosis are usually given insulin as a first step in their treatment. The insulin restores normal glucose metabolism and reduces the rate of formation of ketone bodies. If the patient is also suffering severe dehydration, the return to fluid and acid–base balance is speeded up by the intravenous administration of solutions containing sodium bicarbonate.

ketonemia
An elevated level of ketone bodies in the blood.

ketonuria
The presence of ketone bodies in the urine.

acetone breath
A condition in which acetone can be detected in the breath.

ketosis
A condition in which ketonemia, ketonuria, and acetone breath exist together.

acidosis
A low blood pH.

ketoacidosis
A low blood pH due to elevated levels of ketone bodies.

14.7 Fatty Acid Synthesis

LEARNING OBJECTIVE

7. Describe the pathway for fatty acid synthesis.

When organisms (including humans) take in more nutrients than are needed for energy requirements, the excesses are not excreted; they are converted first to fatty acids and then to body fat. Most of the conversion reactions take place in the liver, adipose tissue, and mammary glands. The mammary glands become especially active in the process during lactation.

As is true in the opposing processes of glycolysis and gluconeogenesis, the pathways for the opposing processes of fatty acid degradation and fatty acid synthesis are not simply the reverse of each other. The processes are separate from each other, utilize different enzyme systems, and are even located in different cellular compartments. Degradation by the β-oxidation pathway takes place in cellular mitochondria, whereas biosynthesis of fatty acids occurs in the cytoplasm.

However, one aspect of fatty acid synthesis and fatty acid degradation is the same. Both processes take place in units of two carbon atoms as a result of involving acetyl CoA. Thus, fatty acid chains are built up two carbons at a time during synthesis and broken down two carbons at a time during β-oxidation.

Acetyl CoA is generated inside mitochondria and so must be transported into the cytoplasm if it is to be used to synthesize fatty acids. This transport is done by first reacting acetyl CoA with oxaloacetate (the first step of the citric acid cycle). The product is citrate:

$$\text{acetyl CoA} + \text{oxaloacetate} + H_2O \rightarrow \text{citrate} + \text{CoA—SH} \qquad (14.8)$$

Mitochondrial membranes contain a citrate transport system that enables excess citrate to move out of the mitochondria into the cytoplasm. Once in the cytoplasm, the citrate reacts to regenerate acetyl CoA and oxaloacetate.

Fatty acid synthesis occurs by way of a rather complex series of reactions catalyzed by a multienzyme complex called the *fatty acid synthetase system*. This system is made up of six enzymes and an additional protein, acyl carrier protein (ACP), to which all intermediates are attached.

The summary equation for the synthesis of palmitic acid from acetyl CoA demonstrates that a great deal of energy, 7 ATP + 14 NADPH (a phosphate derivative of NADH), is required:

$$\underset{\text{acetyl CoA}}{8CH_3-\overset{\overset{\large O}{\|}}{C}-S-CoA} + 14NADPH + 13H^+ + 7ATP \longrightarrow$$

$$\underset{\text{palmitate}}{CH_3(CH_2)_{14}COO^-} + 8CoA-SH + 6H_2O + 14NADP^+ + 7ADP + 7P_i \qquad (14.9)$$

This large input of energy is stored in the synthesized fatty acids and is one of the reasons it is so difficult to lose excess weight due to fat. After synthesis, the fatty acids are incorporated into triglycerides and stored in the form of fat in adipose tissues.

The liver is the most important organ involved in fatty acid and triglyceride synthesis. It is able to modify body fats by lengthening or shortening and saturating or unsaturating the fatty acid chains. The only fatty acids that cannot be synthesized by the body are those that are polyunsaturated. However, linoleic acid (two double bonds) and linolenic acid (three double bonds) from the diet can be converted to other polyunsaturated fatty acids. This utilization of linoleic and linolenic acids as sources of other polyunsaturated fatty acids is the basis for classifying them as essential fatty acids for humans (Section 8.2).

The human body can convert glucose to fatty acids, but it cannot convert fatty acids to glucose. Our cells contain no enzyme that can catalyze the conversion of acetyl CoA to pyruvate, a compound required for gluconeogenesis (Figure 13.13). However, plants and some bacteria do possess such enzymes and convert fats to carbohydrates as part of their normal metabolism.

14.8 Amino Acid Metabolism

LEARNING OBJECTIVE

8. Describe the source and function of the body's amino acid pool.

In terms of amount used, the most important function of amino acids is to provide building blocks for the synthesis of proteins in the body. It is estimated that about 75% of amino acid utilization in a normal, healthy adult is for this function. The maintenance of body proteins must occur constantly in the body because tissue proteins break down regularly from normal wear and tear, as well as from diseases and injuries. The amino acids used in this maintenance come from proteins that are eaten and hydrolyzed during digestion, from the body's own degraded tissue, and from the synthesis in the liver of certain amino acids. The amino acids from these three sources constitute what is called the **amino acid pool** of the body. This cellular supply of amino acids is constantly being restocked to allow the synthesis of new proteins and other necessary metabolic processes to take place as needed.

ThomsonNOW™ Go to Chemistry Interactive to explore **protein catabolism.**

amino acid pool The total supply of amino acids in the body.

The dynamic process in which body proteins are continuously hydrolyzed and resynthesized is called **protein turnover.** The turnover rate, or life expectancy, of body proteins is a measure of how fast the proteins are broken down and resynthesized. The turnover rate is usually expressed as a half-life. The use of radioactive amino acids has enabled researchers to estimate half-lives by measuring the exchange rate between body proteins and the amino acid pool. For example, the half-life of liver proteins is about 10 days. This means that over a 10-day period, half the proteins in the liver are hydrolyzed to amino acids and replaced by equivalent proteins. Plasma proteins also have a half-life of about 10 days, hemoglobin about 120 days, and muscle protein about 180 days. The half-life of collagen, a protein of connective tissue, is considerably longer—some estimates are as high as 1000 days. Other proteins, particularly enzyme and polypeptide hormones, have much shorter half-lives of only a few minutes. Once it is released from the pancreas, insulin has a half-life estimated to be only 7–10 minutes.

protein turnover The continuing process in which body proteins are hydrolyzed and resynthesized.

Thus, the stability of body protein is more apparent than real because, as we have seen, synthesized proteins do not become a permanent part of the body structure or function. The frequent turnover of proteins is advantageous, for it allows the body to continually renew important molecules and to respond quickly to its own changing needs.

In addition to tissue protein synthesis, there is a constant draw on the amino acid pool for the synthesis of other nitrogen-containing biomolecules. These molecules include the purine and pyrimidine bases of nucleic acids; heme structures for hemoglobin and myoglobin; choline and ethanolamine, which are the building blocks of phospholipids; and neurotransmitters such as acetylcholine and dopamine. Like proteins, these compounds are also constantly being degraded and replaced. ■ Table 14.2 lists some of the important nitrogen-containing compounds that are synthesized from amino acids.

Unlike carbohydrates and fatty acids, amino acids in excess of immediate body requirements cannot be stored for later use. They are degraded and, depending on the organism, their nitrogen atoms are converted to either ammonium ions, urea, or uric acid and excreted. Their carbon skeletons are converted to pyruvate, acetyl CoA, or one of the intermediates of the citric acid cycle and

TABLE 14.2 Compounds derived from amino acids

Amino acid	Product	Function
Tyrosine	dopamine	Neurotransmitter
	norepinephrine	Neurotransmitter, hormone
	epinephrine	Hormone
	melanin	Skin pigmentation
Tryptophan	serotonin	Neurotransmitter
Histidine	histamine	Involved in allergic reactions
Serine	ethanolamine	Required in cephalin synthesis
Cysteine	taurine	A compound of bile salts

used for energy production, the synthesis of glucose through gluconeogenesis (Section 13.10), or conversion to triglycerides. The various pathways involved in amino acid metabolism are summarized in ■ Figure 14.7.

14.9 Amino Acid Catabolism: The Fate of the Nitrogen Atoms

LEARNING OBJECTIVES

9. Write equations for transamination and deamination reactions.
10. Explain the overall results of the urea cycle.

As previously noted, excess amino acids cannot be stored in the body but are catabolized for energy production, and the nitrogen is either excreted or used to synthesize other nitrogen-containing compounds. In this section, we will study the chemical pathways involved with the nitrogen. The chemical fate of the carbon skeleton of the amino acids is discussed in Section 14.10.

In essence, there are three stages in nitrogen catabolism (see ■ Figure 14.8). These processes (which occur in the liver) are transamination, deamination, and urea formation.

Stage 1: Transamination

In the tissues, amino groups freely move from one amino acid to another under the catalytic influence of enzymes called aminotransferases, or more commonly,

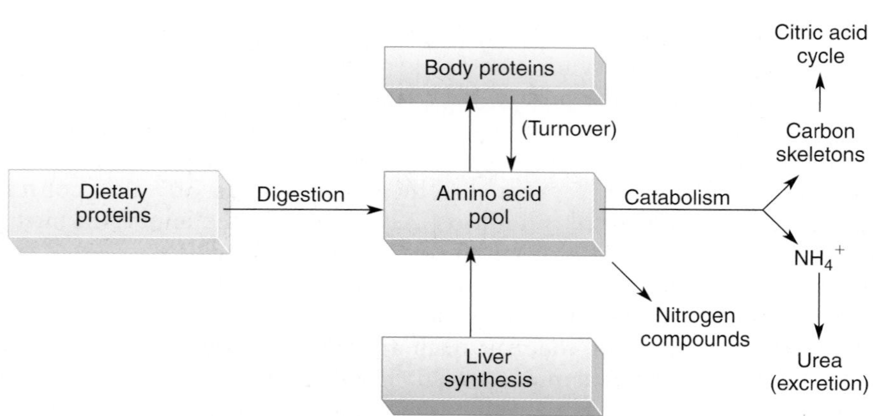

■ **FIGURE 14.7** The major metabolic pathways of amino acids.

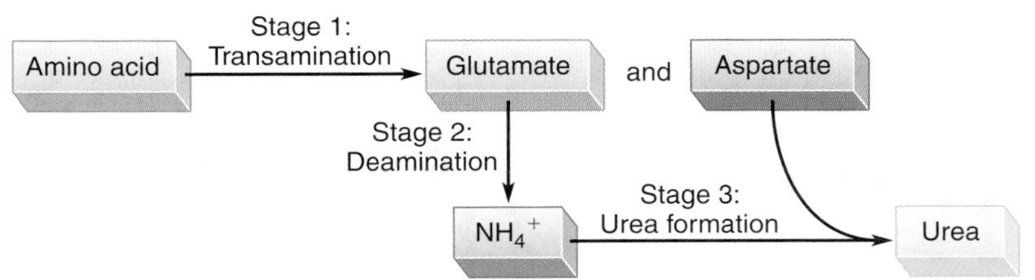

FIGURE 14.8 An overview of the catabolic fate of nitrogen atoms in amino acids.

transaminases. A key reaction for amino acids undergoing catabolism is a **transamination** involving the transfer of amino groups to α-ketoglutarate. The carbon skeleton of the amino acid remains behind as an α-keto acid:

transaminase
An enzyme that catalyzes the transfer of an amino group.

transamination
The enzyme-catalyzed transfer of an amino group to a keto acid.

General reaction:

$$\underset{\substack{\text{a donor}\\ \alpha\text{-amino acid}}}{R-\overset{\overset{\displaystyle NH_3^+}{|}}{C}H-COO^-} + \underset{\substack{\alpha\text{-ketoglutarate}\\ \text{an }\alpha\text{-keto acid}}}{{}^-OOC-CH_2CH_2-\overset{\overset{\displaystyle O}{\|}}{C}-COO^-} \underset{}{\overset{\text{transaminase}}{\rightleftharpoons}}$$

$$\underset{\substack{\text{a new}\\ \alpha\text{-keto acid}}}{R-\overset{\overset{\displaystyle O}{\|}}{C}-COO^-} + \underset{\substack{\text{glutamate}\\ \text{the new amino acid}}}{{}^-OOC-CH_2CH_2-\overset{\overset{\displaystyle NH_3^+}{|}}{C}H-COO^-} \quad (14.10)$$

Specific example:

$$\underset{\text{alanine}}{CH_3-\overset{\overset{\displaystyle NH_3^+}{|}}{C}H-COO^-} + \underset{\alpha\text{-ketoglutarate}}{{}^-OOC-CH_2CH_2-\overset{\overset{\displaystyle O}{\|}}{C}-COO^-} \overset{\substack{\text{alanine}\\ \text{transaminase}}}{\rightleftharpoons}$$

$$\underset{\text{pyruvate}}{CH_3-\overset{\overset{\displaystyle O}{\|}}{C}-COO^-} + \underset{\text{glutamate}}{{}^-OOC-CH_2CH_2-\overset{\overset{\displaystyle NH_3^+}{|}}{C}H-COO^-} \quad (14.11)$$

Notice that in the reaction, the $-NH_3^+$ group trades places with a carbonyl oxygen.

A second important example of transamination is the production of aspartate, which is used in Stage 3, urea formation. In this case, the amino group is transferred to oxaloacetate:

Specific example:

$$\underset{\text{valine}}{CH_3-\underset{\underset{\displaystyle CH_3}{|}}{C}H-\overset{\overset{\displaystyle NH_3^+}{|}}{C}H-COO^-} + \underset{\text{oxaloacetate}}{{}^-OOC-CH_2-\overset{\overset{\displaystyle O}{\|}}{C}-COO^-} \rightleftharpoons$$

$$CH_3-\underset{\underset{\displaystyle CH_3}{|}}{C}H-\overset{\overset{\displaystyle O}{\|}}{C}-COO^- + \underset{\text{aspartate}}{{}^-OOC-CH_2-\overset{\overset{\displaystyle NH_3^+}{|}}{C}H-COO^-} \quad (14.12)$$

The net effect of the transamination reactions in the catabolism of amino acids is to use nitrogen from a variety of amino acids to form glutamate and aspartate. In Section 14.11, we'll see that transamination is also an important method in the biosynthesis of amino acids.

Learning Check 14.3 Complete the following transansamination reaction:

$$\text{HO}-\text{C}_6\text{H}_4-\text{CH}_2-\underset{\text{tyrosine}}{\overset{\overset{\displaystyle NH_3^+}{|}}{\text{CH}}}-\text{COO}^- + {}^-\text{OOC}-\text{CH}_2\text{CH}_2-\overset{\overset{\displaystyle O}{\|}}{\text{C}}-\text{COO}^- \rightleftharpoons$$

Stage 2: Deamination

This phase of amino acid catabolism uses the glutamate produced in Stage 1. The enzyme glutamate dehydrogenase catalyzes the removal of the amino group as an ammonium ion and regenerates α-ketoglutarate, which can again participate in the first stage (transamination). This reaction is the principal source of NH_4^+ in humans. Because the deamination results in the oxidation of glutamate, it is called **oxidative deamination:**

oxidative deamination
An oxidation process resulting in the removal of an amino group.

$$\underset{\text{glutamate}}{{}^-\text{OOC}-\text{CH}_2\text{CH}_2-\overset{\overset{\displaystyle NH_3^+}{|}}{\text{CH}}-\text{COO}^-} + \text{NAD}^+ + H_2O \rightleftharpoons$$

$$NH_4^+ + \underset{\alpha\text{-ketoglutarate}}{{}^-\text{OOC}-\text{CH}_2\text{CH}_2-\overset{\overset{\displaystyle O}{\|}}{\text{C}}-\text{COO}^-} + \text{NADH} + H^+ \qquad (14.13)$$

CHEMISTRY AND YOUR HEALTH 14.1 The Magic Bean

Black bean salad, white bean chili, rice and beans . . . just a few of the many dishes that contain beans as an ingredient. Just as there are many recipes that include beans, there are also many different varieties of beans, though all of them have a common denominator. All beans have amazing nutritional value—they are high in fiber, high in protein, and low in fat and contain no cholesterol. In addition, they can be stored for extended periods of time or canned. Dried beans should be soaked before use, and the soaking water should be changed several times. When canned beans are used, they should be rinsed well to remove excess salt.

Our bodies need a balanced diet containing protein, carbohydrates, and fats. The proteins provide needed amino acids, including some that cannot be produced in sufficient amounts within the body. For good health, these essential amino acids must be included in the diet. Meat is a source of all essential amino acids. No type of bean contains all the essential amino acids, but beans combined with nuts, grains, and seeds will provide sufficient amounts of the essential amino acids needed for good health.

Beans come out a clear winner as a source of lean protein. One cup of cooked beans provides one-third of the U.S. Recommended Dietary Allowance (RDA) of protein for adult males. Meats, on the other hand, are normally high in fat, especially the saturated kind identified as a leading contributor to coronary disease. Twenty percent of the calories in a T-bone steak come from protein and 80% from fat, whereas in one cup of kidney beans, 25% of the calories come from protein, 70% from carbohydrates, and only 5% from fat. In addition, the small amount of fat in beans is unsaturated, and beans are cholesterol-free. Unlike meats and dairy products, beans do not contribute to increased cholesterol levels in the body.

Beans also supply significant amounts of complex carbohydrates in the diet; approximately 40 grams for one cup of cooked beans. The starch of complex carbohydrates is slowly digested into simple sugars in the digestive system. This slow release of sugars makes it an ideal food for diabetics and hypoglycemics as the blood sugar level does not undergo rapid fluctuations.

Another nutritional benefit of beans is that they have high fiber content. The dietary fiber in beans binds with some cholesterol in the digestive system and reduces the amount absorbed into the bloodstream. In this way, beans reduce cholesterol levels in the body. Fiber is also beneficial in terms of acting as a laxative agent. The inclusion of fiber in the diet can help prevent problems such as hemorrhoids, diverticulosis, colon cancer, and appendicitis.

The NADH produced in this second stage enters the electron transport chain and eventually produces 2.5 ATP molecules. Other amino acids in addition to glutamate may also be catabolized by oxidative deamination. The reactions are catalyzed by enzymes in the liver called amino acid oxidases. In each case, an α-keto acid and NH_4^+ are the products.

Stage 3: Urea Formation

The ammonium ions released by the glutamate dehydrogenase reaction are toxic to the body and must be prevented from accumulating. In the **urea cycle**, which occurs only in the liver, ammonium ions are converted to urea (see ■ Figure 14.9). This compound is less toxic than ammonium ions and can be allowed to concentrate until it is convenient to excrete it in urine.

urea cycle
A metabolic pathway in which ammonium ions are converted to urea.

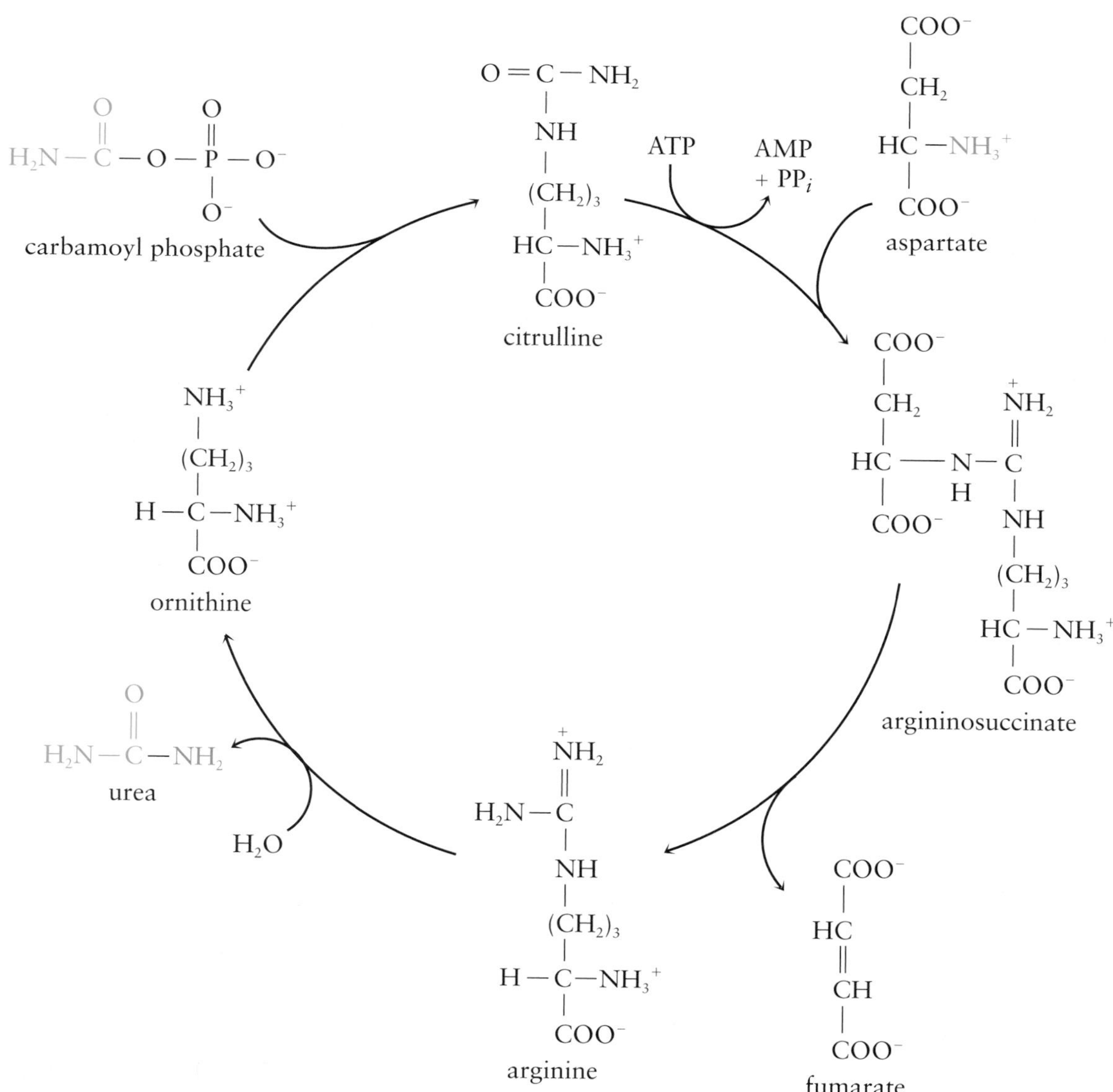

■ **FIGURE 14.9** The urea cycle. Nitrogen atoms enter the cycle from carbamoyl phosphate and aspartate.

The urea cycle processes the ammonium ions in the form of carbamoyl phosphate, the fuel for the urea cycle. This compound is synthesized in the mitochondria from ammonium ions and bicarbonate ions.

$$NH_4^+ + HCO_3^- + 2ATP + H_2O \xrightarrow{\text{carbamoyl phosphate synthetase}} \underset{\text{carbamoyl phosphate}}{H_2N-\overset{\overset{\displaystyle O}{\|}}{C}-O-\overset{\overset{\displaystyle O}{\|}}{\underset{\underset{\displaystyle O^-}{|}}{P}}-O^-} + 2ADP + P_i + 3H^+ \quad (14.14)$$

As we see, the energy for the synthesis of one molecule of carbamoyl phosphate is derived from two molecules of ATP. Figure 14.9 summarizes the reactions of the urea cycle. The net reaction for carbamoyl phosphate formation and the urea cycle is

$$NH_4^+ + HCO_3^- + 3ATP + 2H_2O + \underset{\text{aspartate}}{\begin{array}{l} COO^- \\ | \\ CH_2 \\ | \\ CH-NH_3^+ \\ | \\ COO^- \end{array}} \longrightarrow$$

$$\underset{\text{urea}}{H_2N-\overset{\overset{\displaystyle O}{\|}}{C}-NH_2} + \underset{\text{fumarate}}{\begin{array}{ccc} H & & COO^- \\ & C=C & \\ {}^-OOC & & H \end{array}} + 2ADP + 2P_i + AMP + PP_i \quad (14.15)$$

After urea is formed, it diffuses out of liver cells into the blood, the kidneys filter it out, and it is excreted in the urine. Normal urine from an adult usually contains about 25–30 g of urea daily, although the exact amount varies with the protein content of the diet. The direct excretion of NH_4^+ accounts for a small but important amount of total urinary nitrogen. Ammonium ions can be excreted along with acidic ions, a mechanism that helps the kidneys control the acid–base balance of body fluids.

Learning Check 14.4

a. What is the primary function of the urea cycle?
b. What is the fuel for the urea cycle?

14.10 Amino Acid Catabolism: The Fate of the Carbon Skeleton

LEARNING OBJECTIVE

11. Describe how amino acids can be used for energy production, the synthesis of triglycerides, and gluconeogenesis.

After removal of the amino group of an amino acid is accomplished by transamination or oxidative deamination (Section 14.9), the remaining carbon skeleton undergoes catabolism and is converted into one of several products. We will not study the various pathways followed, but we will focus on the final forms the skeleton assumes and the metabolic fate of these forms.

CHEMISTRY AROUND US 14.1 Phenylketonuria (PKU)

At least 35 different hereditary changes in amino acid metabolism have been identified. One of the more well known is an inherited disease associated with abnormal aromatic amino acid metabolism. In *phenylketonuria (PKU),* there is a lack of the enzyme phenylalanine hydroxylase. As a result, phenylalanine cannot be converted to tyrosine, leading to the accumulation of phenylalanine and its metabolites (phenylpyruvate and phenylacetate) in the tissues and blood:

$$C_6H_5-CH_2-\overset{\overset{\displaystyle O}{\|}}{C}-COO^-$$

phenylpyruvate

$$C_6H_5-CH_2-COO^-$$

phenylacetate

Abnormally high levels of phenylpyruvate and phenylacetate damage developing brain cells, causing severe mental retardation. It is estimated that 1 out of every 20,000 newborns is afflicted with PKU. A federal regulation requires that all infants be tested for this disease, which can be detected very easily by analyzing blood or urine samples for phenylalanine or phenylpyruvic acid.

The disease is treated by restricting the amount of phenylalanine in the diet of the infant starting 2–3 weeks after birth. The diet is also supplemented with tyrosine because it is not being formed in the PKU patient. Strict dietary management is not as critical after the age of 6 because the brain has matured by this time and is not as susceptible to the toxic effects of phenylalanine metabolites.

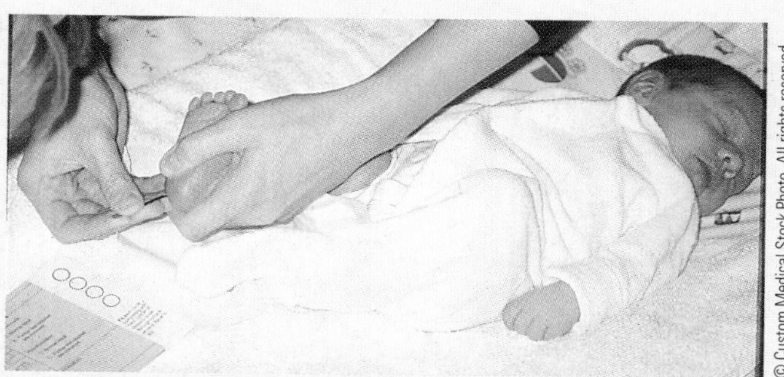

Infants with PKU have an excellent chance for a normal healthy life, if the disease is detected and treated early. Here a nurse is taking blood samples from an infant's foot to be used in a simple chemical test for PKU.

After the amino group is gone, the skeletons of all 20 amino acids are degraded into either pyruvate, acetyl CoA, acetoacetyl CoA (which is degraded to acetyl CoA), or various substances that are intermediates in the citric acid cycle. As shown in ■ Figure 14.10, all these degraded forms of the carbon skeletons are

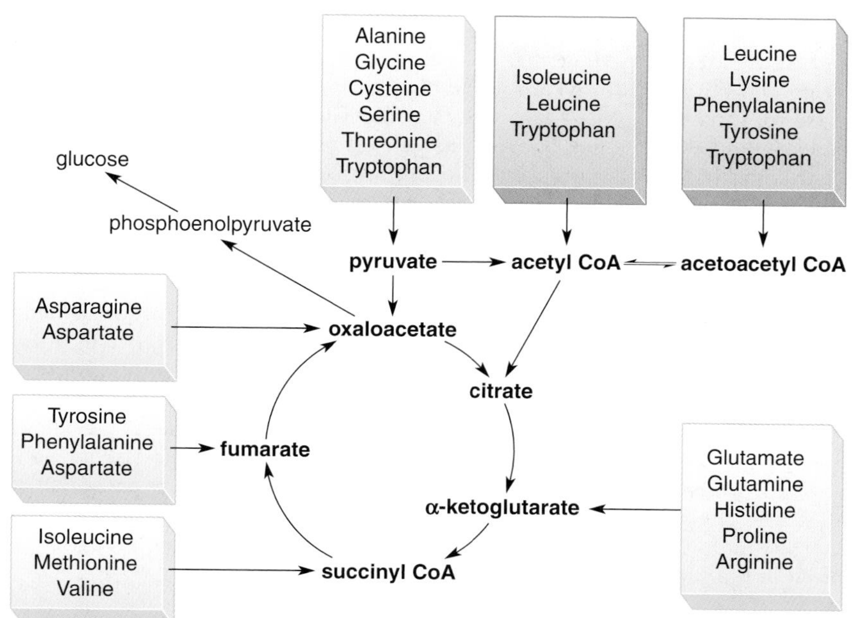

■ **FIGURE 14.10** Fates of the carbon atoms of amino acids. Glucogenic amino acids are shaded yellow, and ketogenic amino acids are shaded red.

a part of or can enter the citric acid cycle and thus may be very important in the production of energy.

Even though the degraded carbon skeletons can be used to produce energy via the citric acid cycle, not all are used this way because cellular needs might dictate other uses. For example, once a carbon skeleton is catabolized to pyruvate, there are two possible uses for it: the production of energy or the synthesis of glucose through gluconeogenesis (Section 13.10). Amino acids that yield a skeleton that is degraded into pyruvate or intermediates of the citric acid cycle can be used to make glucose and are classified as **glucogenic amino acids.** Amino acids with skeletons that are degraded into acetyl CoA or acetoacetyl CoA cannot be converted into glucose but can, in addition to energy production, be used to make ketone bodies and fatty acids. Such amino acids are classified as **ketogenic.** Figure 14.10 shows this catabolic classification of the 20 amino acids. Notice that some amino acids appear in more than one location in the diagram.

glucogenic amino acid
An amino acid whose carbon skeleton can be converted metabolically to an intermediate used in the synthesis of glucose.

ketogenic amino acid
An amino acid whose carbon skeleton can be converted metabolically to acetyl CoA or acetoacetyl CoA.

Glucogenic amino acids play a special role in the metabolic changes that occur during starvation. To maintain the blood glucose level, the body first uses the glycogen stored in the liver. After 12–18 hours without food, the glycogen supplies are exhausted. Glucose is then synthesized from glucogenic amino acids through the process of gluconeogenesis. Hydrolysis of proteins from body tissues, particularly from muscle, supplies the amino acids for gluconeogenesis.

CHEMISTRY AROUND US 14.2 Steroids in High Schools

The word *steroid* means different things to different people. A biochemist might think of a molecule with a specific ring structure. A physician might think of a hormone used to treat inflammation in the body. A sports fan or athlete might think of a substance that will help build muscle tissue in the body. Each of these meanings describes a known characteristic of chemical compounds classified as steroids, but it is the last one that has received the most media attention recently and is best known to the general public.

The muscle-building steroids, known as anabolic steroids, can be administered orally or by injection, and they greatly accelerate the buildup of muscle tissue and body strength when they are included in a program of rigorous physical training. This characteristic makes the use of anabolic steroids attractive to weight lifters, body builders, and other athletes who want to improve their performance quickly.

The use of steroids has become prevalent in professional sports despite bans against their use and knowledge of well-known adverse side effects that accompany their use. In addition to professionals, high school athletes have begun using steroids because of the pressure to do well in order to obtain college athletic scholarships or even professional contracts. Studies indicate that many of these high school adolescents use steroids even though they know such use is illegal and has the potential to cause serious harm.

The use among high school students is growing at a disturbing rate. According to a survey done by the Center for Disease Control and Prevention, the use of steroids by high school students more than doubled between 1991 and 2003. More than 6% of the 15,000 students involved in the survey admitted to trying steroid pills or injections. A 2002 study done at Texas A&M University indicated that up to 42,000 Texas high school students were involved in steroid use.

In another very disturbing study, it was found that not only were male football players and weight lifters using steroids to enhance performance, but that some females in the 9- to 11-year-old age group were using steroids to enhance their "build." This is extremely disturbing because it indicates that parents or other adults are buying the steroids for children. It is also disturbing because it means that at least some people in our society are communicating to children the attitude that it is acceptable to put your health at risk in order to have a specific body type.

Many high school athletic coaches know what is happening but feel unable to stop the practice. In a few unfortunate instances, coaches have been found to be not only encouraging steroid use but actually helping athletes obtain them. Sometimes the "win" is as important to the coach as it is to the athlete.

Unfortunately, many young people have an attitude of invincibility, or at least they are quite shortsighted. They know steroid use is illegal and dangerous, but they think the present benefits outweigh the long-term adverse consequences. Until tragedy strikes them, they think it will always strike someone else. However, tragedies, including some deaths, are known to have accompanied steroid use. Some well-known results that should give users pause for concern include acne, liver tumors, testicular atrophy, decrease in sperm count, and occasional temporary infertility.

TABLE 14.3 Essential and nonessential amino acids

Essential	Nonessential	Essential	Nonessential
Histidine	Alanine	Threonine	Glutamine
Isoleucine	Arginine	Tryptophan	Glycine
Leucine	Asparagine	Valine	Proline
Lysine	Aspartate		Serine
Methionine	Cysteine		Tyrosine
Phenylalanine	Glutamate		

Because each protein in the body has an important function, the body can meet its energy needs for only a limited time by sacrificing proteins.

14.11 Amino Acid Biosynthesis

LEARNING OBJECTIVE

12. State the relationship between intermediates of carbohydrate metabolism and the synthesis of nonessential amino acids.

As we have already seen, the liver is highly active biochemically. In biosynthesis, it is responsible for producing most of the amino acids the body can synthesize. Amino acids that can be synthesized in the amounts needed by the body are called **nonessential amino acids** (see Table 14.3). As explained in Chapter 12, essential amino acids either cannot be made or cannot be made in large enough amounts to meet bodily needs and must be included in the diet.

nonessential amino acid An amino acid that can be synthesized within the body in adequate amounts.

The key starting materials for the synthesis of 10 nonessential amino acids are intermediates of the glycolysis pathway and the citric acid cycle (see Figure 14.11). Tyrosine, the only nonessential amino acid with an aromatic side chain, is produced from the essential amino acid phenylalanine in a conversion that requires a single oxidation step catalyzed by the enzyme phenylalanine hydroxylase:

$$\underset{\text{phenylalanine}}{\text{C}_6\text{H}_5{-}\text{CH}_2{-}\underset{|}{\overset{\text{NH}_3^+}{\text{CH}}}{-}\text{COO}^-} + \text{O}_2 + \text{NADPH} + \text{H}^+ \xrightarrow{\text{phenylalanine hydroxylase}} \underset{\text{tyrosine}}{\text{HO}{-}\text{C}_6\text{H}_4{-}\text{CH}_2{-}\overset{\text{NH}_3^+}{\text{CH}}{-}\text{COO}^-} + \text{NADP}^+ + \text{H}_2\text{O} \tag{14.16}$$

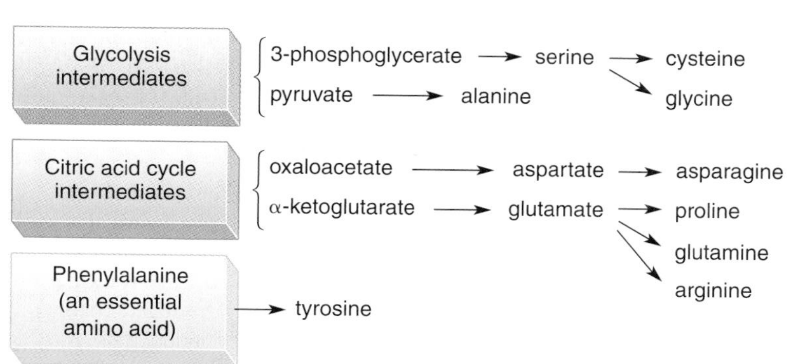

FIGURE 14.11 A summary of amino acid biosynthesis.

Three nonessential amino acids (glutamate, alanine, and aspartate) are synthesized from α-keto acids via reactions catalyzed by transaminases. For example, alanine is produced from pyruvate and glutamate (which furnishes the amino group):

$$\underset{\text{pyruvate}}{CH_3-\overset{\overset{\displaystyle O}{\|}}{C}-COO^-} + \text{glutamate} \xrightleftharpoons{\text{glutamic pyruvic transaminase}} \underset{\text{alanine}}{CH_3-\overset{\overset{\displaystyle NH_3^+}{|}}{CH}-COO^-} + \alpha\text{-ketoglutarate} \quad (14.17)$$

Thus, we see that transaminases perform two vital functions in amino acid metabolism. By taking part in the biosynthesis of nonessential amino acids, they provide a means to help readjust the relative proportions of amino acids to meet the particular needs of the body. This is a vital function because most of our diets do not contain amino acids in the exact proportions the body requires. Also, as we noted in Section 14.9, transamination reactions allow the nitrogen atoms of all amino acids to be transferred to α-keto acids to form glutamate and aspartate when disposal of nitrogen is necessary.

Two other nonessential amino acids, asparagine and glutamine, are formed from aspartate and glutamate by reaction of the side-chain carboxylate groups with ammonium ions:

$$\underset{\text{aspartate}}{^-O-\overset{\overset{\displaystyle O}{\|}}{C}-CH_2-\overset{\overset{\displaystyle NH_3^+}{|}}{CH}-COO^-} + NH_4^+ \xrightarrow[\text{ATP} \quad \text{ADP} + P_i]{} \underset{\text{asparagine}}{H_2N-\overset{\overset{\displaystyle O}{\|}}{C}-CH_2-\overset{\overset{\displaystyle NH_3^+}{|}}{CH}-COO^-} \quad (14.18)$$

$$\underset{\text{glutamate}}{^-O-\overset{\overset{\displaystyle O}{\|}}{C}-CH_2-CH_2-\overset{\overset{\displaystyle NH_3^+}{|}}{CH}-COO^-} + NH_4^+ \xrightarrow[\text{ATP} \quad \text{ADP} + P_i]{} \underset{\text{glutamine}}{H_2N-\overset{\overset{\displaystyle O}{\|}}{C}-CH_2CH_2-\overset{\overset{\displaystyle NH_3^+}{|}}{CH}-COO^-} \quad (14.19)$$

So far, we have shown how 6 of the 11 nonessential amino acids are synthesized. The syntheses of the remaining 5 are more complex, and we will not consider them here.

Concept Summary

ThomsonNOW™ Sign in at **www.thomsonedu.com** to:

- Assess your understanding with Exercises keyed to each learning objective.
- Check your readiness for an exam by taking the **Pre-test** and exploring the modules recommended in your **Personalized Learning Plan.**

Blood Lipids. Fatty acids, glycerol, and monoglycerides that result from lipid digestion are absorbed into the lymph, recombined into triglycerides, and deposited into the bloodstream as lipoprotein aggregates. **OBJECTIVE 1 (Section 14.1), Exercises 14.2 and 14.4**

Fat Mobilization. When stored fats are needed for energy, hydrolysis reactions liberate the fatty acids and glycerol. These component substances of fat then enter the bloodstream and travel to tissues, where they are utilized. **OBJECTIVE 2 (Section 14.2), Exercise 14.8**

Glycerol Metabolism. The glycerol available from fat mobilization is first phosphorylated and then oxidized to dihydroxyacetone phosphate, an intermediate of glycosis. By having an entry point into the glycolysis pathway, glycerol can ultimately be converted into glucose or oxidized to CO_2 and H_2O. **OBJECTIVE 3 (Section 14.3), Exercise 14.12**

The Oxidation of Fatty Acids. The catabolism of fatty acids begins in the cytoplasm, where they are activated by combining with CoA—SH. After being transported into mitochondria, the degradation of fatty acids occurs by the β-oxidation pathway. The β-oxidation pathway produces reduced coenzymes ($FADH_2$ and NADH) and cleaves the fatty acid chain into two-carbon fragments bound to coenzyme A (acetyl CoA). **OBJECTIVE 4 (Section 14.4), Exercise 14.22**

The Energy from Fatty Acids. The energy available from a fatty acid for ATP synthesis can be calculated by determining the amount of acetyl CoA, $FADH_2$, and NADH produced. **OBJECTIVE 5 (Section 14.5), Exercise 14.26.** On an equal-mass basis, fatty acids contain more than twice the energy of glucose.

Ketone Bodies. Acetoacetate, β-hydroxybutyrate, and acetone are known as ketone bodies. They are synthesized in the liver from acetyl CoA. During starvation and in unchecked diabetes, the level of ketone bodies becomes very high, leading to several conditions: ketonemia, ketonuria, acetone breath, and ketosis. **OBJECTIVE 6 (Section 14.6), Exercise 14.28**

Fatty Acid Synthesis. This process occurs in the cytoplasm and uses acetyl CoA as a starting material. Two-carbon fragments are added to growing fatty acid molecules as energy is supplied by ATP and NADPH. **OBJECTIVE 7 (Section 14.7), Exercise 14.34**

Amino Acid Metabolism. Amino acids and proteins are not stored within the body. Amino acids absorbed from digested food or synthesized in the body become a part of a temporary supply or pool that may be used in metabolic processes. **OBJECTIVE 8 (Section 14.8), Exercise 14.38.** The turnover, or life expectancy, of proteins is usually expressed in half-lives. Protein half-lives range from several minutes to hundreds of days. Some amino acids in the body are converted into relatively simple but vital nitrogen-containing compounds. The catabolism of amino acids produces intermediates used for energy production, synthesis of glucose, or formation of triglycerides.

Amino Acid Catabolism: The Fate of the Nitrogen Atoms. As an amino acid undergoes catabolism, the amino group may be transferred to an α-keto acid through a process called transamination. In the process of deamination, an amino acid is converted into a keto acid and ammonia. **OBJECTIVE 9 (Section 14.9), Exercise 14.44.** The urea cycle converts toxic ammonia molecules into urea for excretion. **OBJECTIVE 10 (Section 14.9), Exercise 14.54**

Amino Acid Catabolism: The Fate of the Carbon Skeleton. Amino acids are classified as glucogenic or ketogenic based on their catabolic products. Glucogenic amino acids are degraded into pyruvate or intermediates of the citric acid cycle and can be used for glucose synthesis. Ketogenic amino acids are degraded into acetoacetyl CoA and acetyl CoA for energy production and triglyceride synthesis. **OBJECTIVE 11 (Section 14.10), Exercise 14.58**

Amino Acid Biosynthesis. The body can synthesize amino acids, known as nonessential amino acids. The key starting materials are intermediates of the glycolysis pathway and the citric acid cycle, and phenylalanine in the case of tyrosine. **OBJECTIVE 12 (Section 14.11), Exercise 14.64**

Key Terms and Concepts

Acetone breath (14.6)
Acidosis (14.6)
Adipose tissue (14.2)
Amino acid pool (14.8)
β-oxidation process (14.4)
Chylomicron (14.1)
Fat mobilization (14.2)
Glucogenic amino acid (14.10)
Ketoacidosis (14.6)
Ketogenic amino acid (14.10)
Ketone bodies (14.6)
Ketonemia (14.6)
Ketonuria (14.6)
Ketosis (14.6)
Nonessential amino acid (14.11)
Oxidative deamination (14.9)
Protein turnover (14.8)
Transaminase (14.9)
Transamination (14.9)
Urea cycle (14.9)

Key Reactions

1. Digestion of triglycerides (Section 14.1):

$$\text{triglyceride} + H_2O \xrightarrow{\text{lipase}} \text{fatty acids} + \text{glycerol} + \text{monoglycerides}$$

2. Fat mobilization (Section 14.2):

$$\text{triglyceride} + 3H_2O \xrightarrow{\text{lipase}} 3\ \text{fatty acids} + \text{glycerol}$$

3. Activation of fatty acids (Section 14.4):

$$\underset{\text{fatty acid}}{R-\overset{\overset{\displaystyle O}{\|}}{C}-OH} + HS-CoA \xrightarrow[\text{ATP} \quad \text{AMP} + PP_i]{} \underset{\text{fatty acyl CoA}}{R-\overset{\overset{\displaystyle O}{\|}}{C}-S-CoA} + H_2O$$ Reaction 14.2

4. Oxidation of fatty acids (Section 14.4):

$$R-CH_2CH_2-\overset{O}{\overset{\|}{C}}-S-CoA + NAD^+ + FAD + H_2O + CoA-SH \longrightarrow$$

Reaction 14.5

$$R-\overset{O}{\overset{\|}{C}}-S-CoA + CH_3\overset{O}{\overset{\|}{C}}-S-CoA + NADH + H^+ + FADH_2$$

5. Fatty acid synthesis (Section 14.7):

$$8CH_3-\overset{O}{\overset{\|}{C}}-S-CoA + 14NADPH + 13H^+ + 7ATP \longrightarrow$$

Reaction 14.9

$$CH_3(CH_2)_{14}COO^- + 8CoA-SH + 6H_2O + 14NADP^+ + 7ADP + 7P_i$$

6. Transamination (Section 14.9):

$$R-\overset{NH_3^+}{\overset{|}{CH}}-COO^- + {}^-OOC-CH_2CH_2-\overset{O}{\overset{\|}{C}}-COO^- \rightleftharpoons R-\overset{O}{\overset{\|}{C}}-COO^- + {}^-OOC-CH_2CH_2-\overset{NH_3^+}{\overset{|}{CH}}-COO^-$$

Reaction 14.10

7. Oxidative deamination (Section 14.9):

$${}^-OOC-CH_2CH_2-\overset{NH_3^+}{\overset{|}{CH}}-COO^- + NAD^+ + H_2O \rightleftharpoons NH_4^+ + {}^-OOC-CH_2CH_2-\overset{O}{\overset{\|}{C}}-COO^- + NADH + H^+$$

Reaction 14.13

8. Urea formation (Section 14.9):

$$NH_4^+ + HCO_3^- + 3ATP + 2H_2O + \underset{\text{aspartate}}{COO^--CH_2-CH(NH_3^+)-COO^-} \longrightarrow \underset{\text{urea}}{H_2N-\overset{O}{\overset{\|}{C}}-NH_2} + \underset{\text{fumarate}}{(H)(^-OOC)C=C(COO^-)(H)} + 2ADP + 2P_i + AMP + PP_i$$

Reaction 14.15

Exercises

SYMBOL KEY

Even-numbered exercises are answered in Appendix B.

Blue-numbered exercises are more challenging.

■ denotes exercises available in ThomsonNow and assignable in OWL.

ThomsonNOW To assess your understanding of this chapter's topics with sample tests and other resources, sign in at **www.thomsonedu.com.**

BLOOD LIPIDS (SECTION 14.1)

14.1 How does the Caloric value of fat compare with the Caloric value of glycogen and starch?

14.2 ■ What are the products of triglyceride digestion?

14.3 What happens to the products of triglyceride digestion as they pass into the cells of the intestinal mucosa?

14.4 What mechanism exists for transporting insoluble lipids?

14.5 Describe what happens to the level of blood lipids after a meal rich in fat.

14.6 ■ List the major types of lipoproteins.

FAT MOBILIZATION (SECTION 14.2)

14.7 What signals the need for mobilization of triglycerides from adipose tissue?

14.8 ■ Describe what is meant by the term *fat mobilization*.

14.9 Explain why fatty acids cannot serve as fuel for the brain.

14.10 ■ Which cells utilize fatty acids over glucose to satisfy their energy needs?

GLYCEROL METABOLISM (SECTION 14.3)

14.11 At what point does glycerol enter the glycolysis pathway?

14.12 Describe the two fates of glycerol after it has been converted to an intermediate of glycolysis.

THE OXIDATION OF FATTY ACIDS (SECTION 14.4)

14.13 Where in the cell does fatty acid catabolism take place?

14.14 How is the fatty acid prepared for catabolism? Where in the cell does fatty acid activation take place?

14.15 Why is the oxidation of fatty acids referred to as β-oxidation?

14.16 ■ What hydrogen-transporting coenzymes play important roles in the oxidation of fatty acids?

14.17 Label the α- and β-positions in the following acids:

a. $CH_3CH_2CH_2CH_2CH_2-\overset{\overset{\displaystyle O}{\|}}{C}-OH$

b. $CH_3(CH_2)_6\overset{\overset{\displaystyle OH}{|}}{CH}-CH_2-\overset{\overset{\displaystyle O}{\|}}{C}-OH$

14.18 Write an equation for the activation of palmitic acid (16 carbons).

14.19 Write formulas for the products that result when activated palmitic acid is taken through the following:

a. One turn of the fatty acid spiral

b. Two turns of the fatty acid spiral

14.20 Why is the oxidation of fatty acids referred to as the fatty acid spiral rather than the fatty acid cycle?

14.21 How many molecules of acetyl CoA are produced in the catabolism of a molecule of stearic acid, an 18-carbon carboxylic acid? How many times does the spiral reaction sequence occur?

14.22 ■ Complete the following equations for one turn of the fatty acid spiral:

a. $CH_3CH_2CH_2CH_2CH_2-\overset{\overset{\displaystyle O}{\|}}{C}-S-CoA + FAD \longrightarrow$ ___ + ___

b. ___ + H_2O → ___

c. ___ + NAD^+ → ___ + ___ + ___

d. ___ + CoA—SH → ___ + ___

14.23 Outline the four reactions that occur in the fatty acid spiral.

THE ENERGY FROM FATTY ACIDS (SECTION 14.5)

14.24 Determine the ATP yield per molecule of acetyl CoA that enters the citric acid cycle.

14.25 What is the energy cost in ATP molecules to activate a fatty acid?

14.26 ■ Calculate the number of ATP molecules that can be produced from the complete oxidation of a ten-carbon fatty acid to carbon dioxide and water.

14.27 Compare the energy available (in ATP molecules) from a 6-carbon fatty acid to the energy from a molecule of glucose.

KETONE BODIES (SECTION 14.6)

14.28 What are ketone bodies?

14.29 Explain how excessive ketone bodies may form in the following:

a. During starvation

b. In patients with diabetes mellitus

14.30 Where are ketone bodies formed? What tissues utilize ketone bodies to meet energy needs?

14.31 ■ Differentiate between (a) ketonemia, (b) ketonuria, (c) acetone breath, and (d) ketosis.

14.32 Why is ketosis frequently accompanied by acidosis?

FATTY ACID SYNTHESIS (SECTION 14.7)

14.33 Where within a liver cell does fatty acid synthesis occur?

14.34 ■ Explain why most fatty acids within the body have an even number of carbon atoms.

14.35 What substances furnish the energy for fatty acid synthesis?

14.36 Why can the liver convert glucose to fatty acids but cannot convert fatty acids to glucose?

AMINO ACID METABOLISM (SECTION 14.8)

14.37 ■ List three vital functions served by amino acids in the body.

14.38 What is the amino acid pool?

14.39 What are the sources of amino acids in the pool?

14.40 What term is used to denote the breakdown and resynthesis of body proteins?

14.41 What percentage of amino acids is used to synthesize proteins?

14.42 Name four important biomolecules other than proteins that are synthesized from amino acids.

14.43 Compare the half-life of plasma proteins with the half-life of enzymes.

AMINO ACID CATABOLISM: THE FATE OF THE NITROGEN ATOMS (SECTION 14.9)

14.44 ■ Match the terms *transamination* and *deamination* to the following:

a. An amino group is removed from an amino acid and donated to an α-keto acid.

b. Ammonium ion is produced.

c. New amino acids are synthesized from other amino acids.

d. A keto acid is produced from an amino acid.

14.45 Complete the following transamination reaction and name the amino acid that is produced:

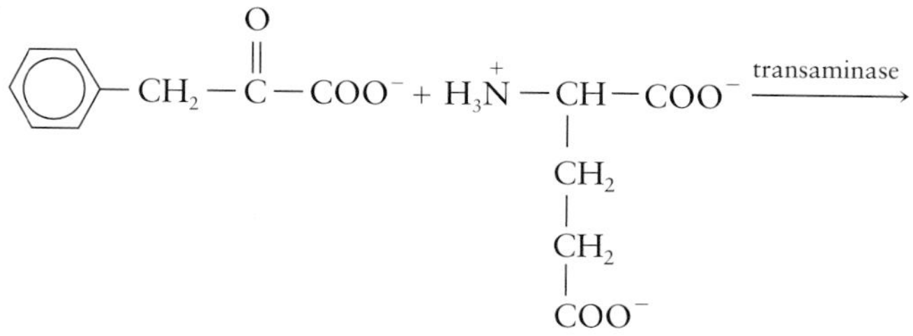

$$C_6H_5-CH_2-\overset{\overset{\displaystyle O}{\|}}{C}-COO^- + H_3\overset{+}{N}-\underset{\underset{\displaystyle \underset{\displaystyle \underset{\displaystyle COO^-}{|}}{CH_2}}{|}}{\underset{\displaystyle CH_2}{|}}{CH}-COO^- \xrightarrow{\text{transaminase}}$$

14.46 Write an equation to illustrate transamination between leucine and pyruvate.

14.47 Complete the following oxidative deamination reaction:

$$H_3\overset{+}{N}-\underset{\underset{\displaystyle \underset{\displaystyle \underset{\displaystyle NH_2}{|}}{C{=}O}}{|}}{\underset{\displaystyle CH_2}{|}}{CH}-COO^- + NAD^+ + H_2O \xrightarrow[\text{oxidase}]{\text{amino acid}}$$

14.48 Write an equation to illustrate the oxidative deamination of glycine.

14.49 By writing two reactions, the first a transamination and the second an oxidative deamination, show how the amino group of alanine can be removed as NH_4^+.

14.50 In what organ of the body is urea synthesized?

14.51 What are the sources of the carbon atom and two nitrogen atoms in urea?

14.52 Name the compound that enters the urea cycle by combining with ornithine.

14.53 How much energy (in number of ATP molecules) is expended to synthesize a molecule of urea?

14.54 ■ Write a summary equation of the formation of urea.

AMINO ACID CATABOLISM: THE FATE OF THE CARBON SKELETON (SECTION 14.10)

14.55 Differentiate between glucogenic and ketogenic amino acids.

14.56 Is any amino acid both glucogenic and ketogenic?

14.57 ■ Identify the five locations at which the carbon skeletons of amino acids enter the citric acid cycle.

14.58 ■ Based on the following conversions, classify each amino acid as glucogenic or ketogenic.

a. aspartate → oxaloacetate

b. leucine → acetyl CoA

c. tyrosine → acetoacetyl CoA

14.59 What special role is played by glucogenic amino acids during fasting or starvation?

AMINO ACID BIOSYNTHESIS (SECTION 14.11)

14.60 Differentiate between essential and nonessential amino acids.

14.61 What essential amino acid makes tyrosine nonessential?

14.62 List two vital functions performed by transaminases.

14.63 ■ What amino acids are synthesized from glutamate?

14.64 List two general sources of intermediates for the biosynthesis of amino acids.

ADDITIONAL EXERCISES

14.65 Trace glycerol metabolism after it enters the glycolytic pathway as dihyroxyacetone phosphate and give an overall net reaction (including ATP, NAD^+, etc.) for the conversion of glycerol to pyruvate.

14.66 A change in the concentration of enzymes that catalyze fatty acid synthesis has been shown to be a major mechanism for controlling synthesis rate. Give two regulatory means of changing enzyme concentrations.

14.67 A ketone body has a molecular weight of 58.1 g/mol and a concentration of 0.050 mg/100 mL in a patient's blood. Which ketone body is it, and how many molecules of it would be found in 1.0 mL of the patient's blood?

14.68 A 7.48-mg sample of a blood lipid was found to contain 1.87 mg of protein. Use Figure 14.4 to classify this lipid.

14.69 The following reaction proceeded in an aqueous solution as the pH was monitored. After a while, the pH quit changing even though there was excess urea present in the solution.

$$\underset{\text{urea}}{NH_2-\overset{\overset{\displaystyle O}{\|}}{C}-NH_2} + H_2O \xrightarrow{\text{urease}} 2NH_{3(aq)} + CO_{2(g)}$$

Was the initial pH when the reaction began acidic, neutral, or basic? How did the pH change over time? Why did the pH quit changing?

ALLIED HEALTH EXAM CONNECTION

Reprinted with permission from Nursing School and Allied Health Entrance Exams, COPYRIGHT 2005 Petersons.

14.70 Which of the following processes must occur before amino acids can be metabolized to release energy?

a. Fermentation

b. Deamination

c. Dehydrogenation

14.71 When in excess, which element is eliminated from the human body through the formation of urea? In what organ of the body is urea synthesized?

14.72 What substances are produced by the digestion of proteins?

14.73 What is the name for amino acids that cannot be manufactured by the body?

14.74 The clinic nurse is evaluating a man wearing a diabetic medic alert band, who appears confused, with hot, dry, flushed skin. His respirations are deep and fast, and he says he is nauseated. His breath smells fruity. What diabetes-related medical condition is this man exhibiting?

CHEMISTRY FOR THOUGHT

14.75 Cellular reactions for the biosynthesis and catabolism of amino acids are not simply the reversals of the same metabolic pathways. Why is this an advantage for the cell?

14.76 Why do you think the half-lives of enzymes are comparatively short?

14.77 Why do you think the half-life of collagen is comparatively long?

14.78 How might HDL levels suggest a partial explanation for the observation that women as a group have fewer heart attacks than do men as a group?

14.79 A student calculates the net yield of ATP molecules from a 12-carbon fatty acid to be 80. What did the student probably overlook?

14.80 If researchers discovered a method of reducing blood cholesterol to extremely low levels, why might it be undesirable to do so?

14.81 Could there be a connection between high sugar intake and high blood levels of triglycerides? Explain.

CHAPTER 15

Body Fluids

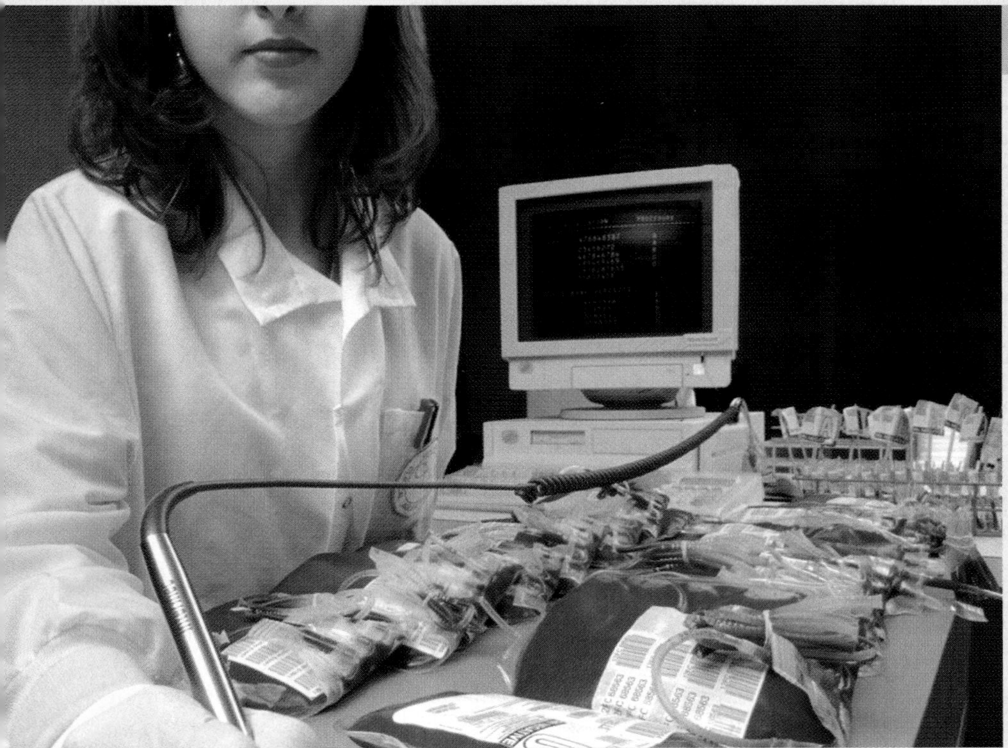

Blood cells of the body are the focus of work done by blood bank technologists and hematology laboratory technicians. Because they perform blood cell counts and identify the kinds of blood cells found in blood, bone marrow, and other body fluids, the technicians are often involved in the diagnosis of leukemias, anemias, infections, and other diseases that affect blood cells. Blood chemistry is one of the major topics of this chapter.

LEARNING OBJECTIVES

When you have completed your study of this chapter, you should be able to:

1. Compare the chemical compositions of plasma, interstitial fluid, and intracellular fluid. (Section 15.1)
2. Explain how oxygen and carbon dioxide are transported within the bloodstream. (Section 15.2)
3. Explain how materials move from the blood into the body cells and from the body cells into the blood. (Section 15.3)
4. List the normal and abnormal constituents of urine. (Section 15.4)
5. Discuss how proper fluid and electrolyte balance is maintained in the body. (Section 15.5)
6. Explain how acid–base balance is maintained in the body. (Section 15.6)
7. Explain how buffers work to control blood pH. (Section 15.7)
8. Describe respiratory control of blood pH. (Section 15.8)
9. Describe urinary control of blood pH. (Section 15.9)
10. List the causes of acidosis and alkalosis. (Section 15.10)

ThomsonNOW™ Throughout the chapter this icon introduces resources on the ThomsonNOW website for this text. Sign in at **www.thomsonedu.com** to:

- Evaluate your knowledge of the material
- Take an exam prep quiz
- Identify areas you need to study with a **Personalized Learning Plan**.

To carry out its complex function, a living cell requires a steady supply of reactants, such as nutrients and oxygen. A cell also requires a reliable system for removing the resulting waste products—carbon dioxide and water, for example. Both of these conditions are met by the simple process of diffusion. Reactants can diffuse into the cell through the thin cell membrane, and the products of cell function can diffuse out of the cell and be absorbed by the surrounding liquid.

However, a complex organism, which consists of many closely packed cells, cannot depend solely on diffusion. Some circulating system must bring necessary materials to the cells and carry away the wastes. Otherwise, the liquid surrounding the cells would soon be depleted of reactants and saturated with wastes.

15.1 A Comparison of Body Fluids

LEARNING OBJECTIVE

1. Compare the chemical compositions of plasma, interstitial fluid, and intracellular fluid.

intracellular fluid
Body fluid located inside cells.

extracellular fluid
Body fluid located outside cells.

interstitial fluid
The fluid surrounding individual tissue cells.

plasma
The liquid portion of whole blood.

ThomsonNOW™ Go to Chemistry Interactive to explore **body fluid composition**.

The average adult body contains about 42 L of fluids, which accounts for two-thirds of the total body weight. Most of these fluids are compartmentalized in three regions of the body: the interior of cells, tissue spaces between cells (including the lymph vessels), and the blood vessels. The majority, about 28 L, is located inside the cells and is called **intracellular fluid.** All body fluids not located inside the cells are collectively known as **extracellular fluids.** Thus, fluids in two of the fluid compartments are extracellular: the **interstitial fluid,** which fills the space between tissue cells and moves in lymph vessels, and the fluid of the bloodstream—**plasma.** Interstitial fluid constitutes about 20% (10.5 L) of the total body fluid, and the plasma (nearly 3.5 L) makes up about 7% of the total. Other body fluids that occur in lesser amounts are urine, digestive juices, and cerebrospinal fluid. The distribution of the various body fluids is illustrated in ■ Figure 15.1. Extracellular fluid provides a relatively constant environment for the cells and transports substances to and from the cells. Intracellular fluid, on the other hand, is the medium in which the vital life-maintaining reactions take place.

Chemically, the two extracellular fluids, plasma and interstitial fluid, are nearly identical. The only difference is that plasma contains an appreciable amount of protein not found in interstitial fluid. Intracellular fluid is chemically different from extracellular fluid. Intracellular fluid contains larger amounts of protein and a much different ionic population (see ■ Figure 15.2). From Figure 15.2, we can see that:

1. The principal cation of plasma and interstitial fluid is Na^+. The principal cation of intracellular fluid is K^+ (actually, about 98% of the body's potassium is found inside cells).
2. The principal anion of extracellular fluid is chloride, Cl^-, whereas phosphate (mainly HPO_4^{2-}) is the main anion inside cells.
3. Intracellular fluid contains more than four times as much protein as plasma, and there is remarkably little protein in interstitial fluid.

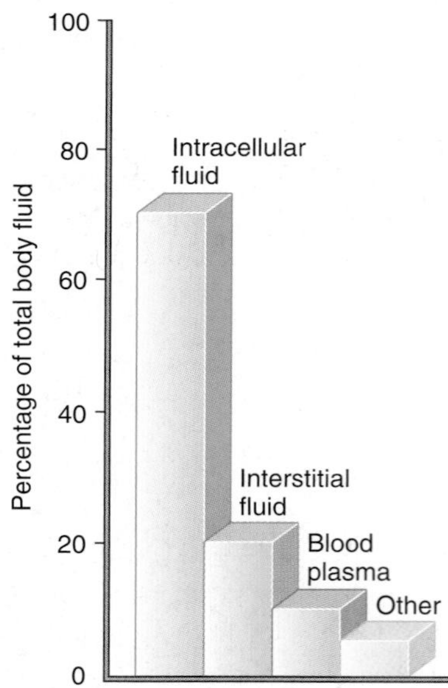

■ **FIGURE 15.1** The distribution of body fluids.

15.2 Oxygen and Carbon Dioxide Transport

LEARNING OBJECTIVE

2. Explain how oxygen and carbon dioxide are transported within the bloodstream.

One of the most important functions of blood is to carry oxygen from the lungs to the tissues. An average adult at rest requires about 350 mL of oxygen per minute.

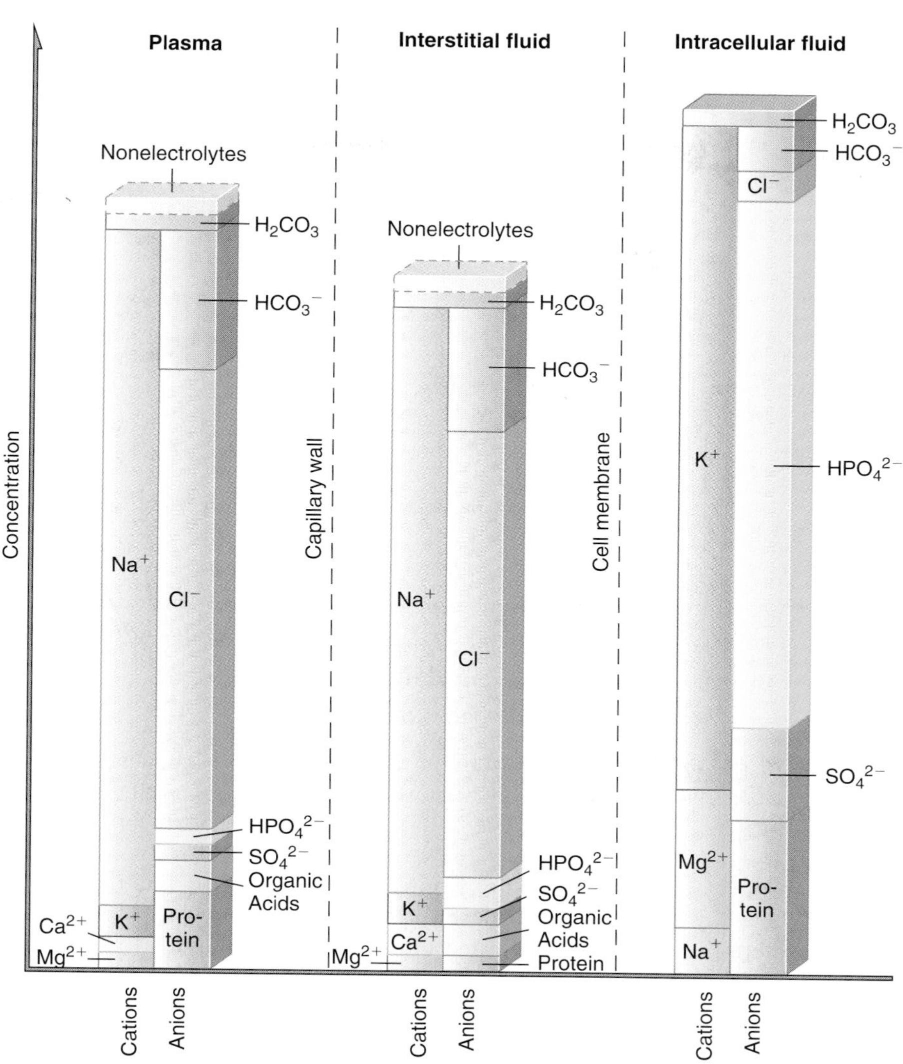

■ **FIGURE 15.2** The relative chemical composition of the major body fluids.

The limited solubility of oxygen in plasma allows only about 2% of this total to be dissolved and transported in solution. About 98% of the needed oxygen is carried by red blood cells (see ■ Figure 15.3) in the form of an oxygen-hemoglobin combination called **oxyhemoglobin.** The nonoxygenated hemoglobin is called **deoxyhemoglobin,** or simply **hemoglobin.** Normal human blood contains about 15 g of hemoglobin per 100 mL.

oxyhemoglobin
Hemoglobin combined with oxygen.

deoxyhemoglobin (hemoglobin)
Nonoxygenated hemoglobin.

Because of its physiological importance, hemoglobin has been studied extensively. It is known to be a conjugated protein containing four chains (the globin portion of the molecule) and four heme groups. Each heme group consists of an Fe^{2+} ion in the center of a nitrogenous heterocyclic structure (see ■ Figure 15.4). The iron in oxyhemoglobin is bonded at six positions, as shown in Figure 15.4. The four nitrogens are part of the heme group, and the histidine is a part of the globin molecule. The sixth bond is between iron and oxygen. Because each hemoglobin molecule contains four heme groups, it can bind a total of four O_2 molecules. When one of the four heme groups binds an oxygen molecule, the affinity of the other heme groups for oxygen is increased. The reversible binding that takes place between hemoglobin (HHb) and molecular oxygen is represented like this:

$$\underset{\text{hemoglobin}}{HHb} + O_2 \rightleftharpoons \underset{\text{oxyhemoglobin}}{HbO_2^-} + H^+ \qquad (15.1)$$

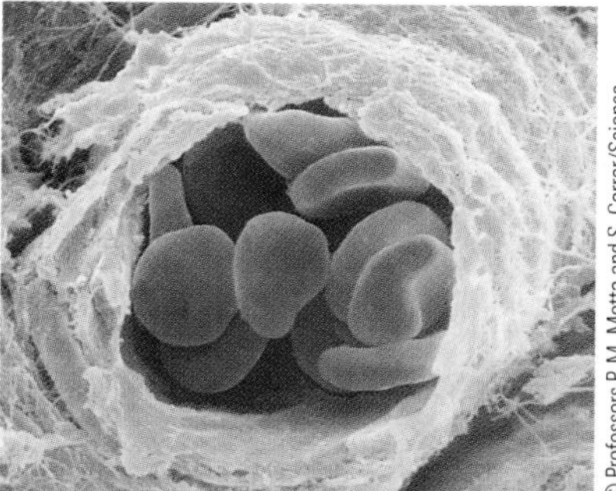

■ **FIGURE 15.3** Red blood cells passing through a narrow capillary.

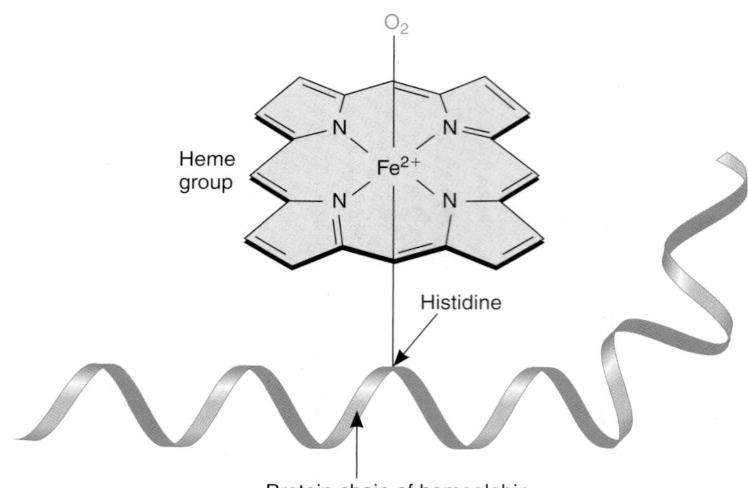

■ **FIGURE 15.4** The iron in oxyhemoglobin.

carbaminohemoglobin Hemoglobin combined with carbon dioxide.

When CO_2 is present, hemoglobin can combine with it to form **carbaminohemoglobin.** This reversible reaction is favored by conditions found in body tissues:

$$\underset{}{HHb} + CO_2 \rightleftharpoons \underset{\text{carbaminohemoglobin}}{HHbCO_2} \qquad (15.2)$$

About 25% of the total CO_2 in the body is carried from body tissues to the lungs in the form of carbaminohemoglobin. Another 5% dissolves in the plasma and travels in solution; the remaining 70% is transported in the form of bicarbonate ions (HCO_3^-).

The transport of oxygen to tissues begins when inhaled air fills the alveoli of the lungs. These clusters of small sacs are surrounded by many tiny blood capillaries, and it is here that oxygen is absorbed into the blood. ■ Figure 15.5 represents the processes that take place as oxygen is absorbed and carbon dioxide is released. The main steps (circled numbers in the figure) are:

ThomsonNOW™ Go to Chemistry Interactive to explore **oxygen transport.**

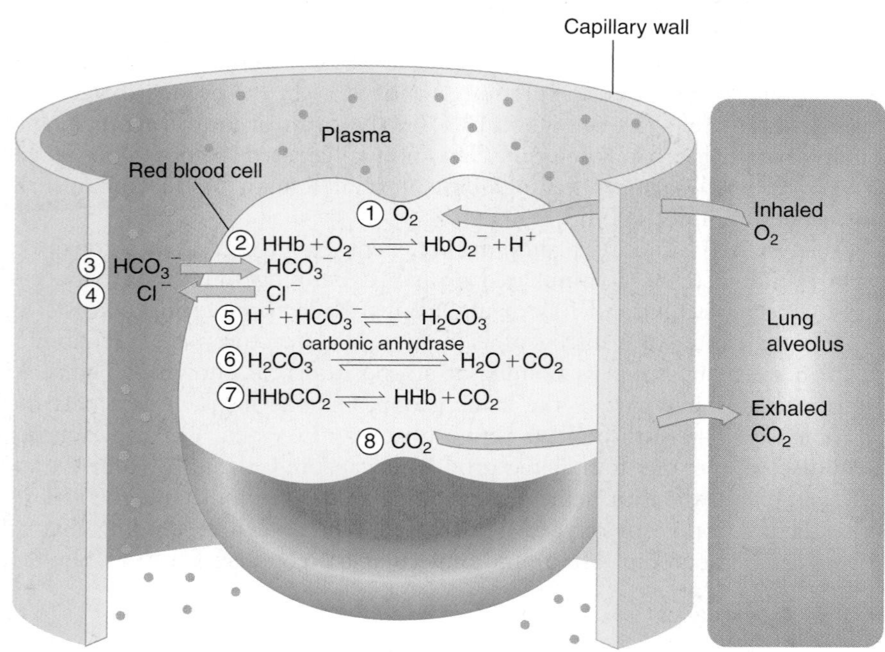

■ **FIGURE 15.5** At the lungs, carbon dioxide of red blood cells is exchanged for oxygen.

1. Oxygen partial pressure is higher in the alveoli than in the red blood cells. As a result, oxygen diffuses from the alveoli through the capillary wall and into the red blood cells.
2. Inside the red blood cells, oxygen reacts with hemoglobin to form oxyhemoglobin and H^+ ions. Oxyhemoglobin formation is favored (according to Le Châtelier's principle) by a high partial pressure of oxygen, a slight cooling of the blood, and an increase in pH (decrease in H^+ ions)—conditions encountered by the blood as it passes through the capillaries of the lungs.
3. Bicarbonate ions (HCO_3^-) diffuse from the plasma into the red blood cells.
4. The bicarbonate ions are replaced in the plasma by chloride ions that diffuse out of the blood cells. This step, called the **chloride shift,** maintains charge balance and osmotic pressure relationships between the plasma and red blood cells.
5. Protons released by hemoglobin as it is oxygenated (Step 2) react with bicarbonate ions to form carbonic acid.
6. An enzyme in the red blood cells, carbonic anhydrase, promotes the breakdown of carbonic acid to water and carbon dioxide. The reaction is reversible, but the low pressure of carbon dioxide in the lungs favors the formation of carbon dioxide and water (Le Châtelier's principle, again).
7. Carbaminohemoglobin ($HHbCO_2$), formed from hemoglobin and CO_2 at the cellular level, breaks apart to yield hemoglobin and carbon dioxide.
8. Because the carbon dioxide pressure is high in the red blood cells but low in the alveoli, carbon dioxide molecules diffuse out of the red blood cells and into the lung. The carbon dioxide and some of the water formed in Step 6 are expelled in the exhaled air.

chloride shift
The movement of chloride ions from one fluid to another to maintain charge neutrality.

OVER THE COUNTER 15.1 Avoiding Food-Drug Interactions

Almost everyone has had the experience of getting a prescription filled or buying an over-the-counter drug only to read the instruction/contraindication sheet and wonder if the drug really should be used because of all the cautions and side effects that are listed. These directions are given for some very good reasons. Both prescription and OTC drugs are chemical compounds that can react with a variety of other compounds or elements, including many that are contained in food. Such interactions with food can influence drugs in a number of ways. The rate of drug absorption from the digestive tract into the body can be influenced. For example, calcium ions found in dairy foods bind to the antibiotic tetracycline and slow down its absorption from the digestive tract. Aspirin is absorbed more slowly when taken with any food. The antibiotic griseofulvin is absorbed better when it is taken with foods that stimulate the release of digestive enzymes. Foods can also change the way drugs are metabolized and thus interfere with the drug's action. The action of phenobarbital is limited by large amounts of folic acid in the diet from foods such as beans, spinach and other greens, and liver.

The following are generally recommended guidelines that will help you avoid food-drug interactions.

1. Read the labels and instruction/contraindication sheets for all medications, both prescription and OTC, and do what they tell you to do. If you need more information or clarification, consult your physician or pharmacist.
2. Unless otherwise directed, take all medications with a full glass of water on an empty stomach to increase the rate of absorption.
3. Don't take medications with food unless directed to do so.
4. Don't take mineral supplements or vitamins with medications without getting approval from your physician or pharmacist.
5. Don't take medications with hot drinks. The heat might decrease the effectiveness of the drug.
6. Never take drugs with alcohol.
7. Always check with your physician or pharmacist before taking a drug that is unfamiliar to you.

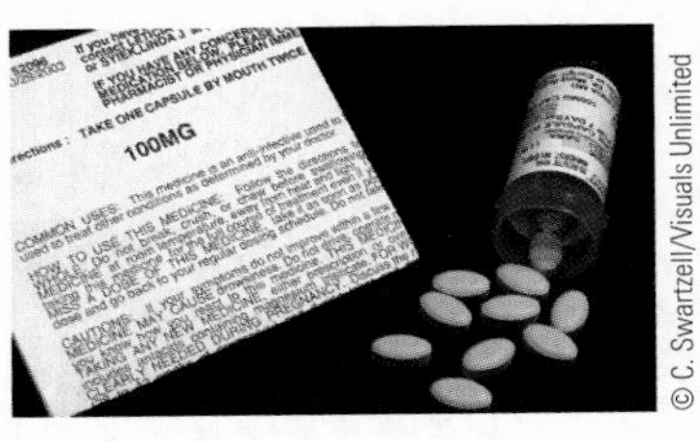

© C. Swartzell/Visuals Unlimited

Always read the instruction/contraindication sheets for your drugs.

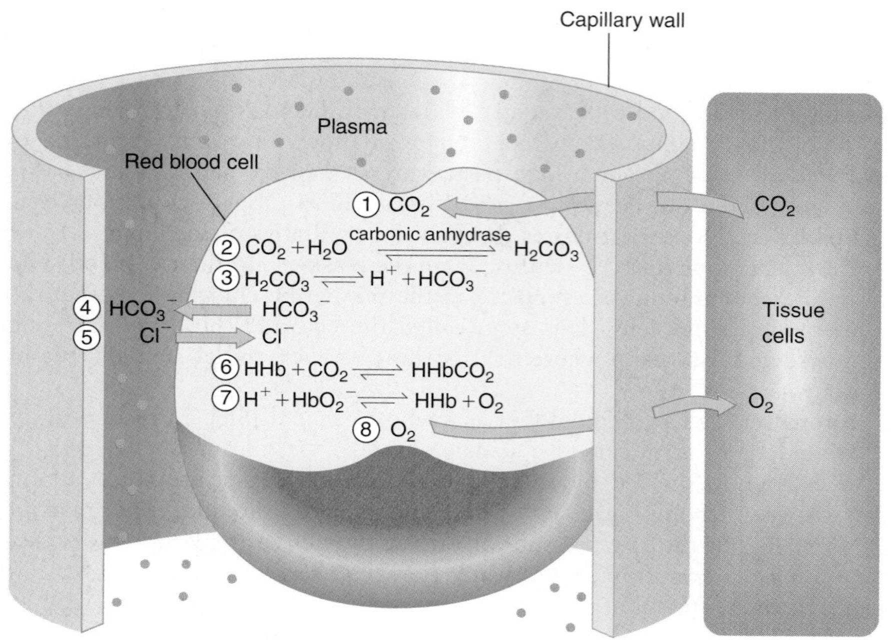

■ **FIGURE 15.6** At the tissue cells, oxygen of red blood cells is exchanged for carbon dioxide.

Oxygenated red blood cells are carried by the bloodstream into the capillaries surrounding the cells of active tissues. ■ Figure 15.6 shows the steps involved in delivering oxygen to the tissues and accepting carbon dioxide for transport to the lungs. These steps are essentially the reverse of those at the lungs:

1. Tissue cells use oxygen to produce energy; carbon dioxide and water are formed as byproducts. The concentration of carbon dioxide is higher in the tissue cells than in the red blood cells, so carbon dioxide diffuses from the tissue cells into the interstitial fluid and then into the red blood cells.
2. Inside the red blood cells, carbon dioxide rapidly reacts with water in the presence of carbonic anhydrase to produce carbonic acid.
3. Carbonic acid dissociates to give hydrogen ions and bicarbonate ions.
4. The bicarbonate ions diffuse into the plasma. The bicarbonate ions in the plasma are the form in which most carbon dioxide moves from tissue cells to the lungs.
5. The chloride shift again takes place (this time in the opposite direction) to maintain electrolyte balance.
6. A smaller amount (25%) of carbon dioxide from Step 1 reacts with hemoglobin to form carbaminohemoglobin ($HHbCO_2$) for transport to the lungs.
7. The increase in concentration of hydrogen ions inside the red blood cells promotes a reaction with oxyhemoglobin, which releases oxygen.
8. The free oxygen diffuses out of the red blood cells through the plasma, capillary membrane, interstitial fluid, and into the tissue cells.

■ **Learning Check 15.1** Decide whether each of the following red blood cell reactions occurs at the lungs or at the site of the tissue cells:

a. $HHbCO_2 \rightleftharpoons HHb + CO_2$
b. $H_2O + CO_2 \rightleftharpoons H_2CO_3$
c. $H^+ + HCO_3^- \rightleftharpoons H_2CO_3$
d. $HbO_2^- + H^+ \rightleftharpoons HHb + O_2$

CHEMISTRY AND YOUR HEALTH 15.1 Exercise Beats Angioplasty

Exercise is not just about looking good and fitting into that new pair of jeans; it is about maintaining a level of health with benefits ranging from a decrease in the chances of having a heart attack or stroke to reducing stress and symptoms of depression. That 20-minute jog around the block is one of the best medicines you can give your body to promote long-term health. Exercise has been identified as providing a surprising number of specific benefits including muscular endurance, increased energy, reduced risk of coronary artery disease, lower blood pressure and cholesterol levels, weight control, increased bone strength, higher metabolism rates, and improved reaction times. After a good workout, the body and mind both experience a state of relaxation that promotes elevated thinking, contributing to a more balanced life. In this fast-paced world driven by work and deadlines, every person should be exercising regularly to promote their mental health as well as their physical health.

Well-rounded physical fitness includes four basic elements that can be improved with regular exercise: cardiovascular endurance, muscular strength, muscular endurance, and flexibility. To be truly fit, each element should be developed, but the most vital is cardiovascular endurance. Physiologically, cardiovascular endurance is the sustained ability of the heart, blood vessels, and blood to carry oxygen to the cells, coupled with the ability of the cells to use oxygen, and the ability of the blood to carry away waste products. Since every cell in the body requires oxygen to function, there is no more basic element of fitness than this—to see that the heart, lungs, and circulatory system function properly.

Cardiovascular endurance is improved through exercises that require the body's systems to deliver even larger amounts of oxygen to working muscles. To achieve this, the exercise should utilize the large muscle groups (such as those in the legs), but, most importantly, it must be sustained. With regular aerobic exercise, your heart will develop the ability to pump more blood and thus deliver more oxygen with greater efficiency. Your muscles will develop a correspondingly greater capacity to use this oxygen. These improvements are part of what is called the aerobic "training effect." Because your heart is stronger, it can pump more blood per beat, and, as a result, your heart rate, both at rest and during exertions, will decrease. Your heart will also acquire the ability to recover from the stress of exercise more quickly.

The next time you are tempted to miss that daily dose of exercise, remember the benefits of getting up and moving. Your body, mind, and heart will thank you.

15.3 Chemical Transport to the Cells

LEARNING OBJECTIVE

3. Explain how materials move from the blood into the body cells and from the body cells into the blood.

For substances to be chemically transported in the body, they must become part of the moving bloodstream. They may dissolve in the water-based plasma (as do sugars, amino acids, ions, and gases to some extent), they may become chemically bonded to cellular components (as do oxygen and carbon dioxide with the hemoglobin of blood cells), or they may form a suspension in the plasma of the blood (as do lipids).

Materials transported by the bloodstream to the cells must move through the capillary wall into the interstitial fluid and through the cell membrane into the cell. The reverse must take place when waste products move from inside tissue cells into the bloodstream. During these processes, capillary walls behave as selectively permeable membranes or filters, allowing water containing dissolved nutrients—including oxygen—to pass in one direction and water containing dissolved wastes to pass in the other direction.

The movement of water and dissolved materials through the capillary walls is governed by two factors—by the pressure of blood against the capillary walls and by the differences in protein concentration on each side of the capillary walls.

The concentration of protein in plasma is much higher than the protein concentration of the interstitial fluid outside the blood vessels. This concentration difference results in an osmotic pressure of about 18 torr, which tends to move fluid into the bloodstream. However, the pumping action of the heart puts the blood under pressure (blood pressure) that is greater at the arterial end of a capillary

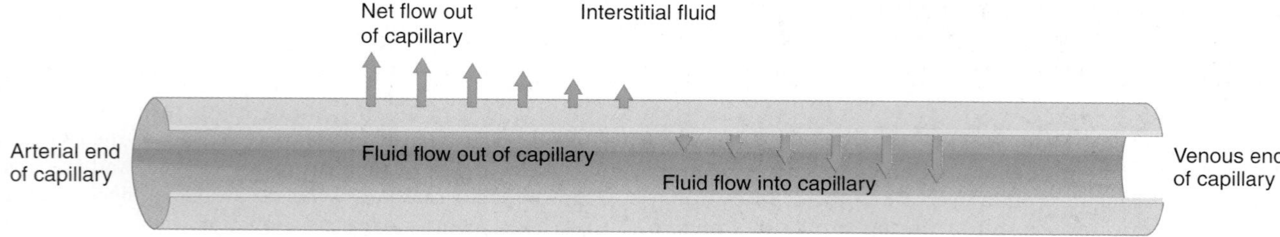

■ **FIGURE 15.7** The flow of fluid between capillaries and interstitial spaces.

(about 32 torr) than at the venous end (about 12 torr). Because the blood pressure at the arterial end is higher than the osmotic pressure, the tendency is for a net flow to occur from the capillary into the interstitial fluid (see ■ Figure 15.7). The fluid that leaves the capillary contains the dissolved nutrients, oxygen, hormones, and vitamins needed by the tissue cells.

As the blood moves through the capillary, its pressure decreases, until at the venous end, it becomes lower than the osmotic pressure, and a net flow of fluid takes place from the interstitial fluid into the capillary. The incoming fluid contains the metabolic waste products, such as carbon dioxide and excess water. This mechanism allows the exchange of nutrients and wastes and maintains fluid balance between the blood and interstitial space.

TABLE 15.1 Major dissolved solids in normal urine

Constituent[a]	Grams per 24 hours[b]
Amino acids	0.80
Urea	25.0
Creatinine	1.5
Uric acid	0.7
H^+	pH 4.5–8.0
Na^+	3.0
K^+	1.7
NH_4^+	0.8
Ca^{2+}	0.2
Mg^{2+}	0.15
Cl^-	6.3
SO_4^{2-}	1.4
$H_2PO_4^-$	1.2
HCO_3^-	0–3
Other compounds	2–3

[a]The 24-hour volume and composition of urine vary widely, depending on the fluid intake and diet.

[b]These data are for an average 24-hour specimen with a total volume of 1400 mL.

15.4 The Constituents of Urine

LEARNING OBJECTIVE

4. List the normal and abnormal constituents of urine.

Urine contains about 96% water and 4% dissolved organic and inorganic waste products. Approximately 40–50 g of dissolved solids are contained in the daily urine output of an adult (see ■ Table 15.1). The urine of a healthy person has a pH of 4.5–8.0, with 6.6 being a reasonable average for a person on an ordinary diet. Fruits and vegetables tend to make the urine alkaline, whereas high-protein foods tend to make the urine acidic.

The composition of urine is an excellent indicator of the state of health. This accounts for the importance of urinalysis as part of a physical checkup or a diagnosis. Normal constituents are considered abnormal when they are eliminated in excess amounts. ■ Table 15.2 lists some conditions that lead to abnormally high

TABLE 15.2 Some abnormal urine constituents

Abnormal constituent	Condition	Causes
Glucose (in large amounts)	Glucosuria	Diabetes mellitus, renal diabetes, alimentary glycosuria
Protein	Proteinuria or albuminuria	Kidney damage, nephritis, bladder infection
Ketone bodies	Ketonuria	Diabetes mellitus, starvation, high-fat diets
Pus (leukocytes)		Kidney or bladder infection
Hemoglobin	Hemoglobinuria	Excessive hemolysis of red blood cells
Red blood cells	Hematuria	Hemorrhage in the urinary tract
Bile pigments (in large amounts)	Jaundice	Blockage of bile duct, hepatitis, cirrhosis

urinary levels of normal constituents; some other substances (such as pus or blood), which are considered abnormal in *any* measurable amount, are also listed.

A urine specimen is often collected from a patient entering a hospital and is checked with a paper test strip that contains bands of different reagents that will react with the different abnormal constituents (see ■ Figure 15.8). This one simple test quickly checks for indications of pathological conditions.

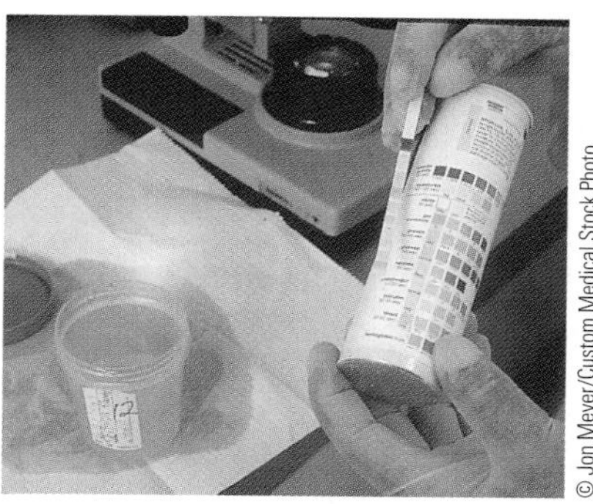

■ **FIGURE 15.8** Urinalysis test strips. A urine sample is tested for several components simultaneously by dipping a test strip into a sample and comparing the reagent areas to a color chart.

15.5 Fluid and Electrolyte Balance

LEARNING OBJECTIVE

5. Discuss how proper fluid and electrolyte balance is maintained in the body.

The body is 45–75% water (by weight), the percentage depending on a number of factors, including age, sex, and especially the amount of body fat. In general, the percentage of water increases as fat content decreases.

Fluid balance within the body is maintained not only by a balance in the total amount but by a normal and stable distribution of fluid in the body's three fluid-containing areas. Thus, the amounts of fluid inside the cells, in the interstitial spaces, and in the blood vessels must remain relatively constant if a fluid balance is to exist.

Fluid and electrolyte balance are interdependent. If one deviates from normal, so does the other. In modern medical practice, a great deal of attention is given to these matters. Many of today's hospital patients receive some form of fluid and electrolyte balance therapy.

An important and obvious principle of fluid balance is that fluid output and intake must be equal. Water from food and drink is absorbed into the body from the digestive tract. Besides this well-known source, water is also derived within the cells from food metabolism. This byproduct water enters the bloodstream along with the water that is absorbed from the digestive tract.

Water intake is regulated in part by the thirst mechanism (see ■ Figure 15.9). When the body loses large amounts of water, salivary secretions decrease, and a dry feeling develops in the mouth. This and other sensations are recognized as thirst, and water is drunk to relieve the condition. The fluid intake compensates for the fluid lost, and balance is reestablished.

Water normally leaves the body through the kidneys (urine), lungs (water vapor in expired air), skin (by diffusion and perspiration), and intestines (feces) (see ■ Table 15.3). Abnormally high fluid losses, and possibly dehydration, can be caused by hyperventilation, excessive sweating, vomiting, or diarrhea.

TABLE 15.3
Typical normal values for water entry and exit

Intake (mL)	
Ingested liquids	1500
Water in foods	700
Water formed by catabolism	200
Total	2400
Output (mL)	
Kidneys (urine)	1400
Lungs (water in expired air)	350
Skin	
By diffusion	350
By perspiration	100
Intestines (in feces)	200
Total	2400

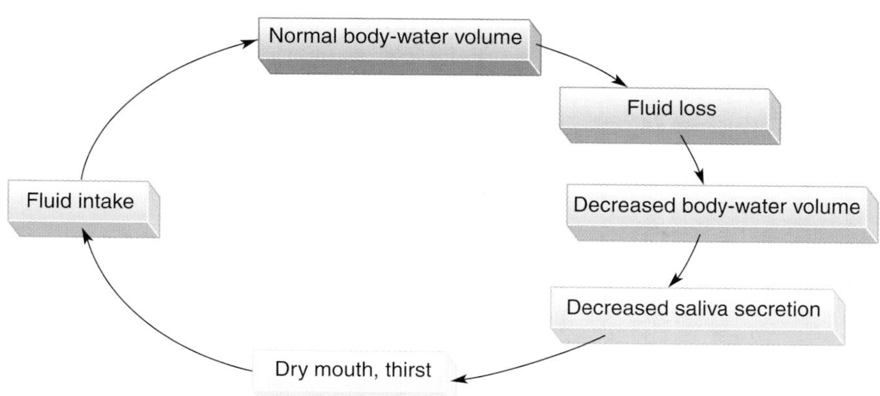

■ **FIGURE 15.9** The thirst mechanism.

Fluid balance in the body is maintained or restored primarily by variations in urine output. Normally, this amounts to about 1400 mL/day, but the total excretion adjusts according to water intake. The primary factor that controls urine production is the rate of water reabsorption from the renal tubules in the kidneys. This rate is regulated chiefly by the pituitary hormone vasopressin and by the adrenal cortex hormone aldosterone.

ThomsonNOW™ Go to Chemistry Interactive to explore **kidney function.**

Vasopressin, also called antidiuretic hormone (ADH), exerts its action by affecting the permeability of the renal tubules to water. In the absence of vasopressin, less water is reabsorbed into the bloodstream, and a large volume of light yellow, dilute urine results. Patients with the rare disease diabetes insipidus do not produce vasopressin; their urine is extremely voluminous and dilute. In the presence of vasopressin, the tubules become more permeable to water that is drawn into the capillaries, thus increasing blood volume. The reabsorption of water from the tubules produces a low volume of dark yellow, concentrated urine. (Diabetes insipidus is very different from diabetes mellitus, mentioned elsewhere in the text.)

With the help of vasopressin, a healthy person can vary the intake of water widely and yet preserve a stable overall concentration of substances in the blood. But sometimes the water level of the body gets dangerously low, from not drinking enough water or excessive water loss caused by diarrhea or excessive perspiration. Under such circumstances, the adrenal gland secretes the steroid hormone aldosterone. At the kidneys, aldosterone stimulates the reabsorption of sodium ions (electrolytes) from the renal tubes into the blood. Chloride ions follow the sodium ions to maintain electrical neutrality, and water follows the sodium chloride. Thus, aldosterone secretion conserves both salt and water in the body. When the fluid level of the body returns to normal, the amount of aldosterone secreted decreases.

15.6 Acid–Base Balance

LEARNING OBJECTIVE

6. Explain how acid–base balance is maintained in the body.

alkalosis
An abnormally high blood pH.

Blood pH normally remains within the rather narrow range of 7.35–7.45. A decrease in pH, called acidosis, or an increase, called **alkalosis,** is serious and requires prompt attention. Death can result when the pH value falls below 6.8 or rises above 7.8 (see ■ Figure 15.10). Acidosis, an abnormally low blood pH, was introduced in Section 14.6.

Large amounts of acids and smaller amounts of bases normally enter the blood. Some mechanisms must neutralize or eliminate these substances if the blood pH is to remain constant. In practice, a constant pH is maintained by the interactive operation of three systems: buffer, respiratory, and urinary. Only when all three parts of this complex mechanism function properly can the acid–base balance be maintained.

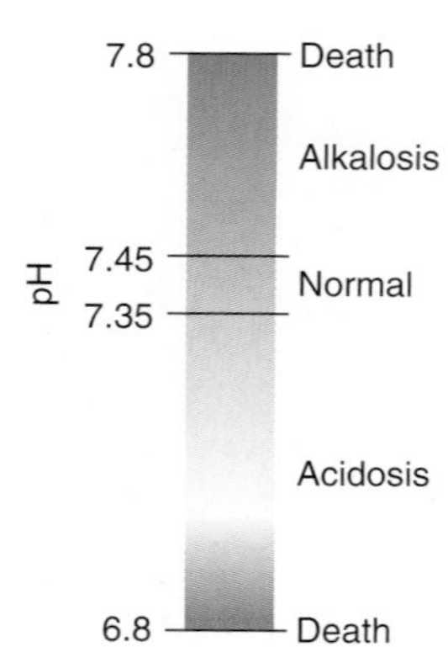

■ **FIGURE 15.10** Human circulatory and respiratory processes operate only within a very narrow range of blood pH.

15.7 Buffer Control of Blood pH

LEARNING OBJECTIVE

7. Explain how buffers work to control blood pH.

A buffer system prevents marked changes in the pH of a solution when an acid or base is added. Three major buffer systems of the blood are the bicarbonate buffer, the phosphate buffer, and the plasma proteins. The most important of these is the bicarbonate buffer system, consisting of a mixture of bicarbonate ions (HCO_3^-) and carbonic acid (H_2CO_3).

Buffer systems work by neutralizing H^+ or OH^- ions that come from the dissociation of acids or bases added to a system. As an example of buffers in action,

CHEMISTRY AROUND US 15.1 Exercise and Altitude

Most of us have experienced the feeling of being "out of breath." We breathe deeply, even gasp, in an attempt to inhale sufficient oxygen. Under normal conditions, the uncomfortable feeling passes rapidly as we restore a balance between the oxygen requirements of our body and the oxygen contained in the air we inhale. But what would happen if the oxygen concentration of the inhaled air was significantly reduced?

The concentration of oxygen in terms of its partial pressure in air decreases significantly with decreasing atmospheric pressure, which decreases with altitude. Physiological symptoms that accompany this decreasing oxygen concentration are relatively mild at lower altitudes. For example, at 8500 feet above sea level, about 25% of individuals not acclimated to that altitude suffer mild to moderate headaches, and a few also experience one or more of the following: loss of appetite, shortness of breath after mild exertion, general fatigue, insomnia, and nausea. More severe symptoms of oxygen deficiency, often called altitude sickness, are not usually encountered except at very high altitudes such as those reached by experienced mountain climbers. These symptoms, which can prove fatal, include very severe headaches, a drunken stagger, excessive laughter, and hallucinations.

People interested in participating in activities at altitudes significantly greater than those in which they normally live should seriously consider training in simulated altitude training environments. These are rooms or chambers in which the air pressure and oxygen concentrations can be adjusted to simulate conditions that would be encountered in any natural environment. Such training is especially important when the contemplated activities push the normal limits.

An extreme example of such an activity is the biannual marathon run that begins near the Mount Everest base camp area in the Himalayas at an altitude of about 16,800 feet above sea level. At this altitude, the atmospheric pressure and partial pressure of oxygen are about 50% of the values at normal sea level. The race finishes 27 miles later in the Sherpa capital of Namche Bazaar at an altitude of 11,200 feet above sea level. This extreme high-altitude challenge was first run in 1987 despite serious objections from altitude experts who believed the risk of running that distance at that extreme altitude was unjustifiable. However, the event has proved to be a safe, but very difficult, race. There have been no fatalities, and only 5–10% of the participants in the seven races have been unable to complete the entire course. These results indicate the tremendous adaptability of the human body if proper training and conditioning are utilized.

Even if a challenging mountain-climbing expedition is not a goal, anyone planning a trip to an area with an altitude significantly greater than their normal environment should use caution in choosing the kinds of activities in which to participate. In addition, training should be done if it is at all possible. At the very minimum, plans should be made for a day or two of low to moderate activity at the higher altitude to allow the body to acclimate before tackling the more strenuous activities.

consider the fate of lactic acid (HLac) entering the bloodstream. When lactic acid dissolves in the blood, it dissociates to produce H^+ and lactate ions (Lac^-). The liberated H^+ ions drive the pH of the blood down. The bicarbonate ions of the bicarbonate buffer system react with the excess H^+ and return the pH to the normal range:

$$\mathrm{HLac} \rightleftharpoons \mathrm{H^+} + \mathrm{Lac^-} \qquad (15.3)$$

$$\mathrm{H^+} + \mathrm{HCO_3^-} \rightleftharpoons \mathrm{H_2CO_3} \qquad (15.4)$$

Reaction 15.4 is a net ionic reaction for the buffering process. When a total reaction is written, we see that the effect of the buffering is to replace lactic acid in the blood by carbonic acid, which is a weaker acid and so releases fewer H^+ ions to the blood than does lactic acid:

$$\underset{\substack{\text{stronger}\\\text{acid}}}{\mathrm{HLac}} + \mathrm{HCO_3^-} \longrightarrow \underset{\substack{\text{weaker}\\\text{acid}}}{\mathrm{H_2CO_3}} + \mathrm{Lac^-} \qquad (15.5)$$

This replacement of a stronger acid with a weaker acid is how biochemists often describe buffering reactions.

Although the bicarbonate system is not the strongest buffer system in the body (the most powerful and plentiful one is the proteins of plasma and cells), it is very important because the concentrations of both bicarbonate and carbonic acid are regulated by the kidneys and by the respiratory system. Without the respiratory and urinary processes, the capacity of buffers would eventually be exceeded.

15.8 Respiratory Control of Blood pH

LEARNING OBJECTIVE

8. Describe respiratory control of blood pH.

The respiratory system helps control the acidity of blood by regulating the elimination of carbon dioxide and water molecules. These molecules, exhaled in every breath, come from carbonic acid as follows (see Figure 15.5):

$$\underset{\text{carbonic acid}}{H_2CO_3} \rightleftharpoons H_2O + CO_2 \quad (15.6)$$

The more CO_2 and H_2O that are exhaled, the more carbonic acid is removed from the blood, thus elevating the blood pH to a more alkaline level.

The respiratory mechanism for controlling blood pH begins in the brain with respiratory center neurons that are sensitive to blood CO_2 levels and pH. A significant increase in the CO_2 of arterial blood, or a decrease below about 7.38 of arterial blood pH, causes the breathing to increase both in rate and depth, resulting in **hyperventilation.** This increased ventilation eliminates more carbon dioxide, reduces carbonic acid and hydrogen-ion concentrations, and increases the blood pH back toward the normal level (see ■ Figure 15.11).

hyperventilation Rapid, deep breathing.

In the opposite situation, an increase in blood pH above normal causes **hypoventilation,** a reduced rate of respiration. Less CO_2 is exhaled, and the higher concentration of carbonic acid remaining in the blood lowers the pH back to normal.

hypoventilation Slow, shallow breathing.

15.9 Urinary Control of Blood pH

LEARNING OBJECTIVE

9. Describe urinary control of blood pH.

Because they can excrete varying amounts of acid and base, the kidneys, like the lungs, play a vital role in pH control. For example, with acidic blood, the excretion of hydrogen ions by the kidneys decreases the urine pH and simultaneously increases the blood pH back toward normal.

■ Figure 15.12 traces the reactions involved in the excretion of hydrogen ions by the kidneys:

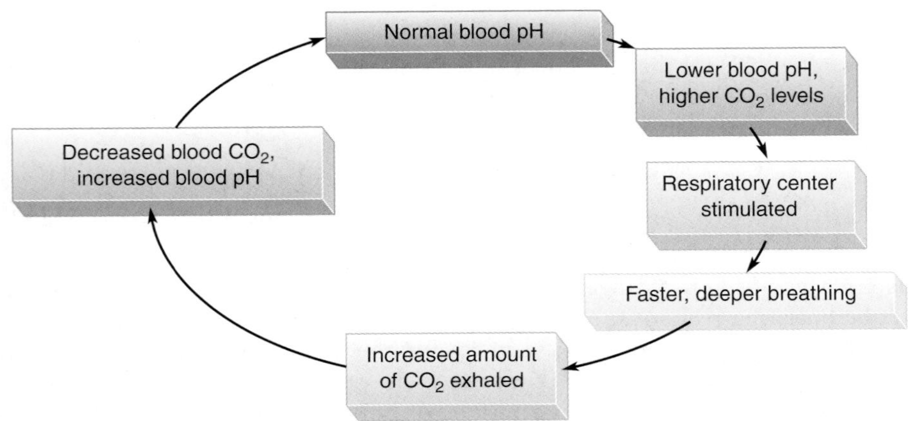

■ **FIGURE 15.11** Respiratory control of blood pH.

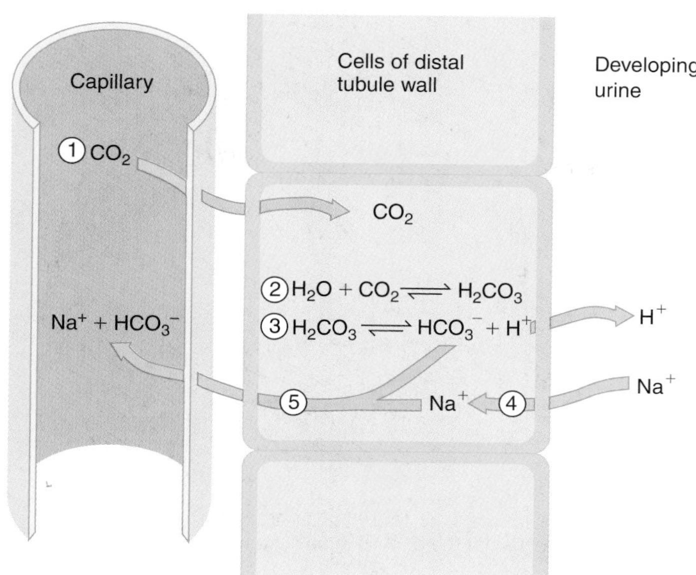

■ **FIGURE 15.12** Urinary control of blood pH in the kidney.

1. Carbon dioxide diffuses from the blood capillaries into the kidney distal tubule cells.
2. Catalyzed by carbonic anhydrase, water and carbon dioxide react to give carbonic acid.
3. The carbonic acid ionizes to give hydrogen ions and bicarbonate ions. The hydrogen ions diffuse into the developing urine.
4. For every hydrogen ion entering the urine, a sodium ion passes into the tubule cells.
5. The sodium ions and bicarbonate ions enter the bloodstream capillaries.

The net result of these reactions is the conversion of CO_2 to HCO_3^- within the blood. Both the decrease in CO_2 and the increase in HCO_3^- tend to increase blood pH levels back to normal. The developing urine has picked up the hydrogen ions, which now react with buffering ions such as phosphate that are present in urine:

$$H^+ + HPO_4^{2-} \rightarrow H_2PO_4^- \qquad (15.7)$$

The presence of the phosphate buffer system ($H_2PO_4^{2-}/HPO_4^{2-}$) usually keeps urine from going much below pH 6, but too great an excess of protons can exceed the buffer capacity of the system and result in quite acidic urine.

15.10 Acidosis and Alkalosis

LEARNING OBJECTIVE

10. List the causes of acidosis and alkalosis.

When blood pH is normal and a state of acid–base balance exists, components of the bicarbonate buffer are present in plasma in a ratio of 20 parts of bicarbonate (HCO_3^-) to 1 part of carbonic acid (H_2CO_3).

When respiratory and urinary systems work together to maintain the blood at pH 7.4, they do so primarily by maintaining the HCO_3^-/H_2CO_3 ratio at 20:1. Any change in the ratio produces a change in pH. An increase in the ratio of bicarbonate to carbonic acid causes the pH to increase (alkalosis); and a decrease, which is much more common, causes the pH to decrease (acidosis) (see ■ Figure 15.13).

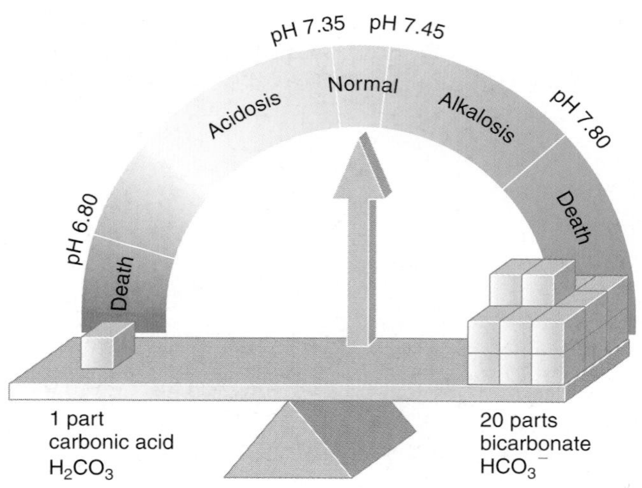

FIGURE 15.13 Blood pH depends on the relative concentrations of carbonic acid and bicarbonate ion.

Two adjectives describe the general cause or origin of pH imbalances in body fluids: metabolic and respiratory. Respiratory acid–base imbalances (respiratory acidosis and respiratory alkalosis) result from abnormal breathing patterns. Metabolic acidosis and metabolic alkalosis are caused by factors other than abnormal breathing.

Respiratory Alkalosis

respiratory alkalosis
A condition of alkalosis caused by hyperventilation.

Respiratory alkalosis is caused by hyperventilating—breathing rapidly and deeply. Hysteria, anxiety, or prolonged crying may result in hyperventilation in which too much carbon dioxide is exhaled, which disturbs the blood's normal carbon dioxide/carbonic acid ratio. According to Le Châtelier's principle, Reaction 15.6 (written again below) shifts to the right as it attempts to restore the equilibrium:

$$\underset{\xrightarrow[\text{restore equilibrium}]{\text{direction of shift to}}}{H_2CO_3 \rightleftharpoons H_2O + \overset{\textit{loss of } CO_2}{CO_2 \uparrow}} \qquad (15.8)$$

The equilibrium shift reduces the amount of carbonic acid present and rapidly increases the HCO_3^-/H_2CO_3 ratio to a value greater than 20:1. As a result, blood pH rises to 7.54 or higher within a few minutes. A corresponding loss of HCO_3^- also occurs, but it is not large enough to maintain the HCO_3^-/H_2CO_3 ratio at 20:1.

The treatment of respiratory alkalosis involves rebreathing one's own exhaled air by breathing into a paper bag (*not* a plastic bag), administering carbon dioxide, and treating the underlying causes of the hyperventilation.

Respiratory Acidosis

respiratory acidosis
A condition of acidosis caused by hypoventilation.

Respiratory acidosis is sometimes the result of slow or shallow breathing (hypoventilation) caused by an overdose of narcotics or barbiturates. Anesthetists need to be particularly aware of respiratory acidosis, because most inhaled anesthetics depress respiration rates. Lung diseases, such as emphysema and pneumonia, or an object lodged in the windpipe can also result in hypoventilation. In respiratory acidosis, too little carbon dioxide is exhaled, and its concentration within the blood increases. The carbon dioxide/carbonic acid equilibrium shifts to the left to restore the equilibrium:

$$H_2CO_3 \rightleftharpoons H_2O + CO_2 \quad \text{(increased } CO_2\text{)} \tag{15.9}$$

$\longleftarrow$ direction of shift of equilibrium

As a result of the shift, the concentration of carbonic acid in the blood increases, the HCO_3^-/H_2CO_3 ratio decreases to less than 20:1, and respiratory acidosis sets in. The treatment of respiratory acidosis involves identifying the underlying causes and possibly the intravenous administration of isotonic sodium bicarbonate solution or hemodialysis.

Metabolic Acidosis

Various body processes—collectively called metabolism—produce acidic substances that liberate H^+ in solution. The diffusion of these substances into the bloodstream causes a shift in the carbonic acid–bicarbonate equilibrium:

$$H_2CO_3 \rightleftharpoons H^+ + HCO_3^- \quad \text{(increased } H^+\text{)} \tag{15.10}$$

$\longleftarrow$ direction of shift to restore equilibrium

The shift decreases the concentration of bicarbonate ions and increases the concentration of carbonic acid. As a result, the HCO_3^-/H_2CO_3 ratio decreases, as does the blood pH. **Metabolic acidosis** is often a serious problem in uncontrolled diabetes mellitus, and it also occurs temporarily after heavy exercise. Other causes include severe diarrhea and aspirin overdose.

metabolic acidosis A condition of acidosis resulting from causes other than hypoventilation.

Symptoms of metabolic acidosis, besides low blood pH and high levels of H_2CO_3, include the following: hyperventilation as the respiratory center tries to get excess CO_2 out of the lungs, increased urine formation as the kidneys try to remove excess acids, thirst (to replace urination water loss), drowsiness, headache, and disorientation. The treatment again depends on the cause; it may involve insulin therapy, intravenous bicarbonate, or hemodialysis. Note carefully that hyperventilation can be a symptom of metabolic acidosis and a cause of respiratory alkalosis, two different conditions requiring very different treatments.

Metabolic Alkalosis

In **metabolic alkalosis,** the body has lost acid in some form, perhaps from prolonged vomiting of the acidic stomach contents, or there has been an ingestion of alkaline substances. Excessive use of sodium bicarbonate (baking soda), for example, a common home remedy for upset stomach, increases the concentration of bicarbonate ions in the blood.

metabolic alkalosis A condition of alkalosis resulting from causes other than hyperventilation.

A decrease in acid concentration within the body (from vomiting) or an increase in HCO_3^- levels (from excessive $NaHCO_3$) causes the HCO_3^-/H_2CO_3 ratio to rise above 20:1, and measurable alkalosis is observed. The respiratory center responds to the higher blood pH by slowing the breathing (hypoventilation). Other symptoms may include numbness, tingling, and headache. ■ Table 15.4 is a summary of causes of acidosis and alkalosis.

TABLE 15.4 Some causes of blood acid–base imbalance

Condition	Causes
Respiratory	
Acidosis: CO_2 ↑ pH ↓	Hypoventilation, blockage of diffusion within lungs, respiratory center depressants
Alkalosis: CO_2 ↓ pH ↑	Hyperventilation, excitement, trauma
Metabolic	
Acidosis: H^+ ↑ pH ↓	Kidney failure, prolonged diarrhea, ketone bodies from diabetes mellitus
Alkalosis: H^+ ↓ pH ↑	Kidney disease, prolonged vomiting, excessive intake of baking soda

Concept Summary

ThomsonNOW™ Sign in at www.thomsonedu.com to:

- Assess your understanding with Exercises keyed to each learning objective.
- Check your readiness for an exam by taking the **Pre-test** and exploring the modules recommended in your **Personalized Learning Plan.**

A Comparison of Body Fluids. The fluids of the body are located inside tissue cells (intracellular fluid) or outside tissue cells (extracellular fluid). Volumetrically, most of the extracellular fluids are of two types—blood and interstitial fluid, the fluid that occupies spaces between cells and moves in the lymph vessels. Other extracellular body fluids that occur in smaller amounts are urine, digestive juices, and cerebrospinal fluid. Chemically, blood (plasma) and interstitial fluid (including lymph) are similar. Intracellular fluid is chemically different from the extracellular fluids. **OBJECTIVE 1 (Section 15.1), Exercise 15.2**

Oxygen and Carbon Dioxide Transport. Nearly all oxygen that is transported from the lungs to tissue cells is transported by the blood in the form of oxyhemoglobin. Most carbon dioxide transported from the cells to the lungs is transported by the blood in the form of bicarbonate ions dissolved in the plasma. Smaller amounts also travel as carbaminohemoglobin and as CO_2 gas dissolved in plasma. The formation of the two hemoglobin compounds and the dissolving of various materials in the plasma are influenced by equilibria related to concentration differences of materials in the blood as it passes through capillaries in the lungs or near tissue cells. **OBJECTIVE 2 (Section 15.2), Exercise 15.10**

Chemical Transport to the Cells. Substances transported by the blood to tissue cells move from the blood, through capillary walls, into the interstitial fluid, then through cell membranes, and finally into the cytoplasm of the cells. Waste products of the cells move in the opposite direction. The movement of fluid through capillary walls is governed by the blood pressure against the capillary walls and by osmotic pressure differences that arise from protein concentration differences between blood and interstitial fluid. **OBJECTIVE 3 (Section 15.3), Exercise 15.14**

The Constituents of Urine. Normal urine is about 96% water and 4% dissolved organic and inorganic wastes. Because urine composition is an excellent indicator of a person's health, urinalysis is a routine procedure in physical checkups. **OBJECTIVE 4 (Section 15.4), Exercise 15.18**

Fluid and Electrolyte Balance. Fluid balance in the body involves a balance between fluid intake and output, and a proper distribution of fluid between the three fluid areas. Fluid and electrolyte balance are interdependent. Water enters the body through food and drink and leaves in the form of urine, water vapor in exhaled air, perspiration, and feces. The amount of fluid in the body is maintained or restored primarily by variations in urine output, which is regulated chiefly by the hormones vasopressin and aldosterone. **OBJECTIVE 5 (Section 15.5), Exercise 15.22**

Acid-Base Balance. Blood pH is normally in the range of 7.35–7.45. Three systems cooperate to maintain the pH in this narrow range. **OBJECTIVE 6 (Section 15.6), Exercise 15.26.** Acidosis and alkalosis exist when the blood pH becomes lower or higher, respectively, than these values.

Buffer Control of Blood pH. The blood contains three major buffer systems: the bicarbonate system, the phosphate system, and the plasma protein system. The most important is the bicarbonate system, which consists of a mixture of bicarbonate ions and carbonic acid. **OBJECTIVE 7 (Section 15.7), Exercise 15.28**

Respiratory Control of Blood pH. The respiratory system influences blood pH by regulating the elimination of carbon dioxide and water. Because carbon dioxide and water are produced during the decomposition of carbonic acid, the greater the elimination of carbon dioxide and water from the body, the greater is the elimination of carbonic acid from the blood. **OBJECTIVE 8 (Section 15.8), Exercise 15.30**

Urinary Control of Blood pH. The kidneys influence blood pH by varying the amount of hydrogen ions excreted in the urine. When the blood is too acidic, carbonic acid is converted to bicarbonate ions and hydrogen ions. The hydrogen ions are excreted, and the bicarbonate ions enter the blood. **OBJECTIVE 9 (Section 15.9), Exercise 15.34**

Acidosis and Alkalosis. Acidosis and alkalosis can result from metabolic or respiratory causes. Respiratory acidosis and alkalosis are the result of abnormal breathing patterns. Alkalosis is caused by hyperventilation that accompanies hysteria, anxiety, or prolonged crying. Acidosis can result from hypoventilation brought on by such factors as an overdose of depressant narcotics or anesthesia. Metabolic acidosis can result from the metabolic production of acids in the body or the elimination of alkaline materials, as during severe diarrhea. Metabolic alkalosis results from the loss of acid in some form such as the prolonged vomiting of acidic stomach contents or the ingestion of alkaline materials. **OBJECTIVE 10 (Section 15.10), Exercise 15.38**

Key Terms and Concepts

Alkalosis (15.6)
Carbaminohemoglobin (15.2)
Chloride shift (15.2)
Deoxyhemoglobin (hemoglobin) (15.2)
Extracellular fluid (15.1)
Hyperventilation (15.8)
Hypoventilation (15.8)
Interstitial fluid (15.1)
Intracellular fluid (15.1)
Metabolic acidosis (15.10)
Metabolic alkalosis (15.10)
Oxyhemoglobin (15.2)
Plasma (15.1)
Respiratory acidosis (15.10)
Respiratory alkalosis (15.10)

Key Reactions

1. Formation of oxyhemoglobin (Section 15.2):

$$HHb + O_2 \rightleftarrows HbO_2^- + H^+$$ Reaction 15.1

2. Formation of carbaminohemoglobin (Section 15.2):

$$HHb + CO_2 \rightleftarrows HHbCO_2$$ Reaction 15.2

3. Decomposition of carbonic acid (Section 15.8):

$$H_2CO_3 \rightleftarrows H_2O + CO_2$$ Reaction 15.6

Exercises

SYMBOL KEY

Even-numbered exercises are answered in Appendix B.

Blue-numbered exercises are more challenging.

■ denotes exercises available in ThomsonNow and assignable in OWL.

ThomsonNOW To assess your understanding of this chapter's topics with sample tests and other resources, sign in at www.thomsonedu.com.

A COMPARISON OF BODY FLUIDS (SECTION 15.1)

15.1 Where is most fluid in the body located? What is it called?

15.2 ■ Which two of the following have nearly the same chemical composition: plasma, intracellular fluid, interstitial fluid?

15.3 In what way does the composition of the two similar fluids in Exercise 15.2 differ?

15.4 Which of the fluids of Exercise 15.2 contains the highest concentration of the following?

a. protein
b. K^+
c. Na^+
d. HPO_4^{2-}
e. HCO_3^-
f. Mg^{2+}
g. Cl^-

15.5 ■ What is the principal cation found in the following?

a. blood plasma
b. interstitial fluid
c. intracellular fluid

15.6 ■ What is the principal anion in the following?

a. blood plasma

b. interstitial fluid

c. intracellular fluid

15.7 What is the principal type of organic compound found in intracellular fluid?

OXYGEN AND CARBON DIOXIDE TRANSPORT (SECTION 15.2)

15.8 Write a reversible chemical equation that shows the reaction of hemoglobin (Hb) with:

a. oxygen b. carbon dioxide

15.9 ■ Hemoglobin exists in several forms—oxyhemoglobin, deoxyhemoglobin, and carbaminohemoglobin. Match each of the following to the type of hemoglobin involved:

a. Formed when CO_2 combines with hemoglobin

b. The oxygen-transporting form of hemoglobin

c. The form of hemoglobin designated as HHb

15.10 ■ What percentages of the total O_2 and CO_2 transported by the blood are in the following forms?

a. O_2 as dissolved gas in plasma

b. O_2 as oxyhemoglobin

c. CO_2 as carbaminohemoglobin

d. CO_2 as dissolved gas in plasma

e. CO_2 as bicarbonate ions

15.11 What term is given to the reversible flow of chloride ion across the red blood cell membrane during the oxygen and carbon dioxide transport process?

15.12 Why are large capillary surface areas critical to O_2 and CO_2 transport processes? What driving force causes O_2 to move from lung alveoli into the bloodstream and CO_2 to move from the bloodstream into lung alveoli?

15.13 ■ Write the equilibrium equation representing the reaction between oxygen and hemoglobin. In which direction will the equilibrium shift when the following occurs?

a. The pH is increased.

b. The concentration (pressure) of O_2 is increased.

c. The concentration of CO_2 is increased (see Figure 15.5).

d. The blood is in a capillary near actively metabolizing tissue cells.

CHEMICAL TRANSPORT TO THE CELLS (SECTION 15.3)

15.14 ■ Which factor, blood pressure or osmotic pressure, has the greater influence on the direction of fluid movement through capillary walls at (a) the venous end of a capillary and (b) the arterial end of a capillary?

15.15 What types of compounds leave the capillaries and enter the interstitial fluid? What types of compounds leave the interstitial fluid and enter the capillaries to be transported away?

THE CONSTITUENTS OF URINE (SECTION 15.4)

15.16 What organic material is normally excreted in the largest amount in urine?

15.17 What cation and anion are normally excreted in the largest amounts in urine?

15.18 ■ List the abnormal constituents of urine.

15.19 Under what circumstances would the presence of a normal constituent of urine be considered abnormal?

FLUID AND ELECTROLYTE BALANCE (SECTION 15.5)

15.20 List four routes by which water leaves the body.

15.21 What are three sources of water that contribute to the fluid intake of a human body?

15.22 ■ What is the primary way fluid balance is maintained in the body?

15.23 Describe how vasopressin and aldosterone function to maintain fluid levels in the blood.

15.24 Trace the events of the regulation of fluid balance by the thirst mechanism.

ACID–BASE BALANCE (SECTION 15.6)

15.25 What is the normal pH range of human blood? Outside what pH range will death occur?

15.26 ■ List three systems that cooperate to maintain blood pH in an appropriate narrow range. Which one of these provides the most rapid response to changes in blood pH?

15.27 What is meant by the terms *acidosis* and *alkalosis?*

BUFFER CONTROL OF BLOOD PH (SECTION 15.7)

15.28 ■ List three major buffer systems of the blood.

15.29 Hemoglobin (HHb) is a weak acid that forms the Hb^- anion. Write equations to show how the HHb/Hb^- buffer would react to added H^+ and added OH^-.

RESPIRATORY CONTROL OF BLOOD PH (SECTION 15.8)

15.30 Describe the respiratory mechanism for controlling blood pH.

15.31 ■ Hyperventilation is a symptom of what type of acid–base imbalance in the body?

15.32 Hypoventilation is a symptom of what type of acid–base imbalance in the body?

URINARY CONTROL OF BLOOD PH (SECTION 15.9)

15.33 Write a short summary description of the role of the kidneys and urine in the control of blood pH.

15.34 ■ What happens to CO_2 in the blood when the kidneys function to control blood pH?

15.35 What buffer system in the urine usually keeps the urine pH from dropping lower than 6?

15.36 What ionic shift maintains electron charge balance when the kidneys function to increase the pH of blood?

ACIDOSIS AND ALKALOSIS (SECTION 15.10)

15.37 What is the normal ratio of bicarbonate to carbonic acid in the blood?

15.38 ■ What is the difference between the causes of respiratory acid–base imbalances and metabolic acid–base imbalances in the body?

15.39 What type of acid–base imbalance in the blood is caused by hypoventilation? By hyperventilation?

15.40 What type of acid–base imbalance might develop in an individual who has taken an overdose of a depressant narcotic? In an individual with uncontrolled diabetes mellitus?

15.41 Explain how prolonged vomiting can lead to alkalosis.

ADDITIONAL EXERCISES

15.42 An average adult excretes about 15.0 g of urea in his or her urine daily. Assuming both nitrogen atoms of the urea molecule come from the deamination of amino acids (one N atom/amino acid), how many amino acid molecules were deaminated to produce this amount of urea?

$$H_2N-\overset{\overset{\displaystyle O}{\|}}{C}-NH_2 \quad \text{urea}$$

15.43 A person inhales O_2 and exhales CO_2. What is the ultimate use for the O_2 in the cell? Is the O_2 that we inhale incorporated into the CO_2 we exhale? What is the source of the CO_2 we exhale?

15.44 When the body is dehydrated, aldosterone causes the active readsorption of Na+ ions from the renal tubules into the blood, which causes water to "follow" the Na+ ions into the blood. Explain what causes this flow of water to occur.

15.45 The body uses the phosphate buffer system shown in the following reaction (a).

$$H^+ + HPO_4^{2-} \leftrightarrows H_2PO_4^- \text{ (a)}$$

$$H^+ + H_2PO_4^- \leftrightarrows H_3PO_4 \text{ (b)}$$

Explain why reaction (b) is not used in the human body as a buffer system. (HINT: What are the pKa values for the dissociations of the individual hydrogen ions?)

15.46 Explain how heavy exercise can cause metabolic acidosis.

ALLIED HEALTH EXAM CONNECTION

Reprinted with permission from Nursing School and Allied Health Entrance Exams, COPYRIGHT 2005 Petersons.

15.47 Rank the following in order from least important to most important in removing CO_2 from the body: carbaminohemoglobin, dissolved in plasma, bicarbonate.

15.48 Movement of respiratory gases into and out of blood is primarily dependent upon the presence of a _____ _____ gradient.

15.49 The most rapid method of pH regulation involves the blood's _____ system.

15.50 Which of the following are normal constituents of urine?

a. Creatinine

b. Urea

c. Water

d. Ammonia

e. Sugar

f. Hormones

g. Blood

15.51 Uncontrolled diabetes mellitus results in _____ acidosis.

CHEMISTRY FOR THOUGHT

15.52 Explain why an hysterical individual might develop alkalosis.

15.53 Explain why a person suffering from metabolic acidosis might develop a thirst.

15.54 Ordinary water may be the most effective "sports drink." Explain why this might be true.

15.55 How does the concentration of hemoglobin of someone living in the Himalayas compare with that of someone living at a lower altitude?

15.56 What happens if the amount of H^+ ions in the urine exceeds urine's buffering capacity?

15.57 Aerobic exercise causes your cells to produce CO_2 and H_2O. Why can you dehydrate during aerobic exercise even if you don't perspire?

15.58 ■ What types of food can make urine acidic? What types of food will make it basic?

15.59 Explain how uncontrolled diarrhea can lead to acidosis.

APPENDIX A

The International System of Measurements

The International System of Units (SI units) was established in 1960 by the International Bureau of Weights and Measures. The system was established in an attempt to streamline the metric system, which included certain traditional units that had historical origins but that were not logically related to other metric units. The International System established fundamental units to represent seven basic physical quantities. These quantities and the fundamental units used to express them are given in Table A.1.

TABLE A.1 Fundamental SI units

Physical quantity	SI unit	Abbreviation
Length	meter	m
Mass	kilogram	kg
Temperature	kelvin	K
Amount of substance	mole	mol
Electrical current	ampere	A
Time	second	s
Luminous intensity	candela	cd

All other SI units are derived from the seven fundamental units. Prefixes are used to indicate multiples or fractions of the fundamental units (Table 1.2). In some cases, it has been found convenient to give derived units specific names. For example, the derived unit for force is kg m/s^2, which has been given the name newton (abbreviated N) in honor of Sir Isaac Newton (1642–1726). Some examples of derived units are given in Table A.2.

TABLE A.2 Examples of derived SI units

Physical quantity	Definition in fundamental units	Specific name	Abbreviation
Volume	m^3 (see NOTE below)	—	—
Force	kg m/s^2	newton	N
Energy	kg m^2/s^3 = N m	joule	J
Power	kg m^2/s^3 = J/s	watt	W
Pressure	kg/m s^2 = N/m^2	pascal	Pa
Electrical charge	A s	coulomb	C
Electrical potential	kg m^2/A s^3 = W/A	volt	V
Electrical resistance	kg m^2/A^2 s^3 = V/A	ohm	Ω
Frequency	1/s	hertz	Hz

NOTE: The liter (L), a popular volume unit of the metric system, has been redefined in terms of SI units as 1 dm^3 or 1/1000 m^3. It and its fractions (mL, μL, etc.) are still widely used to express volume.

APPENDIX B

Answers to Even-Numbered End-of-Chapter Exercises

CHAPTER 1

1.2 Hair, skin, clothing, paper, plastic, carpet

1.4 They all contain carbon.

1.6 Covalent

1.8
- a. Organic; most organic compounds are flammable.
- b. Inorganic; a high-melting solid is typical of ionic compounds.
- c. Organic; many organic compounds are insoluble in water.
- d. Organic; most organic compounds can undergo combustion.
- e. Organic; a low-melting solid is characteristic of covalent materials which are usually organic.

1.10 Organic compounds do not readily accept electrons to transport them, thus making the flow of electrical current impossible.

1.12 The ability of carbon to bond to itself repeatedly; isomerism.

1.14 The *s* orbital of the hydrogen overlaps with the sp^3 orbital of carbon.

1.16 A *p* orbital and an sp^3 orbital both have a two-lobed shape. A *p* orbital has equal-sized lobes, whereas an sp^3 orbital has different-sized lobes.

1.18

```
      H   H   H                 H   O
      |   |   |                 |   ||
a. H—C — C — C—H        c. H—C — C—H
      |   |   |                 |
      H   H   H                 H

      H   H   H                 H   H
      |   |   |                 |   |    H
b. H—C — C = C—H        d. H—C — C—N/
      |                         |   |   \
      H                         H   H    H
```

1.20 b, d, e

1.22
- a. Incorrect; H cannot have two bonds.
- b. Correct
- c. Incorrect; the carbon with the C=O bond has five bonds.
- d. Incorrect; both carbons of the C=C double bond have three bonds.
- e. Correct

1.24
- a. $CH_3CH_2CH_2CH{=}CH-\overset{\overset{\displaystyle O}{\|}}{C}-CH_2CH_3$
- b. $CH_3CH_2-\overset{\overset{\displaystyle O}{\|}}{C}-NH_2$

1.26

```
      O   H   H
      ||  |   |    H
a. H—C — C — C—N/
          |   |    \
          H   H     H

      H         H         H
      |         |         |
b. H—C ——————— C — O — C—H
      |         |         |
      H      H—C—H        H
                |
                H
```

1.28
- a. Branched
- b. Normal
- c. Normal
- d. Normal
- e. Branched
- f. Branched

1.30
- a. Same
- b. Same
- c. Isomers
- d. Same

1.32
- a. Six
- b. Six
- c. Eight

1.34
- a. 3-methylpentane
- b. 2-methylpropane
- c. 3,6-diethyl-4-methylnonane
- d. 4,7-diethyl-5-methyldecane
- e. 3-ethyl-5-methylnonane

1.36

a.
```
CH3—CH2—CH—CH2—CH3
        |
        CH2CH3
```

b.
```
        CH3
        |
CH3—C—CH2—CH3
        |
        CH3
```

c.
```
             CH3    CH2CH2CH3
              |      |
CH3—CH2—C—CH—CH—CH2—CH2—CH2—CH2—CH3
              |  |
            CH3 CH2CH3
```

d.
```
CH3—CH2—CH2—CH2—CH—CH2—CH2—CH2—CH2—CH3
                  |
              CH3CHCH2CH3
```

1.38

```
                        CH3
                        |
CH3CH2CH2CH2CH3     CH3CHCH2CH3
    pentane         2-methylbutane

            CH3
            |
          CH3CCH3
            |
            CH3
      2,2-dimethylpropane
```

1.40

Structure	Name
$CH_3CH_2CH_2CH_2CH_2CH_3$	hexane
$CH_3CH(CH_3)CH_2CH_2CH_3$	2-methylpentane
$CH_3CH_2CH(CH_3)CH_2CH_3$	3-methylpentane
$CH_3C(CH_3)_2CH_2CH_3$	2,2-dimethylbutane
$CH_3CH(CH_3)CH(CH_3)CH_3$	2,3-dimethylbutane

1.42

a.
```
CH3
|
CH2CHCH3
   |
   CH3
```

The longest carbon chain is four carbon atoms rather than three. The correct name is 2-methylbutane.

b.
```
       CH3
       |
CH3CH2CHCHCH3
          |
          CH3
```

The chain should be numbered from the right to give 2,3-dimethylpentane.

c.
```
   CH2CH3
   |
CH3CHCH2CHCH3
        |
        CH3
```

The longest carbon chain is six carbon atoms and should be numbered from the right to give 2,4-dimethylhexane.

d.
```
   Br
   |
CH3CHCHCH3
      |
      CH2CH3
```

The longest carbon chain is five carbon atoms rather than four. The correct name is 2-bromo-3-methylpentane.

1.44

a. cyclopentane

b. 1,2-dimethylcyclobutane

c. 1,1-dimethylcyclohexane

d. 1,2,3-trimethylcyclobutane

1.46

a. CH_2CH_3 (on cyclobutane ring)

b. CH_3, CH_3, CH_3, CH_3 (on cyclohexane ring)

c. $CH_2CH_2CH_2CH_3$ and $CHCH_3$ with CH_3 (on cyclopentane ring)

1.48 b, c, and d

1.50 So that the carbon atoms with attached hydrogens can assume a tetrahedral shape

1.52

a. none

b. none

c.

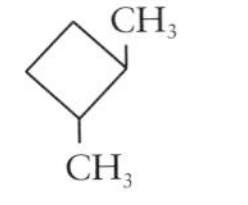

trans-1,2-dimethylcyclobutane

CH_3 CH_3

cis-1,2-dimethylcyclobutane

d.

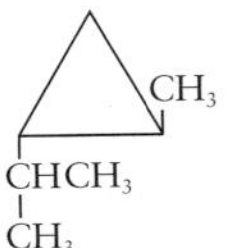

trans-1-isopropyl-2-methylcyclopropane

$CHCH_3$ CH_3 / CH_3

cis-1-isopropyl-2-methylcyclopropane

1.54 a. *trans*-1-ethyl-2-methylcyclopropane

b. *cis*-1-bromo-2-chlorocyclopentane

c. *trans*-1-methyl-2-propylcyclobutane

d. *trans*-1,3-dimethylcyclohexane

1.56 a. Liquid

b. No

c. Yes

d. Less dense

1.58 2-methylheptane

1.60 a. $2C_4H_{10} + 13O_2 \rightarrow 8CO_2 + 10H_2O$

b. $C_5H_{12} + 8O_2 \rightarrow 5CO_2 + 6H_2O$

c. $C_4H_8 + 6O_2 \rightarrow 4CO_2 + 4H_2O$

1.62 $2C_6H_{14} + 13O_2 \rightarrow 12CO + 14H_2O$

1.64 Because of the rigid structure of cycloalkanes, stronger dispersion forces can develop between molecules. This increase of intermolecular attraction causes an increase in boiling point.

1.66 pentane: 414.5 torr; hexane: 113.9 torr; heptane: 37.2 torr

1.68 13 L of air

1.70 a. C_3H_8 $CH_3—CH_2—CH_3$

b. C_8H_{18}
$CH_3—CH_2—CH_2—CH_2—CH_2—CH_2—CH_2—CH_3$

c. C_4H_{10} $CH_3—CH_2—CH_2—CH_3$

CHAPTER 2

2.2 An alkene contains one or more C = C bonds.

An alkyne contains one or more C ≡ C bonds.

An aromatic hydrocarbon contains the characteristic benzene ring or similar feature.

2.4 a. 2-butene

b. 3-ethyl-2-pentene

c. 4,4-dimethyl-2-hexyne

d. 4-methylcyclopentene

e. 6-bromo-2-methyl-3-heptyne

f. 1-ethyl-2,3-dimethylcyclopropene

g. 6-methyl-1,4-heptadiene

2.6

a. 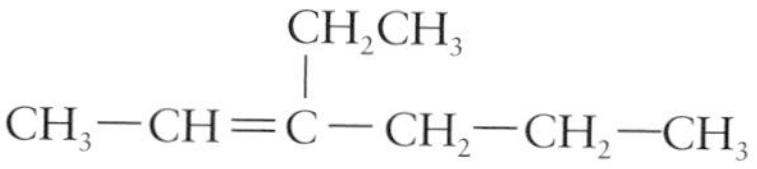
```
              CH2CH3
              |
CH3—CH=C—CH2—CH2—CH3
```

b.
```
            CH3  CH3
            |    |
CH2=CH—CH—CH—CH3
```

c.
```
            CH3
            |
CH2=CH—C=CH—CH3
```

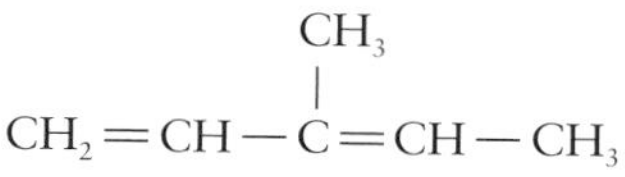

d. 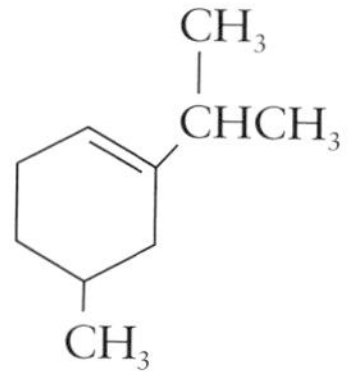
```
          CH3
          |
          CHCH3
   ______/
  /      \\
  |      |
  \\______/
     |
     CH3
```
(cyclohexene ring bearing a $CHCH_3$–CH_3 (isopropyl) group on a double-bonded carbon and a CH_3 group on the ring)

e. 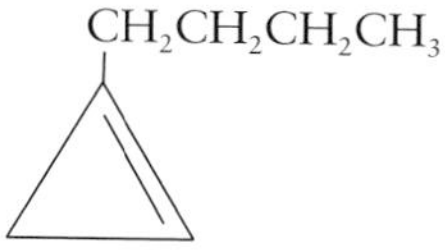
```
   CH2CH2CH2CH3
   |
   /\\
  /  \\\\
 /____\\
```
(cyclopropene bearing a $CH_2CH_2CH_2CH_3$ group)

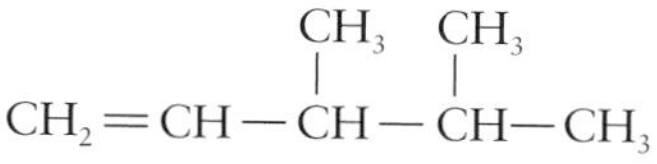

2.8 Several possibilities exist including these:

$CH_2{\equiv}CCH_2CH_2CH_3$ 1-pentyne

$CH_2{=}CHCH{=}CHCH_3$ 1,3-pentadiene

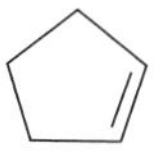

(cyclopentene ring) cyclopentene

2.10 3,7,11-trimethyl-1,3,6,10-dodecatetraene

2.12 a. 
```
    CH3
    |
CH3CH—CH2CH=CHCH3
```
5-methyl-2-hexene

b. $CH_3CH_2CH{=}CHCH{=}CHCH_3$

2,4-heptadiene

c. 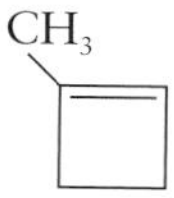 (cyclobutene ring bearing a CH_3 group)

1-methylcyclobutene

2.14 The overlap of two *p* orbitals forms a π (pi) bond containing two electrons.

2.16 Structural isomers have a different order of linkage of atoms. Geometric isomers have the same order of linkage of atoms but different three-dimensional arrangements of their atoms in space.

2.18 a. and c. cannot have cis- and trans- isomers.

b. 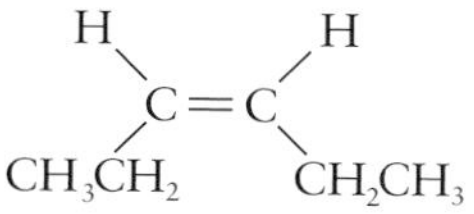
```
  H          H
   \\        /
    C = C
   /        \\
CH3CH2      CH2CH3
```
cis-3-hexene

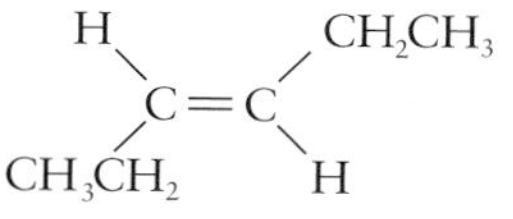

```
  H          CH2CH3
   \\        /
    C = C
   /        \\
CH3CH2       H
```
trans-3-hexene

2.20 a. 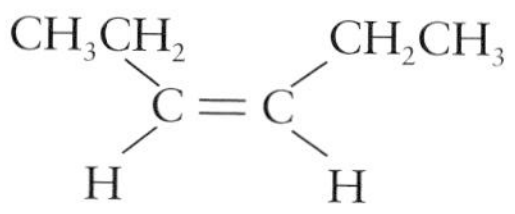
```
CH3CH2      CH2CH3
   \\        /
    C = C
   /        \\
  H          H
```

b. 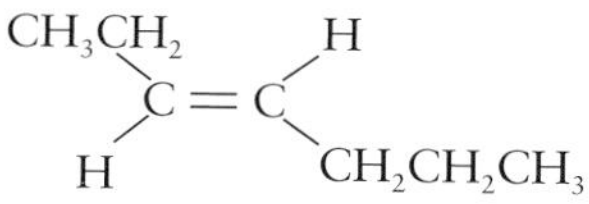
```
CH3CH2      H
   \\        /
    C = C
   /        \\
  H          CH2CH2CH3
```

2.22 Both alkanes and alkenes are nonpolar compounds that are insoluble in water, less dense than water, and soluble in nonpolar solvents.

2.24 When H—X adds to an alkene, the hydrogen becomes attached to the carbon atom that is already bonded to more hydrogens.

$$CH_2{=}CH{-}CH_3 + HCl \longrightarrow CH_3\overset{\displaystyle Cl}{\overset{|}{C}}HCH_3$$

2.26

a. $C_6H_{11}{-}\overset{\displaystyle Br}{\overset{|}{C}}H{-}\overset{\displaystyle Br}{\overset{|}{C}}HCH_3$ b. $C_6H_{11}{-}OH$

c. $CH_3\overset{\displaystyle CH_3}{\overset{|}{C}}HCH_2CH_3$ d. $C_5H_8(CH_3)(Cl)$

2.28 a. Br_2

b. H_2 with Pt catalyst

c. HCl

d. H_2O with H_2SO_4 catalyst

2.30 The addition of Br_2 to these samples will cause the cyclohexane solution to become light orange with unreacted Br_2, whereas the 2-hexene will react with the Br_2 and remain colorless.

2.32 A monomer is a starting material used in the preparation of polymers, long-chain molecules made up of many repeating units. Addition polymers are long-chain molecules prepared from alkene monomers through numerous addition reactions. A copolymer is prepared from two monomer starting materials.

2.34 A carbon–carbon double bond

2.36
$$nCH_2{=}\underset{\displaystyle Cl}{\underset{|}{C}}H \xrightarrow{\text{catalyst}} \left(CH_2{-}\underset{\displaystyle Cl}{\underset{|}{C}}H \right)_n$$

2.38 *sp*; two

2.40 Linear

2.42 Acetylene is used as a fuel for welding torches and in the synthesis of monomers for plastics and fibers.

2.44 $H{-}C{\equiv}C{-}CH_2CH_2CH_3$
1-pentyne

$CH_3{-}C{\equiv}C{-}CH_2CH_3$
2-pentyne

$CH_3\overset{\displaystyle CH_3}{\overset{|}{C}}H{-}C{\equiv}C{-}H$
3-methyl-1-butyne

2.46 Unhydridized *p* orbitals

2.48 Aromatic refers to compounds containing the benzene ring. Aliphatic substances do not contain a benzene ring.

2.50 In cyclopentane, attached groups are located above and below the plane of the ring. In benzene, attached groups are located in the same plane as the ring.

2.52 a. 1,3,5-trimethylbenzene

b. 1,4-diethylbenzene

2.54 a. 3-phenyl-1-pentene

b. 2,5-diphenylhexane

2.56 a. *m*-ethylphenol

b. *o*-chlorophenol

2.58 a. 2-ethyl-3-propyltoluene

b. 3-bromo-5-chlorobenzoic acid

2.60

a. OH, CH_2CH_3 (on benzene ring)

b. HO—C=O, Cl (on benzene ring)

c. $CH_3CH_2\overset{\displaystyle CH_3}{\overset{|}{C}}CH_2CH_3$ (with benzene ring on central C)

2.62 Nonpolar and insoluble in water

2.64 Cyclohexene readily undergoes addition reactions. Benzene resists addition reactions and favors substitution reactions. Both undergo combustion.

2.66 a. phenol

b. styrene

c. aniline

d. phenol

2.68 Heat increases the kinetic energy of the ethylene molecules and increases the number of molecules having the required activation energy.

Pressure on a gas increases the temperature and concentration of the gas.

Catalysts lower the activation energy requirements.

2.70

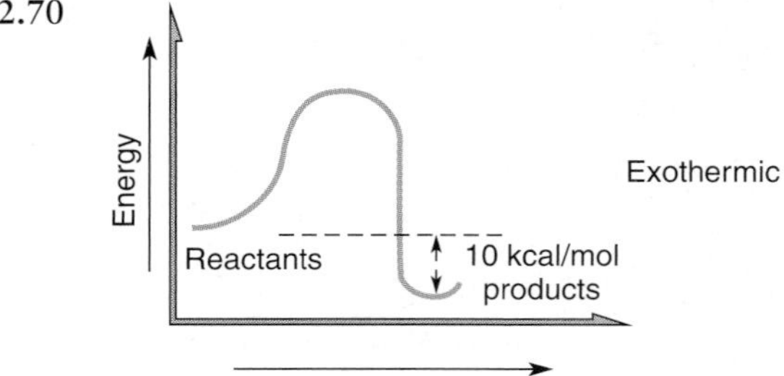

2.72 a and d

CHAPTER 3

3.2 R—O—R

3.4 a. 1-butanol

b. 3,4-dibromo-2-butanol

c. 3-ethyl-3-pentanol

d. 2-phenyl-1-ethanol
e. 2-bromo-6-methylcyclohexanol
f. 2,4-pentanediol

3.6 a. methanol
b. 2-propanol
c. ethanol
d. 1,2-ethanediol
e. 1,2,3-propanetriol

3.8 a. $\mathrm{CH_3\overset{\overset{\textstyle OH}{|}}{\underset{\underset{\textstyle CH_3}{|}}{C}}CH_2CH_2CH_3}$

b. $\mathrm{\overset{\overset{\textstyle OH}{|}}{C}H_2CH_2\overset{\overset{\textstyle OH}{|}}{C}HCH_3}$

c. OH
CH₂CH₃

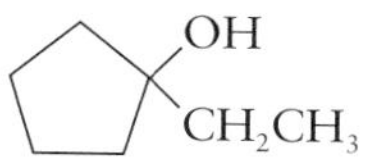

3.10 a. 2-ethylphenol or *o*-ethylphenol
b. 2-ethyl-4,6-dimethylphenol

3.12 a. OH
Br

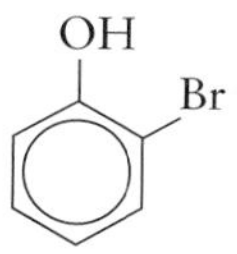

b. OH
$\mathrm{CH_2CH_3}$
$\mathrm{CH_3CH_2}$
$\mathrm{CH_2CH_3}$

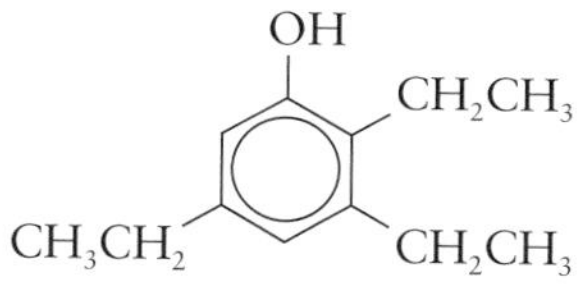

3.14 a. Secondary
b. Primary
c. Primary

3.16

$\mathrm{CH_3CH_2CH_2CH_2-OH}$	1-butanol	primary
$\mathrm{CH_3\overset{\overset{\textstyle OH}{|}}{C}HCH_2CH_3}$	2-butanol	secondary
$\mathrm{CH_3\underset{\underset{\textstyle CH_3}{|}}{C}HCH_2-OH}$	2-methyl-1-propanol	primary
$\mathrm{CH_3\overset{\overset{\textstyle OH}{|}}{\underset{\underset{\textstyle CH_3}{|}}{C}}CH_3}$	2-methyl-2-propanol	tertiary

3.18 a. methanol, ethanol, 1-propanol
b. butane, 1-propanol, ethylene glycol

3.20 a. $\mathrm{CH_3CH_2CH_2CH_2-O-H\cdots O(H)-CH_2CH_3}$

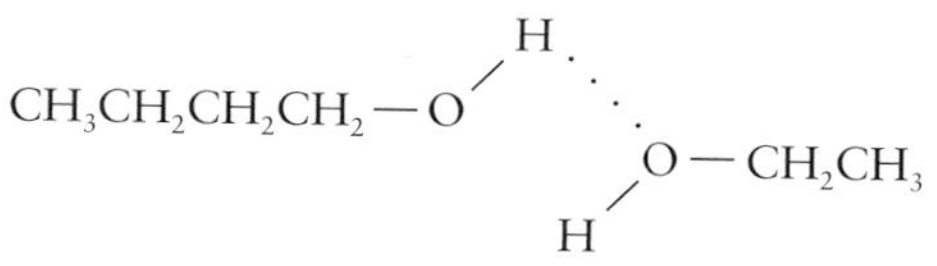

b. $\mathrm{C_6H_{11}-O(H)\cdots H-O-H}$

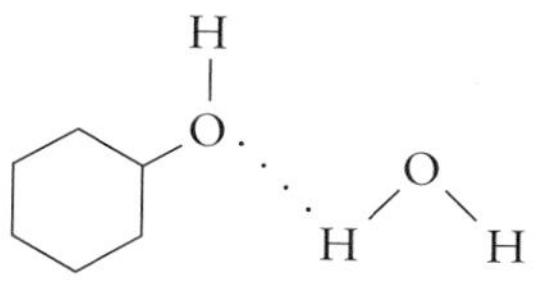

3.22 a. $\mathrm{CH_3CH{=}CHCH_3}$
b. $\mathrm{CH_3CH{=}\overset{\overset{\textstyle CH_3}{|}}{C}CH_3}$

3.24 a. $\mathrm{CH_3CH_2CH_2-O-CH_2CH_2CH_3}$
b. $\mathrm{C_6H_5-O-C_6H_5}$
c. $\mathrm{C_6H_5CH_2CH_2-O-CH_2CH_2C_6H_5}$

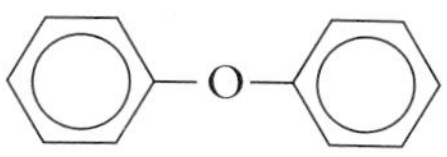

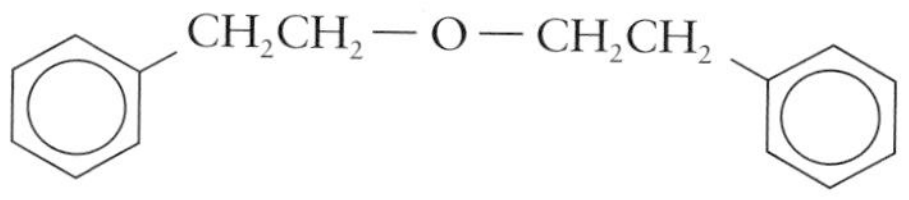

3.26 a. OH (cyclopentanol)

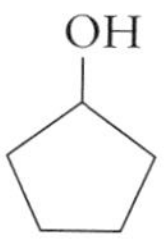

b. $\mathrm{CH_3CH_2\overset{\overset{\textstyle OH}{|}}{C}H\underset{\underset{\textstyle CH_3}{|}}{C}HCH_3}$

c. $\mathrm{CH_3CH_2CH_2\overset{\overset{\textstyle OH}{|}}{C}H_2}$

3.28 a. $\mathrm{CH_3\overset{\overset{\textstyle OH}{|}}{\underset{\underset{\textstyle CH_3}{|}}{C}}CH_2CH_3 + (O) \longrightarrow}$ no reaction

b. $\mathrm{2CH_3CH_2CH_2-OH \xrightarrow[140^\circ C]{H_2SO_4} CH_3CH_2CH_2-O-CH_2CH_2CH_3 + H_2O}$

c. $\mathrm{CH_3CH_2\overset{\overset{\textstyle OH}{|}}{C}HCH_2CH_3 + (O) \longrightarrow CH_3CH_2\overset{\overset{\textstyle O}{\|}}{C}CH_2CH_3 + H_2O}$

d. $\mathrm{CH_3CH_2\overset{\overset{\textstyle OH}{|}}{C}HCH_2CH_3 \xrightarrow[180^\circ C]{H_2SO_4} CH_3CH{=}CHCH_2CH_3 + H_2O}$

e. $\mathrm{CH_3CH_2CH_2CH_2CH_2CH_2-OH + 2(O) \longrightarrow CH_3CH_2CH_2CH_2CH_2-\overset{\overset{\textstyle O}{\|}}{C}-OH + H_2O}$

3.30

a. $CH_3CH_2CH{=}CH_2 + H_2O \xrightarrow{H_2SO_4} CH_3CH_2CH(OH)CH_3$

$CH_3CH_2CH(OH)CH_3 + (O) \rightarrow CH_3CH_2C(=O)CH_3 + H_2O$

b. 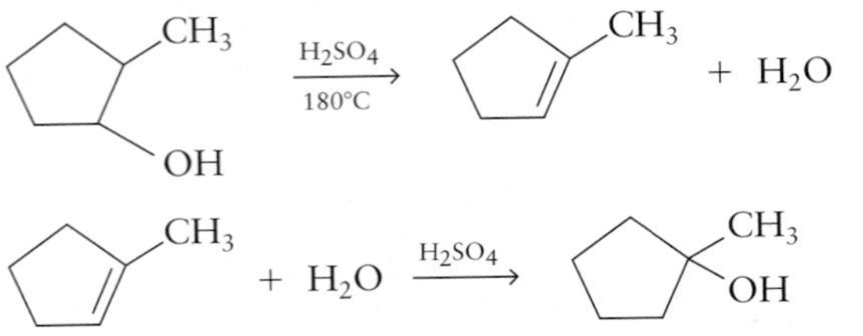 2-methylcyclopentanol $\xrightarrow[180°C]{H_2SO_4}$ 1-methylcyclopentene $+ H_2O$

1-methylcyclopentene $+ H_2O \xrightarrow{H_2SO_4}$ 1-methylcyclopentanol

c. $CH_3CH_2CH_2CH_2OH \xrightarrow[180°C]{H_2SO_4} CH_3CH_2CH{=}CH_2 + H_2O$

$CH_3CH_2CH{=}CH_2 + H_2O \xrightarrow{H_2SO_4} CH_3CH_2CH(OH)CH_3$

$CH_3CH_2CH(OH)CH_3 + (O) \rightarrow CH_3CH_2C(=O)CH_3 + H_2O$

3.32 $CH_3C(=O)-C(=O)-OH$

3.34 Methanol is used as an automobile fuel and in the preparation of formaldehyde, a starting material for certain plastics.

3.36 a. glycerol
b. ethanol
c. ethylene glycol
d. isopropyl alcohol
e. menthol
f. ethanol

3.38 a. *o*-phenylphenol or 2-benzyl-4-chlorophenol
b. 4-hexylresorcinol or phenol
c. BHA or BHT

3.40 a. ethyl isopropyl ether
b. butyl methyl ether
c. diphenyl ether

3.42 a. 1-ethoxypropane
b. 2-ethoxypropane
c. ethoxybenzene
d. 1,2-dimethoxycyclopentane

3.44 a. $CH_3-O-CH_2CH_2CH_3$

b. 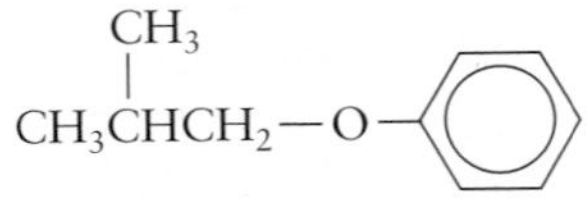 $CH_3CH(CH_3)CH_2-O-C_6H_5$

c. $CH_3-O-CH(CH_3)CH_2CH_3$

d. 1,2,3-trimethoxycyclopropane (CH_3-O groups on each ring carbon)

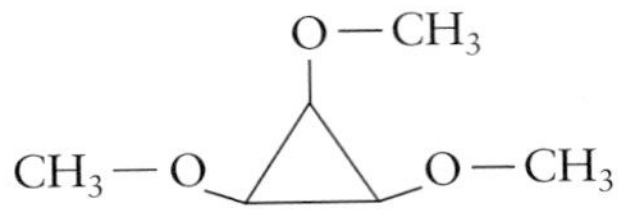

e. $(CH_3)_3C-O-C(CH_3){=}CHCH_2CH_3$

3.46 Inertness

3.48 $CH_3CH_2CH_2-OH$, $CH_3CH_2-O-CH_3$, $CH_3CH_2CH_2CH_3$

This order is based on the decreasing ability to form hydrogen bonds with water.

3.50 $CH_3CH_2-O(CH_2CH_3)\cdots H-O-H$

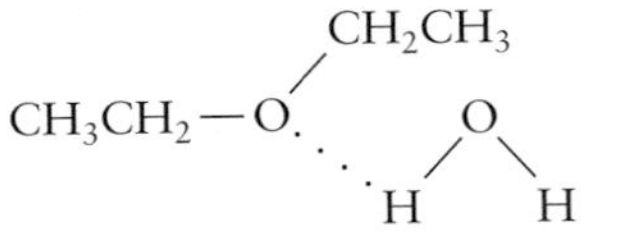

3.52 a. $CH_3CH_2CH_2-S-Hg-S-CH_2CH_2CH_3 + 2H^+$
b. $CH_3CH_2CH_2-S-S-CH_2CH_2CH_3 + H_2O$
c. 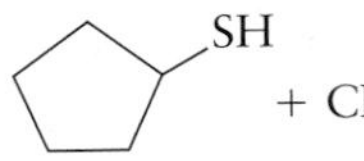 cyclopentanethiol $+ CH_3-SH$

3.54 Alcohols are oxidized to aldehydes and ketones, whereas thiols are oxidized to disulfides.

3.56 aromatic, alkene, aldehyde

3.58 The electronegativity difference between S and H is less than between O and H. This means the S—H bond is less polar, leading to a decreased ability to hydrogen bond. The decrease in hydrogen bonding ability would decrease a thiol's boiling point and water solubility in comparison to the alcohol.

3.60 The negative charge that forms on the oxygen atom when phenol acts as an acid can be delocalized around the ring structure of benzene.

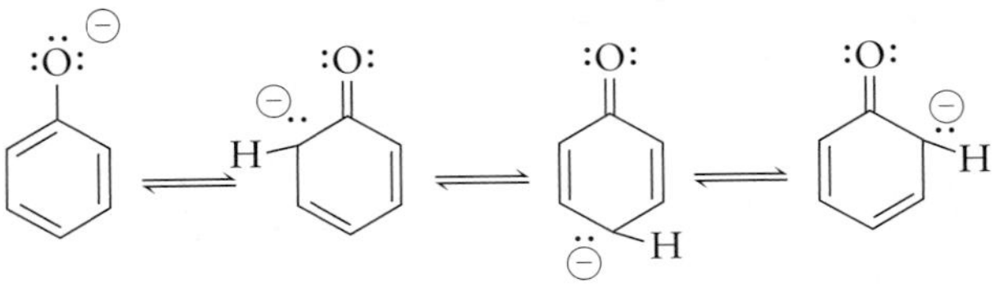

(Resonance structures of phenoxide ion: $:\ddot{O}:^{\ominus}$ on ring ⇌ :O: with ⊖ on ring carbon (H) ⇌ :O: with ⊖ at para carbon (H) ⇌ :O: with ⊖ at other ortho carbon (H))

3.62 a. ethanol
b. isopropanol
c. methanol

CHAPTER 4

4.2

Simplest aldehyde: $H-C(=O)-H$

Simplest ketone: $H_3C-C(=O)-CH_3$

4.4 a. aldehyde

b. neither

c. ketone

d. neither

e. ketone

f. neither

4.6 a. butanal

b. 2,4-dibromobutanal

c. 4-methyl-2-pentanone

d. 2,5-dichlorocyclopentanone

e. 3-methyl-1-phenyl-2-butanone

4.8

a. 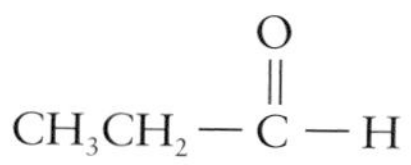 $CH_3CH_2-\overset{O}{\overset{\|}{C}}-H$

b. $CH_3-\overset{O}{\overset{\|}{C}}-\overset{CH_3}{\overset{|}{C}H}-CH_3$

c. O, CH_3, CH_3

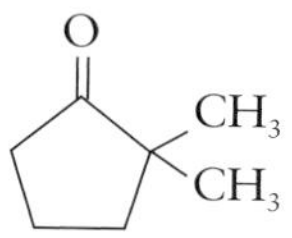

d. 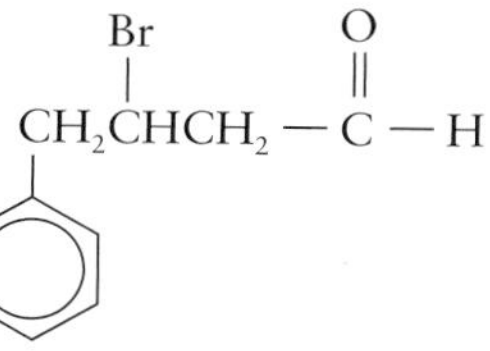 $\overset{Br}{\overset{|}{CH_2CHCH_2}}-\overset{O}{\overset{\|}{C}}-H$ (with a benzene ring attached to the first CH_2)

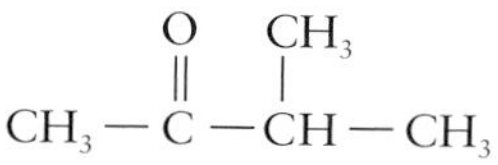

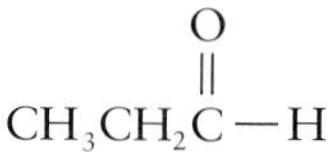

4.10 $CH_3CH_2\overset{O}{\overset{\|}{C}}-H$ and $CH_3-\overset{O}{\overset{\|}{C}}-CH_3$

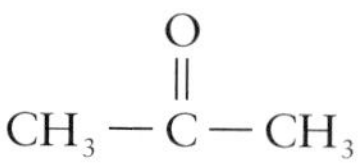

4.12 a. 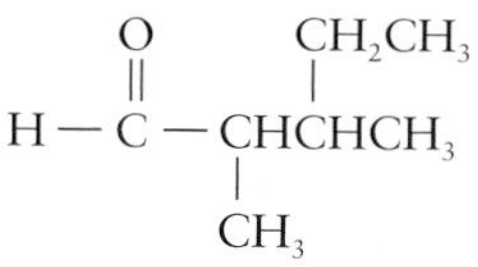 $H-\overset{O}{\overset{\|}{C}}-\overset{CH_2CH_3}{\overset{|}{C}H}\underset{\underset{CH_3}{|}}{C}HCH_3$

The longest chain contains five carbon atoms. The correct name is 2,3-dimethylpentanal.

b. The compound is an aldehyde. Number from the carbonyl group.

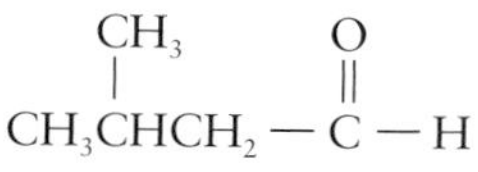

$CH_3\overset{CH_3}{\overset{|}{C}}HCH_2-\overset{O}{\overset{\|}{C}}-H$

The correct name is 3-methylbutanal.

c. 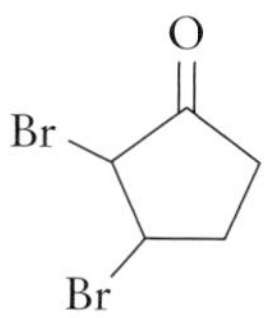 O, Br, Br

Number the ring so that the bromines have lower numbers. The correct name is 2,3-dibromocyclopentanone.

4.14 Acetone dissolves remaining traces of water, and together they are discarded. The remaining small amounts of acetone evaporate.

4.16 Propane is nonpolar. Ethanal is a polar molecule and exhibits greater interparticle forces.

4.18

a. 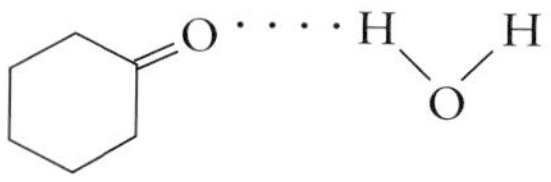 $O\cdots H-O-H$ (cyclohexanone C=O hydrogen-bonded to water)

b. 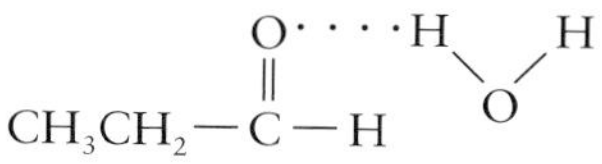 $CH_3CH_2-\overset{O\cdots H-O-H}{\overset{\|}{C}}-H$

4.20 The alcohol will have the higher melting point because of hydrogen bonding. Menthol is a solid, and menthone is a liquid.

4.22

a. 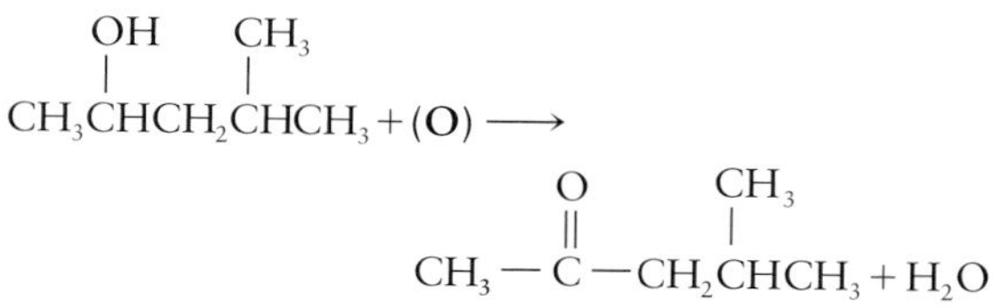 $CH_3\overset{OH}{\overset{|}{C}}HCH_2\overset{CH_3}{\overset{|}{C}}HCH_3 + (O) \longrightarrow CH_3-\overset{O}{\overset{\|}{C}}-CH_2\overset{CH_3}{\overset{|}{C}}HCH_3 + H_2O$

b. (cyclopentyl)$-\overset{OH}{\overset{|}{C}H_2} + (O) \longrightarrow$ (cyclopentyl)$-\overset{O}{\overset{\|}{C}}-H + H_2O$

c. (cyclopentane with OH and CH_2CH_3) $+ (O) \longrightarrow$ (cyclopentanone with CH_2CH_3) $+ H_2O$

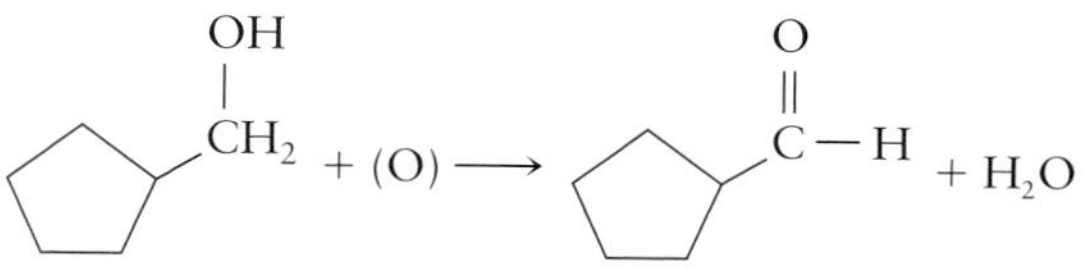

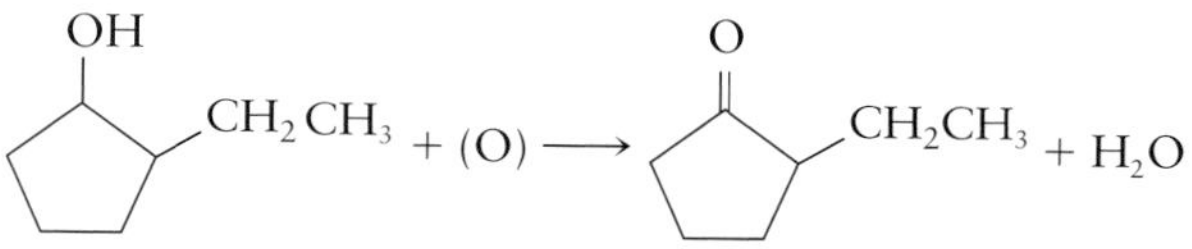

4.24 a. hemiacetal

b. hemiketal

c. neither

d. hemiketal

4.26 a. neither

b. ketal

c. acetal

4.28 a. ketal

b. none of these

c. hemiacetal

4.30 They are unstable, splitting into an alcohol and aldehyde.

4.32 The formation of a silver precipitate or mirror

4.34 b. 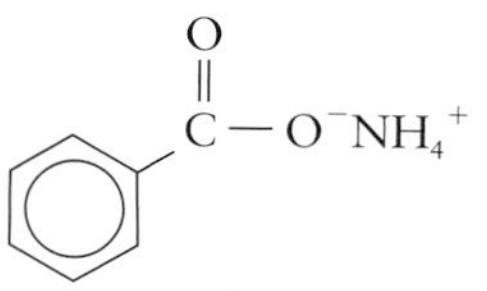 (benzene ring)$-\overset{O}{\overset{\|}{C}}-O^-NH_4^+$

d. 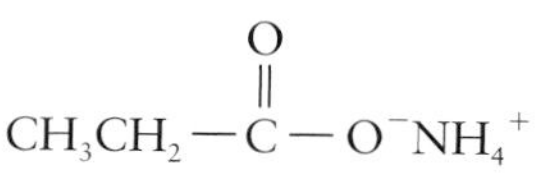 $CH_3CH_2-\overset{O}{\overset{\|}{C}}-O^-NH_4^+$

4.36 a. no
b. no
c. yes
d. yes
e. no

4.38
$$\begin{array}{ccccccccccc} OH & & OH & & OH & & OH & & OH & & O \\ | & & | & & | & & | & & | & & \| \\ CH_2 & - & CH & - & CH & - & CH & - & CH & - & C-H \end{array}$$

4.40
$$\begin{array}{ccccccccccc} OH & & OH & & OH & & OH & & OH & & O \\ | & & | & & | & & | & & | & & \| \\ CH_2 & - & CH & - & CH & - & CH & - & CH & - & C-H \end{array}$$

4.42

a.
$$\begin{array}{c} OH \\ | \\ \text{(cyclohexyl)}-CH-\text{(cyclohexyl)} \end{array}$$

b.
$$\begin{array}{c} OCH_2CH_2CH_3 \\ | \\ CH_3CH_2-C-OCH_3 \\ | \\ CH_3 \end{array} \quad + H_2O$$

c.
$$\begin{array}{c} CH_3 \\ | \\ O-CHCH_3 \\ | \\ CH_3CH_2-C-H \\ | \\ O-CHCH_3 \\ | \\ CH_3 \end{array} \quad + H_2O$$

d.
$$\begin{array}{c} O \\ \| \\ \text{(cyclopentyl)}-CH_2-C-OH \end{array}$$

e.
$$\begin{array}{c} CH_3 \\ | \\ CH_3CH_2CHCH_2-OH \end{array}$$

4.44 Alkenes give alkanes.
Alkynes give alkenes and alkanes.
Aldehydes give primary alcohols.
Ketones give secondary alcohols.

4.46

a. $\text{(cyclohexanone)} + 2CH_3CH_2OH$

b.
$$\begin{array}{c} O \\ \| \\ CH_3CH_2-C-CH_2CH_3 + 2CH_3OH \end{array}$$

c.
$$\begin{array}{c} O \\ \| \\ H-C-H + 2CH_3CH_2CH_2OH \end{array}$$

d.
$$\begin{array}{c} O \\ \| \\ CH_3CH_2C-H + 2CH_3OH \end{array}$$

4.48

a.
$$\begin{array}{ccc} OH & & O \\ | & & \| \\ CH_2 & \quad\quad & C-CH_3 + HO-CH_2CH_3 \\ \diagdown & \text{\_\_\_} & \diagup \end{array}$$

b.
$$\begin{array}{c} CH_2-OH \\ C=O + HO-CH_2CH_2CH_3 \\ \diagdown \\ H \end{array}$$

4.50

a.
$$\begin{array}{c} OH \\ | \\ CH_3CH_2CH_2 + (O) \longrightarrow CH_3CH_2-\overset{\overset{O}{\|}}{C}-H \end{array}$$

$$CH_3CH_2-\overset{\overset{O}{\|}}{C}-H + 2CH_3CH_2-OH \overset{H^+}{\rightleftharpoons}$$

$$\begin{array}{c} OCH_2CH_3 \\ | \\ CH_3CH_2-CH \\ | \\ OCH_2CH_3 \end{array} \quad + H_2O$$

b.
$$\begin{array}{c} CH_3 \quad O \\ | \quad\ \ \| \\ CH_3CH-C-CH_3 + H_2 \end{array} \xrightarrow{Pt} \begin{array}{c} CH_3 \quad OH \\ | \quad\quad | \\ CH_3CH-CH-CH_3 \end{array}$$

$$\begin{array}{c} CH_3 \quad OH \\ | \quad\quad | \\ CH_3CH-CH-CH_3 \end{array} \xrightarrow[180^\circ]{H_2SO_4} \begin{array}{c} CH_3 \\ | \\ CH_3C=CH-CH_3 + H_2O \end{array}$$

4.52 a. menthone
b. biacetyl
c. cinnamaldehyde
d. vanillin

4.54 461 g

4.56
$$\begin{array}{c} O \\ \| \\ CH_3CH_2CCH_2CH_3 + 2CH_3CH_2CH_2-OH \end{array} \xrightarrow{H^+}$$

$$\begin{array}{c} O-CH_2CH_2CH_3 \\ | \\ CH_3CH_2CCH_2CH_3 \\ | \\ O-CH_2CH_2CH_3 \end{array}$$

4.58 a. aldehyde
b. alcohol
c. ketone

CHAPTER 5

5.2 Carboxylic acids with more than ten carbon atoms are called fatty acids. They were originally produced by the saponification of fats.

5.4 lactic acid

5.6 a. 2-methylpropanoic acid
b. 4-bromobutanoic acid
c. 3-isopropylbenzoic acid or *m*-isopropylbenzoic acid
d. 3-methoxypentanoic acid
e. 4-phenylbutanoic acid

5.8 a. $CH_3CH_2CH_2CH_2-\overset{O}{\overset{\|}{C}}-OH$

b. $CH_3CH_2CH_2\underset{CH_3}{\underset{|}{C}}H\overset{Br}{\overset{|}{C}}H-\overset{O}{\overset{\|}{C}}-OH$

c. $HO-\overset{O}{\overset{\|}{C}}-C_6H_4-CH_2CH_2CH_3$ (para)

5.10 a. Acetic acid because it forms dimers
b. Propanoic acid because it forms dimers
c. Butyric acid because it has a higher molecular weight

5.12
$$\begin{array}{l} \quad\quad\quad\;\; O \cdot\cdot\cdot\cdot HO-\underset{\|}{C}-CH_2CH_3 \\ CH_3CH_2-\overset{\|}{C}-OH\cdot\cdot\cdot\cdot O \end{array}$$

5.14 The carboxyl group is hydrophilic; the carbon chain is hydrophobic.

5.16 pentane < ethoxyethane < 1-butanol < propanoic acid

5.18 $CH_3CH_2-\overset{O}{\overset{\|}{C}}-O^-$

5.20 Under cellular pH conditions, lactic acid ionizes to produce the anion lactate.

5.22 $CH_3CH_2CH_2\overset{O}{\overset{\|}{C}}-OH$

5.24 a. $CH_3CH_2-\overset{O}{\overset{\|}{C}}-O^-Na^+ + H_2O$

b. $CH_3\overset{OH}{\overset{|}{C}}H-\overset{O}{\overset{\|}{C}}-O^-K^+ + H_2O$

5.26 a. $CH_3-\overset{O}{\overset{\|}{C}}-OH + NaOH \longrightarrow CH_3-\overset{O}{\overset{\|}{C}}-O^-Na^+ + H_2O$

b. $CH_3-\overset{O}{\overset{\|}{C}}-OH + KOH \longrightarrow CH_3-\overset{O}{\overset{\|}{C}}-O^-K^+ + H_2O$

c. $2CH_3-\overset{O}{\overset{\|}{C}}-OH + Ca(OH)_2 \longrightarrow (CH_3-\overset{O}{\overset{\|}{C}}-O^-)_2Ca^{2+} + 2H_2O$

5.28 a. sodium 3-bromobutanoate
b. calcium methanoate
c. potassium 3-phenoxypropanoate

5.30 a. $CH_3-\overset{O}{\overset{\|}{C}}-O^-K^+$

b. $CH_3-C_6H_4-\overset{O}{\overset{\|}{C}}-O^-Na^+$ (meta)

c. $CH_3CH_2\overset{CH_3}{\overset{|}{C}}H-\overset{O}{\overset{\|}{C}}-O^-Na^+$

5.32 a. sodium stearate
b. acetic acid
c. sodium benzoate
d. zinc 10-undecylenate
e. sodium propanoate
f. citric acid/sodium citrate

5.34 b. $CH_3-\overset{O}{\overset{\|}{C}} \underset{\uparrow}{-} O-CH_2CH_3$

d. $C_6H_5-O\underset{\uparrow}{-}\overset{O}{\overset{\|}{C}}-CH_3$

f. $C_6H_5-\overset{O}{\overset{\|}{C}}\underset{\uparrow}{-}O-CH_2CH_3$

5.36 a. $CH_3-\overset{O}{\overset{\|}{C}}-O-CH_3 + CH_3-\overset{O}{\overset{\|}{C}}-OH$

b. $CH_3CH(CH_3)-C(=O)-O-CH_2-C_6H_5 + H_2O$

c. $CH_3CH(CH_3)-C(=O)-O-C_6H_5 + HCl$

5.38 a. $CH_3CH_2-C(=O)-O-CH_3$

b. $CH_3CH_2-C(=O)-O-C_6H_5$

c. $CH_3CH_2-C(=O)-O-CH_2CH(CH_3)CH_3$

5.40 (structure with $N-CH_3$, ring O, and two acetate groups: $CH_3-C(=O)-O$ and $O-C(=O)-CH_3$)

5.42 $-(O-C(=O)-C(=O)-O-CH_2CH_2CH_2)_n-$

5.44 a. ethyl lactate

b. propyl butyrate

5.46 a. methyl 2-methylpropanoate

b. ethyl 3,5-dichlorobenzoate

5.48 a. methyl propionate

b. methyl butyrate

c. methyl lactate

5.50 a. $H-C(=O)-O-C_6H_5$

b. $NO_2-C_6H_4-C(=O)-O-CH_3$

c. $CH_3CH(Cl)-C(=O)-O-CH_2CH_3$

5.52 $CH_3-C(=O)-OCH_2CH_3 + H_2O \xrightarrow{H^+} CH_3-C(=O)-OH + CH_3CH_2-OH$

$CH_3-C(=O)-OCH_2CH_3 + NaOH \longrightarrow CH_3-C(=O)-O^-Na^+ + CH_3CH_2-OH$

5.54 a. $CH_3(CH_2)_{16}-C(=O)-O^+Na^+ + CH_3CH_2-OH$

b. $CH_3CH(CH_3)-C(=O)-OH + HO-C_6H_5$

5.56 $CH_2(OH)-C(=O)-CH_2-O-P(=O)(OH)-OH + H_2O$

5.58 When $NaHCO_3$ reacts with an acid, it forms CO_2 gas, water, and a salt:

$RCO_2H + NaHCO_3 \rightarrow RCO_2^- \; Na^+ + CO_2 + H_2O$

There would be no reaction with an alcohol. Therefore, the presence of effervescence ("bubbling") indicates an acid.

5.60 $CH_3CH_2CH_2CH(CH_2CH_3)CH_2C(=O)-O-C_6H_5$

phenyl 3-ethylhexanoate

5.62 1. Use an excessive quantity of ethanol.

2. Remove the water as it is being formed.

5.64 Polarity: d > a > b > c

Boiling Points: c < b < a < d

CHAPTER 6

6.2 $R-NH_2$ (primary), R_2NH (secondary), and R_3N (tertiary).

6.4 a. secondary

b. tertiary

c. primary

d. secondary

6.6 (1) $CH_3-CH_2-CH_2-NH_2$ primary

(2) $CH_3-CH(CH_3)-NH_2$ primary

(3) $CH_3-CH_2-NH-CH_3$ secondary

(4) $CH_3-\underset{\displaystyle CH_3}{\underset{|}{N}}-CH_3$ tertiary

6.8 **a.** isopropylmethylamine
b. ethylmethylphenylamine
c. butylamine

6.10 **a.** 2-methyl-1-butanamine
b. cyclobutanamine
c. *N*-methylcyclobutanamine

6.12 **a.** *N*-propylaniline
b. *N*,*N*-dimethylaniline

6.14

a. $CH_3\overset{\displaystyle NH_2}{\overset{|}{C}}H\underset{\displaystyle CH_2CH_3}{\underset{|}{C}}HCH_2CH_3$

b. 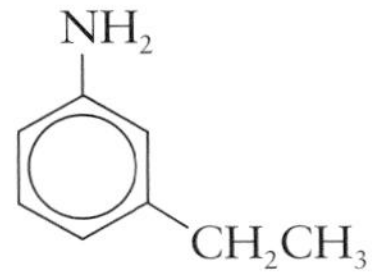 NH_2 (on benzene ring), CH_2CH_3

c. 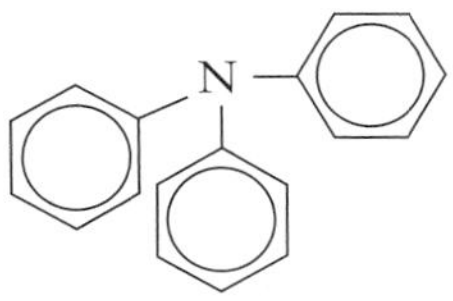 N (bonded to three benzene rings)

6.16 Amines form hydrogen bonds with water.

6.18 Tertiary amines do not form hydrogen bonds among themselves.

6.20 **a.** 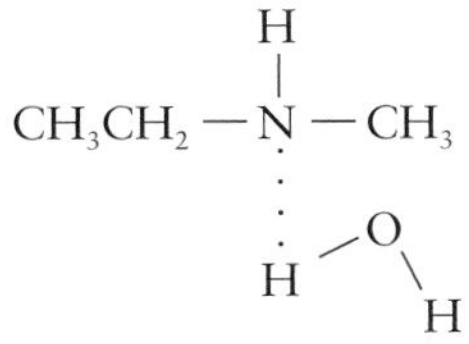 $CH_3CH_2-\overset{\displaystyle H}{\overset{|}{N}}-CH_3$ ⋮ H–O–H

b. 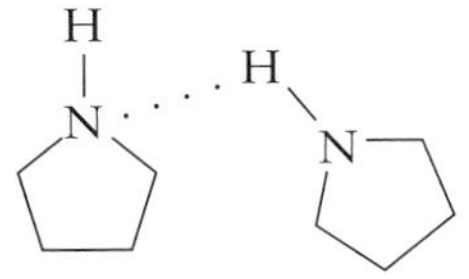 H–N (pyrrolidine ring) · · · H–N (pyrrolidine ring)

6.22 B < C < A

6.24 $CH_3CH_2-NH-CH_2CH_3+H_2O \rightleftharpoons$

$$CH_3CH_2-\overset{\displaystyle H}{\overset{|}{\underset{\displaystyle H}{\underset{|}{N^+}}}}-CH_2CH_3+OH^-$$

6.26 **a.** $CH_3CH_2CH_2CH_2-NH_3^+Cl^-$

b. 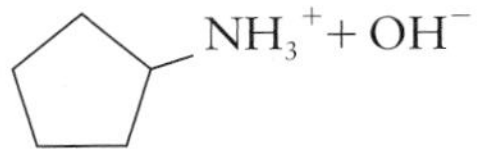 (cyclopentyl)$-NH_3^+ + OH^-$

c. 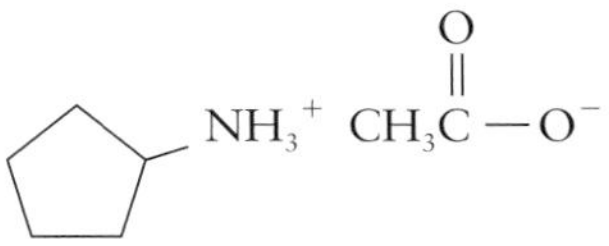 (cyclopentyl)$-NH_3^+\ CH_3\overset{\displaystyle O}{\overset{\|}{C}}-O^-$

d. $CH_3\overset{\displaystyle NH_2}{\overset{|}{C}}HCH_3+NaCl+H_2O$

e. $CH_3\underset{\displaystyle CH_3}{\underset{|}{C}}H-\overset{\displaystyle O}{\overset{\|}{C}}-NH-CH_3+HCl$

f. 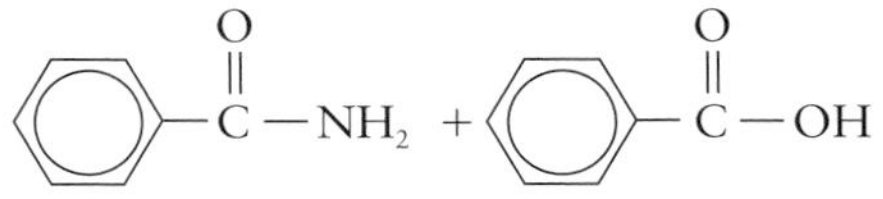 (phenyl)$-\overset{\displaystyle O}{\overset{\|}{C}}-NH_2 +$ (phenyl)$-\overset{\displaystyle O}{\overset{\|}{C}}-OH$

6.28 The amine salts are much more soluble in water (or the aqueous fluids in the body).

6.30 (1) $CH_3-\overset{\displaystyle O}{\overset{\|}{C}}-Cl+CH_3-NH-CH_3 \longrightarrow$

$$CH_3-\overset{\displaystyle O}{\overset{\|}{C}}-\underset{\displaystyle CH_3}{\underset{|}{N}}-CH_3+HCl$$

(2) $CH_3-\overset{\displaystyle O}{\overset{\|}{C}}-O-\overset{\displaystyle O}{\overset{\|}{C}}-CH_3+CH_3-NH-CH_3 \longrightarrow$

$$CH_3-\overset{\displaystyle O}{\overset{\|}{C}}-\underset{\displaystyle CH_3}{\underset{|}{N}}-CH_3+HO-\overset{\displaystyle O}{\overset{\|}{C}}-CH_3$$

6.32 **a.** NaOH
b. NH_3
c. HCl

6.34 Synapse

6.36 Norepinephrine and serotonin

6.38 Norepinephrine, serotonin, acetylcholine, and dopamine

6.40 Epinephrine is used clinically in local anesthetics to constrict blood vessels in the area of injection.

6.42 Plants

6.44 **a.** caffeine
b. atropine
c. nicotine
d. codeine
e. quinine
f. morphine

6.46 **a.** *N*-methylbutanamide
b. *N*,*N*-dimethylethanamide
c. 2-methylbenzamide
d. 2-methylbutanamide

6.48 a. $CH_3CH_2—\overset{\overset{O}{\|}}{C}—NH_2$

b. $CH_3CH_2CH_2CH_2—\overset{\overset{O}{\|}}{C}—NH—CH_2CH_3$

c. $H—\overset{\overset{O}{\|}}{C}—\underset{\underset{CH_3}{|}}{N}—CH_3$

6.50 a. (cyclopentyl)$—\overset{\overset{O}{\|}}{C}—\underset{\underset{H}{|}}{N}—H \cdots\cdots O(H)(H)$

b. $CH_3\underset{\underset{CH_3}{|}}{CH}—\overset{\overset{O}{\|}}{C}—\underset{\underset{H}{|}}{N}—H \cdots\cdots O{=}\underset{\underset{CHCH_3 \atop | \atop CH_3}{|}}{C}—NH_2$

6.52 a. $CH_3—\overset{\overset{O}{\|}}{C}—O^-Na^+ + CH_3CH_2—NH_2$

b. $CH_3CH_2—\overset{\overset{O}{\|}}{C}—OH + NH_4^+Cl^-$

6.54 (CH_3-substituted benzene ring)$—\overset{\overset{O}{\|}}{C}—O^-Na^+ + CH_3CH_2—NH—CH_2CH_3$

6.56 a. saturated

b. one C=C bond

c. two C=C bonds

6.58 $Na^{+-}O_2C(CH_2)_4CO_2^-Na^+$ and $NH_2(CH_2)_6NH_2$

6.60 $CH_3\overset{\overset{O}{\|}}{C}—OH$ and $\left[HO—CH_2CH_2—\underset{\underset{CH_3}{|}}{\overset{\overset{CH_3}{|}}{N}}—CH_3\right]^+$

acetic acid choline

6.62 Primary: $R—NH_2$

Secondary: $R_2—NH$

Tertiary: R_3N

CHAPTER 7

7.2 a. A structural material in plants

b. An energy source in our diet

c. A stored form of energy in animals

d. A stored form of energy in plants

7.4 a. carbohydrate, disaccharide

b. carbohydrate, monosaccharide

c. carbohydrate, polysaccharide

d. carbohydrate, monosaccharide

e. carbohydrate, polysaccharide

f. carbohydrate, disaccharide

g. carbohydrate, polysaccharide

h. carbohydrate, polysaccharide

7.6 They don't have four different groups attached; C number 1 has only three groups; C number 3 has four groups, but two are hydrogens.

7.8 a. No chiral carbon atoms

b. $CH_3CH_2—\underset{\underset{OH}{|}}{\overset{*}{C}H}—\overset{\overset{O}{\|}}{C}—CH_3$ Yes, can have enantiomers. * denotes the chiral carbon

c. (cyclopentyl)$—\underset{\underset{CH_3}{|}}{\overset{\overset{OH}{|}}{C^*}}—CH_2CH_3$ Yes, can have enantiomers. * denotes the chiral carbon

7.10 The chiral carbon is at the intersection of the lines. The H and OH are above the plane of the paper, and the CHO and CH_2CH_3 are below the plane.

7.12 a. The D form is given; the L enantiomer is

CHO
H—+—OH
H—+—OH
HO—+—H
CH_2OH

b. The L form is given; the D enantiomer is

CH_2OH
|
C=O
HO—+—H
H—+—OH
H—+—OH
CH_2OH

7.14 a.

COOH H_2N—+—H CH_3	COOH H—+—NH_2 CH_3
L-alanine	D-alanine

b.

COOH H_2N—+—H $CH_2CH(CH_3)_2$	COOH H—+—NH_2 $CH_2CH(CH_3)_2$
L-leucine	D-leucine

7.16 a. Two chiral carbon atoms; four stereoisomers

b. Three chiral carbon atoms; eight stereoisomers

7.18 8 aldopentoses; 4 are in the D form, and 4 are in the L form.

7.20 Optically active molecules rotate the plane of polarized light.

7.22 a. ketopentose
b. aldopentose

7.24 Some carbohydrates have a sweet taste; these are sugars.

7.26 * denotes sites where the hydrogens of a water molecule can hydrogen-bond.

CHO*
H—OH*
HO*—H
H—OH*
H—OH*
CH₂OH*

7.28 a. The α form is given in the text; the β form is

CH₂OH, HO, O, OH, OH, OH

b. The β form is given in the text; the α form is

HOCH₂, O, OH, OH

7.30 CH₂OH, O, OH, OH, HO, OH
α-mannose

CH₂OH, O, OH, OH, OH, HO
β-mannose

7.32 Nonreducing sugar

7.34 a.

COOH
H—OH
H—OH + Cu_2O
H—OH
CH_2OH

b. CH_2OH, O, OH, HO, OCH_2CH_3, OH

+ CH_2OH, O, OCH_2CH_3, OH, HO, OH + H_2O

c. $HOCH_2$, O, OCH_3, HO, CH_2OH, OH

+ $HOCH_2$, O, CH_2OH, HO, OCH_3, OH + H_2O

7.36 Ribose is found is RNA and deoxyribose is found in DNA.

7.38 Dextrose; blood sugar

7.40 The hexoses glucose and galactose differ in the positions of H and OH on carbon number 4.

7.42 Fructose can be used as a low-calorie sweetener because the caloric content per gram is about the same as for sucrose. However, since fructose is so much sweeter, less is needed to provide the same degree of sweetness.

7.44 a. Sucrose
b. Maltose
c. Lactose
d. Maltose
e. Sucrose
f. Sucrose

7.46 The glucose ring opens at the hemiacetal group, exposing an aldehyde group which then reacts with the Cu^{2+} of the Benedict's reagent.

7.48 Perform a Benedict's test. Sucrose is not a reducing sugar, but honey contains reducing sugars.

7.50 The right ring in lactose has a hemiacetal group that can open to form the aldehyde and alcohol; hence, it is a reducing sugar. Sucrose has no hemiacetal bonds, so it can't reduce Cu^{2+}.

7.52 a. Galactose and glucose
b. Yes; there is a hemiacetal on the glucose ring.
c. $\alpha(1 \rightarrow 6)$

7.54 a. Amylose
b. Cellulose
c. Amylopectin
d. Cellulose
e. Glycogen

7.56 The boiling point of the glucose solution will be higher because the molarity of the solution is higher.

7.58 50.0 g, the molecular weights of glucose and fructose are identical.

7.60 As hydrolysis occurred, the molarity of the solution would increase causing an increase in osmotic pressure, and the solution level in the tube would rise.

7.62 a, b, and d

7.64 Hydrolysis

CHAPTER 8

8.2 Energy storage, structural components, some hormones

8.4 a. Nonsaponifiable

b. Saponifiable

c. Saponifiable

d. Saponifiable

e. Saponifiable

f. Nonsaponifiable

8.6 1. They have a straight-chain carbon segment with a carboxylic acid group.

2. The straight-chain carbon segment is 10–20 carbons long.

3. There is an even number of total C atoms.

4. The carbon chain may be saturated or unsaturated.

8.8 Linoleic acid and linolenic acid are essential fatty acids because they are needed by the body but are not synthesized within the body.

8.10 Unsaturated fatty acids have a bent structure that prevents the molecules from packing as tightly as saturated fatty acids.

8.12 Fats and oils are both triglycerides (esters). Fats are solid at room temperature; oils are liquid at room temperature. Fats tend to have more saturated fatty acids, whereas oils tend to have more unsaturated fatty acids.

8.14
$$\begin{array}{l} CH_2-O-\overset{O}{\overset{\|}{C}}-(CH_2)_{12}CH_3 \\ | \\ CH-O-\overset{O}{\overset{\|}{C}}-(CH_2)_7CH{=}CH(CH_2)_5CH_3 \\ | \\ CH_2-O-\overset{O}{\overset{\|}{C}}-(CH_2)_6(CH_2CH{=}CH)_2(CH_2)_4CH_3 \end{array}$$

(arrows point to the O–C bonds) ester linkages

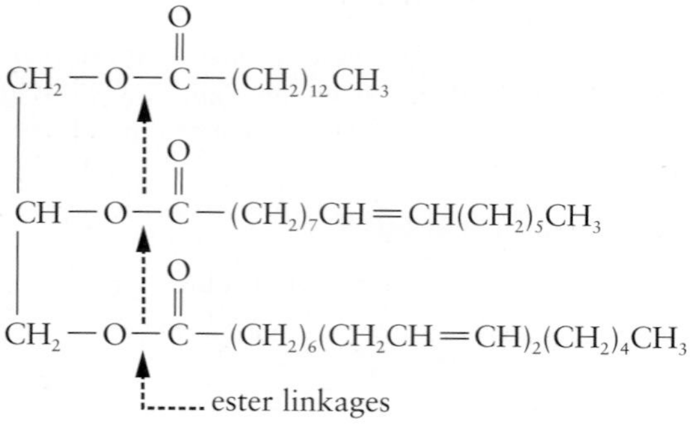

8.16 Triglyceride B

8.18 Hydrogenation

8.20 A number of commercially useful margarines and cooking shortenings are prepared by the hydrogenation of vegetable oils.

8.22 a.
$$\begin{array}{l} CH_2-O-\overset{O}{\overset{\|}{C}}-(CH_2)_{16}CH_3 \\ | \\ CH-O-\overset{O}{\overset{\|}{C}}-(CH_2)_{16}CH_3 + 3H_2O \xrightarrow{H^+} \\ | \\ CH_2-O-\overset{O}{\overset{\|}{C}}-(CH_2)_{16}CH_3 \end{array}$$

$$\begin{array}{l} CH_2-OH \\ | \\ CH-OH \\ | \\ CH_2-OH \end{array} + 3HO-\overset{O}{\overset{\|}{C}}-(CH_2)_{16}CH_3$$

glycerol stearic acid

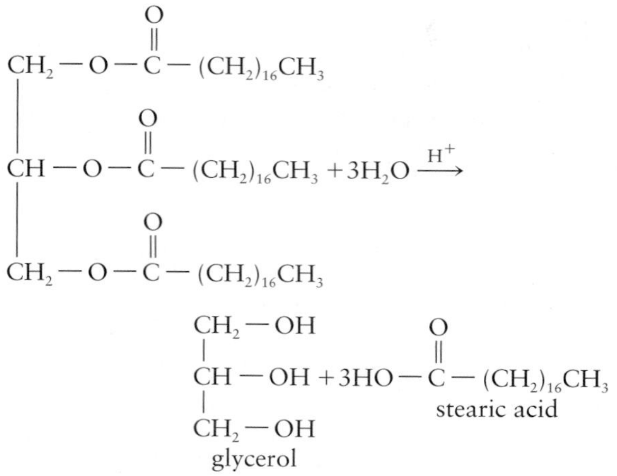

b. Same reaction as in part a.

c.
$$\begin{array}{l} CH_2-O-\overset{O}{\overset{\|}{C}}-(CH_2)_6(CH_2-CH{=}CH)_2(CH_2)_4CH_3 \\ | \\ CH-O-\overset{O}{\overset{\|}{C}}-(CH_2)_6(CH_2-CH{=}CH)_3CH_2CH_3 + 5H_2 \xrightarrow{Ni} \\ | \\ CH_2-O-\overset{O}{\overset{\|}{C}}-(CH_2)_6(CH_2-CH{=}CH)_2(CH_2)_4CH_3 \end{array}$$

$$\begin{array}{l} CH_2-O-\overset{O}{\overset{\|}{C}}-(CH_2)_{16}-CH_3 \\ | \\ CH-O-\overset{O}{\overset{\|}{C}}-(CH_2)_{16}CH_3 \\ | \\ CH_2-O-\overset{O}{\overset{\|}{C}}-(CH_2)_6(CH_2-CH{=}CH)_2(CH_2)_4CH_3 \end{array}$$

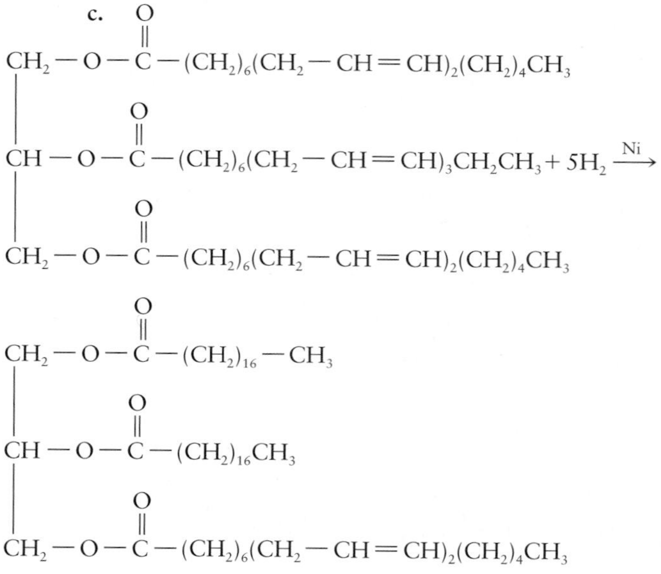

d.
$$\begin{array}{l} CH_2-O-\overset{O}{\overset{\|}{C}}-(CH_2)_{16}CH_3 \\ | \\ CH-O-\overset{O}{\overset{\|}{C}}-(CH_2)_{16}CH_3 + 3NaOH \longrightarrow \\ | \\ CH_2-O-\overset{O}{\overset{\|}{C}}-(CH_2)_{16}CH_3 \end{array}$$

$$\begin{array}{l} CH_2-OH \\ | \\ CH-OH \\ | \\ CH_2-OH \end{array} + 3Na^{+}\,{}^{-}O-\overset{O}{\overset{\|}{C}}-(CH_2)_{16}CH_3$$

glycerol soap

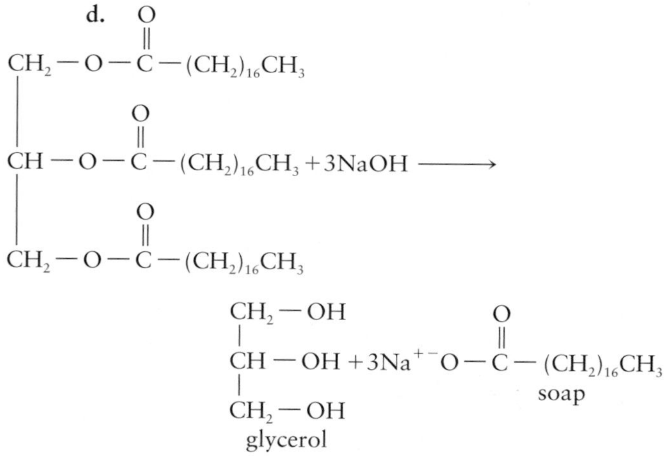

8.24 Fats are a triester between glycerol and fatty acids. Waxes are a monoester between a long-chain alcohol and a fatty acid.

8.26 In nature, waxes often occur as protective coatings.

8.28 g — fatty acid
l
y
c — fatty acid
e
r
o — phosphoric acid
l | alcohol

8.30 Lecithins serve as components in cell membranes and aid in lipid transport.

8.32 They differ in the identity of the amino alcohol portion. Lecithins contain choline, and cephalins contain ethanolamine or serine.

8.34 s
p
h
i
n — fatty acid
g
o
s
i — phosphoric acid
n | choline
e

8.36 Tay-Sachs, Gaucher's, and Niemann-Pick

8.38 Cerebrosides, brain tissue

8.40 Phosphoglycerides, sphingomyelin, and cholesterol

8.42 The model is composed of a flexible bilayer structure with protein molecules floating in the bilayer. The lipid molecules are free to move in a lateral direction.

8.44 The fused ring system:

8.46 Bile salts, sex hormones, and adrenocorticoid hormones. Cell membranes.

8.48 Cholesterol

8.50 The two groups are glucocorticoids and mineralocorticoids. Cortisol, a glucocorticoid, acts to increase the glucose and glycogen concetrations in the body. Aldosterone, a mineralocorticoid, regulates the retention of Na^+ and Cl^- during urine formation.

8.52 Testosterone; estradiol, estrone, and progesterone

8.54 Estrogens are involved in egg development; progesterone causes changes in the wall of the uterus to prepare it to accept a fertilized egg and maintain the resulting pregnancy.

8.56 Arachidonic acid

8.58 Induction of labor, treatment of asthma, treatment of peptic ulcers

8.60 The lipid is unsaturated (see Figure 12.5).

8.62 The phosphate and amino alcohol groups found in complex lipids are much more hydrophilic and capable of interacting with water than simple lipids.

8.64 Ester, ester

8.66 a. Estrogens and progesterone

b. Testosterone

8.68 C explanation: The gallbladder stores bile from the liver, which is used to emulsify fats to aid in their digestion in the intestines.

CHAPTER 9

9.2 The amine group ($—NH_2$) and the carboxylic acid group ($—COOH$), or the corresponding protonated $—NH_3^+$ group and unprotonated $—COO^-$ group.

9.4 a. (H) above C; $(H_3\overset{+}{N})—C—(COO^-)$; side chain (CH — CH_3 | OH)

b. (H) above C; $(H_3\overset{+}{N})—C—(COO^-)$; side chain ($CH_2$ | COO^-)

c. (H) above C; $(H_3\overset{+}{N})—C—(COO^-)$; side chain ($CH_2$ | OH)

d. (H) above C; $(H_3\overset{+}{N})—C—(COO^-)$; side chain ($CH_2$ | benzene ring)

9.6

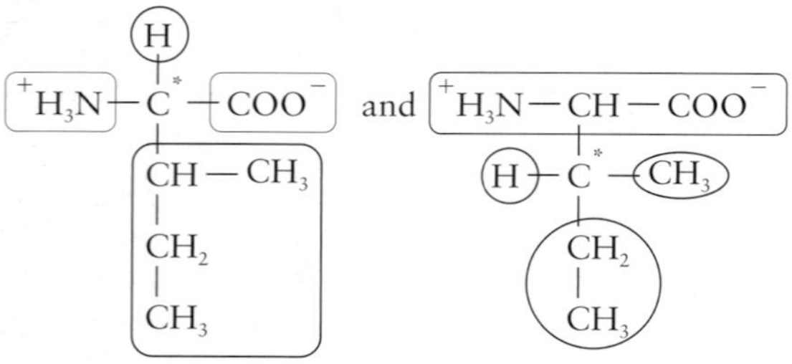

$$\begin{array}{c} \text{(H)} \\ | \\ \boxed{{}^{+}H_3N}-\overset{*}{C}-\boxed{COO^-} \\ | \\ \boxed{\begin{array}{l} CH-CH_3 \\ | \\ CH_2 \\ | \\ CH_3 \end{array}} \end{array} \quad \text{and} \quad \begin{array}{c} \boxed{{}^{+}H_3N-CH-COO^-} \\ | \\ \text{(H)}-\overset{*}{C}-\text{(}CH_3\text{)} \\ | \\ \text{(}CH_2-CH_3\text{)} \end{array}$$

9.8 a.

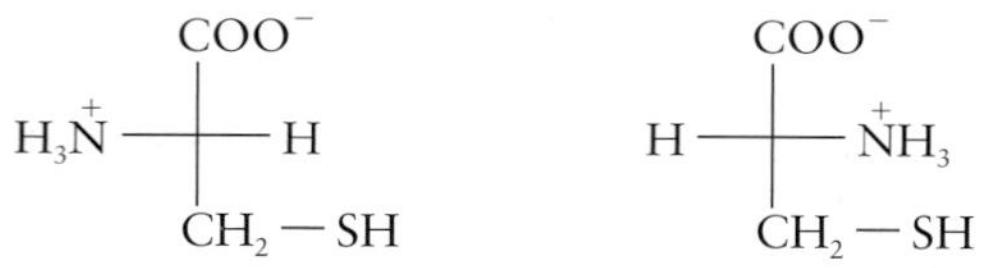

$$\begin{array}{c} COO^- \\ {}^{+}H_3N-\!\!\!+\!\!\!-H \\ CH_2-SH \end{array} \qquad \begin{array}{c} COO^- \\ H-\!\!\!+\!\!\!-\overset{+}{N}H_3 \\ CH_2-SH \end{array}$$

L-cysteine D-cysteine

b.

$$\begin{array}{c} COO^- \\ {}^{+}H_3N-\!\!\!+\!\!\!-H \\ CH_2 \\ CH_2 \\ | \\ COO^- \end{array} \qquad \begin{array}{c} COO^- \\ H-\!\!\!+\!\!\!-\overset{+}{N}H_3 \\ CH_2 \\ CH_2 \\ | \\ COO^- \end{array}$$

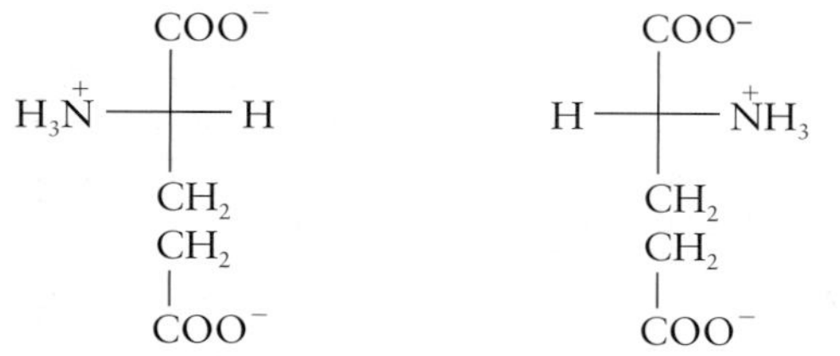

L-glutamate D-glutamate

9.10 They are crystalline solids with relatively high melting points and high water solubilities.

9.12

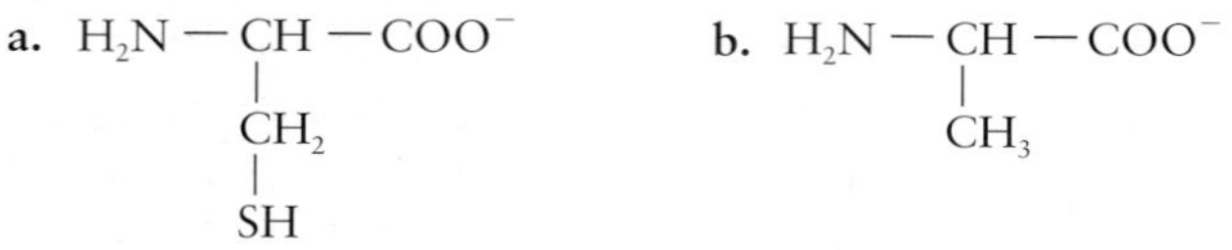

a. $H_2N-\underset{\underset{\underset{SH}{|}}{CH_2}}{\underset{|}{CH}}-COO^-$ b. $H_2N-\underset{CH_3}{\underset{|}{CH}}-COO^-$

9.14 a.

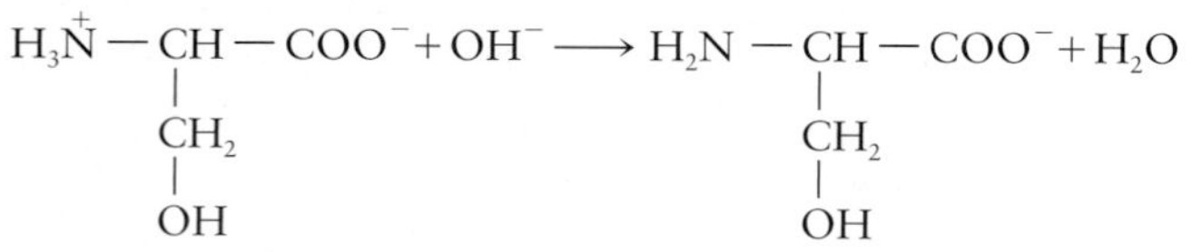

$$H_3\overset{+}{N}-\underset{\underset{OH}{|}}{\underset{CH_2}{\underset{|}{CH}}}-COO^- + OH^- \longrightarrow H_2N-\underset{\underset{OH}{|}}{\underset{CH_2}{\underset{|}{CH}}}-COO^- + H_2O$$

b.

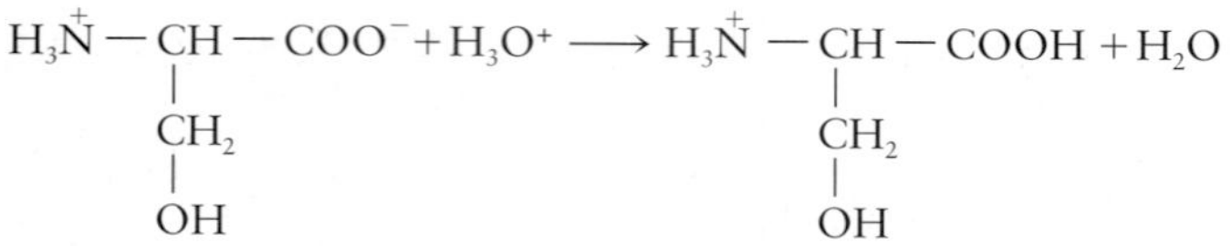

$$H_3\overset{+}{N}-\underset{\underset{OH}{|}}{\underset{CH_2}{\underset{|}{CH}}}-COO^- + H_3O^+ \longrightarrow H_3\overset{+}{N}-\underset{\underset{OH}{|}}{\underset{CH_2}{\underset{|}{CH}}}-COOH + H_2O$$

9.16

$$H_3\overset{+}{N}-\underset{\underset{CH_3}{|}}{\underset{CHCH_3}{\underset{|}{CH}}}-\overset{\overset{O}{\|}}{C}-NH-\underset{\underset{COO^-}{|}}{\underset{CH_2}{\underset{|}{CH}}}-\overset{\overset{O}{\|}}{C}-NH-\underset{\underset{SH}{|}}{\underset{CH_2}{\underset{|}{CH}}}-\overset{\overset{O}{\|}}{C}-O^-$$

Val-Asp-Cys

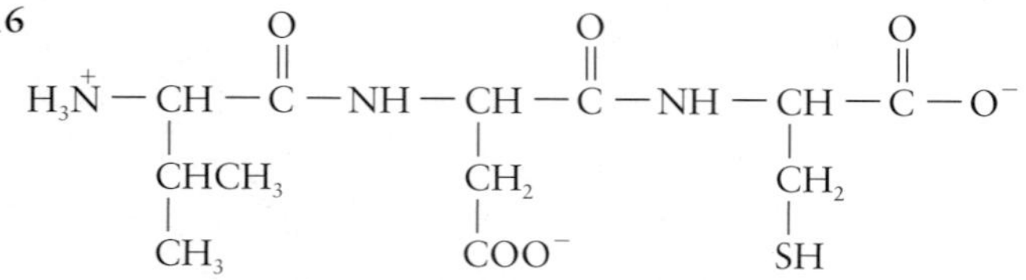

9.18 Ala-Phe-Arg, Phe-Arg-Ala, Arg-Ala-Phe

Ala-Arg-Phe, Phe-Ala-Arg, Arg-Phe-Ala

9.20 Six

9.22 Each peptide contains a disulfide bridge in which two cysteines have been linked.

9.24 Normally proteins do not pass through membranes, and the appearance of an abnormal protein in blood or urine indicates cellular damage, which has released those substances.

9.26 When a protein is at a pH corresponding to the isoelectric point, the net charge on the protein is 0. The protein is not as soluble with a 0 charge as when it possesses a net negative or positive charge.

9.28 Catalysis, structural, storage, protection, regulatory, nerve impulse transmission, motion, transport

9.30 a. Fibrous; collagen is a structural protein.

b. Globular; lactate dehydrogenase is an enzyme and needs to be water soluble.

9.32 Simple proteins contain only amino acid residues, whereas conjugated proteins contain additional components.

9.34 The sequence of the amino acid residues in the protein

9.36

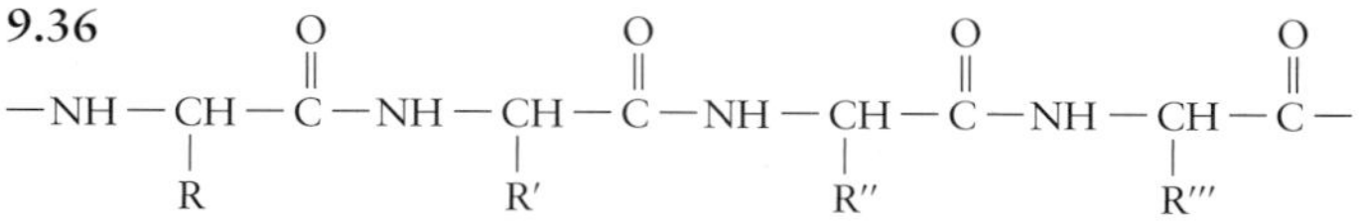

$$-NH-\underset{R}{\underset{|}{CH}}-\overset{\overset{O}{\|}}{C}-NH-\underset{R'}{\underset{|}{CH}}-\overset{\overset{O}{\|}}{C}-NH-\underset{R''}{\underset{|}{CH}}-\overset{\overset{O}{\|}}{C}-NH-\underset{R'''}{\underset{|}{CH}}-\overset{\overset{O}{\|}}{C}-$$

9.38 Hydrogen bonding between the amide hydrogen and the carboxyl oxygen along the backbone

9.40 In secondary structures, the hydrogen bonding is between the atoms in the backbone; in tertiary structures, the hydrogen bonding is between atoms in the side chains.

9.42 a. Hydrogen bonding

b. Salt bridges

c. Hydrophobic interactions

d. Hydrophobic interactions

9.44 The nonpolar R groups in phenylalanine, alanine, and methionine will point inward; the polar R groups in glutamate and lysine will point toward the water.

9.46 Ionic attractions, disulfide bridges, hydrogen bonds, and hydrophobic forces

9.48 A subunit is a polypeptide chain that is part of a larger protein.

9.50 The products of hydrolysis are the separate, individual amino acid molecules. The product of denaturation has the same primary protein backbone with different secondary, tertiary, and quaternary structure.

9.52 The primary structure of the protein is the same.

9.54 Cooking egg white is an irreversible denaturation. Whipping air into them is a reversible denaturation.

9.56 The heavy-metal ions denature the protein in the egg white by interacting with the —SH groups and the carboxylate (—COO^-) groups. Vomiting should then be induced to remove the denatured proteins so they are not digested, releasing the metal ions.

9.58 valine

9.60

Conjugate acid: $H_3\overset{+}{N}-\underset{R}{\overset{H}{C}}-COOH$

Conjugate base: $H_2N-\underset{R}{\overset{H}{C}}-COO^-$

9.62 Oxytocin

9.64 a and c

9.66 Hemoglobin is nearly spherical with four chains (subunits) held together.

CHAPTER 10

10.2 Enzymes are highly specific in the reactions they catalyze, and their activity as catalysts can be regulated.

10.4 Each one is specific for a single type of reaction and often is specific for a single substrate.

10.6 Enzymes that are specific for a single reactant exhibit absolute specificity.

10.8 -ase

10.10 **a.** sucrose

b. amylose

c. lactose

d. maltose

e. arginine

10.12 **a.** catalyzes the dehydrogenation (oxidation) of succinate

b. catalyzes the reduction of L-amino acids

c. catalyzes the oxidation of cytochrome

d. catalyzes the isomerization of glucose 6-phosphate

10.14 An enzyme is a catalytically active structure made up of a protein portion called an apoenzyme and a nonprotein portion called a cofactor.

10.16 Mg^{2+}, Zn^{2+}, Fe^{2+}, Ca^{2+}

10.18 E = enzymes, S = substrate, ES = enzyme–substrate complex, and P = product of reaction. Thus $E + S \rightleftarrows ES \rightarrow E + P$ represents the reaction of an enzyme with a substrate molecule to form an enzyme–substrate complex, which then gives the reaction product and regenerates the enzyme.

10.20 The shape of the enzyme active site allows only specific substrates having the correct shape to bind and be acted upon.

10.22 Induced-fit theory, because the enzyme catalyzes reactions of several related compounds.

10.24 Any observation that allows the rate of product formation or reactant depletion to be measured will allow enzyme activity to be determined.

10.26 One international unit (IU) is the amount of enzyme that will convert 1 micromol (μ mol) of substrate to product in 1 minute. The international unit is useful in medical diagnosis because it represents a standard quantity of enzyme (or enzyme activity level) against which the level present in a patient can be compared.

10.28 **a.** The enzyme activity (rate of reaction) increases with increased substrate concentration until all the enzyme is utilized, at which point no further increase in rate occurs.

b. The activity (rate) increases with an increase in enzyme concentration.

c. There is an optimum pH for the reaction, so at a pH above or below the optimum value the rate will decrease.

d. The activity (rate) increases up to a maximum as the temperature increases and then decreases with further temperature increase.

10.30 Prepare a series of reactions with constant enzyme amount and increasing substrate amounts. Plot reaction rate versus substrate concentration. A constant rate at higher substrate concentrations is V_{max}.

10.32 Enzymes could be destroyed at pH values far from the optimum pH.

10.34 In competitive inhibition another substance, an inhibitor, competes with the substrate in binding to the active site of the enzyme. In noncompetitive inhibition the inhibitor binds to another site on the enzyme, and in doing so alters the shape of the enzyme so that it can no longer bind to the substrate, essentially prohibiting a reaction.

10.36 **a.** Cyanide poisoning is treated with $Na_2S_2O_3$ (sodium thiosulfate), which oxidizes CN^- to SCN^-. The thiocyanate does not bind to the iron in the enzyme.

b. Heavy-metal poisoning is treated by administering chelating agents such as EDTA. The chelating agents bond to the metal ions to form water-soluble complexes that are excreted from the body in the urine.

10.38 1. Zymogens or proenzymes

2. Allosteric regulation

3. Genetic control over the synthesis of enzymes

10.40 So the blood doesn't clot as it circulates, but only when clotting is needed to stop bleeding

10.42 Modulators may increase activity (activators) or decrease the activity (inhibitors).

10.44 Genetic control of enzyme activity means that a cell synthesizes an enzyme when conditions temporarily require it. Such a controlled synthesis is called enzyme induction.

10.46 **a.** CK assays are used in diagnosing heart attack.

b. ALP assays are useful in diagnosing liver or bone disease.

c. Amylase assays are useful in diagnosing diseases of the pancreas.

10.48 An LDH assay is a good initial diagnostic test because the isoenzyme forms of LDH are so widespread in the human system. Thus, the locations of diseases can be pinpointed.

10.50 An ionic bond is formed:

$$H_3\overset{+}{N}-\text{Gly}-\text{Ala}-C{\overset{O^-}{\underset{O}{\lesssim}}}\quad \overset{+}{H_2N}\!\!\gtrsim C-\overset{H}{\overset{|}{N}}-CH_2-CH_2-CH_2-\boxed{\text{Enzyme}}$$

sample substrate arginine side chain

10.52 A covalent ester linkage is formed between the acid chloride and the —OH of the serine:

$$CH_3CH_2\overset{O}{\overset{\|}{C}}-Cl + HO-CH_2-\boxed{\text{Enzyme}} \longrightarrow CH_3CH_2\overset{O}{\overset{\|}{C}}-O-CH_2-\boxed{\text{Enzyme}} + HCl$$

10.54 There will be no shift in the equilibrium. Enzymes are catalysts and cannot change the position of equilibrium, only the speed of which an equilibrium is reached.

10.56 a.

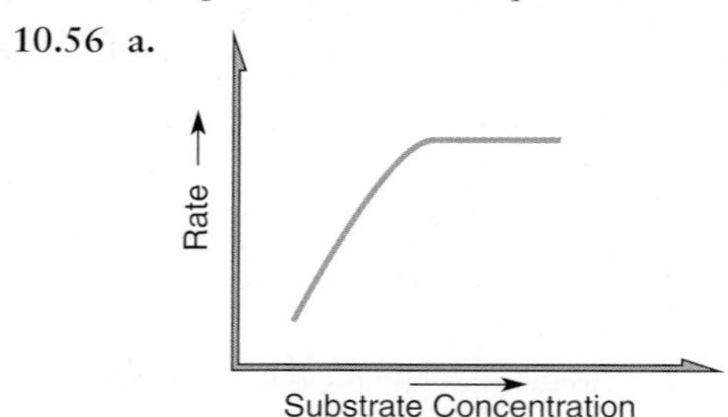

b.

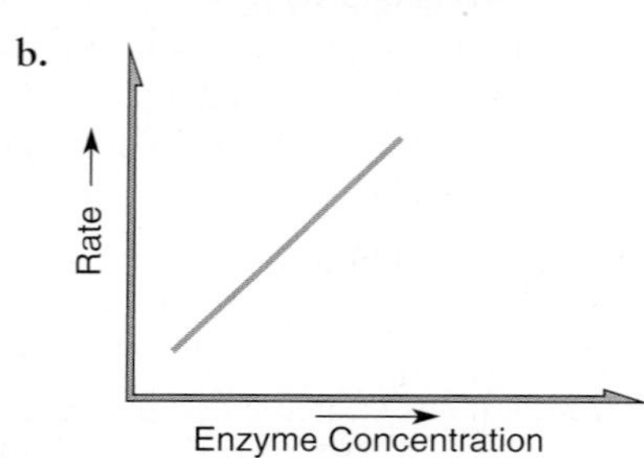

c.

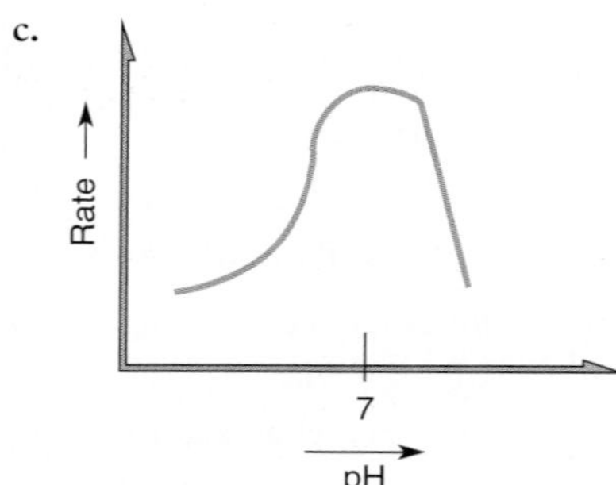

d.

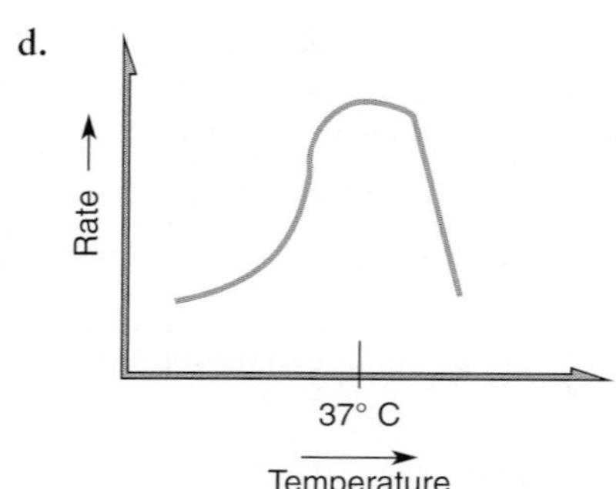

10.58 b

CHAPTER 11

11.2 Deoxyribose in DNA, ribose in RNA

11.4 a. Purine

b. Pyrimidine

c. Pyrimidine

d. Pyrimidine

e. Purine

11.6

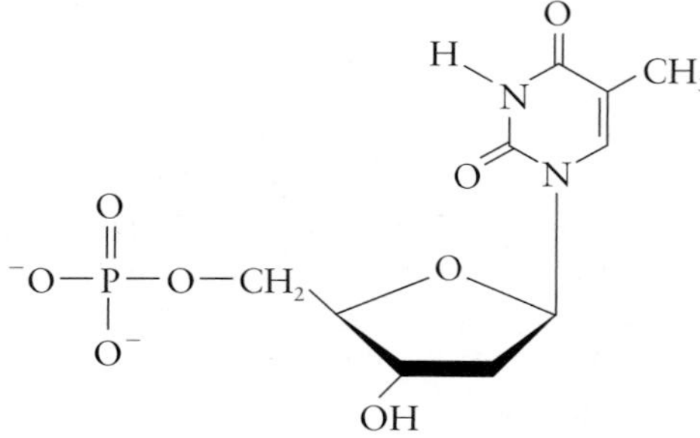

11.8 By convention, in the DNA segment AGTCAT, the 5′ end is the A and the 3′ end is the T.

11.10 Each DNA has two complementary nucleic acid strands arranged in a double helix with the bases protruding into the interior of the helix and the 5′ end of one strand next to the 3′ end of the other. The bases on one strand form hydrogen bonds with the bases on the other.

11.12 a. 15

b. 14

11.14 GATGCAT

11.16 A chromosome is a tightly packed bundle of DNA and protein that is involved in cell division. A human cell normally contains 46 chromosomes with about 25,000 genes.

11.18 The point in the double helix where "unwinding" begins

11.20 1. The double helix unwinds.

2. The new DNA segments are synthesized.

3. The "nicks" in the new strand are closed.

11.22 The daughter strand forms from its 5′ end toward the 3′ end.

11.24 a. ACGTCT

TGCAGA

b.

Old strand	5′ ACGTCT 3′	daughter
New strand	3′ TGCAGA 5′	DNA molecule
New strand	5′ ACGTCT 3′	daughter
Old strand	3′ TGCAGA 5′	DNA molecule

11.26 The sugar in the RNA backbone is ribose, and that in the DNA backbone is deoxyribose.

11.28 1. mRNA functions as a messenger. It consists of a base sequence that is complementary to a DNA strand. It is formed in the nucleus and then moves into the cytoplasm where protein synthesis occurs. It carries the genetic code to control protein synthesis.

2. Ribosomal RNA functions as the site of protein synthesis in the ribosome.

3. tRNA delivers an amino acid to the site of protein synthesis.

11.30 The two important regions of a tRNA molecule are the anticodon, which binds the tRNA to mRNA during protein synthesis, and 3′ end, which binds to an amino acid by an ester linkage.

11.32 Transcription is the transferring of genetic DNA information to messenger RNA. Translation is the expression of messenger RNA information in the form of an amino acid sequence in a protein.

11.34 5′ CGAUCUUAG 3′

11.36 In the DNA of eukaryotic cells, segments called introns carry no amino acid coding, whereas exons are coded segments. Both introns and exons are copied (transcribed) and produce hnRNA. A series of enzyme-catalyzed reactions cut out the introns and splice the remaining segments to produce mRNA.

11.38 **a.** 5′ AUA 3′, tyrosine

b. 5′ AUG 3′, histidine

c. 5′ UGA 3′, serine

d. 5′ AGA 3′, serine

11.40 **a.** False; should be three bases.

b. True

c. False; 61 of 64 codons represent amino acids.

d. False; each living species is thought to use the same genetic code.

e. True

f. False; stop signals for protein synthesis are represented by three codons: UAA, UAG, and UGA.

11.42 A complex of mRNA and several ribosomes; also called polysome.

11.44 Initiation, elongation, termination

11.46 The A site (aminoacyl site) is where an incoming tRNA carrying the next amino acid will bond. The P site (peptidyl site) is where the tRNA carrying the peptide chain is located.

11.48 A change in the DNA of a cell wherein a wrong base sequence is produced on the DNA.

11.50 The protein synthesized would contain a proline rather than an alanine residue at that site in the protein.

11.52 Recombinant DNA is DNA of one organism that contains segments of DNA from another organism.

11.54 Human insulin will become widely available for treating diabetes. Human growth hormone will be used for treating pituitary disorders. Recombivax Hb, a vaccine for hepatitis B, will be marketed.

11.56 Hydrolysis of the phosphodiester linkage would break the nucleic acid backbone and allow the removal of intron segments. Esterification (ester formation) of the $5'$—PO_4^{2-} group of one exon segment to the 3′ —OH group of another exon segment would be used to join exons

11.58 **a.** Glucose and fructose

b. Enzyme induction

c. Somehow the presence of sucrose activates the bacterial cell's DNA to produce more copies of mRNA that codes for enzyme Q.

11.60 DNA → RNA → protein

11.62 **a.** Small circular self-replicating DNA molecules found in bacterial cells that are used as vectors to incorporate recombinant DNA

b. Enzymes used to fragment chromosomes to be incorporated into a plasmid, also used to open plasmids.

c. Enzyme used to join together DNA strands

CHAPTER 12

12.2 Macronutrients are needed in large quantities per day (1 gram or more). Micronutrients are needed in small quantities (milligrams or less).

12.4

Carbohydrates	Energy source Provide materials for cell and tissue synthesis
Proteins	Provide amino acids for cell repair and maintenance Provide ingredients for the synthesis of enzymes and hormones
Lipids	Energy source; solvent for fat-soluble vitamins Source of essential fatty acids

12.6 **a.** Carbohydrates, lipids, proteins

b. Carbohydrates, lipids, proteins

c. Carbohydrates, proteins

d. Carbohydrates, lipids, proteins

e. Proteins, lipids

f. Lipids, proteins, carbohydrates

12.8 Fat, 65 g; saturated fat, 20 g; total carbohydrate, 300 g; fiber, 25 g

12.10 **a.** Fat soluble

b. Water soluble

c. Water soluble

d. Fat soluble

12.12 Water-soluble vitamins can be excreted in water-based fluids much more readily than fat-soluble vitamins. Thus, large doses of fat-soluble vitamins accumulate in the body.

12.14 **a.** vitamin C

b. vitamin B_1

c. vitamin B_{12}

d. niacin

12.16 Ca: component of bones and teeth; regulation of muscle activity

P: component of bones and teeth, nucleic acids, phospholipids, and enzymes

K: charge balance in cells; regulates heartbeat

S: component of the amino acid cysteine; charge balance in cells (as SO_4^{2-})

Na: charge balance in cells

Cl: charge balance in cells

Mg: component of enzymes; regulation of muscle activity

12.18 The general functions of trace minerals in the body are to serve as components of vitamins (Co), enzymes (Zn, Se), hormones (I), or specialized proteins (Fe, Cu).

12.20 Hydrogen

12.22 $6CO_2 + 6H_2O + \text{energy} \rightarrow C_6H_{12}O_6 + 6O_2$

12.24 Energy from the sun powers the process of photosynthesis in plants. As the carbohydrates of plants are converted to CO_2 and H_2O, the released energy is trapped in molecules of ATP.

12.26 Catabolism includes all of the breakdown processes. Anabolism includes all of the synthesis processes.

12.28 a. Stage III

b. Stage I

c. Stage III

d. Stage II

12.30 The production of ATP

12.32 ATP is involved in both.

12.34 ATP molecules store the energy released during the oxidation of food and deliver the energy to the parts of the cell where energy is needed.

12.36 P_i is the symbol for the phosphate ion. PP_i represents the pyrophosphate ion.

12.38 The hydrolysis of phosphoenolpyruvate liberates 14.8 kcal/mol, while the hydrolysis of glycerol 3-phosphate liberates only 2.2 kcal/mol.

12.40 The triphosphate portion

12.42 Mitochondria are the sites for most ATP synthesis in the cells.

12.44 Mitochondria are cigar-shaped organelles with both an outer and an inner membrane. The inner membrane has many folds and contains the enzymes for the electron transport chain and oxidative phosphorylation. The enzymes for the citric acid cycle are attached to or near to the surface of the inner membrane.

12.46 a. Coenzyme A is part of acetyl coenzyme A, which is formed from all foods as they pass through the common metabolic pathway. Acetyl coenzyme A is the fuel for the citric acid cycle.

b. and c. FAD and NAD^+ carry hydrogen atoms from the citric acid cycle to the electron transport chain.

12.48 $^-OOC - CH_2CH_2 - COO^-$ is oxidized. FAD is reduced.

12.50 a. $HO{-}CH_2{-}COOH + NAD^+ \rightarrow O{=}CH{-}COOH + NADH + H^+$

b. $HO{-}CH_2CH_2{-}OH + FAD \rightarrow HO{-}CH{=}CH{-}OH + FADH_2$

12.52

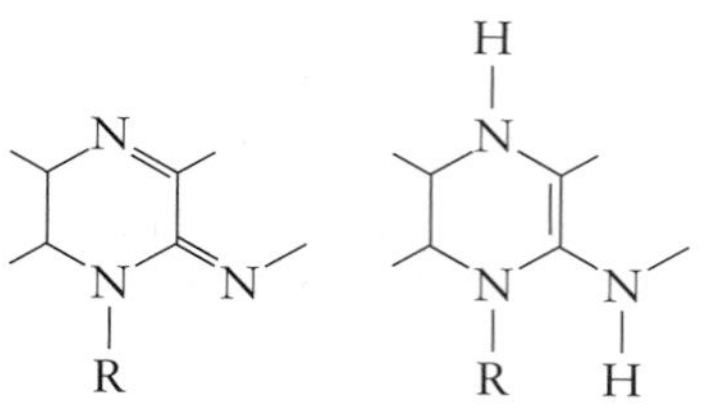

FAD

$FADH_2$

12.54 $CH_4 + O_2 + 2\ NAD^+ \rightarrow CO_2 + 2\ NADH + 2\ H^+$

A total of four electrons are transferred to the 2NADH molecules.

12.56 Protein synthesis, translation

12.58 a. A, D, E, and K

b. B_1

c. C

d. K

e. B_{12}

12.60 a

12.62 Lipids or fats

CHAPTER 13

13.2 a. Glucose

b. Glucose and galactose

c. Glucose and fructose

d. Glucose

13.4 Blood sugar level is the concentration of glucose in the blood. The normal fasting level is the blood glucose concentration after a fast of 8–12 hours.

13.6 a. A blood sugar level below the normal range

b. A blood sugar level above the normal fasting level

c. A blood sugar level at which glucose is not reabsorbed by the kidneys

d. An elevated blood sugar level resulting in the excretion of glucose in the urine

13.8 If the blood sugar is too low, severe hypoglycemia can cause convulsions and shock.

13.10 Glucose gives pyruvate.

13.12 Steps 1 and 3 require ATP, Steps 7 and 10 produce ATP. A net of 2 ATP molecules are produced for each glucose molecule.

13.14 The third control point is in the last step of glycolysis. The enzyme for this step, pyruvate kinase, is inhibited by high ATP concentration. If the cell does not need ATP, the ATP concentration will inhibit glycolysis and slow the production of ATP. As the need for ATP increases, the inhibition of the enzyme no longer occurs and glycolysis proceeds to make more ATP.

13.16 Glucose 6-phosphate is a feedback inhibitor of the enzyme hexokinase and causes a decrease in the enzymatic turnover rate (activity) of that enzyme.

13.18 Aerobic denotes the presence of oxygen, whereas anaerobic means without oxygen.

13.20 Under aerobic conditions, the pyruvate is converted to acetyl CoA. Under anaerobic conditions, the pyruvate is converted to lactate.

13.22 It is released as CO_2. Ethanol formation allows the yeast to regenerate NAD^+ in the absence of oxygen and still generate sufficient ATP to sustain life.

13.24 The citric acid cycle is the principal process for generating NADH and $FADH_2$, which are necessary for ATP synthesis. The citric acid cycle is also important as a source of intermediates in the biosynthesis of amino acids.

13.26 **a.** acetyl CoA

b. CO_2

c. NADH and $FADH_2$

13.28 **a.** Three

b. One

c. One

d. Two

13.30

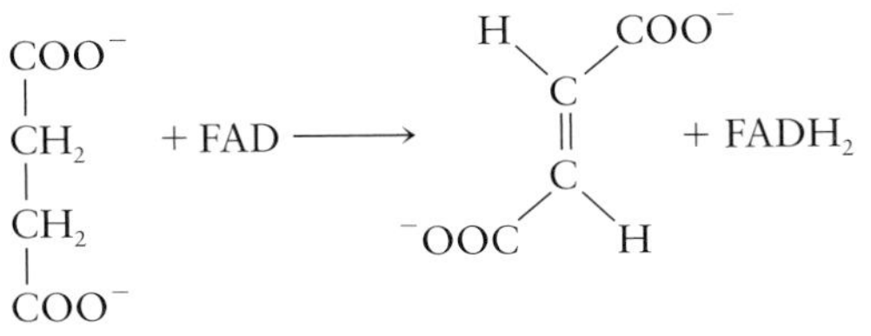

$$\begin{array}{c} COO^- \\ | \\ CH_2 \\ | \\ CH_2 \\ | \\ COO^- \end{array} + FAD \longrightarrow \begin{array}{c} H \quad\; COO^- \\ \diagdown\;\diagup \\ C \\ \| \\ C \\ \diagup\;\diagdown \\ {}^-OOC \quad\; H \end{array} + FADH_2$$

13.32 **a.** Step 1, citrate synthetase
Step 3, isocitrate dehydrogenase
Step 4, α-ketoglutarate dehydrogenase

b. When supplies of NADH and ATP are abundant, the cell has sufficient energy for its needs, so the operation of the citric acid cycle is inhibited.

c. As the cell uses ATP and makes ADP, the enzymes controlling the cycle are activated to make more ATP.

13.34 NADH and $FADH_2$

13.36 Cytochromes pass electrons along the electron transport chain to oxygen, which combines with H^+ ions to form water.

13.38 NADH and $FADH_2$ are oxidized to NAD^+ and FAD, respectively. ADP is phosphorylated to ATP.

13.40 1.5 molecules of ATP are formed during a sequence of reactions.

13.42 100 ATP molecules

13.44 94 mol of ATP

13.46 ATP synthesis results from the flow of protons (H^+) across the inner mitochondrial membranes.

13.48 **a.** 2 **b.** 2 **c.** 28

13.50 Glycogenesis is the synthesis of glycogen, while glycogenolysis is the breakdown of glycogen.

13.52 UTP

13.54 Liver cells have the essential enzyme glucose 6-phosphatase, which is lacking in muscle tissue.

13.56 The liver

13.58 Lactate, glycerol (derived from the hydrolysis of fats), and certain amino acids

13.60 There would be no lactate produced, so the Cori cycle wouldn't operate.

13.62 **a.** Insulin promotes the utilization of blood glucose, causing the level to decrease. Glucagon activates the breakdown of glycogen, thereby raising blood glucose levels.

b. Insulin promotes glycogen formation, while glucagon stimulates the breakdown.

13.64 1.00×10^{24} ATP molecules

13.66 The removal of lactate from muscle cells shifts the equilibrium to the right, ensuring the regeneration of NAD^+ and the production of pyruvate by glycolysis. In the liver, pyruvate is used to make glucose, thus decreasing the pyruvate concentration, which shifts the equilibrium to the left and prevents the accumulation of lactate.

13.68

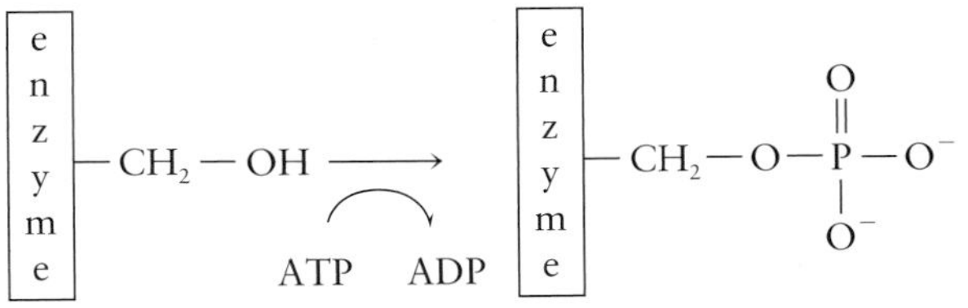

$$\boxed{\text{enzyme}}-CH_2-OH \xrightarrow[\text{ATP}\;\curvearrowright\;\text{ADP}]{} \boxed{\text{enzyme}}-CH_2-O-\overset{\overset{\displaystyle O}{\|}}{\underset{\underset{\displaystyle O^-}{|}}{P}}-O^-$$

An ester linkage is formed.

13.70 a and c

13.72 b, c, and d

13.74 liver and muscles

CHAPTER 14

14.2 Glycerol, fatty acids, and monoglycerides

14.4 Lipids are transported as lipoprotein complexes.

14.6 Chylomicrons, very low-density lipoproteins, low-density lipoproteins, high-density lipoproteins

14.8 Fat mobilization is the process of hydrolyzing triglycerides to produce fatty acids and glycerol, both of which can be used for energy in muscle cells.

14.10 Resting muscle and liver cells

14.12 Glycerol can be converted to pyruvate and contribute to cellular energy production or be converted to glucose.

14.14 A fatty acid is first converted to a fatty acyl CoA, a high-energy compound. This occurs in the cytoplasm of the cell.

14.16 FAD and NAD^+

14.18

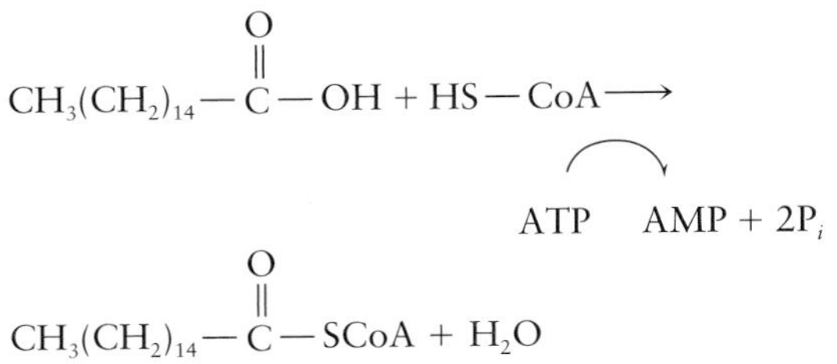

$$CH_3(CH_2)_{14}-\overset{\overset{\displaystyle O}{\|}}{C}-OH + HS-CoA \xrightarrow[\text{ATP}\;\curvearrowright\;\text{AMP} + 2P_i]{}$$

$$CH_3(CH_2)_{14}-\overset{\overset{\displaystyle O}{\|}}{C}-SCoA + H_2O$$

14.20 The word *cycle* implies that the product is identical to the starting material, which is not the case. The fatty acyl product of a trip through the sequence has two fewer carbon atoms than the starting material.

14.22 a. $CH_3CH_2CH_2CH{=}CH{-}C({=}O){-}S{-}CoA + FADH_2$

b. $CH_3CH_2CH_2CH{=}CH{-}C({=}O){-}S{-}CoA + H_2O \longrightarrow CH_3CH_2CH_2CH(OH)CH_2{-}C({=}O){-}S{-}CoA$

c. $CH_3CH_2CH_2CH(OH)CH_2{-}C({=}O){-}S{-}CoA + NAD^+ \longrightarrow CH_3CH_2CH_2C({=}O)CH_2{-}C({=}O){-}S{-}CoA + NADH + H^+$

d. $CH_3CH_2CH_2C({=}O)CH_2{-}C({=}O){-}S{-}CoA + CoA{-}SH \longrightarrow CH_3CH_2CH_2{-}C({=}O){-}S{-}CoA + CH_3{-}C({=}O){-}S{-}CoA$

14.24 10 ATP

14.26 64 ATP molecules

14.28 Ketone bodies are three compounds formed from acetyl CoA: acetoacetate, β-hydroxybutyrate, and acetone.

14.30 The liver; the brain, heart, and skeletal muscles

14.32 Two of the ketone bodies are carboxylic acids, which lower blood pH.

14.34 Fatty acids are built two carbons at a time and broken down two carbons at a time.

14.36 The required enzymes are missing.

14.38 The amino acid pool is a supply of amino acids located within the blood and cellular spaces.

14.40 Protein turnover

14.42 Purines and pyrimidines, heme, choline, ethanolamine

14.44 a. Transamination

b. Deamination

c. Transamination

d. Transamination and deamination

14.46 $\underset{\text{pyruvate}}{CH_3{-}C({=}O){-}COO^-} + \underset{\text{leucine}}{H_3\overset{+}{N}{-}CH(CH_2{-}CH(CH_3){-}CH_3){-}COO^-} \longrightarrow CH_3{-}CH(NH_3^+){-}COO^- + {}^{-}OOC{-}C({=}O){-}CH_2{-}CH(CH_3){-}CH_3$

14.48 $H_3\overset{+}{N}{-}CH_2{-}C({=}O){-}O^- + NAD^+ \longrightarrow H{-}C({=}O){-}C({=}O){-}O^- + NADH + NH_4^+$

14.50 The liver

14.52 Carbamoyl phosphate

14.54 $NH_4^+ + HCO_3^- + 3ATP + 2H_2O + \underset{\text{aspartate}}{{}^{-}OOC{-}CH_2{-}CH(NH_3^+){-}COO^-} \longrightarrow \underset{\text{urea}}{H_2N{-}C({=}O){-}NH_2} + \underset{\text{fumarate}}{{}^{-}OOC{-}CH{=}CH{-}COO^-} + 2H^+ + 2ADP + 2P_i + AMP + PP_i$

14.56 Yes; isoleucine, tryptophan, phenylalanine, and tyrosine

14.58 a. Glucogenic

b. Ketogenic

c. Ketogenic

14.60 Amino acids that can be synthesized by the body are called nonessential. Those amino acids that must be obtained from the diet because the body cannot produce them are called essential amino acids.

14.62 Transaminases are used in the biosynthesis of amino acids and in transfer of nitrogen atoms from amino acids when the disposal of nitrogen is necessary.

14.64 The glycolysis pathway and the citric acid cycle

14.66 Activation of zymogens and the production of new enzymes through induction

14.68 It is an LDL.

14.70 b

14.72 Amino acids

14.74 ketoacidosis

CHAPTER 15

15.2 Plasma and interstitial fluid

15.4 a. Intracellular fluid

b. Intracellular fluid

c. Plasma and interstitial fluid

d. Intracellular fluid

e. Plasma and interstitial fluid

f. Intracellular fluid

g. Plasma and interstitial fluid

15.6 a. Cl^-

b. Cl^-

c. HPO_4^{2-}

15.8 a. $HHb + O_2 \rightleftarrows HbO_2^- + H^+$

b. $HHb + CO_2 \rightleftarrows HHbCO_2$

15.10 a. 2% b. 98% c. 25% d. 5% e. 70%

15.12 A larger surface area increases the amount of O_2 and CO_2 that can diffuse through the surface. The partial pressure gradient is the force that causes O_2 and CO_2 movement.

15.14 a. Osmotic pressure

b. Blood pressure

15.16 Urea

15.18 Glucose (in large amounts), protein, ketone bodies, pus, hemoglobin, red blood cells, bile pigments (in large amounts)

15.20 Kidneys (urine), lungs (water vapor), skin (perspiration), and intestines (feces)

15.22 Variations in urine output

15.24 As fluid level decreases, saliva output decreases, giving the "thirst" feeling. Drinking water raises fluid level.

15.26 Blood buffers, urinary system, and respiratory system. The blood buffer system is the fastest to react.

15.28 Bicarbonate, phosphate, and protein

15.30 Exhaled H_2O and CO_2 are derived from H_2CO_3. The more CO_2 and H_2O that are exhaled, the more carbonic acid is removed from the blood, thus elevating the blood pH.

15.32 Respiratory alkalosis

15.34 CO_2 diffuses into the distal tubule cells to form carbonic acid, which then ionizes to H^+ and HCO_3^-. Thus, CO_2 concentration is lowered in the blood.

15.36 As H^+ enters the urine, Na^+ is reabsorbed into the blood from the urine.

15.38 Respiratory acid-base imbalances are the result of abnormal breathing patterns. Acid-base imbalances caused by any other factor are metabolic acid-base imbalances.

15.40 The depressant narcotic often causes a lower breathing rate, causing respiratory acidosis. Diabetes mellitus can cause metabolic acidosis.

15.42 5.01×10^{23} amino acid molecules

15.44 The increased concentration of Na^+ (and Cl^-) ions in the blood increases the osmotic pressure gradient and water flows from the region of low solute concentration to the region of high solute concentration.

15.46 Heavy exercise would cause anaerobic conditions to occur in some cells, which would then release lactate from glycolysis into the bloodstream.

15.48 Partial pressure

15.50 a, b, c, and d

Solutions to Learning Checks

CHAPTER 1

1.1 Look for the presence of carbon atoms.

a. Inorganic　　d. Inorganic
b. Organic　　e. Organic
c. Organic　　f. Inorganic

1.2 Refer to Table 1.1 for help.

a. Organic
b. Organic
c. Inorganic

1.3 Compound (a) is a structural isomer because it has the same molecular formula but a different structural formula.

1.4 a.
$$\begin{array}{c} \quad\;\; OH \\ \quad\;\; | \\ CH_3-CH-CH_2-CH_2-CH_3 \end{array}$$

b.
$$\begin{array}{c} \qquad\qquad\qquad\quad O \\ \qquad\qquad\qquad\quad \| \\ CH_3-CH-CH_2-C-OH \\ \;\;| \qquad\qquad\qquad \\ \;\;CH_3 \qquad\qquad\qquad \end{array}$$

1.5 a. The number of hydrogen atoms is twice the carbon atoms plus two: C_8H_{18}.

b. $CH_3—CH_2—CH_2—CH_2—CH_2—CH_2—CH_2—CH_3$

1.6 a. Same molecule: In both molecules, the five carbons are bonded in a continuous chain.

b. Same molecule: In both molecules, there is a continuous chain of four carbons with a branch at position 2.

$$\begin{array}{c} \quad\; CH_3 \\ \quad\; | \\ \underset{1}{CH_3}-\underset{2}{CH}-\underset{3}{CH_2}-\underset{4}{CH_3} \end{array} \qquad \begin{array}{c} \quad\; \overset{3}{CH_2}-\overset{4}{CH_3} \\ \quad\; | \qquad\quad \\ \underset{1}{CH_3}-\underset{2}{CH}-CH_3 \end{array}$$

c. Structural isomers: Both molecules have a continuous chain of five carbons, but the branch is located at different positions.

1.7 a.
$$\begin{array}{l} \overset{1}{CH_3} \\ | \\ \underset{2}{CH_2}-\underset{3}{CH_2}-\underset{4}{CH_3} \end{array} \quad \text{a butane}$$

b.
$$\begin{array}{c} \overset{1}{CH_3} \qquad\qquad \overset{5}{CH_3} \\ | \qquad\qquad\quad | \\ \underset{2}{CH_2}-\overset{3}{CH}-\underset{4}{CH_2} \\ | \\ CH_3 \end{array} \quad \text{a pentane}$$

c.
$$\begin{array}{c} \underset{1}{CH_3}-\overset{2}{CH}-CH_3 \\ | \\ CH_3-\underset{3}{CH}-\underset{4}{CH_3} \end{array} \quad \text{or}$$

$$\begin{array}{c} CH_3-\overset{3}{CH}-\overset{4}{CH_3} \\ | \\ \underset{1}{CH_3}-\underset{2}{CH}-CH_3 \end{array} \quad \text{a butane}$$

1.8 a.
$$\begin{array}{l} CH_3-\overset{3}{CH}-\overset{4}{CH_2}-\overset{5}{CH_2}-\overset{6}{CH_3} \\ \qquad\quad | \\ \qquad\quad \overset{2}{CH_2} \\ \qquad\quad | \\ \qquad\quad \overset{1}{CH_3} \end{array}$$

A CH_3 group is located at position 3. If the chain had been numbered beginning at the right, the CH_3 group would have been at position 4.

b. Numbering from the left, groups are located at positions 5, 5, 7. From the right, the groups are at positions 2, 4, 4. The first difference occurs with the first number, so numbering from the right (2, 4, 4) is correct.

1.9 Proceeding from the left, the groups are a methyl, propyl, isopropyl, *sec*-butyl, and ethyl.

1.10 a. The chain is numbered from the right to give 2-methylhexane.

b.
$$\begin{array}{l} \qquad\qquad\qquad\qquad\quad CH_2-CH_3 \\ \qquad\qquad\qquad\qquad\quad | \\ \underset{8}{CH_3}-\underset{7}{CH_2}-\underset{6}{CH_2}-\underset{5}{CH}-\underset{4}{CH_2}-\overset{3}{CH}-CH_3 \\ \qquad\qquad\qquad\qquad\qquad\qquad\qquad | \\ \qquad\qquad\qquad\qquad\qquad\qquad\qquad \underset{2}{CH_2}-\underset{1}{CH_3} \end{array}$$

The chain is numbered from the right to give 5-ethyl-3-methyloctane.

c. The chain is numbered from the right to give 4-isopropyl-2,3-dimethylheptane.

1.11

CH_3 CH_3

a. $CH_3-C-CH_2-CH-CH_3$

CH_3

CH_3

$CH-CH_3$

b. $CH_3-CH_2-CH-CH_2-CH_2-CH_3$

CH_2-CH_3

c. $CH_3-CH-CH-CH-CH_2-CH_2-CH_3$

CH_3 CH_3

1.12 a. 1,4-dimethylcyclohexane

b. The correct name is ethylcyclopropane. When only one group is attached to a ring, the position is not designated.

c. The name 1-ethyl-2-methylcyclopentane is correct, whereas 2-ethyl-1-methylcyclopentane is incorrect because the ring numbering begins with the carbon attached to the first group alphabetically.

1.13 a. (1) *Trans* because the two Br's are on opposite sides of the ring.

(2) *Cis*; both Cl's are on the same side.

(3) *Cis*; the two groups are on the same side.

b. In showing geometric isomers of ring compounds, it helps to draw the ring in perspective:

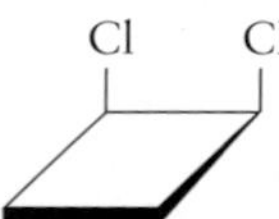

CHAPTER 2

2.1 In each of these alkenes, the double-bonded carbons occur at positions 1 and 2.

a. 3-bromo-1-propene

b. 2-ethyl-1-pentene

c. 3,4-dimethylcyclohexene

2.2 a. The chain is correctly numbered from the right to give 2-methyl-1,3-butadiene.

b. The chain is correctly numbered from the left to give 2-methyl-1,3,6-octatriene.

c. 7-bromo-1,3,5-cycloheptatriene

2.3 a. This structure does not exhibit geometric isomerism because there are two H's attached to the carbon at position 1.

b.

H H H CH_3

$C=C$ $C=C$

CH_3-CH CH_3 CH_3-CH H

CH_3 CH_3

cis *trans*

c. This structure does not exhibit geometric isomerism because there are two methyl groups attached to the left double-bonded carbon:

2.4

Br Br

a. CH_3-C-CH_2

CH_3

b. Cl

Cl

2.5 a. $CH_3-CH-CH_2-CH_3$

CH_3

b. CH_3 CH_3

H

H

=

2.6 a. The major product will be the one where H attaches to the CH carbon:

CH_3

$CH_3-CH_2-C-CH_2-CH_3$

Br

b. Position 1 on the ring has an attached hydrogen, whereas position 2 does not have any attached hydrogens:

CH_3

CH_2CH_3

2

1

H

Thus, H attaches at position 1 to give

CH_3 CH_3

CH_2CH_3 CH_2CH_3

Br = Br

H

H

2.7 a. Markovnikov's rule predicts that H will attach at position 1 to give:

OH H OH

$CH_3CH_2CH_2CH-CH_2 = CH_3CH_2CH_2CHCH_3$

b. H

OH OH

=

2.8 The double bond becomes a single bond:

CH_3 CH_3 CH_3 CH_3

$-CH_2-CH-CH_2-CH-CH_2-CH-CH_2-CH-$

1 2 3 4

2.9 Each chain is correctly numbered from the right.

a. 2-pentyne b. 5-methyl-2-hexyne

2.10 a. Numbers or the term *meta* may be used:

1,3-diethylbenzene or *m*-diethylbenzene

b. The compound must be named as a derivative of cyclopentane. The correct name is 1-chloro-3-phenylcyclopentane.

c. Numbers must be used when there are three groups: 1,2,3-tribromobenzene.

d. If the compound is named as a derivative of benzoic acid, then the methyl group is at position 2. The name is 2-methylbenzoic acid.

CHAPTER 3

3.1 a. $CH_3CH(CH_3)CH_2-OH$ (carbons numbered 3, 2, 1) 2-methyl-1-propanol

b. $CH_3CH(OH)CH(CH_2CH_3)CH_2CH(CH_3)CH_3$ (main chain numbered 5 4 3 2 1, with CH_2CH_3 as 6, 7) 2,5-dimethyl-4-heptanol

c. (cyclopentane ring with CH_3 on C-2 and OH on C-1) 2-methylcyclopentanol

3.2 a. 1,4-butanediol b. 1,2-cyclohexanediol

3.3 Because the compound is named as a phenol, the OH group is at position 1. The name is 2-ethylphenol or *o*-ethylphenol.

3.4 In each case, count the number of carbon atoms attached to the hydroxy-bearing carbon.

a. Primary b. Secondary c. Secondary

3.5 b, c, a

3.6 Methanol molecules form hydrogen bonds, whereas molecules of propane do not.

3.7 a. There are two ways of removing water from this structure, but they both produce the same product:

$$CH_3CH{=}CHCH_2CH_3$$

b. Dehydration will occur in the direction that gives the highest number of carbon groups attached to the double-bonded carbons:

(1-methylcyclohexene: cyclohexene ring with CH_3 on a double-bonded carbon)

3.8 a. Ether formation is favored by H_2SO_4 at 140°C.

b. Alkene formation predominates when H_2SO_4 is used at 180°C.

3.9 (benzene ring)–$\overset{O}{\overset{\|}{C}}-H$ first product; (benzene ring)–$\overset{O}{\overset{\|}{C}}-OH$ second product

3.10 (cyclohexanone) $+ H_2O$

3.11 The alcohols in a and c react because they are secondary and primary, respectively. Alcohol b is tertiary and does not react.

3.12 Working backward, you could prepare an ether product from an alcohol. Thus, the starting alkene must first be converted to an alcohol.

$$CH_2{=}CH_2 + H_2O \xrightarrow{H_2SO_4} CH_3CH_2-OH$$

$$2CH_3CH_2-OH \xrightarrow[140°C]{H_2SO_4} CH_3CH_2-O-CH_2CH_3 + H_2O$$

3.13 Use Table 3.2 to answer these questions:

a. Solvent

b. Alcoholic beverages

c. Rubbing alcohol

d. Antifreeze

e. Food moisturizer

f. Cough drops

3.14 a. Antioxidant in gasoline

b. Throat lozenges

3.15 a. ethyl phenyl ether

b. diisopropyl ether

3.16 a. 1,2-dimethoxycyclopentane

b. 3-isopropoxypentane

3.17 a. Ether molecules do not form hydrogen bonds, whereas alcohol molecules do form hydrogen bonds.

b. They are quite unreactive.

3.18 Oxidation of a thiol produces a disulfide:

$$CH_3CH(CH_3)-S-S-CH(CH_3)CH_3 + H_2O$$

3.19 a. $2CH_3CH(CH_3)-SH$

b. 2 (cyclopentane ring)–SH

3.20 $CH_3CH(CH_3)-S-Pb-S-CH(CH_3)CH_3 + 2H^+$

3.21 Ether, phenolic OH, and benzene ring

CHAPTER 4

4.1 a. The aldehyde contains five carbon atoms, and the name is pentanal.

b. The aldehyde carbon is position 1. The name is 3,4-dimethylpentanal.

4.2 a. The chain is numbered from the right. The name is 4,6-dibromo-2-heptanone.

b. The carbonyl carbon atom is position 1. The name is 2,5-dimethylcyclopentanone.

4.3 a. $CH_3CH(CH_3)CH_2-C(=O)-OH$

b. This is a ketone, and it does not undergo oxidation.

4.4 a. $CH_3CH_2CH(OH)CH_2CH_3$

b. $C_6H_5-CH_2-OH$

4.5 a. $C_6H_5-CH(OH)(OCH_2CH_3) \xrightarrow{H^+,\ CH_3CH_2-OH} C_6H_5-CH(OCH_2CH_3)_2 + H_2O$

hemiacetal → acetal

b. $C_6H_5-C(OH)(OCH_3)-CH_3 \xrightarrow{H^+,\ CH_3-OH} C_6H_5-C(OCH_3)_2-CH_3 + H_2O$

hemiketal → ketal

4.6 a. $CH_3CH_2CH_2-C(=O)-H + 2CH_3CH_2-OH$

b. cyclopentanone $+ 2CH_3-OH$

4.7 a. This compound is not a hemiacetal or hemiketal because there is no OH group. The presence of the ring hydrogen makes it an acetal.

Hydrolysis breaks these bonds

cyclic acetal (ring O, OCH_3, H) $+ H_2O \xrightarrow{H^+}$ open-chain compound with OH and $C=O$ (H) $+ CH_3-OH$

b. An OH group is missing, so this compound is an acetal or ketal. Because the carbon with two attached oxygens does not have a hydrogen present, the compound is a ketal.

cyclic ketal (ring O, CH_3, OCH_3) $+ H_2O \xrightarrow{H^+}$ open-chain compound with OH and $C=O$ (CH_3) $+ CH_3-OH$

Hydrolysis breaks these bonds

CHAPTER 5

5.1 a. The longest chain beginning with the carboxyl group has four carbon atoms. The name is 2-methylbutanoic acid.

b. The carboxyl group is at position 1. The name is 3,4-dibromobenzoic acid.

5.2 $CH_3-C(=O)-C(=O)-O^-$

5.3 $CH_3CH_2CH_2-C(=O)-O^-Na^+ + H_2O$

5.4 a. This is the lithium salt of 2-methylbenzoic acid. The name is lithium 2-methylbenzoate.

b. This is the sodium salt of butanoic acid. The name is sodium butanoate.

5.5 Each of these reactions requires H^+ as a catalyst. The products are as follows:

a. $CH_3CH_2-C(=O)-OCH_3 + H_2O$

b. $CH_3-C(=O)-O-C_6H_5 + H_2O$

c. $C_6H_5-C(=O)-O-CH(CH_3)CH_3 + H_2O$

5.6 a. $C_6H_5-C(=O)-O-C(=O)-C_6H_5 + CH_3-OH \longrightarrow$

$C_6H_5-C(=O)-O-CH_3 + C_6H_5-C(=O)-OH$

b. $CH_3CH_2-C(=O)-Cl + CH_3CH(CH_3)-OH \longrightarrow$

$CH_3CH_2-C(=O)-O-CH(CH_3)CH_3 + HCl$

5.7 a. The common name for the acid used to prepare this ester is formic acid. The alkyl group attached to oxygen is isopropyl. Thus, the common name is isopropyl formate. The IUPAC name for formic acid is methanoic acid. Thus, the IUPAC name for the ester is isopropyl methanoate.

b. The common and IUPAC names are identical when the carboxylic acid is benzoic acid. The name of the ester is methyl benzoate.

5.8 $C_6H_5-\overset{O}{\overset{||}{C}}-OH + HO-CH_2CH_2CH_3$

5.9 **a.** $CH_3CH_2-\overset{O}{\overset{||}{C}}-O^-Na^+ + HO-\underset{CH_3}{\underset{|}{CH}}CH_3$

b. $C_6H_5-\overset{O}{\overset{||}{C}}-O^-K^+ + HO-CH_2CH_3$

CHAPTER 6

6.1 In each case, count the number of carbon atoms directly attached to the nitrogen.

a. Secondary **b.** Secondary **c.** Tertiary

6.2 Identify the groups attached to the nitrogen.

a. propylamine

b. methylphenylamine

c. trimethylamine

6.3 **a.** 1-butanamine

b. *N*-methyl-1-butanamine

6.4 **a.** *N*-ethylaniline

b. *N*-ethyl-*N*-methylaniline

c. 3-ethyl-2-methylaniline

6.5 **a.** $CH_3-\underset{H}{\underset{|}{N}}-H\cdots\underset{H}{\underset{|}{O}}-H$ or $CH_3-\underset{H}{\underset{|}{N}}(\cdots H-O-H)-H$

b. $CH_3CH_2-\underset{CH_3}{\underset{|}{N}}(\cdots H-O-H)-CH_3$

6.6 **a.** $CH_3-\overset{H}{\overset{|}{\underset{CH_3}{\underset{|}{N^+}}}}-CH_2CH_3 + OH^-$

b. $C_6H_5-NH_3^+ + OH^-$

6.7 **a.** $C_6H_5-\underset{CH_3}{\underset{|}{NH_2^+}}Cl^-$

b. $C_5H_9-NH_3^+CH_3CH_2COO^-$

6.8 **a.** $CH_3-\overset{O}{\overset{||}{C}}-NH-CH_2CH_2CH_3 + CH_3-\overset{O}{\overset{||}{C}}-OH$

b. $CH_3CH_2-\overset{O}{\overset{||}{C}}-NH_2 + HCl$

6.9 **a.** This amide is derived from butanoic acid. The name is butanamide.

b. hexanamide

6.10 **a.** This amide is a derivative of the two-carbon acid acetic acid (common name) or ethanoic acid (IUPAC name). The amide is named *N,N*-diethylacetamide (common name) or *N,N*-diethylethanamide (IUPAC name).

b. This amide is a derivative of butyric acid (common name) or butanoic acid (IUPAC name). The amide is named *N*-isobutylbutyramide (common name) or *N*-isobutylbutanamide (IUPAC name).

c. The acid portion of this amide came from benzoic acid. The amide is named *N*-ethyl-*N*-phenylbenzamide.

d. The acid portion of this amide came from 2-methylbenzoic acid. The amide is named 2,*N*-dimethylbenzamide.

6.11 **a.** $CH_3-\overset{O}{\overset{||}{C}}-\underset{H}{\underset{|}{N}}-CH_3 \cdots O=\overset{}{C}(CH_3)-\underset{H}{\underset{|}{N}}-CH_3$ (the N–H of one amide hydrogen-bonded to the C=O oxygen of the other)

b. $H-O-H\cdots O=C(CH_3)-\underset{H}{\underset{|}{N}}-CH_3$ or $CH_3-\overset{O}{\overset{||}{C}}-\underset{H}{\underset{|}{N}}-CH_3$ with $N-H\cdots O(H)H$

6.12 **a.** Under acidic conditions, amide hydrolysis produces a carboxylic acid and the salt of an amine.

$C_6H_5-\overset{O}{\overset{||}{C}}-OH + CH_3CH_2-NH_3^+Cl^-$

b. Under basic conditions, amide hydrolysis produces an amine and the salt of a carboxylic acid.

$$C_6H_5-\overset{\displaystyle O}{\overset{\|}{C}}-O^-Na^+ \;+\; CH_3CH_2-NH_2$$

CHAPTER 7

7.1 In each case, look for the presence of four different groups attached to the colored atom.

a. The carbon is not chiral because of the two attached CH_2OH groups.

b. Yes

c. The carbon is not chiral because it is attached to two H's.

d. Yes

7.2 Each of the atoms marked with an asterisk has four different attached groups and is chiral.

a.

CH_2OH
$C{=}O$
$HO-C^*-H$
$H-C^*-OH$
CH_2OH

b.

CHO
$^*CH-OH$
$^*CH-OH$
$^*CH-OH$
CH_2OH

c. $HOCH_2$ (ring: O; OH; H, H; H, H; OH, OH; chiral carbons marked *)

7.3 The chiral carbon atoms (three of them) are marked with an asterisk. Thus, there are 2^3, or 8, possible stereoisomers:

$$\underset{|}{\overset{OH}{}}CH_2-\overset{OH}{\underset{*}{CH}}-\overset{OH}{\underset{*}{CH}}-\overset{OH}{\underset{*}{CH}}-CHO$$

7.4 The chiral carbon atom is the one with the attached nitrogen. The chiral carbon atom is placed at the intersection of the two lines, and the COOH is placed at the top:

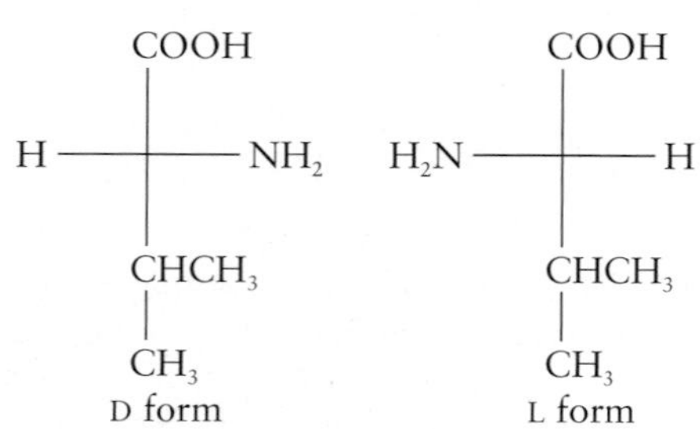

7.5 In each case, look at the direction of the OH group on the bottom chiral carbon.

a. L b. D c. L

7.6 a. aldopentose

b. ketotetrose

c. aldotetrose

7.7

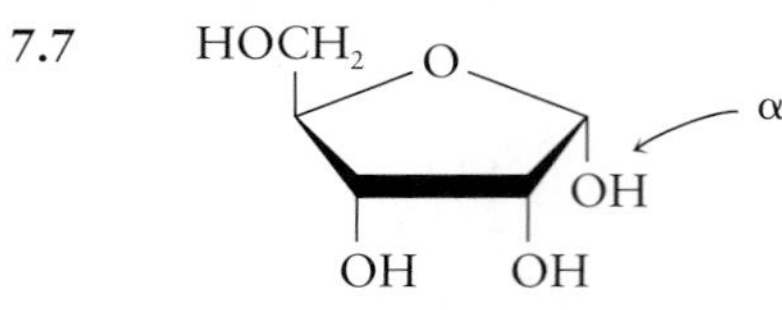

7.8 a.

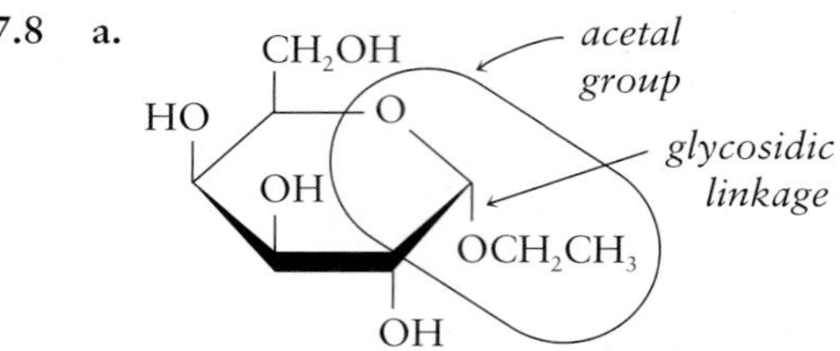

b.

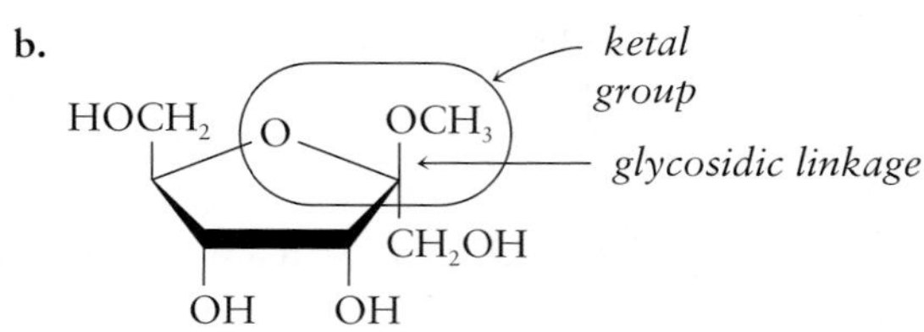

CHAPTER 8

8.1 One of several possibilities is the following structure:

$CH_2-O-\overset{O}{\overset{\|}{C}}-(CH_2)_{16}CH_3$ (stearic acid component)

$CH-O-\overset{O}{\overset{\|}{C}}-(CH_2)_7CH{=}CH(CH_2)_7CH_3$ (oleic acid component)

$CH_2-O-\overset{O}{\overset{\|}{C}}-(CH_2)_{14}CH_3$ (palmitic acid component)

8.2

$CH_2-OH \quad HOOC-(CH_2)_7CH{=}CH(CH_2)_7CH_3$
$CH-OH \;+\; HOOC-(CH_2)_{14}CH_3$
$CH_2-OH \quad HOOC-(CH_2)_{16}CH_3$

8.3

$CH_2-OH \quad Na^{+-}OOC-(CH_2)_{16}CH_3$
$CH-OH + 2Na^{+-}OOC-(CH_2)_7CH{=}CH(CH_2)_7CH_3$
CH_2-OH

8.4 During complete hydrogenation, all CH=CH groups become CH_2—CH_2 groups:

$CH_2-O-\overset{O}{\overset{\|}{C}}-(CH_2)_7CH_2CH_2(CH_2)_7CH_3$

$CH-O-\overset{O}{\overset{\|}{C}}-(CH_2)_7CH_2CH_2CH_2CH_2CH_2CH_2CH_2CH_2CH_2CH_3$

$CH_2-O-\overset{O}{\overset{\|}{C}}-(CH_2)_7CH_2CH_2CH_2CH_2CH_2(CH_2)_4CH_3$

Combining all the CH_2 groups gives

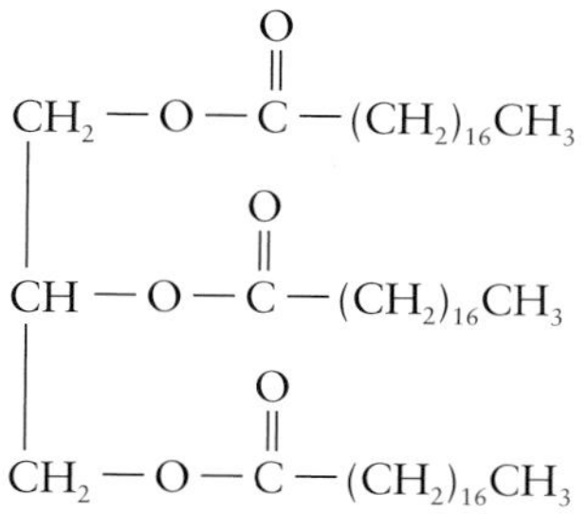

$$CH_2-O-\overset{O}{\overset{\|}{C}}-(CH_2)_{16}CH_3$$
$$CH-O-\overset{O}{\overset{\|}{C}}-(CH_2)_{16}CH_3$$
$$CH_2-O-\overset{O}{\overset{\|}{C}}-(CH_2)_{16}CH_3$$

8.5

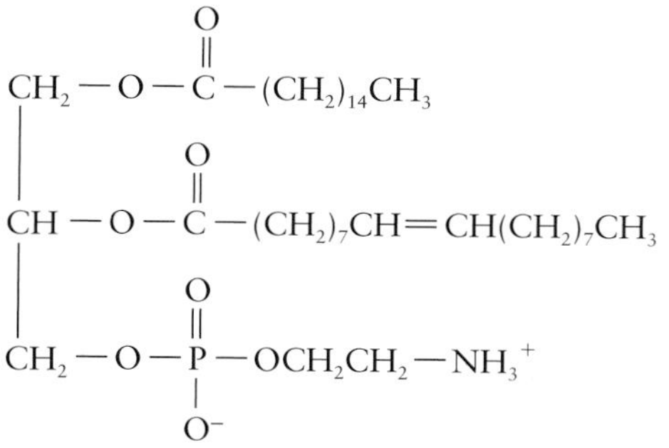

$$CH_2-O-\overset{O}{\overset{\|}{C}}-(CH_2)_{14}CH_3$$
$$CH-O-\overset{O}{\overset{\|}{C}}-(CH_2)_7CH{=}CH(CH_2)_7CH_3$$
$$CH_2-O-\overset{O}{\overset{\|}{P}}(-O^-)-OCH_2CH_2-NH_3^+$$

CHAPTER 9

9.1 a.

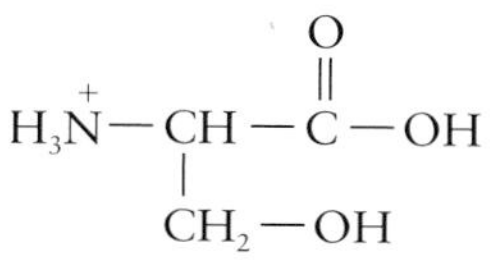

$$H_3\overset{+}{N}-\underset{CH_2-OH}{CH}-\overset{O}{\overset{\|}{C}}-OH$$

b.

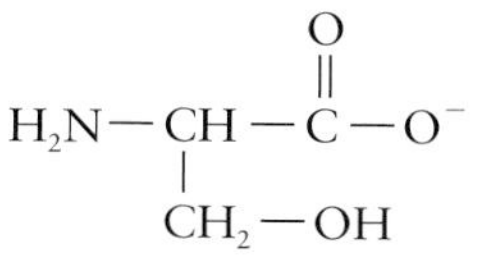

$$H_2N-\underset{CH_2-OH}{CH}-\overset{O}{\overset{\|}{C}}-O^-$$

9.2

$$H_3\overset{+}{N}-\underset{CH_2C_6H_5}{CH}-\overset{O}{\overset{\|}{C}}-NH-\underset{CH_2-SH}{CH}-\overset{O}{\overset{\|}{C}}-NH-$$

N-terminal residue

$$-\underset{CH_2-OH}{CH}-\overset{O}{\overset{\|}{C}}-NH-\underset{CHCH_3-CH_2CH_3}{CH}-\overset{O}{\overset{\|}{C}}-O^-$$

C-terminal residue

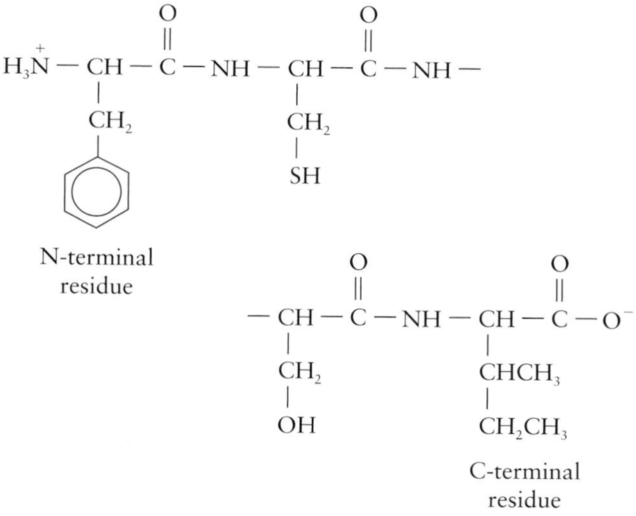

9.3 a. Hydrophobic attractions between nonpolar groups

b. Hydrogen bonding

c. Salt bridges between the ionic groups

CHAPTER 10

10.1 a. maltose

b. peptides

c. glucose 6-phosphate

10.2 Turnover number is the number of substrate molecules acted on per minute by one molecule of enzyme. One enzyme international unit is the amount of enzyme that catalyzes 1 μmol of substrate to react per minute.

10.3 a. An increase in enzyme concentration increases the rate of an enzyme-catalyzed reaction.

b. An increase in substrate concentration increases the rate until the enzyme is saturated. Above the saturation point, rate of reaction is constant.

c. An increase in temperature increases the rate of the enzyme-catalyzed reaction until the optimum temperature is reached. Above the optimum temperature, the rate decreases.

d. An increase in pH increases the reaction rate until the optimum pH is reached. Above the optimum pH, the rate decreases.

10.4 a. The structure of a competitive inhibitor resembles that of the substrate. The structure of a noncompetitive inhibitor bears no resemblance to that of the substrate.

b. A competitive inhibitor binds at the active site. A noncompetitive inhibitor binds at some other region of the enzyme.

c. Increasing substrate concentration reverses the effect of a competitive inhibitor but has no effect on a noncompetitive inhibitor.

10.5 a. Several enzymes are synthesized in inactive forms called zymogens. Under the proper reaction conditions, zymogens are activated to carry out their catalytic role.

b. The catalytic ability of allosteric enzymes may be either increased or decreased by the binding of modulators.

c. Certain enzymes are synthesized in greater amounts when needed and in smaller amounts when not needed.

CHAPTER 11

11.1

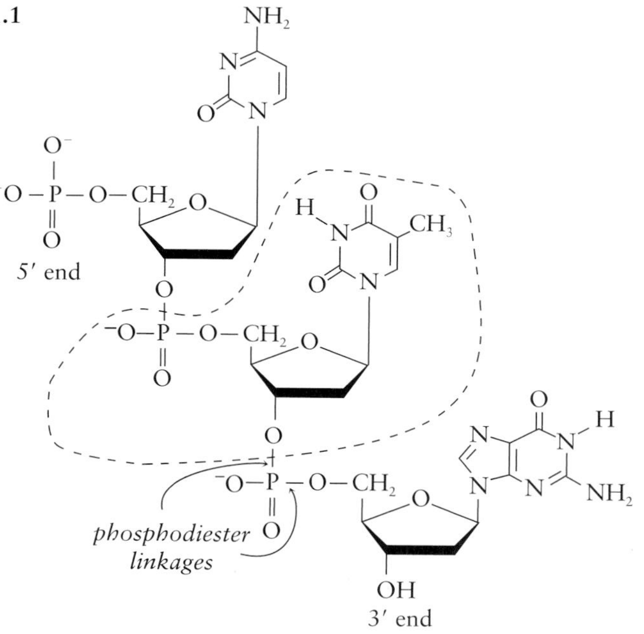

5′ end

phosphodiester linkages

3′ end

11.2 5′ T-T-A-C-G 3′ original strand
3′ A-A-T-G-C 5′ complementary strand

The complementary strand written in the 5′ to 3′ direction is C-G-T-A-A.

11.3 5′ A-T-T-A-G-C-C-G 3′ DNA template
3′ U-A-A-U-C-G-G-C 5′ mRNA

The mRNA written in the 5′ to 3′ direction is C-G-G-C-U-A-A-U.

11.4 fMet-His-His-Val-Leu-Cys

CHAPTER 12

12.1

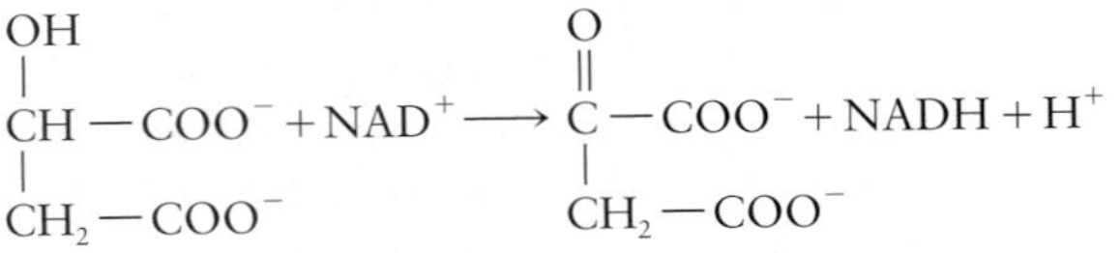

$$\begin{array}{l} OH \\ | \\ CH-COO^- + NAD^+ \\ | \\ CH_2-COO^- \end{array} \longrightarrow \begin{array}{l} O \\ \| \\ C-COO^- + NADH + H^+ \\ | \\ CH_2-COO^- \end{array}$$

12.2 a. $^-OOC-CH=CH-COO^- + FADH_2$

b.

$$\begin{array}{c} \quad O \\ \quad \| \\ CH_3-C-COO^- + NADH + H^+ \end{array}$$

CHAPTER 13

13.1 a. Hexokinase regulates the first step of the glycolysis pathway. The enzyme is controlled by concentrations of glucose 6-phosphate through feedback inhibition.

b. Phosphofructokinase is an allosteric enzyme which regulates the production of fructose 1,6-bisphosphate. The enzyme is inhibited by high concentrations of ATP and citrate, and it is activated by high concentrations of ADP and AMP.

c. Pyruvate kinase controls the last step of glycolysis. It is inhibited by higher concentrations of ATP.

13.2 NAD^+ produced in the various fates of pyruvate can be used in the glycolysis pathway, enabling further glycolysis to take place.

13.3 a. Three: Each $FADH_2$ produces 1.5 molecules of ATP.

b. Five: Each NADH produces 2.5 molecules of ATP.

c. 20: Each acetyl CoA produces 10 molecules of ATP.

13.4 Glucose 6-phosphatase is present in the liver but not in the muscles.

CHAPTER 14

14.1 a. Each acetyl CoA contains two carbon atoms from capric acid. Thus, five acetyl CoA molecules are formed.

b. Ten carbon atoms require four trips through the fatty acid spiral. Thus, four molecules of $FADH_2$ are formed.

c. Four trips through the spiral also produce four NADH molecules and four H^+ ions.

14.2 A 16-carbon fatty acid would yield eight molecules of acetyl CoA. Seven trips through the spiral are necessary and would produce 7 $FADH_2$ and 7 NADH + 7 H^+.

8 acetyl CoA × 10 ATP/acetyl CoA	=	80 ATP
7 $FADH_2$ × 1.5 ATP/$FADH_2$	=	10.5 ATP
7 NADH × 2.5 ATP/NADH	=	17.5 ATP
		108 ATP
Two high-energy bonds in ATP are consumed in the activation step.		−2
		106 ATP

14.3

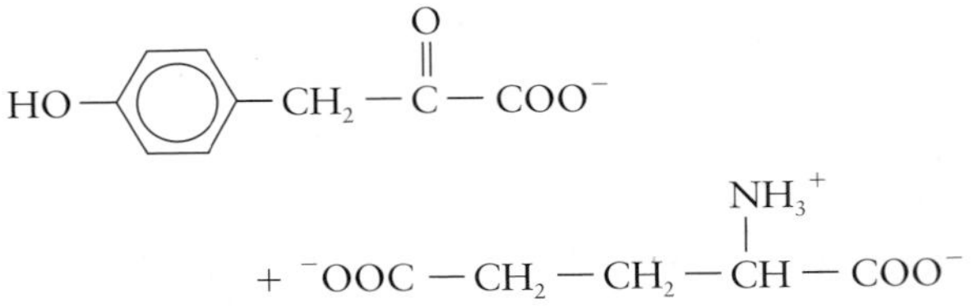

$$HO-C_6H_4-CH_2-\overset{\overset{\large O}{\|}}{C}-COO^-$$

$$+\ ^-OOC-CH_2-CH_2-\underset{}{\overset{\overset{\large NH_3^+}{|}}{CH}}-COO^-$$

14.4 a. The urea cycle converts toxic ammonia into less toxic urea for excretion from the body.

b. carbamoyl phosphate

CHAPTER 15

15.1 a. Lungs

b. Tissue cells

c. Lungs

d. Tissue cells

Glossary

absolute specificity The characteristic of an enzyme that it acts on one and only one substrate.

acetal A compound that contains the $-\overset{\displaystyle OR}{\underset{\displaystyle OR}{\overset{|}{\underset{|}{C}}}}-H$ arrangement of atoms.

acetone breath A condition in which acetone can be detected in the breath.

acidosis A low blood pH.

activator A substance that binds to an allosteric enzyme and increases its activity.

active site The location on an enzyme where a substrate is bound and catalysis occurs.

addition polymer A polymer formed by the linking together of many alkene molecules through addition reactions.

addition reaction A reaction in which a compound adds to a multiple bond.

adipose tissue A kind of connective tissue where triglycerides are stored.

aerobic In the presence of oxygen.

alcohol A compound in which an —OH group is connected to an aliphatic carbon atom.

alcoholic fermentation The conversion of glucose to ethanol.

aldehyde A compound that contains the $-\overset{\displaystyle O}{\overset{\|}{C}}-H$ group; the general formula is $R-\overset{\displaystyle O}{\overset{\|}{C}}-H$.

aliphatic compound Any organic compound that is not aromatic.

alkaloids A class of nitrogen-containing organic compounds obtained from plants.

alkalosis An abnormally high blood pH.

alkane A hydrocarbon that contains only single bonds.

alkene A hydrocarbon that contains one or more double bonds.

alkoxy group The —O—R functional group.

alkyl group A group differing by one hydrogen from an alkane.

alkyl halide See **haloalkane.**

alkyne A hydrocarbon that contains one or more triple bonds.

allosteric enzyme An enzyme with quaternary structure whose activity is changed by the binding of modulators.

alpha-amino acid An organic compound containing both an amino group and a carboxylate group, with the amino group attached to the carbon next to the carboxylate group.

α-helix The helical structure in proteins that is maintained by hydrogen bonds.

amide An organic compound having the functional group $-\overset{\displaystyle O}{\overset{\|}{C}}-\underset{|}{N}-$.

amide linkage The carbonyl carbon-nitrogen single bond of the amide group.

amine An organic compound derived by replacing one or more of the hydrogen atoms of ammonia with alkyl or aromatic groups, as in RNH_2, R_2NH, and R_3N.

amino acid pool The total supply of amino acids in the body.

amino acid residue An amino acid that is a part of a peptide, polypeptide, or protein chain.

amphetamines A class of drugs structurally similar to epinephrine, used to stimulate the central nervous system.

anabolism All reactions involved in the synthesis of biomolecules.

anaerobic In the absence of oxygen.

anomeric carbon An acetal, ketal, hemiacetal, or hemiketal carbon atom giving rise to two stereoisomers.

anomers Stereoisomers that differ in the three-dimensional arrangement of groups at the carbon of an acetal, ketal, hemiacetal, or hemiketal group.

antibiotic A substance produced by one microorganism that kills or inhibits the growth of other microorganisms.

antibody A substance that helps protect the body from invasion by foreign materials known as antigens.

anticodon A three-base sequence in tRNA that is complementary to one of the codons in mRNA.

antioxidant A substance that prevents another substance from being oxidized.

apoenzyme A catalytically inactive protein formed by removal of the cofactor from an active enzyme.

aromatic hydrocarbon Any organic compound that contains the characteristic benzene ring or similar feature.

Benedict's reagent A mild oxidizing solution containing Cu^{2+} ions used to test for the presence of aldehydes.

β-oxidation process A pathway in which fatty acids are broken down into molecules of acetyl CoA.

β-pleated sheet A secondary protein structure in which protein chains are aligned side by side in a sheetlike array held together by hydrogen bonds.

biochemistry A study of the compounds and processes associated with living organisms.

biomolecule A general term referring to an organic compound essential to life.

blood sugar level The amount of glucose present in blood, normally expressed as milligrams per 100 mL of blood.

branched alkane An alkane in which at least one carbon atom is not part of a continuous chain.

carbaminohemoglobin Hemoglobin combined with carbon dioxide.

carbocation An ion of the form $-\overset{+}{\underset{|}{C}}-$.

carbohydrate A polyhydroxy aldehyde, ketone, or substance that yields such compounds on hydrolysis.

carbonyl group The $-\overset{O}{\overset{\|}{C}}-$ group.

carboxyl group The $-\overset{O}{\overset{\|}{C}}-OH$ group.

carboxylate ion The $R-\overset{O}{\overset{\|}{C}}-O^-$ ion that results from the dissociation of a carboxylic acid.

carboxylic acid An organic compound that contains the $-\overset{O}{\overset{\|}{C}}-OH$ functional group.

carboxylic acid anhydride An organic compound that contains the $-\overset{O}{\overset{\|}{C}}-O-\overset{O}{\overset{\|}{C}}-$ functional group.

carboxylic acid chloride An organic compound that contains the $-\overset{O}{\overset{\|}{C}}-Cl$ functional group.

carboxylic ester A compound with the $-\overset{O}{\overset{\|}{C}}-OR$ functional group.

catabolism All reactions involved in the breakdown of biomolecules.

cellular respiration The entire process involved in the use of oxygen by cells.

central dogma of molecular biology The well-established process by which genetic information stored in DNA molecules is expressed in the structure of synthesized proteins.

cephalin A phosphoglyceride containing ethanolamine or serine.

chemiosmotic hypothesis The postulate that a proton flow across the inner mitochondrial membrane during operation of the electron transport chain provides energy for ATP synthesis.

chiral A descriptive term for compounds or objects that cannot be superimposed on their mirror image.

chiral carbon A carbon atom with four different groups attached.

chloride shift The movement of chloride ions from one fluid to another to maintain charge neutrality.

chromosome A tightly packaged bundle of DNA and protein that is involved in cell division.

chylomicron A lipoprotein aggregate found in the lymph and the bloodstream.

cis- On the same side (as applied to geometric isomers).

citric acid cycle A series of reactions in which acetyl CoA is oxidized to carbon dioxide and reduced forms of coenzymes FAD and NAD^+ are produced.

codon A sequence of three nucleotide bases that represents a code word on mRNA molecules.

coenzyme An organic molecule required by an enzyme for catalytic activity.

cofactor A nonprotein molecule or ion required by an enzyme for catalytic activity.

common catabolic pathway The reactions of the citric acid cycle plus those of the electron transport chain and oxidative phosphorylation.

competitive inhibitor An inhibitor that competes with substrate for binding at the active site of an enzyme.

complementary DNA strands Two strands of DNA in a double-helical form such that adenine and guanine of one strand are matched and hydrogen bonded to thymine and cytosine, respectively, of the second strand.

complete protein Protein in food that contains all essential amino acids in the proportions needed by the body.

complex carbohydrates The polysaccharides amylose and amylopectin, collectively called starch.

complex lipid An ester-containing lipid with more than two types of components: an alcohol, fatty acids, plus other components.

condensation polymerization The process by which monomers combine with the simultaneous elimination of a small molecule.

condensed structural formula A structural molecular formula showing the general arrangement of atoms but without showing all the covalent bonds.

conformations The different arrangements of atoms in space achieved by rotation about single bonds.

conjugated protein A protein made up of amino acid residues and other organic or inorganic components.

copolymer An addition polymer formed by the reaction of two different monomers.

Cori cycle The process in which glucose is converted to lactate in muscle tissue, lactate is reconverted to glucose in the liver, and glucose is returned to the muscle.

C-terminal residue An amino acid on the end of a chain that has an unreacted or free carboxylate group.

cycloalkane An alkane in which carbon atoms form a ring.

cytochrome An iron-containing enzyme located in the electron transport chain.

daily reference values (DRV) A set of standards for nutrients and food components (such as fat and fiber) that have important relationships with health; used on food labels as part of the Daily Values.

daily values (DV) Reference values developed by the FDA specifically for use on food labels. The Daily Values represent two sets of standards: Reference Daily Intakes (RDI) and Daily Reference Values (DRV).

dehydration reaction A reaction in which water is chemically removed from a compound.

denaturation The process by which a protein loses its characteristic native structure and function.

deoxyhemoglobin (hemoglobin) Nonoxygenated hemoglobin.

deoxyribonucleic acid (DNA) A nucleic acid found primarily in the nuclei of cells.

dextrorotatory Describes substances that rotate plane-polarized light to the right.

dimer Two identical molecules bonded together.

dipeptide A compound formed when two amino acids are bonded by an amide linkage.

disaccharide A carbohydrate formed by the combination of two monosaccharides.

disulfide A compound containing an —S—S— group.

disulfide bridge A bond produced by the oxidation of —SH groups on two cysteine residues. The bond loops or holds two peptide chains together.

electron transport chain A series of reactions in which protons and electrons from the oxidation of foods are used to reduce molecular oxygen to water.

elimination reaction A reaction in which two or more covalent bonds are broken and a new multiple bond is formed.

enantiomers Stereoisomers that are mirror images.

enzyme A protein molecule that catalyzes chemical reactions.

enzyme activity The rate at which an enzyme catalyzes a reaction.

enzyme induction The synthesis of an enzyme in response to a cellular need.

enzyme inhibitor A substance that decreases the activity of an enzyme.

enzyme international unit (IU) A quantity of enzyme that catalyzes the conversion of 1 μmol of substrate per minute under specified conditions.

essential amino acid An amino acid that cannot be synthesized within the body at a rate adequate to meet metabolic needs.

essential fatty acid A fatty acid needed by the body that cannot be synthesized within the body.

ester A compound in which the —OH of an acid is replaced by an —OR.

ester linkage The carbonyl carbon–oxygen single bond of the ester group.

esterification The process of forming an ester.

ether A compound that contains a $-\overset{|}{\underset{|}{C}}-O-\overset{|}{\underset{|}{C}}-$ functional group.

eukaryotic cell A cell containing membrane-enclosed organelles, particularly a nucleus.

exon A segment of a eukaryotic DNA molecule that is coded for amino acids.

expanded structural formula A structural molecular formula showing all the covalent bonds.

extracellular fluid Body fluid located outside cells.

extremozyme A nickname for certain enzymes isolated from microorganisms that thrive in extreme environments.

fat A triglyceride that is a solid at room temperature.

fat mobilization The hydrolysis of stored triglycerides followed by the entry of fatty acids and glycerol into the bloodstream.

fatty acid A long-chain carboxylic acid found in fats.

feedback inhibition A process in which the end product of a sequence of enzyme-catalyzed reactions inhibits an earlier step in the process.

fermentation A reaction of sugars, starch, or cellulose to produce ethanol and carbon dioxide.

fiber Indigestible plant material composed primarily of cellulose.

fibrous protein A protein made up of long rod-shaped or stringlike molecules that intertwine to form fibers.

Fischer projection A method of depicting three-dimensional shapes for chiral molecules.

fluid-mosaic model A model of membrane structure in which proteins are embedded in a flexible lipid bilayer.

functional group A unique reactive combination of atoms that differentiates molecules of organic compounds of one class from those of another.

furanose ring A five-membered sugar ring system containing an oxygen atom.

gene An individual section of a chromosomal DNA molecule that is the fundamental unit of heredity.

genome A summation of all the genetic material (chromosomes) of a cell; a person's genetic blueprint.

geometric isomers Molecules that differ in the three-dimensional arrangements of their atoms in space and not in the order of linkage of atoms.

globular protein A spherical protein that usually forms stable suspensions in water or dissolves in water.

glucogenic amino acid An amino acid whose carbon skeleton can be converted metabolically to an intermediate used in the synthesis of glucose.

gluconeogenesis The synthesis of glucose from noncarbohydrate molecules.

glucosuria A condition in which elevated blood sugar levels result in the excretion of glucose in the urine.

glycogenesis The synthesis of glycogen from glucose.

glycogenolysis The breakdown of glycogen to glucose.

glycolipid A complex lipid containing sphingosine, a fatty acid, and a carbohydrate.

glycolysis A series of reactions by which glucose is oxidized to pyruvate.

glycoside Another name for a carbohydrate containing an acetal or ketal group.

glycosidic linkage The carbon-oxygen-carbon linkage that joins the components of a glycoside to the ring.

haloalkane A derivative of an alkane in which one or more hydrogens are replaced by halogens; also called alkyl halide.

Haworth structure A method of depicting three-dimensional carbohydrate structures.

hemiacetal A compound that contains the functional group

$$\begin{array}{c} \text{OH} \\ | \\ -\text{C}-\text{H}. \\ | \\ \text{OR} \end{array}$$

hemiketal A compound that contains the $\begin{array}{ccccc} | & & \text{OH} \\ & & | & & | \\ -\text{C} & - & \text{C} & - & \text{C}- \\ | & & | & & | \\ & & \text{OR} \end{array}$ group.

heterocyclic ring A ring in which one or more atoms is an atom other than carbon.

heterogeneous nuclear RNA (hnRNA) RNA produced when both introns and exons of eukaryotic cellular DNA are transcribed.

high-energy compound A substance that on hydrolysis liberates a great amount of free energy.

homeostasis A condition of chemical balance in the body.

homologous series Compounds of the same functional class that differ by a $—CH_2—$ group.

hormone A chemical messenger secreted by specific glands and carried by the blood to a target tissue, where it triggers a particular response.

hybrid orbital An orbital produced from the combination of two or more nonequivalent orbitals of an atom.

hydration The addition of water to a multiple bond.

hydrocarbon An organic compound that contains only carbon and hydrogen.

hydrogenation A reaction in which the addition of hydrogen takes place.

hydrolysis Bond breakage by reaction with water.

hydrophobic Characterizing molecules or parts of molecules that repel (are insoluble in) water.

hydroxy group The —OH functional group.

hyperglycemia A higher-than-normal blood sugar level.

hyperventilation Rapid, deep breathing.

hypoglycemia A lower-than-normal blood sugar level.

hypoventilation Slow, shallow breathing.

inborn error of metabolism A disease in which a genetic change causes a deficiency of a particular protein, often an enzyme.

induced-fit theory The theory of enzyme action proposing that the conformation of an enzyme changes to accommodate an incoming substrate.

inorganic chemistry The study of the elements and all noncarbon compounds.

international Unit (IU) A measure of vitamin activity, determined by biological methods.

interstitial fluid The fluid surrounding individual tissue cells.

intracellular fluid Body fluid located inside cells.

intron A segment of a eukaryotic DNA molecule that carries no codes for amino acids.

invert sugar A mixture of equal amounts of glucose and fructose.

isoelectric point The characteristic solution pH at which an amino acid has a net charge of zero.

isoenzyme A slightly different form of the same enzyme produced by different tissues.

isomerism A property in which two or more compounds have the same molecular formula but different arrangements of atoms.

ketal A compound that contains the $\begin{array}{ccccc} & & \text{OR} \\ | & & | & & | \\ -\text{C} & - & \text{C} & - & \text{C}- \\ | & & | & & | \\ & & \text{OR} \end{array}$ group.

ketoacidosis A low blood pH due to elevated levels of ketone bodies.

ketogenic amino acid An amino acid whose carbon skeleton can be converted metabolically to acetyl CoA or acetoacetyl CoA.

ketone A compound that contains the $\begin{array}{ccccc} & & \text{O} \\ | & & \| & & | \\ -\text{C} & - & \text{C} & - & \text{C}- \\ | & & & & | \end{array}$ group; the general formula is $\begin{array}{c} \text{O} \\ \| \\ \text{R}-\text{C}-\text{R}' \end{array}$.

ketone bodies Three compounds—acetoacetate, β-hydroxybutyrate, and acetone—formed from acetyl CoA.

ketonemia An elevated level of ketone bodies in the blood.

ketonuria The presence of ketone bodies in the urine.

ketosis A condition in which ketonemia, ketonuria, and acetone breath exist together.

lactate fermentation The production of lactate from glucose.

lactose intolerance The inability to digest milk and other products containing lactose.

lecithin A phosphoglyceride containing choline.

levorotatory Describes substances that rotate plane-polarized light to the left.

lipid A biological compound that is soluble only in nonpolar solvents.

lipid bilayer A structure found in membranes, consisting of two sheets of lipid molecules arranged so that the hydrophobic portions are facing each other.

lock-and-key theory The theory of enzyme specificity proposing that a substrate has a shape fitting that of the enzyme's active site, as a key fits a lock.

macronutrient A substance needed by the body in relatively large amounts.

major mineral A mineral found in the body in quantities greater than 5 g.

Markovnikov's rule In the addition of H—X to an alkene, the hydrogen becomes attached to the carbon atom that is already bonded to more hydrogens.

messenger RNA (mRNA) RNA that carries genetic information from the DNA in the cell nucleus to the site of protein synthesis in the cytoplasm.

metabolic acidosis A condition of acidosis resulting from causes other than hypoventilation.

metabolic alkalosis A condition of alkalosis resulting from causes other than hyperventilation.

metabolic pathway A sequence of reactions used to produce one product or accomplish one process.

metabolism The sum of all reactions occurring in an organism.

micelle A spherical cluster of molecules in which the polar portions of the molecules are on the surface and the nonpolar portions are located in the interior.

micronutrient A substance needed by the body only in small amounts.

mineral A metal or nonmetal used in the body in the form of ions or compounds.

mitochondrion A cellular organelle where reactions of the common catabolic pathway occur.

moderator A material capable of slowing down neutrons that pass through it.

modulator A substance that binds to an enzyme at a location other than the active site and alters the catalytic activity.

monomer The starting material that becomes the repeating unit of polymers.

monosaccharide A simple carbohydrate most commonly consisting of three to six carbon atoms.

mutagen A chemical that induces mutations by reacting with DNA.

mutation Any change resulting in an incorrect base sequence on DNA.

native state The natural three-dimensional conformation of a functional protein.

neurotransmitter A substance that acts as a chemical bridge in nerve impulse transmission between nerve cells.

noncompetitive inhibitor An inhibitor that binds to the enzyme at a location other than the active site.

nonessential amino acid An amino acid that can be synthesized within the body in adequate amounts.

normal alkane Any alkane in which all the carbon atoms are aligned in a continuous chain.

N-terminal residue An amino acid on the end of a chain that has an unreacted or free amino group.

nucleic acid A biomolecule involved in the transfer of genetic information from existing cells to new cells.

nucleic acid backbone The sugar-phosphate chain that is common to all nucleic acids.

nucleotide The repeating structural unit or monomer of polymeric nucleic acids.

nutrition An applied science that studies food, water, and other nutrients and the ways living organisms use them.

oil A triglyceride that is a liquid at room temperature.

Okazaki fragment A DNA fragment produced during replication as a result of strand growth in a direction away from the replication fork.

optically active molecule A molecule that rotates the plane of polarized light.

optimum pH The pH at which enzyme activity is highest.

optimum temperature The temperature at which enzyme activity is highest.

organelle A membrane-enclosed structure in the interior of a cell.

organic chemistry The study of carbon-containing compounds.

organic compound A compound that contains the element carbon.

oxidative deamination An oxidation process resulting in the removal of an amino group.

oxidative phosphorylation A process coupled with the electron transport chain whereby ADP is converted to ATP.

oxyhemoglobin Hemoglobin combined with oxygen.

peptide An amino acid polymer of short chain length.

peptide linkage (peptide bond) The amide linkage between amino acids that results when the amino group of one acid reacts with the carboxylate group of another.

phenol A compound in which an —OH group is connected to a benzene ring. The parent compound is also called phenol.

phenyl group A benzene ring with onse hydrogen absent, C_6H_5—.

phosphoglyceride A complex lipid containing glycerol, fatty acids, phosphoric acid, and an alcohol component.

phospholipid A phosphorus-containing lipid.

phosphoric anhydride A compound that contains the

$$-O-\overset{\overset{\displaystyle O}{\|}}{\underset{\underset{\displaystyle O^-}{|}}{P}}-O-\overset{\overset{\displaystyle O}{\|}}{\underset{\underset{\displaystyle O^-}{|}}{P}}-O- \text{ group.}$$

plasma The liquid portion of whole blood.

plasmid Circular, double-stranded DNA found in the cytoplasm of bacterial cells.

polycyclic aromatic compound A derivative of benzene in which carbon atoms are shared between two or more benzene rings.

polyfunctional compound A compound with two or more functional groups.

polymer A very large molecule made up of repeating units.

polymerization A reaction that produces a polymer.

polypeptide An amino acid polymer of intermediate chain length containing up to 50 amino acid residues.

polyribosome A complex of mRNA and several ribosomes; also called polysome.

polysaccharide A carbohydrate formed by the combination of many monosaccharide units.

polysome See **polyribosome**.

polyunsaturated A term usually applied to molecules with several double bonds.

primary alcohol An alcohol in which the carbon bearing the —OH group is attached to one other carbon atom.

primary amine An amine having one alkyl or aromatic group bonded to nitrogen, as in $R—NH_2$.

primary protein structure The linear sequence of amino acid residues in a protein chain.

prokaryotic cell A simple unicellular organism that contains no nucleus and no membrane-enclosed organelles.

prostaglandin A substance derived from unsaturated fatty acids, with hormonelike effects on a number of body tissues.

prosthetic group The non–amino acid parts of conjugated proteins.

protein An amino acid polymer made up of more than 50 amino acids.

protein turnover The continuing process in which body proteins are hydrolyzed and resynthesized.

pyranose ring A six-membered sugar ring system containing an oxygen atom.

quaternary ammonium salt An ionic compound containing a positively charged ion in which four alkyl or aromatic groups are bonded to nitrogen, as in R_4N^+.

quaternary protein structure The arrangement of subunits that form a larger protein.

recombinant DNA DNA of an organism that contains genetic material from another organism.

reducing sugar A sugar that can be oxidized by Cu^{2+} in solution.

reference daily intakes (RDI) A set of standards for protein, vitamins, and minerals used on food labels as part of the Daily Values.

relative specificity The characteristic of an enzyme that it acts on several structurally related substances.

renal threshold The blood glucose level at which glucose begins to be excreted in the urine.

replication The process by which an exact copy of a DNA molecule is produced.

replication fork A point where the double helix of a DNA molecule unwinds during replication.

respiratory acidosis A condition of acidosis caused by hypoventilation.

respiratory alkalosis A condition of alkalosis caused by hyperventilation.

restriction enzyme A protective enzyme found in some bacteria that catalyzes the cleaving of all but a few specific types of DNA.

retrovirus A virus in which RNA directs the synthesis of DNA.

ribonucleic acid (RNA) A nucleic acid found mainly in the cytoplasm of cells.

ribosomal RNA (rRNA) RNA that constitutes about 65% of the material in ribosomes, the sites of protein synthesis.

ribosome A subcellular particle that serves as the site of protein synthesis in all organisms.

saponification The basic cleavage of an ester linkage.

satiety A state of satisfaction or fullness.

saturated hydrocarbon Another name for an alkane.

secondary alcohol An alcohol in which the carbon bearing the —OH group is attached to two other carbon atoms.

secondary amine An amine having two alkyl or aromatic groups bonded to nitrogen, as in R_2NH.

secondary protein structure The arrangement of protein chains into patterns as a result of hydrogen bonds between amide groups of amino acid residues in the chain. The common secondary structures are the α-helix and the β-pleated sheet.

semiconservative replication A replication process that produces DNA molecules containing one strand from the parent and a new strand that is complementary to the strand from the parent.

simple carbohydrates Monosaccharides and disaccharides, commonly called sugars.

simple lipid An ester-containing lipid with just two types of components: an alcohol and one or more fatty acids.

simple protein A protein made up entirely of amino acid residues.

soap A salt of a fatty acid often used as a cleaning agent.

sphingolipid A complex lipid containing the aminoalcohol sphingosine.

stem cell A cell with the ability to divide for indefinite periods in culture and to give rise to specialized cells.

stereochemical specificity An enzyme's ability to distinguish between stereoisomers.

stereoisomers Compounds with the same structural formula but different spatial arrangements of atoms.

steroid A compound containing four rings fused in a particular pattern.

structural isomers Compounds that have the same molecular formula but in which the atoms bond in different patterns.

substrate The substance that undergoes a chemical change catalyzed by an enzyme.

subunit A polypeptide chain having primary, secondary, and tertiary structural features that is a part of a larger protein.

sulfhydryl group The —SH functional group.

tertiary alcohol An alcohol in which the carbon bearing the —OH group is attached to three other carbon atoms.

tertiary amine An amine having three alkyl or aromatic groups bonded to nitrogen, as in R_3N.

tertiary protein structure A specific three-dimensional shape of a protein resulting from interactions between R groups of the amino acid residues in the protein.

thioester bond The carbon–sulfur bond of the

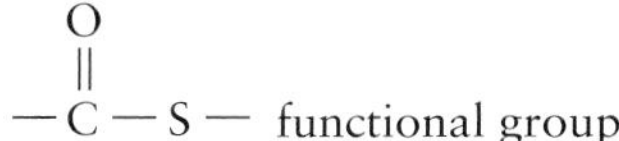

functional group.

thiol A compound containing an —SH group.

Tollens' reagent A mild oxidizing solution containing silver ions used to test for aldehydes.

trace mineral A mineral found in the body in quantities smaller than 5 g.

***trans*-** On opposite sides (as applied to geometric isomers).

transaminase An enzyme that catalyzes the transfer of an amino group.

transamination The enzyme-catalyzed transfer of an amino group to a keto acid.

transcription The transfer of genetic information from a DNA molecule to a molecule of messenger RNA.

transfer RNA (tRNA) RNA that delivers individual amino acid molecules to the site of protein synthesis.

translation The conversion of the code carried by messenger RNA into an amino acid sequence of a protein.

triglyceride (triacylglycerol) A triester of glycerol in which all three alcohol groups are esterified.

turnover number The number of molecules of substrate acted on by one molecule of enzyme per minute.

unsaturated hydrocarbon A hydrocarbon containing one or more multiple bonds.

urea cycle A metabolic pathway in which ammonium ions are converted to urea.

vector A carrier of foreign DNA into a cell.

virus An infectious particle composed of only proteins and DNA or RNA.

vitamin An organic nutrient that the body cannot produce in the small amounts needed for good health.

vitamin international unit A measure of vitamin activity, determined by biological methods.

wax An ester of a long-chain fatty acid and a long-chain alcohol.

zwitterion A dipolar ion that carries both a positive and a negative charge as a result of an internal acid-base reaction in an amino acid molecule.

zymogen or proenzyme The inactive precursor of an enzyme.

Index

A site, 320
Absolute specificity, 276
Acesulfame-K, 200
Acetaldehyde, 104, 364
Acetals, 112–118
Acetaminophen, 167
Acetic acid, 130, 131
Acetoacetate, 392
Acetone, 103, 104, 119, 392–393
Acetone breath, 393
Acetyl CoA. *See* Acetyl coenzyme A
Acetyl coenzyme A (acetyl CoA), 343, 349–350, 363, 365–368
Acetylcholine, 169
Acetylene. *See* Ethyne
Acetylsalicylic acid (aspirin), 144
Acid-base balance, 418
Acidosis, 393, 418, 421–424
  metabolic, 423
  respiratory, 422–423
Acids:
  carboxylic, 130–142
  trans fatty, 229
Actin, 257
Activated tRNA, 312
Activator, 290
Active site, 279
Activity, enzyme, 281. *See also* Enzyme activity
Acyl carrier protein (ACP), 394
Addition polymers, 51–53
Addition reactions:
  of aldehydes and ketones, 111–117
  of alkenes, 47–53
Adenine, 299–300, 344
Adenosine, 344–345
Adenosine diphosphate. *See* ADP
Adenosine monophosphate. *See* AMP
Adenosine triphosphate. *See* ATP
Adipose tissue, 386
ADP (adenosine diphosphate), 148–149, 345
ADP, hydrolysis of, 345
Adrenal glands, 237
Adrenaline. *See* Epinephrine
Adrenocorticoid hormones, 237–238
Adrenocorticotropic hormone (ACTH), 254–255
Aerobic, 363
Alanine, 247, 397, 401, 403
Alanine transaminase (ALT), 293
Albinism, 278
Alcohol dehydrogenase, 351, 364
Alcoholic fermentation, 364
Alcohols, 71–84
  classification of, 74–75
  nomenclature of, 72–74
  physical properties of, 75–77
  reactions of, 77–82
  uses of, 82–84
Aldehydes, 102–120
  nomenclature of, 102–105
  physical properties of, 105–107
  reactions of, 108–118
  uses of, 119–120
Aldoses, 195–196
Aldosterone, 238, 418
Aliphatic compound, 57
Alkaloids, 172–174
Alkalosis, 418
metabolic, 423
respiratory, 422
Alkanes (saturated hydrocarbons), 10–29
  branched, 13
  conformations of, 13–15
  cycloalkanes, 21–25
  nomenclature of, 15–21
  normal, 13
  physical properties of, 26–27
  reactions of, 28–29
  structures of, 10–13
Alkenes, 10, 39–53
  geometry of, 42–46
  nomenclature of, 39–42
  properties of, 46–47
  reactions of, 47–53
Alkoxy group, 88
Alkyl groups, 17
Alkyl halides, 47
Alkynes, 10, 39, 53–56
  nomenclature of, 55
Allosteric enzymes, 290
Allosteric regulation, 290–291
Alpha (α)-amino acids, 246. *See Also* Amino acids
Alpha (α)-helix, 260–261
Alpha hydroxy acids, 134
Alpha (α)-keratin, 257, 258, 261
Alzheimer's disease, 258
Amide linkage, 164, 251
Amides, 164–168, 174–178
  chemical properties of, 176–178
  formation of, 164–168
  nomenclature of, 174–175
  physical properties of, 176
  reactions of, 176–178
  uses of, 178
Amine salts, 161–164
Amines, 158–174
  classification of, 158
  nomenclature of, 158–159
  physical properties of, 160–161
  reactions of, 161–168
  uses and occurrences of, 168–174
Amino acid catabolism, 396–403
Amino acid metabolism, 395–404
  biosynthesis in, 403–404
  fates of the carbon atoms in, 400–403
  fates of the nitrogen atoms in, 396–400
  oxidative deamination in, 398–399
  transamination in, 396–398
  urea cycle in, 399–400
Amino acid oxidase, 399
Amino acid pool, 395, 396
Amino acid residue, 252
Amino acid synthesis, 403–404
Amino acids, 246–253. *See also* names of individual amino acids
  abbreviations of, 247
  essential, 336–337, 403
  fates of carbon atoms in, 400–403
  formation of amides from, 251–252
  glucogenic, 402–403
  ionic forms of, 247–250
  ketogenic, 402–403
  metabolism of, 395–404
  nonessential, 403–404
  reactions of, 250–252
  side chains of, 246–247
  stereochemistry, 248
  structure of, 246
  symbols of, 247
  zwitterions, 248–250
Amino group, 246
AMP (adenosine monophosphate), 301, 346
Amphetamines, 172
Amylase, 293
Amylopectin, 208
Amylose, 207
Amytal, 178
Anabolism, 343
Anaerobic, 363
Analgesic, 167
Androgens, 238
Anemia:
  pernicious, 338
  sickle-cell, 263
Anesthetics, 91
Angiotensin II, 255
Aniline, 59, 63
Animal fats, 224, 334–336

Anomeric carbon, 197–199
Anomers, 197–199
Antibiotics, 286
Antibodies, 256
Anticodons, 311
Antiinflammatory agent, 144, 167
Antioxidants, 44, 87
Antipyretic agent, 167
Antiseptics, 83
Apoenzyme, 278
Arachidonic acid, 222, 239
Arginine, 247, 399
Argininosuccinate, 399
Aromatic hydrocarbons, 10, 39, 56–63
  nomenclature of, 58–61
  polycyclic, 61
  properties of, 61–63
  structure of, 56–58
  uses of, 61–63
Artificial sweeteners, 200
Ascorbic acid. *See* Vitamin C
Asparagine, 247, 403
Aspartame, 200
Aspartate, 247, 397, 399–400
Aspartate transaminase (AST), 293
Aspirin (acetylsalicylic acid), 144, 167
Atherosclerosis, 144, 234
ATP (adenosine triphosphate), 148–149, 344–348
  from complete oxidation of glucose, 370–372
  hydrolysis of, 344–348
  production of, 369–370
ATP-ADP cycle, 347–348
Atropine, 173
Avian flu, 315
Axon, 168–169

B-vitamins, 338. *See also* names of individual B-vitamins
Base pairing (in DNA), 302–304
Bases:
  in nucleic acids, 299–301
Benedict's reagent, 109, 199–201
Benzo [a] pyrene, 61
Benzedrine (amphetamine), 172
Benzene, 56–58, 61, 62
Benzoic acid, 59
Benzoic acid salts, 137
Beriberi, 338
Beta (β)-carotene, 41
Beta (β)-keratin, 257
Beta (β)-oxidation process, 388–391
  cellular location of, 388
  energetics of, 391
Beta (β)-pleated sheet, 261
Bentoquatam, 86
BHA (butylated hydroxyanisole), 87
BHT (butylated hydroxytoluene), 87
Bile, 235–237
Bile salts, 235–237
Biochemistry, 186
Bioengineering, 326
Biomolecules, 186, 205
Bioprocesses, 346
Biotechnology, 326
Biotin, 279, 338
Birth control, 110
Birth defects, 119
Bismuth:
  1,3 -Bisphosphoglycerate, 361
Bladder hyperreflexia, 118
Blood, 410, 411
  buffers in, 418–420
Blood glucose level, 359
Blood lipids, 384–386
Blood sugar level, 359
BMI, 339
Body fluids, 410
  blood, 410, 411
  extracellular, 410
  interstitial, 410, 411
  intracellular, 410, 411
  urine, 416–417
Branched alkane, 13
Buffers:
  in blood, 418–420
Butane, 12–13
  1-Butanol (butanol), 84
Butylated hydroxyanisole (BHA), 87
Butylated hydroxytoluene (BHT), 87

C-terminal residue, 252
Caffeine, 173
Calories, 333
Camphor, 121
Cancer, 44, 59, 61–62, 106, 144
Carbaminohemoglobin, 412
Carbamoyl phosphate, 399–400
Carbamoyl phosphate synthetase, 400
Carbocation, 79
Carbohydrate loading, 376
Carbohydrate metabolism, 359–377
  energy from glucose in, 370–372
  fates of pyruvate in, 363–364
  gluconeogenesis in, 374–375
  glycogen breakdown in, 373–374
  glycogen synthesis in, 372–373
  glycolysis in, 360–363
  regulation of, 375–377
Carbohydrates, 186–212
  as nutrients, 333–334
  classes of, 186–187
  disaccharide, 204–207
  Fischer projections of, 192–195
  metabolism of, 359–377
  monosaccharide, 195–203
  polysaccharide, 207–212
  stereochemistry of, 187–191
Carbolic acid, 85
Carbon, 2
Carbon cycle, 342
Carbon dioxide:
  formation of, by citric acid cycle, 365–367
  in acidosis, 422–423
  in alkalosis, 422
  transport of, 410–414
Carbonic acid, 275, 412–414, 421–422
Carbonic anhydrase, 275
Carbon monoxide poisoning, 29
Carbonyl group, 102
  polarity of, 107
Carboxyl group, 130
Carboxylate ions, 135–136
Carboxylic acid anhydride, 140, 165–166
Carboxylic acid chloride, 140, 165–166
Carboxylic acid dimers, 133
Carboxylic acid salts, 136–138
  nomenclature of, 136–137
  uses of, 137–138
Carboxylic acids, 130–143
  acidity of, 135–136
  ester formation from, 138–142
  nomenclature of, 130–132
  physical properties of, 132–135
  salt formation from, 136
Carboxylic ester, 138. *See also* Esters
Carboxypeptidase, 290
Cardiovascular disease, 254
Cardiovascular endurance, 415
Catabolism, 343
Cells, 232
  eukaryotic, 232
  membrane of, 232–234
  organelles of, 232
  prokaryotic, 232
Cellular respiration, 342
Cellular work, 342
Cellulose, 209–212
Central dogma of molecular biology, 313
Cephalins, 230
Cerebrosides, 231
Chemical carcinogens, 62
Chemiosmotic hypothesis, 369
Chiral carbon, 188
Chiral compounds, 188–191
Chloride shift, 413
Cholesterol, 71, 224, 234, 384–386, 387
Choline, 229
Chromosomes, 305–306
Chylomicrons, 384–386
Chymotrypsin, 290
Chymotrypsinogen, 290
Cigarette smoking. *See* Smoking
Cis isomer, 24, 44–45
Citrate, 366, 368. *See also* Citric acid
Citrate synthetase, 367
Citric acid, 131, 138. *See also* Citrate
Citric acid cycle, 365–368
  entry points for amino acid skeletons in, 401–402
  reactions in, 365–367

regulation of, 367–368
summary and major features of, 365–367
Citrulline, 399
Cloning, 310
Clostridium botulinum, 267
Codeine, 173–174
Codons, 316
Coenzyme A, 349–350
Coenzyme Q (CoQ), 368–369
Coenzymes, 278–279, 349–353. *See also* names of individual coenzymes
Cofactors, 277–279
Collagen, 257, 258
Common catabolic pathway, 343
Competitive inhibitor, 286
Complementary DNA strands, 304
Complete combustion of alkanes, 28–29
Complete proteins, 336
Complex carbohydrates, 334
Complex lipids, 219
Condensation polymerization, 139
Condensed structural formulas, 8–10
Conformation of alkanes, 13–15
Conjugated proteins, 258
Contraception, oral, 110
Copolymer, 53
Corey, Robert, 260
Cori cycle, 375
Corticosteroids, 86
Cortisol, 237
Cortisone, 237
Cosmetics, 134
C-reactive protein, 254
Creatine, 351
Creatine kinase (CPK), 293
Crick, Francis H. C., 302–303
Cristae, 348
Cyanide ions, reaction of:
with cytochrome oxidase, 284
with enzymes, 284
Cycloalkanes, 21–25
nomenclature of, 21–23
shape of, 23–25
Cysteine, 247, 250
Cystine, 250
Cytochrome oxidase, 284
Cytochromes, 368
Cytoplasm, 232
Cytosine, 299–300

Dacron, 140
Daily reference values (DRV), 333
Daily values (DV), 333
Daughter DNA, 307, 309
Deamination, 398–399
Dehydration:
of alcohols, 77
of humans, 417
Dehydration reaction, 77–78
Dehydroepiandrosterone, 236
Denaturation:
of proteins, 267–268
Deoxyhemoglobin, 411
Deoxyribonucleic acid. *See* DNA
Deoxyribose, 202–203, 300
Dextrins, 208
Dextrorotary, 194
Diabetes insipidus, 418
Diabetes mellitus, 200
Dianabol, 238
Diazepam (Valium), 178
Dibenz [*a,h*] anthracene, 61
Diethyl ether, 88, 91
Digestion, 343–344
of carbohydrates, 359
of fats, 384–386
of proteins, 343–344
Dihydroxyacetone (DHA), 106
Dihydroxyacetone phosphate, 361, 388
Dimers, 133
Dipeptide, 252
Dipolar ions (zwitterions), 248–250
Disaccharides, 204–207. *See also* Lactose; Maltose; Sucrose
Disulfide, 91
Disulfide bridges, 253, 262–263
Divinyl ether, 91
DNA, 299
base pairing of, 299–301
fingerprinting (profiling, typing), 324
mutations, 322
recombinant, 322–326
replication of, 305–309
structure of, 301–305
DNA ligase, 308, 325
DNA polymerase, 307
Dopamine, 169
Double helix, 302–305
Drugs:
in sports, 238
interactions, 413
sulfa, 287–288
Dynamite, 148

E, 85
Elastin, 257
Electrolytes:
maintaining balance of, 417–418
Electron transport chain, 368–369
Electrophoresis, 324
Elimination reaction, 77
Enantiomers, 187–191
Endoplasmic reticulum, 232
Energetics:
of fatty acid oxidation, 391
of glycolysis, 370–372
of the citric acid cycle, 370–372
Energy:
and the biosphere, 340–342
free, 345–347
from carbohydrates, 370–372
from fatty acids, 391
from food, 343–344
from proteins, 335–336
Energy conservation from glucose oxidation, 370–372
Enflurane, 91
Enzyme activity, 282–284
effect of enzyme concentration on, 282
effect of pH on, 283
effect of substrate concentration on, 282–283
effect of temperature on, 283
regulation of, 289–292
Enzyme catalysis, 275
Enzyme induction, 291
Enzyme inhibition, 284–289
Enzyme inhibitors, 284
Enzyme international unit, 282
Enzyme-substrate complex, 279–281
Enzymes, 275–294. *See also specific names of enzymes*
activity of, 281–284
characteristics of, 275–276
cofactors, 277–279
inhibition of, 284–289
mechanism of action of, 279–280
medical applications of, 292–294
nomenclature of, 276–277
regulation of activity of, 289–292
specificity, 275–276
Epinephrine (adrenaline), 171–172, 376–377
Essential amino acids (table), 336–337, 403
Essential fatty acids, 335
Ester linkage, 138
Esterification, 138
Esters, 138–143
formation of from carboxylic acids, 138–139
formation of from inorganic acids, 147–149
nomenclature of, 142–143
polyesters, 139, 140
reactions of, 143–147
Estradiol, 238
Estrogens, 238
Estrone, 238
Ethane, 11
1,2-Ethanediol (ethylene glycol), 84
Ethanol, 83, 84, 85, 113, 364
Ethanolamine, 229
Ethene (ethylene), 43
Ethers, 71, 88–90
nomenclature of, 88–89
preparation of, 78
properties of, 89–90
Ethylene. *See* Ethene
Ethylene glycol. *See* Ethanediol
Ethyne (acetylene), 54–55
Eukaryotic cells, 232

Exon, 316
Expanded structural formula, 8–10
Extracellular fluid, 410
Extremozyme, 285

FAD (flavin adenine dinucleotide, oxidized), 350–353
FADH (flavin adenine dinucleotide, reduced), 350–353
Fat mobilization, 386–387
Fat-soluble vitamins, 337–338. *See also* individual names of fat-soluble vitamins
Fats, 223–228
- animal, 224
- caloric content of, 384
- nutrients, 334–335
- reactions of, 225–228
- saturated, 224
- structure of, 223–224

Fatty acid activation, 388
Fatty acid oxidation, 388–391
- cellular location of, 388
- energetics of, 391

Fatty acid spiral, 388–390
Fatty acid synthesis, 394–395
Fatty acid synthetase, 394
Fatty acids, 130, 220–223
- in fats and oils, 334
- omega, 223
- polyunsaturated, 221–222
- saturated, 221–222, 224
- trans, 229
- unsaturated, 221–222, 224

Fatty acyl CoA, 388–390
Fatty acyl group, 388
Feedback inhibition, 291
Fermentation, 83
- alcoholic, 364
- lactate, 364

Ferritin, 257
Fiber, dietary, 210, 332
Fibrin, 259
Fibrinogen, 257
Fibrous protein, 258
Fischer, Emil, 192
Fischer projection, 192–195
Fish, 286
Flavin adenine dinucleotide, oxidized. *See* FAD
Flavin adenine dinucleotide, reduced. *See* FADH
Flavin monocucleotide. *See* FMN
Fleming, Alexander, 286
Fluid mosaic model, 233
Fluids. *See* Body fluids
FMN (flavin mononucleotide), 368
Folacin. *See* Folic acid
Folic acid (folacin), 279, 288, 338
Follicle stimulating hormone (FSH), 255
Food disorders, 347
Food guide pyramide, 334
Food label, 333
Formaldehyde, 103, 104, 119
Formic acid, 131
Formulas:
- condensed structural, 8–10
- expanded structural, 8–10

Free energy change (ΔG), under cellular conditions, 345–347
Fructose, 198, 203
Fructose, -bisphosphate, 362
Fructose -phosphate, 201, 362
Fumarate, 287, 366, 368, 399
Functional groups, 7–9. *See also* names of individual functional groups
Furanose ring, 198–199

Galactose, 203
Galactosemia, 278
Galactosidase, 291–292
Gallstones, 235–237
Gasoline, 29
Gastrin, 255
Gaucher's disease, 232, 278
General anesthetics, 91
Genes, 305
Genetic code, 316–318
Genetic control of enzymes, 291–292
Genetic engineering, 322–325
Genetic foods, 326
Genetic mutations, 322
Geometric (*cis-trans*) isomers, 24–25, 43–45
Globular proteins, 258
Glucagon, 255, 376–377
Glucocorticoid, 237
Glucogenic amino acids, 402–403
Gluconeogenesis, 374–375
Gluconic acid, 201
Glucose, 198, 203, 340–342, 359–363, 364
Glucose -phosphate, 373
Glucose -phosphatase, 374
Glucose -phosphate, 148, 362, 373–374
Glucosuria, 359, 416
Glutamate, 247, 397, 398, 403
Glutamic acid. *See* Glutamate
Glutamine, 247, 403
Glyceraldehyde, 188, 192
Glyceraldehyde 3-phosphate, 361
Glycerol. *See*, 1,2,3-Propanetriol
Glycerol 3-phosphate, 388
Glycine, 247
Glycogen, 209
- breakdown of, 373–374, 375
- synthesis of, 372–373, 375

Glycogen metabolism, 372–375
Glycogen synthesis, 372–373
Glycogenesis, 372–373, 375
Glycogenolysis, 373–374, 375
Glycolipids, 231
Glycolysis, 360–363, 375
- aerobic, 363
- anaerobic, 363
- energy yield of, 370–372
- reactions in, 361
- regulation of, 362

Glycosides, 201–202
Glycosidic linkage, 202
Growth hormone, 255, 256, 257
GTP (guanosine triphosphate), 365–367
Guanine, 300
Guanosine triphosphate. *See* GTP

Half-life:
- of proteins, 395

Haloalkane, 47
Halogenated hydrocarbons, 47–48
Halogenation of alkenes, 47–48
Halothane, 91
Haworth structures, 197–199
Heart attack, 144, 293
Heavy-metal ions, reaction of:
- with enzymes, 284–285
- with thiols, 92

Heme, 266, 412
Hemiacetals:
- from monosaccharides, 196–198
- from simple aldehydes, 112–118

Hemiketals:
- from monosaccharides, 196–198
- from simple ketones, 113–118

Hemoglobin, 266, 410–414
- carbon monoxide poisoning and, 29
- heme in, 266, 412
- oxygen binding in, 410–414
- quaternary structure of, 266
- transport of carbon dioxide by, 410–412
- transport of oxygen by, 410–414
- types of, 410–412

Herbal supplements, 280
Heroin, 173–174
Heterocyclic ring, 89
Heterogeneous nuclear RNA (hnRNA), 316
Hexokinase, 362
Hexylresorcinol, 85–86
High-density lipoproteins (HDL), 385–387
High-energy compound, 346
Histidine, 247
Homocystinuria, 278
Homologous series, 26
Hormones, 237–240. *See also* names of individual hormones
- adrenocorticoid, 237–238
- peptide, 253–254
- sex, 238
- steroid, 237–238

Hybrid orbitals, 5, 42–43, 55, 57–58

Hydration:
- of alkenes, 51
- of fatty acids, in the β-oxidation cycle, 388–390

Hydrocarbons, 10–13. *See also* Alkanes; Alkenes; Alkynes; Aromatic hydrocarbons
Hydrocortisone, 86
Hydrogen bonding:
- in alcohols, 75–76
- in aldehydes and ketones, 107
- in amides, 176
- in amines, 160–161
- in base pairs, 304
- in carboxylic acids, 132–135
- in ethers, 89–90
- in tertiary protein structure, 263–264
- in the α-helix, 260–261
- in the β-pleated sheet, 261
- in the DNA double helix, 302–303
- in tRNA, 310

Hydrogen phosphate ion (inorganic phosphate, P*i*), 345
Hydrogenation:
- of aldehydes and ketones, 111
- of alkenes, 48
- of vegetable oils, 226–228

Hydrolases, 277
Hydrolysis, 116
- of acetals and ketals, 116–118
- of amides, 177–178
- of ATP, 345–348
- of esters, 144–145
- of fats and oils, 225
- of proteins, 266–267
- of pyrophosphate, 347
- of urea, 276–277

Hydrophobic, 26–27, 220
Hydrophobic interactions, 264
β-Hydroxybutyrate, 392–393
Hydroxy group, 71
Hyperglycemia, 359
Hyperventilation, 422
Hypoglycemia, 359
Hypoventilation, 422

Ibuprofen, 167
Inborn errors of metabolism, 278
Incomplete combustion of alkanes, 29
Induced-fit theory, 280
Inhibitors:
- of enzymes, 284

Inorganic chemistry, 2
Inorganic phosphate (Pi), 345
Insulin, 255, 260, 376–377
Interferon, 323
International Union of Pure and Applied Chemistry (IUPAC), 16
International units, 282
Interstitial fluid, 410, 411
Intracellular fluid, 410, 411
Intron, 316
Invert sugar, 206
Iodine, 208
Ionizing radiation, 322
Irreversible inhibitor, 284–286
Isocitrate, 366
Isocitrate dehydrogenase, 367–368
Isoelectric point, 249
Isoenzymes, 293
Isoleucine, 247, 290
Isomerases, 277
Isomerism, 5–7. *See also* Isomers
Isomers, 5–7
- enantiomers, 187–191
- geometric (*cis-trans*), 24–25, 43–46
- of alkanes, 5–7
- stereo-, 24, 187–191
- structural, 5–7

IUPAC (International Union of Pure and Applied Chemistry), 23

Jaundice, 416

Kekulé, Friedrich August, 57
Keratin, 257, 261
- α-, 257, 261
- β-, 257

Ketals, 113–118
Ketoacidosis, 393
Ketogenic amino acids, 402–403
α-Ketoglutarate, 366, 367, 397, 398
Ketone bodies, 392
Ketonemia, 393
Ketones, 102–120
- nomenclature of, 102–105
- physical properties of, 105–107
- reactions of, 108–118
- uses of, 119–120

Ketonuria, 393, 416
Ketoprofen, 167
Ketoses, 195
Ketosis, 393
Kidneys:
- in acid-base balance, 420–421
- in electrolyte balance, 417–418
- in fluid balance, 417–418

Krebs cycle. *See* Citric acid cycle
Krebs, Hans A., 365

Lactase, 360
Lactate, 363–364, 374. *See also* lactic acid
Lactate accumulation, 367
Lactate dehydrogenase (LDH), 278, 293–294–363
Lactate fermentation, 363–364
Lactic acid, 131. *See also* Lactate
Lactose, 204–205, 292
Lactose intolerance, 360
Lecithin (phosphatidylcholine), 229
Leucine, 247
Leukemia, 293
Levorotatory, 194
Ligases, 277
Linoleic acid, 222, 223, 335
Linolenic acid, 222, 223, 335
Lipid bilayers, 233–234
Lipid metabolism, 384–395
- acetyl CoA in, 388–391
- blood lipids in, 384–386
- cholesterol in, 384–386, 387
- fat mobilization in, 386–387
- fatty acid synthesis in, 394–395
- glycerol in, 388
- ketone bodies in, 392–393
- oxidation of fatty acids in, 388–391
- Lipids, 219–240
- caloric value of, 384
- classification of, 219
- complex, 219
- fats and oils, 223–228
- fatty acids, 220–223
- membranes, 232–234
- nutrients, 334–336
- phosphoglycerides, 228–230
- prostaglandins, 239–240
- simple, 219
- sphingolipids, 231–232
- steroids, 234–237
- waxes, 228

Lipoproteins, 359, 384–387
Lister, Joseph, 85
Lock-and-key theory, 280
Low-density lipoproteins (LDL), 385–386
Lyases, 277
Lycopene, 44
Lymph system, 384
Lysine, 247
Lysosomes, 232

Macronutrients, 333–336
- carbohydrates, 333–334
- lipids, 334–335
- proteins, 335–336

Major minerals, 339
Malate, 366, 368
Malonate, 287
Maltose, 204
Maple syrup urine disease, 278
Marathons, energy for, 376
Margarine, 227
Markovnikov's rule, 49–51
Maximum velocity (Vmax), 282–283
Mechanisms, 56, 79, 114, 146, 279–280
Melanin, 59
Melatonin, 236
Membranes, 232–234
- composition of, 232–234
- fluid mosaic model of, 233
- lipid bilayers of, 233

Menthol, 84
Mercury, 286
Messenger RNA. *See* mRNA

Metabolic acidosis, 423
Metabolic alkalosis, 423
Metabolic pathways, 343. *See also* names of individual metabolic pathways
Metabolism, 343
- of amino acids, 395–404
- of carbohydrates, 359–377
- of glycogen, 372–374
- of lipids, 384–395

Methandrostenolone (Dianabol), 238
Methane, 11
Methanol, 82, 84
Methedrine, 172
Methemoglobinemia, 278
Methionine, 247
Methyl salicylate (oil of wintergreen), 139
Micelles, 220
Micronutrients, 337–340
- fat-soluble vitamins, 337–338
- minerals, 339–340
- water-soluble vitamins, 337–338

Mineralocorticoid, 237
Mineral oil, 27
Minerals, 339–340
- major, 339
- trace, 339

Mitochondria, 232, 348
Mitochondrial matrix, 348
Mitochondrial membranes, 348
Modulators, 290
Moisturizers, 27
Monomer, 52
Monosaccharides, 195–203. *See also* names of individual monosaccharides
- physical properties of, 196
- reactions of, 196–202
- uses of, 202–203

Morphine, 163, 173–174
mRNA, 311
Mutagens, 322
Mutations, genetic, 322
Mylar, 140
Myoglobin, 257
Myosin, 257

N-terminal residue, 252
NAD, 350–353
NADH, 350–353
NADP, 394
NADPH, 394
Naphthalene, 61
Native state, 267
Natural gas, 11, 28
Neurotransmitter, 168
Niacin (vitamin B), 279, 338
Nicotinamide, 350
Nicotinamide adenine dinucleotide, oxidized. *See* NAD
Nicotinamide adenine dinucleotide phosphate, oxidized. *See* NADP
Nicotinamide adenine dinucleotide, reduced. *See* NADH
Nicotinamide adenine dinucleotide phosphate, reduced. *See* NADPH
Nicotine, 173
Niemann-Pick disease, 232, 278
Night blindness, 338
Nirenberg, Marshall, 316
Nitroglycerin, 148
Nobel, Alfred, 148
Nomenclature:
- of alcohols, 72–74
- of aldehydes and ketones, 102–105
- of alkanes, 15–21
- of alkenes, 39–42
- of alkynes, 55
- of amides, 174–175
- of amines, 158–159
- of benzene derivatives, 58–61
- of carboxylic acid salts, 136–137
- of carboxylic acids, 130–132
- of enzymes, 276–277
- of esters, 142–143
- of ethers, 88–89
- of phenols, 72–74

Noncompetitive inhibitor, 289
Nonessential amino acids, 403–404
Nonnutritive sweeteners, 200
Norepinephrine, 169
Norethynodrel, 110
Normal alkane, 13
Nucleic acid backbone, 301
Nucleic acids, 299–326
- components of, 299–301
- DNA replication, 305–309
- DNA structure, 301–305
- genetic code, 316–318
- mutations, 322
- recombinant DNA, 322–326
- RNA structure, 309–312
- RNA synthesis, 314–316
- supplements, 305
- translation and protein synthesis, 318–321
- viruses, 315

Nucleotides, 299–301
Nucleus, cell, 232
Nutrients, 332–333
- macro-, 333–336
- micro-, 337–340

Nutrition, 332–340
- and macronutrients, 333–336
- and micronutrients, 337–340

Nutritional minerals, 339–340
Nutritional requirements, 332–333
Nylon, 166–168

Oils, 223–224
- reactions of, 225–228
- structure of, 223–224
- vegetable, 224, 334–336

Okazaki fragments, 308
Okazaki, Reiji, 308
Oleic acid, 222
Oligosaccharides, 187
Omega fatty acids, 223
Optically active molecule, 194
Optimum pH, 283
Optimum temperature, 283
Oral contraception, 110
Orbitals:
- hybrid, 5, 42–43, 55, 57–58

Organelles, 232
Organic chemistry, 2–3
Organic compounds, 2
- classes of, 7–9
- properties of, 3–4

Organic foods, 11
Ornithine, 399
Ovalbumin, 257
Oxalic acid, 131
Oxaloacetate, 366, 397
Oxidant. *See* Oxidizing agent
Oxidation:
- of alcohols, 78–80
- of aldehydes, 108–111
- of cysteine, 250
- of fatty acids, 388–391
- of glucose, 201, 370–372

Oxidative deamination, 398–399
Oxidative phosphorylation, 369–370
Oxido-reductases, 277
Oxygen, transport of, 410–414
Oxyhemoglobin, 411
Oxytocin, 253, 255

P site, 319
p-Aminobenzoic acid (PABA), 288
Palmitic acid, 222
Palmitoleic acid, 222
Pantothenic acid, 279, 338, 349
Paraffin, 28
Partially hydrogenated, 227
Pauling, Linus, 5, 260
Pellagra, 338
Penicillins, 286, 287
Pentothal, 187
Pepsin, 283, 290
Pepsinogen, 290
Peptide bonds, 252
Peptide hormones, 253–254
Peptide linkage, 252
Peptides, 253–254
Peptidyl transferase, 320
Pernicious anemia, 338
Petroleum, 26
Petroleum jelly, 27
pH:
- buffer control of, 418–420
- effect on enzymes, 283
- effect on proteins, 266–268

respiratory control of, 420
urinary control of, 420–421
Pharmacist, 1
Phenols, 59, 63, 71
characteristics and uses, 85–87
nomenclature of, 72–74
Phenylacetate, 401
Phenylalanine, 63, 247, 401, 403
Phenylalanine hydroxylase, 401, 403
Phenyl group, 60
Phenylketonuria (PKU), 278, 401
Phenylpyruvate, 401
Phosphate, 300
Phosphate esters, 147–148, 201, 301
Phosphatidylcholine. *See* Lecithin
Phosphodiester linkages, 301
3-Phosphoenolpyruvate, 361
Phosphofructokinase, 362
Phosphoglucomutase, 373
2-Phosphoglycerate, 361
3-Phosphoglycerate, 361
Phosphoglycerides, 228–230, 384–386
Phospholipids, 228
Phosphoric acid, 300
Phosphoric anhydride, 148
Phosphorylation, oxidative, 369–370
Photosynthesis, 340
P*i*. *See* Inorganic phosphate
Plasma, 410, 411
Plasmids, 323–324
Poison ivy, 86
Polyamide, 166
Polycyclic aromatic compounds, 61–62
Polyester, 139, 140
Polyethylene, 52–53
Polyfunctional compounds, 92–93
Polymer, 52
Polymerization of alkenes, 51–53
Polypeptides, 252
Polypropylene, 54
Polyribosome, 321
Polysaccharides, 207–212. *See also* individual names of polysaccharides
Polysome, 321
Polyunsaturated, 48
Poly (vinyl acetate), 54
Poly (vinyl chloride), 53, 54
PP*i*. *See* Pyrophosphate
Prednisolone, 237
Primary (1°) alcohols, 74
Primary (1°) amines, 158
Primary structure:
of nucleic acids, 301–302
of proteins, 259–260
Procarboxypeptidase, 290
Proenzymes (zymogens), 289
Progesterone, 110, 120, 238
Prokaryotic cell, 232
Prolactin, 255
Proline, 247
Propane, 11
1,2,3-Propanetriol (glycerol), 83–84, 223, 388
2-Propanol (isopropyl alcohol), 83, 84
Prostaglandins, 239–240
Prosthetic group, 258
Proteins, 246–268
amino acid components of, 246–248
characteristics of, 255–259
classification of, 258
denaturation of, 267–268
function of, 256–258
glycoproteins, 257
hydrolysis of, 266–267
nutrients, 335–336
primary structure of, 259–260
quaternary structure of, 265–266
reference daily intake, 335
secondary structure of, 260–261
synthesis, 318–321
tertiary structure of, 262–265
turnover, 395
Prothrombin, 290
Purines, 299
Pyranose ring, 197
Pyridoxal, 279
Pyridoxal phosphate, 279
Pyridoxamine, 279
Pyridoxine, 279
Pyrimidines, 299
Pyrophosphate (PP*i*), 346
Pyruvate, 361, 363–365, 374–375, 397, 403
conversion of, to acetyl CoA, 363
reduction of, to ethanol, 364
reduction of, to lactate, 363–364
Pyruvate decarboxylase, 364
Pyruvate fates, 363–365
Pyruvate kinase, 362

Quaternary ammonium salts, 164
Quaternary structure of proteins, 265–266
Quinine, 173

Radiation:
ultraviolet, 59
Reaction mechanism, 56, 79, 114, 146
Recombinant DNA, 322–325
Reducing sugars, 201
Reference Daily Intakes (RDI), 332
Relative specificity, 276
Renal threshold, 359
Replication, 305–309
Replication fork, 307
Residue, amino acid, 252
Respiration, cellular, 342
Respiratory acidosis, 422–423
Respiratory alkalosis, 422
Restriction enzymes, 323
Retinal, 129
Reversible inhibitor, 286–289
Rhodopsin, 257
Riboflavin (vitamin B), 63, 279, 338, 352
Ribonucleic acid. *See* RNA
Ribose, 187, 202–203, 300, 301
Ribosomal RNA. *See* rRNA
Ribosome, 311, 319
Rickets, 338
RNA, 299, 309–312
RNA polymerase, 314
mRNA, 311
rRNA, 311
tRNA, 311–312

Saccharin, 200
Salt bridges, 263
Salts:
from amines, 161–164
from carboxylic acids, 136–138
quaternary ammonium, 164
Saponification:
of esters, 145, 146
of fats and oils, 226
Saran wrap, 53
Satiety, 335
Saturated hydrocarbons. *See* Alkanes
Scurvy, 338
Secondary (2°) alcohols, 74
Secondary (2°) amines, 158
Secondary structure:
of DNA, 302–305
of proteins, 260–261
Semiconservative replication, 307
Serine, 247
Serotonin, 169
Sex hormones, 238
Shortening, 227
Sickle-cell disease, 263
Sickle-cell hemoglobin (HbS), 263
Simple carbohydrates, 334
Simple lipids, 219
Simple proteins, 258
Smoking, 62
Soaps, 226
Sodium benzoate, 137
Sodium glycocholate, 235
Somatostatin, 255
Sorbitol, 200
Sphingolipids, 231–232
Sphingomyelins, 231–232
Sphingosine, 231
Standard free energy change (ΔG^o):
in the electron transport chain, 369–370
of ADP hydrolysis, 345
of ATP hydrolysis, 345–347
Starch, 207–208
Start signal, 317, 318
Stearic acid, 137, 222
Stem cell, 319
Stereochemical specificity, 276
Stereoisomers, 24–25, 43–46, 187–191
Steroid hormones, 237–238

Steroids, 234, 402
Stop signals, 317, 318
Structural formulas, 9–10
Structural isomers, 5–7
Styrene, 63
Styrofoam, 54
Substrates, 276
Subunits, 265
Succinate, 287, 366, 368
Succinate dehydrogenase, 287
Succinyl CoA, 366, 368
Sucrose, 205–207
Sugars:
  foods free of, 200
  in blood, 359
  in the diet, 334, 359
  sweetnesses scale for, 197
Sulfa drugs, 287–288
Sulfanilamide, 287–288
Sulfhydryl group, 90
Supplements, 305
Sweeteners, artificial, 200
Synaptic terminals, 168–169
Systematic enzyme names, 277

Tanning, 106
Taxol, 71
Tay-Sachs disease, 232, 278
Teflon, 54
Temperature:
  effect on enzymes, 283
  effect on proteins, 266–268
Tertiary (3°) amines, 158
Tertiary (3°) alcohols, 74
Tertiary structure (of proteins), 262–265
Testosterone, 120, 238
Tetrahedron, 5
Tetrahydrocannabinol, 93
Thiamin (vitamin B), 279, 338
Thioester bond, 350
Thiol, 90–92
Thirst mechanism, 417
Threonine, 247, 290
Threonine deaminase, 290
Thrombin, 257, 290
Thymine, 300
Tollens' reagent, 109
Toluene, 59, 63
Trace minerals, 339
Trans fatty acid, 229
Trans isomer, 25
Tranquilizers, 178
Transaminases, 396–398
Transaminations, 396–398
Transcription, 313, 314–316
Transfer RNA. *See* tRNA
Transferases, 277
Transferrin, 257
Translation, 313, 318–321
Translocation, 320
Triacylglycerols. *See* Triglycerides
Triglycerides (triacylglycerols), 223
Triple bond, 55
Triplet code, 316
Trypsin, 290
Trypsinogen, 290
Tryptophan, 247
Turnover number, 281
Tylenol, 167
Tyrosine, 247, 403

UDP, 373
Unsaturated hydrocarbon, 39–63. *See also* Alkenes, Alkynes, and Aromatic hydrocarbons
Uracil, 300
Urea, 2, 276, 399–400, 416
  formation of, 399–400
Urea cycle, 399–400
Urease, 276
Uridine diphosphate. *See* UDP
Uridine triphosphate. *See* UTP
Urinary system:
  fluid and electrolyte balance by, 417–418
  pH control by, 420–421
Urine, composition of, 416–417
UTP, 373

Valine, 247, 397
Valium, 178
Vanillin, 118, 121
Vanilloids, 118
Vasopressin, 253, 255, 418
Vector, 323
Vegetable oils, 224, 334–336
Vinegar, 130
Viruses, 315
Vitamin A, 41, 119
Vitamin B_1. *See* Thiamin
Vitamin B_{12}, 279, 338
Vitamin B_2. *See* Riboflavin
Vitamin B_3. *See* Niacin
Vitamin B_6, 338
Vitamin C (ascorbic acid), 338
Vitamin D, 338
Vitamin E, 93, 338
Vitamin K, 338
Vitamin international unit, 119
Vitamins, 62, 280, 337–338. *See also* names of individual vitamins

Water:
  balance, 417–418
  distribution of, in the human body, 410
Water-soluble vitamins, 337–338. *See also* names of individual water-soluble vitamins
Watson, James D., 302–303
Waxes, 228
Wöhler, Friedrich, 2
Wool, 261

Zephiran chloride, 164
Zinc -undecylenate, 137
Zwitterions, 248–250
Zymogens (proenzymes), 289